September 9–13, 2012
Dublin, Ireland

**Association for
Computing Machinery**

Advancing Computing as a Science & Profession

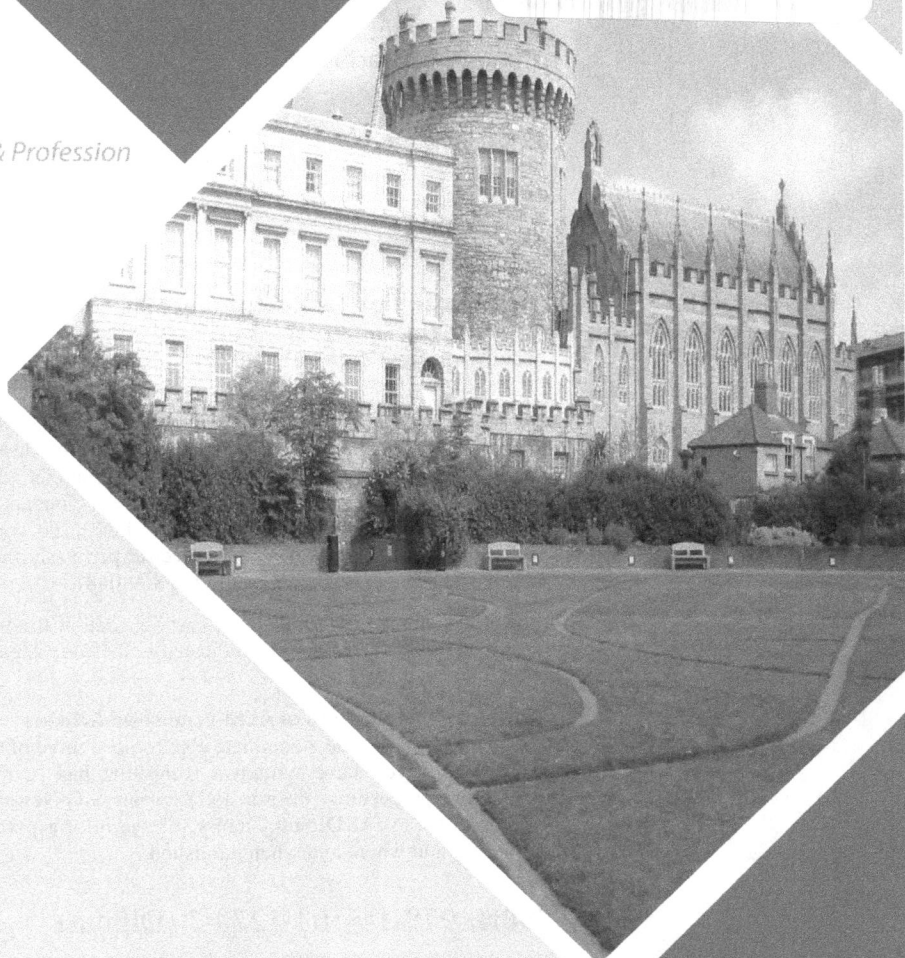

I0060976

RecSys'12

Proceedings of the Sixth ACM Conference on

Recommender Systems

Sponsored by:
ACM SIGWEB

In cooperation with:
ACM SIGART, SIGIR, ACM KDD.org, SIGWEB, & ACM SIGecom

Supported by:
SFI, Fáilte Ireland, IBM Research, Netflix, Technicolor, & LinkedIn

Association for
Computing Machinery

Advancing Computing as a Science & Profession

The Association for Computing Machinery
2 Penn Plaza, Suite 701
New York, New York 10121-0701

Notice to Past Authors of ACM-Published Articles

ACM intends to create a complete electronic archive of all articles and/or other material previously published by ACM. If you have written a work that has been previously published by ACM in any journal or conference proceedings prior to 1978, or any SIG Newsletter at any time, and you do NOT want this work to appear in the ACM Digital Library, please inform permissions@acm.org, stating the title of the work, the author(s), and where and when published.

ISBN: 978-1-4503-1270-7 (Digital)

ISBN: 978-1-4503-1882-2 (Print)

Additional copies may be ordered prepaid from:

ACM Order Department
PO Box 30777
New York, NY 10087-0777, USA

Phone: 1-800-342-6626 (USA and Canada)
+1-212-626-0500 (Global)
Fax: +1-212-944-1318
E-mail: acmhelp@acm.org
Hours of Operation: 8:30 am – 4:30 pm ET

Printed in the USA

Foreword

It is our great pleasure to welcome you to the 6th ACM Recommender Systems Conference (ACM RecSys 2012), held in Dublin, Ireland on September 9-13. As RecSys enters the second half of its first decade, it has clearly established itself as the premier international venue for research and development in the field of Recommender Systems, where leading researchers and practitioners from around the world meet and discuss their latest results and solutions.

The growing maturity of the conference is reflected through several changes in the program as compared to previous years: the tutorial program includes four exciting tutorials, which, for the first time, were not invited, but rather submitted and rigorously reviewed. We also put a special emphasis on soliciting demonstrations, leading to a record number of 10 demos accepted to the conference. In addition, this year continues the tradition of attractive collocated workshops, with a total of 8. Finally, 8 PhD students have been accepted to participate in the Doctoral Symposium. In parallel to growing and diversifying participation opportunities, the selectivity of the short paper track was substantially increased, while the competitive acceptance rate of the long paper track was maintained.

Overall, the number of paper submissions went up almost 10% as compared to last year, with a total of 185 submissions. Out of 119 full paper submissions, 24 were accepted (20.2%) for oral presentation at the conference. Out of 66 short paper submissions, 21 were accepted (31.8%) for poster presentation at the conference. 5 of these 21 have been identified as industry short papers, highlighting the fact that their main contribution lies in the description of a real system, typically already in wide use.

The conference program includes two keynotes, one from an academic perspective by Jure Leskovec (Stanford University) and one from an industrial perspective by Ron Kohavi (Microsoft). The Industry program also includes a rich set of talks by Ralf Herbrich (Facebook), Ronny Lempel (Yahoo! Research), Sumanth Kolar (StumbleUpon), Anmol Bhasin (LinkedIn), Thore Graepel (Microsoft Research), and Paul Lamere (The Echo Nest).

Many people have contributed to make ACM RecSys 2012 a success: we would like to thank the workshop chairs Jill Freyne and Pearl Pu; the doctoral symposium chairs Alexander Felfernig and Michael O'Mahony; the tutorial chairs Alexander Tuzhilin and Werner Geyer; the demo chairs Elizabeth Daly and Michael Ekstrand; the publicity chairs Karen Church and Alan Said; and Asia liaison Li Chen. We would like to thank all the senior and regular Program Committee members as well as additional reviewers for their diligent work. We would also like to thank our sponsors for their generous support.

We hope that you will find this program interesting and thought-provoking and that the conference will provide you with a valuable opportunity to share ideas and have interesting conversations with leading researchers and practitioners in the field. In addition to attending the conference, we hope you will use *recommendations* from local attendees or past travelers to explore other opportunities to enjoy the lovely city of Dublin.

Padraig Cunningham
and Neil Hurley
ACM RecSys 2012 General Chairs

Ido Guy and
Sarabjot Singh Anand
ACM RecSys 2012 Program Chairs

Yehuda Koren
and Daniel Tunkelang
ACM RecSys 2012 Industry Chairs

Table of Contents

Industry Session 1

Industry Session 2

Short Papers

Workshop Outlines

6th ACM Conference on Recommender Systems Organization

General Chairs:	Pádraig Cunningham *(University College Dublin, Ireland)*
	Neil Hurley *(University College Dublin, Ireland)*
Program Chairs:	Ido Guy *(IBM Haifa Research Laboratory, Israel)*
	Sarabjot Singh Anand *(University of Warwick, England)*
Workshop Chairs:	Jill Freyne *(CSIRO, Australia)*
	Pearl Pu *(EPFL, Switzerland)*
Doctoral Symposium Chairs:	Alexander Felfernig *(TU Graz, Austria)*
	Michael O'Mahony *(University College Dublin, Ireland)*
Industry Chairs:	Yehuda Koren *(Google, Israel)*
	Daniel Tunkelang *(LinkedIn, USA)*
Tutorial Chairs:	Alexander Tuzhilin *(Stern School of Business, New York University, USA)*
	Werner Geyer *(IBM Research, USA)*
Demo Chairs:	Elizabeth Daly *(IBM Research, Ireland)*
	Michael Ekstrand *(University of Minnesota, USA)*
Publicity Chairs:	Karen Church *(Telefonica Research, Spain)*
	Alan Said *(Technische Universität Berlin, Germany)*
Asia Liaison:	Li Chen *(Hong Kong Baptist University, China)*
Senior Program Committee:	Gediminas Adomavicius *(University of Minnesota, USA)*
	Xavier Amatriain *(Netflix, USA)*
	Liliana Ardissono *(Università di Torino, Italy)*
	Lawrence Bergman *(IBM T.J. Watson Research Center, USA)*
	Shlomo Berkovsky *(CSIRO, Australia)*
	Derek Bridge *(University College Cork, Ireland)*
	Peter Brusilovsky *(University of Pittsburgh, USA)*
	Robin Burke *(DePaul University, USA)*
	Martin Ester *(Simon Fraser University, Canada)*
	Alexander Felfernig *(TU Graz, Austria)*
	Jill Freyne *(CSIRO, Australia)*
	Gerhard Friedrich *(Alpen-Adria-Universität Klagenfurt, Austria)*
	Matthias Fuchs *(Mid Sweden University, Sweden)*
	Werner Geyer *(IBM Research, USA)*
	Jennifer Golbeck *(University of Maryland, USA)*
	Andreas Hotho *(University of Würzburg, Germany)*
	Alejandro Jaimes *(Yahoo! Research, Spain)*
	Dietmar Jannach *(TU Dortmund, Germany)*
	Alfred Kobsa *(University of California, Irvine, USA)*
	Yehuda Koren *(Google, Israel)*

Senior Program Committee (continued):

Benjamin Marlin *(University of British Columbia, Canada)*
Bamshad Mobasher *(DePaul University, USA)*
Michael Paul O'Mahony *(University College Dublin, Ireland)*
Paul Resnick *(University of Michigan, USA)*
Francesco Ricci *(Free University of Bozen-Bolzano, Italy)*
John Riedl *(University of Minnesota, USA)*
Lars Schmidt-Thieme *(University of Hildesheim, Germany)*
Giovanni Semeraro *(Università di Bari, Italy)*
Barry Smyth *(University College Dublin, Ireland)*
Domonkos Tikk *(Gravity R&D/Budapest University of Technology and Economics, Hungary)*
Marc Torrens *(Strands, Spain)*
Alexander Tuzhilin *(New York University, USA)*
Hannes Werthner *(TU Vienna, Austria)*
Markus Zanker *(Alpen-Adria-Universität Klagenfurt, Austria)*

Program Committee:

Linas Baltrunas *(Free University of Bozen-Bolzano, Italy)*
Xinlong Bao *(Google Inc., USA)*
Mathias Bauer *(Mineway, Germany)*
Smriti Bhagat *(Technicolor, USA)*
Iván Cantador *(Universidad Autónoma de Madrid, Spain)*
Pablo Castells *(Universidad Autónoma de Madrid, Spain)*
Oscar Celma *(Gracenote, Emeryville, USA)*
Li Chen *(Hong Kong Baptist University, China)*
Wen-Huang Cheng *(Academia Sinica, Taiwan)*
Zunping Cheng *(Oracle EMEA, Ireland)*
Paolo Cremonesi *(Politecnico die Milano, Italy)*
Elizabeth Daly *(IBM Cambridge Research Center, USA)*
Marco De Gemmis *(University of Bari, Italy)*
Ernesto De Luca *(Technical University of Berlin, Germany)*
Christian Desrosiers *(École Polytechnique de Montréal, Canada)*
Hendrik Drachsler *(Open University of the Netherlands, The Netherlands)*
Zeno Gantner *(Nokia)*
Derek Greene *(UCD, Ireland)*
Asela Gunawardana *(Microsoft Research, USA)*
Conor Hayes *(NUI Galway, Ireland)*
Tim Hussein *(University of Duisburg-Essen, Germany)*
Mohsen Jamali *(Simon Fraser University, Canada)*
Robert Jäschke *(University of Kassel, Germany)*
Alexandros Karatzoglou *(Telefonica Research, Spain)*
George Karypis *(University of Minnesota, USA)*
Benjamin Keller *(Eastern Michigan University, USA)*
Georgia Koutrika *(IBM Almaden Research Center, USA)*
Antonio Krüger *(Saarland University, Germany)*
Tsvi Kuflik *(University of Haifa, Israel)*
Paul Lamere *(The Echo Nest, USA)*
Daniel Lemire *(Université du Québec à Montréal, Canada)*

Program Committee (continued):   Pasquale Lops *(Università di Bari, Italy)*
Nikos Manouselis *(Agricultural University of Athens, Greece)*
Kevin McCarthy *(University College Dublin, Ireland)*
Sean McNee *(Attenex Corp., USA)*
Bashkar Meta *(Google Research, Zurich, Switzerland)*
Stuart Middleton *(University of Southampton, UK)*
Alexandros Nanopoulos *(University of Hildesheim, Germany)*
Olfa Nasraoui *(University of Louisville, USA)*
Wolfgang Nejdl *(L3S and University of Hannover, Germany)*
Barry O'Sullivan *(University College Cork, Ireland)*
Seung-Taek Park *(Samsung Electronics Co. Ltd, South Korea)*
Istvan Pilaszy *(Gravity R&D Ltd, Hungary)*
Naren Ramakrishnan *(Virginia Tech, USA)*
James Reilly *(Google Inc., USA)*
Steffen Rendle *(University of Konstanz, Germany)*
Hagee Roitman *(IBM Haifa Research Laboratory, Israel)*
Lior Rokach *(Ben-Gurion University, Israel)*
Inbal Ronen *(IBM Haifa Research Lab, Israel)*
Alan Said *(TU Berlin, Germany)*
Olga C. Santos *(aDeNu Research Group – UNED, Spain)*
Shilad Sen *(Macalester College, USA)*
Guy Shani *(Ben-Gurion University, Israel)*
Bracha Shapira *(Ben-Gurion University, Israel)*
Kostyantyn Shchekotykhin *(Alpen-Adria-Universität Klagenfurt, Austria)*
Markus Stolze *(Hochschule für Technik Rapperswil, Switzerland)*
Panagiotis Symeonidis *(Aristotle University Thessaloniki, Greece)*
Nava Tintarev *(University of Aberdeen, UK)*
Jun Wang *(University College London, UK)*
Mi Zhang *(Fudan University, China)*
Jiyong Zhang *(EPFL, Switzerland)*
Daniel Xiaodan Zhou *(University of Michigan, USA)*
Tingshao Zhu *(Graduate University of Chinese Academy of Sciences, Beijing, China)*

Additional reviewers:

Panagiotis Adamopoulos
Hyung-Il Ahn
David C. Anastasiu
Pierpaolo Basile
Sanat Kumar Bista
Annalina Caputo
Jilin Chen
Yi-Ling Chen
Evangelia Christakopoulou
Maciej Dabrowski
Tommaso Di Noia
Ernesto Diaz-Aviles
Eyal Dim
Stephan Doerfel
Lucas Drumond
Asmaa Elbadrawy
Philipp Fleiss
Christoph Freudenthaler
Arik Friedman
Mouzhi Ge
Fatih Gedikli
Jonathan Gemmell
Michele Gorgoglione
Josif Grabocka
Mark Graus
Negar Hariri
Benjamin Heitmann
Martin Hochmeister
Min-Chun Hu
Leo Iaquinta
Santosh Kabbur

Rasoul Karimi
Gilad Katz
Benjamin Kille
Bart Knijnenburg
Youngok Kwon
Jui-Hsin Lai
Carlos Lamsfus
Yin-Tzu Lin
Corrado Mencar
Piero Molino
Donn Morrison
Cataldo Musto
Fedelucio Narducci
Julia Neidhardt
Xia Ning
Owen Phelan
Nataliia Pobiedina
Rachael Rafter
Fergal Reid
Patrick Rodler
Shaghayegh Sahebi
Tom Schimoler
Jianqiang Shen
Yun-Chung Shen
Stefan Siersdorfer
Gabor Takacs
Eleftherios Tiakas
Amit Tiroshi
Violeta Vogel
Jingjing Zhang

RecSys 2012 Sponsors & Supporters

Sponsor:

In Cooperation with:

Supporters:

IBM Research

Online Controlled Experiments:
Introduction, Learnings, and Humbling Statistics

Ron Kohavi
Microsoft
One Microsoft Way
Redmond, WA 98052
ronnyk@microsoft.com

ABSTRACT

The web provides an unprecedented opportunity to accelerate innovation by evaluating ideas quickly and accurately using controlled experiments (e.g., A/B tests and their generalizations). Whether for front-end user-interface changes, or backend recommendation systems and relevance algorithms, online controlled experiments are now utilized to make data-driven decisions at Amazon, Microsoft, eBay, Facebook, Google, Yahoo, Zynga, and at many other companies. While the theory of a controlled experiment is simple, and dates back to Sir Ronald A. Fisher's experiments at the Rothamsted Agricultural Experimental Station in England in the 1920s, the deployment and mining of online controlled experiments at scale—thousands of experiments now—has taught us many lessons. We provide an introduction, share real examples, key learnings, cultural challenges, and humbling statistics.

This presentation will be available at http://www.exp-platform.com/Pages/2012RecSys.aspx

Categories and Subject Descriptors: G.3 [Probability and Statistics]: Experimental design

General Terms: Experimentation, Measurement

Keywords: A/B Testing, Controlled Experiments, Online Experiments

Conducting User Experiments in Recommender Systems

Bart P. Knijnenburg
Department of Informatics
University of California, Irvine
bart.k@uci.edu

ABSTRACT

There is an increasing consensus in the field of recommender systems that we should move beyond the offline evaluation of algorithms towards a more user-centric approach. This tutorial teaches the essential skills involved in conducting user experiments, the scientific approach to user-centric evaluation. Such experiments are essential in uncovering how and why the user experience of recommender systems comes about.

Categories and Subject Descriptors

H.1.2. [**Models and principles**]: User/Machine Systems–*software psychology*; H.5.2 [**Information Interfaces and Presentation**]: User Interfaces–*evaluation/methodology*; H.4.2. [**Information Systems Applications**]: Types of Systems–*decision support*

General Terms

Measurement, Experimentation, Human Factors, Standardization.

Keywords

User experiments, recommender systems, user experience, user-centric evaluation.

1. INTRODUCTION

From a methodological perspective, the evaluation of recommender systems has undergone an interesting development [8]. The recent focus on "user-centric" evaluation [4, 11] is inspired by the suggestion that higher accuracy does not always mean higher user satisfaction [9], and that the algorithm accounts for only a small part of the real-world relevance of a recommender system. Other aspects such as the presentation and interaction have a signification impact on the user experience [2, 5, 10].

To make inferences about the users' experience, we need to move beyond measuring their behavior, and measure their *subjective valuations* as well [6]. Moreover, as users' interaction with recommender systems is highly context-dependent [1, 3], personal and situational characteristics also need to be taken into account.

In Knijnenburg et al. [6] we present a framework for the user-centric evaluation of recommender systems that takes all these aspects into consideration (Figure 1). This framework can be used as a guideline for *user experiments* to reveal how and why the user experience of recommender systems comes about. However, conducting such experiments is a complex endeavor. How does one test whether a certain system aspect has a significant influence on e.g. users' satisfaction with the system? How does one measure a subjective concept like "user satisfaction" to begin with?

As recommender systems evaluation is becoming more user-centric, an increasing number of recommender systems researchers have to deal with these tricky questions. We thus often find papers without clearly defined hypotheses, lacking proper experimental manipulations, and/or testing a large number of seemingly unrelated effects using simple t-tests. Although it is encouraging to see user-centric evaluation efforts bloom, these evaluations are with notable exceptions not up to par with state-of-the-art research methods and statistical analyses. Whereas our RecSys 2011 short paper provides a pragmatic yet curtailed approach to user-centric evaluation [7] that fits such budding research efforts, this tutorial provides a more thorough treatment of user experiments as a mature scientific approach to the user-centric evaluation of recommender systems.

For the intended audience of recommender systems researchers wanting to get serious about user-centric evaluation, the tutorial covers all aspects involved in conducting user experiments: developing testable hypotheses, sampling participants from the right population, constructing useful experimental manipulations, robustly measuring behavior and subjective valuations, and analyzing the results using modern statistical methods.

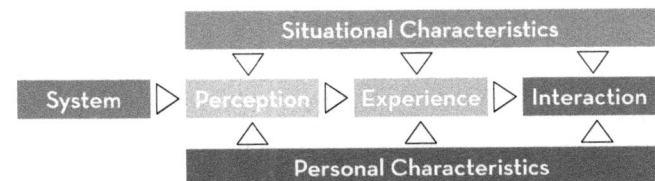

Figure 1. Framework for the user-centric evaluation of recommender systems, adapted from Knijnenburg et al. [6]

2. TOPICS

2.1 Hypotheses

Trying to find out whether your recommender system is "good" for users is not a workable goal in user-centric evaluation. Instead, researchers need to operationalize a set of hypotheses: testable predictions about how a recommender system influences the user. This part of the tutorial teaches the principle of *ceteris paribus* as a way to single out the effects of specific system aspects on the user experience.

2.2 Participants

Many researchers believe that a recommender system can be evaluated with a handful of willing colleagues or students. This part of the tutorial instead makes the case that systems should be tested with participants sampled from the target population (i.e. the target audience of the system). It also discusses the typical sample size needed to allow for statistical inferences (which is usually much larger than a handful).

2.3 Testing A vs. B

Although it may make intuitive sense to test the user experience of a recommender system in a holistic fashion, such "test everything at once" evaluations cannot discern the specific causes of the user experience. Good user experiments instead try to single out the effects of specific aspects of the system. To single out the effect of an aspect, one needs to *manipulate* that aspect by creating two or more *conditions* (versions of the aspect). The tutorial covers adequate manipulations, and discusses the pros and cons of between-subjects (i.e. each participants gets to see only one condition) and within-subjects (i.e. each participant gets to see all conditions) experiments.

2.4 Measurement

Measuring user behavior is insufficient to make inferences about the user experience. Behavior is highly context-dependent and difficult to interpret. Subjective valuations, gathered through questionnaires, typically provide a more robust measurement of the users' experience with the recommender system. Moreover, subjective evaluations are better predictors of longer-term system goals such as adoption and user retention.

This part of the tutorial teaches the art of creating questionnaire items, typically presented as statements to which users can agree or disagree on a 5- or 7-point scale. Currently, researchers typically use one such questionnaire item for each concept (e.g. satisfaction, perceived control, understandability) that they want to measure. This tutorial instead makes the case for creating multi-item measurements for each concept. It presents *factor analysis* as a statistical method to turn such multi-item measurements into robust unidimensional scales.

2.5 Analysis

Statistical analysis of user-centric research typically involves correlations, t-tests and linear regressions. This part of the tutorial presents *structural equation models* as a more sophisticated and modern statistical method to make causal inferences. Structural equation models can test complex causal structures, such as whether a certain manipulation (e.g. a different algorithm) has a significant influence on users' perceptions (e.g. perceived recommendation quality), and whether this perception in turn influences their experience (e.g. system effectiveness), and behavior (e.g. item ratings).

2.6 Evaluation framework

The tutorial concludes by returning to the Knijnenburg et al. [6] evaluation framework. This framework for the user-centric evaluation of recommender systems can be used to develop causal hypotheses, to select and construct subjective measures, and to integrate new and existing user-centric research on recommender systems.

3. CONCLUSION

If you work on recommender systems—as a system developer, an algorithms researcher, or a user interface designer—user-centric evaluations are the way to go. This tutorial presents user experi-

ments as an essential skill in uncovering *how* and *why* the user experience of recommender systems comes about.

4. REFERENCES

[1] Adomavicius, G. and Tuzhilin, A. 2011. Context-Aware Recommender Systems. *Recommender Systems Handbook*. F. Ricci, L. Rokach, B. Shapira, and P.B. Kantor, eds. Springer US. 217–253.

[2] Bollen, D., Knijnenburg, B.P., Willemsen, M.C. and Graus, M. 2010. Understanding choice overload in recommender systems. *Proceedings of the fourth ACM conference on Recommender systems* (Barcelona, Spain, 2010), 63–70.

[3] Knijnenburg, B.P., Reijmer, N.J.M. and Willemsen, M.C. 2011. Each to his own: how different users call for different interaction methods in recommender systems. *Proceedings of the fifth ACM conference on Recommender systems* (Chicago, IL, 2011), 141–148.

[4] Knijnenburg, B.P., Schmidt-Thieme, L. and Bollen, D.G.F.M. 2010. Workshop on user-centric evaluation of recommender systems and their interfaces. *Proceedings of the fourth ACM conference on Recommender systems* (New York, NY, USA, 2010), 383–384.

[5] Knijnenburg, B.P. and Willemsen, M.C. 2009. Understanding the effect of adaptive preference elicitation methods on user satisfaction of a recommender system. *Proceedings of the third ACM conference on Recommender systems* (New York, NY, 2009), 381–384.

[6] Knijnenburg, B.P., Willemsen, M.C., Gantner, Z., Soncu, H. and Newell, C. 2012. Explaining the user experience of recommender systems. *User Modeling and User-Adapted Interaction*. 22, 4-5 (2012), 441–504.

[7] Knijnenburg, B.P., Willemsen, M.C. and Kobsa, A. 2011. A pragmatic procedure to support the user-centric evaluation of recommender systems. *Proceedings of the fifth ACM conference on Recommender systems* (New York, NY, USA, 2011), 321–324.

[8] Konstan, J. and Riedl, J. 2012. Recommender systems: from algorithms to user experience. *User Modeling and User-Adapted Interaction*. 22, 1 (2012), 101–123.

[9] McNee, S.M., Riedl, J. and Konstan, J.A. 2006. Being accurate is not enough. *CHI '06 extended abstracts on Human factors in computing systems - CHI '06* (Montreal, Quebec, Canada, 2006), 1097–1101.

[10] Pu, P., Chen, L. and Hu, R. 2012. Evaluating recommender systems from the user's perspective: survey of the state of the art. *User Modeling and User-Adapted Interaction*. 22, 4 (2012), 317–355.

[11] Willemsen, M., Bollen, D. and Ekstrand, M. 2011. UCERSTI 2: second workshop on user-centric evaluation of recommender systems and their interfaces. *Proceedings of the fifth ACM conference on Recommender systems* (New York, NY, USA, 2011), 395–396.

Personality-based Recommender Systems: An Overview

Maria Augusta S. N. Nunes
Universidade Federal de Sergipe
Departamento de Computação
C. universitária- São Cristovão-SE-Brazil
+557988555676

gutanunes@gmail.com

Rong Hu
Human Computer Interaction Group
Swiss Federal Institute of Technology in Lausanne
CH-1015, Lausanne, Switzerland
+41 216931327

rong.hu@epfl.ch

ABSTRACT
Personality is a critical factor which influences people's behavior and interests. There is a high potential that incorporating users' characteristics into recommender systems could enhance recommendation quality and user experience. The goal of this tutorial is to give an overview of personality-based recommender systems and discuss challenges and possible research directions in this topic.

Categories and Subject Descriptors
H.3.3 [**Information Storage and Retrieval**]: Information Search and Retrieval – *Information filtering*; H.1.2 [**Models and Principles**]: User/Machine Systems – *human information processing*

General Terms
Algorithms, Experimentation, Performance, Human Factors

Keywords
Personality-based recommender system

1. INTRODUCTION
Over the last 20 decade, recommender systems have obtained great success as an intelligent information system to help deal with the information overload problem, especially in the field of e-commerce [1]. Prior studies on recommender systems mainly consider leveraging user preference information (e.g., user ratings, users' past behavior), item properties (e.g., price) , or user demographic information (e.g., gender) [1]. For example, collaborative filtering approaches first build a model from a user's past behavior (e.g., items previously purchased and/or ratings given to those items), then use that model to predict items (or ratings for items) that the user may have an interest in by considering the opinions of other like-minded users. Content-based filtering approaches utilize a series of discrete characteristics of an item to recommend items with similar properties. In recent years, other information (e.g., contexts, tags and social information) has also taken into account in the implementation of recommender system [5]. However, few studies have considered addressing the recommendation problem from the angel of users' psychological characteristics.

Personality can be defined as a set of characteristics possessed by a person that uniquely influences his or her cognitions, emotions, motivations, and behaviors in various situations. That is, personality is a critical factor which influences

how people make their decisions. People with similar personality characteristics are more likely to have similar interests and preferences. For example, extraverted people usually like the music genre with positive emotions [4]. The effect of personality has on human behavior has been widely studied in psychology, marketing and behavioral economics.

Personality could help us explain why we prefer one option to the other. It is implied that incorporating personality into recommender systems could help understand the reasons that essentially determine user preferences. Currently, some researchers have considered incorporating personality aspects into recommender systems to personalize recommendations and enhance both recommendation quality and user experience [2, 3]. Furthermore, it could be noticed that commercial recommenders have been starting to implement personality-based recommendation engines in their systems. The successful applications of personality-based recommendation technologies include social matching systems (e.g., online dating systems), gift recommenders, music recommenders and movie recommenders. It is an emerging research field. There are still many challenges, which have not been addressed yet. This tutorial will help catch more attentions on this topic from both academia and industry and form a platform for their discussion and cooperation.

2. OUTLINE
This tutorial will give an overview of personality-based recommender systems (PBRS) covering theories and practices, as well as discuss challenges and possible solutions in implementing personality-based recommender systems.

First, the tutorial will give an introduction to the related theories in psychology, human decision making and affective computing. In particular, the tutorial will present in detail personality theory as well as how exactly personality data look like, how to extract them, how to represent them and how to standardize those definitions to be used as input recommender data towards to improve recommendations.

Next, the tutorial will present the state-of-the-art technologies and applications of personality-based recommender systems and show the examples of existing tools and systems through live demos. Furthermore, user experience issues in personality-based recommender systems will be discussed by showing the results of real user experiments.

The tutorial will conclude with a discussion of challenges in research and practice and an outline of research directions.

The detailed outline of this tutorial is given as follows.

1. Basic Theories and Knowledge
   - A brief introduction to affective computing.

- Human decision making & Computer decision making.
- A brief introduction to personality theory.
- Computational personality acquisition methods.
- Personality profile representation and standardization.

2. Personality-based Recommender Systems

- A review of the state-of-the-art of personality-based recommender technologies and systems.
- Advantages and disadvantages of personality-based recommendation technologies.
- User perception issues.
- Challenges and new research directions.

3. INSTRUCTORS' BIOGRAPHIES

Maria Augusta S. N. Nunes: in 2008 she finished her PhD in France, her thesis [3] was the starting point to the Personality-based Recommender Systems (as described in many papers published in 2011 at ACM, IEEE, UMUAI). (Available as a book at Amazon or for download at http://professores.dcomp.ufs.br/~gutanunes/hp/publications/Tese.zip. Since her PhD her research focus are Affective Computing and how to model and represent the Human Psychological aspects, mainly personality, in computers aiming improve the personalization of information, products and services for humans during their interaction in e-commerce environment, for instance. From 2009 she is an associate professor and researcher at DCOMP/Universidade Federal de Sergipe (http://professores.dcomp.ufs.br/~gutanunes/). Her more recent projects include how to extract and store human Personality in order to motivate and personalize services in Recommender System considering mainly the user Psychological aspects. In the last years she has been writing many books, book chapters and papers about the use of Affective Computing in order to motivate and personalize information for people. In 2011 we received 3 awards in projects which considering aspects such as accessibility, Recommendation and Personality Traits. Curriculum available at (http://lattes.cnpq.br/9923270028346687). The Personality Portal with her research is available at (http://www.personalityresearch.com.br/).

Rong Hu: She obtained her PhD from the Human Computer Interaction (HCI) group at École Polytechnique Fédérale de Lausanne (EPFL) in 2012. Her research interests concentrate on how to incorporate psychological factors (in particular, personality) into recommender systems to improve recommendation quality and user experience. She has published many papers related to personality-based recommender systems at top-tier international conferences, such as IUI, RecSys and UMAP. She is working as a post-doc in the HCI group at EPFL since March 2012.

4. REFERENCES

[1] Adomavicius, G. and Tuzhilin, A. 2005. Toward the Next Generation of Recommender Systems: A Survey of the State-of-the-Art and Possible Extensions. *IEEE Trans. on Knowl. and Data Eng. 17*, 6, 734-749.

[2] Hu, R. and Pu, P. 2010. A Study on User Perception of Personality-Based Recommender Systems. *User Modeling, Adaptation, and Personalization*, De, B., Kobsa, A. and Chin, D., eds. 6075, Springer Berlin / Heidelberg, 291-302.

[3] Nunes, M.A.S.N. 2009. *Recommender Systems based on Personality Traits: Could human psychological aspects influence the computer decision-making process?* VDM Verlag, Berlin.

[4] Rentfrow, P.J. and Gosling, S.D. 2003. The do re mi's of everyday life: the structure and personality correlates of music preferences. *J Pers Soc Psychol. 84*, 6 (Jun), 1236-1256.

[5] Ricci, F., Rokach, L., Kantor, P.B. and Shapira, B. 2011. *Recommender Systems Handbook*. Springer.

Building Industrial-scale Real-world Recommender Systems

Xavier Amatriain
Netflix
xamatriain@netflix.com

Categories and Subject Descriptors: H.3 [Information Search and Retreival]: Information Filtering

General Terms: Algorithms, Experimentation

Keywords: Recommender Systems, Applications

1. INTRODUCTION

In 2006, Netflix announced the Netflix Prize, a machine learning and data mining competition for movie rating prediction. We offered $1 million to whoever improved the accuracy of our existing system called Cinematch by 10%. We conducted this competition to find new ways to improve the recommendations we provide to our members, which is a key part of our business. However, we had to come up with a proxy question that was easier to evaluate and quantify: the root mean squared error (RMSE) of the predicted rating.

A year into the competition, the Korbell team won the first Progress Prize with an 8.43% improvement. They reported more than 2000 hours of work in order to come up with the final combination of 107 algorithms that gave them this prize. And, they gave us the source code. We looked at the two underlying algorithms with the best performance in the ensemble. To put these algorithms to use, we had to work to overcome some limitations, for instance that they were built to handle 100 million ratings, instead of the more than 5 billion that we have, and that they were not built to adapt as members added more ratings. But once we overcame those challenges, we put the two algorithms into production, where they are still used as part of our recommendation engine.

You might be wondering what happened with the final Grand Prize ensemble that won the $1M two years later. This is a truly impressive compilation and culmination of years of work, blending hundreds of predictive models to finally cross the finish line. We evaluated some of the new methods offline but the additional accuracy gains that we measured did not seem to justify the engineering effort needed to bring them into a production environment.

This example highlights the fact that, besides improving offline metrics such as the RMSE, recommender systems need to take into account other practical issues such as scalability or deployment. In this tutorial, we go over some of those practical issues that many times are as important as the theory, if not more, in order to build an industrial-scale real-world recommender system.

2. BEYOND RATING PREDICTION

We have discovered through the years that there is tremendous value in incorporating recommendations to personalize as much as possible. Personalization starts on our homepage, which consists of groups of videos arranged in horizontal rows. Each row has a title that conveys the intended meaningful connection between the videos in that group. Most of our personalization is based on the way we select rows, how we determine what items to include in them, and in what order to place those items.

Take as a first example the Top 10 row: this is our best guess at the ten titles you are most likely to enjoy. Of course, when we say "you", we really mean everyone in your household. It is important to keep in mind that Netflix' personalization is intended to handle a household that is likely to have different people with different tastes. That is why when you see your Top10, you are likely to discover items for dad, mom, the kids, or the whole family. Even for a single person household we want to appeal to your range of interests and moods. To achieve this, in many parts of our system we are not only optimizing for accuracy, but also for **diversity**.

Another important element in personalization is **awareness**. We want members to be aware of how we are adapting to their tastes. This not only promotes trust in the system, but encourages members to give feedback that will result in better recommendations. A different way of promoting trust with the personalization component is to provide **explanations** as to why we decide to recommend a given movie or show. We are not recommending it because it suits our business needs, but because it matches the information we have from you: your explicit taste preferences and ratings, your viewing history, or even your friends' recommendations.

Some of the most recognizable personalization in our service is the collection of "genre" rows. Each row represents 3 layers of personalization: the choice of genre itself, the subset of titles selected within that genre, and the ranking of those titles. As with other personalization elements, **freshness** and diversity is taken into account when deciding what genres to show from the thousands possible.

Similarity is also an important source of personalization in our service. Think of similarity in a very broad sense; it can be between movies or between members, and can be in multiple dimensions such as metadata, ratings, or viewing data. Furthermore, these similarities can be blended and used as features in other models. Similarity is used in multiple contexts, for example in response to a member's action such as searching or adding a title to the queue.

The goal of recommender systems is to present a number

of attractive items for a person to choose from. This is usually accomplished by selecting some items and sorting them in the order of expected enjoyment (or utility). Since the most common way of presenting recommended items is in some form of list, such as the various rows on Netflix, we need an appropriate **ranking model** that can use a wide variety of information to come up with an optimal ranking of the items for each of our members.

If you are looking for a ranking function that optimizes consumption, an obvious baseline is item popularity. The reason is clear: on average, a member is most likely to watch what most others are watching. However, popularity is the opposite of personalization: it will produce the same ordering of items for every member. Thus, the goal becomes to find a personalized ranking function that is better than item popularity, so we can better satisfy members with varying tastes. One obvious way to approach this is to use the member's predicted rating of each item as an adjunct to item popularity. At this point, we are ready to build a ranking prediction model using these two features. For example, we could use a simple linear function of the form form $rank(u,v) = w1p(v) + w2r(u,v) + b$, where u=user, v=video item, p=popularity and r=predicted rating

Once we have such a function, we can pass a set of videos through our function and sort them in descending order according to the score. You might be wondering how we can set the weights w1 and w2 in our model. In other words, in our simple two-dimensional model, how do we determine whether popularity is more or less important than predicted rating? There are at least two possible approaches to this. You could sample the space of possible weights and let the members decide what makes sense after many A/B tests. This procedure might be time consuming and not very cost effective. Another possible answer involves formulating this as a machine learning problem: select positive and negative examples from your historical data and let a machine learning algorithm learn the weights that optimize your goal. This family of machine learning problems is known as "Learning to rank" and is central to application scenarios such as search engines or ad targeting.

As you might guess, apart from popularity and rating prediction, we have tried many other features. Some have shown no positive effect while others have improved our ranking accuracy tremendously. On the other hand, many supervised classification methods can be used for ranking. There is no easy answer to choose which model will perform best in a given ranking problem. The simpler your feature space is, the simpler your model can be. But it is easy to get trapped in a situation where a new feature does not show value because the model cannot learn it. Or, the other way around, to conclude that a more powerful model is not useful simply because you don't have the feature space that exploits its benefits.

3. SYSTEMS & ARCHITECTURE

When we design a Recommender System, we need to take into account under what conditions it will be operated and deployed. As we saw with the outcome of the Netflix Prize, issues such as scalability need to be considered. Another important factor that will determine the feasibility of an approach is the overall system **latency** measured as the time elapsed since the user gives us some feedback to the time the UI will present a different recommendation that is influ-

enced by that input. The best algorithm will be useless if the system is unable to respond to user actions in a timely fashion.

In Netflix, we approach scalability and latency by optimizing our systems in several ways. We take advantage of the scalability of the cloud using Amazon Web Services. We also use Hadoop for distributed data processing, and Cassandra for efficient distributed storage. Our architecture follows an "offline-online" pattern to maximize data throughput while minimizing latency. Heavy data computation jobs that are not very sensitive to latency are processed offline. These jobs can be triggered periodically, or in response to user events. On the other hand, jobs that depend on real-time signals are processed online by taking advantage of the previously computed results. A simple an effective for some of these cases is to design filters that immediately filter out items such as a movie the user just watched. But, not everything can be solved with a simple filter. And, it is important to understand how things like filters can impact algorithms.

4. CONSUMER DATA SCIENCE

The abundance of source data, measurements and associated experiments allow us to operate a data-driven organization. Netflix has embedded this aproach into its culture since the company was founded, and we have come to call it Consumer (Data) Science. We strive for an innovation culture that allows us to evaluate ideas rapidly, inexpensively, and objectively. And, once we test something we want to understand why it failed or succeeded.

So, how does this work in practice? It is a slight variation over the traditional scientific process called A/B testing (or bucket testing): (1) Start with a hypothesis: Algorithm/feature/design X will increase member engagement with our service and ultimately member retention. (2) Design a test: Develop a solution or prototype. Think about dependent & independent variables, control, and significance. (3) Execute the test. (4) Let data speak for itself

When we execute A/B tests, we track many different metrics. But we ultimately trust member engagement (e.g. hours of play) and retention. Tests usually have thousands of members and anywhere from 2 to 20 cells exploring variations of a base idea. We typically have scores of A/B tests running in parallel. A/B tests let us try radical ideas or test many approaches at the same time, but the key advantage is that they allow our decisions to be data-driven.

An interesting follow-up question that we have faced is how to integrate our machine learning approaches into this data-driven A/B test culture at Netflix. We have done this with an offline-online testing process that tries to combine the best of both worlds. The offline testing cycle is a step where we test and optimize our algorithms prior to performing online A/B testing. To measure model performance offline we track multiple metrics used in the machine learning community: from ranking measures such as normalized discounted cumulative gain, mean reciprocal rank, or fraction of concordant pairs, to classification metrics such as accuracy, precision, recall, or F-score. We also use the famous RMSE from the Netflix Prize or other more exotic metrics to track different aspects like diversity. We keep track of how well those metrics correlate to measurable online gains in our A/B tests. However, since the mapping is not perfect, offline performance is used only as an indication to make informed decisions on follow up tests.

The Challenge of Recommender Systems Challenges

Alan Said
Technische Universität Berlin
Berlin, Germany
alan@dai-lab.de

Domonkos Tikk
Gravity R&D
Budapest, Hungary
domonkos.tikk@
gravityrd.com

Andreas Hotho
Universität Würzburg
Würzburg, Germany
hotho@informatik.
uni-wuerzburg.de

ABSTRACT

Recommender System Challenges such as the Netflix Prize, KDD Cup, etc. have contributed vastly to the development and adoptability of recommender systems. Each year a number of challenges or contests are organized covering different aspects of recommendation. In this tutorial and panel, we present some of the factors involved in successfully organizing a challenge, whether for reasons purely related to research, industrial challenges, or to widen the scope of recommender systems applications.

Categories and Subject Descriptors

D.2.8 [**Software Engineering**]: Metrics - complexity measures, performance measures; H.3.3 [**Information Storage and Retrieval**]: Information Search and Retrieval - Information filtering, Retrieval models, Selection process; H.3.4 [**Information Technology and Systems Applications**]: Decision support; H.3.5 [**Online Information Services**]: Data Sharing; H.5.1 [**Multimedia Information Systems**]: Evaluation/methodology

General Terms

Algorithms, Design, Experimentation, Human Factors, Measurement, Performance

Keywords

benchmarks, evaluation, challenge, contest

1. INTRODUCTION

With the launch of the Netflix Prize[1] in 2006, recommender systems challenges, or contests, allowed the everyman to access large-scale, real-world user data, and try to beat a commercial recommender system. Thousands of people participated in the challenge over its three year run.

---

[1] http://www.netflixprize.com

Subsequently, other challenges have been organized, both by the industry[2], as well as academia[3], and today there are several challenges (large and small) organized on a yearly basis. This tutorial gives an insight into the organization of a challenge, from planning and preparation of data, to the execution and final evaluation of submitted approaches.

2. OVERVIEW

The tutorial is organized into three parts, each covering a different topic in the concept of recommender systems challenges. In the first section (Introduction), an introduction to previous and current recommender systems challenges is given. Challenge-related topics are introduced and discussed, e.g. datasets, evaluation details with specific focus on which type of aspects should be evaluated, e.g.

- Qualitative requirements - include recommendation-related features such recommendation quality, satisfaction, etc.

- Technical requirements - include aspects related to the recommendation setting, such as scalability, robustness and data constraints

The second part of the tutorial (Preparation, Execution & Evaluation of a challenge) covers the practicalities involved in the process of organizing a challenge. Topics such a evaluation types (offline vs. online), anonymization of data are discussed. In the final part of the tutorial (Conclusion) the concept of the recommender system challenge is revisited to sum up the most important aspects of the realization of a recommender systems challenge, e.g. benchmarking, data, user-centricity.

After the tutorial, a panel on recommender system challenges is held to discuss relevant topics and answer any question from the audience.

3. OBJECTIVES

The objectives of this tutorial is to elicit more interest in the concept of challenges, whether as an organizer or participant. Recommender systems challenges have undoubtedly brought the related research forward. By organizing a challenge, any problem can be brought to light and entice interest among those who would have not approached the problem otherwise.

---

[2] http://overstockreclabprize.com
[3] e.g. the CAMRa Challenge http://www.camrachallenge.com or the ECML/PKDD Discovery Challenge - specifically http://www.kde.cs.uni-kassel.de/ws/rsdc08/ and http://www.kde.cs.uni-kassel.de/ws/dc09/

4. OUTLINE

The tutorial starts with a presentation of challenge-related concepts and concludes with a panel discussion.

- Introduction
  - History
  - Why recommender challenges are important?
  - Requirements of recommender systems to be evaluated
- Execution
  - Evaluation methods: offline and online
  - Dataset preparation
  - Organization
  - Evaluation
    * Qualitative assessment: error rate, accuracy, ranking-based metrics
    * Technical assessment: scalability, robustness, real-time responding
- Conclusion
- Panel discussion

5. ORGANIZERS

Alan Said is a postgraduate researcher at the Competence Center for Information Retrieval and Machine Learning at the Distributed Artificial Intelligence (DAI) Lab of Technische Universität Berlin. He is working in the field of recommender systems, focusing on recommender system evaluation, context-aware and hybrid recommender systems. He has been a co-chair of the Challenges on Context-Aware Movie Recommendation (CAMRa) held in conjunction with ACM RecSys in 2010 and 2011, and the 2012 RecSysChallenge held in conjunction with ACM RecSys 2012.

Domonkos Tikk is the Chief Scientific Officer at Gravity R&D Inc., a recommender solution vendor company. Domonkos obtained his PhD in 2000 in computer science from Budapest University of Technology and Economics. He has been working on machine learning and data and text mining topics in the last decade. His team, Gravity, participated at the Netflix Prize challenge, and was a leader of the The Ensemble team finished tied at the first position of the challenge. The team members founded the company Gravity to exploit the results achieved in Netflix Prize. Domonkos published actively in the field of recommender systems, co-authored about 20 papers in the last years. He also acted as the co-chair of at the recommender system related KDD-Cup 2007, RecsysChallange 2012 and RecSys Doctoral Symposium in 2011.

Andreas Hotho is a professor at the University of Würzburg. He holds a Ph.D. from the University of Karlsruhe, where he worked from 1999 to 2004 at the Institute of Applied Informatics and Formal Description Methods (AIFB) in the areas of text, data, and web mining, semantic web and information retrieval. He earned his Master's Degree in information systems from the University of Braunschweig in 1998. From 2004 to 2009 he was a senior researcher at the University of Kassel. He joined the L3S in 2011. Since 2005 he has been leading the development of the social bookmark and publication sharing platform BibSonomy. Andreas Hotho has published over 90 articles in journals and at conferences, co-edited several special issues and books, and co-chaired several workshops, e.g. the Workshop on Recommender Systems and the Social Web in 2011 and 2012 held in conjunction with ACM RecSys and the ECML PKDD Discovery Challenge in 2008 and 2009. He worked as a reviewer for journals and was a member of many international conferences and workshops program committees. His research focuses on the combination of data mining, information retrieval and the semantic web. Further, he is interested in the analysis of social media systems, in particular folksonomies, tagging, and sensor data emerging trough ubiquitous and social activities. As the World Wide Web is one of his main application areas, his research contributes to the field of web science.

Multiple Objective Optimization in Recommender Systems

Mario Rodriguez
LinkedIn Corp.
Mountain View, CA
mrodriguez@linkedin.com

Christian Posse
LinkedIn Corp.
Mountain View, CA
cposse@linkedin.com

Ethan Zhang
LinkedIn Corp.
Mountain View, CA
ezhang@linkedin.com

ABSTRACT

We address the problem of optimizing recommender systems for multiple relevance objectives that are not necessarily aligned. Specifically, given a recommender system that optimizes for one aspect of relevance, *semantic matching* (as defined by any notion of similarity between source and target of recommendation; usually trained on CTR), we want to enhance the system with additional relevance signals that will increase the utility of the recommender system, but that may simultaneously sacrifice the quality of the semantic match. The issue is that semantic matching is only one relevance aspect of the utility function that drives the recommender system, albeit a significant aspect.

In talent recommendation systems, job posters want candidates who are a good match to the job posted, but also prefer those candidates to be open to new opportunities. Recommender systems that recommend discussion groups must ensure that the groups are relevant to the users' interests, but also need to favor active groups over inactive ones. We refer to these additional relevance signals (job-seeking intent and group activity) as *extraneous features*, and they account for aspects of the utility function that are not captured by the semantic match (i.e. post-CTR downstream utilities that reflect engagement: time spent reading, sharing, commenting, etc). We want to include these extraneous features into the recommendations, but we want to do so while satisfying the following requirements: 1) we do not want to drastically sacrifice the quality of the semantic match, and 2) we want to quantify exactly how the semantic match would be affected as we control the different aspects of the utility function. In this paper, we present an approach that satisfies these requirements.

We frame our approach as a general constrained optimization problem and suggest ways in which it can be solved efficiently by drawing from recent research on optimizing non-smooth rank metrics for information retrieval. Our approach features the following characteristics: 1) it is model and feature agnostic, 2) it does not require additional la-

beled training data to be collected, and 3) it can be easily incorporated into an existing model as an additional stage in the computation pipeline. We validate our approach in a revenue-generating recommender system that ranks billions of candidate recommendations on a daily basis and show that a significant improvement in the utility of the recommender system can be achieved with an acceptable and predictable degradation in the semantic match quality of the recommendations.

Categories and Subject Descriptors

H.3.3 [**Information Search and Retrieval**]: Information Filtering

Keywords

Recommender systems, multiple objective optimization

1. INTRODUCTION

In designing recommender systems, we often have to balance multiple competing objectives. An example scenario can be drawn from a revenue-generating product at LinkedIn called *TalentMatch*, in which the recommender system, triggered by a job posted on the site, scours the entire member database to find the best candidates for the job. Those receiving the recommendations, job posters, want the candidates recommended to be a good fit for the job, but also prefer that the candidates be open to pursuing new opportunities. More specifically, a job poster would rather be recommended a candidate who is a *great* match for the job and also happens to be looking to change jobs, than the *best* match who happens to not be interested in exploring new opportunities. On the other hand, recommending a candidate who will certainly take the job if the offer was made, but who is not a good match for the job, will negatively affect the experience of the job poster. Therefore, given a ranking of candidates according to how well they match a given job and a ranking of candidates with regards to their job-seeking intent, the challenge is to combine both rankings into a final ranking that is optimal with regards to a given utility function.

In most recommender systems, there is a utility function to be maximized: relevant engagement. In *TalentMatch*, relevant engagement is a multi-faceted objective: a) the job poster decides to purchase the set of candidate recommendations based on a snippet of information for each of the candidates (see Figure 1), b) the job poster decides to initiate communication with each of the recommended candidates in

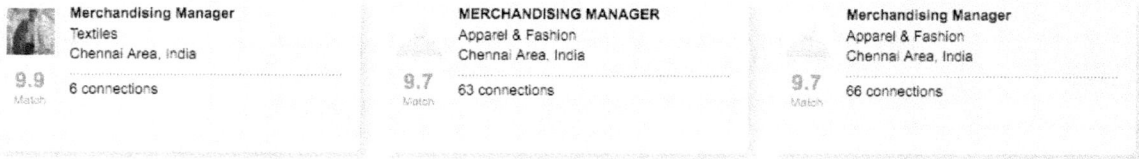

Merchandising Manager	MERCHANDISING MANAGER	Merchandising Manager
Textiles	Apparel & Fashion	Apparel & Fashion
Chennai Area, India	Chennai Area, India	Chennai Area, India
9.9 Match 6 connections	**9.7** Match 63 connections	**9.7** Match 66 connections

Figure 1: Sample *TalentMatch* recommendation snippet shown to a job poster. We only present the result set snippet to a job poster if we have at least 6 candidates in the result set.

the purchased set, and c) each of the candidates contacted respond in a favorable fashion to the job poster.

In the *TalentMatch* system, the semantic model computes the probability that the feature vector representing the member and the feature vector representing the job are a good match. The model does this by computing similarities between subsets of the member's feature vector and semantically related subsets of the job's feature vector. The various similarities in this vector are then weighted by training against a given CTR metric using a supervised learning algorithm. We refer to the features used to generate the similarity vector as *semantic features*, a concrete example being the job description in the job posting and the job description of the member's current position. In this case, the semantic features being compared are explicit, however, it may also be the case that the semantic features are latent (as in matrix factorization approaches to recommender systems). *Extraneous features*, on the other hand, exist only in the entity being recommended, not in the entity being recommended to. An example of an extraneous feature would be the job-seeking intent.

The job-seeking intent of each candidate is generated using another model, which for purposes of this paper, can be treated as a black-box that takes as input a candidate member and based on that member's data (e.g. activity on the site and profile information), outputs a job-seeking propensity score and a probabilistic assignment to each of the job-seeking intent categories: *active, passive, and non-job-seeker*. Many members who do not self-identify as job-seekers on the site actually display job-seeking behavior and characteristics. Therefore, we can estimate job-seeking intent for every member of the site. Though only a proxy, job-seeking intent is a very good indicator of the likelihood with which a member contacted regarding a job opportunity will respond favorably (which is one of the aspects of the utility function in *TalentMatch*). The other two aspects of the utility function (purchase rate of candidate recommendations and likelihood that a job poster will communicate with the purchased recommendations) are accounted for by the semantic model. Intuitively, increasing the number of individuals with high job-seeking intent (those classified as *active* or *passive*) in the top-K recommendations, without drastically sacrificing the semantic match of the recommendations, should increase the utility of the *TalentMatch* recommender system by connecting job posters with candidates who will engage with them. Figure 2 gives a high-level overview of the relevant system components.

Compounding the issue of multi-faceted objectives is the fact that in live production systems, models often need to evolve in a progressive fashion: the model may have initially optimized for only one aspect of relevant engagement (e.g. purchase rate of candidate recommendations) and it would

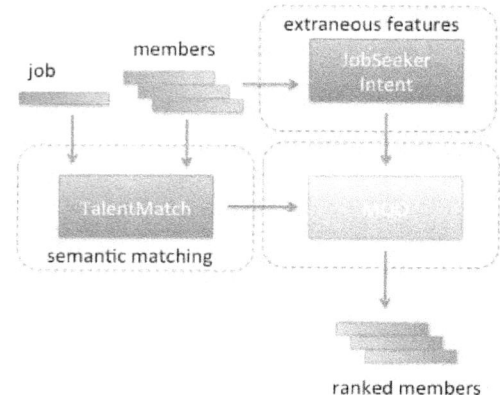

Figure 2: *TalentMatch* ranks all members according to their semantic match to a given job. The semantic score is combined with extraneous features, e.g. job-seeking intent, in the Multiple Objective Optimization (MOO) component, which outputs the final recommendation ranking.

be preferable to improve the model incrementally (as soon as a new feature like job seeking intent becomes available), rather than waiting for a complete redesign and development lifecycle of a new model. The incrementally improved model then bridges the old and the new models, and allows for additional analysis on the performance of the new feature, which may in turn influence how it is incorporated in the new model.

In this paper we describe a general approach for incorporating extraneous features into a semantic model, which result in the need to optimize for objectives which are not necessarily aligned. More specifically, given a model which outputs recommendations ranked according to some notion of semantic relevance, we want to add certain features which contribute to the overall utility of the users of the recommender system, but that may negatively affect the semantic relevance of the recommendations. This approach is model and feature agnostic, does not require additional labeled training data to be collected, and can be easily incorporated into an existing model as an additional stage in the computation pipeline. We validate our approach by A/B testing it on *TalentMatch* system, which currently ranks billions of job-member pairs on a daily basis, and show that a significant improvement in the utility of the recommender system (**42% increase on email reply rate**) can be achieved with an *acceptable* and *predictable* degradation in the original relevance of the recommendations.

2. PROBLEM FORMULATION

In this section we discuss a general template for framing the kinds of problems we are targeting in this paper and in Section 4.1 we discuss the instantiation of this general template for the specific *TalentMatch* scenario.

We start with a model which is optimized for semantic relevance. We then want to enhance this model with additional features which will increase the utility of the recommender system, but at a potential loss in the semantic relevance of the recommendations. Adding these additional features to the model will result in an enhanced model with additional objectives to be optimized. These additional objectives will be optimized conditionally on the semantic relevance objective having already been optimized.

2.1 Adding a single competing objective

In the simplest case, we would have only one feature to add to the semantic model, which equates to one additional objective to be increased (adding more features that map to only one additional objective can be handled similarly). We also want to penalize enhanced models in a manner that is correlated with the distance between the semantic relevance score distribution of the items in the top-K ranking as output by the semantic model, and the semantic relevance score distribution of the items in the top-K ranking as output by the enhanced model (note that the enhanced model outputs a ranking based on the enhanced scores, but we need to map those scores back to their semantic counterparts to compare the two distributions). These requirements are expressed by the following loss function:

$$L(w) = -g(f(\boldsymbol{Y}, \boldsymbol{X}, w)) + \lambda \Delta(\pi(\boldsymbol{Y}), \pi(f(\boldsymbol{Y}, \boldsymbol{X}, w))) \quad (1)$$

Where:

- $\boldsymbol{X}$ is the single feature to be added, a matrix of dimensions $m \times n$;

- m is the number of targets of the recommender system;

- n is the number of recommendations per target;

- $\boldsymbol{Y}$ represents semantic relevance scores, also a matrix of dimensions $m \times n$;

- w is the parameter associated with the feature to be added;

- f is the enhanced model which perturbs the semantic relevance score $\boldsymbol{Y}$ according to the new feature $\boldsymbol{X}$ and parameter w;

- g is an objective to be maximized which contributes the overall utility of the recommender system but is not accounted for in the semantic model;

- Δ is a non-negative measure which is indicative of the distance between the semantic match score distribution and the enhanced score distribution;

- π is a function which returns a top-K ranking, and n >> K;

- λ is a positive trade-off parameter.

Alternatively, it may be easier to visualize the objective as a constrained optimization problem where we have a limit on how much we are allowed to deviate from the top-K score distribution based on the semantic model:

$$\text{maximize } g(f(\boldsymbol{Y}, \boldsymbol{X}, w))$$

$$\text{s.t. } \Delta_i(\pi(\boldsymbol{Y}), \pi(f(\boldsymbol{Y}, \boldsymbol{X}, w))) \leq c_i, i = 1, \ldots, l \quad (2)$$

Where:

- c_i is the *ith* constraint on the top-K distribution distance between the semantic match score distribution and the enhanced score distribution;

- l is the number of constraints on the semantic score distribution deviation.

Given the constrained optimization perspective, in the simple case where $l = 1$ we could analyze how g trends as a function of various values for c, from which we could extract the Pareto frontier [6] and which we could use to make a data-driven decision on what value of c is appropriate. Note that if, in the Pareto frontier, g turns out to be a linear function of c, then the slope of the line would be the value of the λ parameter in Equation 1.

2.2 Adding multiple competing objectives

In the case where the additional features with which we will enhance the semantic model lead to multiple additional objectives, we have the following general version of the problem:

$$\text{maximize } G(F(\boldsymbol{Y}, \boldsymbol{X}, \boldsymbol{w}))$$

$$\text{s.t. } \Delta_i(\pi(\boldsymbol{Y}), \pi(f(\boldsymbol{Y}, \boldsymbol{X}, w))) \leq c_i, i = 1, \ldots, l \quad (3)$$

Where:

- G is the set of additional objectives to be maximized $\{g_1, g_2, \ldots, g_t\}$;

- t is the number of additional objectives to be maximized;

- $\boldsymbol{X}$ are the features to be added, a matrix of dimensions $m \times n \times p$;

- p is the number of features being added;

- $\boldsymbol{w}$ is the parameter vector associated with the features to be added with dimension $1 \times p$;

- F is the enhanced model which perturbs the semantic match score $\boldsymbol{Y}$ according to the new features $\boldsymbol{X}$ and parameter $\boldsymbol{w}$.

3. COMPUTATIONAL STRATEGY

The functions g and Δ described in Section 2 are non-smooth since they depend on a ranking which in turn depends on a sort operation. Therefore, traditional optimization approaches which leverage the gradient of a function are not directly applicable.

In very small parameter spaces (one or two parameters), grid (exhaustive) search is an acceptable and very simple to implement computational strategy. For larger parameter spaces, we can devise smoothed approximations to g and Δ that are amenable to traditional gradient-based methods and therefore able to handle parameter spaces where grid search would be unfeasible. In this Section we discuss using such approximations in our problem formulation and in Section 4.2 we discuss the computational strategy followed in the *TalentMatch* case study.

Recent research on "learning to rank" for information retrieval addresses the need to optimize non-smooth rank-based metrics. There are two approaches that are particularly interesting in this direction: *SoftRank* [9] and *SmoothRank* [3]. These approaches develop smooth approximations to IR metrics such as the Normalized Discounted Cumulative Gain (NDCG) and the Average Precision (AP). We can formulate our g and Δ functions so that they have a similar form to those IR metrics and then we can employ the techniques described in [9, 3] for optimizing them.

For example, we can consider the original semantic relevance score from the *TalentMatch* model to be the ground-truth measure of relevance of each candidate member given a job. In an IR setting, we would have queries and documents, where documents have a measure of relevance to a particular query. In the *TalentMatch* model, a *job* is equivalent to a *query* and a *candidate member* is equivalent to a *document*.

One possible instantiation of g and Δ would be as follows: assume we do not wish to distinguish between *active* and *passive* candidates; we would then have a binary notion of relevance for a given candidate, {*job-seeker = 1*, *non-job-seeker = 0*} that we want to maximize in the top-K results. This is a good match for the AP measure. We then need a constraint function which penalizes how much deviation there is from the original relevance-based ranking. It turns out that an adapted form of the NDCG measure would be appropriate here.

This leaves us with the following smooth approximation to our objective function, using the approximation in equation 9 from [3]:

$$g_j(f) \approx (A)AP_j(f, \sigma) =$$

$$\frac{1}{\sum_{i=1}^{n} l_i} \sum_{i,k=1}^{n} l_i l_k \frac{1 + exp((f(\boldsymbol{x_i}) - f(\boldsymbol{x_k}))/\sigma)^{-1}}{\hat{r}_i} \quad (4)$$

Where:

- $(A)AP_j$ is the Approximate Average Precision for a given job j;

- f is the enhanced model which perturbs the semantic match score, originally defined in Equation 1;

- σ is the smoothing parameter as described in [3];

- n is the number of recommendations per job j;

- l_i is the label of the i_{th} recommended candidate, either *job-seeker = 1* or *non-job-seeker = 1*;

- l_k is the label of the recommended candidate at rank k, either *job-seeker = 1* or *non-job-seeker = 1*;

- $\hat{r}_i$ is the smooth rank, as defined in equation 5 in [3].

And the following smooth approximation to our constraint function, using the approximation from equation 8 in [3]:

$$\Delta_j(f) \approx (A)NDCG_j(f, \sigma) =$$

$$\sum_{i,k=1}^{n} S_i D(k) h_{ik} \quad (5)$$

Where:

- $(A)NDCG_j$ is the Approximate Normalized Discounted Cumulative Gain for a given job j;

- S_i is the semantic score for the i_{th} candidate, obtained from using the *TalentMatch* model;

- n is the number of recommendations per job j;

- $D(k)$ is the discounting associated with ranking k, which could be defined as $D(k) = 1/log_2(1 + k)$;

- h_{ik} is defined in equation 7 in [3] to be a soft version of an indicator variable that indicates the probability that the j_{th} recommendation is ranked at the k_{th} position.

f, which is the enhanced model that perturbs the semantic match score, originally defined in Equation 1, enters Equation 5 through h_{ik}:

$$h_{ik} =$$

$$exp(-\frac{(f(\boldsymbol{x_i}) - f(\boldsymbol{x_{d(k)}}))^2}{\sigma}) / \sum_{j=1}^{n} exp(-\frac{(f(\boldsymbol{x_i}) - f(\boldsymbol{x_{d(k)}}))^2}{\sigma})$$

$$(6)$$

Where $d(k)$ is the index of the recommendation which was ranked at position k by f.

There are many other ways to formulate our approach using these smoothed approximations. For example, if instead of the job-seeking categories (*active, passive, and non-job-seeker*) we wished to use the job-seeking intent score, we could formulate g using $(A)NDCG$ instead of $(A)AP$. Additionally, if the functional form of f in equation 1 is such that a parameter vector $\boldsymbol{w}$ of $\boldsymbol{0}$ in the enhanced model yields the equivalent of the semantic model, an Euclidean norm constraint on $\boldsymbol{w}$ could be used instead of $(A)NDCG$ for the Δ function.

4. TALENT MATCH CASE STUDY

We illustrate our approach with the *TalentMatch* system, where given a job posted on the site, we generate a ranked list of candidates with regards to how well the candidates match the job. This semantic model outputs the probability that the candidate is a good match to the job. We want to enhance this model with the job-seeking intent of the candidate so that the candidates being recommended are both good matches for the job, as well as open to new job opportunities. Our hypothesis is that this will contribute to increased engagement between the job poster and the recommended candidates.

There are many ways to incorporate the job-seeking intent signal into the *TalentMatch* model. As discussed in section

1, the job-seeking intent model ouputs, for each member, a job-seeking propensity score and a probabilistic assignment to each of the job-seeking intent categories: *active, passive, and non-job-seeker*. Our objective is to increase the average number of *active* and *passive* candidates in the top-K recommendations. We want to achieve this objective by perturbing *slightly* the semantic ranking so that if there is a candidate C_x with a semantic score of 0.9 in rank 1 who has a low job-seeking intent (classified as *non-job-seeker*), and another candidate, C_y, in rank 2 with a match score of 0.88, but that happens to have a high job-seeking intent (classified as *active* or *passive*), then we would like to bump C_y up to rank 1 and bump C_x down to rank 2. We do not necessarily want to eliminate C_x from the final ranking, nor do we want to excessively bump the candidate down the ranking. More importantly, we want a systematic way to perform this re-ranking perturbation.

A simple strategy would be to remove from the ranked list based on semantic matching scores all those recommended candidates with a job-seeking intent score below a certain threshold t, backfilling if needed to make sure we have K recommendations (we discuss below how this specific heuristic is a special case of our suggested approach). This approach still requires us to estimate the threshold t, but more crucially, it also incurs the risk of completely eliminating high-quality matches from the final ranking, an outcome we do not want.

4.1 TalentMatch Problem Formulation

In order to come up with a strategy for re-ranking that satisfies our requirements, we frame our problem using the template described in section 2. The average number of *active* and *passive* candidates in the top-K recommendations is actually an instance of a familiar metric: *mean precision at K*, where our binary relevance measure is an indicator function that returns 1 if the member is *active* or *passive* and 0 otherwise, $l_i \in \{0, 1\}$. For a given job posted to the site, precision at K is:

$$Prec@K = \frac{1}{K} \sum_{i=1}^{n} l_i \mathbf{1}\{r(i) \leq K\} \qquad (7)$$

Where $\mathbf{1}\{A\}$ is the indicator function applied to A, and $\mathbf{1}\{A\} = 1$ is A is true and 0 otherwise, $r(i)$ is the ranking of the i_{th} candidate, and n is the number of candidates in the result set. Our objective to be maximized, the mean precision at K, which maps to the g function in equations 1 and 2 is:

$$g = MeanPrec@K = \frac{1}{m} \sum_{i=1}^{m} Prec@K(i) \qquad (8)$$

The functional form of f, the enhanced model in equations 1 and 2, can also be specified in a variety of ways. One possible option is to use a a linear combination of the *TalentMatch* semantic and job-seeking intent scores. This would not be ideal: we want both, good matches *and* likely to be job-seeking candidates; therefore, a multiplicative feature interaction is what we seek. We settled on the following formulation:

$$f(y, x, w = [\alpha, \beta]) =$$

$$y \times (\alpha^{\mathbf{1}\{x == active\}}) \times (\beta^{\mathbf{1}\{x == passive\}}) \qquad (9)$$

This is equivalent to applying a small boost to the semantic match score (y), and allowing for the boost to be different for actives (α) and passive (β) candidates. Solving the optimization problem defined in equations 1 and 2, with the specific functional forms defined here will yield appropriate values for α and β.

Given our chosen functional for f, it can be seen that the simple heuristic suggested earlier is actually a special case in our approach, where α and/or β are set to large enough values so as to effectively rank all members with a job-seeking intent score above the threshold t over those members with a score below t. Section 5 discusses how this strategy is suboptimal (it causes an unacceptable loss in semantic relevance).

Finally, we need to specify how we will measure the deviation of the enhanced model distribution from the semantic model distribution, that is, the functional form for Δ in equations 1 and 2. There are various histogram distance functions to choose from [2], examples of which include Euclidean distance and Kullback-Leibler divergence. We settled on using the Euclidean distance between the two histograms, or more specifically, the sum of squared errors of the histogram buckets, each histogram having b buckets:

$$\Delta = \Delta_{SSE}(H_s, H_e) = \sum_{i=1}^{b} (H_s[i] - H_e[i])^2 \qquad (10)$$

Where H_s is the histogram of semantic match scores of the top-K candidates ranked by the semantic match score and H_e is the histogram of semantic match scores of the top-K candidates ranked by the enhanced score.

4.2 TalentMatch Computational Strategy

Since we only have two parameters: α and β, and given our intuition that the optimal parameters will probably lie in the interval $[1.0, 2.0]$, a grid search turns out to be an acceptable computational strategy in this scenario. We break up the grid search into 2 runs: a coarse run (to see what region of the search space we should focus on) and a fine run (to zero in on the desired values). In each run we generate all the plans to be tested (a plan being an assignment of values to α and β) and evaluate our g and Δ functions for each plan generated.

5. EVALUATION AND RESULTS

For estimating the α and β parameters to be used in Equation 9, we created a sample dataset of jobs recently posted to the site and computed a maximum of 9000 recommendations for those jobs using the *TalentMatch* model. We filtered all recommendations with a threshold of 0.6 on the *TalentMatch* semantic score, and then removed all jobs which did not have at least 6 recommendations (we do not show results on the site unless there are at least 6 relevant matches and we include only the top-24 candidates in the recommendation set). This left us with a total of 760 jobs, each with anywhere from 6 recommendations to 9000 recommendations. We then generated the plans as per Section 4.2 and evaluated our g and Δ functions.

Our g function is the mean precision at K, as defined in Equation 8, where $K = 12$ since that is how many snippets of candidate recommendations we show in a single page.

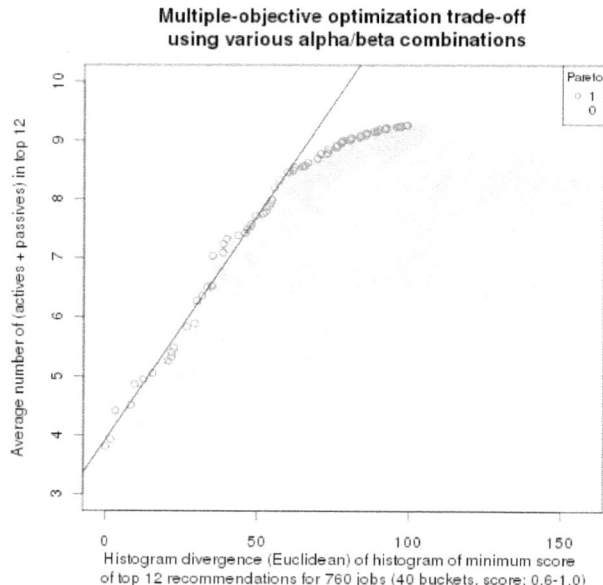

Figure 3: Multiple Objective Optimization trade-off between the objective being maximized and the penalty incurred. Up until a histogram divergence of a little over 60, the relationship is strongly linear ($\mathbf{R}^2 = 0.985$) on the Pareto front, with a slope of 0.076.

Table 1: Sample plans from figure 3. Plan 1 is the original plan, where the relevance score of active and passive candidates is not boosted.

Plan	Avg. Count	Hist. Divergence	α	β	Pareto
1	3.8	0.0	1.0	1.0	Yes
2	7.0	35.7	1.07	1.07	Yes
3	8.1	64.0	1.15	1.15	Yes
4	6.5	141.1	1.3	1.0	No

Also, for the measure of divergence, our Δ function, we compared the distribution of the minimum score of the top-12 ranking, given that we want to ensure relevant recommendations in the worst case on the first results page.

Figure 3 shows the result of the fine grid search run, which illustrates the risk-reward trade-off in our experiment: up until a histogram divergence of a little over 60, we pay a penalty that is linear with regards to the increase in the average number of *active* and *passive* members in the top-12 result set. Table 1 shows a few of the points used in the plot, including the original plan (equivalent to setting α and β to 1.0), which also indicates the average number of *active* and *passive* candidates in the top-12 result set of the original plan to be nearly 4. Figure 3 tells us that we can double that number if we are willing to pay a penalty of about 64 in the histogram divergence, and also tells us what to set α and β to: 1.15 (see table 1). Figures 4(a)-4(d) give an idea of how good/bad a histogram divergence of 64 is. For reference, as per table 1, setting α to 1.3 and α to 1.0 causes an unacceptable loss in relevance (the histogram divergence is too high and the gain in the objective does not justify it).

All of the plans on the Pareto front in figure 3 have similar coefficients for α and β, which is tied to the fact that the goal we are trying to maximize is the combined number of *active/passive* candidates in the top-12, and presumably the values for the weights would diverge more had we favored one category or the other. These results point to reasonable strategies that should be evaluated using A/B testing: a plan where $\alpha = \beta = 1.07$ and a plan where $\alpha = \beta = 1.15$. A/B testing turns out to be a crucial component to the methodology described in this paper. Our approach provides the tools for generating reasonable values for α and β: no matter what the desired risk-reward trade-off of a specific application, only plans in the Pareto frontier should be chosen. However, our choice for what is an acceptable histogram divergence will only be meaningful if once in production, the rate with which job posters purchase candidate set recommendations and the rate with which job posters contact the purchased recommended candidates does not decrease substantially.

As mentioned in Section 1, we would like to increase the likelihood of relevant engagement for *TalentMatch*. If the job-seeking intent is to be of any use to us, we would expect the rate of replies to *InMails* (LinkedIn e-mails) from job posters about job opportunities to be higher for members with high job-seeking intent. We determined that members classified as having a high job-seeking intent (*actives/passives*) are **16**× more likely to reply to an InMail regarding a career opportunity, with a 95% confidence interval of 15-17x (intervals computed by the method of E. C. Fieller [4]). These numbers are based on InMail activity that took place over a period of 10 days, during which time the number of *non-job-seekers* contacted was nearly the same as the number of *actives/passives* contacted.

Assuming that all members in the top-12 ranking are contacted about the job opportunity, our results suggest that we can double the desired relevant engagement. Given that the probability of positively replying for *non-job-seekers* is $p_n(reply) = 0.028$, and the probability of replying for *actives/passives* is $p_{a/p}(reply) = 0.45$ (as computed using the 10-day sample), and given that our analysis shows that we can double the average number of *actives/passives* in the top-12 from 4 to 8 at an acceptable relevance loss, we expect to double the expected relevant engagement: $0.028 \times 8 + 0.45 \times 4 \approx 2$ versus $0.028 \times 4 + 0.45 \times 8 \approx 4$.

We deployed the A/B test experiment and let it run for a couple of weeks before we started collecting data for analysis (until the "novelty effect" often caused by a new feature being deployed live had subsided). To measure the change in *InMail* reply rate, we looked at all emails created in a period of three weeks and observed how many were replied to. Table 2 shows the increase in response rate for each A/B test treatment bucket along with their confidence intervals. The actual increase follows the expected linear relationship as expected from the Pareto frontier, and though there is high variance, the 95% confidence intervals contain the expected values.

We now turn our attention to the effect of the re-ranking perturbation on the booking rate. More specifically, we want to quantify the effect of the histogram divergence. For this, during the same time period, we looked at the booking rate in each of A/B test buckets. Table 3 summarizes our findings. It shows a slight degradation in booking rate for the most extreme treatment bucket. However, looking at the

(a) $\alpha = \beta = 1, g = 3.8, \Delta = 0$

(b) $\alpha = 1.02, \beta = 1.04, g = 5.8, \Delta = 27$

(c) $\alpha = 1.1, \beta = 1.11, g = 7.9, \Delta = 54$

(d) $\alpha = 1.3, \beta = 1.3, g = 9.3, \Delta = 99$

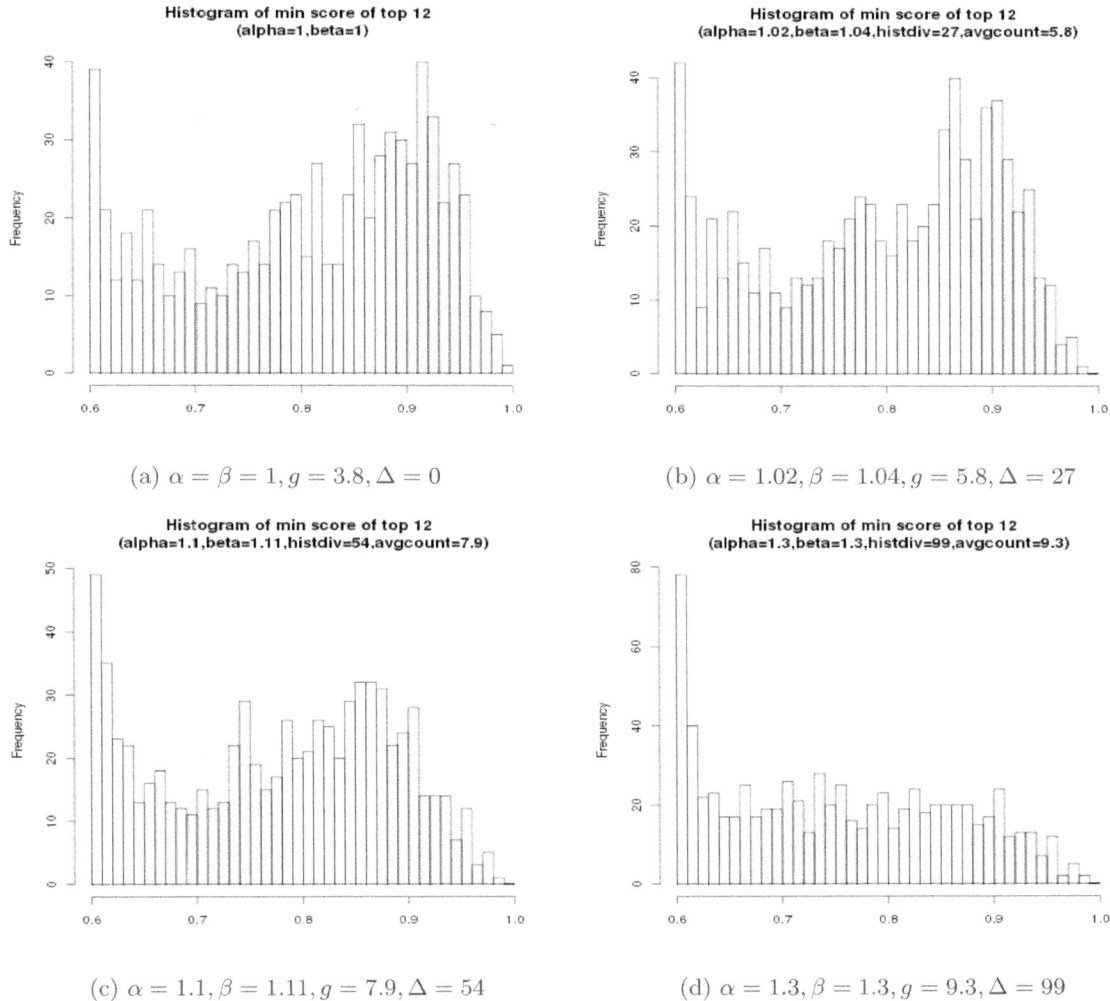

Figure 4: Histogram of the minimum score in the top-12 ranking for 760 jobs using various α, β plans. The top-left plan depicts the original *TalentMatch* semantic model, and as scores of job-seekers are boosted with increasing strength, a gradual degradation in the score distribution can be observed.

Table 2: InMail response rate ratio of treatment group over control group. Data collected over a period of three weeks around May/27/2012-June/17/2012.

Treatment	Treatment/Control	95% CI
$\alpha = \beta = 1.15$	1.42	0.95 - 2.01
$\alpha = \beta = 1.07$	1.22	0.94 - 1.59

rate witch which job posters email candidates in the purchased set (*InMails* per booking) shows a different story.

Table 4 shows that on average, job posters choose to email candidates more often in the treatment group than in the control group. This is evidence that job posters do not email all candidates in the purchased set, but rather pick and choose who they will contact. This fact explains why the actual average increase in InMail response rate was not as high as expected: we had assumed that job posters contact all of the candidates in the result set, but this turns out not to be the case. The control group shows that job

posters only contact an average of approximately 2 members from the purchased candidate result set. More importantly, since job posters do not have access to the job-seeking intent of each candidate (nor to most of the information we use to determine job-seeking intent, such as job searches), there must be something else in the candidate's profile which compels job posters to email candidates we've identified as job-seekers more often than those we have not. Perhaps job-seeking candidates have more complete or more curated profiles. Nevertheless, this means that the snippet we show job posters, based on which they make the decision of whether or not to purchase, is not well representative of the value of the candidate result set. This finding is something that we plan on exploring further, as it suggests that, given the right snippet (one which better conveys the value of the candidate result set to the job poster), **the booking rate for the treatment groups should be higher than for the control group.**

Table 3: Booking rate ratio of treatment group over control group. Data collected over a period of three weeks around May/27/2012-June/17/2012.

Treatment	Treatment/Control	95% CI
$\alpha = \beta = 1.15$	0.96	0.81 - 1.15
$\alpha = \beta = 1.07$	1.00	0.85 - 1.20

Table 4: Inmails/booking ratio of treatment group over control group. Data collected over a period of three weeks around May/27/2012-June/17/2012.

Treatment	Treatment/Control	95% CI
$\alpha = \beta = 1.15$	1.25	1.07 - 1.51
$\alpha = \beta = 1.07$	1.31	1.11 - 1.57

6. FUTURE WORK

An area we would like to expand this work is in incorporating tighter bounds into the additional objectives. For example, we were able to say that we could double the average number of candidates with high job-seeking intent into the top-12 *TalentMatch* ranking, but we have not been specific enough regarding the variance of that average, which was evident in our results.

7. RELATED WORK

In [1], the authors present an approach to optimize jointly for clicks and post-click downstream utilities (time-spent, revenue, etc). Similar to this paper, they propose a multi-objective programming approach in which multiple objectives are modeled in a constrained optimization framework. However, our approach is different: we want to optimize for downstream utilities *given* that the recommendations are of high-quality, whereas they manage the trade-off by showing a portion of their customers results optimized for relevance, while another portion of their customers are shown results optimized for another downstream utility measure. Their approach would not be suitable for a system like *Talent-Match*, since that would mean that some job-posters would see a ranking of candidates which is optimized for semantic relevance (which would not increase the desired relevant engagement), while other job posters would see a ranking of candidates based on the job-seeking intent score (resulting in an excessive loss in relevance which in turn would mean a decrease in our desired relevant engagement, since job-posters would most likely not purchase or contact the candidates provided).

Another related approach is presented in [5], where the authors get much closer to our requirement of explicitly controlling any potential loss in relevance as a result of optimizing for other utility aspects. We expand on their formulation in that we consider objective and constraint functions which can be nonlinear. So, if the system performance will be measured with metrics such as NDCG@K and Precision@K, then the system can be optimized precisely on those metrics.

Researchers from the field of Information Retrieval have also considered the problem of "Learning to Rank" with multiple objective functions [8, 7]. In [8], the authors show how multiple measures can be combined into what they refer to as a *graded measure* that can be learned. This notion of a *graded measure* allows them to formalize some sources of relevance labels as being more important than others, and so once an ordering is imposed, they can optimize for a given

metric provided all other more important metrics have already been optimized. The authors experiment with their suggested approach in the field of IR, where the typical notions of relevance have a low cardinality, a common grading scheme being: {*perfect, excellent, good, fair, bad*}. With this coarse notion of relevance, collisions or ties are likely to occur in a given ranking, and so it seems there is more room to optimize on a secondary metric if a primary one has already been optimized. In the *TalentMatch* system, however, relevance is determined as the probability that a given candidate is a good match for the specified job (the model does indeed allow for a very fine granularity in the relevance measure), and unless the scores or probabilities can somehow be discretized, collisions or ties are unlikely to occur, leaving little room for optimizing for a secondary label source such as job-seeking intent.

In [7], the authors do not have a predefined *graded measure*, but instead learn the relative importance of each label source from data, which implies having to collect additional labelled data, which is a potentially costly and time-consuming proposition.

8. CONCLUSION

We describe an approach for framing multiple objective optimization problems in recommender systems. The approach allows for fine control over any potential loss in relevance as additional aspects of the system's overall utility function are optimized. We illustrate the approach with a detailed analysis of how it was used to improve *TalentMatch*, a revenue-generating product which ranks billions of recommendations on a daily basis at LinkedIn. We show that we increased relevant engagement with an acceptable and predictable degradation in the relevance of the recommendations.

9. REFERENCES

[1] D. Agarwal, B.-C. Chen, P. Elango, and X. Wang. Click shaping to optimize multiple objectives. In *KDD*, 2011.

[2] S.-H. Cha. Comprehensive survey on distance/similarity measures between probability density functions. *International Journal of Mathematical Models and Methods in Applied Sciences*, 2007.

[3] O. Chapelle and M. Wu. Gradient descent optimization of smoothed information retrieval metrics. *Information Retrieval Journal*, June 2010.

[4] E. C. Fieller. Some problems in interval estimation. *Journal of the Royal Statistical Society*, 1954.

[5] T. Jambor and J. Wang. Optimizing multiple objectives in collaborative filtering. In *RecSys*, 2010.

[6] Y. Jin and B. Sendhoff. Pareto-based multiobjective machine learning: An overview and case studies. *IEEE Transactions on Systems, Man, and Cybernetics*, 2008.

[7] C. Kang, X. Wang, Y. Chang, and B. Tseng. Learning to rank with multi-aspect relevance for vertical search. In *WSDM*, 2012.

[8] K. M. Svore, M. N. Volkovs, and C. J. Burges. Learning to rank with multiple objective functions. In *WWW*, 2011.

[9] M. J. Taylor, J. Guiver, S. Robertson, and T. Minka. Softrank: optimizing non-smooth rank metrics. In *WSDM*, 2008.

Pareto-Efficient Hybridization for Multi-Objective Recommender Systems

Marco Tulio Ribeiro[1,2]   Anisio Lacerda[1,2]   Adriano Veloso[1]   Nivio Ziviani[1,2]

[1]Universidade Federal de Minas Gerais
Dept. of Computer Science
Belo Horizonte, Brazil
{marcotcr, anisio, adrianov, nivio}@dcc.ufmg.br

[2]Zunnit Technologies
Belo Horizonte, Brazil
{marcotcr, anisio, nivio}@zunnit.com

ABSTRACT

Performing accurate suggestions is an objective of paramount importance for effective recommender systems. Other important and increasingly evident objectives are novelty and diversity, which are achieved by recommender systems that are able to suggest diversified items not easily discovered by the users. Different recommendation algorithms have particular strengths and weaknesses when it comes to each of these objectives, motivating the construction of hybrid approaches. However, most of these approaches only focus on optimizing accuracy, with no regard for novelty and diversity. The problem of combining recommendation algorithms grows significantly harder when multiple objectives are considered simultaneously. For instance, devising multi-objective recommender systems that suggest items that are simultaneously accurate, novel and diversified may lead to a conflicting-objective problem, where the attempt to improve an objective further may result in worsening other competing objectives. In this paper we propose a hybrid recommendation approach that combines existing algorithms which differ in their level of accuracy, novelty and diversity. We employ an evolutionary search for hybrids following the Strength Pareto approach, which isolates hybrids that are not dominated by others (i.e., the so called Pareto frontier). Experimental results on two recommendation scenarios show that: (i) we can combine recommendation algorithms in order to improve an objective without significantly hurting other objectives, and (ii) we allow for adjusting the compromise between accuracy, diversity and novelty, so that the recommendation emphasis can be adjusted dynamically according to the needs of different users.

Categories and Subject Descriptors

H.3.3 [**Information Storage and Retrieval**]: Information Filtering

Keywords

Hybridization, Pareto-Optimality, Diversity, Novelty

1. INTRODUCTION

Recommender systems are increasingly emerging as enabling mechanisms devoted to overcoming problems that are inherent to

information overload, providing intelligent information access and delivery, and thus potentially improving browsing and consumption experience. Historically, the typical goal of a recommender system is to maximize accuracy as much as possible in predicting and matching user information needs, often by considering individual delivered items in isolation [12]. More recently, however, it has become a consensus that the success of a recommender system depends on other dimensions of information utility, notably the diversity and novelty of the suggestions performed by the system [9, 19, 25, 33]. More specifically, even being accurate, obvious and monotonous recommendations are generally of little use, since they do not expose users to unprecedent experiences.

Increasing novelty and diversity by completely giving up on accuracy is straightforward - and meaningless, since the system will not meet the users needs anymore. In fact, there is an apparent trade-off between these dimensions, which becomes evident by inspecting the performance of existing top-N recommendation algorithms. An easy conclusion is that different algorithms may perform distinctly depending on the dimension of interest (i.e., the best performer in terms of accuracy is not the best one in terms of novelty and diversity), and thus it is hard to point to a best performer if all the dimensions are considered simultaneously. A conclusion which is harder to reach is whether these algorithms are indeed complementary, so that the strengths of an algorithm may compensate the weaknesses of others. The potential synergy between different recommendation algorithms is of great importance to multi-objective recommender systems, since they must achieve a proper level of each dimension (i.e., objective).

In this paper we hypothesize that it is possible to properly aggregate different recommendation algorithms, so that the resulting hybrids balances the level of accuracy, diversity and novelty in its suggestions. In this case, each potential hybrid is given as a weighted combination of well-established recommendation algorithms (e.g., simple algorithms as well as representative of the state-of-the-art). Our proposed hybridization approach consists in finding appropriate weights for the constituent algorithms. By considering each dimension (i.e., accuracy, novelty and diversity) as a separate objective, we reduce the hybridization task to a multi-objective optimization problem, in which we search for the optimal combination of weights that maximizes accuracy, diversity and novelty.

Since the considered objectives are potentially conflicting, we employ an evolutionary search for optimal hybrids. Evolutionary algorithms denote a class of optimization methods that are characterized by a set of candidate solutions (aka individuals) called a population, which is maintained during the entire optimization process. The population of individuals evolves towards better (and potentially optimal) solutions by employing genetic operators, such as reproduction, mutation and crossover. In our context, each individual represents a possible combination of weights (i.e., a possible hybrid). Optimal hybrids lie in the so-called Pareto frontier [37], and are optimal in the sense that no hybrid in the frontier can be improved upon without hurting at least one of its objectives. Therefore, the evolutionary algorithm evolves the population towards producing hybrids that are located closer to the Pareto frontier, and then a linear search returns the most dominant hybrid [37], which is likely to balance accuracy, novelty and diversity. Alternatively, hybrids in the Pareto frontier can be selected according to a certain need, allowing the recommender system to adjust the compromise between accuracy, novelty and diversity, so that the recommendation emphasis can be adapted dynamically according to the needs of each user (i.e., new users may benefit more from more accurate suggestions, whereas older users may require more novel and diversified suggestions).

We conducted a systematic evaluation involving different recommendation scenarios, with explicit user feedback (i.e., movies from the MovieLens dataset), as well as implicit user feedback (i.e., artists from the LastFM dataset). The experiments showed that it is possible to (i) combine different algorithms in order to produce better recommendations and (ii) control the desired balance between accuracy, novelty and diversity. In order to evaluate the baseline algorithms and our hybrids, we used the methodology for top-N evaluation proposed in [12] and measured novelty and diversity using the framework proposed in [33].

2. PRELIMINARIES

In this section we review the main concepts about evolutionary algorithms and multi-objective optimization. Finally, we discuss related work on hybrid and multi-objective recommender systems.

2.1 Evolutionary Algorithms

Evolutionary algorithms are meta-heuristic optimization techniques that follow processes such as inheritance and evolution as key components in the design and implementation of computer-based problem solving systems [15, 20]. In evolutionary algorithms, a solution to a problem is represented as an individual in a population pool. The individuals may be represented as different data structures, such as vectors, threes, or stacks [26]. If the individual is represented as a vector, for example, each position in the vector is called a gene.

Typically, evolutionary algorithms employ a training and a validation set, as described in Algorithm 1. Initially, the population starts with individuals created randomly (line 6). The evolutionary process is composed of a sequence of solution generations. The process evolves generation by generation through genetic operations (lines 7-12). The goal of this process is to obtain better solutions after some generations. A fitness function is used to assign a fitness value to each individual (line 9), which represents its performance on the training set or in a cross validation set. To produce a new generation, genetic operators are applied to individuals with the aim of creating more diverse and better individuals (line 12). Typical operators include reproduction, mutation, and crossover.

2.2 Multi-Objective Optimization

Since we are interested in maximizing three different objectives for the sake of recommender systems (i.e. accuracy, novelty, and diversity), we use a multi-objective evolutionary algorithm. In multi-objective optimization problems there is a set of solutions that are superior to the remainder when all the objectives are considered together. In general, traditional approaches to multi-objective optimization problems are very limited because they become too ex-

---

Algorithm 1 Evolutionary Algorithm.

1 Let $\mathcal{M}$ be a training set
2 Let $\mathcal{V}$ be a validation set
3 Let $\mathcal{N}_g$ be the number of generations
4 Let $\mathcal{N}_I$ be the number of individuals
5 $\mathcal{S} \leftarrow \emptyset$
6 $\mathcal{P} \leftarrow$ initial random population of individuals
7 For each generation g of $\mathcal{N}_g$ do
8   For each individual $i \in \mathcal{P}$ do
9      $fitness \leftarrow fitness(i, \mathcal{M}, \mathcal{V})$
10   $\mathcal{S}_g \leftarrow \mathcal{N}_I$ top-ranked individuals of generation g according to their fitness
11   $\mathcal{S} \leftarrow \mathcal{S} \cup \mathcal{S}_g$
12   $\mathcal{P} \leftarrow$ New population created by applying genetic operators to individuals in $\mathcal{S}_g$
13 BestIndividual$\leftarrow$ SelectionMethod($\mathcal{S}$)

---

pensive as the size of the problem grows [8]. Multi-objective evolutionary algorithms are a suitable option to overcome such an issue.

Typically, multi-objective evolutionary algorithms are classified as Pareto or non-Pareto [37]. In the non-Pareto optimization case, the objectives are combined into a single evaluation value that is used as fitness value (i.e., average of the objectives). In Pareto algorithms, on the other hand, a vector of objective values is used (i.e., the individual is given as an objective vector). The evaluation of Pareto approaches follows the *Pareto dominance* concept. An individual dominates another if it performs better in at least one of the objectives considered. Given two arbitrary individuals, the result of the dominance operation has two possibilities: (i) one individual dominates another, or (ii) the two individuals do not dominate each other. An individual is denoted as *non-dominated* if it is not dominated by any other individual in the population, and the set of all non-dominated individuals compose the *Pareto frontier*.

In this work we use a second version of the strength Pareto evolutionary algorithm (SPEA-2) [36, 37]. The aim is to find or approximate the Pareto-optimal set for multi-objective problems. The main features of this algorithm are: (i) the fitness assignment scheme takes into account how many individuals each individual dominates or is dominated by, (ii) it uses a nearest neighbour density estimation technique to break ties in solutions with the same fitness, (iii) the size of the population of non-dominated solutions is a fixed value η. Thus, we have two situations. First, when the actual number of non-dominated solutions is lower than η, the population is filled with dominated solutions; second, when the actual number of non-dominated solutions exceeds η, some of them are discarded by a truncation operator which preserves boundary conditions, even though we always keep the current Pareto Frontier in a list separate from the population, so we can later retrieve the individuals in it.

2.3 Related Work

Traditionally, hybrid recommender strategies are the combination of two different families of algorithms - namely, content-based and collaborative filtering [1]. In this work, we combine many (up to 8) recommendation algorithms - different content-based and collaborative filtering algorithms that deal with explicit and implicit feedback, etc. We treat each recommendation algorithm as a black-box, so adding or removing recommendation algorithms is easy. Different hybridization strategies have been proposed to combine recommender methods, such as weighted approaches [10], voting mechanisms [30], switching between different recommenders [6, 24], and re-ranking the results of one recommender with another [7]. A prominent use of hybridization in recommender systems is the Belkor system that won the Netflix competition [4, 5]. Their method is a statically weighted linear combination of 107 collaborative filtering engines. There are important differences between their work and ours: (i) their solution is single-objective (accuracy), (ii) they combine only collaborative filtering information, and (iii) the recommendation task is rating prediction, focused on RMSE - which makes the aggregation simpler, since all of the ratings are on the same scale and consist of the same items.

There has been an increasing consensus in the recommender systems community about the importance of proposing algorithms and methods to enhance novelty and diversity [17, 33]. As showed in [35], user satisfaction does not always correlate with high recommender accuracy. Thus, different multi-objective algorithms have been proposed to improve user experience considering either diversity or novelty. For instance, in [35], the authors define a greedy re-ranking algorithm that diversifies baseline recommendations. Another approach to improve diversity is presented in [34], where they suggest an optimization method to improve two objective functions reflecting preference similarity and item diversity.

On the other hand, novelty has been understood as recommending long-tail items, i.e., those items which few users have accessed. In [33], the authors present hybrid strategies that combine collaborative filtering with graph spreading techniques to improve novelty. The authors in [9] take an alternative approach: instead of assessing novelty in terms of the long-tail items that are recommended, they follow the paths leading from recommendations to the long tail using similarity links. As far as we know, this is the first work that proposes a hybrid method that is multi-objective in terms of the three metrics, i.e., accuracy, diversity and novelty.

Extensive research has also been performed exploiting the robust characteristics of genetic algorithms in recommender systems. For instance, in [28] the authors build a content-based recommender system and use genetic algorithms to assign proper weights to the words. Such weights are combined using the traditional IR *vector space model* [2] to produce recommendations. In [23] the authors use a genetic algorithm to build a recommender method that considers the browsing history of users in real-time. In contrast to our approach (which uses a GA to combine multiple recommender methods), they use GA to build a single-method.

In [22], the authors present an implementation of GA for optimal feature weighting in the multi-criteria scenario. Their application of GA consists in selecting features that represent users' interest in a collaborative filtering context, in contrast to our method, which focuses on assigning weights to different recommendation algorithms in order to improve the overall performance in terms of accuracy, novelty and diversity.

3. PARETO-EFFICIENT HYBRIDIZATION

In this section we introduce our search approach for Pareto-Optimal hybrids. We start by discussing how different recommendation algorithms are combined, so that potential hybrids are created. Then we describe the evolutionary search for Pareto-Optimal hybrids. Finally, we discuss an approach to deal with the compromise between accuracy, novelty and diversity, so that the system is able to adjust itself for different user perspectives.

3.1 Weighted Hybridization

Our hybridization approach is based on assigning weights to each constituent algorithm. We denote the set of constituent algorithms as A and the score given by algorithm A_j for an item i is represented by $A_j(i)$. As the constituent algorithms may output scores in drastically different scales, a simple normalization proce-

dure is necessary to ensure that all algorithms in A operate in the same scale. The aggregated score for each item i is calculated as:

$$S(i) = \sum_{j=1}^{|A|} A_j(i) * W_j \qquad (1)$$

where W is a vector that represents the weight assigned to each constituent algorithm. The assignment of weights to each algorithm is formulated as a search problem which we discuss next.

3.2 Searching for Pareto-Optimal Hybrids

Finding a suitable vector of weights W can be viewed as a search problem in which possible solutions are given as a combination of weights $\{w_1, w_2, \dots, w_{|A|}\}$, such that each w_i is selected in a way that optimizes an established criterion. We consider the application of evolutionary algorithms for searching optimal solutions. These algorithms iteratively evolve a population of individuals towards optimal solutions by performing operations based on reproduction, mutation, recombination, and selection [18]. This approach is interesting because we have no knowledge of the search space, since any number of different algorithms may be used, in different domains. Next, we precisely define an individual.

Definition 1: *An individual is a candidate solution, which is encoded as a sequence of $|A|$ values $[w_1, w_2, \dots, w_{|A|}]$, where each w_i indicates the weight associated with algorithm $A_i \in A$.*

Each algorithm A_i assigns scores to items using a cross-validation set. Finally, weights are assigned to each algorithm and their scores are aggregated according to Equation 1, producing an individual. A fitness function is computed for each individual in order to make them directly comparable, so that the population can evolve towards optimal solutions.

Definition 2: *An optimal solution is a sequence of weights $W = \{w_1, w_2, \dots, w_{|A|}\}$, satisfying:*

$$\text{maximize } \phi(o_i) \ \forall \ o_i \in \{\text{accuracy, novelty, diversity}\} \qquad (2)$$

where $\phi(o_i)$ is a metric used to measure an objective, which can be either accuracy, novelty or diversity. These metrics are better discussed in Section 4. For now it suffices to notice that the performance of each individual is given by a 3-dimensional objective vector, containing the average accuracy, novelty and diversity over the users in the cross validation set (since different metrics may operate in different scales, we normalize each $\phi(o_i)$ to the 0-1 interval). Searching for optimal solutions, therefore, is a multi-objective optimization problem, in which the value of $\phi(o_i)$ must be maximized for each of the 3 objectives that compose an optimal solution. Therefore, multiple optimal individuals are possible. It is worth noticing that different datasets and combinations of algorithms and A will generate different optimal individuals.

A general strategy for solving a multi-objective optimization problem is to exploit the concept of Pareto dominance, which may be used to find solutions that are not dominated by others. These non-dominated solutions lie in the so-called Pareto frontier, and are optimal in the sense that no solution in the frontier can be improved upon without hurting at least one of its objectives. Therefore, the evolutionary algorithm evolves the population towards producing individuals that are located closer to the Pareto frontier, and then a linear search returns the individual which simply maximizes the average (or some other combination, as we see on the next section) of the three objectives. Under this strategy, we follow the well-known Strength Pareto Evolutionary Algorithm approach [36], which has shown to be highly effective and also because it provides more diverse results when compared to existing approaches [11, 13, 32] for many problems of interest. The Strength Pareto approach isolates individuals that achieve a compromise between maximizing the competing objectives by evolving individuals that are likely to be non-dominated by other individuals in the population.

It is worth noticing that our approach does not depend on which recommendation algorithms are being aggregated, nor does it depend on the data domain. This makes adding or removing algorithms trivial, and allows the data to determine how each algorithm contributes to each of the objectives - an algorithm may be the most accurate when ratings are available, but not so accurate when only implicit feedback is used.

The Pareto-Optimal search is computationally expensive. However, it can be performed in an off-line manner, and with low frequency. After the Pareto-Optimal weights are discovered, there is no need to perform the search repeatedly, unless a recommendation algorithm is added or removed, or a lot of new feedback data enters the system. Therefore, using this approach would not hinder the system's online performance.

3.3 Adjusting the System Priority

It is well recognized that the role that a recommender system plays may vary depending on the target user. For instance, according to [19], the suggestions performed by a recommender system may fail to appear trustworthy to a new user because it does not recommend items the user is sure to enjoy but probably already knows about. Based on this, a recommender system might prioritize accuracy instead of novelty or diversity for new users, while prioritizing novelty for users that have already used the system for a while. This is made possible by our hybridization approach, by searching which individual in the *Pareto frontier* better solves the user's current needs.

The choice of which individual in the Pareto frontier is accomplished by performing a linear search on all of the individuals, in order to find which one maximizes a simple weighted mean on each of the three objectives in the objective vector, where the weights in the weighted mean represent the priority given to each objective. It is worth noting that fitness values are always calculated using the cross-validation set. Therefore, considering a 3-dimensional priority vector Q, which represents the importance of each objective j, the individual in the Pareto frontier P is chosen as:

$$arg \max_{i \in P} \sum_{j=1}^{|O|} Q_j O_{ij} \quad (3)$$

4. EVALUATION METHODOLOGY

The testing methodology we adopted in this paper is similar to the one described in [12], which is appropriate for the top-N recommendation task. For each dataset, ratings are split into two subsets: the training set $\mathcal{M}$ and the test set $\mathcal{T}$. The training set $\mathcal{M}$ can (if necessary) be split into two subsets: the cross-validation training set $\mathcal{C}$ and the cross-validation test set $\mathcal{V}$, which is used in order to tune parameters or adjust models. The test set $\mathcal{T}$ and the cross-validation test set $\mathcal{V}$ only contain items that are considered relevant to the users in the set. For explicit feedback (i.e., MovieLens), this means that the sets $\mathcal{T}$ and $\mathcal{V}$ only contain 5-star ratings.

In the case of implicit feedback (i.e., Last.fm), we normalized the observed item access frequencies of each user to a common rating scale [0,5], as used in [33]. Namely, $r(u, i) = n * F(frec_{u,i})$, where $frec_{u,i}$ is the number of times u has accessed i and $F(frec_{u,i}) = |j \in \mathbf{u}|f_{u,j} < f_{u,i}|/|\mathbf{u}|$ is the cumulative distribution function of $frec_{u,i}$ over the set of items accessed by the user u, denoted as **u**. In this case, the test set and the cross validation test set only

contain ratings such that $r(u, i) >= 4$, since the number of 5 star ratings is very small using this mapping of implicit feedback into ratings. It is worth noting that all the sets have a corresponding implicit feedback set, used by the recommendation algorithms that can deal with implicit feedback.

The detailed procedure to create $\mathcal{M}$ and $\mathcal{T}$ is the same used in [12], in order to maintain compatibility with their results. Namely, for each dataset we randomly sub-sampled 1.4% of the ratings from the dataset in order to create a probe set. The training set $\mathcal{M}$ contains the remaining ratings, while the test set $\mathcal{T}$ contains all the 5-star ratings in the probe set (in the case of explicit feedback) or 4+ star ratings (in the case of implicit feedback mapped into explicit feedback). We further divided the training set in the same fashion, in order to create the cross-validation training and test sets $\mathcal{C}$ and $\mathcal{V}$. The ratings in the probe sets were not used for training.

In order to evaluate the algorithms, we first train the models using $\mathcal{M}$. Then, for each item in $\mathcal{T}$ that is relevant to user u:

- We randomly select 1,000 additional items unrated by user u. The assumption is that most of them will not be interesting to u.

- The algorithm in question forms a ranked list by ordering all of the 1,001 items. The most accurate result corresponds to the case where the test item i is in the first position.

Since the task is top-N recommendation, we form a top-N list by picking the N items out of the 1,001 that have the highest rank. If the test item i is among the top-N items, we have a *hit*. Otherwise, we have a *miss*. Recall and precision are calculated as follows:

$$recall(N) = \frac{\#hits}{|T|} \quad (4)$$

$$precision(N) = \frac{\#hits}{N * |T|} = \frac{recall(N)}{N} \quad (5)$$

In order to measure the novelty of the recommendations, we used a popularity-based item novelty model proposed in [33], so that the probability of an item i being seen is estimated as:

$$P(seen|i_k) = \frac{|u \in U|r(u, i) \neq \emptyset|}{|U|} \quad (6)$$

where U denotes the set of users. Since the testing methodology supposes that most of the 1,000 additional unrated items are not relevant, we used the metrics in the framework proposed in [33] without relevance awareness. The novelty of a top-N recommendation list from R presented to user u is therefore given by:

$$nov(R(N)) = EPC(N) = C \sum_{i_k \in R}^{i_N} rd(k)(1 - p(seen|i_k)) \quad (7)$$

where $rd(k)$ is a rank discount given by $rd(k) = 0.85^{k-1}$[33] and C is a normalizing constant given by $1/\sum_{i_k \in R}^{i_N} rd(k)$. Therefore, this metric is rank-sensitive (i.e. the novelty of the top-rated items counts more than the novelty of other items). As is the case with precision and recall, we average the EPC@N value of the top-N recommendation lists over the test set.

We used a distance based model in order to measure the diversity of the recommendation lists. Once again, we used the metrics from [33] without relevance-awareness. The recommendation diversity, therefore, is given by:

$$div(R(N)) = EILD(N) = \sum_{\substack{i_k \in R \\ i_l \in R \\ l \neq k}}^{i_N, l_N} C_k rd(k) rd(l|k) d(i_k, i_l) \quad (8)$$

where $rd(l|k) = rd(max(1, l-k))$ reflects a relative rank discount between l and k, and $d(i_k, i_l)$ is the cosine similarity between two items, given by:

$$d(i, j) = \frac{|\mathbf{U_i} \cap \mathbf{U_j}|}{\sqrt{|\mathbf{U_i}|}\sqrt{|\mathbf{U_j}|}} \qquad (9)$$

such that $\mathbf{U_i}$ denotes the users that liked item i, and $\mathbf{U_j}$ denotes the users that liked item j.

5. EXPERIMENTAL EVALUATION

We apply the methodology presented in Section 4 to two different scenarios, in order to evaluate our hybrid approach: movie and music recommendation. For movie recommendation, we used the MovieLens dataset [27]. This dataset contains 1,000,209 ratings from 6,040 users on 3,883 movies. For music recommendation, we used an implicit preference dataset from [9], which consists of 19,150,868 user accesses to music tracks on the website Last.fm[1]. This dataset involves 176,948 artists and 992 users, and we considered the task of recommending artists to users. Mapping the implicit feedback into user-artist ratings yielded a total of 889,558 ratings, which were used by the algorithms that cannot deal with implicit feedback, and to separate the dataset into the training and test sets $\mathcal{M}$ and $\mathcal{T}$.

5.1 Recommendation Algorithms

We selected eight recommendation algorithms to provide the base for our hybrids. To represent latent factor models, we selected PureSVD with 50 and 150 factors (**PureSVD50** and **PureSVD150**), described in [12]. These were the only algorithms we used that are based on explicit feedback. To compute the scores for the items in the Last.fm dataset, we used the mappings of implicit feedback into ratings explained in Section 5.3.

As for recommendation algorithms that use implicit feedback, we used algorithms available in the MyMediaLite package [16]. We used **WeightedItemKNN** (WIKNN) and **WeightedUserKNN** (WUKNN) as representative of neighbourhood models based on collaborative data [14] (we only used **WeightedItemKNN** on the MovieLens dataset, as MyMediaLite's implementation cannot yet handle datasets where the number of items is very large, which is the case in the Last.fm dataset). As a baseline, and to allow for comparison with [12], we used MyMediaLite's **MostPopular** implementation, which is the same as TopPop in [12]. We also used **WRMF** − a weighted matrix factorization method based on [21, 29], which is very effective for data with implicit feedback. In order to represent content-based algorithms, we used **ItemAttributeKNN**(IAKNN), a K-nearest neighbor item-based collaborative filtering using cosine-similarity over the movie genres for MovieLens (we could not use this method in the Last.fm dataset, because it does not contain content data). Finally, we used **UserAttributeKNN**(UAKNN), a K-nearest neighbor user-based collaborative filtering using cosine-similarity over the user attributes, such as sex, age, etc. (which both datasets provide).

5.2 Hybrid Approaches

As a baseline, we used a voting-based hybrid based on Borda-Count (BC) which is similar to [30], where each constituent algorithm gives n points to each item i such that $n = |R| - p_i$, where $|R|$ is the size of the recommendation list and p_i is the position of i in R. We also used STREAM as baseline, a stacking-based approach with additional meta-features [3]. We used the same additional meta-features as [3], namely, the number of items that a

[1] www.Last.fm

certain user has rated and the number of users that has rated a certain item (denoted as RM_1 and RM_2 in [3]). We tried the learning algorithms proposed in [3], and Linear Regression yielded the best results, so the results presented for STREAM are generated using Linear Regression as the meta-learning algorithm. Our last baseline is the weighted hybrid we proposed in Section 3.1, using equal weights for each constituent algorithm. We called this baseline Equal Weights (EW).

As for our genetic approach, we combined all of the the recommendation algorithms cited in the last subsection. We used an open-source implementation of SPEA2 [36, 37] from DEAP [31].We used a two points crossover operator [20], and a uniform random mutation operator with probability 0.05. SPEA-2 was configured with the following parameters:

Parameters	Datasets	
	Movielens	Last.fm
Population Size	100	100
Gene length	8 algorithms	6 algorithms
# of Objectives	3	3
# of Generations	300	800
Mutation Rate	0.2	0.2
Crossover Rate	0.5	0.5

5.3 Results and Discussion

The results achieved by each of the constituent recommendation algorithms can be seen in Tables 1 and 2. We show the accuracy results (recall and precision) over different values of N. Since both EPC(novelty) and EILD(diversity) are rank-sensitive metrics, we only presented their values for $N = 20$. There is a clear compromise between accuracy, novelty and diversity of these algorithms. For the MovieLens dataset (Table 1), the constituent algorithm that provides the most accurate recommendations is PureSVD50. The constituent algorithm that provides the most novel recommendation with an acceptable degree of accuracy is PureSVD150, but its accuracy is much worse than the accuracy obtained by PureSVD50, and its diversity is much worse than the other algorithms. TopPop provided the most diverse recommendations, although it performs significantly worse in accuracy and novelty. It is worth noting that ItemAttributeKNN is based only on genres, which explains its poor accuracy results.

On the Last.fm dataset, the constituent algorithm that provides the most accurate recommendations is WRMF. This is expected, as Last.fm is originally an implicit feedback dataset, to which WRMF is more suitable. Once again, PureSVD150 proved its capacity to suggest novel items, being the algorithm with the most novel recommendations. WeightedUserKNN proved to be the algorithm that provided the most diverse recommendations, while maintaining a reasonable accuracy degree. In this dataset the compromise between the three objectives is once again illustrated by the fact that there is no algorithm that dominates the others in every objective.

Regarding the performance of the baselines in the MovieLens dataset, STREAM performs worse then PureSVD50 on accuracy, maintaining the same level of novelty and performing better in terms of diversity. Borda Count performed poorly on accuracy and reasonably well in terms of novelty and diversity. Equal Weights performed poorly on accuracy and novelty and well on diversity. On the Last.fm dataset, STREAM performed slightly worse than WRMF in accuracy, while maintaining the same level of diversity and improving slightly on novelty. Once again, Borda Count performed poorly on accuracy and reasonably well on novelty and di-

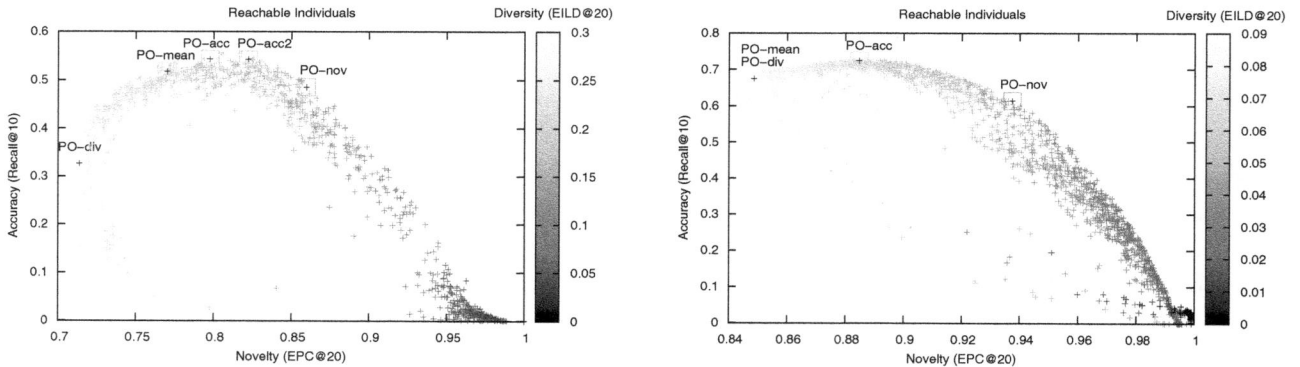

Figure 1: Individuals found in the Movielens (left) and Last.fm (right) datasets.

versity. Finally, Equal Weights performed poorly on accuracy and novelty, while performing well on diversity.

Now, with our evolutionary approach, we could reach any of the individuals in Figure 1, which represent the accuracy (in this case, Recall@10) and novelty (EPC@20) of the recommendations in x and y axes, and diversity (EILD@20) with a color scale. These graphics show the results in the test set for the individuals that represented the Pareto frontier in the cross-validation. It is clear that there is a compromise between the three objectives: the individuals with the most novel recommendations provide less accurate and diverse lists, and so on. This compromise can be adjusted dynamically with little extra cost, since the cost of reaching these individuals is as low as a linear search (for the individual that maximizes a weighted mean, as described on Section 3.2) over the Pareto frontier individuals' scores on the cross validation set. The Pareto frontier consists of 1,418 individuals in the MovieLens dataset and of 1,995 individuals in the Last.fm dataset, so a linear search can be done very quickly. We chose to demonstrate a few of these individuals in Tables 1 and 2. First, Pareto-Optimal-mean (PO-mean) represents the individual that optimizes the mean of the three normalized objectives, assuming each of them are equally important. This would be an option if personalization was not desired, or if the designers of the recommender systems did not know which combination of the three objectives would result in higher user satisfaction. However, in a more realistic situation, the recommender system would most likely want to select different individuals for different users. We selected as examples the following individuals, which were found by the process explained in Section 3.2 with the represented associated weighted vectors:

- PO-acc: [Accuracy:0.85, Novelty:0.1, Diversity:0.05]
- PO-acc2: [Accuracy:0.7, Novelty:0.3, Diversity:0]
- PO-nov: [Accuracy:0.15, Novelty:0.85, Diversity:0]
- PO-div: [Accuracy:0.15, Novelty:0.15, Diversity:0.7]

We compared PO-acc and PO-acc2 with PureSVD50, which is the stand-alone algorithm with the most accurate recommendations. Both perform as well as PureSVD50 on accuracy, but PO-acc performs much better on diversity (and equally well on novelty), and PO-acc2 performs better on novelty while maintaining the diversity level. We compared PO-nov with Pure-SVD150, which presented the most novel recommendations to the users, with reasonable accuracy. PO-nov performs slightly better on novelty than

PureSVD150, but performs much better in terms of accuracy, and slightly on diversity. Finally, we compared PO-div with MostPopular, the algorithm with the most diverse recommendations. PO-div loses very slightly on diversity, while improving on accuracy and novelty. We were able, therefore, to find individuals in the Pareto frontier that performed close or better than the best algorithms in each individual objective, but better on the other objectives. Once again, we could have chosen to compromise more accuracy and diversity if we desired more novelty, as is shown by Figure 1 (left).

As for the Last.fm dataset, we selected the following individuals:

- PO-acc: [Accuracy:0.7, Novelty:0.3, Diversity:0]
- PO-nov: [Accuracy:0.15, Novelty:0.85, Diversity:0]
- PO-div: [Accuracy:0.45, Novelty:0.05, Diversity:0.5]

For the Last.fm dataset, we compared PO-acc with WRMF, which is the most accurate stand-alone algorithm on this dataset. PO-acc is much more accurate than WRMF, while also improving on novelty and maintaining the diversity level. The individual PO-nov was compared with PureSVD150, and it performed equally well on accuracy, while delivering a much higher novelty, and only a slightly worse diversity. PO-div was compared against WeightedUserKNN, and it faired equally well on diversity and novelty, while slightly improving on accuracy. It is worth noticing that the individual represented by PO-div is the same individual that maximizes the mean with equal weight (PO-mean). Once again, we were able to find interesting individuals in the Pareto frontier, but we could have reached any of the individuals in Figure 1 (right) by tweaking the weight value for each objective.

6. CONCLUSIONS

In this paper, we propose a hybridization technique for combining different recommendation algorithms, following the Strength Pareto approach. We show that different recommendation algorithms do not perform uniformly well when evaluated in accuracy, novelty and diversity, but our technique allows for the dynamic adjustment of the compromise between these three aspects of user satisfaction. This can be very useful in different scenarios, one example being the personalization of recommendations according to the users. According to [25], "New users have different needs from experienced users in a recommender. New users may benefit from an algorithm which generates highly ratable items, as they need

	Algorithm	Accuracy								Novelty	Diversity
		R@1	R@5	R@10	R@20	P@1	P@5	P@10	P@20	EPC@20	EILD@20
Constituent Algorithms	PSVD50 †	**0.1900**	**0.4155**	**0.5402**	**0.6643**	**0.1900**	**0.0831**	**0.0540**	**0.0332**	0.8070	0.1667
	PSVD150 •	0.1237	0.3203	0.4450	0.5658	0.1237	0.0641	0.0445	0.0283	0.8519	0.1375
	TopPop ◇	0.0722	0.2061	0.2895	0.3994	0.0722	0.0412	0.0289	0.0200	0.7079	**0.2598**
	WRMF	0.1513	0.3453	0.4545	0.5674	0.1513	0.0691	0.0455	0.0284	0.7847	0.1993
	WIKNN	0.1529	0.3564	0.4624	0.5806	0.1529	0.0713	0.0462	0.0290	0.7744	0.2160
	WUKNN	0.1510	0.3364	0.4437	0.5707	0.1510	0.0673	0.0444	0.0285	0.7560	0.2215
	UAKNN	0.0614	0.1762	0.2504	0.3387	0.0614	0.0352	0.0250	0.0169	0.7386	0.2270
	IAKNN	0.0000	0.0072	0.0105	0.0144	0.0000	0.0014	0.0011	0.0007	**0.9723**	0.0106
Baselines	STREAM	**0.1792**	**0.3961**	**0.5169**	**0.6426**	**0.1792**	**0.0792**	**0.0517**	**0.0321**	0.8078	0.1914
	BC	0.0473	0.1657	0.2639	0.4352	0.0473	0.0331	0.0264	0.0218	**0.8210**	0.1609
	EW	0.1562	0.3574	0.4752	0.5980	0.1562	0.0715	0.0475	0.0299	0.7441	**0.2284**
Our Hybrids	PO-mean †	0.1894	0.3991	0.5176	0.6400	0.1894	0.0798	0.0518	0.0320	0.7700	0.2096
	PO-acc †	**0.1999**	**0.4227**	**0.5432**	**0.6705**	**0.1999**	**0.0845**	**0.0543**	**0.0335**	0.7977	0.1897
	PO-acc2 †	0.1946	0.4188	0.5425	0.6659	0.1946	0.0838	0.0543	0.0333	0.8223	0.1685
	PO-nov •	0.1513	0.3725	0.4854	0.6111	0.1513	0.0745	0.0485	0.0306	**0.8597**	0.1411
	PO-div ◇	0.0820	0.2271	0.3269	0.4470	0.0820	0.0454	0.0327	0.0223	0.7138	**0.2563**

Table 1: Results for Recommendation Algorithms on the MovieLens dataset, with the three objectives (i.e., accuracy, novelty, and diversity). The recommender methods variants are grouped into: (i) constituent algorithms, (ii) hybrid baselines, and (iii) our proposed hybrids. We used the symbols: †, •, ◇ to point out our method and the respective baseline. For instance, PSVD150 is the baseline with respect to the selected PO-nov individual. For each group, the best results for each metric are in bold. Underlined values means that the selected individual and the respective baseline are statistically different (95%).

to establish trust and rapport with a recommender before taking advantage of the recommendations it offers." Therefore, our approach could be used to provide new users with the most accurate recommendations as possible, even if the recommendations are not novel at all - so the users would have items to rate, and build trust in the system. The costly part of our technique (the evolutionary algorithm) is performed off-line, and the online cost of choosing an individual in the pareto frontier and weighting the results for different algorithms is very small, since the pareto frontier is comprised of few individuals.

We performed highly reproducible experiments on public datasets of implicit and explicit feedback, using open-source implementations. In our experiments, we demonstrated our technique's ability to balance each of the objectives according to the desired compromise, and we showed some examples of reached solutions that are competitive with the best algorithms according to each objective and almost always better on the other objectives.

7. ACKNOWLEDGMENTS

This work was partially sponsored the Brazilian National Institute of Science and Technology for the Web (grant MCT/CNPq 573871/2008-6), and by the authors' individual grants from CNPq.

8. REFERENCES

[1] G. Adomavicius and A. Tuzhilin. Toward the next generation of recommender systems: A survey of the state-of-the-art and possible extensions. *IEEE Transactions on Knowledge and Data Engineering*, 17(6):734–749, 2005.

[2] R. Baeza-Yates and B. Ribeiro-Neto. *Modern information retrieval*. Addison-Wesley, 2011 (second edition).

[3] X. Bao, L. Bergman, and R. Thompson. Stacking recommendation engines with additional meta-features. In *ACM RecSys*, pages 109–116, 2009.

[4] R. Bell, Y. Koren, and C. Volinsky. Chasing $1,000,000: How we won the netflix progress prize. *ASA Statistical and Computing Graphics Newsletter*, 18(2):4–12, 2007.

[5] J. Bennett, S. Lanning, and N. Netflix. The netflix prize. In *KDD Cup and Workshop in conjunction with KDD*, 2007.

[6] D. Billsus and M. Pazzani. User modeling for adaptive news access. *User modeling and user-adapted interaction*, 10(2):147–180, 2000.

[7] R. Burke. Hybrid recommender systems: Survey and experiments. *User modeling and user-adapted interaction*, 12(4):331–370, 2002.

[8] R. Cecchini, C. Lorenzetti, A. Maguitman, and N. Brignole. Multiobjective evolutionary algorithms for context-based search. *JASIST*, 61(6):1258–1274, 2010.

[9] O. Celma and P. Herrera. A new approach to evaluating novel recommendations. In *ACM RecSys*, pages 179–186, 2008.

[10] M. Claypool, A. Gokhale, T. Miranda, P. Murnikov, D. Netes, and M. Sartin. Combining content-based and collaborative filters in an online newspaper. In *ACM SIGIR Workshop on Recommender Systems*, pages 40–48, 1999.

[11] D. Corne, J. Knowles, and M. Oates. The pareto envelope-based selection algorithm for multi-objective optimisation. In *Parallel Problem Solving from Nature*, pages 839–848, 2000.

[12] P. Cremonesi, Y. Koren, and R. Turrin. Performance of recommender algorithms on top-n recommendation tasks. In *ACM RecSys*, pages 39–46, 2010.

[13] K. Deb. Multi-objective genetic algorithms: Problem difficulties and construction of test problems. *Evolutionary Computation*, 7(3):205–230, 1999.

[14] C. Desrosiers and G. Karypis. A comprehensive survey of neighborhood-based recommendation methods. In F. Ricci,

	Algorithm	Accuracy								Novelty	Diversity
		R@1	R@5	R@10	R@20	P@1	P@5	P@10	P@20	EPC@20	EILD@20
Constituent Algorithms	PSVD50	**0.3859**	0.5997	0.6649	0.7178	**0.3859**	0.1199	0.0665	0.0359	0.8878	0.0557
	PSVD150 •	0.3265	0.5241	0.6055	0.6667	0.3265	0.1048	0.0605	0.0333	**0.8998**	0.0484
	TopPop	0.1879	0.4114	0.5198	0.6224	0.1879	0.0823	0.0520	0.0311	0.8508	0.0755
	WRMF †	0.3834	**0.6148**	**0.7073**	**0.7858**	0.3834	**0.1230**	**0.0707**	**0.0393**	0.8735	0.0669
	WUKNN ◇	0.3272	0.5662	0.6562	0.7340	0.3272	0.1132	0.0656	0.0367	0.8481	**0.0821**
	UAKNN	0.1922	0.3790	0.4712	0.5328	0.1922	0.0758	0.0471	0.0266	0.8605	0.0726
Baselines	STREAM	**0.3898**	**0.6022**	**0.6685**	**0.7185**	**0.3898**	**0.1204**	**0.0668**	**0.0359**	**0.8882**	0.0555
	BC	0.2973	0.5346	0.6026	0.6692	0.2973	0.1069	0.0603	0.0335	0.8606	0.0742
	EW	0.3017	0.5850	0.6785	0.7595	0.3017	0.1170	0.0679	0.0380	0.8473	**0.0807**
Our Hybrids	PO-mean †	<u>0.3387</u>	<u>0.5857</u>	<u>0.6749</u>	<u>0.7487</u>	<u>0.3387</u>	<u>0.1171</u>	<u>0.0675</u>	<u>0.0374</u>	<u>0.8487</u>	**<u>0.0831</u>**
	PO-acc †	**<u>0.4269</u>**	**<u>0.6501</u>**	**<u>0.7239</u>**	**<u>0.7822</u>**	**<u>0.4269</u>**	**<u>0.1300</u>**	**<u>0.0724</u>**	**<u>0.0391</u>**	<u>0.8839</u>	<u>0.0643</u>
	PO-nov •	0.3143	0.5320	0.6141	0.6800	0.3143	0.1064	0.0614	0.0340	**<u>0.9374</u>**	<u>0.0392</u>
	PO-div ◇	<u>0.3387</u>	<u>0.5857</u>	<u>0.6749</u>	<u>0.7487</u>	<u>0.3387</u>	<u>0.1171</u>	<u>0.0675</u>	<u>0.0374</u>	<u>0.8487</u>	**<u>0.0831</u>**

Table 2: Results for Recommendation Algorithms on the Last.fm dataset, with the three objectives (i.e., accuracy, novelty, and diversity). The recommender methods variants are grouped into: (i) constituent algorithms, (ii) hybrid baselines, and (iii) our proposed hybrids. We used the symbols: †, •, ◇ to point out our method and the respective baseline. For each group, the best results for each metric are in bold. Underlined values means that the selected individual and the respective baseline are statistically different (95%).

L. Rokach, B. Shapira, and P. Kantor, editors, *Recommender Systems Handbook*, pages 107–144. Springer, 2011.

[15] A. Eiben and J. Smith. *Introduction to evolutionary computing*. Springer Verlag, 2003.

[16] Z. Gantner, S. Rendle, C. Freudenthaler, and L. Schmidt-Thieme. MyMediaLite: A free recommender system library. In *ACM RecSys*, 2011.

[17] M. Ge, C. Delgado-Battenfeld, and D. Jannach. Beyond accuracy: evaluating recommender systems by coverage and serendipity. In *ACM RecSys*, pages 257–260, 2010.

[18] E. Goldberg. *Genetic Algorithms in Search, Optimization and Machine Learning*. Addison-Wesley, 1989.

[19] J. L. Herlocker, J. A. Konstan, L. G. Terveen, and J. T. Riedl. Evaluating collaborative filtering recommender systems. *ACM Trans. on Information Systems*, 22(1):5–53, 2004.

[20] J. Holland. *Adaptation in natural and artificial systems*. Number 53. University of Michigan press, 1975.

[21] Y. Hu, Y. Koren, and C. Volinsky. Collaborative filtering for implicit feedback datasets. In *ICDM*, pages 263–272, 2008.

[22] C. Hwang. Genetic algorithms for feature weighting in multi-criteria recommender systems. *JCIT*, 5(8), 2010.

[23] M. Jung, J. Oh, and E. Lee. Genetic recommend generating method with real-time fitness function adaption. *International Journal of u-and e-Service, Science and Technology*, 1(1):9–16, 2008.

[24] G. Lekakos and P. Caravelas. A hybrid approach for movie recommendation. *Multimedia tools and applications*, 36(1):55–70, 2008.

[25] S. McNee, J. Riedl, and J. Konstan. Being accurate is not enough: how accuracy metrics have hurt recommender systems. In *CHI Extended Abstract*, pages 1097–1101. ACM, 2006.

[26] Z. Michalewicz. *Genetic algorithms+ data structures*. Springer Verlag, 1996.

[27] B. N. Miller, I. Albert, S. K. Lam, J. A. Konstan, and J. Riedl. Movielens unplugged: experiences with an occasionally connected recommender system. In *IUI*, pages 263–266, 2003.

[28] J. Pagonis and A. Clark. Engene: A genetic algorithm classifier for content-based recommender systems that does not require continuous user feedback. In *UKCI*, pages 1–6, 2010.

[29] R. Pan, Y. Zhou, B. Cao, N. N. Liu, R. Lukose, M. Scholz, and Q. Yang. One-class collaborative filtering. In *ICDM*, pages 502–511, 2008.

[30] M. J. Pazzani. A framework for collaborative, content-based and demographic filtering. *Artificial Intelligence Review*, 13(5-6):393–408, 1999.

[31] F. Rainville, F. Fortin, M. Gardner, M. Parizeau, and C. Gagne. Deap: A python framework for evolutionary algorithms. In *EvoSoft Workshop (GECCO 2012)*, 2012.

[32] N. Srinivas and K. Deb. Multiobjective optimization using nondominated sorting in genetic algorithms. *Evolutionary Computation*, 2(3):221–248, 1994.

[33] S. Vargas and P. Castells. Rank and relevance in novelty and diversity metrics for recommender systems. In *ACM RecSys*, pages 109–116, 2011.

[34] M. Zhang and N. Hurley. Avoiding monotony: improving the diversity of recommendation lists. In *ACM RecSys*, pages 123–130, 2008.

[35] C. Ziegler, S. McNee, J. Konstan, and G. Lausen. Improving recommendation lists through topic diversification. In *WWW*, pages 22–32, 2005.

[36] E. Zitzler, M. Laumanns, and L. Thiele. Spea2: Improving the strength pareto evolutionary algorithm. Technical Report 103, ETH, Zurich, 2001.

[37] E. Zitzler and L. Thiele. Multi-objective evolutionary algorithms: a comparative case study and the strength pareto approach. *IEEE Transactions on Evolutionary Computation*, 3(4):257–271, 1999.

User Effort vs. Accuracy in Rating-based Elicitation

Paolo Cremonesi
Politecnico di Milano
p.zza L.da Vinci 32, Milano
Italy
paolo.cremonesi@polimi.it

Franca Garzotto
Politecnico di Milano
p.zza L.da Vinci 32, Milano
Italy
franca.garzotto@polimi.it

Roberto Turrin
Moviri
via Schiaffino 11, Milano
Italy
roberto.turrin@moviri.com

ABSTRACT

One of the unresolved issues when designing a recommender system is the number of ratings – i.e., the profile length – that should be collected from a new user before providing recommendations. A design tension exists, induced by two conflicting requirements. On the one hand, the system must collect "enough" ratings from the user in order to learn her/his preferences and improve the accuracy of recommendations. On the other hand, gathering more ratings adds a burden on the user, which may negatively affect the user experience. Our research investigates the effects of profile length from both a subjective (user-centric) point of view and an objective (accuracy-based) perspective. We carried on an offline simulation with three algorithms, and a set of online experiments involving overall 960 users and four recommender algorithms, to measure which of the two contrasting forces influenced by the number of collected ratings – recommendations relevance and burden of the rating process – has stronger effects on the perceived quality of the user experience. Moreover, our study identifies the potentially optimal profile length for an explicit, rating based, and human controlled elicitation strategy.

Categories and Subject Descriptors

H.5.2 [**Information Interfaces and Presentation**]: User Interfaces; H.3.3 [**Information Storage and Retrieval**]: Information Search and Retrieval

General Terms

Design, Experimentation, Human Factors

Keywords

Elicitation, Profile Length, New User Problem, User-Centric Evaluation, Perceived Relevance, Perceived Quality, Accuracy

1. INTRODUCTION

Whenever a *new user* joins a Recommender System (RS), the system tries first to learn her preferences in order to provide personalized recommendations as soon as possible. This *preference elicitation process* is fundamental both at cold-start time (i.e., when bootstrapping a new RS) and during the normal operational life of the system, and has effects along multiple dimensions. The preference elicitation strategy can affect the "new user utility" (how well the system can make good

recommendations to the new user who is undergoing the elicitation process) and the "system or community utility" (how well the system can provide good recommendations to *all* users, given what it learns from the new user)[12][18][27]. In addition, the elicitation process represents the user's initial experience with the recommender and is crucial to shape her attitude towards the system and her decision process and behavior (i.e., what she will do with the recommendations).

A wide amount of studies have explored different techniques for preferences elicitation (considering, for example, which questions to ask a new user, which and how many items to propose, in which form and order). A number of design criteria have been identified for making this process more effective in terms of both (new user and community) utility and the quality of the user interaction with the RS. Maximizing both utility and quality of use are somehow conflicting requirements. Obviously, the system needs to learn from new users and to collect enough preferences to generate good and satisfying recommendations; not gathering enough information can result in a poor user model, which may lead to limited accuracy of recommendations and in turn may negatively affect the quality of the user interaction with the RS. Still, requiring users to spend too much time and energy with the system before they receive any recommendation can be annoying, and cause some users to give up the sign up process. Hence the developers of elicitation strategies must face a potential *design tension*: to raise utility by increasing the amount of information gathered from new users, and to make the elicitation process smooth from a user interaction perspective, limiting complexity and user effort during sign-up tasks.

Finding a compromise that solves this tension is not obvious, and represents an unsolved issue in current research on RSs. This paper investigates this challenge for a specific category of elicitation strategies, which can be referred to as explicit, rating based, and human controlled. An *explicit* elicitation process means that the system learns from specific facts provided by the new users about their taste and preferences. In explicit *rating based* elicitation processes, such facts are user's opinions, i.e., binary or multi-scale ratings, on a set of items that, in *human controlled* methods, are selected by the users themselves. In the context of this kind of strategies, one possible measure of the user effort during the elicitation process is the *profile length* - the number of ratings the new user must provide to the system before starting receiving recommendations. Hence we focus on the trade-off that exists between *maximizing* the user utility and *minimizing* the rating effort. More precisely, we explore the following research question: "*Which of the two potentially contrasting "forces" that depend on profile length – user utility and user effort – have stronger effects on the perceived quality of the user interaction?*"

There are two implicit assumptions in the above research question, which are intuitive but not always confirmed by prior studies. The first assumption is that profiles length positively affects user utility. Some works show that profile length of new users is positively correlated to the accuracy of recommendations in term of user utility [12][13][4]. However, this result cannot be easily generalized, as its supporting experiments are limited to item-based collaborative algorithms, and accuracy is measured only in terms of error metrics: RMSE [13] and MAE [4]. Moreover, [28] finds that the correlation between *profile length* and utility is not always present, but it depends on the elicitation strategy adopted. These studies instill some doubts on the general assumption that a longer profile corresponds to more accurate recommendations. We may wonder, for example, to what extent we can claim that the fallout of a content-based recommender algorithm improves with the profile length.

The second assumption in the above research question is that profile length negatively affects the perceived quality of the user interaction because of the rating burden. As we discuss in the next section, several studies have explored the relationship between the characteristics of the preferences elicitation strategy, the user effort during sign-up, and the user interaction with the RS. Still, existing empirical findings reveal some discrepancies and provide different results in different experimental contexts. According to some authors, there is a negative force induced by the user effort which seems to dominate in the user interaction with respect to the force originated from increased utility. In contrast, other studies suggest that as users become aware of the better quality of the recommendations that result from a richer amount of preferences, they somehow feel that the system best understand their taste, and tend to not perceive the extra burden.

For all these reasons, before exploring the general question we need to address two *preliminary* research questions:

"Does the accuracy of a recommender algorithm increase with the profile length?"

"Does the increased burden of ratings collection affect perceived quality of user interaction?"

The three research questions have been addressed by carrying on *three main studies*, involving off-line and on-line experiments that have been replicated in different experimental conditions, involving overall *four recommender algorithms* and *960* users.

The rest of the paper is organized as it follows. The next section provides an overview of the state of the art which is more relevant for the scope of our research, pinpointing the contrasting results that emerge from the current literature. Section 3 describes the design of the three studies. Section 4 presents the findings and discusses the key results. Section 5 draws the conclusions and outlines directions for future work.

2. RELATED WORK

It is generally acknowledged that the strategy and the interface designed to elicit new users' preferences influences the perceived quality of the user interaction with the RS, and has impact on users' decision accuracy and the intention to return [5]. Hence a wide amount of studies have explored the elicitation process, trying to understand how the construction of preferences process takes place [26], which questions to ask a new user [27], which items to propose [26], in which form and order [26]. For more exhaustive reviews of these issues, the reader is referred to [5]. In this section, we shortly outline the works that are more relevant for the context of our research.

In their review [24], the authors discuss the tradeoff between accuracy versus effort, and suggest "minimizing preferences elicitation in the profile initialization". The arguments for this design guideline are both theoretical and empirical. According to behavioral decision theories [14], users are likely to settle on the immediate benefit of saving effort over the delayed gratification

of higher accuracy. A number of works discussed in [28] support this principle (e.g., [12][16]), which is also confirmed by a more recent online study presented in [13], which pinpoints that, in a content-based recommender, a higher perceived system effectiveness is related to reduced effort in the elicitation activity, measured in terms of amount of browsing before receiving recommendations.

Still, not all studies confirm the above guideline. While some works show that the risk in requiring users to provide too much information is to annoy them [19], or to have them give up the sign up process [22][28], other researches show that users are willing to face a more complex elicitation if they feel they are rewarded with useful recommendations [28].

Most authors, such as [27] and [28], consider *system controlled* elicitation methods and explore different measures to select items for the user to rate (e.g., popularity, entropy). These authors conducted a set of off line experiments to evaluate, in terms of accuracy, the strength and weakness of the different item selection strategies. They then compared these results with those ones emerging from online studies where they collected users' opinion about the perceived effort of the signup process in the different experimental conditions. Their findings show that, even if different item selection measures caused an objective variation of effort (measured by the number of pages the user must see before starting to get recommendations), users seemed not to notice the extra burden; hence the authors suggest that the initial recommendation quality (i.e., accuracy) and not the user effort should be considered as the deciding factor to judge (and choose) the desired elicitation strategy.

Similar results are outlined in [29], where experiments show that more elicited ratings do not necessarily imply more perceived effort. However, these findings have a different motivation with respect to [27] and [28]. Users in [29] perceive a low effort with poor quality recommender algorithms, even in the case of a very long elicitation process, as they feel the need to provide more ratings to the RS in order to improve quality.

The experiments described in [22] explore the design tradeoff between user effort and the benefits it brings either to the system (who needs to learn about the user) and the user (who needs to receive useful or convincing recommendations) in different conditions of user control during the elicitation process. The authors compare three interfaces to elicit information from new users, respectively using a system-control method, where the system proposes the lists of items to evaluate, a user-control method, where the user herself selects the items to rate, and a hybrid, mixed-initiative method (a combination of the other two methods). For each interface, they measured the quality of the user models, using a common measure of recommendation accuracy (MAE), and, through a survey, users' perception of the complexity and burden (time/effort) of the sign-up process. They found that the two "pure" interfaces both provide accurate user models. Still, users in the user controlled elicitation group who completed the sign up process were 8-10% less than the other groups and spent twice more time, which indicates that the extra burden had a significantly negative effect on these subjects. On the other hand, the persons who completed the process thought that the system best understood their taste, felt more motivated being in charge of the process, and did not feel the extra effort. This result is confirmed by a study reported in [13], which tests the effects on the user interaction of explicit vs. implicit elicitation methods. These findings show that explicit control over preferences elicitation, in spite of the extra burden on the user, leads to a slightly higher perceived recommendation quality. In addition, in the same study the system effectiveness is judged

Table 1.Studies at-a-glance

Study	Type	Research scope	Metric (dependent variable)	Research question	Algorithms	Users	Profile lengths
1	Off-line simulation	Profile length vs. new user utility	Accuracy (recall and fallout)	*does the accuracy of a recommender algorithm increase with the profile length?*	PureSVD AsySVD DirectContent	60,000 (simulated)	5 – 40
2	On-line user experiment	Burden of the rating process	Global satisfaction	*does the increased burden of ratings collection affect perceived quality of user interaction?*	TopPop	60 (total)	5,10
3	On-line user experiment	Design tension between utility and burden	Global satisfaction Perceived relevance	*which of the two potentially contrasting "forces" that are created by the profile length – user utility and user effort – have stronger effects on the perceived quality of the user interaction?*	PureSVD DirectContent	900 (total)	5,10,20

higher by participants who rate more items (as they noticed an increase of accuracy due to a wider system's knowledge about them).

A number of works have analyzed and compared the cognitive effort related to non-rating-based sign-up processes (tagging items [12], elicitation of user preferences on product features [2], personality quiz [15], and affective feedback [26]). In the experiments described in [26], for example, cognitively less demanding elicitation methods were perceived low in effort and high in liking. Still, follow-up studies reported in the same paper, which explored the trade-off between giving detailed preference feedback and effort, show that users are willing to spend more effort if the feedback mechanism enables them to be more expressive. This provides some insights on the intrinsic motivational factors that lead people to spend more effort to give more detail about a preference.

Few works have studied the impact of the *profile length*, i.e., the number of collected ratings, on the accuracy of recommendations, and results are sometimes contrasting. Some authors highlight that *profile length* is positively correlated to the accuracy of recommendations, both in term of:

(i) *new user* utility [13][4], measured on collaborative RSs with error metrics such as MAE and RMSE;

(ii) *community* utility [11][12], such as eliciting ratings for movies that don't have many, or committing users to do more valuable work for the community (e.g., tagging content and posting comments).

Surprisingly, [28] finds that the correlation between *profile length* is not always present, but depends on the elicitation strategy adopted.

3. THE DESIGN OF THE STUDIES

The three research questions presented in Section 1 (Introduction) have been explored in three main (sub)studies - one off-line simulation and two on-line experiments – summarized in Table 1.

3.1 Study 1: Accuracy

The first study analyzes the accuracy of three recommender algorithms as a function of the new user profile length.

For the evaluation, we used a subset of the Netflix dataset. Our subset consisted of 6,500 items and about 8.8 million ratings given by 250,000 users. In addition, for the purpose of using a content-based algorithm, the dataset was integrated with metadata collected online (e.g., genre, actors, director). The subset was created by extracting movies for which we were able to find the complementary data. The data was added automatically, yet their

quality was manually checked to cleanup any possible redundancy[1].

As recent research founds that improvements in MAE and RMSE are not necessarily the path to improvements in the user experience [18], accuracy has been measured by using information retrieval metrics that are wider adopted in the evaluation of commercial RSs [10]. In particular we focused our attention on *recall* (the percentage of relevant items that are recommended to a user) and *fallout* (the percentage of non-relevant items that are recommended to a user). We did not include precision as accuracy metric because it cannot be estimated in a reliable way unless all ratings are known for all users and all items. Most datasets contains a large number of unrated items: as these are considered irrelevant, they miss a fraction of unknown positive relevance, and lead to precision underestimation [3].

The study considers accuracy vs. profile length for *three algorithms*: two collaborative algorithms (PureSVD and AsySVD) and one content-based algorithm (DirectContent). PureSVD and AsySVD are based on matrix-factorization and previous research shown that their accuracy is one of the best [7][10]. DirectContent recommends items whose content is similar to the content of items the user has positively rated in the past [21]. For instance, in the case of movies the content can be the title, the playing actors, the director, the genre, and the summary. DirectContent is a simplified version of the LSA algorithm described in [1].

The testing methodology adopted in this study is a modified version of the technique described in [10]. Users in the dataset are randomly split into two subsets: *training* set (70% of the users) and *test* set (30% of the users). The test set is further modified by randomly removing 30% of ratings from each user's profile. The removed ratings are the *probe* set, while the modified test set is used to simulate new users. The test set contains 75,000 users and 60,000 of them have a user profile with less than 40 ratings (after removing the probe set).

In order to measure recall, we first trained the algorithm using the ratings in the training set. Then, for each user in the test set and for each item in the probe set that was rated 5-stars by the user, we followed these steps:

- We randomly selected 1,000 additional items that were not rated by the user. We assumed that the user was not interested in most of them.
- We predicted the ratings for the 5-stars rated item and for the additional 1,000 items.

---

[1] The dataset is available for free download at the following address:

http://home.dei.polimi.it/cremones/recsys/Enriched_Netflix.zip

When using the dataset, please cite this paper.

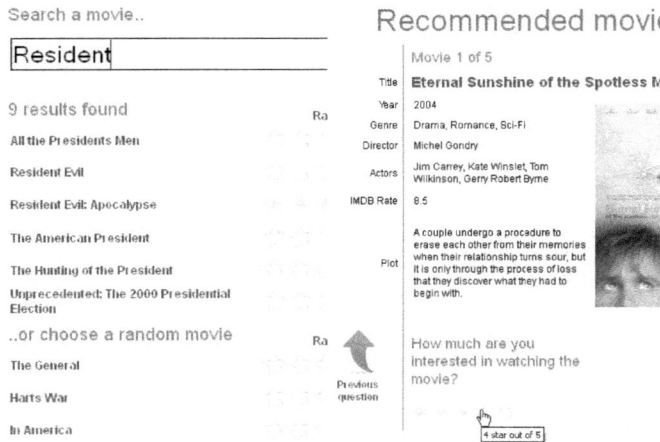

Figure 1 area

Search a movie..

Resident

9 results found

All the Presidents Men

Resident Evil

Resident Evil: Apocalypse

The American President

The Hunting of the President

Unprecedented: The 2000 Presidential Election

..or choose a random movie

The General

Harts War

In America

Recommended movie

Movie 1 of 5

Title	Eternal Sunshine of the Spotless M
Year	2004
Genre	Drama, Romance, Sci-Fi
Director	Michel Gondry
Actors	Jim Carrey, Kate Winslet, Tom Wilkinson, Gerry Robert Byrne
IMDB Rate	8.5
Plot	A couple undergo a procedure to erase each other from their memories when their relationship turns sour, but it is only through the process of loss that they discover what they had to begin with.

Ra

Ra

Previous question

How much are you interested in watching the movie?

4 star out of 5

(a) Providing explicit ratings during navigation

(b) Providing relevance scores for a recommended item

Figure 1. PoliRec framework

- We formed a top-5 recommendation list by picking the 5 items with the largest predicted ratings.

Overall, we generated a number of recommendation lists equal to the number of 5-stars ratings in the probe set. For each list we had a *hit* (e.g., a successful recommendation) if the 5-star rated item was in the list, because we can reasonably state that the item was relevant to the user. Therefore, the overall recall was computed by counting the number of hits (i.e., the number of successful recommendations) over the total number of recommendations

$$recall = \frac{\text{\# times the removed 5 stars item is in the list}}{\text{\# recommendation lists}}$$

Recall is defined as the percentage of items interesting for the user that have been effectively recommended by the system.

A similar approach was used to measure fallout, with the only difference being that we selected 1-stars ratings from the probe set, as we can reasonably state that these ratings refer to items not relevant to the users. The fallout was computed as

$$fallout = \frac{\text{\# times the removed 1 stars item is in the list}}{\text{\# recommendation lists}}$$

Fallout is defined as the percentage of items uninteresting for the user that have been erroneously recommended. Recall and fallout range from 0% to 100%. An ideal algorithm should be able to recommend all interesting items (i.e., recall equals 100%) and to discard all uninteresting items (i.e., fallout equals 0%).

3.2 Study 2 (Burden)

The second experiment investigates the impact of the profile length on the *perceived quality* of the interaction with a RS, measured in term of *global satisfaction*. Global satisfaction is an indicator of how users feel about the overall experience with the system, and represents an important quality factors in user centric approaches to RS evaluation [6].
We measured perceived quality of a RS in the movie domain in *two* different experimental conditions. In each experimental condition we used a recommender system having the same dataset, the same algorithm, and the same user interface, but a *different rating process*, asking users to rate a *different* number of movies. In other words, the two experimental conditions were characterized by different profile lengths (*independent variable*), respectively 5 and 10. As we wanted the only difference in the two experimental conditions to be the objective user effort

(measured by the profile length), we considered a *non-personalized* algorithm, which recommends the same predefined list of items to everybody, regardless his or her user profile, hence not sensitive to profile length. Specifically, we used a simple, non-personalized algorithm (*TopPop*), which recommends top-N items with the highest popularity (largest number of ratings) [7]. As the accuracy of recommendations generated by TopPop does not depend on the user profile, the only measured force playing in the two experimental conditions – long and short profile – is the rating burden. Hence, in this study, we expect the perceived quality to decrease with the profile length.

3.2.1 Instruments
We used the web-based recommender and evaluation framework PoliRec, shown in Figure 1, and powered by the ContentWise[2] recommendation engine. PoliRec supports users with a wide range of functionalities that are common in on-line DVD rental services such as Netflix and Lovefilm. Users can browse a catalog of 2137 movies, retrieving the detailed description of each item, rating it, and getting recommendations. In each experimental condition, the modularization and customization features of PoliRec allowed us to select and apply a specific recommender algorithm among the three that we considered, and to set the desired profile length, i.e., the minimum number of ratings a user has to provide before receiving recommendations. PoliRec also embeds an on-line questionnaire system that allows researchers to collect quantitative and qualitative from the user in a relatively easy way.

3.2.2 Participants
This empirical research involved 60 subjects, who were split in two groups of the same size, and randomly assigned to either experimental condition 1 (short profile) or experimental condition 2 (long profile). The same demographic characteristics were maintained in each subgroup: subjects aged between 20 and 50, evenly distributed into three age categories: 20-30, 30-40, 40-50. None of them had been previously exposed to the system of our study, and none of them had any technical knowledge about RSs.

3.2.3 Procedure
Each participant was initially asked to provide his/her personal information (age, gender, education, nationality, and how many movies they watched per month). Afterwards, users were invited to browse the movie catalog using PoliRec, rating his/her degree of appreciation or interest for the movies encountered at any point during navigation (Fig. 1a), using a 1 to 5 scale (1 = low interest for or appreciation of the movie; 5 = high). Recommendations were generated once X ratings (X = profile length) were collected.

All the users were told they were receiving personalized recommendations on the basis of their input ratings, although all users were receiving exactly the same list of 5 top popular movies. Finally, each user was invited to explore the recommendations, to score perceived relevance for each recommended item on a 1 to 5 scale (Fig. 1b) and to reply to questions regarding global satisfaction.

Each user session lasted between 15 and 20 minutes, and took place in informal environments, such as the university, the interviewer's place, and the interviewee's place. Test results did not present significant differences that can be referred to the execution context. Recruitment and data collection was carried out by a PhD student of the Computer Science Engineering Department at Politecnico di Milano, as part of his PhD research.

---

[2] www.polirec.org – www.contentwise.tv

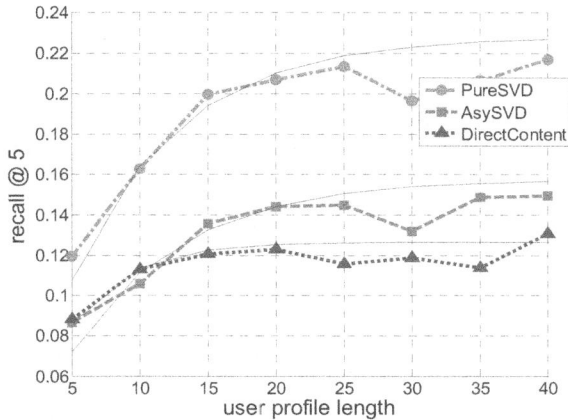

Figure2. Study 1: Off-line simulation: Recall and Fallout of new users on the Netflix dataset as a function of the profile length

3.3 Study 3 (Tension)

In this final study we analyze the *combined* effect on the user interaction of variations in utility (accuracy) and rating burden.

The third research was designed as *two replicated* between-subjects studies. In each study, we measured user's *perceived quality* of a RS in the movie domain in *three* different experimental conditions. Similarly to the previous study, the three experimental conditions were characterized by different profile lengths (*independent variable*), respectively 5, 10 and 20.

User's perceived quality has been operationalized in terms of two measurable factors (*dependent variables*): *perceived relevance* and *global satisfaction* (defined in the previous section). Perceived relevance measures how well the user believes that recommendations match his or her interests, preferences, and taste, and, similarly to global satisfaction, is acknowledged as an important quality factors in user centric evaluation framework [6].

3.3.1 Participants

The overall empirical research involved 900 subjects over a period of two months (December 2011 - January 2012). In *each* of the two replicated studies, we involved 450 subjects who were split in three groups of the same size, and randomly assigned to either experimental condition 1 (profile length = 5 ratings), experimental condition 2 (profile length = 10 ratings) or experimental condition 3 (profile length = 20 ratings). The same demographic characteristics were maintained in each subgroup: subjects aged between 20 and 50, evenly distributed into three age categories: 20-30, 30-40, 40-50. Overall, 52% of the subjects were male and 48% female. None of them had been previously exposed to the system used in our study, and none of them had any technical knowledge about RSs.

3.3.2 Procedure

The third study was executed using the same system used in Study 2 (PoliRec and ContentWise), but two personalized recommender algorithms were used: PureSVD and DirectContent. PureSVD has been chosen because, according to Study 1, it exhibits the best accuracy in term of recall. DirectContent has been chosen as representative of content-based algorithms.

The procedure adopted for this study was identical to the one adopted for Study 2 (Burden), but recruitment and data collection were carried out by a team of 45 master students, organized in 6 groups (2 groups per each experimental condition). They were selected among the best students attending the "Interactive TV" course at our School of Information Engineering. They were trained to perform the study, were given written instructions on the evaluation procedure, and were regularly supervised by a

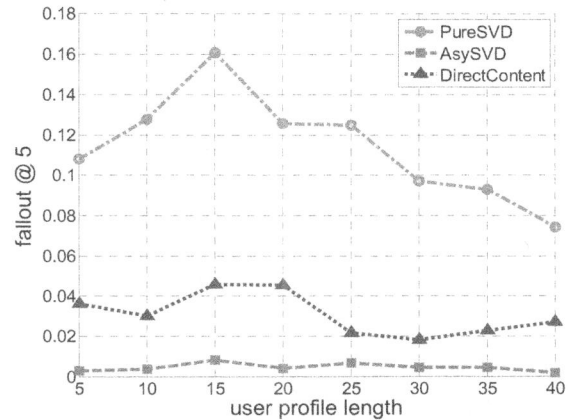

teaching assistant during their activities. They were motivated to perform the evaluation to the best of their capabilities, as the work was constantly monitored and accounted for 20% of their grade in the course.

4. DISCUSSION ON RESULTS

In this section we analyze and discuss the results of the three studies.

4.1 Study 1: Accuracy

From the analysis of Figure 2, we can observe that accuracy in terms of recall clearly improves with the number of ratings in the new user profile for the three tested algorithms. On the contrary, accuracy in terms of fallout does not change significantly, with the exception of PureSVD, where we observe an improvement (i.e., a decrease) in the fallout values.

We can also observe that the increase in recall seems to be bounded by an asymptotic limit, this limit being different for different algorithms. In order to find this asymptotic value, for each of the tested algorithms we have fitted recall with the exponential function below

$$r(l) = a(1 - e^{-bl})$$

where r is the recall, l is the number of ratings in the new user's profile, and a and b are two unknown parameters. Parameter a represents the asymptotic value of the recall, i.e., the maximum recall achievable by an algorithm in the hypothesis of having a very large number of ratings in the user's profile. Parameter b represents the speed at which recall increases towards its maximum limit.

By using least squares, we have fitted the exponential model to the recall data. The resulting parameters are listed in Table 2 and the corresponding exponential functions are plotted in Fig. 2 as continuous lines. The same table shows the profile length values for which recall reaches 80% of its maximum value. *This maximum value is close to 10 ratings for all of the three algorithms.* Hence we should expect that *profile lengths longer than 10 ratings do not increase user perceived relevance in the recommendations* – a hypothesis that we will further analyze in light of the results of Study 3.

Table 2. Study 1: Exponential fitting of recall

	a	b	number of ratings at 80% max recall
PureSVD	0.2281	0.1275	13
AsySVD	0.1579	0.1223	13
DirectContent	0.1265	0.2291	7

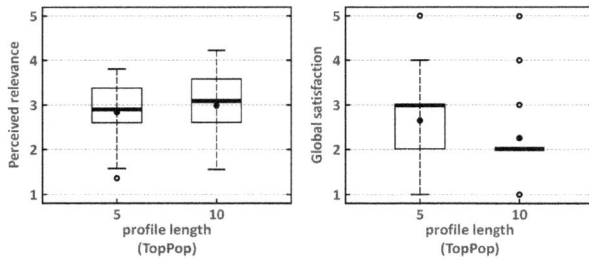

Figure 3.Study 2: Burden of the rating process (non personalized algorithm).

4.2 Study 2: Burden

Figure 3 shows the box plot of perceived relevance and global satisfaction in the two experimental conditions (respectively asking users to rate 5 and 10 movies, but proposing them exactly the same list of 5 top popular movies). Upper and lower ends represent 25th and 75th percentiles. Whiskers extend to the most extreme data point, which is no more than 1.5 times the interquartile range. The median is depicted with a solid line, while the mean with a dot. Outliers are represented with empty circles.

As highlighted by Figure 3, there is *no significant difference* in *perceived relevance* between users with short profile and users with long profile. Moreover, in both cases only 50% of users are moderately satisfied with the quality of the recommendations (perceived relevance greater than 3) and almost no user is greatly satisfied (perceived relevance greater than 4). These findings are somehow expected, as all users were receiving the same list of recommended movies, without any attempt to match their preferences.

Still, our findings show that there is a *significant difference* in *global satisfaction* between users with short profile and users with long profile ($p < 0.1$). This finding is not surprising, and answers to research question 2: in the absence of extra benefits, an increased burden of ratings collection *negatively* affects perceived quality of user interaction, as users with a longer profile have a longer sign-up process with respect to the short profile users but no increase of quality benefits in terms of better recommendations.

4.3 Study 3: Tension

Figure 4 highlights the key results of our investigation on the combined effect on the user interaction of the positive force induced by better accuracy (as confirmed by Study 1) and the negative force induced by increased rating burden (as confirmed by Study 2). Figure 4 shows the box plot of the perceived relevance for both algorithms considered in Study 3. Both of them have a mean relevance between 3 and 4 (the median for all algorithms is greater than or equal to 3). This shows that, on average, users were satisfied with the quality of the recommendations generated by both algorithms in all the three experimental conditions (profile length 5, 10, and 20).

The first notable result is that *perceived relevance changes with the profile length, reaching its maximum when users rated 10 items*. This results is true for both the collaborative (PureSVD and content (DirectContent) algorithms. According to Figure 5, a similar result does not hold for *global satisfaction*, which is substantially unchanged for all algorithms and all profile lengths.

In order to compare the results on perceived relevance more analytically, we ran pair-wise comparison tests using Tukey's method. All tests were run using a significance level $\alpha = 0.1$. When looking at the PureSVD algorithm, *the perceive relevance for new users with 10 ratings in their profile is significantly better*

than the relevance perceived by users with either 5 or 20 ratings ($p < 0.01$). The same applies for the DirectContent algorithm ($p < 0.1$).

4.4 Discussion

The results of the first two studies provide a somehow expected answer to their respective research questions: both accuracy of recommendations and perceived user effort do increase with the number of ratings elicited during the sign-up of new users.

More specifically, the study on accuracy (Study 1) partially supports previous results that investigated the relationship between accuracy and profile length in terms of error metrics (MAE and RMSE) for two item-based recommender systems [12][13]. We have provided empirical evidence that recall improves with the profile length, and this correlation exists for different algorithms. However, the same does not happen for *fallout, which is not correlated with the profile length.*

It is interesting to compare the results of Study 2 with the findings of Study 3. According to Study 2, an increased user effort during the sign-up process negatively affects the perceived quality of user interaction, in terms of global satisfaction, if the extra burden is not compensated by an increased utility (i.e., improved relevance of recommendations). This effect is visible in our experiment because of the low quality (relevance) of the recommendations provided by the non-personalized algorithm we adopted. Users were little rewarded by useful recommendations, regardless of the higher number of ratings, and they were more susceptible to feel the additional burden.

The same phenomenon does not occur in Study 3, where users receive good-quality (relevant) recommendations. If a more demanding rating process is balanced by significantly better recommendations, the global satisfaction is not affected negatively by the increased effort. It is as if the two contrasting forces (accuracy and user effort) generated by profile length on user interaction quality mutually compensate. The potentially positive effects of increased accuracy resulting from a longer profile is eroded by the burden of a more demanding rating process, but this effect is not strong enough to decrease the global opinion of the users towards the recommender system.

Still, when comparing the values of a different indicator of perceived user interaction quality – perceived relevance – in the different profile length conditions, a different phenomenon can be noticed (see Figure 5). With 5 and 10 ratings, we can observe that perceived relevance *increases* with the number of ratings – this was expected, because of the more accurate recommendations. However, when comparing profile lengths with *10 and 20 ratings*, we observe a somehow surprising behavior: *perceived relevance decreases*. This finding confirms our intuition that, as the relevance of recommendations does not increase indefinitely with the profile length, there should be a maximum number of ratings that can be elicited from a new user without having the negative force induced by increased burden overcome the positive force of relevance. The result is also coherent with the findings of Study 1, which pinpoint (see Table 2) that the maximum value for recall is close to 10 ratings for the algorithms considered in Study 3, and therefore suggest that profile lengths longer than 10 do not increase perceived relevance of the recommendations. Still, the motivation of this result from a user interaction perspective is not obvious. A possible interpretation is that the two different quality factors – global satisfaction and perceived relevance – concern different spheres of the user experience. Global satisfaction is a form of "perception", which denotes, in the terminology of [17], whether certain objective aspects of the interaction with a system register with the user at all; perceived relevance is a form of

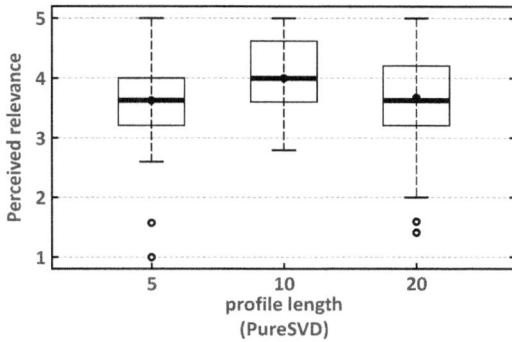

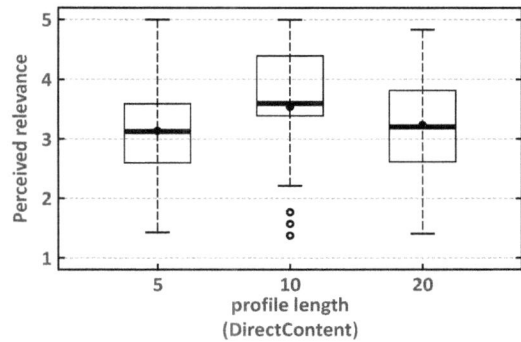

Figure 4. Study 3: Perceived relevance per algorithm and experimental condition (profile length 5-10-20).

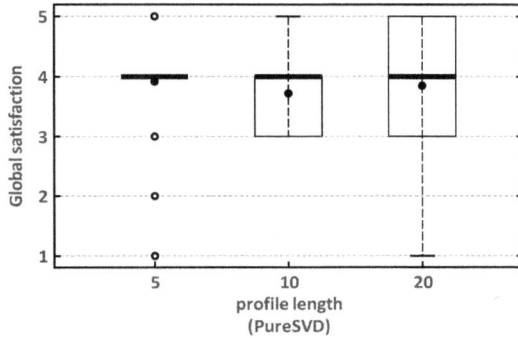

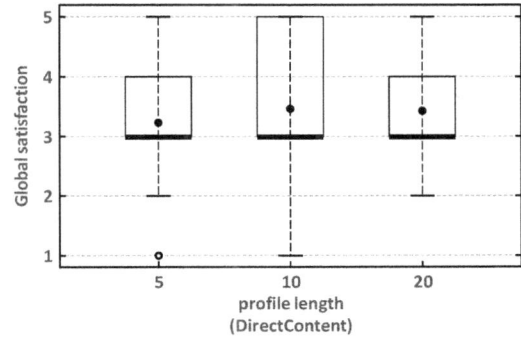

Figure 5. Study 3: Global satisfaction per algorithm and experimental condition (profile length 5-10-20).

"evaluation", which denotes whether a perceived aspect has any personal value for the user. An hypothesis could be that when the users' judgment is related to perception - represented by global satisfaction - all perceived aspects of the experience with a system are integrated into a more holistic perspective, and mutually compensate, unless some forces are significantly stronger or weaker than others, as it happens in Study 2. When the user's judgment moves towards a more value-oriented reflection – as expressed by perceived relevance – the mechanism pinpointed by [24] comes into play, and users are likely to focus on the immediate benefit of saving effort over the less (for them) measurable benefit of higher accuracy.

5. CONCLUSIONS

The studies reported in this paper shed a light on some unresolved issues in RS design: how can the system collect "enough" information (ratings) from the user, in order to learn her/his preferences and improve the quality of recommendations without adding an excess of burden on the user, which may dissolve the perceived benefits of good personalized recommendations? Our work has investigated this problem from multiple perspectives, in the context of rating based user controlled elicitation techniques, thru a set of vast offline and online studies. The validity of our findings is restricted to the actual algorithms and experimental conditions considered. In addition, a weakness of our work is the limited number of user-centric attributes considered for RS quality with respect of the spectrum of user interaction factors proposed by emerging frameworks for the evaluation of RS user experiences [6][17]. Still, in a field where empirical work is particularly complex and resource demanding, our research represents a wide and articulated study that bridges user-centric online evaluation with offline evaluation, and provides contributions for both RS research and design practice.

From a design perspective, our findings (Studies 1 and 3) suggest that the optimal number of ratings in the movie domain is between 5 and 20 ratings (more likely 10 ratings): it is within this range that the contrasting forces induced by profile length achieve a better balance, from a user interaction quality perspective. This result can be distilled into the heuristic: "10 ratings are enough", which can help designers to prioritize design decisions and suggests that there is no real need of building systems that collect extremely long profiles.

From a theoretical perspective, our findings (Study 1) show that it cannot be given for granted that the objective (i.e., statistically measured) quality of recommendations improves with the increase of profile length. We have provided empirical evidence, for different algorithms, that a positive correlation exists between recall and profile length. Still, in our experiments the same phenomenon does not happen for fallout, which is not correlated with the profile length; this infers that recommendations errors depend on the algorithm only, and are largely independent of the number of ratings collected from new users. Finally, the findings of Study 3 confirm and extend some of our previous results [7][8][9]: accuracy metrics measure "weak" user interaction forces, which are less crucial than we may expect in improving the user's perception of a recommender quality. In addition, our experiments indicate that also the burden of the rating process is, in absolute terms, a weak force, not able to strongly decrease the overall satisfaction of users if they are provided with useful recommendations. This may suggest that other "forces" exist, yet to be fully investigated, that intervene in the complex trajectory from the experience of the system to users' quality perception and evaluation of RS interaction [23]. This whole topic deserves a wider and more systematic exploration and calls for the definition of appropriate conceptual frameworks to help us building deeper and more coherent interpretations of empirical results and increasing our understanding of RS elicitation.

6. REFERENCES

[1] Bambini, R., Cremonesi, P. and Turrin, R., 2011. A Recommender System for an IPTV Service Provider: a Real Large-Scale Production Environment. *Recommender Systems Handbook*, F. Ricci, L. Rokach, B. Shapira, and P. B. Kantor, Eds. Springer US, 299-331.

[2] Adomavicius, G. and Tuzhilin, A., 2005. Toward the Next Generation of Recommender Systems: A Survey of the State-of-the-Art and Possible Extensions. *IEEE Trans. on Knowl. and Data Eng.* 17(6) June 2005, 734-749.

[3] Bellogn, A., Castells, P., and Cantador, I., 2011. Precision-oriented evaluation of recommender systems: An algorithmic comparison. In *Proc. of RecSys'11.* ACM, New York, NY, USA, 333–336.

[4] Berkovsky, S., Eytani, Y., Kuflik, T., Ricci, F., 2007. Enhancing privacy and preserving accuracy of a distributed collaborative filtering. In *Proc. of RecSys'07.* ACM, New York, NY, USA, 9-16.

[5] Chen,L. and Pu, P., 2009. Interaction design guidelines on critiquing-based recommender systems. *User Modeling and User-Adapted Interaction* 19(3) August 2009, 167-206.

[6] Hu, R., and Pu, P., 2011. Enhancing collaborative filtering systems with personality information. In *Proc. of RecSys'11.* ACM, New York, NY, USA, 197-204.

[7] Cremonesi, P., Garzotto, F., and Turrin, R., 2012. Investigating the Persuasion Potential of Recommender Systems from a Quality Perspective: an Empirical Study, *ACM Transactions on Interactive Intelligent Systems*, 2(2) June 2012, 1-41.

[8] Cremonesi, P., Garzotto, F. Negro, S. Papadopoulos, A. Turrin, R., 2011. Looking for "good" recommendations: a comparative evaluation of recommender systems. In *Proc. of Human-computer interaction (INTERACT'11).* Vol. Part III. Springer-Verlag, 152-168.

[9] Cremonesi, P., Garzotto, F. Negro, S. Papadopoulos, A. Turrin, R., 2011. Comparative evaluation of recommender system quality. In *Proc. of annual conf. extended abstracts on Human factors in computing systems.* CHI EA '11. ACM, New York, NY, USA, 1927–1932.

[10] Cremonesi, P., Koren, Y. and Turrin, R., 2010. Performance of recommender algorithms on top-n recommendation tasks. In *Proc.of RecSys'10.* ACM, New York, NY, USA, 39–46.

[11] Cremonesi, P., Lentini, E., Matteucci, M., Turrin, R., 2008. An evaluation methodology for recommender systems. In *Proc. of Int. Conf. on Automated Solutions for Cross Media Content and Multi-channel Distribution*, 224–231.

[12] Drenner, S., Sen, S., and Terveen, L., 2008. Crafting the initial user experience to achieve community goals. In *Proc. of RecSys'08.* ACM, New York, NY, USA, 187-194.

[13] Golbandi, N., Koren, Y., and Lempel, R., 2010. On bootstrapping recommender systems. In *Proc. of the 19th ACM Int. Conf. on Information and knowledge management.* ACM, New York, NY, USA, 1805-1808.

[14] Haubl, G., Trifts, V., 2000. Consumer decision making in online shopping environments: the effects of interactive decision aids. *Mark. Sci.*19, 4–21.

[15] Hu, R. and Pu., P., 2009. A comparative user study on rating vs. personality quiz based preference elicitation methods. In *Proc. of Int.l Conf. on Intelligent user interfaces* (IUI '09). ACM, New York, NY, USA, 367-372.

[16] Jones, N., Pu, P., 2007. User technology adoption issues in recommender systems. In *Proc. of Networking and Electronic Commerce Research Conf.* (NAEC '07), 379–394.

[17] Knijnenburg, B.P., Willemsen, M.C., Gantner, Z., Soncu, H. and Newell, C., 2012. Explaining the user experience of recommender systems, *User Modeling and User-Adapted Interaction*, 22(4-5) March 2012, 441-504.

[18] Konstan, J.A., Riedl, J., 2012. Recommender systems: from algorithms to user experience. *User Model. User-Adapt. Interact.* 22, 101-123.

[19] Lekakos, G., Giaglis, G. M., 2007. A hybrid approach for improving predictive accuracy of collaborative filtering algorithms. *User Modeling and User-Adapted Interaction* 17(1-2) March 2007, 5–40.

[20] Liu, N.N., Meng, X, Liu, C., and Yang, Q., 2011. Wisdom of the better few: cold start recommendation via representative based rating elicitation. In *Proc. of RecSys'11.* ACM, New York, NY, USA, 37-44.

[21] Lops, P., De Gemmis, M., and Semeraro, G., 2011. Content-based recommender systems: State of the art and trends. In *Recommender Systems Handbook*, F. Ricci, L. Rokach, B. Shapira, and P. B. Kantor, Eds. Springer US, 73–105.

[22] Mcnee, S., Lam, S., Konstan, J., and Riedl, J., 2003. Interfaces for eliciting new user preferences in recommender systems. In *Proc. of Int. Conf. on User Modeling*, 178–188.

[23] Mcnee, S. M., Riedl, J., and Konstan, J. A., 2006. Being accurate is not enough: how accuracy metrics have hurt recommender systems. In *CHI '06 extended abstracts on Human factors in computing systems*, ACM, New York, NY, USA, 1097–1101.

[24] Pu,.P., Chen, L., Hu, R., 2012. Evaluating recommender systems from the user's perspective: survey of the state of the art. *User Modeling and User-Adapted Interaction*, 22(4), 317-355.

[25] Pu P. Chen L, Hu R., 2011. A User-Centric Evaluation Framework for Recommender Systems. In *Proc. of RecSys'11*, ACM, New York, NY, USA, 157-164.

[26] Pommeranz, A., Broekens, J., Wiggers, P., Brinkman, W.P.,Jonker, C.M., 2012. Designing interfaces for explicit preference elicitation: a user-centered investigation of preference representation and elicitation process, *User Modeling and User-Adapted Interaction*, 22(4-5) 15 March 2012, 357-397.

[27] Rashid, A.M., Albert, I., Cosley, D., Lam, S.K., McNee, S.M., Konstan, J.A., and Riedl, J., 2002. Getting to know you: learning new user preferences in recommender systems. In *Proc. of Int. conf. on Intelligent user interfaces* (IUI '02). ACM, New York, NY, USA, 127-134.

[28] Rashid, A.M., Karypis, G., and Riedl, J., 2008. Learning preferences of new users in recommender systems: an information theoretic approach. *SIGKDD Explorer Newsletter 10*, 90–100.

[29] Swearingen, K. and Sinha, R., 2000. Interaction design for recommender systems. *Designing Interactive Systems.* London, 25-28.

TasteWeights:
A Visual Interactive Hybrid Recommender System

Svetlin Bostandjiev, John O'Donovan, Tobias Höllerer
Department of Computer Science
University of California, Santa Barbara
{alex, jod, holl}@cs.ucsb.edu

ABSTRACT

This paper presents an interactive hybrid recommendation system that generates item predictions from multiple social and semantic web resources, such as Wikipedia, Facebook, and Twitter. The system employs hybrid techniques from traditional recommender system literature, in addition to a novel interactive interface which serves to explain the recommendation process and elicit preferences from the end user. We present an evaluation that compares different interactive and non-interactive hybrid strategies for computing recommendations across diverse social and semantic web APIs. Results of the study indicate that explanation and interaction with a visual representation of the hybrid system increase user satisfaction and relevance of predicted content.

Categories and Subject Descriptors

H.3.3 [**Information Storage and Retrieval**]: Information Search and Retrieval–*relevance feedback*; H.5.2 [**Information Interfaces and Presentation**]: User Interfaces–*graphical user interfaces (GUI), user-centered design*

Keywords

User Interfaces, Visual Knowledge Representation, Hybrid Recommender Systems, Data Integration, Social Web

General Terms

Algorithms, Design, Experimentation, Human Factors

1. INTRODUCTION

The social web has become the dominant modality for distribution of media and collection of user-provided content such as text articles, feedback ratings, and comments for instance. Recommendation systems play an increasingly important role in this domain as they serve to filter and refine a user's information space according to their personal tastes and current requirements. However, social web APIs and

other data sources are constantly evolving, and traditional recommender system techniques such as automated collaborative filtering (CF) [9, 16] need to adapt to the changing data environment on the social web. For example, the traditional approach of pre-processing a large, static database of user ratings to produce a correlation matrix (i.e: the Netflix approach) to finding recommendation partners, can not be applied to user preference data on Facebook because of privacy restrictions in their API. However, we demonstrate that with some adaptation to the CF algorithm, Facebook data can still be effectively harnessed to produce useful personalized recommendation in a collaborative manner.

We make the following contributions addressing the challenges of evolving and emergent data sources for recommender systems: We present two enhancements of traditional recommendation processes. First, a novel and synergistic approach to combining predictions from multiple sources on the social web, such as social (Facebook), content-based (Wikipedia) and expert-based (Twitter) recommendations. Second, a novel interactive user interface which serves to both explain the provenance of recommended content in a transparent manner, and to elicit preference data and relevance feedback from users at recommendation time.

To evaluate our approaches, we introduce *TasteWeights*, a hybrid music recommendation system with an interactive interface, allowing users to both understand and control aspects of the recommendation process that would otherwise go unnoticed. Figure 1 shows a snapshot of the interface, highlighting three social web context sources with a variety of weighting options, along with item recommendations on the right side of the interface. A video demonstration of the system can be watched at [1]. Using this system, a user evaluation was performed with 32 participants. The evaluation used participants' own social connections and music preference data. The study addressed the following core questions:

- What (if any) is the benefit of explaining a hybrid recommendation process through a user interface?

- How does interaction at recommendation time affect accuracy and user experience?

- Can a hybrid strategy combining different social APIs provide better recommendations than traditional CF (over Facebook music preferences)?

While the *TasteWeights* system (Figure 1) is capable of recommending any media content listed in a Facebook profile, such as books, TV shows, and movies, recommendations

[1]TasteWeights video demo: `http://bit.ly/TasteWeights`

Figure 1: Screenshot of *TasteWeights* illustrating the main interaction features: (a) changing the weight of an item (b) restoring the default value of an item (c) removing an item (d) changing the weight of a context source (e) changing the visible portion of a context source (f) navigating context sources

described in this paper were restricted to music items in order to reduce complexity in our evaluations.

2. TASTEWEIGHTS OVERVIEW

Figure 1 shows a screenshot of the *TasteWeights* music recommender system. The system is organized into three distinct layers and computational steps.

1. *Profile Layer*: This leftmost layer contains the user's profile. In this case, the profile consists of "liked" music items sourced from Facebook. This layer supports re-rating of profile items through an array of sliders.

2. *Context Layer*: The central or "context" layer contains items coming from different sources that can be used to produce recommendations, in this case, Wikipedia, Facebook and Twitter. This layer also contains sliders for weighting of these items, and control of hybridization through weighting of sources.

3. *Recommendation Layer*: The rightmost layer contains the combined recommendations from each source, ranked by relevance. Edges are displayed to illustrate the provenance of each recommended item.

As the system's name, *TasteWeights*, implies, users are encouraged to adjust their tastes via interactive slider-weights and other UI components. While a user drags a slider,

weights of the items connected via outgoing links change accordingly in real time. For example, in Figure 1, as the user drags a slider for "Pink Floyd" to the right, the value of "English Rock Music Groups" increases simultaneously and so also do the values of "Beatles", "Rolling Stones", "Radiohead", and "Oasis". Section 4 describes our design decisions and methodologies in detail.

3. RELATED WORK

Research related to this work falls into the categories of hybrid recommender systems and the role of interaction and visualization for recommendation systems in general.

3.1 Hybrid Recommender Systems

Traditional recommender system techniques such as collaborative filtering (CF) [9, 16], content-based [11, 6], and knowledge-based filtering [17], each have unique strengths and limitations. For example, CF suffers from sparsity and cold start problems [16], while content-based approaches suffer from narrowness and require descriptions. However, a hybrid approach can use one approach to make predictions where the other fails, resulting in a more robust recommender system [12]. Burke [4] proposes a taxonomy of hybrid systems. For example, recommendation algorithms can work in parallel before combining their results, may be pipelined such that the output of one algorithm is the input of the next, or may be combined into one monolithic algorithm. *TasteWeights* falls into the parallelized design class, since our approach firstly generates predictions from individual recommender system techniques, then applies a hybridization strategy afterwards.

3.2 Visualization for Recommender Systems

The focus of this paper is on a visual, interactive interface that supports control of a hybrid recommender system. Through visualization we are creating an "explanation interface" for our recommender system, and, moreover, allowing the end user to control aspects of the hybridization and other elements in the recommendation process through a simple informative and interactive interface [19]. A prominent work in this area is Herlocker's study of recommendation explanations [9]. Herlocker et al. evaluate a "white box" conceptual model of recommendation as opposed to the run-of-the-mill black box approach. They present a user study where 21 different recommendation interfaces are presented to users, explaining various types of internal information from the recommender algorithm. Their general findings agree with Middleton's [13], in that "explanation interfaces lead to improved acceptance of a predicted rating." Herlocker's work highlights justifications for explaining recommendations through some form of interface, and those justifications also apply in our design decisions for the *TasteWeights* recommender system. According to Herlocker, explanatory interfaces:

- help users justify and understand the reasoning behind a recommendation, so that confidence can be decided
- increase users' sense of involvement. (i.e. keep the user "in the loop")
- educate users about the recommendation process
- increase users' acceptance of recommendations

In addition to these roles of an explanation interface, we posit that interaction can further aid in the recommendation process, namely by:

- allowing users to dynamically update their preference profile during a recommendation session
- enabling users to provide ratings directly on the entities used to produce recommendations
- supporting exploration of "what-if" scenarios based on different profile configurations

Previous work focused on interactive visualization of genre information to elicit preference-feedback from users to enhance the quality of movie recommendations generated from a large scale data set [15, 14]. Gretarsson et al.'s Small-Worlds system [7] explored the effect of interactive visualization for a movie recommendation system. They found that an interactive interface helped produce more accurate recommendations and increase user acceptance of the predictions. The unique contribution in this work is an analysis of factors across both hybrid recommendation systems and interactive explanatory interfaces.

4. SYSTEM DESIGN & INTERACTION

Following Herlocker's guidelines in [9], *TasteWeights* was designed to improve the user's understanding of how the hybrid recommender system works under the hood. Burke

Figure 2: Additional info shown when an item gets clicked: (a) profile, recommendation, or Wikipedia item (b) Facebook friend (c) Twitter expert

[4] suggests that recommender systems have three distinct parts: input, background, and suggestions. *TasteWeights* follows a similar design structure, as shown by the three columns in Figure 1. Multiple UI controls allow users to fine-tune their preferences and receive real-time feedback on how their actions affect the output. Users are able to tweak the underlying algorithms by changing weights associated with individual items (Figure 1(a)). As the user moves a slider associated with a weight, they can see how that change affects the system as a whole. Individual items are enhanced by additional information when clicked, in a detail-on-demand fashion [18]: profile, recommendation, and Wikipedia items are accompanied by an image and abstract, Facebook friends are shown with their profile photos and music profiles, and Twitter experts are accompanied by their items of expertise (Figure 2). On a larger scale, users are able to express their relative trust in each context source by manipulating a slider for each context source (Figure 1(d)). The system provides dynamic recommendation feedback in real time while these interactions are being performed.

To emphasize the hybridity of the system, distinct colors for each context source are used as visual cues. The opacity of each context source box changes proportionally with the weight of the source expressed through its source slider. Any edges connected to a context source item inherit the context source's color. The context column usually cannot fit within the screen. To handle this, we have developed two UI features: a slider allowing to resize the visible portion of the context source (Figure 1(e)), and a scrollbar revealing current position within the column and expressing the relative source size through color coding (Figure 1(f)).

5. CONTEXT SOURCES

TasteWeights is a general solution that can be applied to a wide range of data sources on the social web. For our evaluation we have chosen three popular social APIs which relate to three different core recommender system techniques: Wikipedia (content-based / semantic), Facebook (collaborative / social), and Twitter (expert-based).

5.1 Wikipedia

Wikipedia is the most popular community-driven online encyclopedia, consisting of millions of user-provided articles, some of which are templatized and contain both free text and more structured, tabular data. We query Wikipedia for articles and categories that are most relevant to the user's music profile. The results are presented in the top part of the middle (context) layer in Figure 1. We find relevant Wikipedia articles indirectly through DBpedia [1, 2], a semantic web resource that crawls structured data from Wikipedia and organizes it into a database that is queryable through a SPARQL endpoint[2]. The database is an RDF store of subject-predicate-object triples. Subjects and objects correspond to Wikipedia articles and each predicate is a labeled link between two articles. For example, the band "U2" is linked to the music genre "Alternative Rock" via a link labeled "genre". *TasteWeights* leverages Wikipedia by mapping music items in a user's profile to actual Wikipedia articles. For example, "Pink Floyd" profile item corresponds to `http://en.wikipedia.org/wiki/Pink_Floyd`).

5.2 Facebook

Facebook is the world's largest online social network with over 800 million active users in April 2012 [3]. Although their API is limited by privacy restrictions, some music preference data is still accessible from direct friends of a user who is authenticated to the API. Facebook music preference data is used to bootstrap the *TasteWeights* system. The user's music profile items all map to specific pages that represent the artists. In the context layer of Figure 1, Facebook items are a user's friends who have at least one liked item in common with the user, i.e. have similar tastes to the user. This data is mined through the Facebook Graph API[4].

5.3 Twitter

Twitter is a popular Social web microblogging service. Users can upload short text "tweets' through a variety of applications and devices. Twitter is commonly used for propagation of news events and for following expertise on various topics. Accordingly, we incorporate this service to produce expert-based recommendations for our *TasteWeights* system. Specifically, a user's music profile items can be mapped to hash tags. For example, "Pink Floyd" corresponds to the twitter hash tag *#pinkfloyd*. In our implementations, an online service from wefollow.com is used to find Twitter experts on the items in the user's music profile. wefollow.com is a user dictionary that curates lists of the most influential Twitter users for a large number of domains.

6. APPROACH

Now we provide a description of the various models used to gather data and predictions from each source. In the context of Figure 1, computations flow from left-to-right across the three columns. Each data item in the system is associated with a "score" (analogous to a weight) from 0 to 1 that is visually encoded in the slider bars.

Step 1: Profile Initialization. To initialize a user profile, music preference information is gathered though the Facebook graph API. The list of music preferences in the user's profile are used as input to each of the source-specific computational models described in Section 6.1. Preference information for music on Facebook is binary, that is, no scaled

preference rating is available. Accordingly, each profile item is initialized with a score of 0.5.

Step 2: Modeling Similarity. The three context sources (middle column in Figure 1) provide different items that can potentially generate recommended items. Each source requires a different model / strategy to extract these source items (i.e: Facebook friends, Wikipedia articles, Twitter experts). Those are described in 6.1 .

Step 3: Generating Predictions. Once relevant items have been collected by each source, the next step is to generate predictions. Individual recommendation scores are calculated as the sum of the weights of all items within the source that are linked to the recommendation. In Section 6.2 we discuss a few different methods for combining the recommendation scores from individual sources.

6.1 Source-specific Models

This section describes the specific modelling and prediction processes for each context source.

Wikipedia Model. Facebook music profile items are mapped to Wikipedia articles through dynamic queries over Google's Search API. For each profile item, a search is performed within the English Wikipedia and the top result is selected. Next, (as we discussed in Section 5) a query is issued to DBpedia's SPARQL endpoint for items (articles and categories) that are linked to at least two music items in the active user's profile. This can be viewed as a content-matching approach to generating recommendations. An overall weight for each Wikipedia item (articles or categories) is calculated as the sum of the individual user-provided weights of the profile items it shares links with, as represented by the slider bars in the interface. To generate recommendations from Wikipedia, a further query is sent to DBPedia, this time to retrieve new (recommendation) items that are linked to at least two of the relevant Wikipedia items that were found in the previous step. The recommendation items are filtered by type, in the context of music: "Musical Artist" or "Band". For example, as shown in Figure 1, the article for "Pink Floyd" has a semantic link to the category "English Rock Music Groups", which in turn is linked to "The Beatles". In this manner, "The Beatles" becomes a candidate recommendation from this source.

Facebook Model. Our recommendation strategy for Facebook is similar to traditional collaborative filtering, in that the opinions of similar friends are used to generate predictions. These friends are ranked according to their similarity with an active user's taste using a Pearson's correlation coefficient. We have adapted the correlation formula to account for the fact that Facebook items in users' music profiles are binary and do not contain scaled ratings. The similarity of each Facebook friend to the active user is given by:

$$W_{friend_i} = \frac{TWCI_{user,friend_i}}{\sqrt{TWI_{user}^2 \cdot TWI_{friend_i}^2}} \qquad (1)$$

where $TWCI_{x,y}$ is the total weight of the items x and y like in common, and TWI_x is the total weight of items liked by user x.

[2] `http://dbpedia.org/sparql`
[3] `http://facebook.com/press/info.php?statistics`
[4] `https://developers.facebook.com`

Twitter Model. In the Twitter domain, the goal is to source users that have expertise in the items listed in the active user's profile. We first map profile items to Twitter hash tags (i.e. Michael Jackson gets mapped to #michaeljackson) and so on. Next, we retrieve the top Twitter experts on those items according to wefollow.com for each hash tag. For example, Pink Floyd experts are found here: `http://wefollow.com/twitter/pinkfloyd`. For each expert found, recommendations are produced using the following equation to compute a score for each candidate item.

$$S_{exp_i, item_j} = \frac{|Exp_{item_j}| - Rank_{exp_i, item_j}}{|Exp_{item_j}|} \qquad (2)$$

where $Rank_{exp, item_j}$ is the expert's ranking for the item and $|Exp_{item_j}|$ is the total number of experts for the item. For example, if an expert is ranked 20th out of 100 experts for a specific item the expert gets a score of 0.8 for that item. The overall weight of a Twitter expert is determined by the linear combination:

$$W_{exp_i} = \sum_{Linked(prof_j, exp_i)} (W_{prof_j} \cdot S_{exp_i, prof_j})$$

All hash tags that resolve to bands or musical artists that the relevant Twitter experts have knowledge in are potentially recommendable.

6.2 Hybrid Strategies

As pointed out in Section 3, *TasteWeights* uses a parallelized design, that is, predictions are made by each source-specific model individually and then combined in a final processing step. Parallelized hybrids are further classified by Burke [4] into mixed, weighted, and switching hybrids. We describe and evaluate the following three strategies used in *TasteWeights*: Weighted, Mixed and Cross-Source. In order to perform the hybrid step we first need to resolve entities across the different context sources. For example, the system needs to know that the Wikipedia article on the band "Asian Dub Foundation" corresponds to a page in the Facebook graph and to the Twitter hash tag #adf. Of the three sources used in this paper, Wikipedia presents the most evolved semantic graph in terms of completeness and non-redundancy [8], and therefore it is the best available resource for entity-resolution. Accordingly, we use Google Search API to map all recommendations to Wikipedia articles to confirm their identities. After this mapping stage we proceed with our different hybrid methods.

Weighted Hybrid. In this approach, a score for each recommended item is simply the weighted sum of the recommendation scores for each source. Weights for each context source are user-configurable through interactive sliders in the *TasteWeights* interface as described earlier.

Automatically optimizing the set of weights for each context source is desirable, but not trivial. Empirical bootstrapping can be used to calculate an optimal weighting scheme [20], however, historical data is needed for this approach. The P-Tango system looks into dynamic optimization of weights of a content-based and a collaborative recommender [5]. In their model, dynamic optimization starts with a uniform distribution of weights and dynamically adjusts the weights to minimize predictive error as users rate

more items. This procedure can be applied on a per item and per user basis and the results can be combined and used for new users of the system. The evaluations presented in this paper do not use dynamic weighting, since the focus is on other interactive aspects of the system. For simplicity, our weights were fixed evenly across the three sources.

Mixed Hybrid. In this approach, recommendations for each source are ranked, and then the top-n are picked from each source, one recommendation at a time by alternating the sources. This approach only considers relative position in a ranked list and does not include individual recommendation scores. In cases where a recommendation is produced by multiple context sources (i.e. was previously picked from another source) the algorithm simply selects the next recommendation from the ranked list for that source.

Cross-Source Hybrid. This approach strongly favors recommendations that appear in more than one source. We believe that if a recommendation is generated from more than one context source / algorithm, i.e. by both collaborative filtering (Facebook) and content-based recommendation (Wikipedia), then it should be considered more important. To compute a final recommendation set, the weighted hybrid approach (Section 6.2) is first applied, then each recommendation's weight is multiplied by the number of sources in which it appeared. The following equation describes the the cross-source hybrid approach:

$$W_{rec_i} = \sum_{s_j \in S} (W_{rec_i, s_j} \cdot W_{s_j}) \cdot |S_{rec_i}|$$

where $|S_{rec_i}|$ is the number of context sources recommendation i was generated by (i.e. 1, 2, or 3).

7. EVALUATION

We evaluated aspects of the *TasteWeights* system in terms of both recommendation accuracy and user experience. We compared nine methods: recommendations generated by the three individual sources (Wikipedia, Facebook and Twitter; cf. Section 7.2.1), recommendations produced by the three hybrid methods (Weighted, Mixed, and Cross-Source; cf. Section 6.2), and recommendations generated by three interaction variants that allowed users to fine-tune their preferences. The interaction variants differed based on how much of the recommendation process users could reflect on:

Profile Interaction. Users could only view and fine-tune items weights in their profile (left column in Figure 1).

Sources Interaction. In addition to profile tuning, users were able to change the weights on context source items (middle column).

Full Interaction. In addition to profile and sources tuning users could see the effects of their tuning actions on the recommendations (all columns were visible). Note, this is the default interface for the system.

The three different interactive methods could potentially use any of the hybrids as their underlying algorithm. To reduce complexity in our study, the best performing hybrid strategy was chosen for use in the three interactive methods.

A pilot study consisting of 7 user trials was performed to find that the cross-source hybrid outperformed the others.

7.1 Setup

We performed a controlled user study (N=32) with the objective of answering the research questions posed in our earlier discussion (Section 1).

To assess the effects of explanation and interaction with the system on user experience and understanding of the recommendation process a qualitative analysis was performed based on a post-study questionnaire. We asked questions on how useful the explanation of hybridity was and how users perceived refining different aspects of the system.

We also performed a quantitative analysis on the performance of the nine recommendation methods. We used a within-subjects experimental design. The independent variable was recommendation method and the dependent variable was accuracy. Each of the nine methods produced a ranked list of recommendations. To compute the overall accuracy of a given recommendation list, we first asked the user to rate the top 15 recommendations in the list in random order and then used Breeze's *R-Score* "utility" metric [3] to determine a utility score for the list. The metric assumes that the value of recommendations decline exponentially based on position in the recommended list. The utility of a given recommendation list for user u is given by:

$$R_u = \sum_j \frac{max(r_{ui_j} - d, 0)}{2^{\frac{j-1}{\alpha-1}}} \qquad (3)$$

where i_j is the item in the jth position, r_{ui} is user u's rating of item i, (1 to 5 stars), d is Breese's "don't care" threshold (experimentally chosen as 2 stars), and α is the half-life parameter, which we set to 1.5, controlling the exponential decline of the value of positions in the ranked list.

We considered measuring accuracy via popular approaches including variants of Root Mean Squared Error (RMSE) and Mean Average Error (MAE). However, we opted against using those for two reasons: first, our system's input is music that is not rated by the user but only "liked", so in a way all Facebook "likes" correspond to 5-star ratings; and second, because in the real world people look mostly at the top-n recommendations than the complete recommendation list.

7.1.1 Participants

32 users participated in the main study (17 male, 15 female) ranging in age from 19 to 35. Participants were recruited through a university-wide experimental program and were paid a nominal amount for their time. Most participants were students and spanned 10 different majors. Pre- and post- study questionnaires were completed by each participant. Most participants reported that they were regular Facebook users (86% daily), and that they frequently used Wikipedia (36% daily, 45% weekly). There was a notable drop-off in reported use of Twitter in the study group, with 5% daily users, 18% weekly users and 63% who had never used the microblog. Since our system is bootstrapped from a participant's Facebook music profile and associated network, probe questions were asked to assess the amount of available data. On average, participants had 416 Facebook friends (notably far larger than the average of 130 for the

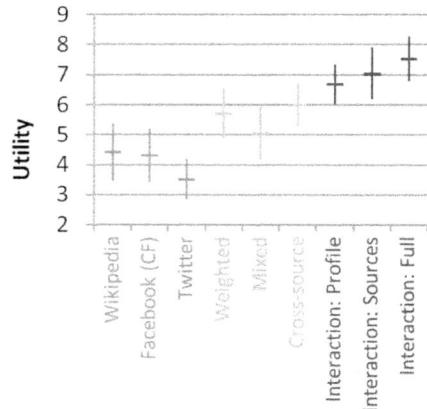

Figure 3: Plot of means of recommendation methods over utility with 95% confidence intervals. The utility metric is described in 7.1.

Method 1	Method 2	Diff	Lower	Upper	P Val
Cross Hybrid	Wikipedia	1.568	0.119	3.017	0.023
Cross Hybrid	Facebook (CF)	1.678	0.229	3.127	0.011
Cross Hybrid	Twitter	2.477	1.028	3.926	0.000
Full Interaction	Cross Hybrid	1.542	0.935	2.991	0.027

Table 1: Results from a Tukey post-hoc analysis of the recommendation methods: multiple comparisons of means with 95% family-wise confidence level

social network[5]), and Dunbar's optimal number of friend associations (150)[10]. Participants reported that they were familiar with recommender systems such as Pandora and Netflix (3.8 out of 5). When asked about their primary methods for discovering new music, participants' top choices were "Friends" (45%), "Pandora" (36%) and "Radio" (23%).

7.1.2 Procedure

After completing the pre-study questionnaire, participants were given an explanation of the system controls and approximately one minute to familiarize themselves with the various UI components. Then, participants were asked to tweak the system using each of the three interactive methods, described in Section 7.1, which were presented in a random order. After that, users were asked to rate a randomized list of output from each of the nine tested methods. The purpose of this task was for participants to rate 15 recommendations produced by each approach on a 5 star scale, 1 being the lowest and 5 being the highest. Ratings were performed in bulk at the end of the study. To rate unknown bands the user was given the chance to look at the band's LastFM[6] page. The page not only contains relevant information about the band but also music samples. After having rated all recommendations, users were asked to answer a post-questionnaire and provide feedback on their perception of the system.

7.2 Recommendation Accuracy

Figure 3 presents a plot of the means of the nine methods over utility with 95% confidence intervals. Overall, the

---

[5]http://facebook.com/press/info.php?statistics
[6]http://www.last.fm

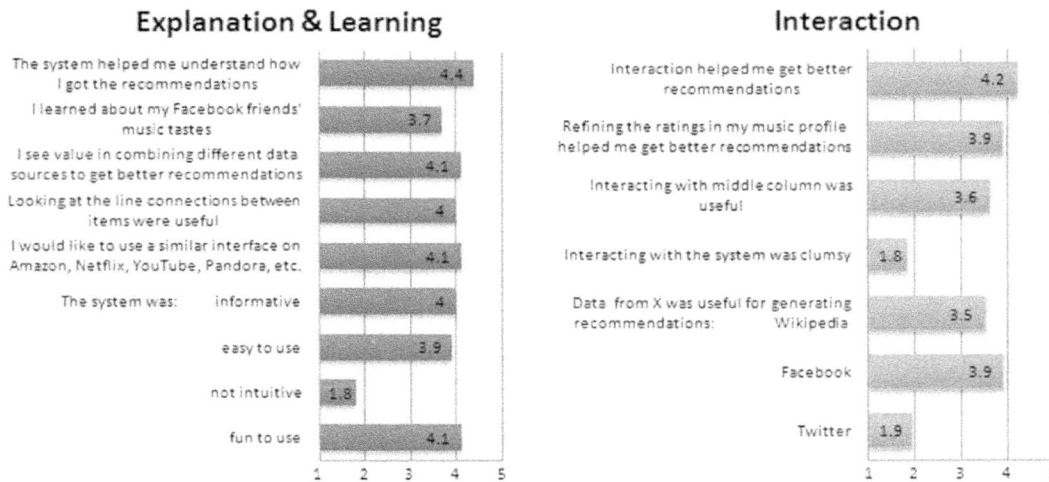

Figure 4: Post-study questionnaire results

full interaction method was found to have the highest utility score, while the twitter method produced the lowest utility score. On average, the hybrid methods performed better than the individual ones, and the interactive methods performed better than the hybrids.

Mauchly's test showed a violation of sphericity against Method ($W(44) = 0.005$, $p = 0.01$). We ran one-way repeated-measure ANOVA and made Greenhouse-Geisser correction ($\varepsilon = 0.49$). It revealed a significant effect of the method variable on utility ($F(3.72, 52.11) = 8.17$, $p < 0.01$). To assess the statistical significance of pair-wise differences within our methods, a Tukey post-hoc analysis was performed and the results are presented in Table 1. Note that not all pair-wise results are shown but only relevant ones.

7.2.1 Single Source Results

Here, we examine the accuracy of predictions generated from each individual context source. To recap, we examined Facebook (collaborative / social filtering), Wikipedia (semantic / content-based filtering) and Twitter (expert-based recommendations). Based on our 32 users, we found no significant difference in the three methods. Wikipedia had the highest average utility of 4.42 and "Expert recommendations" sourced from Twitter exhibited the lowest average utility of 3.52. Based on our observations, it appeared that recommendations derived from Twitter were more obscure than the other two sources. The authors note that this is likely a result of the particular recommendation technique used, and not necessarily a reflection on the quality of the underlying data in Twitter.

7.2.2 Hybrid Results

Our second analysis focuses on a comparison of the three hybrid recommendation approaches. The middle portion of Figure 3 shows the utility score for each approach (*weighted*, *mixed*, and *cross-source* hybrid). Only the cross-source hybrid approach, in which we favor recommendations coming from more than one source, performed better than all three single-source methods. The Tukey pair-wise test showed significant differences between the cross-source hybrid method and Wikipedia, Facebook (CF), and Twitter, with p=0.023, p=0.011, and p<0.001 respectively. This is a strong indi-

cation that hybridization across social web APIs can help increase predictive accuracy in recommender systems.

7.2.3 Interaction Results

The three methods of interaction described in Section 7 were tested in the study and the results are shown in Figure 3. The full interaction method is the standard use case for the *TasteWeights* system and it exhibited improved performance over the best hybrid approach (p=0.027) indicating that interaction with the full system helped the user get better recommendations. As expected, out of all interactive methods the full interaction one achieved the highest accuracy score of 7.54. However, we note that this is clearly not a fair comparison since in this method, participants could see the recommendations change as they interacted with the system, meaning that recommendation feedback could inform their interactions. While this is not a fair scientific comparison, we posit that a mechanism which allows such informed, interactive feedback can be beneficial in real world recommender applications.

7.3 Explanation & Transparency

To assess the effects of interactive visualization on the perceived quality of recommendations, a post-questionnaire was completed by all participants. The study also analyzed the role of the interface as an explanatory mechanism for the underlying algorithms, and as a mechanism to help users learn about the underlying data. Looking at factors affecting explanation and learning, the left side of Figure 4 shows that users generally viewed the system as informative. The highest agreement was for the "helped understand how I got my recommendations" question, indicating that the system is performing well as an explanation interface.

The graph on the right side of Figure 4 shows results for the perceived usefulness of interaction with the system and the quality of each prediction strategy. Users felt that interaction helped them get better recommendations. Facebook was reported as the most useful source for generating recommendations, followed by Wikipedia, with Twitter reported as by far the least useful. Interestingly, perceived usefulness shows a relative improvement of 9.7% for Facebook over Wikipedia, while accuracy from Figure 3 indicates the

Wikipedia slightly outperforming Facebook. This increase in perceived utility of Facebook may be a result of participants favoring recommendations that come from real people who they trust and have prior information about.

8. CONCLUSION

This paper presented *TasteWeights*, an interactive hybrid recommendation system. The system employs new models for sourcing recommendations from a range of web APIs and presents hybridization strategies for combining those recommendations. The *TasteWeights* explanatory interface educates users about hybrid recommendation systems and enables them to tweak the underlying algorithms in real-time. A supervised user study was performed using the system to explore research questions relating to visual interactive recommendation systems. The study results indicate that:

- Explaining a hybrid recommendation process through a user interface can increase user satisfaction.

- Interaction at recommendation time can improve recommendation accuracy and user experience.

- Hybrid strategies combining different social APIs can provide better recommendations than traditional CF (over Facebook music preferences).

9. ACKNOWLEDGEMENTS

This work was partially supported by the U.S. Army Research Laboratory under Cooperative Agreement No. W911NF-09-2-0053; by NSF grant IIS-1058132; and by the U.S. Army Research Laboratory under MURI grant No. W911NF-09-1-0553; The views and conclusions contained in this document are those of the authors and should not be interpreted as representing the official policies, either expressed or implied, of ARL, NSF, or the U.S. Government. The U.S. Government is authorized to reproduce and distribute reprints for Government purposes notwithstanding any copyright notation here on. The authors would also like to thank Bart Knijnenburg for his valuable advice.

10. REFERENCES

[1] S. Auer, C. Bizer, G. Kobilarov, J. Lehmann, R. Cyganiak, and Z. Ives. Dbpedia: A nucleus for a web of open data. In *Proceedings of 6th International Semantic Web Conference, 2nd Asian Semantic Web Conference (ISWC+ASWC 2007)*, 2008.

[2] S. Bostandjiev, J. O'Donovan, C. Hall, B. Gretarsson, and T. Höllerer. Wigipedia: A tool for improving structured data in wikipedia. In *Proceedings of the 5th IEEE International Conference on Semantic Computing (ICSC 2011)*, pages 328–335.

[3] J. S. Breese, D. Heckerman, and C. Kadie. Empirical analysis of predictive algorithms for collaborative filtering. pages 43–52. Morgan Kaufmann, 1998.

[4] R. Burke. Hybrid recommender systems: Survey and experiments. *User Modeling and User-Adapted Interaction*, 12(4):331–370, 2002.

[5] M. Claypool, A. Gokhale, T. Miranda, P. Murnikov, D. Netes, and M. Sartin. Combining content-based and collaborative filters in an online newspaper. In *Proceedings of ACM SIGIR Workshop on Recommender Systems*, 1999.

[6] K. Goldberg, T. Roeder, D. Gupta, and C. Perkins. Eigentaste: A constant time collaborative filtering algorithm. *Information Retrieval*, 4(2):133–151, 2001.

[7] B. Gretarsson, J. O'Donovan, S. Bostandjiev, C. Hall, and T. Höllerer. Smallworlds: Visualizing social recommendations. *Comput. Graph. Forum*, 29(3):833–842, 2010.

[8] A. Halavais and D. Lackaff. An analysis of topical coverage of wikipedia. *Journal of Computer-Mediated Communication*, 13(2):429–440, 2008.

[9] J. L. Herlocker, J. A. Konstan, and J. Riedl. Explaining collaborative filtering recommendations. In *Proceedings of ACM CSCW'00 Conference on Computer-Supported Cooperative Work*, 2000.

[10] A. Hernando, D. Villuendas, C. Vesperinas, M. Abad, and A. Plastino. Unravelling the size distribution of social groups with information theory in complex networks. *The European Physical Journal B - Condensed Matter and Complex Systems*, 2010.

[11] P. Melville, R. Mooney, and R. Nagarajan. Content-boosted collaborative filtering. In *In Proceedings of the Eighteenth National Conference on Artificial Intelligence*, 2002.

[12] P. Melville, R. J. Mooney, and R. Nagarajan. Content-boosted collaborative filtering for improved recommendations. In *Eighteenth national conference on Artificial intelligence*, Menlo Park, CA, USA, 2002. American Association for Artificial Intelligence.

[13] S. E. Middleton, H. Alani, and D. D. Roure. Exploiting synergy between ontologies and recommender systems. *CoRR*, cs.LG/0204012, 2002.

[14] J. O'Donovan, B. Gretarsson, S. Bostandjiev, T. Höllerer, and B. Smyth. A visual interface for social information filtering. In *CSE (4)*, pages 74–81. IEEE Computer Society, 2009.

[15] J. O'Donovan, B. Smyth, B. Gretarsson, S. Bostandjiev, and T. Höllerer. Peerchooser: visual interactive recommendation. In *CHI '08: Proceeding of the twenty-sixth annual SIGCHI conference on Human factors in computing systems*, pages 1085–1088, New York, NY, USA, 2008. ACM.

[16] P. Resnick, N. Iacovou, M. Suchak, P. Bergstrom, and J. Riedl. Grouplens: An open architecture for collaborative filtering of netnews. In *Proceedings of ACM CSCW'94 Conference on Computer-Supported Cooperative Work*, pages 175–186, 1994.

[17] F. Ricci, L. Rokach, B. Shapira, and P. B. Kantor, editors. *Recommender Systems Handbook*. Springer, 2011.

[18] B. Shneiderman. The eyes have it: A task by data type taxonomy for information visualizations. In *Proceedings of the 1996 IEEE Symposium on Visual Languages*.

[19] N. Tintarev and J. Masthoff. A survey of explanations in recommender systems. In *Data Engineering Workshop, 2007 IEEE 23rd International Conference on*, pages 801 –810, april 2007.

[20] M. Zanker and M. Jessenitschnig. Case-studies on exploiting explicit customer requirements in recommender systems. *User Model. User-Adapt. Interact.*, 19(1-2):133–166, 2009.

Inspectability and Control in Social Recommenders

Bart P. Knijnenburg
Department of Informatics
University of California, Irvine

bart.k@uci.edu

John O'Donovan
Department of Computer Science
University of California, Santa Barbara

jod@cs.ucsb.edu

Svetlin Bostandjiev
Department of Computer Science
University of California, Santa Barbara

alex@cs.ucsb.edu

Alfred Kobsa
Department of Informatics
University of California, Irvine

kobsa@uci.edu

ABSTRACT

Users of social recommender systems may want to inspect and control how their social relationships influence the recommendations they receive, especially since recommendations of social recommenders are based on friends rather than anonymous "nearest neighbors". We performed an online user experiment (N=267) with a Facebook music recommender system that gives users control over the recommendations, and explains how they came about. The results show that inspectability and control indeed increase users' perceived understanding of and control over the system, their rating of the recommendation quality, and their satisfaction with the system.

Categories and Subject Descriptors

H.1.2. [**Models and principles**]: User/Machine Systems–*software psychology*; H.4.2. [**Information Systems Applications**]: Types of Systems–*decision support*; H.5.2 [**Information Interfaces and Presentation**]: User Interfaces–*evaluation/methodology, interaction styles, user centered design*

General Terms

Measurement, Design, Experimentation, Human Factors, Theory.

Keywords

Social recommender systems, human-computer interaction, usability, user experience, user interfaces, control, inspectability, visualization, explanations, novelty, understandability, satisfaction

1. INTRODUCTION

Collaborative recommender systems compare users' preferences to those of all other users, and recommend items that are liked by those users who have similar preferences [21]. Social recommenders limit the set of other users to your friends, thereby leveraging personal connections [37, 39, 52]. We suspect that users of social recommenders may not be satisfied with only a static list of recommendations. Rather, they may want to *inspect* and *control* the way in which the system uses their social network to select this list of recommendations, for at least two reasons:

1. Users seem to appreciate it when recommender systems explain their recommendations [9, 12, 21, 47, 48, 50]. In social recommenders, where users know the people on which the recommendations are based, the system can provide such explanation by showing how the overlap between one's preferences and those of one's friends resulted in a set of recommendations. Such a "recommendation graph" increases the *inspectability* (or transparency [9, 47]) of a system, which could have a positive effect on users' experience [44].

2. Users seem to appreciate *control* in their interaction with recommender systems [30, 35]. Recommenders have to somehow gather users' preferences, and different types of preference elicitation methods provide different levels of control [6, 29]. In a recommender system that leverages social networks such as Facebook, the system can use users' "likes" to construct a preference model, and the overlap with their friends' "likes" to compute recommendations. However, users may want some control over this process, because they may not like each item equally well, or they may value a friend's preferences beyond (or short of) the amount of mutual overlap in "likes". Users may therefore want to give additional (or lower) weight to some of the items and/or some of their friends.

Although this reasoning may seem intuitive, little research has been done to establish the effect of inspectability and control on the users' experience with social recommender systems ([17, 18] are notable exceptions). This paper describes the results of an online user experiment (N=267) with a Facebook music recommender in which we independently manipulated the level of inspectability and control. The results show that the versions of our system that offer high inspectability and control indeed provide a better user experience. The structural model in which we present these results allows us to explain *why* inspectability and control are important qualities of social recommender systems.

2. RELATED WORK

Before discussing the experiment we first survey related work on the effects of inspectability and control in recommender systems in general and in social recommenders specifically. We also dis-

This work was partially supported by the U.S. Army Research Laboratory under Cooperative Agreement No. W911NF-09-2-0053 and by NSF grant IIS-1058132; The views an conclusions contained in this document are those of the authors and should not be interpreted as representing the official policies, either expressed or implied, of ARL, NSF, or the U.S. Government. The U.S. Government is authorized to reproduce and distribute reprints for Government purposes notwithstanding any copyright notation here on.

cuss related work on the user-centric evaluation of recommender systems, and on personal characteristics of the user that may influence our results.

2.1 Inspectability

Many aspects of the explanation process have been studied in the recommender systems literature. For example, an earlier approach in [22] discusses a simple glass-box paradigm that provides only minimal information to the user. In this paper our notion of inspectability is similar to Tintarev and Masthoff's concept of transparency in [44], which is to "explain how the system works", and it is treated separately from control in our study. [46] also introduces the concept of scrutability with the (more interactive) aim to "allow users to tell the system it is wrong", while Czarkowski and Kay [10] examine scrutability and control as separate mechanisms, in the context of a student model application. Kay and Lum [26] also focus on scrutability, but in terms of providing explanations of why individual elements and relations in the underlying model have particular values.

Herlocker [19] argues that explanation provides transparency, "exposing the reasoning behind a recommendation". The reasoning and insight into the recommendation process exposed by an explanation interface can also increase the *inspectability* of the system as a whole. Tintarev and Masthoff [48] show that explanations make it easier to judge the quality of recommendations. Consequently, such explanations increase users' trust in the recommendations and, in turn, the perceived competence of the system ([9, 12], see also [19, 50]). Sinha and Swearingen [44] demonstrate that users rate systems that provide detailed information about items as more useful and easier to use.

In the realm of social recommenders, Groh et al. [18] present a study that outlines the "extensive need" for explanation, and Gretarsson et al. [17] present a small-scale study of an explanation interface, finding that the explanation process has a positive effect on satisfaction with recommendations.

2.2 Control

Researchers have implemented various aspects of control in recommender systems, ranging from simple preference elicitation at recommendation time [7] to more complex iterative processes such as dynamic critiquing which allows users to tweak ordered numerical attributes during the recommendation phase [6, 32, 33, 42]. More recent work [17, 37, 39] discusses interactive graphical representations of the recommendation process, to enable control over both item- and user-level preferences in collaborative recommender systems.

Multiple studies highlight the benefits of interactive interfaces that support control over the recommendation process. In a general comparison of user-controlled versus static recommendation interfaces, McNee et al. [35] found that study participants preferred user-controlled interfaces because these systems "best understood their tastes". McNee et al. also showed that participants had higher retention rates with the controlled interface. Knijnenburg et al. [30] found that controllable recommendations are typically deemed more varied than automatic ones, and Willemsen et al. [51] show that diversifying recommendations can be useful to overcome choice overload.

Other common methods of control include rating items [14, 41] and assigning weights to item attributes [20, 31]. Research shows that the choice between these different methods has a substantial impact on the user experience [6, 29, 31].

2.3 Social Recommender Systems

Previous studies have attempted to explain recommendations by showing the link between the recommendations and the "nearest neighbors" on which the they are based [21, 35, 47]. An interesting aspect of social recommenders is that recommendations are based on users' similarity with their *friends* rather than a set of anonymous nearest neighbors. In effect, social recommenders can leverage users' acquaintance with the source of the recommendation, which instantly attaches a wealth of established social information to the recommendations that can be further explored and exploited in the processes of inspection and control [18].

Specifically, we hypothesize that visualizing the link between recommendations and the nearest neighbors on which they are based increases the inspectability of social recommenders beyond regular recommenders, because the neighbors are known.

Furthermore, in social recommenders one could allow the user to not only rate items, but also their friends [5]. This would give certain friends additional (or less) weight beyond the weight computed based on preference overlap. Arguably, this method allows users to indicate how much they "trust" their friends' preferences in the recommendation domain. Several researchers have investigated this idea of assigning trust scores to friends in collaborative recommenders, through explicit mechanisms such as in Golbeck's FilmTrust system [15] which can support propagation of trust scores around a network of peers, and through automated mechanisms for modeling trust such as [11, 38]. Several recent studies have extended these ideas to prediction of personality, and by derivation, behavior of a user within the system [1] in terms of both trust and distrust [16].

In this paper we take a step beyond existing work on social recommenders by explicitly testing the effect of leveraging users' knowledge about their friends to improve inspectability and control. By allowing a user to inspect and control the elements of trust on which the recommendations are based, we can gain an understanding of the effect of inspectability and control on the user experience with the recommender system.

2.4 User Experience

In order to determine the impact of inspectability and control, we need to measure users' *experience* with the recommender system. Specifically, we are not only interested in the quality of the recommendations, but also in users' satisfaction with the system as a whole. Moreover, we need to consider concepts like understandability and perceived control to explain how inspectability and control influence the users' experience.

Knijnenburg et al. [30] developed a framework for user-centric evaluation of recommender systems through user experiments. It describes how certain manipulations (in our case: inspectability and control) influence subjective system aspects (i.e. understandability, perceived control and recommendation quality), which in turn influence user experience (i.e. system satisfaction).

2.5 Personal Characteristics

The framework also allows us to include the effect of *personal characteristics* on users' experience. In social recommenders, the degree of trustfulness (or "trusting propensity") could play an important role, because in order to accept the recommendations, users will need to trust their friends' preferences [15, 38]. Users who are not trustful usually want to take matters in their own hands and therefore demand more control over the system [49].

Figure 1. The TasteWeights system as used in the online user experiment. This is the inspection phase of the "full graph" condition. Users can click on items, friends and recommendations to see the links between them. The inspection phase of the "list only" condition shows the rightmost list (recommendations) only.

Another personality trait that can have an influence on users' experience is choice persistence, a characteristic that divides users into satisficers and maximizers. Satisficers will stop their search when they encounter an item that meets their criteria, while maximizers continue their efforts until they find the best possible option [43]. Maximizers may thus show more appreciation for systems that allow extensive control and inspection (although we do not find evidence for this in [29]).

Finally, users' domain knowledge can also have a significant impact on their experience with a recommender system. Kamis and Davern show that domain experts perceive personalized recommenders as less useful than novices [24], while other researchers have consistently found that experts have a higher appreciation for the recommendations as well as the recommender system itself [3, 30, 51]. In order to be satisfied with a recommender, however, domain experts want more control over their recommendations than novices [13, 29, 40, 45]. Novices, on the other hand, require a simple and understandable recommender system [31], and they may even prefer to give up control in return for simplicity [29].

3. ONLINE USER EXPERIMENT

The related work shows that inspectability and control have a positive influence on users' experience with recommender systems in general, and we suspect that these benefits may be even more pronounced in social recommenders. A user study is needed to investigate the nature and extent of these benefits, and the factors that influence them. We therefore conducted an online user experiment with 267 participants employing a modified version of the TasteWeights [4, 17] social recommender.

3.1 System

The TasteWeights system recommends new artists/bands based on the music "likes" of the user and her Facebook friends.

3.1.1 Recommendation algorithm

The TasteWeights system calculates its recommendations in two steps. First, weights are computed for each friend based on their similarity to the user. Specifically, the similarity of the user to a friend is given by the overlap in music "likes" between them, as defined by Pearson's correlation coefficient:

$$W_{friend_i} = \frac{TWCI_{user,friend_i}}{\sqrt{TWI^2_{user} \cdot TWI^2_{friend_i}}}$$

where $TWCI_{x,y}$ is the total weight of the items ("likes") that users x and y have in common, and TWI_x is the total weight of items liked by user x. As Facebook users can only "like" artists/bands without specifying how much, item weights are initialized to 0.5.

Once all friend weights are computed, recommendations are generated by assigning weights to all friends' music items (excluding the items that the user already "likes"):

$$W_{rec_i} = \sum_{friend_j \ likes \ rec_i} W_{friend_j}$$

Here, the weight of a recommendation i is the sum of the weights of all friends that like i. The recommendations are displayed in decreasing order of recommendation weight.

3.1.2 Inspectability and Control

In terms of inspectability, the TasteWeights system displays a graph (Figure 1) that shows the users' items, their friends, and the recommendations. By clicking at the graph, the connections between these entities can be explored. The system also shows a short description for each recommended band/artist with a link to their LastFM information page.

The system allows two types of control over the recommendations: users can adjust the weights of their items (initialized at 0.5) and their friends (initially weighted by similarity). Changing the weight of an item will influence the similarity scores, and thus the recommendation weights. Changing the weight of a friend will add or subtract a proportion from that friend's similarity score, and thus also influence the recommendation weights.

3.2 Study setup

3.2.1 Experimental conditions

The original TasteWeights system allows users to interactively inspect and control the recommendation graph (i.e. change the weights and inspect the graph simultaneously and iteratively). However, to investigate the effects of inspectability and control independently, we let participants in our experiment interact with the system in two stages: a control stage and an inspection stage.

In the control stage, participants are assigned to one of three conditions (Figure 2): they either skip the control stage altogether (the "no control" condition), they are asked to adjust the weights of the items they "like" (the "item control" condition), or they are asked to adjust the weights of their friends (the "friend control" condition). Our primary interest is to see how these control conditions compare against the no control condition, but we are also interested in differences between the two control conditions.

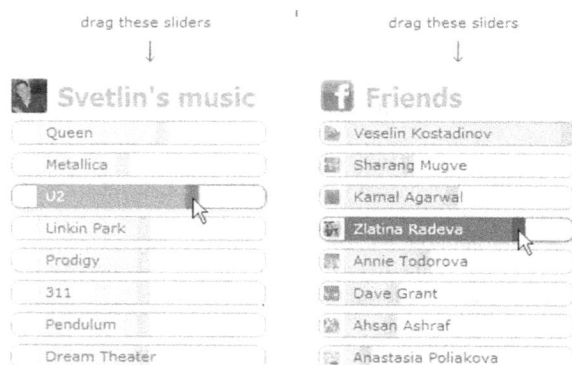

Figure 2. The control phase of item control (left) and friend control (right) conditions.

In the inspection stage, participants are assigned to one of two conditions: the system either shows only the list of recommendations (the "list only" condition), or the full recommendation graph (the "full graph" condition). To give each participant a comparable experience, we limited the number of music likes and friends to be considered by the recommender to 10 each (with a minimum of 5 each). The number of recommendations was fixed to 10.

3.2.2 Participants and Procedure

267 participants were recruited using Craigslist and Amazon Mechanical Turk[1]. Only adult Facebook users living in the U.S. were allowed to participate. 156 participants were female, and 130 were between the ages of 18 and 25, 114 between 25 and 40, and 23 older than 40[2]. In order to provide a meaningful experience, we only allowed users to participate if their recommendation graph would show at least 5 music "likes", showing overlap with at least 5 friends, and resulting in at least 10 recommendations. Denied

---

[1] In [27] we found no substantial differences between these two participant populations.

[2] These numbers reflect the general Facebook population, with a slight underrepresentation of the older demographic. See http://bit.ly/insidefacebook-20100104.

participants were given the suggestion to populate their Facebook profile with more music "likes" and then try again.

Eligible participants were then asked to answer 15 questions about their personal characteristics (music expertise, trusting propensity and choice persistence). Questions were statements to which participants could agree or disagree on a five point scale. They subsequently completed the control stage (unless they were assigned to the "no control" condition), in which they were asked to adjust the weights of either their items or their friends (depending on the control condition). Next, they completed the inspection stage, where they were asked to carefully inspect the list of recommendations or the recommendation graph (depending on the inspectability condition). Finally, they were asked to indicate whether they already knew each of the recommended band/artist or not, and subsequently to rate the recommendation on a 5-star scale. Users were encouraged to click on the provided LastFM link to improve their judgment of unknown bands/artists. After the experiment, participants answered another 29 questions about their user experience. Full questionnaires can be found in [28].

3.2.3 Questionnaires

The answers to the 44 questionnaire items were submitted to a confirmatory factor analysis[3] with categorical indicators and a weighted least squares estimator, estimating 7 factors:

- *Music expertise*: 4 items adapted from [3], e.g. "Compared to my peers I listen to a lot of music.", α: .74, AVE: .627
- Trusting propensity: 3 items adapted from [29], e.g. "In general, people really do care about the well-being of others.", α: .80, AVE: .657
- *Understandability*: 3 new items, e.g. "The recommendation process is clear to me.", α: .92, AVE: .877
- *Perceived control*: 4 items adapted from [29], e.g. "Compared to how I normally get recommendations, Taste-Weights was very limited." (reversed), α: .84, AVE: .643
- *Perceived recommendation quality*: 5 items adapted from [30], e.g. "I liked the bands/artists recommended by the TasteWeights system.", α: .90, AVE: .738
- *System satisfaction*: 7 items adapted from [30], e.g. "I can find better music using TasteWeights.", α: .92, AVE: .708
- *Familiarity with recommenders*: 2 new items, e.g. "I am familiar with online recommender systems.", α: .86, AVE: .794

10 questionnaire items were excluded from the analysis due to low communality, high cross-loadings and/or high residual correlations. Additionally, 5 items on choice persistence (taken from [36]) failed to converge to a stable factor. For the remaining factors the values of Cronbach's α and average variance extracted (AVE) were high[4], indicating convergent validity. Moreover, the square root of the AVE is higher than the factor correlation for all factors, indicating discriminant validity.

3.3 Results

We subjected the 7 factors, the experimental conditions, and selected behaviors to structural equation modeling, which simultaneously fits the factor measurement model and the structural rela-

---

[3] Factor analysis and structural equation modeling as applied in this paper are explained in Appendix A of [30].

[4] For alpha, >.70 is acceptable, >.80 is good, >.90 is excellent. AVE should be >.50 for convergent validity.

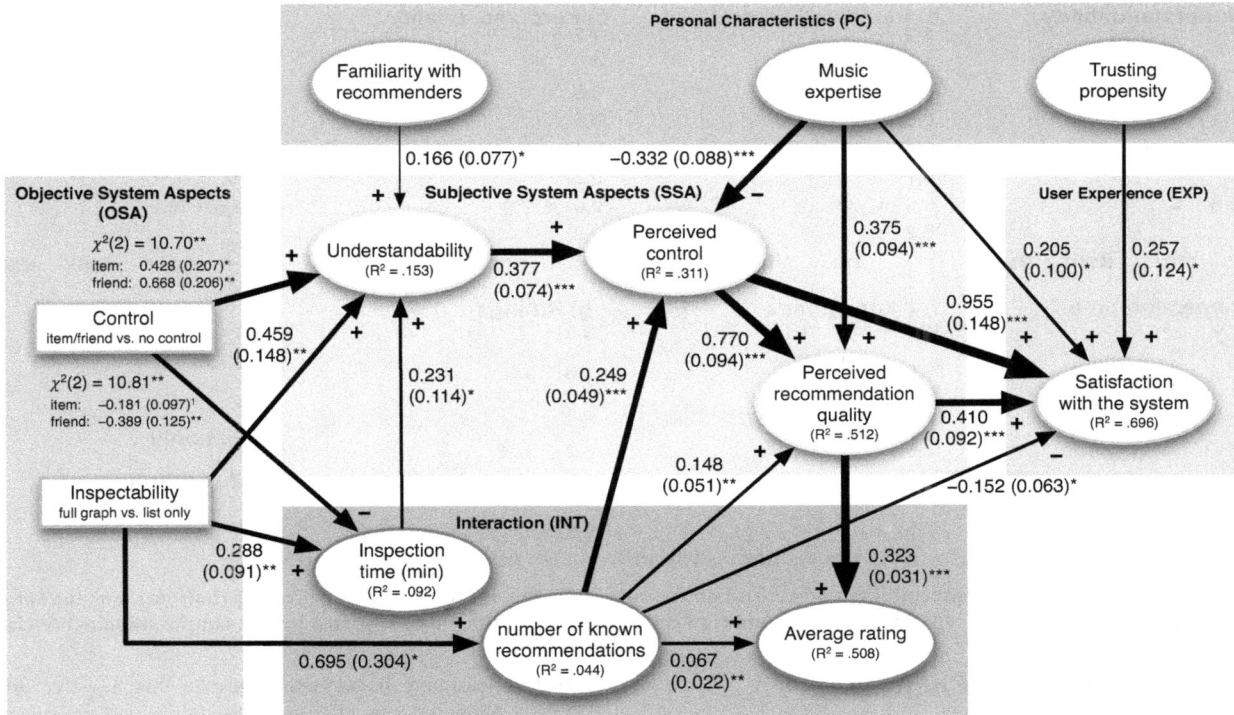

Figure 3. The structural equation model for the data of the experiment. Significance levels: *** p < .001, ** p < .01, 'ns' p > .05. R^2 is the proportion of variance explained by the model. Numbers on the arrows (and their thickness) represent the β coefficients (and standard error) of the effect. Factors are scaled to have an SD of 1.

tions between factors and other variables. The model (Figure 3) has a good[5] model fit: $\chi^2(537) = 639.22, p < .01$; $RMSEA = 0.027$, 90% CI: [0.017, 0.034], $CFI = 0.993$, $TLI = 0.992$.

3.3.1 Subjective Experience

The model shows that the inspectability and control manipulations each have an independent positive effect on the understandability of the system: the full graph condition is more understandable than the list only condition, and the item control and friend control conditions are more understandable than the no control condition (see also Figure 4a). Understandability is in turn related to users' perception of control, which is in turn related to the perceived quality of the recommendations. The perceived control and the perceived recommendation quality finally determine participants' satisfaction with the system (for the marginal effects of control and inspectability on these factors, see Figure 4b,c,d).

3.3.2 User Behavior

There exist additional effects of inspectability and control on understandability, which are mediated by the inspection time (the amount of time users take to inspect the recommendations, see Figure 4e). In the full graph condition, participants take more time to inspect the recommendations (about 7.3 seconds more), and this results in an additional increase of understandability. For the two control conditions, however, the inspection time is shorter (about 10.9 seconds less in the item control condition and about

23.3 seconds less in the friend control condition), which counters the positive effect on understandability.

In the full graph condition, participants indicate that they already know more of the recommendations than in the list only condition (see Figure 4f). In turn, the more recommendations the participant already knows, the higher is the perceived control and perceived recommendation quality, but the lower is the satisfaction.

The perceived recommendation quality and the number of known recommendations determine the average rating participants give to the recommendations. The marginal effects of the inspectability and control manipulations on the average rating (Figure 4g) indicate that the ratings in the item control condition are somewhat lower (mean: 3.146) than the no control condition (mean: 3.267), whereas the ratings in the friend control condition are somewhat higher (mean: 3.384). The difference between the two control conditions is small but significant (p = .031).

3.3.3 Personal Characteristics

Participants who are familiar with recommenders find the system more understandable. Participants with music expertise perceive less control over the system, but perceive a higher recommendation quality and system satisfaction. Finally, trusting propensity influences participants' satisfaction with the system.

4. Discussion

Based on the results of our experiment, we can describe in detail how the benefits of inspectability and control in social recommenders come about. We can also describe these results in the light of users' personal characteristics. Finally, we can provide some preliminary suggestions on the relative effectiveness of controlling items versus friends.

---

[5] A model should not have a non-significant χ^2, but this statistic is regarded as too sensitive [2]. Hu and Bentler [23] propose cut-off values for other fit indices to be: $CFI > .96$, $TLI > .95$, and $RMSEA < .05$, with the upper bound of its 90% CI below 0.10.

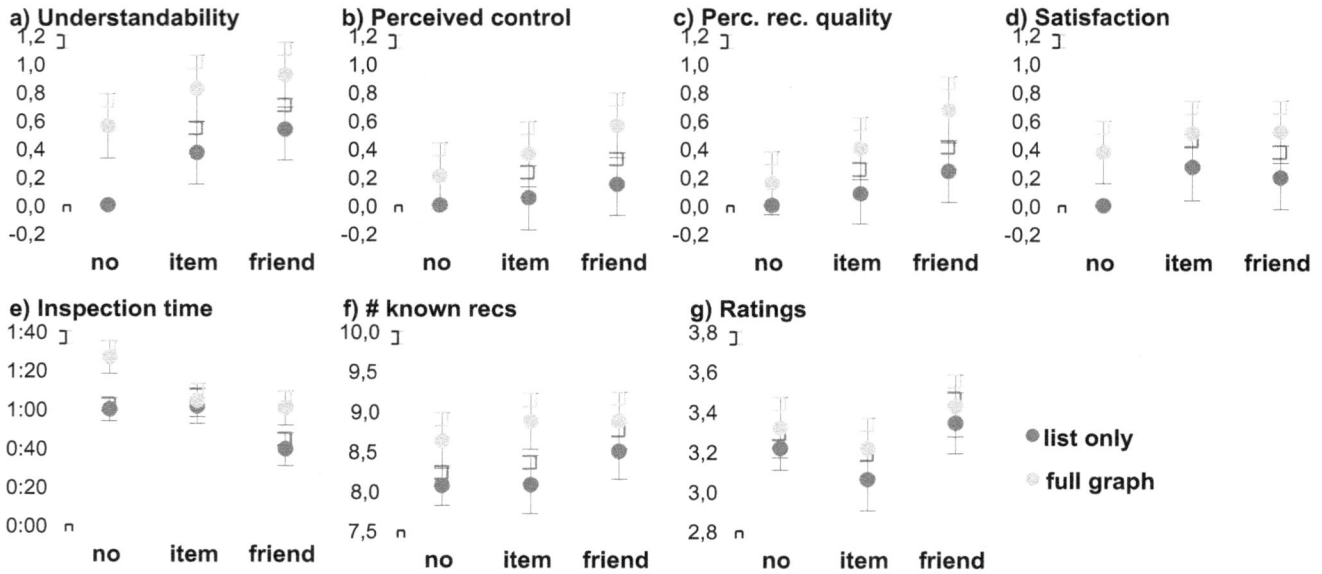

Figure 4. Marginal effects of inspectability and control on the subjective factors (top) and on behaviors (bottom). For the subjective factors, the effects of the "no control, list only" condition is set to zero, and the y-axis is scaled by the sample standard deviation.

4.1 Inspectability and Control

Both inspectability and control have a positive effect on the user experience, primarily because an inspectable and controllable recommender system is easier to understand. The increased understandability causes users to feel more in control over the system, and this in turn increases the perceived quality of the recommendations, also indicated by increased ratings. Finally, the higher perceived control and recommendation quality cause users to be more satisfied with the system.

Inspectability works partially due to a direct effect on understandability, and partially due to its influence on user behavior. Specifically, users take more time for inspection in the "full graph" condition (which increases understandability), and users in this condition already know more of the recommendations (which increases perceived control and recommendation quality, but decreases system satisfaction). The effect of inspectability on the number of recommendations that the participant already knows may seem counterintuitive, because the inspectability conditions do not influence the actual recommendations. However, in the "full graph" condition users can see which friends are connected to the recommendations, and this may allow users to *recognize* more of the recommendations as already known (e.g. "I remember John playing this band's album for me")[6].

Arguably, this recognition effect is an important aspect of inspectability, because knowing recommendations may raise users' trust in the recommender [8, 44]. In our experiment, known recommendations increase users' perceived control (total effect: $\beta = 0.372$, $p = .001$) and the perceived recommendation quality (total effect: $\beta = 0.389$, $p = .002$). On the other hand, known recommendations are less useful, as they contain no novelty, which explains the decrease in system satisfaction (McNee at al. [34] show that users are happy with a set of recommendations as long as it con-

tains at least one novel item). Despite this negative effect of known items, the total effect of inspectability on system satisfaction is however still statistically significant: $\beta = 0.409$, $p = .001$.

Item control and friend control result in a more understandable system despite the shorter inspection time (total effects: $\beta = 0.386$, $p = .063$ and $\beta = 0.578$, $p = .004$, respectively). Note that although inspection time is shorter, participants in these conditions spend additional time controlling the recommendations.

4.2 Personal Characteristics

Several personal characteristics have an effect on users' experience when using our system. Trusting propensity has a positive effect on system satisfaction, which may be due to the fact that users with a higher general trusting propensity seem more likely to trust their friends' music preferences. Arguably, then, trustfulness is an important precondition for a social recommender to work for a user.

Moreover, users with some expertise about music feel less in control, but they view the recommendations and the system itself more positively. Music experts may feel that bands/artists are too crude of a building block for recommendations (for them, bands may have both amazing and terrible albums), which could have caused the reduced perception of control (this effect is consistent with findings in [24]). On the other hand, music experts are more capable of judging the quality of the recommendations, which may be the reason for the increased perceived recommendation quality and satisfaction with the system (these effects are consistent with findings in [3, 30, 51]).

4.3 Which Type of Control?

Besides comparing the control conditions against the "no control" condition, we are also interested in comparing the control conditions against each other, to determine which type of control users prefer. Figure 4 shows that the understandability, perceived control and perceived recommendation quality are consistently higher for the "friend control" condition than for the "item control" condition, but the difference between these two conditions is not sta-

---

[6] Conformity bias could be an alternative explanation: "If all my friends know this band, I ought to know it too!"

tistically significant. The only significant difference between the two control conditions is in users' ratings of the recommendations: "friend control" results in higher ratings, but the difference is again very small (3.146 vs. 3.384 on a 5-star scale).

On the other hand, the "friend control" condition results in slightly more known recommendations, which may be one reason why the system satisfaction is also slightly worse in the "friend control" condition. With a note of caution due to the lack of significant results, we can interpret these trends to conclude that the "friend control" condition may give the user more accurate control, but the "item control" condition may lead to a perception of more novel recommendations.

5. CONCLUSION AND FUTURE WORK

Our results show that social recommender systems (and arguably recommender systems in general) indeed benefit from facilities that improve their inspectability and control.

Recommenders that display a recommendation graph result in a better user experience, partially because they encourage users to spend more time on inspecting the recommendations. Moreover, by inspecting the links between friends and recommendations, users get hints about their previous encounters with the recommended items, which allows them to more accurately evaluate the quality of the recommendations. In effect, the recommendation graph increases the understandability, perceived control, perceived recommendation quality, and system satisfaction.

Recommenders that give users control over the item weights and friend weights are more understandable, which leads to higher perceived control, perceived recommendation quality and overall system satisfaction. As to which control method is preferred, the results suggest that giving users control over the friend weights results in slightly higher quality recommendations, but that controlling the item weights may heighten users' perception of recommendation novelty. Arguably, making both control mechanisms available may preserve the best aspects of both methods.

In fact, as we find that the effects of inspectability and control are additive, the best user experience may arise when users can control items and friends simultaneously and directly in the recommendation graph. Such interactive control could arguably result in *scrutability*: allowing users to find and correct mistakes in the recommendation process [25, 47]. In this paper we purposefully disentangled inspectability and control to isolate their respective effects; in future work we purpose to investigate the benefits of scrutability in a fully interactive social recommender.

6. ACKNOWLEDGEMENTS

We thank Nikhil Rao for his extensive help conducting the study and Tobias Höllerer for his feedback throughout the process.

7. REFERENCES

[1] Adali, S., Escriva, R., Goldberg, M.K., Hayvanovych, M., Magdon-Ismail, M., Szymanski, B.K., Wallace, W.A. and Williams, G. Measuring behavioral trust in social networks. *2010 IEEE Intl. Conference on Intelligence and Security Informatics* (Vancouver, Canada, 2010), 150–152.

[2] Bentler, P.M. and Bonett, D.G. Significance Tests and Goodness of Fit in the Analysis of Covariance Structures. *Psychological Bulletin*. 88, 3 (1980), 588–606.

[3] Bollen, D., Knijnenburg, B.P., Willemsen, M.C. and Graus, M. Understanding choice overload in recommender systems. *Proceedings of the fourth ACM conference on Recommender systems* (Barcelona, Spain, 2010), 63–70.

[4] Bostandjiev, S., O'Donovan, J. and Höllerer, T. Taste-Weights: A Visual Interactive Hybrid Recommender System. *Proceedings of the sixth ACM conference on Recommender Systems* (Dublin, Ireland, 2012).

[5] Bourke, S., McCarthy, K. and Smyth, B. Power to the people: exploring neighbourhood formations in social recommender system. *Proceedings of the fifth ACM conference on Recommender systems* (New York, NY, 2011), 337–340.

[6] Chen, L. and Pu, P. Critiquing-based recommenders: survey and emerging trends. *User Modeling and User-Adapted Interaction*. 22, 1-2 (Oct. 2011), 125–150.

[7] Chen, L. and Pu, P. *Survey of Preference Elicitation Methods*. Technical Report #IC/2004/67 (2004).

[8] Cooke, A.D.J., Sujan, H., Sujan, M. and Weitz, B.A. Marketing the Unfamiliar: The Role of Context and Item-Specific Information in Electronic Agent Recommendations. *Journal of Marketing Research*. 39, 4 (Nov. 2002), 488–497.

[9] Cramer, H., Evers, V., Ramlal, S., Someren, M., Rutledge, L., Stash, N., Aroyo, L. and Wielinga, B. The effects of transparency on trust in and acceptance of a content-based art recommender. *User Modeling and User-Adapted Interaction*. 18, 5 (Aug. 2008), 455–496.

[10] Czarkowski, M. and Kay, J. Bringing Scrutability to Adaptive Hypertext Teaching. *Intelligent Tutoring Systems*. Springer (2000). 423–432.

[11] DuBois, T., Golbeck, J. and Srinivasan, A. Predicting Trust and Distrust in Social Networks. *IEEE 3rd intl. conference on social computing* (Boston, MA, Oct. 2011), 418–424.

[12] Felfernig, A. Knowledge-Based Recommender Technologies for Marketing and Sales. *Intl. J. of Pattern Recognition and Artificial Intelligence*. 21, 2 (2007), 333–354.

[13] Fitzsimons, G.J. and Lehmann, D.R. Reactance to Recommendations: When Unsolicited Advice Yields Contrary Responses. *Marketing Science*. 23, 1 (Jan. 2004), 82–94.

[14] Gena, C., Brogi, R., Cena, F. and Vernero, F. The Impact of Rating Scales on User's Rating Behavior. *User Modeling, Adaption and Personalization*. J.A. Konstan, R. Conejo, J.L. Marzo, and N. Oliver, eds. Springer (2011). 123–134.

[15] Golbeck, J. Generating Predictive Movie Recommendations from Trust in Social Networks. *Trust Management*. K. Stølen, W.H. Winsborough, F. Martinelli, and F. Massacci, eds. Springer (2006). 93–104.

[16] Golbeck, J., Robles, C. and Turner, K. Predicting personality with social media. *Proceedings of the 2011 annual conference extended abstracts on Human factors in computing systems* (Vancouver, Canada, 2011), 253–262.

[17] Gretarsson, B., O'Donovan, J., Bostandjiev, S., Hall, C. and Höllerer, T. SmallWorlds: Visualizing Social Recommendations. *Computer Graphics Forum*. 29, 3 (2010), 833–842.

[18] Groh, G., Birnkammerer, S. and Köllhofer, V. Social Recommender Systems. *Recommender Systems for the Social Web*. Springer (2012). 3–42.

[19] Guy, I., Ronen, I. and Wilcox, E. Do you know?: recommending people to invite into your social network. *Proceedings of the 14th international conference on Intelligent user interfaces* (New York, NY, 2009), 77–86.

[20] Haubl, G. and Trifts, V. Consumer Decision Making in Online Shopping Environments: The Effects of Interactive Decision Aids. *Marketing Science*. 19, 1 (Jan. 2000), 4–21.

[21] Herlocker, J.L., Konstan, J.A. and Riedl, J. Explaining collaborative filtering recommendations. *Proc. of the 2000 ACM*

conference on Computer supported cooperative work (Philadelphia, PA, 2000), 241–250.

[22] Höök, K., Karlgren, J., Wærn, A., Dahlbäck, N., Jansson, C., Karlgren, K. and Lemaire, B. A glass box approach to adaptive hypermedia. *User Modeling and User-Adapted Interaction.* 6, 2-3 (Jul. 1996), 157–184.

[23] Hu, L. and Bentler, P.M. Cutoff criteria for fit indexes in covariance structure analysis: Conventional criteria versus new alternatives. *Structural Equation Modeling: A Multidisciplinary Journal.* 6, 1 (Jan. 1999), 1–55.

[24] Kamis, A. and Davern, M.J. Personalizing to product category knowledge: exploring the mediating effect of shopping tools on decision confidence. *37th an. Hawaii intl. conf. on System Sciences* (Big Island, HI, 2004), 10 pp.

[25] Kay, J. Stereotypes, Student Models and Scrutability. *Intelligent Tutoring Systems.* G. Gauthier, C. Frasson, and K. VanLehn, eds. Springer (2000). 19–30.

[26] Kay, J. and Lum, A. Ontology-based user modelling for the Semantic Web. *Workshop on Personalisation on the Semantic Web* (Edinburgh, UK, 2005), 15–23.

[27] Knijnenburg, B.P. and Kobsa, A. Making Decisions about Privacy: Information Disclosure in Context-Aware Recommender Systems. *submitted to the ACM Transactions on Intelligent Interactive Systems, Special Issue on Human Decision Making and Recommender Systems* (2012). http://www.isr.uci.edu/tech_reports/UCI-ISR-12-1.pdf

[28] Knijnenburg, B.P., Rao, N. and Kobsa, A. *Experimental Materials Used in the Study on Inspectability and Control in Social Recommender Systems.* Technical Report #UCI-ISR-12-4 (2012).

[29] Knijnenburg, B.P., Reijmer, N.J.M. and Willemsen, M.C. Each to his own: how different users call for different interaction methods in recommender systems. *Proceedings of the fifth ACM conference on Recommender systems* (Chicago, IL, 2011), 141–148.

[30] Knijnenburg, B.P., Willemsen, M.C., Gantner, Z., Soncu, H. and Newell, C. Explaining the user experience of recommender systems. *User Modeling and User-Adapted Interaction.* 22, 4-5 (2012), 441–504.

[31] Kramer, T. The Effect of Measurement Task Transparency on Preference Construction and Evaluations of Personalized Recommendations. *Journal of Marketing Research.* 44, 2 (May. 2007), 224–233.

[32] McCarthy, K., Salem, Y. and Smyth, B. Experience-Based Critiquing: Reusing Critiquing Experiences to Improve Conversational Recommendation. *Case-Based Reasoning. Research and Development.* I. Bichindaritz and S. Montani, eds. Springer (2010). 480–494.

[33] McGinty, L. and Reilly, J. On the Evolution of Critiquing Recommenders. *Recommender Systems Handbook.* F. Ricci, L. Rokach, B. Shapira, and P.B. Kantor, eds. Springer (2011). 419–453.

[34] McNee, S.M., Albert, I., Cosley, D., Gopalkrishnan, P., Lam, S.K., Rashid, A.M., Konstan, J.A. and Riedl, J. On the recommending of citations for research papers. *Proceedings of the 2002 ACM conference on Computer supported cooperative work* (New Orleans, LA, 2002), 116–125.

[35] McNee, S.M., Lam, S.K., Konstan, J.A. and Riedl, J. Interfaces for Eliciting New User Preferences in Recommender Systems. *User Modeling 2003.* P. Brusilovsky, A. Corbett, and F. Rosis, eds. Springer Heidelberg. 178–187.

[36] Nenkov, G.Y., Morrin, M., Ward, A., Schwartz, B. and Hul-

land, J. A short form of the Maximization Scale: Factor structure, reliability and validity studies. *Journal of Judgment and Decision Making.* 3, (June 2008), 371–388.

[37] O'Donovan, J., Gretarsson, B., Bostandjiev, S., Hollerer, T. and Smyth, B. A Visual Interface for Social Information Filtering. *Intl. Conf. on Computational Science and Engineering* (Vancouver, Canada, 2009), 74–81.

[38] O'Donovan, J. and Smyth, B. Trust in recommender systems. *Proceedings of the 10th international conference on Intelligent user interfaces* (San Diego, CA, 2005), 167–174.

[39] O'Donovan, J., Smyth, B., Gretarsson, B., Bostandjiev, S. and Höllerer, T. PeerChooser: visual interactive recommendation. *Proc. of the 26th SIGCHI conf. on Human factors in computing systems* (Florence, Italy, 2008), 1085–1088.

[40] Pereira, R.E. Optimizing human-computer interaction for the electronic commerce environment. *Journal of Electronic Commerce Research.* 1, 1 (2000), 23–44.

[41] Pommeranz, A., Broekens, J., Wiggers, P., Brinkman, W.-P. and Jonker, C.M. Designing interfaces for explicit preference elicitation: a user-centered investigation of preference representation and elicitation process. *User Modeling and User-Adapted Interaction.* 22, 4-5 (Mar. 2012), 357–397.

[42] Pu, P., Chen, L. and Kumar, P. Evaluating product search and recommender systems for E-commerce environments. *Electronic Commerce Research.* 8, 1-2 (2008), 1–27.

[43] Schwartz, B., Ward, A., Monterosso, J., Lyubomirsky, S., White, K. and Lehman, D.R. Maximizing versus satisficing: Happiness is a matter of choice. *Journal of Personality and Social Psychology.* 83, 5 (2002), 1178–1197.

[44] Sinha, R.R. and Swearingen, K. Comparing Recommendations Made by Online Systems and Friends. *DELOS Workshop on Personalisation and Recommender Systems in Digital Libraries* (2001), 64–67.

[45] Spiekermann, S. and Paraschiv, C. Motivating human–agent interaction: Transferring insights from behavioral marketing to interface design. *Electronic Commerce Research.* 2, 3 (2002), 255–285.

[46] Tintarev, N. and Masthoff, J. A Survey of Explanations in Recommender Systems. *Data Engineering Workshop* (Istanbul, Turkey, Apr. 2007), 801–810.

[47] Tintarev, N. and Masthoff, J. Designing and Evaluating Explanations for Recommender Systems. *Recommender Systems Handbook.* F. Ricci, L. Rokach, B. Shapira, and P.B. Kantor, eds. Springer (2011). 479–510.

[48] Tintarev, N. and Masthoff, J. Evaluating the effectiveness of explanations for recommender systems. *User Modeling and User-Adapted Interaction.* 22, 4-5 (Feb. 2012), 399–439.

[49] Vries, P. *Trust in systems: effects of direct and indirect information.* Thesis Technische Universiteit Eindhoven (2004).

[50] Wang, W. and Benbasat, I. Recommendation agents for electronic commerce: Effects of explanation facilities on trusting beliefs. *Journal of Management Information Systems.* 23, 4 (2007), 217–246.

[51] Willemsen, M.C., Graus, M.P., Knijnenburg, B.P. and Bollen, D. Not just more of the same: Preventing choice overload in recommender systems by offering small diversified sets. *Submitted for journal publication.* (2012).

[52] Zhao, S., Zhou, M.X., Yuan, Q., Zhang, X., Zheng, W. and Fu, R. Who is talking about what: social map-based recommendation for content-centric social websites. *Proceedings of the fourth ACM conference on Recommender system* (Barcelona, Spain, 2010), 143–150.

Spotting Trends: The Wisdom of the Few

Xiaolan Sha
EURECOM
Sophia Antipolis
France
xiaolan.sha@eurecom.fr

Daniele Quercia
The Computer Laboratory
University of Cambridge
United Kingdom
daniele.quercia@cl.cam.ac.uk

Pietro Michiardi
EURECOM
Sophia Antipolis
France
pietro.michiardi@eurecom.fr

Matteo Dell'Amico
EURECOM
Sophia Antipolis
France
matteo.dell-amico@eurecom.fr

ABSTRACT

Social media sites have used recommender systems to suggest items users might like but are not already familiar with. These items are typically movies, books, pictures, or songs. Here we consider an alternative class of items - pictures posted by design-conscious individuals. We do so in the context of a mobile application in which users find "cool" items in the real world, take pictures of them, and share those pictures online. In this context, temporal dynamics matter, and users would greatly profit from ways of identifying the latest design *trends*. We propose a new way of recommending trending pictures to users, which unfolds in three steps. First, two types of users are identified - those who are good at uploading trends (trend makers) and those who are experienced in discovering trends (trend spotters). Second, based on what those "special few" have uploaded and rated, trends are identified early on. Third, trends are recommended using existing algorithms. Upon the *complete* longitudinal dataset of the mobile application, we compare our approach's performance to a traditional recommender system's.

Categories and Subject Descriptors

H.4 [**Information Systems Applications**]: Miscellaneous

General Terms

Experimentation

Keywords

Mobile, Social Media, Trend Detection

1. INTRODUCTION

Good user experience is what makes online services attractive. To be outstanding, many services not only try to provide easy access to the content of what users are looking for, but also attempt to help them discover new information which they might be interested in. Mostly, they use recommender systems to give personalized suggestions to each user. Depending on the context of the applications, recommended items are of all kinds [2, 6, 9, 10, 20, 21, 25, 32, 33, 34].

We here consider a social mobile application called iCoolhunt [1]. Its users are encouraged to take pictures of objects they think "cool" and share these pictures with friends online. The idea of "hunting cool" items has attracted a very special community - users are mainly technology-savvy and design-conscious individuals. In such a context, helping users to discover trends becomes important. As we shall see in Section 2, in the literature, there have been studies on what trends are and who creates them. But no study has been yet conducted to show how such knowledge can be leveraged to make personalized recommendation on trends.

To fill the gap, we propose a trend-aware recommender system (Section 4). We build upon the insight offered by Amatriain *et al.* [2]: that is, the recommendation process should not always rely on many (crowd) ratings but might also benefit from *few* expert ratings. By exploiting this insight, we make two main contributions:

- We propose a method that detects trends by paying attention to the activities of two types of users: trend makers (those who upload items that become trends) and trend spotters (those who rate items before they become trends).

- We construct a preference matrix based on the identified trends, and test the extent to which a state-of-the-art matrix factorization algorithm (*Implicit SVD* [14]) effectively recommends trends.

We evaluate the effectiveness of our trend-aware recommender system at helping users discover trends. To this end, we compare our approach to a traditional item-based recommender system (Section 5). We then conclude by discussing how such a trends-aware recommender approach could be applied to a real-life working system (Section 6).

2. RELATED WORK

This paper builds upon existing studies in two main research areas: personalized recommendations, and analyses of trends in social networks.

---

[1] www.icoolhunt.com

Personalized Recommendations. Recommender systems are used in different online services. Traditionally, studies have been focused on recommending books, CDs [21], movies [2, 34], songs [18, 32], news [9], and videos [10]. With the advent of mobile applications, many applications have been able to know where users are, and some services have thus started to recommend location-based events [25], activities, and *POIs* (Points of Interests) [33]. Adding the users' social connections to their geographic information has been found to improve the quality of recommendations [15, 19]. Also, new social connections have themselves become "items to recommend" [6, 24]. There has been a lot of work on algorithms over the last few years (a useful categorization of them can be found in Adamic *et al.*'s work [1]), and effective techniques such as matrix factorization have emerged [14, 15, 33].

Analyses of trends in social networks. "Trends" are items (e.g., pictures, videos) that receive abrupt attention and are of two types: endogenous and exogenous depending on the sources that triggered them. Endogenous trends are triggered by collective activities *within* the user community, while exogenous ones result from activities *outside* the community. In Youtube, researchers have been able to distinguish between endogenous and exogenous trends by simply looking at temporal patterns (more specifically, at the user response rates) [8]. In Twitter, instead, researchers identified trends based on a richer set of features - content, user interactions and social networks [23]. In addition to identify trends, researchers have also looked at the human beings behind trends and have tried to answer the question of who creates them. In his popular book "The Tipping Point", Malcolm Gladwell argued that the creators belong to the "special few". Those are often called "influentials" [12] and are believed to be characterized by large social networks [17] and disproportionally high social standing [27, 29, 31]. By contrast, Watts argues that trends are created by accidental activities, that is, by adopting the right item at the right time [28]. More recently, researchers are coming round to the idea that trends in a networked environment are actually generated by a combination of accidental activities and presence of influential individuals [3, 13].

These two lines of research (i.e., recommender systems and trend analysis) are here brought together to study whether one could build simple ways of facilitating discovery and recommendation of trends. We experimentally do so in the context of a mobile social-networking application with which users discover, organize and share pictures of "cool" design items.

3. APPLICATION

iCoolhunt users are encouraged to take pictures of objects that they think 'cool', upload them and share them with friends online. If their devices are GPS-enabled, pictures are automatically tagged with *(longitude, latitude)* points corresponding to where pictures are uploaded. Uploaders are asked to tag each of their pictures with one of the predefined categories, which include technology, lifestyle, music, design and fashion. They must also add a brief textual description to each of their pictures. By following each other, users can then vote others' pictures using 'like' and 'dislike' buttons and can comment on them. Users are automatically assigned to different levels of expertise depending on the quality of their uploads and votes, and the number of followers they attract. Our anonymized dataset is complete, in that, it covers all user activities from February, 2010 (its launch) to August 2010 (before the time when iCoolhunt launched their web application).

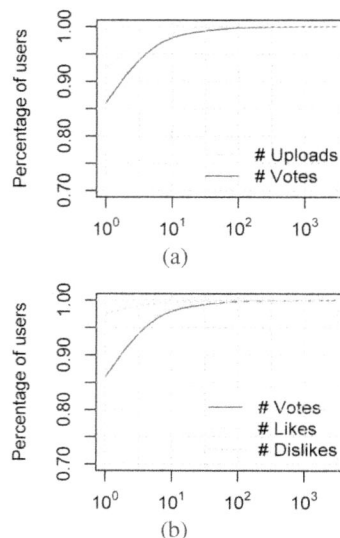

Figure 1: Empirical CDF of 1(a) #uploads and #votes per user; 1(b) #likes, #dislikes and #votes per user.

3.1 Dataset Analysis

Our dataset includes 9,316 iCoolhunt mobile application users, 6,395 photos, and 21,252 votes. To filter away inactive users and lurkers (which, based on our data, are indistinguishable), we focus on users who have uploaded or voted at least once.

Uploads and Votes. In the mobile application, uploading and voting pictures are two main user activities. From the distributions of uploads and votes per user in Figure 1(a), we see that, as one expects, most of the users (92%) have uploaded only once, or voted (85%) only once. The remaining active minority have contributed 83% of pictures and 94% of votes. This results in data sparsity: the fraction of non-zero values in the user-item matrix is around 1%.

Users can vote others' pictures using 'like' and 'dislike' buttons. Figure 1(b) shows the distribution of the number of 'likes', 'dislikes' and the total number of votes per user. One can see that the distribution of number of 'likes' is very similar to the total number of votes per user, suggesting that users tend to frequently express their opinions.

Social Connectivities. Users can also create social contacts and follow each other. We depict the distribution of the number of followers/followees per user in Figure 2, and find that only a very small portion of users follow others, suggesting that using social ties to recommend content would not work in this specific application.

From this brief analysis, two main points emerge: 1) there is a core group of users who contributed most of the content; and 2) relying only on the social graph to recommend trends would not be beneficial. For these two reasons, we propose a way of recommending trends based on the idea of identifying the "special few".

4. TREND-AWARE RECOMMENDATION

To recommend trends, we perform three steps (Figure 3):

1. Identify trend makers and trend spotters (Section 4.1);

2. Identify trends (Section 4.2);

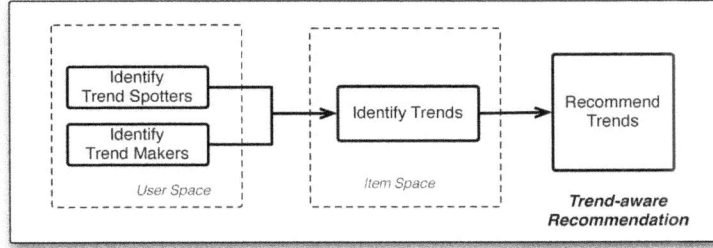

Figure 3: Trend-aware Recommender System

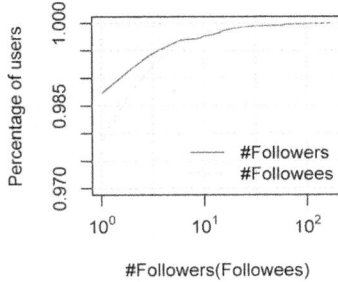

Figure 2: Empirical CDF of #followers and #followees per user.

3. Recommend the previously identified trends (Section 4.3).

In Section 6 on "Discussion", we will explain why it is not a good idea to identify trends directly and, instead, it is beneficial to identify trend spotters (makers) first and then trends later on.

4.1 Trend makers and trend spotters

In every social application, there are large behavioral differences among users [22]: some are able to identify trends early on, and some are leisure laggards. To identify the former type, we focus on two user categories – trend makers and trend spotters.

Trend makers are those who tend to *upload* items that then become trends, and trend spotters are those who tend to *vote* items that then become trends early on. To quantitatively identify those users, we consider the following features:

Activity. The main activities on the application are two and are voting and uploading. From them, we compute three activity features: *daily votes*, *daily uploads* and *lifetime*. This last feature reflects whether users are early adopters (i.e., are those who shape social norms [4]) or not [11].

Content. Users vary in how diverse their interests are: one could have a wide variety of interests, while another one could "focus" on very specific and limited set of interests. In Twitter, for example, it has been shown that influential users focus on very specific topics [5, 29]. To differentiate users based on their interest diversity, we consider two measures of content diversity. Both use the Shannon Index and are called *upload diversity* and *vote diversity*.

Social Network. Since information might partly propagate along social connections, we also account for how well a user is connected by considering *number of followers*, *number of followees*, and the user's *clustering coefficient* (computed on the social graph in which each node represents a user, and an edge links two users with at least one following relationship [28]).

Geography. We finally consider: 1) how much and how often one user is *wandering* in the real world by using the radius of gyration [7]; and 2) a user's *geographical span of followers* computed as the average distance between where the user is and where his/her followers are.

Having these features, we now perform three steps. For each user, we:

Step 1 Compute the user's spotter score and maker score;

Step 2 Discretize the user's scores.

Step 3 Predict the user's discretized scores on input of the previous features of activity, content, social network, and geography.

Let us now spell out each of the steps.

Step 1. To begin with, we introduce two metrics that reflect the extent to which a trend maker (spotter) u is successfully uploading (spotting) trends. User u's $makerScore(u)$ is:

$$makerScore(u) = \frac{\sum_{i \in \mathcal{I}_u} I(i \ is \ a \ trend)}{|\mathcal{I}_u|}, \qquad (1)$$

where $\mathcal{I}_u$ is the set of trends that u has uploaded, and $I(i \ is \ a \ trend)$ is an indication function which equals to 1, if i is a trend; otherwise, it is 0. To establish whether an item is a trend or not, we use a metric similar to the one proposed in [23]. That is, for each time unit t, each item i is assigned with a $trendScore(i,t)$ computed as:

$$trendScore(i,t) = \frac{|v_{i,t}| - \mu_i}{\sigma_i}, \qquad (2)$$

where $|v_{i,t}|$ is the number of votes item i has received within time unit t, μ_i is the mean number of votes it received per time unit, and σ_i is its standard deviation. A high trend score tells that the item have received more attention than expected within the time unit. In each time unit, items are sorted according to their trend scores, and top-N items are extracted and identified as trends. From our analysis, we found that the temporal resolution (one week or two weeks) and the length of the recommended list do not significantly change the scores. In Figure 4, one observes that the trend spotter score does not change as the list length (top-10 vs. top-50) changes.

To add the spotter score to the maker score, we observe that the ability of spotting trends is largely determined by three factors – how many, how early, and how popular one's spotted trends become. To incorporate the factor of how early and how popular trends become, for each trend i that u has spotted (voted), we compute the following gain $g_{u,i}$ score:

$$g_{u,i} = v_i \times \alpha^{-p_{u,i}}, \qquad (3)$$

53

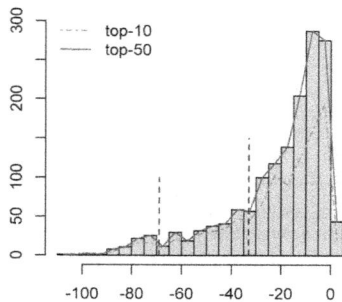

Figure 4: Trend spotter score (log). We split trend spotters into three classes using a proportional 3-interval discretization, as the two vertical lines show.

in which v_i is the total votes i received, $p_{u,i}$ captures that u is the pth user who spotted i (the lower p, the better), and α is a decay factor. Combining a user's gains all together, we obtain a cumulative $spotterScore$ for user u (which is normalized by user u's total number of votes):

$$spotterScore(u) = \frac{\sum_{i \in \mathcal{I}_u} g_{u,i}}{v_u}, \qquad (4)$$

Step 2. Based on their maker scores and trend scores, we are able to cluster users into K classes, which indicate their ability of uploading (spotting) trends. To do so, we apply a proportional k-interval discretization [30] over the whole range of maker (spotter) scores and assign each user to one of the three classes (with $k = 3$). We chose three because the distribution of trend spotter(maker) scores in Figure 4 shows three distinct increments in score and thus lends itself to the identification of three classes of users.

Step 3. Based on a user's (activity, content, social network, and geographic) features, machine learning models predict to which class the user belongs. We tried a variety of models - Naive Bayesian, linear regression, and *SVM* - and found that *SVM* works best (Section 5).

4.2 Identify Trends

Trend makers and trend spotters are the source of trends, but not all items uploaded and voted by those users become trends - there is a certain probability that they will be so. More generally, an item is likely to become a trend depending on: the extent to which the item's uploader is a trend maker; and the extent to which the item's voters are trend spotters. We model these insights in the following logistic regression:

$$Pr(y_i = 1) = logit^{-1}(X_i \beta) \qquad (5)$$

where $Pr(y_i = 1)$ is the probability of an item y_i is a trend ($Pr(y_i = 1)$), and X_i are a set of predictors, which are the uploader's trend maker class and the voters' trend spotter classes. Results are again presented in Section 5.

4.3 Recommend Trends

Having identified items that are likely to be trends, we are now able to build a trend-aware preference matrix $\mathcal{P}'$, in which $p'_{u,t}$ is 1 or 0 depending on whether u liked item t that has then become a trend. Upon this matrix, we apply two existing recommender systems algorithms: *Implicit SVD* [14] and item-based collaborative filtering [21]. We compare how these algorithms perform on input of the trend-aware matrix and on input of a traditional preference

matrix $\mathcal{P}$, in which $p_{u,t}$ is 1 or 0 depending, again, on whether u voted on item t that has then become a trend. The difference between the two preference matrices is that the trend-aware one is less sparse because, at the columns, it does not have all items but only those that we have predicted to be trends.

5. EVALUATION

We now evaluate the effectiveness of each of the three steps we have just introduced.

Classifying users into trend spotter(maker) classes. We first evaluate the extent to which *SVM* is able to classify each user into one of the three maker/spotter classes on input of the user's features (introduced in Section 4.1). To this end, upon our dataset that has 209 unique trends, 50 trend makers, and 531 trend spotters, we run a 10-fold cross validation and test three algorithms: linear regression, Naive Bayesian, and *SVM*. The best accuracies are returned by *SVM*: 83.80% of trend spotters and 60.7% of trend makers are correctly identified.

Determining whether an item is a trend or not. After ascertaining that *SVM* is able to reasonably identify trend spotters and trend makers, we now need to test whether the logistic regression in Section 4.2 is able to identify trends based on information about uploaders and voters. The regression predicts whether an item is a trend or not based on four features: the uploader's trend maker class (first feature), and the number of votes from users who belong to: the low spotter class (second feature), the medium spotter class (third feature), and high spotter class (fourth feature). To test the logistic regression, we build a balanced dataset that contains our 209 trends plus 209 (randomly extracted) non-trends and obtain the results in Table 1. The statistically significant coefficients suggest that an item is more likely to become a trend, if its uploader is a good trend maker and its voters are in the upper (trend spotter) class.

To avoid overfitting in Equation 5, we add a Tikhonov regularization term. The problem of learning β now translates into the following optimization problem:

$$\beta' = \arg\min_{\beta} \sum_i log(1 + exp(-y_i \beta X_i)) + \lambda ||\beta||_2^2 \qquad (6)$$

We split the dataset of trends into two subsets: the first subset consists of 80% of the entire dataset and is used to *train* the model, while the remaining 20% is used to *test* the model. With a 10-fold cross validation, we first fix the value of λ and then fit the model with the training set. To measure the accuracy of the regularized logistical regression model, we apply the trained model to the test set. We obtain the ROC curve plot that reflects both the model's TPR (true positive rate) and FPR (false positive rate). In Figure 5, one sees that the regression returns accurate results, whereby the point (0,1) corresponds to the best classification performance (the diagonal reflects the baseline of random guess).

Recommending trends. At this point, we have ascertained our ability to identify trends. Now the question is: if we were to build a *user-by-trend* matrix out of the predicted trends, what would be the performance of an existing collaborative filtering algorithm? To answer that question, we need to establish three things:

1. We need to select which existing algorithm to use. For now, we pick a simple item-based collaborative filtering algorithm [26]. Later, we will see whether we can improve performance with *Implicit SVD* [14].

Predictors	Coefficient
Uploader's trend maker class	6.21 * * *
#Voters from low trend spotter class	-1.30
#Voters from medium trend spotter class	-1.17 *
#Voters from high trend spotter class	0.64 * * *

Table 1: Coefficients of the logistic regression (a correlation coefficient within 2 standard errors is statistically significant. The significance levels are marked with ∗'s: $p < 0.001(∗ ∗ ∗)$, $p < 0.01(∗∗)$, $p < 0.05(∗)$))

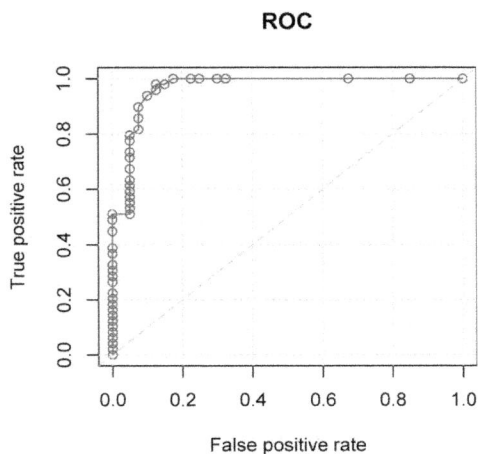

Figure 5: ROC curve for the logistic regression that predicts whether an item is a trend or not.

2. We need to determine which metrics reflect performance. To be in line with the literature, we compute precision and recall [14]:

$$recall(N) = \frac{\#hits}{|T|} \qquad (7)$$

$$precision(N) = \frac{\#hits}{N * |T|} \qquad (8)$$

where T is the test set, and N is number of items to recommend.

3. Finally, we need to determine the baseline against which our trend-aware approach will be compared. To ease interpretability of the results, we again select item-based collaborative filtering but, this time, the algorithm would take in input the original user-item preference matrix $\mathcal{P}$, in which:

$$p_{u,i} = \begin{cases} 1 & \text{if } u \text{ likes } i \ \& \ i \text{ is a trend} \\ 0 & \text{otherwise} \end{cases}$$

Figure 6 shows precision and recall for the traditional and trend-aware item-based collaborative filtering as a function of the recommended list size (top-N recommendations). For both systems, precision and recall improve linearly as N increases, but the trend-aware's increase is faster, suggesting that a traditional item-based recommender system would not be able to recommend trends (precision/recall is only 0.05 for top-10 recommendations), while a trend-aware system would (precision/recall is 0.2). These results also suggest that, in the presence of data sparsity, relying on few expert ratings is an effective way of recommending trends.

So far, we have analyzed how an item-based collaborative filtering algorithm would perform. Next, we test whether a matrix

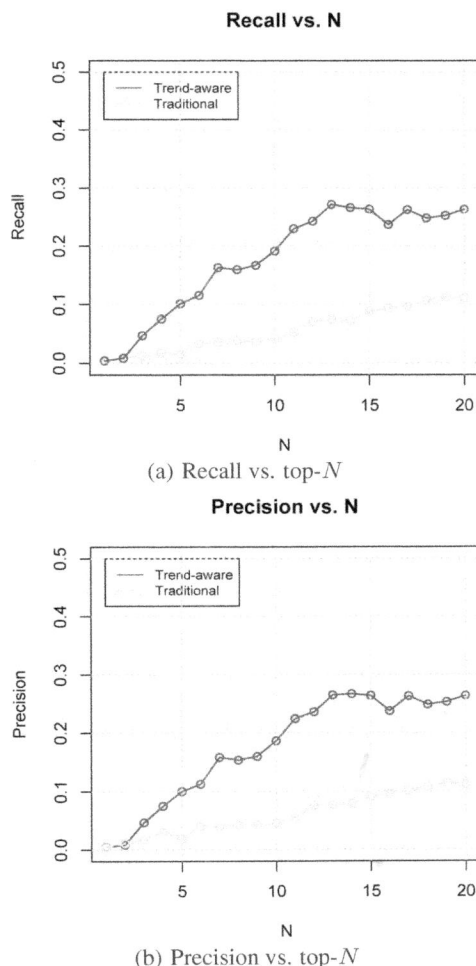

(a) Recall vs. top-N

(b) Precision vs. top-N

Figure 6: Precision and Recall. Results for trend-aware recommender vs. item-based recommender. The size of the recommended list is N.

factorization approach - *Implicit SVD* - would improve the performance. Figure 7 shows this to be the case. As the size of the recommended list increases, *Implicit SVD* consistently returns better precision and recall than item-based's.

Interestingly, if one were to recommend only popular trends, precision and recall would be worst. This suggests that even for trends - items that one expects to be non-long tail - personalization makes sense. But up to a point - precision and recall results are limited, and that is largely because of the very nature of trends.

To sum up, the results on precision and recall should be interpreted in comparative terms, while the other results answer a more fundamental question - whether trend detection helps the recommendation process; and the answer is a definite 'Yes'.

6. DISCUSSION

We now discuss some open questions.

Why not detecting activity bursts directly? Since a burst detection algorithm could be easily applied to identify trends (in a way similar to expression (2)), one might wonder why we are going to the trouble of having the intermediate step of identifying trend

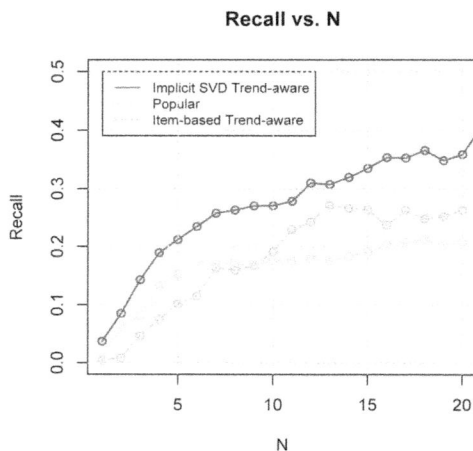

Recall vs. N

(a) Recall vs. top-N

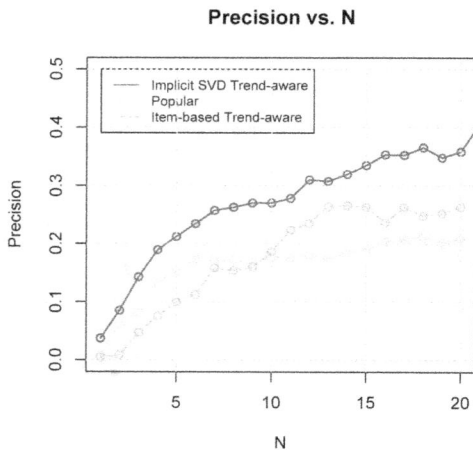

Precision vs. N

(b) Precision vs. top-N

Figure 7: Precision and Recall. Results for two trend-aware recommenders (item-based and Implicit SVD) and for recommendations of most popular trends.

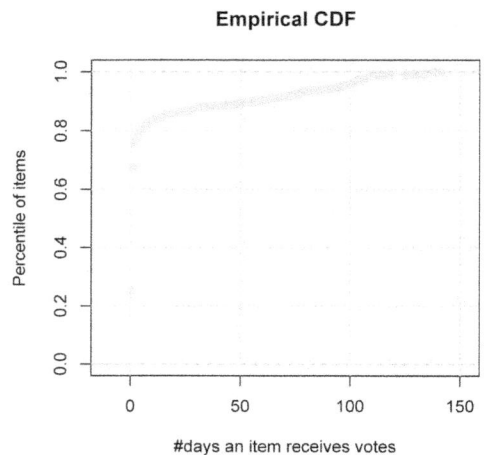

Empirical CDF

(a) Item Lifetime

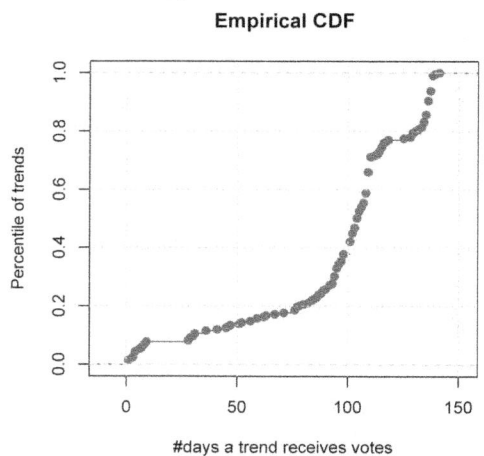

Empirical CDF

(b) Trend Lifetime

Figure 8: Number of days an item (*a*) vs. a trend (*b*) receives votes for.

spotters (makers) and not, instead, identifying trends directly. The main reasons for this choice are efficiency and time:

1. *Efficiency.* Monitoring a limited number of users who are trend spotters and trend makers is more efficient than requiring the complete and up-to-date view of the system: a typical burst detection algorithm would require to monitor *all* votes on *all* items.

2. *Time.* This is the most important reason and comes from the temporal dynamics typical of trends. The average item is generally short-lived and dies off the first or second day (Figure 8(a)); by contrast, a trend persists for a longer period of time (as one would expect based on preferential attachment[2]), yet it also takes off only after two weeks or so (Figure 8(b)). As such, burst detection would miss trends for a long period of time, while monitoring key individuals - trend spotters (makers) - can be done quickly and efficiently. One contribution of this paper has been to show that monitoring

<hr />

[2]http://en.wikipedia.org/wiki/Preferential_attachment

key individuals not only is quick and efficient but is also an accurate way of identifying trends.

Online Updating. In our analysis, we have not registered the frequent emergence of new trend spotters and trend makers. However, in a system with a larger user base, that might be the case, and ways of updating the pool of key users - trend spotters (makers) - would be needed. To decide when and how to run such updates, we are currently exploring the use of controllers that automatically and accurately estimate frequency of updates. These techniques have been recently introduced with the idea of ensuring stable and high-quality recommendations in dynamically evolving environments [16].

7. CONCLUSION

We have shown that, upon activity, network, and geographic attributes, a machine learning approach can identify key users - trend spotters and trend makers. A simple logistic regression can then reliably infer whether an item (a picture in this case) will be a trend or not based on whether the item has been uploaded by a successful trend maker and voted by trend spotters. We have then seen that existing recommender systems can profit from this ability of identifying these special users. One promising future research direction is to simultaneously model the two processes here treated separately - collaborative filtering and trend detection. This could be done by, for example, combing Amatriain *et al.*'s work [2] with existing models of temporal dynamics.

8. ACKNOWLEDGEMENTS

We thank Neal Lathia for his comments on earlier drafts of the paper. The work was funded by the French National Research Agency (ANR) under the PROSE project (ANR-09-VERS-007), by an industrial scholarship from PlayAdz [3], and by RCUK through the Horizon Digital Economy Research grant (EP/G065802/1).

9. REFERENCES

[1] G. Adomavicius and A. Tuzhilin. Toward the Next Generation of Recommender Systems: A Survey of the State-of-the-art and Possible Extensions. *IEEE Transactions on Knowledge and Data Engineering*, 2005.

[2] X. Amatriain, N. Lathia, J. Pujol, H. Kwak, and N. Oliver. The Wisdom of the Few: A Collaborative Filtering Approach Based on Expert Opinions from the Web. In *Proceedings of the 32nd ACM International Conference on Research and Development in Information Retrieval (SIGIR)*, 2009.

[3] S. Aral and D. Walker. Identifying Influential and Susceptible Members of Social Networks. *Science*, June 2012.

[4] D. Boyd. The Future of Privacy: How Privacy Norms Can Inform Regulation. Invited Talk at the 32nd International Conference of Data Protection and Privacy Commissioners, October 2010.

[5] M. Cha, H. Haddadi, F. Benevenuto, and K. Gummadi. Measuring User Influence in Twitter: The Million Follower Fallacy. In *Proceedings of the 4th International AAAI Conference on Weblogs and Social Media (ICWSM)*, 2010.

[6] J. Chen, W. Geyer, C. Dugan, M. Muller, and I. Guy. Make New Friends, But Keep the Old: Recommending People on Social Networking Sites. In *Proceedings of the 27th ACM International Conference on Human Factors in Computing Systems*. ACM, 2009.

[7] Z. Cheng, J. Caverlee, K. Lee, and D. Sui. Exploring Millions of Footprints in Location Sharing Services. In *Proceedings of the 5th International AAAI Conference on Weblogs and Social Media (ICWSM)*, 2011.

[8] R. Crane and D. Sornette. Robust Dynamic Classes Revealed by Measuring the Response Function of a Social System. In *Proc. of PNAS Vol.105*, 2008.

[9] A. Das, M. Datar, A. Garg, and S. Rajaram. Google News Personalization: Scalable Online Collaborative Filtering. In *Proceedings of the 16th ACM International Conference on World Wide Web*, 2007.

[10] J. Davidson, B. Liebald, J. Liu, P. Nandy, T. Van Vleet, U. Gargi, S. Gupta, Y. He, M. Lambert, B. Livingston, et al. The youtube video recommendation system. In *Proceedings of the 4th ACM Conference on Recommender Systems*, 2010.

[11] C. Droge, M. Stanko, and W. Pollitte. Lead Users and Early Adopters on the Web: The Role of New Technology Product Blogs. *Journal of Product Innovation Management Vol.27*, 2010.

[12] M. Gladwell. *The Tipping Point: How Little Things Can Make a Big Difference*. Little, Brown and Company, 2000.

[13] S. Gonzalez-Bailon, J. Borge-Holthoefer, A. Rivero, and Y. Moreno. The Dynamics of Protest Recruitment through an Online Network. December 2011.

[14] Y. Hu, Y. Koren, and C. Volinsky. Collaborative Filtering for Implicit Feedback Datasets. In *Proceedings of the 8th IEEE International Conference on Data Mining*, 2008.

[15] M. Jamali and M. Ester. A Matrix Factorization Technique with Trust Propagation for Recommendation in Social Networks. In *Proceedings of the 4th ACM conference on Recommender systems*, 2010.

[16] T. Jambor, J. Wang, and N. Lathia. Using Control Theory for Stable and Efficient Recommender Systems. In *Proceedings of the 21st International Conference on World Wide Web (WWW)*, 2012.

[17] D. Kempe, J. Kleinberg, and E. Tardos. Influential Nodes in a Diffusion Model for Social Networks. In *Automata, Languages and Programming Vol.3580*. Springer Berlin / Heidelberg, 2005.

[18] N. Koenigstein, G. Dror, and Y. Koren. Yahoo! Music Recommendations: Modeling Music Ratings with Temporal Dynamics and Item Taxonomy. In *Proceedings of the 5th ACM conference on Recommender systems (RecSys)*. ACM, 2011.

[19] I. Konstas, V. Stathopoulos, and J. Jose. On Social Networks and Collaborative Recommendation. In *Proceedings of the 32nd International ACM SIGIR Conference on Research and Development in Information Retrieval*, 2009.

[20] H. Li, Y. Wang, D. Zhang, M. Zhang, and E. Chang. Pfp: Parallel FP-growth for Query Recommendation. In *Proceedings of the 2nd ACM Conference on Recommender Systems*, 2008.

[21] G. Linden, B. Smith, and J. York. Amazon.com Recommendations: Item-to-item Collaborative Filtering. *IEEE Internet Computing*, 2003.

[22] M. Muller, N. S. Shami, D. R. Millen, and J. Feinberg. We are all Lurkers: Consuming Behaviors among Authors and Readers in an Enterprise File-Sharing Service. In *Proceedings of the 16th ACM International Conference on Supporting Group Work (GROUP)*, 2010.

[23] M. Naaman, H. Becker, and L. Gravano. Hip and Trendy: Characterizing Emerging Trends on Twitter. *Journal of the American Society for Information Science and Technology Vol.65*, 2011.

[24] D. Quercia and L. Capra. FriendSensing: Recommending Friends Using Mobile Phones. In *ACM RecSys*, 2009.

[25] D. Quercia, N. Lathia, F. Calabrese, G. Di Lorenzo, and J. Crowcroft. Recommending Social Events from Mobile Phone Location Data. In *Proceedings of the 10th IEEE International Conference on Data Mining (ICDM)*, 2010.

[26] B. Sarwar, G. Karypis, J. Konstan, and J. Reidl. Item-based collaborative filtering recommendation algorithms. In *Proceedings of the 10th International ACM Conference on World Wide Web (WWW)*, 2001.

[27] J. Tang, J. Sun, C. Wang, and Z. Yang. Social Influence Analysis in Large-scale Networks. In *Proceedings of the 15th International Conference on Knowledge Discovery and Data Mining (KDD)*, 2009.

[28] D. Watts and S. Strogatz. Collective Dynamics of Small-world Networks. *Nature Vol.393*, 1998.

[29] J. Weng, E.-P. Lim, J. Jiang, and Q. He. TwitterRank:

---

[3] http://www.playadz.com

Finding Topic-sensitive Influential Twitterers. In *Proc. of the 3rd ACM WSDM*, 2010.

[30] I. H. Witten and E. Frank. *Data Mining: Practical Machine Learning Tools and Techniques, Second Edition* . Morgan Kaufmann Publishers Inc., San Francisco, CA, USA, 2005.

[31] L. Yu, S. Asur, and B. A. Huberman. What Trends in Chinese Social Media. *The 5th SNA-KDD Workshop*, 2011.

[32] Y. Zhang, D. Séaghdha, D. Quercia, and T. Jambor. Auralist: Introducing Serendipity into Music Recommendation. In th

Proceedings of the 5 ACM International Conference on Web Search and Data Mining. ACM, 2012.

[33] V. Zheng, Y. Zheng, X. Xie, and Q. Yang. Collaborative Location and Activity Recommendations with GPS History Data. In *Proceedings of the 19th ACM International Conference on World Wide Web*, 2010.

[34] Y. Zhou, D. Wilkinson, R. Schreiber, and R. Pan. Large-scale Parallel Collaborative Filtering for the Netflix Prize. *Algorithmic Aspects in Information and Management*, 2008.

Real-Time Top-N Recommendation in Social Streams

Ernesto Diaz-Aviles[1], Lucas Drumond[2], Lars Schmidt-Thieme[2], and Wolfgang Nejdl[1]

[1]L3S Research Center / University of Hannover, Germany
{diaz, nejdl}@L3S.de

[2]Information Systems and Machine Learning Lab / University of Hildesheim, Germany
{ldrumond, schmidt-thieme}@ISMLL.de

ABSTRACT

The Social Web is successfully established, and steadily growing in terms of users, content and services. People generate and consume data in real-time within social networking services, such as Twitter, and increasingly rely upon continuous streams of messages for real-time access to fresh knowledge about current affairs. In this paper, we focus on analyzing social streams in real-time for personalized topic recommendation and discovery. We consider collaborative filtering as an online ranking problem and present *Stream Ranking Matrix Factorization* – RMFX –, which uses a pairwise approach to matrix factorization in order to optimize the personalized ranking of topics. Our novel approach follows a selective sampling strategy to perform online model updates based on active learning principles, that closely simulates the task of identifying relevant items from a pool of mostly uninteresting ones. RMFX is particularly suitable for large scale applications and experiments on the *476 million Twitter tweets* dataset show that our online approach largely outperforms recommendations based on Twitter's global trend, and it is also able to deliver highly competitive Top-N recommendations faster while using less space than Weighted Regularized Matrix Factorization (WRMF), a state-of-the-art matrix factorization technique for Collaborative Filtering, demonstrating the efficacy of our approach.

Categories and Subject Descriptors: H.3.3 [**Information Storage and Retrieval**]—*Information Filtering*

General Terms: Algorithms, Experimentation, Measurement, Performance

Keywords: Collaborative Filtering; Matrix Factorization; Online Learning; Ranking; Selective Sampling; Twitter

1. INTRODUCTION

The amount of user generated content in social media applications and the rate at which such content is made available poses a challenge to state-of-the-art recommender system algorithms. For instance, the number of users of the

popular micro-blog service *Twitter*[1] is estimated to have surpassed 300 million generating more than 200 million *tweets* (micro-blog posts) per day [18].

When dealing with user generated content in social media applications, it is crucial that finding good patterns and making inference about them is done in a reasonable time. One scenario where this point becomes particularly critical is that of filtering a continuous stream of incoming tweets in order to recommend topics that match user interests at a specific moment, given the huge scale of this kind of data. The challenge here is to account for dynamic short term information needs of the users. As a surrogate for a user's topic interests one could use *hashtags*. For example, in Twitter, if user *Alice* often tags her tweets with the hashtag *#TechCrunch* and never uses the hashtag *#fashion*, we can exploit this information and use it as a good indicator for her preferences. We can infer that, currently, *Alice* is more interested in *technology news* than, for instance, in *fashion*. Thus the task can be cast as that of recommending hashtags to users.

Collaborative filtering (CF) has been shown to be an effective approach to recommender systems. The essence of CF lies in analyzing past user and item interactions to generate personalized recommendations based on the preferences of other users with similar behavior. One of CF's most successful techniques are low dimensional linear factor models, that assume user preferences can be modeled by only a small number of *latent* factors [11].

Although latent factor models are able to generate high quality recommendations, coping with fast changing trends in the presence of large scale data might be a challenge, since retraining such models is costly. One alternative is to learn the parameters online, updating the decision function for each new observation [2]. Unfortunately, the gain in processing time achieved by online learning algorithms comes at the cost of reduced prediction quality and work has been done on closing this gap. The main issue with online approaches is their short-term "memory", i.e., since the updates based only on the most recent data point do not take into account past observations, the model quickly "forgets" them. Recently Zhao et al. [23] proposed an online learning algorithm for maximizing Area Under the ROC Curve (AUC), a metric that is widely used for measuring the classification performance for imbalanced data distributions, that addresses this problem by keeping a representative sample of the data set in a reservoir and using only this sample plus the new observation for retraining.

---

[1]**Twitter**: http://twitter.com

One issue that arises with the reservoir approach is how to perform the model updates. In this paper we propose to selectively sample the most informative instances from a reservoir using personalized small buffers and perform stochastic gradient descent updates based on *active learning* principles [6, 22]. We elaborate on the notion of "informative elements" and illustrate the application of our approach to learning factorization models. We demonstrate its usefulness on the task of recommending hashtags to Twitter users based on real world data.

To summarize, the main contributions of this work are as follows:

- We introduce a novel framework for online collaborative filtering. The novelty of our approach lies in a selective sampling strategy to update the model based on personalized small buffers.

- We propose Stream Ranking Matrix Factorization (RMFX), an online leaning algorithm based on a pairwise ranking approach for matrix factorization that is intended for streaming data, and is well founded in stochastic gradient descent.

- For unpersonalized learning to rank, many studies have been made in the field of information retrieval. This paper presents an innovative personalized ranking perspective to matrix factorization for social media streams, which has not been reported before in the literature.

- Finally, this paper provides an example of integrating large-scale collaborative filtering with the real-time nature of Twitter.

The reminder of the paper is organized as follows: In Section 2, we present background material and notation. In Section 3, we present our Stream Ranking Matrix Factorization approach. Section 4 discusses related work. In Section 5, we show the results of our approach by analyzing real-world data consisting of millions of tweets. Finally, in Section 6, we conclude and present directions for future work.

2. BACKGROUND

First we introduce some notation that will be useful in our setting. Let $U = \{u_1, \ldots, u_n\}$ and $I = \{i_1, \ldots, i_m\}$ be the sets of all users and all items, respectively. We reserve special indexing letters to distinguish users from items: for users u, v, and for items i, j. Suppose we have interactions between these two entities, and for some user $u \in U$ and item $i \in I$, we observe a *relational score* x_{ui}.

Thus, each instance of the data is a tuple (u, i, x_{ui}). For example in the movie recommendation case, the tuple might correspond to an explicit "rating" given by user u to movie i or, in the case of hashtag/topic recommendation, to a "weight" that is implicitly derived from user u's interaction patterns, e.g., how many times the user u has used hashtag i. Typical CFs organize these tuples into a sparse matrix $\mathbf{X}$ of size $|U| \times |I|$, using (u, i) as index and x_{ui} as entry value. The task of the recommender system is to estimate the score for the missing entries. The relational scores themselves are ordinal and need not be numbers. Thus, we assume a total order between the possible score values. We distinguish predicted scores from the known ones, by using $\hat{x}_{ui}$. The set S of all observed scores is defined as follows:

$$S := \{(u, i, x_{ui}) \mid (u, i, x_{ui}) \in U \times I \times \mathbb{N}\} .$$

For convenience, we also define for each user the set of all items with an observed score, denoted by B_u^+:

$$B_u^+ := \{i \in I \mid (u, i, x_{ui}) \in S\} .$$

Low dimensional linear factor modeling are popular collaborative filtering approaches [11]. These models consider that only a small number of *latent* factors can influence the preferences. Their prediction is a real number, $\hat{x}_{ui}$, per user item pair (u, i). Some of the most successful realizations of latent factor models are based on matrix factorization (MF). In its basic form, matrix factorization estimates a matrix $\mathbf{X} : U \times I$ by the product of two low-rank matrices $\mathbf{W} : |U| \times k$ and $\mathbf{H} : |I| \times k$:

$$\hat{\mathbf{X}} := \mathbf{W}\mathbf{H}^\mathsf{T} , \qquad (1)$$

where k is a parameter corresponding to the rank of the approximation.

The factorization process is performed by minimizing a loss function that measures the quality of the reconstruction $\hat{\mathbf{X}}$. One alternative to learn the optimal parameters of the model is to use a Stochastic Gradient Descent (SGD) approach [2].

Even though the *squared loss* has been successfully used for MF in the context of rating prediction (e.g., [11]) and item prediction [8], we are interested in a *ranking* approach to MF, and therefore require an *ordinal loss* to guide the factorization process. In particular we are interested in a *pairwise* approach, similar to the one used by RankSVM [9], a popular ranking method in the field of learning to rank. We present the details of our approach in the next section and discuss related work in Section 4.

3. STREAM RANKING MATRIX FACTORIZATION

In the presence of a continuous stream of incoming tweets, arriving at a high rate, our objective is to process the incoming data in bounded space and time and recommend a short list of interesting topics that meet users' individual taste.

The high rate makes it harder to: (i) capture the information transmitted, (ii) compute sophisticated models on large pieces of the input, and (iii) store the input data, which can be significantly larger than the algorithm's available memory.

This problem setting fits a streaming model of computation by Muthukrishnan [12], which establishes that, by imposing a space restriction on algorithms that process streaming data, we may not be able to store all the data we see. The impact is that the data generated in real-time carries high-dimensional information which is difficult to extract and process. Any time lag in modeling the data could render the outcome of the modeling obsolete and useless.

We assume that topics of interest are captured by the hash-tagging behavior in Twitter. Hashtags are words or phrases prefixed with the symbol #, e.g., *#recsys*, a form of metadata tag used to mark keywords or topics in a tweet. Hashtags evolve over time, reflecting the dynamics of user preferences in the social stream. Our approach seeks to incorporate these dynamics to produce a short list of interesting recommendations based on a matrix factorization model for CF, which is learned online.

In this section, we formally define the problem and introduce our approach Stream Ranking Matrix Factorization or RMFX, and develop our model in steps discussing the rationale behind them. Such steps are illustrated in Figure 1.

Figure 1: **Main steps of our approach Stream Ranking Matrix Factorization (RMFX).**

3.1 Pairwise Approach for Personalized Rank Learning

We focus on learning a matrix factorization model for collaborative filtering in presence of streaming data. To this end, we will follow a pairwise approach to minimize an ordinal loss. Our formalization extends the work of Scully [15] for unpersonalized learning to rank, to an online collaborative filtering setting.

With slight abuse of notation, we also use S to represent the input stream $s_1, s_2, \ldots$ that arrives sequentially, instance by instance. Let $p_t = ((u,i), (u,j))_t$ denote a pair of training instances sampled at time t, where $(u,i) \in S$ has been observed in the stream and $(u,j) \notin S$ not.

Formally, we define the set P as the set of tuples $p = ((u,i), (u,j))$ selected from the data stream S, as follows:

$$P := \{((u,i),(u,j)) \mid i \in B_u^+ \ \wedge \ j \notin B_u^+\} .$$

We require pairs that create a *contrast* in the preferences for a given user u over items i and j. Since we are dealing with implicit, positive only feedback data (i.e. the user never explicitly states a negative preference for an item) we follow the rationale from Rendle et al. [13] and assume that user u prefers item i over item j. We will restrict the study to a binary set of preferences $x_{ui} = \{+1, -1\}$, e.g., *observed* and *not-observed*, represented numerically with $+1$ and -1, respectively. For example, if a user u in Twitter posts a message containing hashtag i, then we consider it as a positive feedback and assign a score $x_{ui} = +1$. More formally, $x_{ui} = +1 \iff i \in B_u^+$. In future work we plan to explore how repeated feedback can be exploited to establish a total order for items in B_u^+.

It is obvious that, in the case of streaming data, we do not compute P explicitly, but instead select pairs from the stream, at each time step, that meet P's membership requirements.

With P defined, we find $\theta = (\mathbf{W}, \mathbf{H})$ that minimizes the pairwise objective function:

$$\underset{\theta=(\mathbf{W},\mathbf{H})}{\operatorname{argmin}} L(P, \mathbf{W}, \mathbf{H}) + \frac{\lambda_W}{2} \|\mathbf{W}\|_2^2 + \frac{\lambda_H}{2} \|\mathbf{H}\|_2^2 . \quad (2)$$

In this paper, we explore the use of the SVM loss, or *hinge-loss*, used by RankSVM for the learning to rank task [9].

RMFX Framework
Input:
    Reservoir representing a sample of the stream at time t: R; Regularization parameters λ_W, λ_{H^+}, and λ_{H^-}; Learning rate η_0; Learning rate schedule α; Number of iterations T_S, and T_θ; Parameter c to control how often to perform the model updates.
Output: $\theta = (\mathbf{W}, \mathbf{H})$
1: initialize $\mathbf{W}_0$ and $\mathbf{H}_0$
2: initialize sample stream $S' \leftarrow \emptyset$
3: counter $\leftarrow 0$
4: **for** $t = 1$ to T_S **do**
5:     $R \leftarrow$ **updateReservoir**(R)
6:     counter $\leftarrow$ counter $+ 1$
7:     **if** $c =$ counter **then**
8:         $\theta \leftarrow$ **updateModel**$(S_t, \lambda_W, \lambda_{H^+}, \lambda_{H^-}, \eta, \alpha, T_\theta)$
9:         counter $\leftarrow 0$
10:    **end if**
11: **end for**
12: **return** $\theta_T = (\mathbf{W}_T, \mathbf{H}_T)$

Figure 2: **RMFX Framework for Real-Time CF.**

Given the predicted scores $\hat{x}_{ui}$ and $\hat{x}_{uj}$, the ranking task is reduced to a pairwise classification task by checking whether the model is able to correctly rank a pair $p \in P$ or not. Thus, $L(P, \mathbf{W}, \mathbf{H})$ is defined as follows:

$$L(P, \mathbf{W}, \mathbf{H}) = \frac{1}{|P|} \sum_{p \in P} \hbar(y_{uij} \cdot \langle \mathbf{w}_u, \mathbf{h}_i - \mathbf{h}_j \rangle) , \quad (3)$$

where $\hbar(z) = max(0, 1-z)$ is the hinge-loss; $y_{uij} = sign(x_{ui} - x_{uj})$ is the $sign(z)$ function, which returns $+1$ if $z > 0$, i.e., $x_{ui} > x_{uj}$, and -1 if $z < 0$. The prediction function $\langle \mathbf{w}_u, \mathbf{h}_i - \mathbf{h}_j \rangle = \langle \mathbf{w}_u, \mathbf{h}_i \rangle - \langle \mathbf{w}_u, \mathbf{h}_j \rangle$ corresponds to the difference of predictor values $\hat{x}_{ui} - \hat{x}_{uj}$.

Please note that in this special case of binary rank values of *observed* and *not-observed*, the optimization problem defined by Eq. (3) is equivalent to the problem of optimizing area under the ROC curve (AUC) for binary-class data [15].

Other convex loss functions can also be applied, e.g., *squared* or *logistic* loss [13, 15], as well as any prediction function besides the dot product $\langle \cdot, \cdot \rangle$ [14].

To conclude this section, we compute the gradient of the pairwise loss at instance $p_t \in P$ with non-zero loss, and model parameters $\theta_t = (\mathbf{w}_u, \mathbf{h}_i, \mathbf{h}_j)$, as follows:

$$-\nabla \hbar(p_t, \theta_t) = \begin{cases} y_{uij} \cdot (\mathbf{h}_i - \mathbf{h}_j) & \text{if } \theta_t = \mathbf{w}_u, \\ y_{uij} \cdot \mathbf{w}_u & \text{if } \theta_t = \mathbf{h}_i, \\ y_{uij} \cdot (-\mathbf{w}_u) & \text{if } \theta_t = \mathbf{h}_j, \\ 0 & \text{otherwise.} \end{cases}$$

Our goal is to develop an algorithm to efficiently optimize the objective function (2).

Based on stochastic gradient descent concepts [2], we present the framework of our algorithm in Figure 2. The main components of this framework are: (i) a sampling procedure done on the streaming data and (ii) a selective model update based on small buffers created per each user.

3.2 Random Sampling the Social Stream with a Reservoir

When processing streams of data, one usually wants to avoid the cost of retraining a model every time new data points arrive; thus online updates are usually used. Unfortunately, the gain in processing time and bounded space achieved by this online learning approach comes at the cost of reduced prediction quality, compared to more accurate models that the large training set could allow. The main issue with online approaches is their short-term "memory", i.e., since the updates based only on the most recent data point do not take into account past observations, the model quickly "forgets" them. In the presence of an abundant source of training examples, a way to reduce complexity of a learning algorithm consists of picking a random subset of training examples and building a model on this subset. In this phase of our model, we employ the technique of random sampling with a reservoir [20], which is widely used in data streaming, and recently has been proposed for online AUC maximization in the context of binary classification [23].

A reservoir sampling algorithm incrementally maintains a random sample of fixed size of the incoming stream of tweets. We represent the reservoir as a list $R := [s_1, s_2 \ldots, s_{|R|}]$ that "remembers" $|R|$ random instances from stream S. Instances can occur more than once in the reservoir, reflecting the distribution of the observed data, thus the reservoir captures an accurate "sketch" of history under the constraint of fixed space.

Let be t the index reflecting the order of arrival of data in the stream, note that until $t = |R|$ all data points enter the reservoir. When $t = |R|$ we have a random sample of size $|R|$ of the stream; indeed the entire dataset so far is in the reservoir. For subsequent t we need to decide whether the newly arrived data should be put in the reservoir and, if so, which data already in the reservoir it should replace.

Vitter shows in [20] that if one includes the tth data instance with probability $|R|/t$ and replaces uniformly at random an instance from the reservoir, the reservoir is a random sample of the current dataset. This reservoir sampling mechanism is implemented by the procedure **updateReservoir**(R) in Figure 2.

3.3 Selective Model Update from Small Buffers

The random sampling with a reservoir allows us to retain a fixed size of observed instances, bounding the space available for the algorithm to a set of $|R|$ randomly chosen samples from the stream and update the model using this history. Although simply updates of the model based on the reservoir may yield better results than single online updates, it is still far from the accuracy achieved by the offline cases. On top of that, in the reservoir we store only user and item pairs observed in the stream, and the question of how to sample the pairs needed for creating the contrasts P still remains.

In order to address this drawback, we need to exploit as much information as possible from the sampled tweets in the reservoir. In particular we propose to perform model updates and retraining on the most informative examples present in the reservoir, then, the question is how to select such examples from this sketch of the stream. This scenario is similar to the one of *active learning*, where the system asks the user to evaluate a minimum set of items which will contribute the most to learning his/her preferences (e.g., [10]).

Consider the case of binary classification using Support Vector Machines (SVM). SVM attempt to find a hyperplane that divides the two classes with the largest margin. From the theoretical foundations of SVM we know that only the support vectors have an effect on the solution. The support vectors are the points that lie closest to the hyperplane, therefore the *most informative* training points, and the goal of training is to discover them [19].

Usually, the training set is chosen to be a random sampling of instances, for example the tweets in our reservoir. However, in many cases principled criteria can be used to sample the training data with the goal to reduce its need for large quantities of labeled data.

Our scenario of dyadic data, i.e., user-item interactions, differs from the one of SVM in two fundamental ways: (i) since we are learning personalized rankings, there are as many hyperplanes as users, unlike an SVM, (ii) we are not just learning a hyperplane per user, but simultaneously also the item feature vectors, in contrast to SVM where the values of the features vectors, defining the training points, are known and given in advance.

Moreover, remember that we are concerned with learning personalized rankings from pairwise comparisons, hence the most informative instances are the ones that have opposite labels but are close to each other in the ranking induced by the user's hyperplane, intuitively they are more difficult to order than the ones away from each other in the ranking [22]. Figure 3 illustrates how user u's feature vector $\mathbf{w}_u$ induces a particular (personalized) ranking at a given iteration in a two dimensional example[2]. $\mathbf{w}_u$ determines the ordering of four item points. For any user *weight* vector $\mathbf{w}_u$, the items are ordered by the projection onto $\mathbf{w}_u$, or equivalently, by their signed distance to a hyperplane with normal vector $\mathbf{w}_u$. The items in the figure are ordered (h_1, h_3, h_2, h_4). We denote as δ the distance between two projections of data points with different labels on the induced ranking, the smaller the δ, the more informative the instances are for training the model.

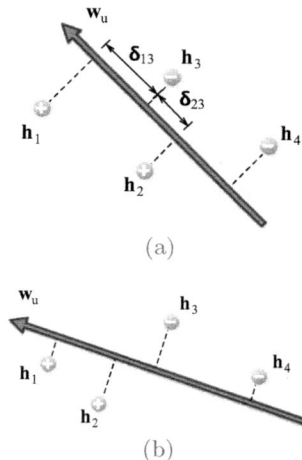

(a)

(b)

Figure 3: **Example of how a user *weight* vector $\mathbf{w}_u$ ranks four item points. (a) The vector $\mathbf{w}_u$ ranks the points as (h_1, h_3, h_2, h_4), erroneously ranking h_3 higher than h_2. (b) The model updates vector $\mathbf{w}_u$ (user features) and the item features iteratively based on pairwise differences and learns the correct ordering (h_1, h_2, h_3, h_4). In this example, the pair (h_2, h_3) with δ_{23} is considered more informative than (h_1, h_2) with δ_{13}, since $|\delta_{23}| < |\delta_{13}|$, i.e., the smaller the δ, the more informative the instances are for training the model.**

---

[2]Observe that Figure 3 can be regarded as a *personalized* adaptation of Figure 2 in [9].

Finally, to answer the question of how to select such examples from the reservoir, we will use an active learning inspired approach. In classical active learning [17], the search for the most informative instance is performed over the entire dataset, which involves the recomputation of each training example's distance to the new hyperplane. This process is prohibitively expensive for large datasets or unbounded data streams. Therefore, we propose a selection method based on the "59 trick" [16, 6], that establishes that randomly sampling only 59 instances, regardless the training set size, is enough to guarantee with 95% probability, that one of them is among the top 5% closest instances to the hyperplane. This approach also simulates the real world scenario of given a pool of items, ranking the positive ones higher than the negatives, modeled into the recommender system evaluation protocol proposed in [3].

At each iteration, we select at random a user-item (u,i) interaction from the reservoir, which represents a positive feedback observation. Next, we construct a small buffer for user u by sampling 59 negative items j's, creating the required *contrast* in the preferences for user u over items i and j's. The user buffer contains exactly 59 pairs of the form $p_b = ((u,i),(u,j_b))$, $b = 1 \ldots 59$. Then, we compute the values δ_{uijb} between the projections on $\mathbf{w}_u$ of each instance in the pair p_b. Finally, we sample a pair p^* with probability proportional to its *informativeness*, which is given by $1/\delta_{uijb}$, and use p^* to perform the matrix factorization model updates. This procedure is shown in Figure 4, which includes three regularization constants: λ_W, λ_{H+}, and λ_{H-}, one for the user factors, the other two for the positive and negative item factors updates. Moreover, we include a learning rate, and a learning rate *schedule* α that adjusts the step size of the updates at each iteration.

4. RELATED WORK

Learning of large-scale recommender systems for dealing with dynamic and fast changing content has been addressed before, for instance in the context of the *Google News* system [4]. However the problem setting in [4] is different from the one addressed here, since their work does not deal with a continuous stream of user generated data, but instead provides recommendations to users based on offline models.

The Fast Online Bilinear Factor Model (FOBFM) [1] addresses the related task of click through rate prediction. They combine offline training with online updates in a principled framework. While FOBFM addresses a regression task, we are concerned here with a learn to rank problem. Also our approach does not need an explicit dimensionality reduction step for the offline learned features.

Online matrix factorization learning methods have also been investigated by Rendle and Schmidt-Thieme for rating prediction [14]. They propose online update rules on a stochastic gradient descent style based on the last example observed. While those update rules take into account only the last observed data point, RMFX uses a *reservoir* with a representative set of previously seen data points from the stream. This idea has been previously explored by Zhao et al. [23] in the context of binary classification, in contrast, a novel idea introduced in this work is the selective sampling based on *personalized* buffers according to the distance of points to the decision boundary which, as shown in our experiments, delivers better results than using exclusively the random sampling technique used by Zhao et al. [23].

RMFX Model Update based on SGD for MF using active learning with small buffers

Input:
> Reservoir representing a sample of the stream at time t: R; Regularization parameters λ_W, λ_{H+}, and λ_{H-}; Learning rate η_0; Learning rate schedule α; Number of iterations T_θ.

Output: $\theta = (\mathbf{W}, \mathbf{H})$

1: **procedure** UPDATEMODEL($S_t, \lambda_W, \lambda_{H+}, \lambda_{H-}, \eta_0, \alpha, T_\theta$)
2:     **for** $t = 1$ to T_θ **do**
3:         Select a user-item pair (u,i) from R uniformly at random
4:         Construct a small buffer for user u by sampling 59 negative items j's from R ("59 trick" [16, 6])
5:         Compute the distances δ_{uijb} for each pair $p_b = ((u,i),(u,j_b)) \in P$, $b = 1 \ldots 59$ in the small buffer
6:         Sample a pair $p^* = ((u,i),(u,j))$ from the buffer with probability proportional to its *informativeness*: $1/\delta_{uijb}$
        // Perform the model updates as follows:
7:         $y_{uij} \leftarrow sign(x_{ui} - x_{uj})$
8:         $\mathbf{w}_u \leftarrow \mathbf{w}_u + \eta \, y_{uij} \, (\mathbf{h}_i - \mathbf{h}_j) - \eta \, \lambda_W \, \mathbf{w}_u$
9:         $\mathbf{h}_i \leftarrow \mathbf{h}_i + \eta \, y_{uij} \, \mathbf{w}_u - \eta \, \lambda_{H+} \, \mathbf{h}_i$
10:        $\mathbf{h}_j \leftarrow \mathbf{h}_j + \eta \, y_{uij} \, (-\mathbf{w}_u) - \eta \, \lambda_{H-} \, \mathbf{h}_j$
11:        $\eta = \alpha \cdot \eta$
12:     **end for**
13:     **return** $\theta = (\mathbf{W}_{T_\theta}, \mathbf{H}_{T_\theta})$
14: **end procedure**

Figure 4: **Matrix factorization model update based on SGD using personalized active learning with small buffers. The procedure minimizes the SVM loss, or hinge-loss, following a pairwise learning to rank approach for dyadic data.**

Yu proposed a selective sampling technique for learning ranking functions in the context of data retrieval applications [22]. Our method, on the other hand, is a learning to rank approach for personalized item prediction.

Since we deal with pairwise classification from positive-only data, negative examples must be sampled. The sampling of the 59 negative examples for each positive one has been proposed, discussed and proved in [6]. Whereas they do it for active learning, we adapt it for our online learning to rank scenario.

5. EXPERIMENTAL STUDY

In this section, we demonstrate our approach by analyzing real-world data consisting of millions of tweets. We present the evaluation protocol and experimental setting, as well as the results of the empirical study.

476 million Twitter tweets Dataset

The dataset corresponds to the *476 million Twitter tweets*[3] [21], which includes over 476 million Twitter posts from 20 million users, covering a 7 month period from June 1, 2009 to December 31, 2009. The number of hashtags present in the dataset is 49,293,684. It is estimated that this is about 20-30% of all public tweets published on Twitter during the particular time frame. For our evaluation we computed a 5-core of the dataset, i.e., every user has used at least 5 different hashtags, and every hashtag has been used

---
[3] http://snap.stanford.edu/data

at least by 5 different users. The 5-core consists of 35,350,508 tweets, 413,987 users and 37,297 hashtags.

Evaluation Methodology

Evaluation of a recommender in the presence of stream data requires a time sensitive split. We evaluated by splitting the dataset S into two sets: a training set S_{train} and a testing set S_{test}. Consider we make the split at time t_{split}, then we put into S_{train} the individual training examples (tweets) with timestamps less that t_{split}. Into S_{test}, we put the user rankings with timestamps greater than t_{split}. The recommenders are trained on S_{train} and then their performance is measured on S_{test}. Note that given the dynamics in Twitter, there might be users in S_{train} not present in S_{test}.

To evaluate the recommenders we used a variant of the *all-but-1* protocol, also known as the leave-one-out holdout method. In particular, we follow a similar schema as the one described in [3].

Our goal is to evaluate the system performance when it suggests *Top-N* topics to a user. For example, recommending the user a few specific hashtags which are supposed to be the most attractive to him. That is, to find the relative position of these interesting items within the total order of items ranked for a specific user.

To this end, for each user $u \in |U_{test}|$ we aggregate his rankings in the test set S_{test} by accumulating the item frequencies across those rankings in order to produce a single total ranking. The items are again sorted in descending order of their accumulated frequencies.

We take one item i at random from the top-10 of the aggregated ranking and hide it. The goal of a recommender system is to help users to discover new items of interest, therefore we impose the additional restriction that the hidden item has to be *novel* for the user, and therefore we remove from the training set all occurrences of the pair (u, i). In total, we have $|U_{test}| = 260,246$ hidden items.

Then, for each hidden item i, we randomly select 1000 additional items from the test set S_{test}. Notice that most of those items selected are probably not interesting to user u.

We predict the scores for the hidden item i and for the additional 1000 items, forming a ranking by ordering the 1001 items according to their scores. The best expected result is that the interesting item i_u to user u will precede the rest 1000 random items.

Finally, we generate a *Top-N* recommendation list by selecting the N items with the highest score. If the test item i_u is in the *Top-N*, then we have a *hit*, otherwise we have a *miss*. We measure the quality by looking at the *recall* metric.

Evaluation Metric: Recall

Traditionally, collaborative filtering algorithms are evaluated by the accuracy of their predicted ratings. One commonly used performance metric for rating accuracy is the Mean Absolute Error (MAE).

However, we are interested in measuring Top-N recommendation performance and not in rating prediction. Therefore, we measure the quality by looking at the *recall* metric, also known as *hit rate*, which is widely used for evaluating Top-N recommender systems (e.g., [5, 3]).

In our recommender systems setting, the *recall* metric is defined as follows:

$$\text{recall} := \frac{\sum_{u \in U_{test}} \mathbb{1}_{[i_u \in \text{Top-N}_u]}}{|U_{test}|}, \qquad (4)$$

where $\mathbb{1}_{[z]}$ is the indicator function that returns 1 if condition z holds, and 0 otherwise. A recall value of 1.0 indicates that the system was able to always recommend the hidden item, whereas a recall of 0.0 indicates that the system was not able to recommend any of the hidden items. Since the precision is forced by taking into account only a restricted number N of recommendations, there is no need to evaluate *precision* or *F1* measures, i.e., for this kind of scenario, precision is just the same as recall up to a multiplicative constant.

Experimental Setting

We implemented our `RMFX`, and evaluated them against the following competing models:

1. RMF-RSV: Reservoir Sampling involves retaining a fixed size of observed instances in a *reservoir*. The reservoir captures an accurate "sketch" of history under the constraint of fixed space. We randomly choose $|R|$ samples from the stream and update the model using this history, i.e., without performing any selective sampling.

2. RMF-SP: Single Pass takes a single pair from the stream and performs an update of the model every iteration. This approach does not "remember" previously seen instances. That is, we sample a pair $p_t \in P$ at iteration t, and directly execute the model updates described in lines 7 to 11 in Figure 4.

3. Trending Topics (TT). This model sorts all hashtags based on their popularity, so that the top recommended hashtags are the most popular ones, which represents the trending topics overall. This naive baseline is surprisingly powerful, as crowds tend to heavily concentrate on few of the many thousands available topics in a given time frame.

4. Weighted Regularized Matrix Factorization (WRMF). This is a state-of-the-art matrix factorization model for item prediction introduced by Hu et al. [8]. Their method outperforms neighborhood based (item-item) models in the task of item prediction for implicit feedback datasets. The model is computed in *batch mode*, assuming that the whole stream is stored and available for training. It is expected that the performance of this offline method, with full access to the user-item interactions, will set an upper bound for the online approaches.

We simulate the stream receiving one instance at the time based on the tweets' publication dates. Tweets without hashtags were ignored.

For `RMFX`, RMF-RSV, and RMF-SP we set regularization constants $\lambda_W = \lambda_{H^+} = \lambda_{H^-} = 0.1$, learning rate $\eta_0 = 0.1$, and a learning rate schedule $\alpha = 1$, and find that the setting gives good performance. We are currently investigating how to efficiently perform a grid search on stream data to tune up the hyperparameters dynamically. Moreover, the number of iterations is set to the size of the reservoir for both `RMFX` and RMF-RSV.

WRMF setup is as follows: $\lambda_{\text{WRMF}} = 0.015$, $C = 1$, $epochs = 15$, which corresponds to a regularization parameter, a confidence weight that is put on positive observations, and to the number of passes over observed data, respectively [8][4].

We divided the six-month Twitter activity of our dataset, by choosing the first five months (from first of June, 2009 to end of November, 2009) for training. We use the remaining

---

[4] We have observed that WRMF is not so sensitive to changes in the hyperparameters, the most important aspect is the number of iterations before early stopping, i.e., epochs=15

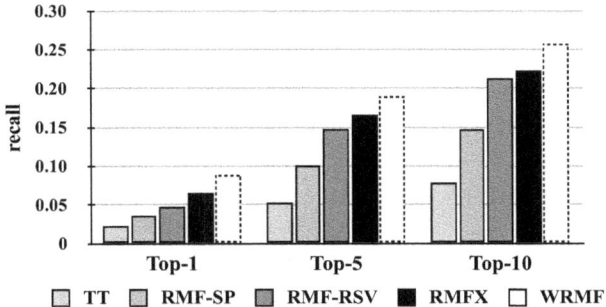

Figure 5: **Recommendation performance in terms of recall for Top–{1, 5, 10}. The reservoir size for RMF-RSV and our method RMFX is set to 8M. The number of factors for the matrix factorization models is set to 128. The batch mode WRMF provides an upper bound reference for the the online methods' performance.**

month, i.e., December, to build 10 independent test sets following the evaluation protocol described previously in this section. The models RMFX and RMF-RSV are built on the sketch of the stream available just before the evaluation period, i.e., end of November, 2009. For TT, we use as predictors the most popular hashtags from the last four weeks before the evaluation, i.e., TT of November, 2009.

We restricted the analysis to short list of recommendations and computed the recall metric for Top-N recommendations, where $N \in \{1, 5, 10\}$. The value of the metric for a particular Top-N is denoted as *recall@N*. The performance is evaluated on the test set only and the reported results are the average over 5 runs. All the differences reported are statistically significant (two-sample t-test, $p < 0.015$).

Reproducibility of Experiments. We will provide an anonymized dataset of the 5-core dataset used in our experiments and a reference implementation of RMFX upon request by email. We used the WRMF implementation provided by *My-MediaLite*, a free software recommender system library [7][5].

Results

Figure 5 summarizes the recommendation performance for RMFX, the baselines and the upper bound given by WRMF. We can observe that RMFX is superior with respect to the online methods RMF-SP and RMF-RSV, and largely outperforms the trending topics (TT). Please note that the trending topics from the previous four weeks achieve a recall@10=7.8%, this performance based on the crowd behavior in Twitter is much better than a random model, whose recall@10 is under 1%.

Table 1 shows that RMFX achieves the best performance over all online methods evaluated with reservoir sizes 2, 4 and 8 million. As expected, the offline method WRMF sets an upper bound for the online approaches achieving a recall@5 of 18.96%, but RMFX is still competitive with a recall@5=16.58% and the advantage of real-time updates.

Finally, with a fixed reservoir size of 8M, we also explored the impact of model dimensionality over the recommendation quality for RMFX. The results are presented in Figure 6 and Table 2. From the figure, we see that RMFX consistently outperforms the baseline TT and the online competitors for 32, 64 and 128 dimensions.

[5]MyMediaLite:
http://www.ismll.uni-hildesheim.de/mymedialite

Method	Reservoir Size	Recall@1	Recall@5	Recall@10
RMFX	8M	6.50%	16.58%	22.25%
RMF-RSV	8M	4.70%	14.72%	21.25%
RMFX	4M	4.16%	11.24%	15.84%
RMF-RSV	4M	2.82%	9.02%	13.70%
RMFX	2M	1.95%	5.59%	8.41%
RMF-RSV	2M	1.68%	4.89%	7.36%
RMF-SP	–	3.57%	10.03%	14.69%
TT	–	2.26%	5.22 %	7.80%
WRMF	(Batch)	8.85%	18.96%	25.73%

Table 1: **Recommendation performance for different sizes of the reservoir. The number of factors is set to 128 for the online methods RMFX, RMF-RSV, as well as for the offline approach WRMF.**

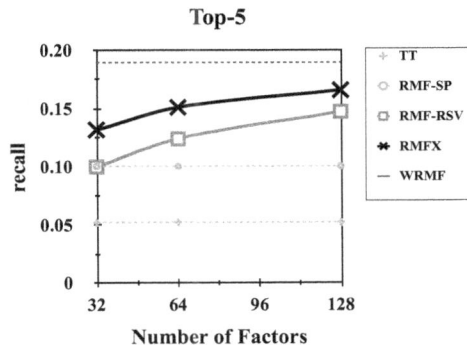

Figure 6: **Recall@5 for 32, 64 and 128 factors.**

Method	Factors	Recall@1	Recall@5	Recall@10
RMFX	64	5.57%	15.10%	20.79%
RMF-RSV	64	3.79%	12.40%	18.89%
RMFX	32	4.32%	13.18%	18.96%
RMF-RSV	32	2.69%	9.98%	16.07%

Table 2: **Recommendation performance for 32 and 64 factors, with a fixed reservoir of size 8M (the information for 128 factors is shown in Table 1).**

Time and Space Savings vs. Recommendation Quality

We report in this section the CPU training times and space required for the best performing variation of our online approach: RMFX, and the ones for the strongest baseline: WRMF. We also discuss the trade-off between time and space savings against the recommendation performance.

RMFX is implemented in the Python programming language using SciPy[6]. We ran RMFX on a Intel Xeon 1.87GHz machine. For WRMF, we used the implementation provided by *MyMediaLite* [7], which is implemented in C#. The baseline WRMF was run on a machine with a slightly faster CPU (Intel Xeon 2.27GHz). The experiments were conducted using GNU/Linux 64-bit as operating system. None of the methods was parallelized and therefore used one single CPU for computations. Please remember that running times heavily depend on platform and implementation, so they should be only taken as relative indicators.

In Table 3, we can observe the gains in speed of our approach over the baseline for all the evaluated reservoir sizes. We can observe that for all reservoir sizes RMFX is faster and more space efficient than WRMF. For example, RMFX with a

[6]SciPy : http://scipy.org

Method (128 factors)	Time (hh:mm:ss)	recall@10	Space	x times faster than WRMF	Recommendation Quality w.r.t WRMF
WRMF (Batch) [Baseline]	96:21:50.093	0.2573	100.00%	–	100%
RMFX 2M	0:33:12.898	0.0841	5.66%	174.07	32.69%
RMFX 4M	1:07:10.570	0.1584	11.32%	86.07	61.52%
RMFX 8M	2:14:32.355	0.2225	22.63%	42.98	86.47%

Table 3: **Time, Space and Recommendation Quality Comparison.** RMFX **is compared against WRMF for different reservoir sizes: 2, 4 and 8 million. The dimensionality of all methods is 128 factors. Time is measured in seconds and corresponds to the training time of 15 epochs in case of WRMF and one single epoch in case of** RMFX, **the gain is computed w.r.t. WRMF's duration. Space is measured as the percentage of all transactions in the dataset, which is 100%=35.35M. Recommendation performance quality is computed with respect to WRMF's recall, which is set as an upper-bound.**

reservoir size 8M is approximately 43 times faster and uses 77% less space than WRMF, and yet it delivers a highly competitive recommendation performance corresponding to 86.47% of the state-of-the-art baseline computed offline.

6. CONCLUSIONS AND FUTURE WORK

We proposed RMFX, an approach for recommending topics to users in presence of streaming data. Our online setting for collaborative filtering captures: "what is interesting to *me* right now within the social media stream". RMFX represents a novel principled approach for online learning from streams, that selects a subsample of the observed data based on the objective function gradients, and uses it to guide the matrix factorization.

RMFX receives instances from a microblog stream and updates a matrix factorization model following a pairwise learning to rank approach for dyadic data. At the core of RMFX is stochastic gradient descent which makes our algorithm easy to implement and efficiently scalable to large-scale datasets. Our empirical study used Twitter as a test bed and showed that models updated using the selective sampling approach proposed here, significantly outperform online methods that use random samples of the data.

In future work, we plan to investigate how the frequency of repeated events (i.e. users using the same hashtag or listening to the same song) can be incorporated to the model to generate more accurate predictions.

Acknowledgments We would like to thank Zeno Gantner and Stefan Siersdorfer for fruitful discussions during the genesis of this paper. This work was funded, in part, by the European Commission FP7/2007-2013 under grant agreement No.247829 for the M-Eco Project, the DFG project Multi-relational Factorization Models, and the NTH School (Niedersächsische Technische Hochschule) for IT Eco-systems. Lucas Drumond is sponsored by a scholarship from CNPq, a Brazilian government institution for scientific development.

7. REFERENCES

[1] D. Agarwal, B.-C. Chen, and P. Elango. Fast online learning through offline initialization for time-sensitive recommendation. In *Proceedings of the ACM KDD conference*, 2010.

[2] L. Bottou. Online Algorithms and Stochastic Approximations. In *Online Learning and Neural Networks*. 1998.

[3] P. Cremonesi, Y. Koren, and R. Turrin. Performance of Recommender Algorithms on Top-N recommendation Tasks. In *Proceedings of the ACM RecSys conference*, 2010.

[4] A. S. Das, M. Datar, A. Garg, and S. Rajaram. Google news personalization: scalable online collaborative filtering. In *Proceedings of the World Wide Web*, 2007.

[5] M. Deshpande and G. Karypis. Item-based top-n recommendation algorithms. *ACM Trans. Inf. Syst.*, 22:143–177, January 2004.

[6] S. Ertekin, J. Huang, L. Bottou, and L. Giles. Learning on the Border: Active Learning in Imbalanced Data Classification. In *Proceedings of the ACM CIKM conference*, 2007.

[7] Z. Gantner, S. Rendle, C. Freudenthaler, and L. Schmidt-Thieme. MyMediaLite: A free recommender system library. In *Proceedings of the ACM RecSys conference*, 2011.

[8] Y. Hu, Y. Koren, and C. Volinsky. Collaborative filtering for implicit feedback datasets. In *Proceedings of the IEEE ICDM conference*, 2008.

[9] T. Joachims. Optimizing Search Engines Using Clickthrough Data. In *Proceedings of the eighth ACM SIGKDD international conference on Knowledge discovery and data mining*, KDD '02, 2002.

[10] R. Karimi, C. Freudenthaler, A. Nanopoulos, and L. Schmidt-Thieme. Towards optimal active learning for matrix factorization in recommender systems. In *IEEE ICTAI conference*, 2011.

[11] Y. Koren, R. Bell, and C. Volinsky. Matrix Factorization Techniques for Recommender Systems. *Computer*, August 2009.

[12] S. Muthukrishnan. *Data streams: algorithms and applications*. Now Publishers, 2005.

[13] S. Rendle, C. Freudenthaler, Z. Gantner, and L. Schmidt-Thieme. BPR: Bayesian Personalized Ranking from Implicit Feedback. In *Proceedings of UAI conference*, 2009.

[14] S. Rendle and L. Schmidt-Thieme. Online-updating regularized kernel matrix factorization models for large-scale recommender systems. In *Proceedings of the ACM RecSys conference*, 2008.

[15] D. Sculley. Combined regression and ranking. In *Proceedings of the ACM KDD conference*, 2010.

[16] A. J. Smola and B. Schölkopf. Sparse greedy matrix approximation for machine learning. In *Proceedings of the International Conference on Machine Learning*, 2000.

[17] S. Tong and D. Koller. Support vector machine active learning with applications to text classification. *J. Mach. Learn. Res.*, 2:45–66, Mar. 2002.

[18] twittereng. 200 million tweets per day. Twitter Blog. http://goo.gl/eybp0, June 2011.

[19] V. N. Vapnik. *The nature of statistical learning theory*. Springer-Verlag New York, Inc., New York, NY, USA, 1995.

[20] J. S. Vitter. Random sampling with a reservoir. *ACM Trans. Math. Softw.*, 11:37–57, March 1985.

[21] J. Yang and J. Leskovec. Patters of temporal variation in online media. In *Proceedings of the ACM WSDM conference*, 2011.

[22] H. Yu. SVM Selective Sampling for Ranking with Application to Data Retrieval. In *Proceedings of the eleventh ACM KDD conference*, 2005.

[23] P. Zhao, S. Hoi, R. Jin, and T. Yang. Online AUC Maximization. In *Proceedings of the International Conference on Machine Learning*, 2011.

On Top-k Recommendation using Social Networks

Xiwang Yang[†], Harald Steck[‡][*], Yang Guo[‡] and Yong Liu[†]
[†]Polytechnic Institute of NYU, Brooklyn, NY, USA 11201
[‡]Bell Labs, Alcatel-Lucent, New Jersey
xyang01@students.poly.edu, hsteck@gmail.com, Yang.Guo@alcatel-lucent.com
yongliu@poly.edu

ABSTRACT

Recommendation accuracy can be improved by incorporating trust relationships derived from social networks. Most recent work on social network based recommendation is focused on minimizing the root mean square error (RMSE). Social network based top-k recommendation, which recommends to a user a small number of items at a time, is not well studied. In this paper, we conduct a comprehensive study on improving the accuracy of top-k recommendation using social networks. We first show that the existing social-trust enhanced Matrix Factorization (MF) models can be tailored for top-k recommendation by including observed and missing ratings in their training objective functions. We also propose a Nearest Neighbor (NN) based top-k recommendation method that combines users' neighborhoods in the trust network with their neighborhoods in the latent feature space. Experimental results on two publicly available datasets show that social networks can significantly improve the top-k hit ratio, especially for cold start users. Surprisingly, we also found that the technical approach for combining feedback data (e.g. ratings) with social network information that works best for minimizing RMSE works poorly for maximizing the hit ratio, and vice versa.

Categories and Subject Descriptors

H.2.8 [**Database Management**]: Database Applications–Data Mining

General Terms

Algorithms, Design, Measurement, Experimentation

Keywords

Recommender System, Social Network, Matrix Factorization

1. INTRODUCTION

The idea of recommender systems (RS) is to automatically suggest items to each user that s/he may find appealing, e.g. see [1] for an overview. Traditional collaborative filtering approaches predict users' interests by mining user rating history data [3, 9, 16, 21–24]. In real life, people often resort to their friends for recommendations in real life. It is therefore tempting to improve RS by incorporating information on the trust relationships between users in social networks. In the literature, e.g. [4, 6–8, 11, 12, 15, 17–20], it was shown that using social network information in addition to feedback data (e.g. ratings) can significantly improve recommendation accuracy. While there is some improvement on the average recommendation accuracy over all users, the improvement is particularly significant for the so-called *cold users*, who have provided only very little feedback (e.g. very few ratings) [5, 7, 8, 15]. Most of the existing social network based recommender systems are optimized for the *root mean square error (RMSE)*, which has enjoyed perhaps the largest popularity among the various accuracy measures in the recommender literature. On the other hand, top-k recommendation, a small number k of items are recommended to a user at a time is pervasive in Real-World recommendation task.

In this paper, we provide a comprehensive study on improving the accuracy of top-k RS using social networks. Towards this goal, we first show that the existing social-trust-enhanced Matrix Factorization (MF) models [7,8,11,12] can be conveniently tailored for top-k recommendation by extending their training objective functions to include both the observed ratings and the missing ratings with the consideration that rating data are missing not at random(MNRA). For the Nearest Neighbor (NN) based top-k recommendation, we propose to combine users' neighborhoods in their trust networks with their neighborhoods in the *user latent feature space* derived from matrix factorization considering MNAR. To generate top-k recommendation from a combined neighborhood, instead of taking the weighted average over only the observed ratings, we propose to use voting-based algorithm as a simple approach to consider both observed and missing ratings.

To assess the performance of the proposed social-network-based top-k RSes, we then undertake a comprehensive comparison study using two publicly available real-world data sets: Epinions and Flixster. The major findings are:

1. Trust information derived from social networks significantly improves top-k hit ratio when incorporated properly both in MF and NN models.

2. Our proposed social network based NN models and RS trained with our modified objective function considerably outperform the only published top-k approach us-

[*]Now at Netflix Inc. Work was done while at Bell Labs.

ing social network information in recent literature [7], an NN approach.

3. Among the various ways of combing feedback data with social network information, the one that was found to be worst with respect to RMSE turns out to be the best concerning the top-k hit ratio. This illustrates that the technical details of minimizing RMSE can be very different from the ones that work to optimize top-k hit ratio.

This paper is organized as follows. Some related work is discussed in section 2. Section 3 outlines the top-k models using social network information. In particular, in Sections 3.1.1 to 3.1.3, we outline various matrix factorization (MF) models for combining feedback data and social trust information that were proposed in the literature. As they are typically geared towards RMSE, we modify their training objectives functions towards the top-k hit ratio. We extend the nearest neighbor (NN) models in Section 3.2. In Section 4, we present the comprehensive comparison study of the various approaches–with respect to the top-k hit ratio. The paper is concluded in Section 5.

2. RECOMMENDATION TASK

We consider the following *Real-World Recommendation Task*: for each user, the recommender system has to recommend a *small* number, say k, of items from among *all available* items. One may distinguish between two slightly different variants of this task: (a) all items are eligible for recommendation, including the items that have been rated[1] by the user in the past (this assumes that a user may consume an item possibly several times, e.g. listen to a song on the online radio); or (b) only those items, which have not been rated by the user in the past, are eligible for recommendation (this assumes that a user consumes an item at most once, e.g. purchase of a movie DVD).

A meaningful offline test ideally should provide a good approximation to the utility function optimized by the deployed system (e.g. user satisfaction, increase in sales). This immediately suggests the corresponding procedures for offline testing on available data:

- **all items:** for each user, all items are considered, whether rated or not by the user (in the training set or the test set).
- **all unrated items:** for each user, only those items are considered that have not been rated by the user in the training set. Note that this contains items with and without ratings in the test data.

Due to the data sparsity, the difference between these two variants is expected to be small, as confirmed by our experiments. We hence will only report results concerning the second variant in this paper.

The user's selection bias causes the observed feedback (e.g. ratings, purchases, clicks) in the data to be missing *not* at random (MNAR). This is an important issue in practice, but largely ignored in the literature (the few exceptions include [2, 13, 14, 25]). Selection bias may result from user's tendency to rate only the items they like or know. Compared to the (unknown) distribution over (a random subset of) all ratings, the distribution of observed ratings is skewed due

---

[1]For simplicity, we use 'ratings' as a synonym for feedback in this paper. As will become clear, the presented approach is applicable to both explicit and implicit feedback data.

to the selection bias. Recent work by Marlin et al. [13, 14] provided empirical evidence that the data typically used for training and testing recommender systems indeed exhibit a significant selection bias, i.e., the ratings are missing not at random: their histograms of the distribution of ratings in the Yahoo!LaunchCast data show that *low* ratings are much more likely to be missing from the observed data than *high* ratings (see Figure 2 in [13]).

RMSE *on the observed data* is agnostic to the selection bias, as the data in the training set and the test set are from the same skewed distribution. In contrast, in top-k recommendation, as the k recommended items have to be chosen from *all* items (that were not rated in the training set), the unknown distribution over all ratings influences the recommendation accuracy, and hence user satisfaction in practice. If the ratings in the available data had been missing at random (MAR), unbiased results could have been expected from the common test procedures using **observed ratings** only. It is shown in [25] that the top-k hit ratio has desirable properties when applied to all (unrated) items in MNAR test data. Note that approaches that perform well with respect to RMSE on the observed ratings may perform poorly with respect to the top-k hit-ratio on all items [2, 10, 25].

As to compute the top-k hit ratio or recall, for each user u, we rank the items i according to the predicted rating $\hat{R}_{i,u}$. An item is defined as *relevant* to a user in the test set if s/he finds it appealing or interesting (e.g., the assigned rating in the test data is above a certain threshold). For instance, in our experiments with Epinions data, the rating values range from 1,..., 5 stars, and we consider 5-star ratings as relevant (i.e. the user definitely liked these items), while other rating values and missing rating values are considered not relevant. Other choices led to similar results. Now the top-k hit ratio or recall can be defined as the fraction of relevant items in the test set that are in the top-k of the ranking list, denoted by $N(k, u)$, from among all relevant items, $N(u)$. For each user u, the top-k hit ratio is given by

$$\mathrm{H}(k, u) = \frac{N(k, u)}{N(u)}, \qquad (1)$$

which can be aggregated over all users to obtain the average top-k hit ratio or recall for the test set. The recall is computed as follows:

$$recall = \frac{\sum_u N(k, u)}{\sum_u N(u)}, \qquad (2)$$

Note that a higher top-k hit ratio or recall is better. In our experiments, the evaluation metric is recall. We noticed that recommender systems that perform well with respect to recall also perform similarly well regarding other ranking measures like precision or nDCG on all items, while the RMSE measure on observed ratings behaves very differently in comparison.

3. TOP-K RECOMMENDER SYSTEMS US-ING SOCIAL NETWORKS

Recommender systems using social network information were mainly developed to optimize RMSE on observed ratings, e.g. [6,8,11,12,15]. Various approaches are used. While neighborhood [4,7] and random walk [6] methods were used on the social network graph, matrix factorization methods were found to be the most accurate models [8,11,12].

For this reason, we start with matrix factorization (MF)

approaches using social network information, and modify them as to optimize the top-k hit ratio (rather than the original RMSE). Each of the three models and its modification is outlined in the following sub-sections. Other than MF approaches, we also consider nearest neighbor (NN) approaches. In recent recommender literature, the only top-k approach using social network information is a NN method [7] to the best of our knowledge. In Section 3.2, we develop several variants of NN approaches by adopting user latent features derived from MF optimized for top-k hit ratios.

3.1 Top-k MF using Social Networks

In the following subsection, we briefly review the existing MF approaches in the literature that combine rating data with social network information [8, 11, 12]. As to optimize the top-k hit ratio (rather than RMSE), we modify their training function as to account for *all* items, rather than the observed ratings only, analogous to *AllRank* proposed in [25] for rating data.

The social network information is represented by a matrix $S \in \mathbb{R}^{u_0 \times u_0}$, where u_0 is the number of users. The directed and weighted social relationship of user u with user v (e.g. user u trusts/knows/follows v) is represented by a positive value $S_{u,v} \in (0, 1]$. An absent or unobserved social relationship is reflected by $S_{u,v} = s_m$, where typically $s_m = 0$.

3.1.1 Social Recommendation (SoRec) Model

Social Recommendation (SoRec) was introduced in [12]. In this model, the social network matrix S (see beginning of Section 3.1) may be slightly modified as follows [12]:

$$S_{u,v}^* = S_{u,v} \sqrt{\frac{d_v^-}{d_u^+ + d_v^-}},$$

where d_u^+ is the out-degree of user u in the social network (i.e. the number of users whom u follows/trusts), and d_v^- is the in-degree of user v in the network (ie the number of users who follow/trust user v). The predicted ratings are obtained from the model as follows:

$$\hat{R} = r_m + QP^\top, \qquad (3)$$

with matrices $P \in \mathbb{R}^{i_0 \times j_0}$ and $Q \in \mathbb{R}^{u_0 \times j_0}$, where $j_0 \ll i_0, u_0$ is the rank; and $r_m \in \mathbb{R}$ is a (global) offset. Besides the rating data, also the social network information is used for training this model. The social relationships are predicted as follows:

$$\hat{S}^* = s_m + QZ^\top, \qquad (4)$$

where $Z \in \mathbb{R}^{u_0 \times j_0}$ is a third matrix in this model, besides P and Q. Note that the matrix Q is shared among the two equations (3) and (4). Due to this constraint, one can expect Q (i.e. the user profiles Q_u for each user u) to reflect information from both the ratings and the social network as to achieve accurate predictions for both. Note that the matrix Z is not needed for predicting rating values, and hence may be discarded after the matrices P and Q have been learned. Both (3) and (4) are combined as follows in the training objective function. Analogous to [25], we modify the training function as to account for *all* items (instead of RMSE on the *observed* ratings) for improved top-k hit-rate on the test data:

$$\sum_{\text{all } u} \sum_{\text{all } i} W_{u,i} \left(R_{u,i}^{\text{o\&i}} - \hat{R}_{u,i} \right)^2 + \sum_{\text{all } u} \sum_{\text{all } v} W_{u,v}^{(S)} \left(S_{u,v}^* - \hat{S}_{u,v}^* \right)^2$$
$$+ \lambda \left(||P||_F^2 + ||Q||_F^2 + ||Z||_F^2 \right), \qquad (5)$$

where $|| \cdot ||_F$ denotes the Frobenius norm of the matrices, and λ is the usual regularization parameter. $R_{u,i}^{\text{o\&i}}$ equals the actual rating value in the training data if observed for user u and item i; otherwise the value $R_{u,i}^{\text{o\&i}} = r_m$ is imputed. The training weights are [25]

$$W_{u,i} = \begin{cases} 1 & \text{if } R_{u,i}^{\text{obs}} \text{ observed} \\ w_m & \text{otherwise} \end{cases}. \qquad (6)$$

The term concerning the social network (in the second line) is analogous to the first term concerning the ratings. In particular, the absent or unobserved social links are treated analogous to the missing ratings in AllRank [25], i.e. we impute the value s_m with weight $w_m^{(S)}$. Like $W_{u,i}$ in (6), $W_{u,v}^{(S)}$ is defined as follows:

$$W_{u,v}^{(S)} = \gamma \cdot \begin{cases} 1 & \text{if } S_{u,v}^* \text{ observed} \\ w_m^{(S)} & \text{otherwise} \end{cases}, \qquad (7)$$

where $\gamma \geq 0$ determines the weight of the social network information compared to the rating data. Obviously, $\gamma = 0$ corresponds to the extreme case where the social network is ignored when learning the matrices P and Q. As γ increases, the influence of the social network increases. The effect is that the user profiles Q_u and Q_v of two users u and v become more similar to each other if they are friends. While only positive social relationships are considered in the original model [12], we note that this model allows also for *negative* values of $S_{u,v}$, representing e.g. distrust among users. This objective function can be optimized using the popular (stochastic) gradient descent method.

3.1.2 Social Trust Ensemble (STE) Model

Recommendation with Social Trust Ensemble (STE) was introduced in [11]. The predicted ratings are obtained from a model comprising the matrices $P \in \mathbb{R}^{i_0 \times j_0}$ and $Q^{u_0 \times j_0}$:

$$\hat{R}_{u,i} = r_m + \alpha Q_u P_i^\top + (1 - \alpha) \sum_v S_{u,v} Q_v P_i^\top, \qquad (8)$$

where we omitted the logistic function, as we found its effect rather negligible in our experiments. The reason is that only the ranking/order of the predicted rating values is important for the top-k hit ratio, while it is irrelevant if the predicted rating values are confined to valid rating values (e.g. the interval $[1, 5]$ stars). The trade-off between the feedback data (ratings) and the social network information is determined by $\alpha \in [0, 1]$. Obviously, the social network information is ignored for $\alpha = 1$, while $\alpha = 0$ assigns the highest possible weight to the social network information. Intermediate values of α result in a weighted combination of the information from both sources. (8) is equivalent to the matrix notation

$$\hat{R} = r_m + S_\alpha QP^\top, \qquad (9)$$

where $S_\alpha = \alpha I + (1 - \alpha) S$, and I is the identity matrix. Analogous to the previous section, we modify the training function to be geared towards the top-k hit ratio as follows:

$$\sum_{\text{all } u} \sum_{\text{all } i} W_{u,i} \cdot \left(R_{u,i}^{\text{o\&i}} - \hat{R}_{u,i} \right)^2 + \lambda \left(||P||_F^2 + ||Q||_F^2 \right), \qquad (10)$$

where $|| \cdot ||_F$ denotes the Frobenius norm. $W_{u,i}$ and $R_{u,i}^{\text{o\&i}}$ are defined as in the previous section. Again, this training objective function can be optimized efficiently using stochastic gradient descent.

3.1.3 Social MF Model

The SocialMF model was proposed in [8], and was found to outperform SoRec and STE with respect to RMSE. Each of the rows of the social network matrix S has to be normalized to 1, resulting in the new matrix S^* with $S^*_{u,v} \propto S_{u,v}$, and $\sum_v S^*_{u,v} = 1$ for each user u.

The predicted ratings are obtained from the model, comprising the matrices $P \in \mathbb{R}^{i_0 \times j_0}$ and $Q^{u_0 \times j_0}$, as follows:

$$\hat{R} = r_m + QP^\top, \tag{11}$$

where we again omitted the logistic function, as we found its effect rather negligible in our experiments. Like before, we modify the training function in [8] as to better optimize the top-k hit ratio (instead of RMSE):

$$\sum_{\text{all } u} \sum_{\text{all } i} W_{u,i} \cdot \left(R^{\text{o\&i}}_{u,i} - \hat{R}_{u,i} \right)^2$$
$$+ \beta \sum_{\text{all } u} \left((Q_u - \sum_v S^*_{u,v} Q_v)(Q_u - \sum_v S^*_{u,v} Q_v)^\top \right)$$
$$+ \lambda \left(||P||^2_F + ||Q||^2_F \right) \tag{12}$$

The tradeoff between the feedback data (ratings) and the social network information is determined by $\beta \geq 0$. Obviously, the social network information is ignored for $\beta = 0$, while increasing values of β shift the tradeoff more and more towards the social network information. The term in the second line constitutes a constraint that a user profile Q_u should be similar to the (weighted) average of his/her friends' profiles Q_v (measured in terms of the square error). We optimize also this modified training function by means of stochastic gradient descent.

3.2 Nearest Neighbor Methods

In a NN method, top-k recommendations are generated not from all items, but only from items "liked" by a subset of users who are "nearest" (under certain distance metric) to the target user. The neighborhood of a user can be calculated using collaborative filtering, or it can be just a set of directly or indirectly connected friends in a social network. This makes it convenient to incorporate social trust into NN based top-k recommendation.

Basically, NN based RS is a different approach from MF based RS. In Real-World systems, there are lots of user's feedbacks every day, e.g., as it is reported that there are billions of the like buttons served daily in facebook. NN based RS enjoys a unique advantage in that it can incrementally integrate user's new feedback into recommendation. Because nearest neighbors of a user is comparably stable within a short period, so a user's new feedbacks influence the recommendation to its neighbors in real time. While, in MF based approach, in order to integrate user's new feedbacks, it requires new matrix factorization which is not so efficient when deployed in real systems.

To the best of our knowledge, [7] is the only work that incorporates social network into NN based top-k recommender system. Two neighborhood based approaches are studied in [7] and their performance are comparable. We select one model, termed as *Trust-cf*, as the baseline for comparison.

In Trust-cf, Breadth First Search (BFS), starting from a source user u, is performed to obtain a set of trusted neighbors, namely trusted neighborhood. Meanwhile, collaborative filtering (CF) neighborhood consists of users who are close to the source user u in terms of Pearson Correlation coefficient. The items rated highly by users in either neighborhoods are considered to be candidates for top-k recommendation. Trust-cf calculates the predicted rating for a candidate item as the weighted average of all observed ratings in the two neighborhoods. The weight for a user in the trusted neighborhood is set to $1/d_v$, where d_v is the depth of user v from user u in the trust network. The weight for a user in the CF neighborhood is the Pearson Correlation coefficient between this user and the source user. If an item has predicted ratings from both neighborhoods, two predicted ratings are combined using weighted average with weights proportional to the neighborhood size for this item. Finally, Trust-cf sorts all the candidate items by their predicted ratings and recommends the top-k to the source user.

We propose a set of social network based NN approaches to achieve high top-k hit ratio by considering both social trust and MNAR. We always denote by k_1 the number of nearest users identified by the Collaborative Filtering (CF) approach, and by k_2 the number of trusted users identified by the social network based approach.

- **CF-ULF approach.** *CF-ULF* uses MF (i.e., *AllRank* [25]) to obtain the user latent features. The users are then clustered in the user latent feature space using the Pearson correlation coefficient. The k_1 users nearest to the source user u are identified. The relevant items of these nearest users are voted to form the top-k recommended items. The voting for the candidate items are computed as follows:

$$Vote_{u,i} = \sum_{v \in N_u} \sum_i sim(u,v) \delta_{i \in I_v}, \tag{13}$$

where δ is the Kronecker delta; I_v denotes the set of *relevant* items of user v; and N_u is the set of k_1 nearest neighbors of user u (as determined by the Pearson correlation). $Vote_{u,i}$ is the voting concerning item i for user u; the k_1 nearest neighbors of user u are weighted according to their similarity $sim(u,v)$ with user u, measured in terms of the Pearson correlation coefficient between user u and v (in user latent feature space).

- **PureTrust approach.** *PureTrust* approach employs the breadth-first search (BFS) in the social network to find k_2 trusted users to the source user u. The voting scheme is similar to the scheme employed in *CF-ULF*.

$$Vote_{u,i} = \sum_{v \in N_u^{(t)}} \sum_i w_t(u,v) \delta_{i \in I_v}, \tag{14}$$

where $N_u^{(t)}$ is the set of trusted users of u, and $w_t(u,v)$ is the voting weight from user v. We set $w_t(u,v) = 1/d_v$, where d_v is the depth of user v in the BFS tree rooted at user u.

- **Trust-CF-ULF approach.** *Trust-CF-ULF approach* is the combination of user latent feature space based collaborative filtering (CF-ULF) approach and social network based approach. We firstly find k_1 closest neighbors from the CF neighborhood, then find k_2 closet neighbors from the trust neighborhood which are not in the k_1 set. Then users in the combined neighborhood vote for their relevant items. $w(u,v)$ is defined as:

$$Vote_{u,i} = \sum_{v \in N_u^{(c)}} \sum_i w(u,v) \delta_{i \in I_v}, \tag{15}$$

where, $N_u^{(c)}$ is the combined neighborhood.

$$w(u,v) = \begin{cases} sim(u,v) & \text{if } v \in N_u \\ w_t(u,v) & \text{if } v \in N_u^{(t)} \end{cases} \tag{16}$$

In the basic version of *Trust-CF-ULF*, we set $k_1 = k_2$.

• **Trust-CF-ULF-best approach.** *Trust-CF-ULF-best* improves upon *Trust-CF-ULF* by dynamically tuning the value of k_1 and k_2 so as to obtain the best recall results.

The main differences between our proposed NN methods and Trust-cf are: 1) Our CF neighbors are derived from user latent features obtained from MF (i.e., *AllRank* [25]), which is expected to have higher top-k hit ratio than the Pearson correlation coefficient based only on observed ratings; 2) We use voting, instead of the weighted averaging of the observed ratings, to construct top-k recommendations. Voting can be treated as the simplest approach to consider all items with and without ratings. As will be shown in Section 4.2.4, our NN models outperform the existing social network based NN models.

4. EXPERIMENTS

In this section, we perform experiments for the proposed top-k RSes on Epinions[2] and Flixster[3] datasets. We focus on the *top-k hit ratio* or *recall* for testing recommendation accuracy (as motivated in Section 2). Concerning the three MF models, SoRec, STE and SocialMF (see Section 3.1), we used rank $j_0 = 10$, like in [8, 11, 12]. We find that Trust information significantly improves top-k hit ratio when incorporated properly both in MF and NN models. We also find that our proposed NN based RS and MF models trained with our modified objective function considerably outperform the existing top-k approach using social network information in recent literature [7]. Moreover, among the three models for combining rating data with social network information, the model with the worst performance concerning RMSE surprisingly turns out to achieve the best top-k hit ratio. This illustrates that approaches that work well for the vastly popular RMSE are not necessarily useful for optimizing the more realistic top-k hit ratio or recall.

4.1 Dealing with High Computation Cost

Training on all items admittedly increases the computation complexity, which is another key performance metric, other than accuracy, in designing recommender systems. Good recommendation algorithms not only need to provide accurate results, but also need to be scalable to large problems. To work with two large real-world datasets, we conducted our experiments on a Linux server with four E5640 Intel Xeon CPUs. Each CPU has four cores, and each core has 12.3 MB cache. The shared memory size is 12 GB. We implemented multi-thread C++ programs to parallelize large-scale matrix operations encountered in model training and parameter optimization. The running times for different models ranges from seconds to hours. For the STE model, we could not afford the computation cost to get the exact optimal solution, and we resorted to approximation methods. We found that the stochastic gradient descend and gradient descend methods easily got stuck in local minima when training with missing ratings, while ALS performed better in many cases.

4.2 Experiments on Epinions Dataset

4.2.1 Dataset

Epinions is a consumer opinion site where users review various items, such as cars, movies, books, software, etc., and assign ratings to the items. The ratings are in the range

of 1(min) to 5(max). Users also assign trust values (i.e. a value of 1) to other users whose reviews and/or ratings they find valuable. No trust value indicates that a user either does not know the other, or distrusts him. We used the Epinions dataset[4] published by the authors of [26].

The Epinions data set consists of 71,002 users with a total number of 104,356 rated items. The total number of reviews is 571,235, and the total number of pairwise, directed trust relationships is 508,960. In our experiments, the data set is divided into two sub-sets: the training set and the test set. For users with less than five ratings, one randomly selected rating is put into the test set. For users with five ratings or more, 10% of the randomly selected ratings are moved to the test set. We define cold user as user who had rated fewer than 5 items. We further split the test set randomly into two disjoint sets of equal size. The first test set is used for cross-validation during training as to determine the tuning parameters in our objective function. The second test set is used as a truly held-out data set for final evaluation of the trained model. We report the result of testing the second test set. We consider 5-star ratings as relevant[5] to a user, i.e. the user definitely likes these items, and report the recall test results mostly for the *top 500* items (as defined in Section 2). The reason we set $k = 500$ is as follows. In the Epinions data set, there is a much larger number of items than users, which is different from many other data sets, e.g. the Netflix dataset. [6] Thus, using a small value for k will produce generally poor results for all the compared methods. Actually, we have performed experiments for $k = 5$. The recall of modified SoRec model on all users and cold users were 2.06% and 2.45% respectively, and the recall of modified *No Trust* on all users and cold users were 1.31% and 0.93% respectively. Nonetheless, we show the results of recall as the value of k changes in Figure 3.

4.2.2 Recall for MF Models

We found the following tuning parameters of the training objective functions for the MF models to result in the highest recall: $\lambda = 0.4$, $r_m = -1$, $w_m = 0.0002$ for all models; the optimal tradeoff between the rating data and the social network information is determined by the parameters β for SocialMF, α for STE, and γ (with $w_m^{(S)} = 0.00003, s_m = 0$) for SoRec. The results are shown in Figure 1. As expected, it is important to find the right tradeoff between the social network information and the rating data.

The recall test results for the optimal tuning parameters are summarized for all these MF models in Table 1. While all three models show an improvement in recall compared to *No Trust* case, it is particularly large for SoRec. The SoRec model with our modified training objective function outperforms *No Trust* by 23.1% in terms of overall recall and 101.8% in terms of cold-user recall. It shows that the social network is very helpful in terms of top-k recommendation, especially for recommendations for cold start users. Moreover, recall is even slightly higher for cold users than it is for all users. This may be explained by the fact that cold users have a slight tendency to rate popular items (i.e. items with a large number of ratings), which can naturally

[2]http://www.epinions.com/
[3]http://www.flixster.com/

[4]http://alchemy.cs.washington.edu/data/epinions/
[5]Considering both 4 and 5 star ratings as relevant, experiments showed similar differences among the various approaches.
[6]http://en.wikipedia.org/wiki/Netflix_Prize

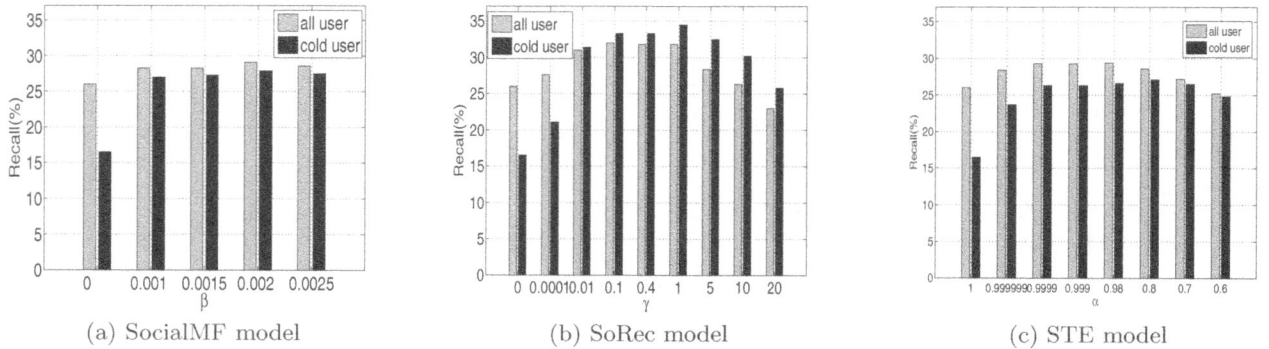

(a) SocialMF model　　　　　　(b) SoRec model　　　　　　(c) STE model

Figure 1: Top-500 Recall for social-network-based matrix-factorization models on Epinions data.

test users	MF models			
	No Trust	SocialMF	STE	SoRec
original training (on observed ratings)				
all	1.9%	3.5%	2.7%	2.6%
cold	1.5%	1.0%	2.8%	2.9%
modified training (on all ratings)				
all	26.0%	29.1%	29.4%	32.0%
cold	16.5%	27.9%	26.6%	33.3%

Table 1: Epinions data: recall (top 500) in percent for three MF models trained with original and modified training objective. 'No Trust' is the baseline MF model that only uses rating data.

be recommended more accurately. In the Epinions data, the average item rated by a cold user has received 102 ratings, while the average item rated by all users has received only 93 ratings. Table 1 shows that the SoRec model with our modified training objective function achieves the best recall. This may be unexpected, as the SoRec model was found to achieve a worse RMSE than STE in [11], and STE was found to have a worse RMSE than SocialMF in [8]. This result illustrates that the best way of combining rating data with social network information concerning the popular RMSE measure is not necessarily the best way to maximize recall.

4.2.3 RMSE for MF Models

Apart from optimizing for recall, we also determined the optimal tuning parameters as to minimize RMSE, and found $\lambda = 0.1$, $r_m = 4$, $w_m = 0$, $j_0 = 10$, which resulted in the following RMSE values:

- $RMSE = 1.174$, if only the rating data is used,

- $RMSE = 1.095$, for SocialMF (with $\beta = 20$),

- $RMSE = 1.157$, for STE (with $\alpha = 0.5$),

- $RMSE = 1.117$, for SoRec (with $\gamma = 50$ and $w_M^{(S)} = 0$).

These results are consistent with RMSE results in the literature [8, 11, 12]. It verifies that social network information is useful for improving RMSE.

4.2.4 Recall for NN Models

As a further comparison, Figures 2 shows the recall test results we obtained for various nearest neighbor models, which are outlined in Section 3.2. To the best of our knowledge, this includes *Trust-cf*, the only top-k approach using social

network information [7]. In the Trust-cf model, k_1 is set to be 5 which leads to the best recall in user-based CF. The top-500 recommendation result on the Epinions dataset of Trust-cf is shown in Figure 2(c).

Among the NN approaches, the one in Figure 2(c) achieves a considerably worse hit ratio than any of the NN approaches in Figure 2(a) and 2(b). This poor performance of the *Trust-cf* approach is due to the following reason: *Trust-cf* predicts the rating value of a user in terms of the average rating values of the user's friends–which is obviously based on the *observed ratings only*. In contrast, the various NN approaches in Figure 2(a) and 2(b) use voting–which is the simplest possible way of accounting for *all items*, i.e. by counting 0 for an absent rating and counting 1 for an observed relevant rating (with weights defines in Section 3.2). As the rating value is ignored, this is the simplest possible approach to account for all items during training. Though recall of NN based RS is not as good as MF based RS. As we mentioned in Section 3.2, NN based approach has the advantage of integrating new feedbacks incrementally while MF based approach not able to.

4.3 Experiments on Flixster Dataset

Flixster is a social network site where users add other users to their friend lists to form a social network. Flixster data has about one million users, who rate movies and share reviews. The ratings in Flixster have ten discrete values from 0.5 to 5, with step size of 0.5. Flixster is different from Epinions in that social relations in Flixster are bi-directional. The Flixster data [7] used here is from [8]. The Flixster social network has 26.7 million connections. The trace consists of 8.2 million movie ratings on 49,000 movies and 1 million users. The number of users who made at least one rating is 150,000. Despite the different properties of the Epinions and Flixster data sets, the results on the Flixster data confirm our results on the Epinions data. The result on Flixster dataset is very similar to Epinions dataset. Due to space limit, we cannot present all results here. The complete results on Flixster dataset is available in our technical report [8].

We split the data into a training set and a disjoint test set. For users with less than 10 ratings, we randomly choose one rating and put it into test set. For users with 10 or more than 10 ratings, we randomly chose 5% as put them into the test set. We further split the test set randomly into two disjoint

---

[7]http://www.sfu.ca/ sja25/datasets/

[8]http://eeweb.poly.edu/faculty/yongliu/docs/topk_tr.pdf

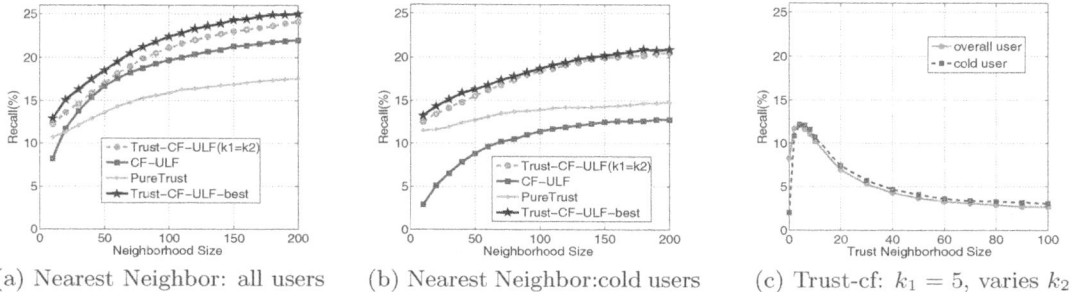

(a) Nearest Neighbor: all users (b) Nearest Neighbor:cold users (c) Trust-cf: $k_1 = 5$, varies k_2

Figure 2: Top-500 Recall by Nearest Neighbor based Models on Epinions dataset

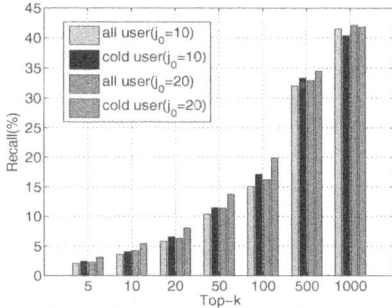

Figure 3: Recall vs Top-k and impact of Dimensionality on Epinions data.

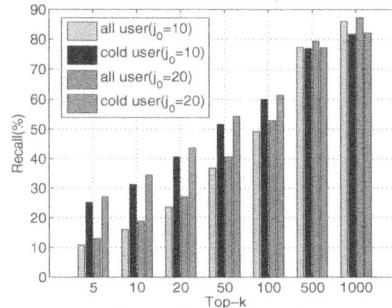

Figure 4: Recall vs Top-k and impact of Dimensionality on Flixster data.

sets of equal size and report results of testing the second test set akin to Epinions case. We defined rating values of 4 or larger as relevant for computing recall on the test set. We report recall for the top 100 items recommendation. The top-k hit ratio of different k value is presented in Figure 4.

The optimal values of tuning parameters are: $\lambda = 0.1$, $r_m = 1.0$ and $w_m = 0.2$. The optimal β is 0.1 in modified SocialMF model, and the optimal α is 0.99 in STE model. The optimal γ is 0.05 (with $w_m^{(S)} = 0.2, s_m = 0$) in modified SoRec model. For these optimal training parameters, the recall test results are summarized for all the MF models in Table 2.

test	MF models			
users	No Trust	SocialMF	STE	SoRec
original training (on observed ratings)				
all	4.4%	4.7%	5.3%	8.2%
cold	6.3%	6.6%	7.2%	15.4%
modified training (on all ratings)				
all	44.3%	45.2%	47.1%	49.1%
cold	30.8%	38.3%	47.6%	59.2%

Table 2: Flixster data: recall (top 100) in percent for three MF models trained with original and modified training objective. 'No Trust' is the baseline MF model that only uses rating data.

As before, SoRec with modified training objective function achieves the largest recall. We can see from Table 2 that SoRec model with our modified training objective function outperforms *No Trust* by 10.8% in terms of overall recall and 92.2% in terms of cold user recall. It again shows that social network is very helpful in terms of top-k recommendation especially for recommendation of cold start users. Note that the improvement for the cold users over all the users is

particularly pronounced for the Flixtser data, as cold users have a rather strong tendency to rate popular items. In the Flixster data, the average item rated by a cold user has received 9,601 ratings, while the average item rated by all users has received only 5,367 ratings.

4.4 Impact of Dimensionality and Top-k

Figure 3 and 4 depict recall vs. the top-k number with dimensionality $j_0 = 10$ and $j_0 = 20$ for Epinions data and Flixster data respectively. We can see from Figure 3 and 4 that dimensionality $j_0 = 20$ performs better than $j_0 = 10$. This is because larger dimensionality captures more latent features of users and items and hence improves recall. It should be noted that top-k hit ratio of Flixster data is much better than Epinions data. Counting that the number of items in Epinions dataset is about twice of Flixster dataset, still, we find that recall of Flixster is more than twice of Epinions for top-5 to top-500 recommendations. This is probably because of the fact that Epinions data is a multi-category data which contains items from many categories (cars, movies, books, software, etc.), while items in Flixster are all movies which makes the recommendation easier in general. Furthermore, users in Flixster dataset averagely have more social connections and item ratings compared to Epinions dataset.

5. CONCLUSIONS

Social recommendation is prevalent in the real-world, but top-k recommendation using online social networks has been insufficiently studied in the recommendation literature. In this paper, we presented a comprehensive study on improving the accuracy of top-k recommendation using trust information derived from social networks. We showed that the existing social-network based recommender-systems can be conveniently tailored for top-k recommendations by modifying their training objective functions to account for both observed ratings and missing ratings. Through experiments on two large-scale data sets, we made three majors findings: 1)Trust information significantly improves the top-k hit-ratio when incorporated properly both in MF and NN models; 2)Our proposed social network based NN models and RS trained with our modified objective function considerably outperform the only published top-k approach using social network information in recent literature [7], an NN approach; 3) surprisingly, we found that among the various ways of combing feedback data with social network information, the one that was found to be worst with respect to RMSE turns out to be the best concerning the top-k hit ratio, and vice versa. This illustrates that the technical details of minimizing RMSE can be very different from the ones that work to optimize the (more realistic) top-k hit ratio. Our work demonstrated that top-k recommendations pose unique challenges, and social trust information, when incorporated properly, can significantly improve the hit ratio of top-k recommendations.

6. ACKNOWLEDGMENTS

The authors would like to thank Debasis Mitra and Shiv Panwar for initiating this collaboration. Harald Steck would also like to thank Tin Kam Ho for her encouragement and support of this work. Xiwang Yang and Yong Liu were partially supported by USA NSF grant CNS-1018032 to conduct this work.

7. REFERENCES

[1] G. Adomavicius and A. Tuzhilin. Toward the next generation of recommender systems: A survey of the state-of-the-art and possible extensions. *IEEE Transactions on Knowledge and Data Engineering*, 17:734–49, 2005.

[2] P. Cremonesi, Y. Koren, and R. Turrin. Performance of recommender algorithms on top-N recommendation tasks. In *ACM Conference on Recommender Systems*, pages 39–46, 2010.

[3] S. Funk. Netflix update: Try this at home, 2006. http://sifter.org/ simon/journal/20061211.html.

[4] J. A. Golbeck. Computing and applying trust in web-based social networks, 2005.

[5] Y. Hu, Y. Koren, and C. Volinsky. Collaborative filtering for implicit feedback datasets. In *IEEE International Conference on Data Mining (ICDM'08)*.

[6] M. Jamali and M. Ester. Trustwalker: a random walk model for combining trust-based and item-based recommendation. In *ACM Conference on Knowledge Discovery and Data Mining (KDD)*, 2009.

[7] M. Jamali and M. Ester. Using a trust network to improve top-n recommendation. In *ACM Conference on Recommender Systems (RecSys)*, 2009.

[8] M. Jamali and M. Ester. A matrix factorization technique with trust propagation for recommendation in social networks. In *ACM Conference on Recommender Systems (RecSys)*, 2010.

[9] R. Keshavan, A. Montanari, and S. Oh. Matrix completion from noisy entries. *Journal of Machine Learning Research*, 11:2057–78, 2010.

[10] Y. Koren. Factorization meets the neighborhood: a multifaceted collaborative filtering model. In *ACM Conference on Knowledge Discovery and Data Mining*, pages 426–34, 2008.

[11] H. Ma, I. King, and M. R. Lyu. Learning to recommend with social trust ensemble. In *ACM conference on Research and development in information retrieval (SIGIR)*, 2009.

[12] H. Ma, H. Yang, M. R. Lyu, and I. King. Sorec: Social recommendation using probabilistic matrix factorization. In *International Conference on Information and Knowledge Management (CIKM'08)*.

[13] B. Marlin and R. Zemel. Collaborative prediction and ranking with non-random missing data. In *ACM Conference on Recommender Systems (RecSys)*, 2009.

[14] B. Marlin, R. Zemel, S. Roweis, and M. Slaney. Collaborative filtering and the missing at random assumption. In *Conference on Uncertainty in Artificial Intelligence (UAI)*, 2007.

[15] P. Massa and P. Avesani. Trust-aware recommender systems. In *ACM Conference on Recommender Systems (RecSys)*, 2007.

[16] R. Pan, Y. Zhou, B. Cao, N. Liu, R. Lukose, M. Scholz, and Q. Yang. One-class collaborative filtering. In *IEEE International Conference on Data Mining (ICDM)*, 2008.

[17] X. Yang, Y. Guo and Y. Liu. "Bayesian-inference based recommendation in online social networks". *The 30th Annual IEEE International Conference on Computer Communications (INFOCOM)*, 2011.

[18] X. Yang, H. Steck and Y. Liu. Circle-based Recommendation in Online Social Networks. *Proc. of 2012 ACM SIGKDD Int. Conf. on Knowledge Discovery and Data Mining (KDD'12)*.

[19] Q. Yuan, S. Zhao, L. Chen, Y. Liu , S. Ding, X. Zhang and W. Zheng. Augmenting collaborative recommender by fusing explicit social relationships. Recommender Systems & the Social Web workshop, RecSys, 2009.

[20] V. Vasuki, N. Natarajan, Z. Lu and I. Dhillon. Affiliation Recommendation using Auxiliary Networks. In *ACM Conference on Recommender Systems (RecSys)*, 2010.

[21] A. Paterek. Improving regularized singular value decomposition for collaborative filtering. In *KDDCup*, 2007.

[22] R. Salakhutdinov, A. Mnih, and G. Hinton. Restricted Boltzmann machines for collaborative filtering. In *International Conference on Machine Learning (ICML)*, 2007.

[23] R. Salakhutdinov and N. Srebro. Collaborative filtering in a non-uniform world: Learning with the weighted trace norm. In *Advances in Neural Information Processing Systems 24 (NIPS)*, 2010.

[24] N. Srebro and T. Jaakkola. Weighted low-rank approximations. In *International Conference on Machine Learning (ICML)*, pages 720–7, 2003.

[25] H. Steck. Training and testing of recommender systems on data missing not at random. In *ACM Conference on Knowledge Discovery and Data Mining*, pages 713–22, 2010.

[26] M. Richardson and P. Domingos. Mining Knowledge-Sharing Sites for Viral Marketing, In ACM Conference on Knowledge Discovery and Data mining, 2002.

Optimal Radio Channel Recommendations with Explicit and Implicit Feedback

Omar Moling
Free University of
Bozen-Bolzano
Piazza Domenicani 3
Bolzano, Italy
omoling@gmail.com

Linas Baltrunas
Telefonica Research
Plaza se E. Lluchi Martin 5
Barcelona, Spain
linas@tid.es

Francesco Ricci
Free University of
Bozen-Bolzano
Piazza Domenicani 3
Bolzano, Italy
fricci@unibz.it

ABSTRACT

The very large majority of recommender systems are running as server-side applications, and they are controlled by the content provider, i.e., who provides the recommended items. This paper focuses on a different scenario: the user is supposed to be able to access content from multiple providers, in our application they offer radio channels, and it is up to a personal recommender installed on the clients' side to decide which channel to select and recommend to the user. We exploit the implicit feedback derived from the user's listening behavior, and we model channel recommendation as a sequential decision making problem. We have implemented a personal RS that integrates reinforcement learning techniques to decide what channel to play every time the user asks for a new music track or the current track finishes playing. In a live user study we show that the proposed system can sequentially select the next channel to play such that the users listen to the streamed tracks for a larger fraction, and for more time, compared to a baseline system not exploiting implicit feedback.

Categories and Subject Descriptors

H.3.3 [**Information Storage and Retrieval**]: Information Search and Retrieval—*information filtering*

General Terms

Design, Experimentation, Human Factors

Keywords

Sequential music recommendations, implicit feedback, reinforcement learning

1. INTRODUCTION

Recommender Systems (RS) target the "information overload" problem, i.e., the difficulty to deal with the increasing amount of available information, e.g., in Internet, when facing an information search problem or a decision task. RSs suggest items that are estimated to be of interest and appropriate for the users in a particular usage context [15]. RSs have become important tools in many application domains, but especially for music information retrieval, since the today popular and large repositories of digital music (Last.fm, Pandora, iTunes) could not be fully exploited without such kind of support. In the last years, several recommendation techniques have been applied to the music domain, such as, collaborative filtering, content-based, social-based approaches and their combination as hybrid systems [6, 5].

1.1 Addressed Limitations of RSs

The large majority of RSs are running as server-side applications, and, more importantly, they are controlled by the content provider, i.e., who provides the recommended items. When a user accesses the system for retrieving new content, e.g., music tracks, the recommender computes a personalized selection of items and delivers them to the user. In [2], and more recently in [11], the authors advocate for new recommendation technologies that can work client-side, are controlled by the user, and can exploit items offered by several independent content providers. In their proposed scenario the recommender selects, among the items offered by the content providers, the most relevant ones for the user in a particular context [1]. Client-side recommenders can still be networked, and can benefit from interaction data collected by observing a community of users, but are designed to optimize performance metrics that are important for the users, and that could be different from those typically considered by the content providers (e.g., conversion rate).

Other two severe limitations of more traditional RSs are: the lack to support sequential consumption of items, and the difficulty to exploit implicit evaluation feedback on items. Regarding the first point, one can observe that users often listen to sequences of tracks during a listening session (playlists), and their preferences at a certain point in time are influenced by the tracks previously listened to. Therefore recommending a sequence of items defines a different problem than recommending individual items. Sequential items selection problems are not found in music applications only. For instance, after having read certain books, one decides what books to read next. These decisions are influenced by what has been read before, the background knowledge of a person, and what is available and preferred at the time the decision is made.

Regarding the second point, i.e., the difficulty to exploit implicit feedback on consumed items, consider the situation where a user is recommended a track, but after listening to it for a few seconds, she requests a new recommendation. That action should be interpreted as a sign that, at that point in time, i.e., after having listened to some particular tracks before, and given the user's preferences, that suggestion was not optimal. In the book recommender scenario we may observe a similar situation; a reader may stop reading a book for a while, or may even start reading a new one. These signs could be interpreted by the recommender as indications that the recommended book was not a convenient choice at that point in time.

Some recent research works have tried to address these two issues: recommending a sequence of items and exploiting implicit feedback in building the user profile and generating recommendations. We will review them in Section 6.

1.2 Recommendation Scenario and Results

This paper focuses on a specific scenario: the user is supposed to be able to access content from multiple providers, in our application they offer music channels, but it is up to a personal recommender to decide what channel to select and play. We show how we have addressed the specific application scenario and the above-mentioned issues, namely, sequential item recommendation, and the usage of implicit feedback in a music RS named RLradio. RLradio plays music streamed by different channels, each one featuring music of a particular genre. Moreover, it learns whether to stay tuned on the current channel or to switch to a different one. This decision is taken, for a target user, when the currently played track finishes, or when the user, using a *next* button of the system GUI, explicitly requests a new track from one of the available channels. We stress that the system has no control on the precise music tracks that are streamed by the available channels, and can only automatize the channel selection action (tune in) that is normally executed manually by a user. The goals are: a) to avoid the need to manually change channel in situations where this could result in a usability problem, e.g., in a car driving scenario; and b) to adapt in real time to the changed preferences of the user, or the group of users listening to the radio system.

In order to implement the above mentioned functionality and requirements, RLradio leverages both explicitly revealed music channel preferences, and implicit feedback produced by the users: the fraction of the track played in the recommended channel that is actually listened to by a user. It exploits a particular type of reinforcement learning (RL) techniques, i.e., R-Learning that is suited for continuous tasks learning [17]. The goal is to learn the optimal policy for selecting the music channel to suggest next.

We first acquired a model of the listening behavior of the users by observing their interaction with RLradio while the system was using a baseline channel selection policy that exploits only the explicitly revealed channel preferences of the users. We call this system variant RLradio P. Then, we computed the optimal policy for the acquired state transition model off-line, and forced RLradio to use it. This new system variant is called RLradio RL. Finally, during the successive interactions of the users with RLradio RL, the system adapted its behavior on-line using R-Learning.

We evaluated RLradio in a live user study. We conducted a within-subjects experiment where the users tested two RL-

radio variants with identical GUI. The first variant (RLradio P) uses only the user's channel preferences elicited at the beginning of the listening session. The second one (RLradio RL) exploits also the implicit feedback derived from the listening behaviors of all the users. The evaluation showed that the learning process is effective: users prefer RLradio RL, and they listen to both a larger fraction of the suggested tracks and a higher daily time span while using RLradio RL. Hence, in conclusion RLradio offers an effective solution for the considered sequential item recommendations problem. It minimizes user effort, and is capable to track the evolving preferences of the user (or users) receiving the recommendations.

The rest of the paper is structured as follows: sections 2 and 3 describe the recommender system functionality and techniques. The evaluation approach and the results are described in sections 4 and 5. Then in section 6 we survey some related work, and finally, conclusions are drawn and future work defined in section 7.

2. RLRADIO RECOMMENDER SYSTEM

RLradio is a music player that offers music channels, each one featuring music of a rather homogeneous genre, such as in a thematic channel. The user can listen to the played music track either fully (when he probably likes it), or can request a new track (presumably when he does not like what is currently played). At that point RLradio decides whether to stay tuned on that channel, or to switch to another one offering music of a different genre. In a normal radio, changing the channel is up to the user, who would search for one, among a set of available channels, which is broadcasting music that he likes. This takes time and effort, and may not be convenient in some situations (e.g., when the user is driving or biking). RLradio automatizes and simplifies this process by providing a *single* button to request a new track (the *next* button in Figure 1(b)). At that point, it is up to the system to decide whether to stay tuned on the current channel, or to switch to another one, with the ultimate goal of providing music that the user, or the users, will like at that moment.

In our system prototype we have considered nine popular Internet radios. These are presented to a user indicating only the music genre played in the channel (Figure 1(a)). At the beginning of each listening session, RLradio collects the user's preferences for the available channels. These preferences are modeled as the percentage of music of that channel that the user would like to listen to (Figure 1(a)). In fact, these explicit preferences have been used by a baseline system, here called RLradio P, that at each channel recommendation step, i.e., each time the currently played track finishes, or when the user clicks on the next button: a) selects a channel with a probability proportional to the user's preference for that channel, and b) plays the next track in the queue that is offered by that channel. The sequence of the channels is therefore (pseudo) random, it exploits the user's explicitly entered preferences, but RLradio P does not take into account neither what channels were previously played nor any knowledge about when the user has used the next button.

We consider this selection policy as a *baseline*. Note that this is not a bad policy, and in fact, in the experimental evaluation, users exposed to this policy reported rather high satisfaction scores (see Section 5). Nevertheless, we hypothesize that *the performance of RLradio, measured as the percent-*

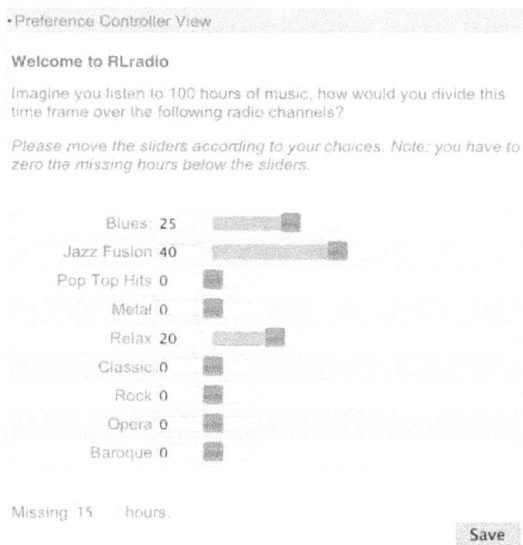

(a) Music channels preferences collection   (b) RLradio main view

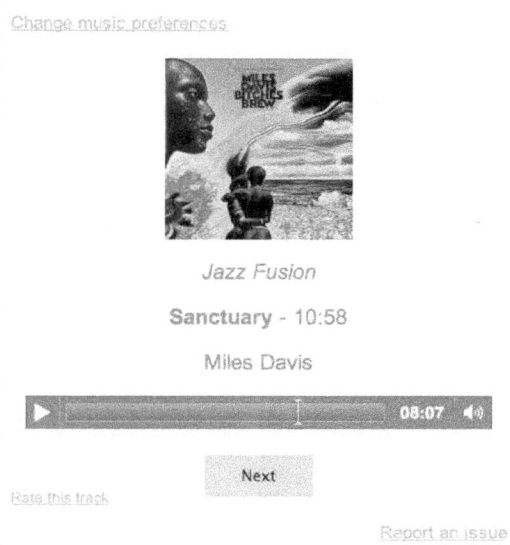

Figure 1: RLradio user interface

age of each proposed track that is actually listened to, and the amount of time, per day, spent by the user listening to the recommended music, can be improved by using also implicit feedback, i.e., the knowledge of when the user clicks next. In other words, we conjectured that the system performance can be improved by observing what channel is played when the user stops listening to a music track in that channel, and requests a new track. We hypothesized that these observations can be exploited to learn a better channel selection policy (an optimal policy).

This policy, for instance, can override explicit preferences made by the user, and adapt to the real listening behavior. This is useful when the user is not entirely sure about his preferences, makes misjudgment, or his preferences change during the session, e.g., in a in-car player when new passengers enters into the car and they contribute to the decisions of requesting new tracks. We also conjectured that the policy can also model the optimal *sequence* of music genres preferred by the user, i.e., how music from different channels should be combined. For instance, the user may not like to switch continuously from one channel that he likes to another that he also likes, but could prefer to listen to more music tracks from the same channel before switching to another.

In our experiments, the nine Internet radio channels that have been selected offer music of a particular genre: blues, jazz fusion, pop, metal, ambient, classic, rock, opera and baroque. For each of the nine selected channels, 24 consecutive hours of music were ripped from the selected Internet radios and the music tracks were split automatically using the track metadata provided by the channels. This resulted in a total of 3,637 music tracks. The number of tracks per channel varies, as the average length of tracks belonging to different music genres is different. We decided to rip the music and then take the music from a local repository to make

the experiments replicable and reduce the possible connection problems to external music servers.

3. THE RECOMMENDATION AGENT

In order to implement the channel recommendation agent used in RLradio, we modeled the recommendation task as a Markov Decision Problem (MDP) [17]. In general, a MDP is defined by:

- A set of states S, modeling the possible different situations that the decision maker can face.

- A set of actions A, listing the possible alternative actions, or decisions, that the decision maker can take.

- A real-valued function $T(s, a, s')$ that defines for each state s and action a combinations, the probability to make a transition to the new state s'. This is called the "environment model" because it describes the reaction of the environment to the decisions taken.

- A real-valued reward function $R(s, a, s')$ that measures the payoff obtained by the decision maker when making a particular state transition after the application of an action.

The solution to a MDP is an optimal policy, i.e., a rule for selecting at each state the best action, i.e., the actions that, transition after transition, will obtain the largest accumulated reward. We used reinforcement learning (RL) to implement the recommendation agent, i.e., the channel selection decision policy at each step. The actions taken by the recommendation agent in RLradio correspond to the channels to play next, and the environment is the "place" where the actions are executed, which in our case is the user listening to the tracks in the suggested channels. The user decides whether to listen, and to what extent, to the played tracks. The agent must learn from the user's (implicit) feedback what action to execute in each possible state in order to maximize the accumulated reward, i.e., as we will show, the percentage of the played tracks that is actually listened to. In the following we will instantiate the model to our particular application.

State Model S**.** The state models the current recommendation situation, i.e., it represents the current channel preferences of the user, and the previous listening behavior. In principle, the full knowledge of the user's listening behavior, i.e., the sequence of all the previously listened channels, and how much the user listened to them should be considered. The main drawback of this solution is the exponential growth of the size of the state space, and the difficulty to estimate the state transition probabilities, and compute the optimal policy. Moreover, one can hypothesize that the most recent listening history plays a stronger role in determining the preferences of a user at a certain point in time.

For these reasons, RLradio records only the two previously recommended channels and the percentages of the music tracks in these two channels that were actually listened to before the user clicked on the next button (if he clicked). These percentages were discretized in three levels that are described later when the Reward function is presented.

Moreover, in order to keep the state model small, the music channel preferences entered by the user were also discretized in four levels: low preference for a channel ($< 15\%$), fair (between 15% and 40%), high (between 40% and 60%) and very high (above 60%). These thresholds were defined heuristically, and we thought that four levels were adequate to score the user preference for a music channel, like in a five star Likert rating schema.

With such a discretization there are 4^9 possible configurations of the user explicit preferences. In fact, we discovered that only 77 configurations were actually entered by the users while using the baseline system. Hence, we constrained the system to use only these configurations or profiles, and when a user entered different preferences we associated this user to the closest profile among these 77 configurations. With these assumptions we produced a state space of 56, 133 states, i.e., $9 * 3 * 9 * 3 * 77$. Where, 9 is the number of possible channels, and 3 is the number of possible (discretized) listening percentages, for the two tracks previously listened by the user. A summary of the state variables is shown in Figure 1.

Actions A**.** The actions correspond to the nine possible channel recommendations that can be selected by the system to pick up the new track to be played. We observe that the recommendation agent cannot recommend a particular track, but only the channel of the next track. The tracks in a channel have a predefined order and are played one after the other. This decision is motivated by the application scenario, i.e., we assume that the recommender can only decide the channel to tune in, and the compilation is defined for all the listeners of that channel by the channel's manager. All the nine actions are available in each state of the state space. Hence, by considering the size of the state space and the available actions, there were $56, 133 * 9 = 505, 197$ state-action combinations.

Reward R**.** The reward is computed observing the user's implicit feedback, i.e., the fraction of the played track's length in the suggested channel that was actually listened

Table 1: State variables

Variable	Values
The last proposed channel	9 channels
Percentage of the track in the last proposed channel that was actually listened to	3 discretized levels
The second-last proposed channel	9 channels
Percentage of the track in the second-last proposed channel that was actually listened to	3 discretized levels
Explicitly entered channel preferences	77 music profiles

to by the user, before he clicked *next*, if he did. This percentage is mapped to a three-levels discrete reward score that measures how successful the action (channel selected) was in that particular state (user preferences and previous listening behavior). 0 reward is assigned to the state transition if the music track is listened to less than 15%, i.e., the user requests another track before the playback reaches 15% of the length of the track. We considered this a strong sign of not liking the currently played track. If the user requests another track while the playback is between 15% and 60%, the reward equals 1, while for more than 60% up to the full length, i.e., the user does not request another track, we decided to assign a double reward of 2. These decisions are somewhat arbitrary, but necessary for defining the MDP problem. Slightly different choices are possible, and they can result in a different optimal policy.

Moreover, we note that here a feedback on a track is interpreted as an implicit evaluation of the channel. In fact, the user is listening to a particular track, and the skipping feedback is related to that specific music and only indirectly to the channel of that music. The rationale of this choice is again related to the application scenario: even though we could learn from the implicit feedback if a user likes or dislikes a track (as in [13]) this is not very useful since the recommendation agent has not full control on the next track that is going to be played, it can only select the channel.

Transition Probabilities. We collected raw data on the users' listening behavior by letting a group of 29 users to use the previously mentioned baseline system (RLradio P) for approximately two weeks. This is a common practice in RL. The baseline system adopts a policy that considers only the explicit music channel preferences entered by the user at the beginning of each listening session, and provides music from a channel with a probability equal to the value of the user preference for the channel. For example, setting 20% preference to channel *classic* results in having, at each step, a probability of 20% that classical music is selected. In addition, 10% of the proposals were chosen randomly in order to better explore the state space, a common practice in RL [17].

From the collected data, several values were computed, in particular we estimated the transition probabilities: i.e., the probability that a user will listen to a certain (discretized) percentage of a track (current state), if it belongs to the suggested channel A (action), given a particular channel preferences profile and the percentage of the two previously played tracks (in channels B and C) that were listened to.

We also estimated the probability that a user will listen to a certain percentage of a track in a channel given his channels' preferences alone. These prior values were used to provide a default state transition model for states that were poorly explored by the users. We combined these priors with the estimated state transition probabilities with Laplace smoothing [17]. The smoothing was such as to assign more weight to the prior values when a given state was visited poorly or not at all, while more weight was set on the experienced transition probabilities for good or fairly good explored states.

Optimal Policy. We computed the optimal policy using Policy Iteration with $\gamma = 0.8$ [17]. γ is a parameter that measures how the future reward is discounted compared with the immediate reward. We tried some other values, but we did not find significant differences; 0.8 is a rather standard value for this parameter. All other variables were set analyzing the data collected during the usage of the baseline system (RLradio P).

The obtained optimal policy served then as the starting policy for the learning-capable agent system variant (RLradio RL), which implemented the *R-Learning for Undiscounted Tasks* [17] technique, since music recommendation is considered a continuous task, as there is no identifiable terminal state. The algorithm is illustrated in Figure 2.

Initialize p and $Q(s, a)$, for all s, a, arbitrarily
repeat
    $s \leftarrow$ current state
    Choose action a in s using ϵ-greedy behavior policy
    Take action a, observe r, s'
    $Q(s, a) \leftarrow Q(s, a) + \alpha \left[r - p + \max_{a'} Q(s', a') - Q(s, a) \right]$
    if $Q(s, a) = \max_a Q(s, a)$ **then**
        $p \leftarrow p + \beta \left[r - p + \max_{a'} Q(s, a) - \max_a Q(s, a) \right]$
    end if
until forever

Figure 2: R-Learning

As explained in more details in [17], R-Learning maintains two policies: a behavior policy, which serves to generate experience, and an estimation policy, which allows a generalized policy iteration process. The behavior policy is the ϵ-greedy action selection with $\epsilon = 0.1$ [17]. The estimation policy only maximizes the action according to action-value function $Q(s, a)$. In addition to the two policies, also the action-value function $Q(s, a)$ and the average reward p are maintained. In Figure 2, r denotes the immediate reward of an action, while the average reward p is set initially for each user to the overall average reward, and then is updated individually for each user. The values of the step-size parameters α and β were set to $\alpha = 0.3$, $\beta = 0.4$, which are common values for these parameters, such that p reacts faster than the state-action value $Q(s, a)$.

4. EXPERIMENTAL STUDY

As we mentioned above, we hypothesized that the proposed approach, which is implemented in RLradio RL, can improve the performance of the baseline system (RLradio P). We measure the system performance as the average percentage of the proposed music tracks that is actually listened to by each user, and as the average number of minutes of music, per day, that the users listen to. We focus on these measures because we aim at building a system that automatically changes the channel such that the played tracks are actually listened to by the users as much as possible, and the system is overall largely used.

We note that RLradio P faithfully uses the channel preferences of the user, while RLradio RL discretizes them, but in addition exploits the dynamically generated implicit feedback provided by the users while interacting with the system. Hence, the expected improvements must be due to the exploitation of the implicit feedback, i.e., the knowledge of when the user requests a new music track while listening to a track in the suggested channel, and having listened previously to other tracks from possibly different channels.

We designed a within group evaluation study. We split the users in two groups: (Group #1) first tested the baseline system (System P), and then the system equipped with the optimal policy and also using R-Learning (System RL); while (Group #2) executed the test in the opposite order. This decision is aimed at balancing a possible order effect. For each group, at least one week of pause was inserted between the evaluations of the two systems. The two groups were formed by 70 users in total and all users were initially informed that the evaluation consisted of two phases, but they had absolutely no knowledge about the systems' differences. The graphical user interfaces of the two systems and the music tracks played in the channels were identical. Hence the two systems differed only in the way the recommender selected the channel to use for playing the next music track.

At the beginning of each phase, all involved users were contacted via email. The initial steps of entering a nickname and defining the channel preferences were explained on the home page. Then, the user interface of the system was deemed as simple, and thus no special explanations were provided to evaluate it. Each evaluation phase lasted for 12 days on average. All data generated by the system usage, i.e. listening sessions, explicit preferences and implicit feedback, was stored on the server running the RS.

In addition to the collection of implicit feedback throughout all the test phases, all users were asked to take a survey at the end of each evaluation phase, i.e., after having used the Systems P and RL (according to their evaluation order). The aim of the survey was to measure the level of satisfaction of the user with the proposed system, and the effectiveness of the music recommendations. The survey included statements, for which the users were asked to express the level of agreement. The statements had possible answers on a five-level Likert scale: *Strongly Disagree*, *Disagree*, *Undecided*, *Agree* and *Strongly Agree*.

The most relevant statements were the following:

S1: I liked the music played by the system.
S2: The system switched the played channel at the right time.
S3: I would use such a player if available.
S4: The system is easy to use.

Statement S1 is related to the degree of satisfaction for the channels and tracks proposed and played by the player. S2 and S3 refer to the behavior of the system, and whether the

Table 2: Number of implicit feedback events collected

System	Feedback items
RLradio P	3423
RLradio RL	4385
Total	7807

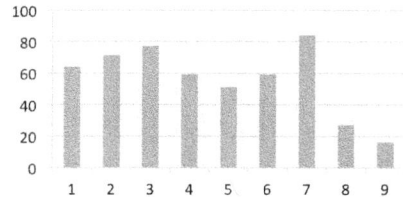

Figure 3: Frequency of sessions with a given number of channels with not null preference

user would like a system that behaves like the proposed one. Finally, S4 refers to the evaluation of the user-interface.

5. EXPERIMENT RESULTS

We compared RL and P system with paired and unpaired data. We considered paired data comparing the average percentage of played tracks listened by each user. In this case we could consider only the users that tested both systems: 17 for Group #1 and 13 for Group #2. We also considered not paired data while comparing the daily system usage, in minutes, by a user. In this case we could consider all the daily recorded sessions and we compared the length of the sessions observed for the two systems. Moreover, in order to increase the reliability of the observations we discarded the data generated by the sessions where the user did not interact with the system during a sequence of five tracks. That could be a sign that the user was not paying attention to the played music.

We note that RLradio RL is not a technology that is supposed to provide a large benefit just after the user has entered the channel preferences. This technology is supposed to adapt to the user while the user is interacting with the system. So in order to detect some differences the users must interact with the application for some time. We also observe that the system P was stable during the evaluation, i.e., its channel selection policy was not changing as new implicit feedback was accumulated. Conversely, the RL system was continuously updating its behavior as new implicit feedback was acquired. Hence, when a user tried the RL system, the implicit feedback provided by all the users that tried the system before was used by the R-Learning algorithm to improve the system's behavior. This is the major benefit of using reinforcement learning: the system a) adapts to the users that are actually using the system, and b) benefits from the feedback produced by previous usage, even by different users.

Table 2 illustrates the overall number of implicit feedback events collected during the evaluation phases, divided per system. We note that 1723 feedback events were collected during the first evaluation phase of RLradio P, and served as training data for the computation of the optimal policy for RLradio RL. A feedback event records: the channel of the track currently listened to, the channel of the two tracks that were listened before, the channels' preferences of the user (his profile configuration), and the time, during the track playback, when the user clicked on the *next* button (if he did it).

Figure 3 represents the frequencies of the observed listening sessions where the user gave a not null preference to $n = 1, \ldots, 9$ channels (e.g., in more than 80 sessions the user gave a positive preference to 7 different channels). This illustrates that typically the users entered a not null pref-

erence for many channels, hence switching channel makes sense for satisfying these users.

We then analyzed the implicit feedback data. Regarding the within-group evaluation, i.e., considering the listening sessions generated by the users that tested both systems. On average each user listened to 64.35% of the length of the proposed music when using RLradio P, with a standard deviation (SD) of 0.209. Instead, when using RLradio RL, the average percentage of track length listened to by each user was 67.41%, with a SD of 0.182. Hence, the improvement in the percentage of the listened track length of RLradio RL over RLradio P is 4.76%. This difference is significant (paired-samples t-test $p = 0.028$).

We also measured that 63.33% *of the users, while using RLradio RL, listened, on average, to a larger percentage of the track length, compared to RLradio P.* In practice the large majority of the users intervened less frequently when using the RL system compared to when they used the baseline P system and listened longer to the proposed tracks.

Considering the daily usage of the system, we observed that the listening sessions with system RL were significantly longer than the sessions performed with P. The average daily usage when using RLradio P was 62.68 minutes per day. That increased to 75.58 minutes per day when using RLradio RL, i.e., the users listened to music while using RL 20% more than while using P. Again this difference is significant to a t-test (unpaired samples p=0.043).

The optimal policy, which was computed by the policy iteration algorithm on the log data of the System P, and served as the starting policy for System RL, was improved online by the R-Learning algorithm during the evaluation. The users visited collectively 1606 states. Actually, this is a small percentage of all the possible states. So, in practice the learning that we could perform with the collected data is still minimal compared to that achievable if the system had been used for a longer period.

We analyzed the changes made by R-Learning to the optimal policy by looking at the state-action values of the best actions, i.e., the actions with the highest value, for each of the possible initial states. The best action, computed by R-Learning, changed for 29.8% of the initial states. This clearly indicates that RLradio RL had a different channel selection strategy. This is confirmed by observing the log files and the state-to-action map of the learned policy. Several different patterns regarding the sequence of the suggested music channels could be detected. For instance, there is a clear preference of the system to keep proposing a channel if the user does not use the *next* button. We also observed the tendency to switch to another channel, which received

Table 3: Mean and standard deviation (SD) values of the responses to the usability survey

Stat.	P (SD)	RL (SD)	p-value
S1	4.13 (0.43)	4.03 (0.56)	0.317
S2	3.33 (0.92)	3.50 (0.82)	0.519
S3	3.70 (0.92)	4.13 (0.82)	**0.012**
S4	4.63 (0.49)	4.70 (0.47)	0.527

Table 4: Evaluation results summary

Result	P	RL	Imprv.
Avg. track's listening time percentage	64.35%	67.41%	+4.76%
Avg. daily listening time	62.6	75.5	+12.9 min
Percentage of users that listened to the suggested tracks longer while using the system vs. the other system	36.67%	63.33%	n.a.
Intention to use the system	3.70	4.13	p-value 0.012

by the user some initial explicit preference, when the user starts using the *next* button.

Finally, we analyzed the answers of the surveys by performing the related-samples Wilcoxon signed-rank test on the results of the (within-group) users' evaluations, as shown in Table 3.

As the user interfaces are absolutely identical for the two compared systems, we did not expect notable changes regarding their usability. Overall, RLradio (P and RL) received a very good evaluation, especially with respect to the quality of the played music (S1) and the ease of use (S4), and, as expected, with no significant difference between the two systems. Regarding the channel switching time (S2), we did measure an improvement, although it is not significant. We measured a statistically significant improvement of RLradio RL for statement S3 ("I would use such a player if available."). This result is very important as it represents the users' likeliness of using the player, and shows that the users had generally a higher satisfaction level when using System RL.

In conclusion, the evaluation results and the improvement of RLradio RL over RLradio P are summarized in Table 4.

6. RELATED WORK

The importance of implicit feedback has been recognized by some researchers who noted that the acquisition of user preferences, or explicit feedback (e.g., ratings), which is the most popular type of input for a RS, requires a remarkable user effort [12, 9]. Moreover, explicit feedback data is not always available, thus implicit feedback can come up as a necessary alternative to build the user profile.

In [9] the authors try to characterize explicit and implicit feedback on Last.fm. The dataset consists of explicit positive feedback (loved tracks) and implicit positive feedback (the number of times a track is played). Their analysis shows

that explicit feedback is very scarce. Moreover, they found that the rate at which a user provides explicit feedback decreases with time, and that overall, leaving explicit feedback has a negative effect on the user's behavior. [14] presents an analysis of the functional dependency between implicit and explicit feedback, coming to the conclusion that the prediction of a user's explicit feedback (ratings), starting from implicit feedback, can be made with an acceptable level of accuracy. [8] presents an extension of matrix factorization to incorporate implicit feedback (whether the user bought or not an item) into the optimization process. They show that a significant improvement of the RS accuracy can be achieved.

Regarding the second main issue tackled in our paper, i.e., the generation of a meaningful sequence of recommendations, [16] was the first to model sequential recommendation as a Markov Decision Process (MDP). They introduced a state model that represents previous system recommendations and user selections. Then, they used a model-based reinforcement learning algorithm to learn an optimal policy, showing the benefit of such approach in a book RS.

[10] applies MDP and reinforcement learning for selecting the next conversational act that the system must follow. The goal is to adapt the recommendation process, and it is shown that in this way the effectiveness of the interaction is improved, as measured by the ratio of successful sessions (when the user actually bought something) over the full set of recommendation sessions. In particular, it is shown that by leveraging user actions (implicit feedback), the system can reduce the usage of explicit feedback, e.g., in that application, the amount of travel specific preferences.

The most similar approach to generate a sequence of music track recommendations is probably presented in [13]. Here, an approach to automatically generate a playlist given a song to start with (seed song), and the user's immediate feedback to the system proposals is presented and evaluated. The authors use an audio-based similarity measure to generate recommendations. The user gives feedback by pressing a skip button if the user dislikes the current song. Songs similar to the skipped songs are removed, while songs similar to the accepted ones are added to the playlist. It is shown that by using audio similarity and simple heuristics it is possible to drastically reduce the number of necessary skips. We stress again that in our scenario the recommender cannot select the next track but only the next channel, hence it cannot choose any track among a set of available ones, as in this system.

RL approaches have been applied to the music domain in the past by [7], but not with the aim of proposing a sequence of music tracks to be listened to. The music sequencing problem was explored in [4], where the authors present a mobile music recommender for a group of users. Finally, in [3] the authors model sequential music recommendation as a case-based reasoning problem. To select each song in the sequence, they search for songs musically associated with the last song of the sequence. Then, the preferences of the audience, expressed as cases, are reused to customize the selection.

In conclusion, although RL techniques have been explored in the past in RSs research, there is no known approach that exploits them to deal with sequential music recommendation and that integrates both explicit and implicit feedback as we propose.

7. CONCLUSIONS

In this paper we have presented RLradio, a novel type of music recommender system, that learns how to automatically switch channel from a collection of radio channels, and hence offering more useful music to the user, i.e., music that the user listen more. RLradio is a system sitting on the user side, not the provider side, as it is more common in currently available recommender systems [11]. RLradio exploits and combines explicit channel preferences and implicit feedback in a MDP that is solved with reinforcement learning. Implicit feedback is generated by the user when he requests a new track, from the current or a different channel, while listening to music in the current channel.

The users tested both a baseline system that chooses the next channel to be played next by considering only the user's channels' preferences, and a system extending the baseline with reinforcement learning that is exploiting implicit feedback. The proposed RL-based approach has shown a significant performance improvement with respect to the baseline. Overall, the users listened to a larger part of the recommended tracks and their daily listening sessions were longer. Hence, the proposed approach can minimize the user intervention, i.e., requesting a new track from the available channels, while the radio is playing music that the user does not like. The subjective evaluation of the two compared system variants also showed a significant improvement of the users' intention to use the reinforcement learning based system over the baseline.

The proposed techniques are not domain specific. In this paper we focussed on a radio player, but we are developing a new application of the same techniques in the domain of adaptive document presentation. The goal is to find a meaningful and effective way to present chunks of a document as the user reads the presented chunks and requests a new one. We are developing a system that adaptively presents information about the patient disease in a hospital information system.

As we have mentioned already, RLradio is not currently designed to choose a channel based on information about the track that is going to be played. But this information could be available in some Internet radios or in a different application scenario. Therefore, in the future we plan to investigate ways to combine reinforcement learning and collaborative filtering techniques (CF) to extend channel selection with actual track selection. Such hybridization would allow to generate accurate user preference estimation using CF, and would effectively solve the exploration/exploitation tradeoff using RL.

8. REFERENCES

[1] G. Adomavicius, B. Mobasher, F. Ricci, and A. Tuzhilin. Context-aware recommender systems. *AI Magazine*, 32(3):67–80, 2011.

[2] G. Adomavicius and A. Tuzhilin. An architecture of e-butler: A consumer-centric online personalization system. *International Journal of Computational Intelligence and Applications*, 2(3):313–327, 2002.

[3] C. Baccigalupo and E. Plaza. A case-based song scheduler for group customised radio. In *7th International Conference on Case-Based Reasoning, ICCBR 2007, Belfast, Northern Ireland, UK, August 13-16, 2007, Proceedings*, pages 433–448, 2007.

[4] L. Baltrunas, M. Kaminskas, B. Ludwig, O. Moling, F. Ricci, A. Aydin, K.-H. Lüke, and R. Schwaiger. Incarmusic: Context-aware music recommendations in a car. In *E-Commerce and Web Technologies - 12th International Conference, EC-Web 2011, Toulouse, France, August 30 - September 1, 2011. Proceedings*, pages 89–100. Springer, 2011.

[5] R. Burke. Hybrid web recommender systems. In *The Adaptive Web*, pages 377–408. Springer Berlin / Heidelberg, 2007.

[6] Ò. Celma and P. Lamere. If you like radiohead, you might like this article. *AI Magazine*, 32(3):57–66, 2011.

[7] N. Collins. The potential of reinforcement learning for live musical agents. In *Proceedings of ICMC2008*, Belfast, 2008.

[8] Y. Hu, Y. Koren, and C. Volinsky. Collaborative filtering for implicit feedback datasets. In *ICDM '08: Proceedings of the 2008 Eighth IEEE International Conference on Data Mining*, pages 263–272. IEEE Computer Society, December 2008.

[9] G. Jawaheer, M. Szomszor, and P. Kostkova. Characterisation of explicit feedback in an online music recommendation service. In *Proceedings of the 2010 ACM Conference on Recommender Systems, RecSys 2010, Barcelona, Spain, September 26-30, 2010*, pages 317–320, 2010.

[10] T. Mahmood, F. Ricci, and A. Venturini. Improving recommendation effectiveness by adapting the dialogue strategy in online travel planning. *Journal of Information Technology and Tourism*, 11:285–302, 2010.

[11] F. J. Martin, J. Donaldson, A. Ashenfelter, M. Torrens, and R. Hangartner. The big promise of recommender systems. *AI Magazine*, 32(3):19–27, 2011.

[12] D. Oard and J. Kim. Implicit feedback for recommender systems. In *Proceedings of the AAAI Workshop on Recommender Systems*, pages 81–83, 1998.

[13] E. Pampalk, T. Pohle, and G. Widmer. Dynamic playlist generation based on skipping behavior. In *ISMIR 2005, 6th International Conference on Music Information Retrieval, London, UK, 11-15 September 2005, Proceedings*, pages 634–637, 2005.

[14] D. Parra and X. Amatriain. Walk the talk - analyzing the relation between implicit and explicit feedback for preference elicitation. In *User Modeling, Adaption and Personalization - 19th International Conference, UMAP 2011, Girona, Spain, July 11-15, 2011. Proceedings*, pages 255–268, 2011.

[15] F. Ricci, L. Rokach, and B. Shapira. Introduction to recommender systems handbook. In F. Ricci, L. Rokach, B. Shapira, and P. Kantor, editors, *Recommender Systems Handbook*, pages 1–35. Springer Verlag, 2011.

[16] G. Shani, D. Heckerman, and R. I. Brafman. An mdp-based recommender system. *Journal of Machine Learning Research*, 6:1265–1295, 2005.

[17] R. S. Sutton and A. G. Barto. *Reinforcement Learning: An Introduction*. MIT Press, 1998.

Alternating Least Squares for Personalized Ranking

Gábor Takács [*]
Széchenyi István University, Dept. of
Mathematics and Computer Science
Egyetem tér 1.
Győr, Hungary
gtakacs@sze.hu

Domonkos Tikk
Gravity Research and Development Ltd.
Expo tér 5–7.
Budapest, Hungary
domonkos.tikk@gravityrd.com

ABSTRACT

Two flavors of the recommendation problem are the explicit and the implicit feedback settings. In the explicit feedback case, users rate items and the user–item preference relationship can be modelled on the basis of the ratings. In the harder but more common implicit feedback case, the system has to infer user preferences from indirect information: presence or absence of events, such as a user viewed an item. One approach for handling implicit feedback is to minimize a ranking objective function instead of the conventional prediction mean squared error. The naive minimization of a ranking objective function is typically expensive. This difficulty is usually overcome by a trade-off: sacrificing the accuracy to some extent for computational efficiency by sampling the objective function. In this paper, we present a computationally effective approach for the direct minimization of a ranking objective function, without sampling. We demonstrate by experiments on the Y!Music and Netflix data sets that the proposed method outperforms other implicit feedback recommenders in many cases in terms of the ErrorRate, ARP and Recall evaluation metrics.

Categories and Subject Descriptors

I.2.6 [**Artificial Intelligence**]: Learning—*parameter learning*

Keywords

alternating least squares, ranking, collaborative filtering

1. INTRODUCTION

The goal of recommender systems is to provide personalized recommendations on items to users [1]. The basis of the recommendation is user event history related to items, and metadata about users and items. Two flavors of the recommendation problem are the explicit and the implicit feedback settings.

In the *explicit feedback* case, users rate items, therefore the preference relationship between a set of user–item pairs is directly known. A famous example of the explicit feedback setting is the movie recommendation problem of the Netflix Prize competition [2]. A typical characteristic of explicit feedback problems is that user–item pairs where no rating is present are not necessarily needed for modeling users' taste.

In many application domains, however, ratings are not available. The recommender system has to infer user preferences from *implicit feedback* [4], such as the presence or absence of purchase, click, or search events. A significant part of the information is contained in the absence of events, therefore the system has to consider each user–item pair. This makes the implicit feedback problem substantially harder than the explicit feedback one. This paper deals with the implicit feedback variant of the recommendation problem.

Many of the known methods for giving recommendation can be classified as prediction based or ranking based. *Prediction based* methods (e.g. [4]) try to predict the presence or absence of interaction between users and items. *Ranking based* methods (e.g. [5]) try to model users' choice between item pairs. In both cases, model building (training) is typically executed as the minimization of an objective function.

In the implicit feedback case, the naive minimization of the objective function is expensive. This difficulty is usually overcome by a trade-off: sacrificing the accuracy to some extent for computational efficiency by sampling the objective function.

In this paper, we propose RankALS, a computationally effective approach for the direct minimization of a ranking objective function without sampling, that is able to cope with the implicit feedback case. We show an efficient implementation of this method; for that we apply a similar but more complex derivation as used in ImplicitALS [4]. We also prove that the time complexity of RankALS is identical with that of ImplicitALS. We perform experiments on two "implicitized" rating data sets, Y!Music and Netflix. In a comparison we show that RankALS outperforms ImplicitALS in most cases in terms of three evaluation metrics, and it is usually better than ranking based methods employing the sampling strategy.

The rest of the paper is organized as follows. First, the notation is introduced. In Section 2, prediction and ranking based objective functions are reviewed. Next, we survey related work and highlight ImplicitALS, the algorithm that inspired RankALS. Section 4 introduces RankALS; first, its

---

[*]Also affiliated with Gravity Research and Development Ltd., Expo tér 5–7, Budapest, Hungary.

main idea and its advantageous computational properties are shown, then pseudocode with explanation is presented. The last part discusses experiments and performance comparisons with other methods on two publicly available large-scale data sets.

1.1 Notation

U and I will denote the number of users and items. $\mathcal{U} = \{1, \ldots, U\}$ and $\mathcal{I} = \{1, \ldots, I\}$ will denote the set of items and users respectively. We use $u \in \mathcal{U}$ as index for users, and $i, j \in \mathcal{I}$ as indices for items.

The implicit rating of user u on item i is r_{ui}, and its prediction is $\hat{r}_{ui}$. Implicit ratings r_{ui} are arranged in the matrix $R \in \mathbb{R}^{U \times I}$. $\mathcal{T}$ denotes the set of (u, i) indexes of R where a positive feedback is provided from user u on item i. We assume that the implicit rating value is 0 for negative feedback (i.e. $r_{ui} = 0$, if $(u, i) \notin \mathcal{T}$) and positive for positive feedback. We assume that users give positive feedback for items at most once.

$T = |\mathcal{T}|$ denotes the number of positive feedbacks. $\mathcal{I}_u = \{i : (u, i) \in \mathcal{T}\}$ denotes the set of items for which a positive feedback is provided by user u. $\mathcal{U}_i = \{u : (u, i) \in \mathcal{T}\}$ denotes the set of user who provided positive feedback for item i. $z_u = |\mathcal{I}_u|$ denotes the number of positive feedbacks of user u and $z \in \mathbb{R}^U$ denotes the vector of $\mathcal{I}_u$ values.

We also employ the usual matrix algebraic notations $\text{diag}(x)$, for the diagonal matrix containing the elements of vector x in its main diagonal, and $\text{tr}(X)$ for the trace of square matrix X. In matrix algebraic expressions, 1 means a vector of all ones of appropriate size.

$X_{[\mathcal{S}]}$ denotes the submatrix of matrix X selected by the row index set $\mathcal{S}$, and $x_{[\mathcal{S}]}$ denotes the subvector of vector x selected by the index set $\mathcal{S}$. $X_{[\mathcal{S}_1, \mathcal{S}_2]}$ denotes the submatrix of X selected by the row index set $\mathcal{S}_1$ and the column index set $\mathcal{S}_2$. Submatrix selection has precedence over the transpose operation, i.e. $X_{[\ldots]}^T$ is evaluated as $(X_{[\ldots]})^T$.

2. OBJECTIVE FUNCTIONS

Two major approaches for the recommendation problem are prediction based and ranking based recommendation. The debate on which approaches is better and when one is preferred to the other is still active. In the next subsections, we define objective functions for the two approaches. We will omit regularization terms in order to keep the discussion shorter.

2.1 Prediction based objective function

The objective function associated with a simple prediction based approach can be defined as

$$f_P(\Theta) = \sum_{u \in \mathcal{U}} \sum_{i \in \mathcal{I}} c_{ui} (\hat{r}_{ui} - r_{ui})^2,$$

where Θ contains the parameters of the prediction model. [1] The $c_{ui} > 0$ values are parameters of the objective function that have to be fixed in advance. The goal of model building is to make the sum of weighted squared differences between the predicted and true ratings small. c_{ui} means the extent to which we penalize the error on user u and item i.

The standard choice for c_{ui} in the explicit feedback case is $c_{ui} = 1$, if $(u, i) \in \mathcal{T}$ and 0 otherwise, which means that

the objective function contains only T terms. In the implicit feedback case, this simplification is not reasonable, because a significant part of the information is contained in the absence of positive feedback.

A typical choice for c_{ui} in the implicit feedback case is $c_{ui} = c^+$, if $(u, i) \in \mathcal{T}$ and c^- otherwise. The values c^+ and c^- can be interpreted as confidence levels associated with positive and negative feedbacks respectively. We will refer to this specific variant of the prediction based objective function as f_I (I stands for implicit). The function f_I contains $U \cdot I$ terms, therefore its naive minimization is expensive.

Usually, $c^+ \gg c^-$, which means that the presence of positive feedback has to be predicted more accurately than the absence of positive feedback. To see the rationale behind this asymmetry, consider a webshop scenario. If user u has bought item i that we can consider this as a positive feedback with high confidence. If i is not present in the purchase history of u, then we can treat this as a negative feedback from user u on item i, but only with low confidence.

Note that the interpretation of implicit feedback data may not necessarily reflect user satisfaction. Even the presence of positive feedback may be misleading: (1) a purchased item could be disappointing for the user but the purchase information is generated before the user could evaluate the product; (2) the user may purchase the item for someone else, therefore her selection only reflects her knowledge and assumption on the preference of another person. The interpretation of the absence of positive feedback is even more speculative: we cannot directly interpret missing navigational or purchase information as an intentional negative feedback; in the implicit setting direct negative feedback is not available.

2.2 Ranking based objective function

The objective function associated with a simple ranking based approach can be defined as

$$f_R(\Theta) = \sum_{u \in \mathcal{U}} \sum_{i \in \mathcal{I}} c_{ui} \sum_{j \in \mathcal{I}} s_j \left[(\hat{r}_{ui} - \hat{r}_{uj}) - (r_{ui} - r_{uj}) \right]^2,$$

where the c_{ui} and s_j values are parameters of the objective function. This function was introduced by [5] in the recommender system literature (although other ranking based objectives appeared earlier). Here we only consider the implicit feedback case and assume that $c_{ui} = 0$ if $r_{ui} = 0$, and 1 otherwise. The role of c_{ui} is to select user–item pairs corresponding to positive feedbacks from all possible pairs. The meaning of s_j is the importance weight of the j-th item in the objective function. Note that the number of terms in f_R is $T \cdot I$, therefore the naive minimization of f_R is expensive.

Note that there is an asymmetric relation between the objective functions f_I and f_R. A good solution with respect to f_I will also be a good solution with respect to f_R, but the opposite is not necessarily true. One can also observe that user-only dependent terms in $\hat{r}_{ui}$ do not change the value of f_R, in other words, f_R is invariant to user biases.

---

[1] Note that $\hat{r}_{ui}$ depends on Θ, but for brevity, we will not write $\hat{r}_{ui}(\Theta)$ explicitly.

3. RELATED WORK

The literature of explicit feedback based recommendation algorithms is rich. We do not highlight any approach here, since this paper focuses on implicit feedback, just direct the interested reader to the survey work [10].

The area of implicit feedback based recommendation is much less explored. One of the earliest solutions for handling implicit feedback is Hu et al.'s ImplicitALS [4] that consists of a matrix factorization model, a prediction based objective function, and an alternating least squares (ALS) optimizer, tuned for the problem (see a more detailed discussion in Section 3.1).

A faster, approximative version of ImplicitALS was proposed by Pilászy et al. [8]. The speedup is achieved by replacing the exact least squares solver by a coordinate descent method. The authors reported on a marginal loss of accuracy compared to the significant decrease in training time in their experiments.

Pan et al. [7] proposed two prediction based frameworks for handling implicit feedback. The first one is similar to ImplicitALS, but it contains a naive ALS optimizer instead of a tuned one. The second one is based on negative example sampling.

Recently, the results of Track 2 at KDD Cup 2011 [3] have shown an evidence that ranking based approaches are able to handle implicit feedback efficiently on large-scale problems. Ranking based methods had significant weight in the winning solution [6], and they were the most important components of the second runner-up solution [5]. In particular, the key idea of the latter is to define a ranking based objective function, and apply a stochastic gradient descent (SGD) optimizer on a sampled approximation of the objective function.

Another ranking based approach for implicit feedback is Rendle et al's Bayesian Personalized Ranking (BPR) [9]. The objective function of BPR is derived from the Bayesian analysis of the problem. The optimization technique used for training is bootstrapping based SGD.

Our proposed RankALS method differs from all of the above approaches in various aspects:

- Unlike [4], [8] and [7], it is ranking based.

- Unlike [9] and the ranking based methods in [5] and [6], it does not approximate the original objective function, and applies ALS for training instead of SGD.

3.1 ALS for prediction

Here we overview the main points of Hu et al.'s ImplicitALS [4] that can be considered as a straight predecessor of our proposed RankALS method. The prediction formula of ImplicitALS for user u and item i is $\hat{r}_{ui} = p_u^T q_i$. The parameters of the model $p_1, \ldots, p_U \in \mathbb{R}^F$ and $q_1, \ldots, q_I \in \mathbb{R}^F$ are called user and item feature vectors; F denotes the number of features. [2]

Let P and Q denote the matrices that contain user and item feature vectors as rows. The objective function associated with the model is f_I, thus the goal of ImplicitALS is to approximate the rating matrix as $R \approx PQ^T$ so that the weighted squared error of the approximation is low.

The key idea of ImplicitALS is that although the objective function consists of $U \cdot I$ terms, its alternating least squares

---

[2] We omitted user and item bias parameters, because they are not necessary for illustrating the ideas.

Algorithm 1: Alternating least squares based training.

> Initialize Q with small random numbers.
> **for** E *times* **do**
>   Compute the P that minimizes f_I for fixed Q.
>   Compute the Q that minimizes f_I for fixed P.
> **end**

(ALS) based minimization can be done efficiently. The outline of the training of ImplicitALS is given in Algorithm 1.

To see why this scheme can be implemented efficiently, let us rewrite the derivative of f_I with respect to p_u as

$$\frac{\partial f_I(P, Q)}{\partial p_u} = \sum_{i \in \mathcal{I}} c_{ui} \left(q_i^T p_u - r_{ui} \right) q_i =$$

$$\underbrace{\left(\sum_{i \in \mathcal{I}} c_{ui} q_i q_i^T \right)}_{\bar{A}_u} p_u - \underbrace{\sum_{i \in \mathcal{I}} c_{ui} r_{ui} q_i^T}_{\bar{b}_u} = \bar{A}_u p_u - \bar{b}_u.$$

Minimizing f_I in P means that we make the derivatives vanish by setting p_u to $\bar{A}_u^{-1} \bar{b}_u$ for all u. Note that $\bar{b}_u$ can be simplified to $\sum_{i \in \mathcal{I}_u}$ and $\bar{A}_u$ can be decomposed into a user-independent and a user-dependent part as $\bar{A}_u = \sum_{i \in \mathcal{I}} c^- q_i q_i^T + \sum_{i \in \mathcal{I}_u} (c^+ - c^-) q_i q_i^T$. As a consequence, the evaluation of $\bar{A}_u^{-1} \bar{b}_u$ for all u can be done in $O(TF^2 + UF^3)$ time.

Applying the same argument for items yields that item feature vectors can be updated in $O(TF^2 + IF^3)$ time. Therefore, the computational cost of one training step is $O(TF^2 + (U+I)F^3)$ which means much less operations than iterating over the terms of the objective function.

A nice property of ImplicitALS is that it does not replace objective function by an approximation, as other approaches often do. It achieves speedup by applying mathematical simplification.

4. ALS FOR RANKING

This section presents RankALS, our proposed method for personalized ranking. The prediction formula of RankALS is $\hat{r}_{ui} = p_u^T q_i$ as in the case of ImplicitALS. The difference is that the function to minimize is now the ranking objective function f_R.

Recall that the ranking objective function contains $T \cdot I$ terms. We will show that even though f_R contains more and more complex terms than f_I, its ALS based minimization can be done at the same asymptotic computational cost. We will use a similar ideas as for ImplicitALS to break down the computational complexity of the naive implementation of RankALS, but the mathematical derivation and the resulting training algorithm will be more complex.

At first, let us rewrite the derivative of f_R with respect to the user feature vector p_u.

$$\frac{\partial f_R(P,Q)}{\partial p_u} =$$

$$\sum_{i\in\mathcal{I}} c_{ui} \sum_{j\in\mathcal{I}} s_j \left[(q_i - q_j)^T p_u - (r_{ui} - r_{uj}) \right] (q_i - q_j) =$$

$$\underbrace{\left(\sum_{j\in\mathcal{I}} s_j\right)}_{\tilde{1}} \underbrace{\left(\sum_{i\in\mathcal{I}} c_{ui} q_i q_i^T\right)}_{\bar{A}} p_u - \underbrace{\left(\sum_{i\in\mathcal{I}} c_{ui} q_i\right)}_{\bar{q}} \underbrace{\left(\sum_{j\in\mathcal{I}} s_j q_j^T\right)}_{\tilde{q}^T} p_u -$$

$$\underbrace{\left(\sum_{j\in\mathcal{I}} s_j q_j\right)}_{\tilde{q}} \underbrace{\left(\sum_{i\in\mathcal{I}} c_{ui} q_i^T\right)}_{\bar{q}^T} p_u + \underbrace{\left(\sum_{i\in\mathcal{I}} c_{ui}\right)}_{\bar{1}} \underbrace{\left(\sum_{j\in\mathcal{I}} s_j q_j q_j^T\right)}_{\tilde{A}} p_u -$$

$$\underbrace{\left(\sum_{i\in\mathcal{I}} c_{ui} q_i r_{ui}\right)}_{\bar{b}} \underbrace{\left(\sum_{j\in\mathcal{I}} s_j\right)}_{\tilde{1}} + \underbrace{\left(\sum_{i\in\mathcal{I}} c_{ui} q_i\right)}_{\bar{q}} \underbrace{\left(\sum_{j\in\mathcal{I}} s_j r_{uj}\right)}_{\tilde{r}} +$$

$$\underbrace{\left(\sum_{i\in\mathcal{I}} c_{ui} r_{ui}\right)}_{\bar{r}} \underbrace{\left(\sum_{j\in\mathcal{I}} s_j q_j\right)}_{\tilde{q}} - \underbrace{\left(\sum_{i\in\mathcal{I}} c_{ui}\right)}_{\bar{1}} \underbrace{\left(\sum_{j\in\mathcal{I}} s_j q_j r_{uj}\right)}_{\tilde{b}} =$$

$$\left(\tilde{1}\bar{A} - \bar{q}\tilde{q}^T - \tilde{q}\bar{q}^T + \bar{1}\tilde{A}\right) p_u -$$

$$\left(\bar{b}\tilde{1} - \bar{q}\tilde{r} - \bar{r}\tilde{q} + \bar{1}\tilde{b}\right).$$

We converted the expression of the derivative to 8 double sums by expanding the inner term. Then, we factorized each double sum in order to make the evaluation of the expression more efficient. The advantage of this rewritten form of the derivative is that computing the statistics $\bar{A}$, $\bar{b}$, $\bar{q}$, $\bar{r}$, $\bar{1}$, $\tilde{A}$, $\tilde{b}$, $\tilde{q}$, $\tilde{r}$, $\tilde{1}$ for all u can be done in $O(TF^2)$ time, therefore the evaluation of $\frac{\partial f_R(P,Q)}{\partial P}$ takes $O(TF^2)$ time. Note that except $\tilde{A}$, $\tilde{q}$ and $\tilde{1}$ the statistics depend on u, but we do not subscript them by u in order to make formulae cleaner.

Now let us rewrite the derivative of f_R with respect to q_i.

$$\frac{\partial f_R(P,Q)}{\partial q_i} =$$

$$\sum_{u\in\mathcal{U}} \sum_{j\in\mathcal{I}} c_{ui} s_j p_u \left[p_u^T (q_i - q_j) - (r_{ui} - r_{uj}) \right] +$$

$$\sum_{u\in\mathcal{U}} \sum_{j\in\mathcal{I}} c_{uj} s_i (-p_u) \left[p_u^T (q_j - q_i) - (r_{uj} - r_{ui}) \right] =$$

$$\underbrace{\left(\sum_{u\in\mathcal{U}} c_{ui} p_u p_u^T\right)}_{\bar{A}} \underbrace{\left(\sum_{j\in\mathcal{I}} s_j\right)}_{\tilde{1}} q_i - \underbrace{\left(\sum_{u\in\mathcal{U}} c_{ui} p_u p_u^T\right)}_{\bar{A}} \underbrace{\left(\sum_{j\in\mathcal{I}} s_j q_j\right)}_{\tilde{q}} -$$

$$\underbrace{\left(\sum_{u\in\mathcal{U}} c_{ui} p_u r_{ui}\right)}_{\bar{b}} \underbrace{\left(\sum_{j\in\mathcal{I}} s_j\right)}_{\tilde{1}} + \underbrace{\left(\sum_{u\in\mathcal{U}} c_{ui} \overbrace{\left(\sum_{j\in\mathcal{I}} s_j r_{uj}\right)}^{\tilde{r}} p_u\right)}_{\bar{\bar{p}}_1} -$$

$$\underbrace{\left(\sum_{u\in\mathcal{U}} p_u p_u^T \overbrace{\left(\sum_{j\in\mathcal{I}} c_{uj} q_j\right)}^{\bar{q}}\right)}_{\bar{\bar{p}}_2} s_i + \underbrace{\left(\sum_{u\in\mathcal{U}} p_u p_u^T \overbrace{\left(\sum_{j\in\mathcal{I}} c_{uj}\right)}^{\bar{1}}\right)}_{\bar{A}} s_i q_i +$$

$$\underbrace{\left(\sum_{u\in\mathcal{U}} \overbrace{\left(\sum_{j\in\mathcal{I}} c_{uj} r_{uj}\right)}^{\bar{r}} p_u\right)}_{\bar{\bar{p}}_3} s_i - \underbrace{\left(\sum_{u\in\mathcal{U}} p_u r_{ui} \overbrace{\left(\sum_{j\in\mathcal{I}} c_{uj}\right)}^{\bar{1}}\right)}_{\bar{\bar{b}}} s_i =$$

$$\left(\bar{A}\tilde{1} + \bar{\bar{A}} s_i\right) q_i + \left(-\bar{A}\tilde{q} - \bar{b}\tilde{1} + \bar{\bar{p}}_1 - \bar{\bar{p}}_2 + \bar{\bar{p}}_3 s_i - \bar{\bar{b}} s_i\right).$$

We applied the same idea as before. The expression of the derivative was converted into 8 double sums, and then the double sums were factorized. Again, computing the statistics $\bar{A}$, $\bar{b}$, $\tilde{1}$, $\tilde{q}$, $\tilde{1}$, $\bar{\bar{A}}$, $\bar{\bar{b}}$, $\bar{\bar{p}}_1$, $\bar{\bar{p}}_2$, $\bar{\bar{p}}_3$ for all i can be done in $O(TF^2)$ time, therefore the evaluation of $\frac{\partial f_R(P,Q)}{\partial Q}$ takes $O(TF^2)$ time.

Note that $\bar{A}$ and $\bar{b}$ are defined differently as in the user-derivative. Also note that except $\bar{\bar{A}}$, $\tilde{q}$ and $\tilde{1}$ the statistics depend on i or u, but we avoided subscripting them.

The previous results imply that the ALS based minimization of f_R with a matrix factorization model can be implemented efficiently. The two main steps of ALS are evaluating the derivatives and setting them to zero. We have seen that the first step can be speeded up by rewriting the derivatives. The second step is relatively cheap, because it consists of solving linear systems.

Now we are ready to put blocks together and specify the model building procedure of RankALS. The pseudocode of RankALS training can be seen in Algorithm 2.

Each iteration consists of a P-step and a Q-step. Most of the pseudocode deals with the calculation of statistics needed for derivative evaluation. The statistic $\tilde{1}$ is constant during the algorithm, therefore it is computed before the main loop.

The P-step starts with calculating the statistics $\tilde{q}$ and $\tilde{A}$ that are common for each user. Then, for each user u, the user-dependent statistics $\bar{1}$, $\bar{r}$, $\bar{q}$, $\bar{b}$, $\bar{A}$, $\tilde{r}$, and $\tilde{b}$ are computed, and p_u is updated by setting the derivative $\frac{\partial f_R(P,Q)}{\partial p_u}$ to zero.

The Q-step starts with calculating the item-independent statistics $\tilde{q}$, $\tilde{A}$ and $\bar{p}$. Next, the more complex item-independent statistics $\tilde{r}$, $\bar{\bar{p}}_2$ and $\bar{\bar{p}}_3$ are calculated by iterating over users. Then, for each item i, the item-dependent statistics $\bar{b}$, $\bar{A}$, $\bar{\bar{b}}$ and $\bar{\bar{p}}_1$ are computed, and q_i is updated by setting the derivative $\frac{\partial f_R(P,Q)}{\partial q_i}$ to zero.

The tunable parameters of the algorithm are as follows:

- $E \in \mathbb{N}$: The number of iterations.

- $\sigma \in \mathbb{R}$: Initialization range parameter.

- $s \in \mathbb{R}^I$: Item importance weights in the objective function f_R. We experimented with two settings:

  (i) $s_i = 1$, i.e. items are equally important,

  (ii) $s_i = |\mathcal{U}_i|$, i.e. popular items are more important.

The computationally most expensive parts of the P-step are calculating $\bar{A}$ and updating p_u. Computing $\bar{A}$ and p_u for

---
Algorithm 2: Alternating least squares for ranking.

Input: $R \in \mathbb{R}^{U \times I}$, $E \in \mathbb{N}$, $\sigma \in \mathbb{R}$, $s \in \mathbb{R}^I$
Output: $P \in \mathbb{R}^{U \times F}$, $Q \in \mathbb{R}^{I \times F}$

$Q \leftarrow$ uniform random numbers from $[-\sigma, \sigma]$
$\tilde{1} \leftarrow \sum_{i \in \mathcal{I}} s_i$

for $e \leftarrow 1, \ldots, E$ **do**
   // P-step
   $\tilde{q} \leftarrow Q^T s$, $\tilde{A} \leftarrow Q^T \text{diag}(s) Q$

   for $u \leftarrow 1, \ldots, U$ **do** ;   // for each user
      $r_u \leftarrow R_{[\{u\}, \mathcal{I}_u]}^T$, $\bar{1} \leftarrow z_u$
      $\bar{r} \leftarrow r_u^T 1$, $\bar{q} \leftarrow Q_{[\mathcal{I}_u]}^T 1$
      $\bar{b} \leftarrow Q_{[\mathcal{I}_u]}^T r_u$, $\bar{A} \leftarrow Q_{[\mathcal{I}_u]}^T Q_{[\mathcal{I}_u]}$
      $\tilde{r} \leftarrow s_{[\mathcal{I}_u]}^T r_u$, $\tilde{b} \leftarrow Q_{[\mathcal{I}_u]}^T \text{diag}(s_{[\mathcal{I}_u]}) r_u$

      $M \leftarrow \tilde{1}\bar{A} - \bar{q}\tilde{q}^T - \tilde{q}\bar{q}^T + \bar{1}\tilde{A}$
      $y \leftarrow \bar{b}\tilde{1} - \bar{q}\tilde{r} - \bar{r}\tilde{q} + \tilde{1}\bar{b}$
      $p_u \leftarrow M^{-1} y$
   end

   -

   // Q-step
   $\tilde{q} \leftarrow Q^T s$, $\bar{\bar{A}} \leftarrow P^T \text{diag}(z) P$, $\bar{p} \leftarrow P^T z$

   $\tilde{r}, \bar{r}, \bar{Q} \leftarrow 0$
   for $u \leftarrow 1, \ldots, U$ **do** ;   // for each user
      $r_u \leftarrow R_{[\{u\}, \mathcal{I}_u]}^T$, $\tilde{r}_u \leftarrow r_u^T s_{[\mathcal{I}_u]}$
      $\bar{r} \leftarrow r_u^T 1$, $\bar{Q}_{[\{u\}]} \leftarrow Q_{[\mathcal{I}_u]}^T 1$
   end
   $\bar{\bar{p}}_2 \leftarrow P^T \text{tr}(P, \bar{Q})$, $\bar{\bar{p}}_3 \leftarrow P^T \bar{r}$

   for $i \leftarrow 1, \ldots, I$ **do** ;   // for each item
      $r_i \leftarrow R_{[\mathcal{U}_i, \{i\}]}$
      $\bar{b} \leftarrow P_{[\mathcal{U}_i]}^T r_i$, $\bar{A} \leftarrow P_{[\mathcal{U}_i]}^T P_{[\mathcal{U}_i]}$
      $\bar{\bar{b}} \leftarrow P_{[\mathcal{U}_i]}^T \text{diag}(z_{[\mathcal{U}_i]}) r_i$, $\bar{p}_1 \leftarrow P_{[\mathcal{U}_i]}^T \tilde{r}_{[\mathcal{U}_i]}$

      $M \leftarrow \bar{A}\tilde{1} + \bar{\bar{A}}s_i$
      $y \leftarrow \bar{A}\tilde{q} + \bar{b}\tilde{1} - \bar{p}_1 + \bar{\bar{p}}_2 - \bar{\bar{p}}_3 s_i + \bar{\bar{b}}s_i$
      $q_i \leftarrow M^{-1} y$
   end
end

all users takes $O(TF^2)$ and $O(UF^3)$ time respectively. The most expensive parts of the Q-step are calculating $\bar{A}$ and updating q_i Computing $\bar{A}$ and q_i for all items takes $O(TF^2)$ and $O(IF^3)$ time respectively. The overall complexity of one iteration is $O(TF^2 + (U + I)F^3)$, which is the same as for ImplicitALS.

RankALS training is a relatively complex algorithm, and implementing it without error is not trivial. We validated our code by also implementing naive derivative evaluation and comparing the two variants. Apart from the very tiny difference due to round-off errors, the two versions produced the same result.

5. EXPERIMENTS

We compared our proposed method with other approaches on "implicitized" versions of the Y!Music [3] and the Netflix [2] data sets. The characteristics of the data sets are as follows:

- **Y!Music**: The Y!Music data set for Track2 of KDD Cup 2011 consists of 62,551,438 ratings ranging from 0 to 100 from 249,012 users on 296,111 items. The items are arranged in a taxonomy: the leaf nodes are tracks, the higher level nodes are albums, artists and genres. We considered a rating as a positive implicit feedback, if the rating value was greater than or equal to 80. Moreover, we considered each node of the hierarchy as a separate item, and did not exploit the parent-child relationship between items. For training and testing we used the official Track2 training and test sets (containing 22,699,314 and 303,516 implicit ratings respectively).

- **Netflix**: The Netflix training set consists of 100,480,507 ratings ranging from 1 to 5 from 480,189 users on 17,770 items. We defined the implicit rating matrix by assigning 1 to user–item pairs with Netflix rating value 5, and 0 to other user–item pairs. As training and test sets, we used the Netflix training set minus the Netflix probe and the Netflix probe set (containing 22,783,659 and 384,573 implicit ratings respectively).

We compared the following methods:

- **Pop**: Item popularity based, non-personalized baseline method. Its prediction for user u and item i is $|\mathcal{U}_i|$, the number of positive feedbacks for i.

- **ImplicitALS**: The approach described in [4]. It uses an MF model, a prediction based objective function and an ALS optimizer.

- **RankSGD**: The approach described in [5] under the name SVD. It uses an MF model, a ranking based objective function and an SGD optimizer.

- **RankSGD2**: The approach described in [5] under the name AFM. It uses an asymmetric factor (NSVD1) model, a ranking based objective function and an SGD optimizer.

- **RankALS**: Our proposed method that uses an MF model, a ranking based objective function, and an ALS optimizer.

We measured the following performance indicators:

- **ErrorRate**: The evaluation metric used in Track 2 of KDD Cup 2011 [3]. For ErrorRate measurement, the test set contains two sets of items $\mathcal{I}_u^+$ and $\mathcal{I}_u^-$ for each user u. $\mathcal{I}_u^+$ ($\mathcal{I}_u^-$) contains items for that u provided (did not provide) positive feedback. The negative item set $\mathcal{I}_u^-$ is created by drawing $|\mathcal{I}_u^+|$ items randomly so that the probability of item i to be chosen is proportional to $|\mathcal{U}_i|$.[3] For each user u, the predicted rating of u for items in $\mathcal{I}_u^+ \cup \mathcal{I}_u^-$ is calculated and items are ranked based on these values. ErrorRate is the relative frequency of items from $\mathcal{I}_u^+$ not being among the top $|\mathcal{I}_u^+|$ items of the ranking. A lower ErrorRate value means better performance.

- **ARP**: The test set contains a set of preferred items $\mathcal{I}_u^+$ for each user u, and another set of items $\mathcal{I}'$ that is shared between users. Normally $\mathcal{I}'$ contains all items, but it can be a random subset, if there are too many items in the data set. For each user u, the elements of $\mathcal{I}'$ are ranked, based on prediction values. ARP is the average relative position of items from $\mathcal{I}_u^+$ in the ranking [8]. A lower ARP value indicates better performance.

- **Recall**: The structure of the test set, and the ranking procedure is the same as for ARP. Recall is the relative frequency of items from $\mathcal{I}_u^+$ being among the top K items of the ranking. A higher Recall value means better performance.

The set $\mathcal{I}'$ of ARP and Recall measurement contained all 17,770 items in the case of the Netflix data set, and 17,770 uniformly drawn items in the case of the Y!Music data set. For Y!Music, the sets $\mathcal{I}_u^-$ were taken from the official test set of KDD Cup 2011, Track 2. For Netflix, the sets $\mathcal{I}_u^-$ were drawn randomly, since no official negative item sets are available. The parameter K of Recall measurement was 50.

The learning rate was $\eta = 0.008$ for RankSGD and $\eta = 0.016$ for RankSGD2 in all experiments. We did not apply regularization for RankSGD, RankSGD, and RankALS.[4] For ImplicitALS and RankALS, training started with a Q-step and not with a P-step as in the pseudocodes. For SGD-based approaches we ran $E = 30$, for ALS-based approaches we ran $E = 10$ iterations. We stopped training, if the performance started to decrease on the test set. The confidence level parameters of the objective function f_I were set to $c^+ = 100$ and $c^- = 1$.

5.1 Support based weighting

The choice of the item weight vector s can greatly influence the behavior of the ranking based approaches (RankSGD, RankSGD2, RankALS). In the first experiment, we tried to find the appropriate variant of the methods for the different evaluation criteria. The results for the Y!Music and the Netflix data set are shown in Table 1 and 2. The column s.w. indicates if support based weighting was applied ($s_i = |\mathcal{U}_i|$) or not ($s_i = 1$).

The results suggest that support based weighting should be turned on if one optimizes for ErrorRate. This is not

---

[3]This rule could be relaxed. We apply it, because it was used in Track2 of KDD Cup 2011.

[4]According to [5], regularizing RankSGD and RankSGD2 models did not help too much in Track 2 of KDD Cup 2011.

method	s.w.	ErrorRate	ARP	Recall
RankSGD	no	0.1874	**0.0362**	0.2153
RankSGD	yes	**0.0617**	0.0387	**0.3574**
RankSGD2	no	0.1163	**0.0275**	0.3815
RankSGD2	yes	**0.0642**	0.0416	0.3333
RankALS	no	0.0961	**0.0220**	0.3719
RankALS	yes	**0.0529**	0.0375	0.2562

Table 1: Comparison of ranking based method variants on the Y!Music data set ($F = 20$)

method	s.w.	ErrorRate	ARP	Recall
RankSGD	no	0.3103	**0.0509**	0.1525
RankSGD	yes	**0.1861**	0.1234	**0.1770**
RankSGD2	no	0.2651	**0.0572**	0.1757
RankSGD2	yes	**0.2011**	0.1291	0.1589
RankALS	no	0.2878	**0.0478**	**0.1731**
RankALS	yes	**0.1842**	0.1278	0.1331

Table 2: Comparison of ranking based method variants on the Netflix data set ($F = 20$)

surprising, since the negative items of the test set are drawn with probability proportional to $|\mathcal{U}_i|$ in the case of ErrorRate. In further experiments with ErrorRate, we will always use the setting s.w. = yes.

If the evaluation criterion is ARP, then support based weighting should not be applied. This result was again expectable, because negative items of the test set are drawn uniformly in the case of ARP. In further experiments with ARP, we will always use the setting s.w. = no.

The results for Recall are diversified: support based weighting helps significantly in the case of RankSGD, but has negative effect for RankSGD2 and RankALS. In further experiments with Recall, we will always use the setting s.w. = no for RankSGD and RankALS, and s.w. = yes for RankSGD.

We also mention for comparison that the ImplicitALS approach with $F = 20$ achieves ErrorRate = 0.1154, ARP = 0.0291, Recall = 0.3181 on the Y!Music, and ErrorRate = 0.2732, ARP = 0.0494, Recall = 0.1835 on the Netflix data set. These numbers suggest that ImplicitALS tends to be less accurate for a given evaluation metric than the best ranking based method variant, but is is less sensitive to the choice of the metric.

5.2 Accuracy versus model size

In this experiment, we ran all methods on all datasets with different number of features. The results are shown in Figure 1–6. It can be observed that RankALS clearly outperforms other methods on the Y!Music data set in terms of ErrorRate and ARP independently from the number of features, while it is on par with SGD-based method according to Recall. On the Netflix data set, RankALS shows similar behavior, with the exception of $F = 50$ and ARP. Interestingly, on this data set ImplicitALS outperforms other methods for Recall.

5.3 Accuracy per iteration

In this experiment, we compared the convergence speed of RankSGD and RankALS. In particular, we investigated how the ErrorRate decreases with the number of iterations on the Y!Music data set. The number of features was set to $F =$

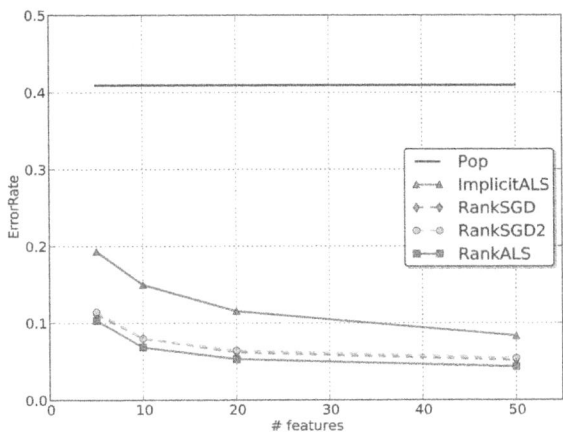

Figure 1: ErrorRate on the Y!Music data set

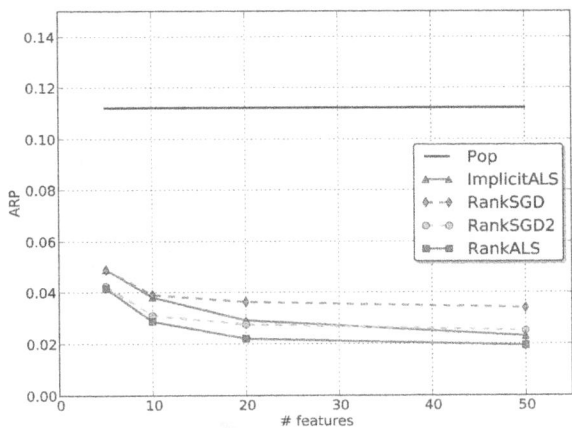

Figure 2: ARP on the Y!Music data set

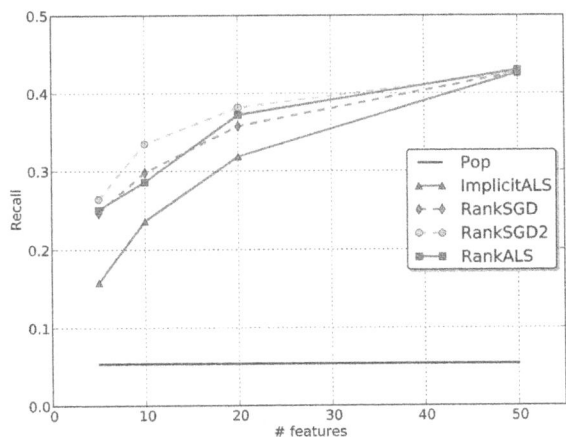

Figure 3: Recall on the Y!Music data set

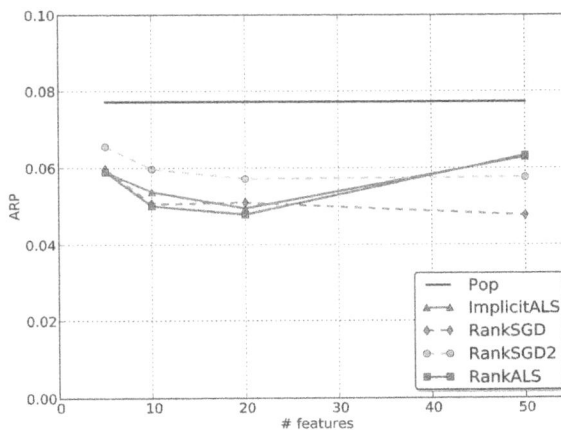

Figure 4: ErrorRate on the Netflix data set

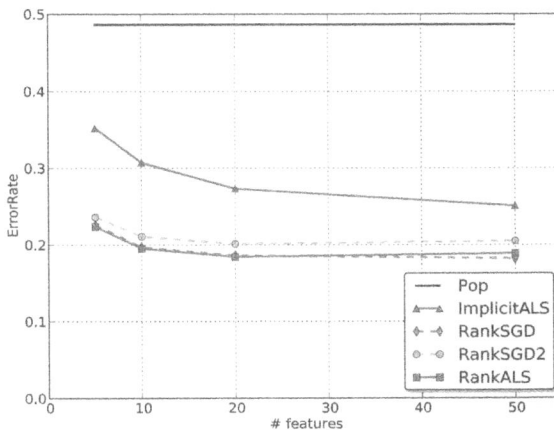

Figure 5: ARP on the Netflix data set

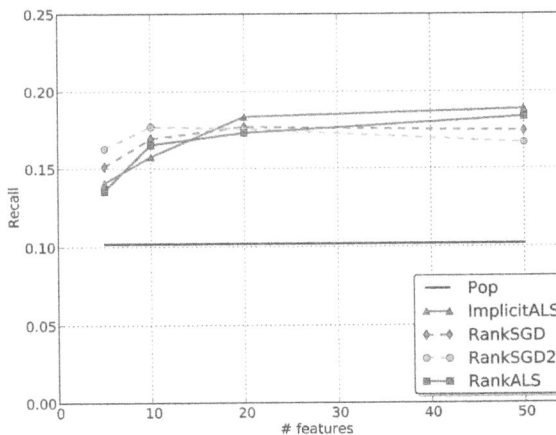

Figure 6: Recall on the Netflix data set

100 for both methods. The results are shown in Figure 7, where it can be observed that RankALS convergences faster.

5.4 Iteration time

We also compared the iteration times of the methods on the Y!Music data set. In the case of ImplicitALS and Rank-ALS an iteration contains both a P- and a Q-step. We used one core of a Intel Xeon E5405 CPU for the measurement. The results are shown in Table 3.

As expected, RankSGD and RankSGD2 methods that use

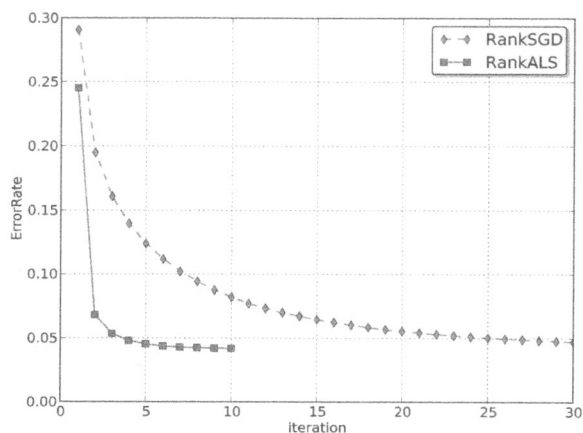

Figure 7: ErrorRate per iteration for RankSGD and RankALS on the Y!Music data set.

method	#5	#10	#20	#50	#100
ImplicitALS	82	102	152	428	1404
RankSGD	21	25	30	42	65
RankSGD2	26	32	40	61	102
RankALS	167	190	243	482	1181

Table 3: Time per iteration in seconds on the Y!Music data set (column labels refer to the number of features, F)

SGD and sampling are faster then ALS-based methods that optimize the original objective function. Note that SGD-based methods have the learning rate as additional parameter, which need to be adjusted appropriately; this introduces another optimization round into the parameter setting. We also mention that our RankSGD and RankSGD2 implementations were more optimized than our ImplicitALS and RankALS implementations.

6. CONCLUSION

Learning user–item preference from implicit feedback is a key problem of recommender systems research. One approach for handling implicit feedback is to minimize a ranking objective function instead of the conventional prediction mean squared error. Ranking objective functions are typically expensive to optimize, and this difficulty is usually overcome by sampling the objective function.

In this paper, we proposed a computationally efficient ranking based method RankALS that optimizes the original objective function, without sampling. RankALS was inspired by the prediction based method ImplicitALS [4]. The key components of RankALS are a matrix factorization model, a ranking based objective function, and an alternating least squares optimizer. Speedup is achieved by the mathematical simplification of the objective function's derivative.

As we have shown, RankALS outperforms the sampling based methods used in the comparison in terms of the ErrorRate and ARP in most cases, and is also on par with SGD-based methods in terms of Recall. Another advantage of the method is its robustness: it is not sensitive to the order of training examples (unlike SGD-based models) and

with the fewer parameters (no learning rate) its adaptation to a new data set can be done with less experiments.

There are various ways to extend the RankALS method. One possible direction is to make RankALS faster by replacing the exact least squares solver by an approximate one [8]. Furthermore, its modeling capability can be improved by adding context-awareness and introducing time-dependent or user/item metadata-dependent terms into the prediction formula.

7. REFERENCES

[1] G. Adomavicius and A. Tuzhilin. Toward the next generation of recommender systems: a survey of the state-of-the-art and possible extensions. *IEEE Transactions on Knowledge and Data Engineering*, 17:734–749, 2005.

[2] J. Bennett and S. Lanning. The Netflix Prize. In *Proc. of KDD Cup Workshop at 13th ACM SIGKDD Int. Conf. on Knowledge Discovery and Data Mining*, KDD '07, pages 3–6, San Jose, CA, USA, 2007.

[3] G. Dror, N. Koenigstein, Y. Koren, and M. Weimer. The Yahoo! Music dataset and KDD-Cup'11. In *Proc. of KDD Cup 2011*, 2011.

[4] Y. Hu, Y. Koren, and C. Volinsky. Collaborative filtering for implicit feedback datasets. In *Proc. of 8th IEEE Int. Conf. on Data Mining*, ICDM '08, pages 263–272, Pisa, Italy, 2008.

[5] M. Jahrer and A. Töscher. Collaborative filtering ensemble for ranking. In *Proc. of KDD Cup Workshop at 17th ACM SIGKDD Int. Conf. on Knowledge Discovery and Data Mining*, KDD '11, San Diego, CA, USA, 2011.

[6] T. G. McKenzie, C. S. Ferng, Y. N. Chen, C. L. Li, C. H. Tsai, K. W. Wu, Y. H. Chang, C. Y. Li, W. S. Lin, S. H. Yu, C. Y. Lin, P. W. Wang, C. M. Ni, W. L. Su, T. T. Kuo, C. T. Tsai, P. L. Chen, R. B. Chiu, K. C. Chou, Y. C. Chou, C. C. Wang, C. H. Wu, H. T. Lin, C. J. Lin, and S. D. Lin. Novel models and ensemble techniques to discriminate favorite items from unrated ones for personalized music recommendation. In *Proc. of KDD Cup Workshop at 17th ACM SIGKDD Int. Conf. on Knowledge Discovery and Data Mining*, KDD '11, San Diego, CA, USA, 2011.

[7] R. Pan, Y. Zhou, B. Cao, N. N. Liu, R. M. Lukose, M. Scholz, and Q. Yang. One-class collaborative filtering. In *Proc. of 8th IEEE Int. Conf. on Data Mining*, ICDM '08, pages 502–511, Pisa, Italy, 2008.

[8] I. Pilászy, D. Zibriczky, and D. Tikk. Fast ALS-based matrix factorization for explicit and implicit feedback datasets. In *Proc. of the 4th ACM conference on Recommender Systems*, RecSys '10, pages 71–78, Barcelona, Spain, 2010.

[9] S. Rendle, C. Freudenthaler, Z. Gantner, and L. Schmidt-Thieme. Bpr: Bayesian personalized ranking from implicit feedback. In *Proc. of the 25th Conf. on Uncertainty in Artificial Intelligence*, UAI '09, pages 452–461, 2009.

[10] X. Su and T. M. Khoshgoftaar. A survey of collaborative filtering techniques. *Advances in Artificial Intellegence*, 2009, 2009. Article ID: 421425, 19 pages.

Local Implicit Feedback Mining for Music Recommendation

Diyi Yang, Tianqi Chen, Weinan Zhang, Qiuxia Lu, Yong Yu
Dept. of Computer Science and Engineering
Shanghai Jiao Tong University
800 Dongchuan Road, Shanghai China, 200240
{yangdiyi, tqchen, wnzhang, luqiuxia, yyu}@apex.sjtu.edu.cn

ABSTRACT

Digital music has experienced a quite fascinating transformation during the past decades. Thousands of people share or distribute their music collections on the Internet, resulting in an explosive increase of information and more user dependence on automatic recommender systems. Though there are many techniques such as collaborative filtering, most approaches focus mainly on users' global behaviors, neglecting local actions and the specific properties of music. In this paper, we propose a simple and effective local implicit feedback model mining users' local preferences to get better recommendation performance in both rating and ranking prediction. Moreover, we design an efficient training algorithm to speed up the updating procedure, and give a method to find the most appropriate time granularity to assist the performance. We conduct various experiments to evaluate the performance of this model, which show that it outperforms baseline model significantly. Integration with existing temporal models achieves a great improvement compared to the reported best single model for Yahoo! Music.

Categories and Subject Descriptors

H.3.3 [**Information Systems**]: Information Search and Retrieval—*Information Filtering*

Keywords

Collaborative Filtering, Recommender System, Local Implicit Feedback, Efficient Training

1. INTRODUCTION

During the past decades, music has experienced a quite fascinating transformation since expensive records and CDs are replaced by enjoying a lot of music free online. People could ask music store staff or just select from existing records to get what they wanted in the past, while nowadays they can surf on the Internet for music service, and turn to an automatic recommender system for a recommendation as the amount of music available expands explosively.

There are many approaches to music recommendation, such as Collaborative Filtering (CF)[12], which has been widely used and proved to be quite effective in handling users' preferences. Latent factor models, like matrix factorization (MF), and neighborhood models are two canonical approaches in CF to capture users' interests. MF, or singular value decomposition, maps users and items into the same low-dimension space, maintaining vectors of items and users. MF predicts a user's rating on an item through the inner product of two vectors. On the other hand, neighborhood models, put more emphasis on detecting the similarities and correlations between users or items, and make prediction based on observed information. Besides, there are other techniques for recommendation, such as graph-based models [26], and content-based approaches[8].

Though effective in maintaining users' overall preferences, many CF methods concentrate more on users' global interests. Such information might be enough to capture users' preferences in some cases, such as recommending academic papers or movies. However, music recommendation differs from those in several ways[1]. Firstly, music has a large item space and a low consumption time, where each song receives attention for a relatively short time. Secondly, a person can listen to songs in many situations, such as working, jogging or resting. However, he or she cannot read papers or watch movies all the time. Finally, people often consume music continually. That is, a user's behaviors in the next time interval are likely to be consistent with his/her recent behaviors.

Many traditional CF approaches focus on users' *global interests* that stay stable for a long time. They do not take the local consistent behaviors into full consideration[18]. That is, things which have happened just now or very recently are likely to influence users' decisions in the near future. We call these *local preferences*. For example, John might suddenly change his favorite Jazz to light music just because he was going to sleep. Or, a person might be blue and turn to some brooding tracks due to a week of rainy days. Such local information is often neglected and not captured. However, better performance may be achieved if local behaviors are considered. In this paper, we propose a simple and effective local implicit feedback model. We also design an efficient training algorithm to speed up the training. Our main contributions are as follows:

---

[1]http://recsys.acm.org/2011/tutorials.shtml, by Oscar Celma and Paul Lamere

- **Local Implicit Feedback Model** We are the first to propose a local implicit feedback model. A combination of local and global information, represented by implicit feedback, can capture users' stable and local changing preferences well.

- **Efficient Training Algorithm** Due to the large amount of data to be processed and the high cost, we design an efficient training algorithm. It greatly decreases the complexity when using the classical stochastic gradient descent method.

- **Time Granularity** We verify our local model by testing with different time granularties, in order to get the most appropriate granularity to maximize performance. We also propose a method to find the optimal time granularity.

Experiments conducted on three different datasets using two evaluation metrics show that our local model outperforms baselines in capturing users' local preferences in both rating and ranking prediction. We also show that our local model is to some extent complementary to other time-aware models. We combine our model with the best single model reported on Yahoo! Music, which achieves an RMSE decrease from 22.346 to 21.879.

The remainder of this paper is organized as follows. In Section 2, we present our local implicit feedback model. Our efficient training algorithm is described in Section 3, and experiments are discussed in Section 4. Related work is in Section 5. We conclude our work and point out some future directions in Section 6.

2. LOCAL IMPLICIT FEEDBACK MODEL

In this section, we present our local implicit feedback model that extracts users' local preferences, and also present the ranking approach we used. To demonstrate that this model is complementary to time-aware models, we integrate our model with some temporal approaches.

2.1 Local Preferences

Local user preferences, as stated above, are mainly used to capture user's varying behaviors in a very local time period. That is, his/her actions during current time interval might have effects on his/her decisions in the next few minutes. His/her actions in the next time period tend to be consistent with his/her recent behaviors, but might be different from his/her global behaviors. For example, if Alice is a music lover keen on Hip Hop and Rock, but something unfortunate has happened so she is depressed at present, then it is likely for her to turn to some sad songs. It is possible that people might change their accustomed habits temporarily and return to them after a short time. Habits can also change quite gradually, and those changes might persist for a long time. Transient events affect users' preferences during a short time period, like a week, a day, an hour or even a minute. On one hand, a user's interests might be affected by his/her established preferences; on the other hand, they could be also influenced by a very local event, like a new record release or a song that rises to fame overnight. Moreover, such very local and transient incidents, tend to be more determined by the characteristic of music that people's consumption is always continuous. However, such contextual information might be unavailable to recom-

Representations	Descriptions
U,I	user set, item set
u,i	user u, item i
p_u,q_i	K dimension vectors of user u and item i
$N(u)$	the set of items rated by user u
$N(u,t)$	the set of items rated by user u during time period t
b_u, b_i	users/item bias
μ	overall average ratings
$bias$	b_u, b_i and μ
ϕ_i	global predictive influence of item i a K dimension vector
φ_j	local predictive influence of item j a K dimension vector

Table 1: Commonly Used Notations

mender systems. To solve this problem, we propose to use users' local behaviors to model their local preferences.

2.2 Local Implicit Feedback

User local preferences are characterized by using users' implicit feedback in a short time period. The implicit feedback is represented by users' behaviors in history, i.e items he/she rated. Through rated/unrated binary information, the implicit feedback model is provided with a non-explicit ability to capture users' potential and global interests. This model was originally proposed by Koren[12], whose formulation was:

$$\widehat{r}_{ui} = \mu + b_u + b_i + {p_u}^T q_i + (|N(u)|^{-\frac{1}{2}} \sum_{j \in N(u)} \phi_j)^T q_i \quad (1)$$

Table 1 gives some commonly used notation for our work. Here, given two items i and j, ϕ_j is an indication of user u's preference, and will be high if j is predictive on item i. Based on MF and implicit feedback, this model focuses more on users' global behaviors. Since implicit feedback works well in characterizing users' global and potential interests, we implement our local preferences idea based on it. Our local implicit feedback model is formulated as this:

$$\widehat{r}_{ui}(t) = b_u + b_i + \left(p_u + \frac{\sum_{j \in N(u)} \phi_j}{\sqrt{|N(u)|}} + \frac{\sum_{j \in N(u,t)} \varphi_j}{\sqrt{|N(u,t)|}}\right)^T q_i$$
$$(2)$$

Here, user' local preferences are characterized by using their very localized rating history represented by $N(u,t)$, which we call **Local Implicit Feedback**. By using users' implicit information during a localized time interval, we assume that user behaviors during that period correlate to their current decisions. The **Time Granularity** is defined as the length of the local time interval, which can be a minute, an hour, a day and even a week. A period larger than a week is beyond what we define as a local time period. For example, if we set the time granularity as day, it means a user's ratings during the current day will be picked as his/her local implicit feedback. The most appropriate time granularity that directly and accurately reflects users' local interests, can be discovered by testing different granularity settings. However, it could be observed that due to differences in various music datasets, the optimal time granularity might be in different datasets. If there are few items rated during a time period, then that granularity might have little ability to capture users' local behaviors. Having implemented our local model with classical SVD++[12] , we could characterize users' global interests with the global implicit feedback, and local preferences by using the local implicit feedback model. We present detailed statistics in the experimental part.

2.3 Temporal Dynamics Integration

To investigate whether our implicit feedback model can further improve the performance of the existing temporal models, we integrate our model with those existing approaches. Some classical temporal models have already been proposed, such as [13, 5]. Incorporating those time-aware models with our local implicit feedback into an integrated model, we get the following formulation.

$$\widehat{r}_{ui} = \mu + b_i + b_i(t) + b_u + b_u(t) + p_u(t)^T q_i(t)$$
$$+ \left(\frac{\sum_{j \in N(u)} \phi_j}{\sqrt{|N(u)|}} + \frac{\sum_{j \in N(u,t)} \varphi_j}{\sqrt{|N(u,t)|}} \right)^T q_i(t) \quad (3)$$

$p_u(t)$, $q_i(t)$ are used to denote the corresponding p_u and q_i that change over time. This integration model is used to test that our implicit feedback model has an unique ability to discover some unheeded information.

2.4 Ranking Optimization

There are two kinds of prediction tasks, rate prediction and top K ranking recommendation. Rate prediction is a general recommendation orientation [13], and ranking based approaches [17] are popularly used on datasets like Pandora[2] and Last.fm. Both rate prediction and ranking task will be studied in our paper.

To recommend the top K items, we have to rank items over the $\widehat{r}_{ui}$ that directly reflects users' interests. Optimization of ranking order is needed to get the updated parameters. Most traditional ranking approaches maximize the area under the ROC curve as follows:

$$AUC(u) := \frac{1}{|N(u)^+||I \setminus N(u)^+|} \sum_{i \in N(u)^+} \sum_{j \in I \setminus N(u)^+} \delta(\hat{r}_{ui} - \hat{r}_{uj})$$

where $N(u)^+$ is the set of users' liked items in $N(u)$. It means user u prefers items in $N(u)^+$ over $I \setminus N(u)^+$. To optimize the rank order, we first define a hard 0-1 function as this:

$$\delta(x) = \begin{cases} 1 & x > 0 \\ 0 & otherwise \end{cases} \quad (4)$$

$\delta(x)$ is non-differentiable, thus we can replace it with surrogate functions $-l(x)$. One widely-used surrogate function is logistic loss, which is adopted later in our ranking approaches.

$$l(x) = \ln(1 + e^{-x}) \quad (5)$$

3. EFFICIENT TRAINING ALGORITHM

In this section, we present our efficient training algorithm that is used to speed up training in our experiments, after which we provide a complexity analysis.

3.1 Traditional Updating

We can characterize implicit feedback models described in Equation 1 and 2 with the following general implicit feed-

back model:

$$\widehat{r}_{u,i} = bias + \left(p_u + \sum_{j \in \Gamma(u)} \alpha_j \psi_j \right)^T q_i \quad (6)$$

Here, $\Gamma(u)$ stands for implicit feedback information that could include global implicit feedback described in Equation 1 as well as local feedback in Equation 2. ψ_j is the implicit feedback term, and could be ϕ_j in global implicit feedback or φ_j in local model. Meanwhile, α_j could be given the value $\beta_j = \frac{1}{\sqrt{|N(u)|}}$ for global implicit feedback terms, and $\gamma_j = \frac{1}{\sqrt{|N(u,t)|}}$ for local terms.

The traditional stochastic gradient descent algorithm [14] updates the above model as follows:

$$p_u = p_u + \eta \left(\hat{e} q_i - \lambda_1 p_u \right) \quad (7)$$
$$q_i = q_i + \eta \left(\hat{e} p_u - \lambda_2 q_i \right) \quad (8)$$
$$\psi_j = \psi_j + \eta \left(\hat{e} \alpha_j q_i - \lambda_3 \psi_j \right), \forall \alpha_j \neq 0, \alpha_j \in \Gamma(u) \quad (9)$$

We focus on p_u, q_i and ψ_j; the rules for updating $bias$ are omitted. The λs are regularization parameters. η is the learning rate, and $\hat{e}$ is the difference between actual and predicted ratings. Obviously, the cost of updating ψ_j is linearly related to the number of non-zero entries in $\Gamma(u)$, i.e. proportional to the number of items the users have rated.

3.2 Efficient Training

The traditional updating procedure becomes expensive when the average number of items rated by users is large. For example, last.fm customers have listened to 3996 songs on average, so their updates would be at a great cost. When we add more implicit feedback information, that problem becomes more prominent. Thus, some optimization methods are needed.

Before turning to our efficient algorithm, let us do some observations on the general model first. Define a derived user implicit feedback factor p^{im} as follows:

$$p^{im} = \sum_j \alpha_j \psi_j \quad (10)$$

The updating rule of ψ_j after one step is like this (omitting regularization terms):

$$\Delta \psi_j = \eta \hat{e} \alpha_j q_i \quad (11)$$

Difference in p^{im} before and after updating is as follows:

$$\Delta p^{im} = \eta \hat{e} (\sum_j \alpha_j^2) q_i \quad (12)$$

With more care, we find that there exists a relation between Δp^{im} and $\Delta \psi_j$, which is formulated as follows:

$$\Delta \psi_j = \frac{\alpha_j}{\sum_k \alpha_k^2} \Delta p^{im} \quad (13)$$

Therefore, to get new Δp^{im}, we do not need to update each ψ_j. For logistic loss or square loss, those update rules and relations also remain valid. However, if the L_2 regularization term is added, update rules changes to this:

$$\Delta \psi_j = \eta (\hat{e} \alpha_j q_i - \lambda \psi_j) \quad (14)$$

Corresponding differences in p^{im} change similarly. Those changes do not affect the use of this relation, since the regularization term is usually small.

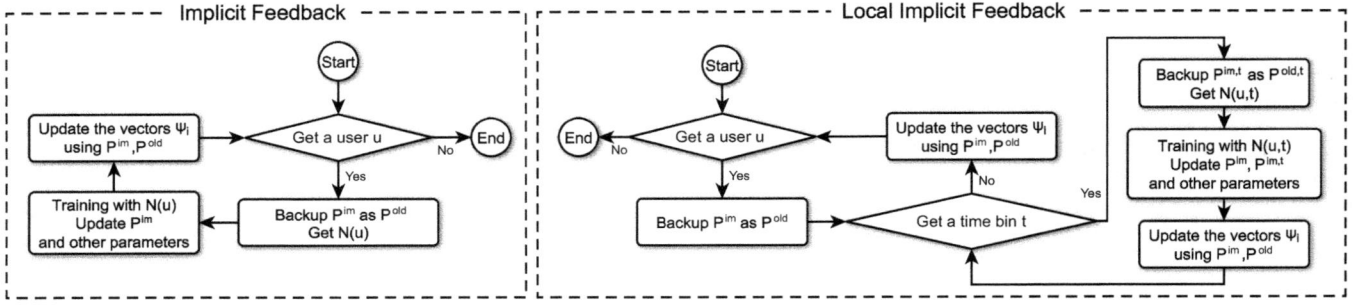

Figure 1: Comparison of Efficient Training Algorithm of Global and Local Implicit Feedback Models

This general relation is used to speed up the stochastic gradient descent training, and can also handle all variants of implicit feedback models, such as SVD++ in Equation 1 and local implicit feedback in Equation 2. We develop an efficient algorithm for our local implicit feedback model described in Equation 2 by using the general relation in Equation 13. The detailed algorithm is presented as Algorithm 1. The general idea is that we do not need to update each ϕ_j or φ_j. Instead, we directly update p^{im} and $p^{im,t}$, and return changes to ϕ_j and φ_j using the relation in Equation 13. Figure 1 presents an intuitive comparison of local and global implicit feedback models using the efficient training algorithm. It is obvious that efficient training of SVD++ is just a special case of Algorithm 1.

Algorithm 1 Efficient Training Algorithm for Training with Local Implicit Feedback

---

for all user u **do**
  $p^{im} \leftarrow \sum_{j \in N(u)} \beta_j \phi_j$
  $p^{old} \leftarrow p^{im}$
  for all time intervals of user u **do**
    $p^{im,t} \leftarrow \sum_{j \in N(u,t)} \gamma_j \varphi_j$
    $p^{old,t} \leftarrow p^{im,t}$
    for all training examples in $N(u,t)$ **do**
      update relevant parameters
      replace $\sum_j \gamma_j \varphi_j$ with $p^{im,t}$, $\sum_j \beta_j \phi_j$ with p^{im}
      update $p^{im,t}$ and p^{im} directly
    end for
    for all i, $\gamma_j \neq 0$ **do**
      $\varphi_j \leftarrow \varphi_j + \frac{\gamma_j}{\sum_k \gamma_j{}^2}(p^{im,t} - p^{old,t})$
    end for
  end for
  for all i, $\beta_j \neq 0$ **do**
    $\phi_j \leftarrow \phi_j + \frac{\beta_j}{\sum_k \beta_j{}^2}(p^{im} - p^{old})$
  end for
end for

3.3 Complexity Analysis

Compared with classical stochastic gradient descent algorithm, our efficient training algorithm achieves a significantly reduction in time complexity. Algorithm 1 has a time complexity of $O(N_e K + U\overline{R})$ for general implicit feedback form, i.e. $O(N_e K)$, usually $K > 1$. N_e stands for all non-zero entries, and K refers to the latent dimension. U is the number of users, while $\overline{R}$ represents the average number of items rated by a user. The cost of updating p^{im} for all users amounts to the product of dimension and all non-zero en-

tries of user-item rating samples. While the complexity of updating all ψ_j for all users is approximately the number of non-zero entries for all users. Thus, the overall complexity is the product of all non-zero entries and dimension. Compared to the cost of updating p^{im}, this can be neglected. However, since original stochastic gradient descent needs to update ψ_j whenever you get a updated p^{im}, its complexity is $O(N_e K\overline{R})$. Obviously, $\xi = \frac{O(N_e K\overline{R})}{O(N_e K)} = O(\overline{R})$. That is, our training algorithm performs $\overline{R}$ times faster than original one, where $\overline{R}$ is larger than 100 in general.

4. EXPERIMENT

In this section, we present our experimental results to evaluate the proposed local implicit feedback models. There are several questions we want to answer.

- Can this local implicit feedback model perform better than a model with only global implicit feedback?

- Which time granularity is the most powerful in reflecting users' local changing preferences? Could we give some methods to find a relatively optimal time granularity?

- Could our local implicit feedback model extract users' potential preferences that complement the existing time-aware models?

4.1 Experimental Setup

Our experiments are conducted on three datasets, Yahoo! Music[3], Last.fm[4], and Douban Music[5]. Yahoo! Music dataset is used by 2011 KDD Cup Workshop track 1. Last.fm is a famous personalized music website. We use the dataset provided by Celma et al.[4]. Douban Music is the largest Chinese music recommendation website. We crawl the dataset from users' recent listening history along with time stamps. Table 2 is a statistical comparison for each dataset. Here, STG stands for the Smallest Time Granularity provided by the datasets or available, and AIR is the Average Items Rated per user. Traditional rating prediction is performed on Yahoo! Music, and is evaluated with RMSE(root mean square error). We train local models on the provided training set and test them on the validation set. Personalized ranking is implemented on the remaining Last.fm and Douban Music datasets. We follow the top-N recommendation evaluation metric proposed by Cremonsei

---

[3]http://kddcup.yahoo.com
[4]http://last.fm
[5]http://music.douban.com

Dataset	User	Item	Ratings	STG	AIR
Yahoo! Music	1000990	624961	253M	Min	252
Last.fm	1001	471997	3.95M	Min	3996
Douban Music	22454	497744	2.94M	Day	88

Table 2: Comparison of DataSets

et al.[6]. First, we randomly 4 : 1 split the original training sets into training and validation sets. Then we get our final testing sets by sampling 1000 not rated items for each high rating example in the validation sets. Recall@K is used as evaluation metric like this:

$$Recall@K = \frac{H}{T} \qquad (15)$$

For each positive example with 1000 negative samples, it is called a hit if the positive example is ranked in the top K among the 1001 items. H is the overall hit number. T is the total number of items that users like in the test set.

4.2 Performance Comparison

We conduct several kinds of experiments on different datasets to evaluate the performances of purely global implicit feedback model and our local implicit feedback models.

4.2.1 Model Names

To fully illustrate the local models later, detailed descriptions of relevant models referred to Equation 2 are given as follows:

- IMFB: is for an implicit feedback model based on basic MF and global information, i.e the classical SVD++.

- MLIF: is minute local implicit feedback, regarding a minute's ratings as local implicit information.

- HLIF: is hour local implicit feedback model, using an hour' ratings as local implicit information.

- DLIF: is day local implicit feedback, treating a day's ratings as local information.

- WLIF: is week local implicit feedback model, taking a week's ratings as local information.

The MLIF, HLIF, DLIF and WLIF models are all implemented with IMFB, a combination of global and local implicit feedback information. IMFB is regarded as a baseline. The five models are all trained with our efficient training algorithm, and are built based on the time granularity available in the corresponding dataset. Through comparisons we set the latent factor as 100 in our experiments.

4.2.2 RMSE Performance on Yahoo! Music

Since Yahoo! Music is a rate prediction task, RMSE is adopted as our evaluation metric. It is appropriate to this specific dataset. The detailed RMSE results of the five models are in Figure 2. From the RMSE results of each of the models with different time granularity, we can get some basic observations. Firstly, we find that our proposed approaches do give significant improvements in this rate prediction task. MLIF, which achieves the best performance in terms of accuracy, attains an RMSE of 22.673. While our baseline model IMFB gives 23.233. It can be also observed that the performances of the day and week models are similar, with RMSE of 22.922 and 22.972 respectively. No matter what

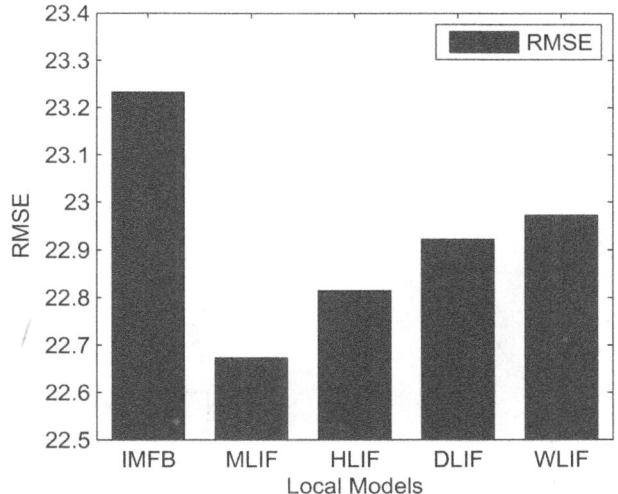

Figure 2: RMSE of Local Models on Yahoo! Music

Models,Recall@K	5	10	15	20
IMFB	0.173	0.240	0.304	0.335
MLIF	0.175	0.250	0.305	0.339
HLIF	0.220	0.288	0.338	0.373
DLIF	**0.224**	**0.301**	**0.345**	**0.383**
WLIF	0.224	0.297	0.341	0.380

Table 3: Recall of Localized Models on Last.fm

time interval is used, a local model always outperforms the baseline, though degrees of improvement are different. Thus, we can conclude that a local implicit feedback model works better than a purely global model. Secondly, it seems possible that a smaller time granularity gives better performance. That is, MLIF works better than DLIF, while DLIF is superior to others except the minute model. However, it might not be true on all datasets. The different performance among various time granularity reflect different descriptions of users' local behaviors and we will discuss this in detail in Section 4.3. We do not present the basic MF method here, since it is generally inferior to SVD++[12].

4.2.3 Recall Performance on Last.fm

In this section, we report the performance of our local implicit feedback models. From Table 2, the STG of Last.fm is minute. Based on time granularity available, minute, hour, day and week local implicit feedback models are constructed. We give the comparison of the best local implicit feedback against the global one in Figure 3. The detailed recall results of all models are shown in Table 3. It is evident that our implicit feedback model achieves a significant improvement over the baseline. Our localized model DLIF achieves a recall of 0.383 when $K = 20$, which means an item that a user likes has a probability of 38.3% to be ranked in the top 20 in DLIF during a large sample collection (1001 items). While the baseline gives a Recall@20 of 0.335. That is, day local implicit feedback has a 14% improvement over the baseline. Such big progress indicates that our model does outperform a purely global model significantly. Table 3 shows detailed recall results for the local implicit feedback model. From Table 3, we know that compared to the

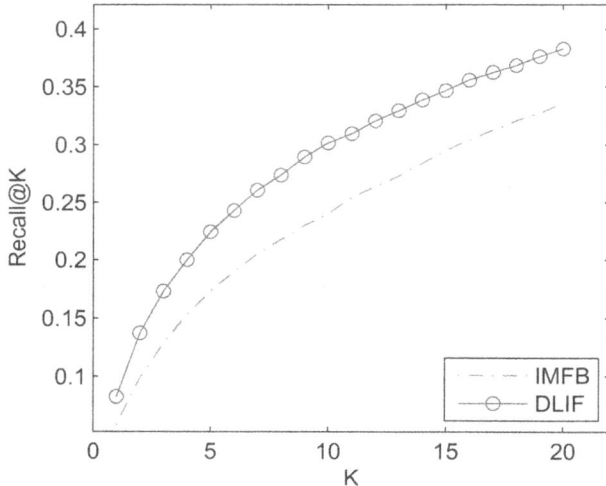

Figure 3: Recall@K of Localized and Baseline Models on Last.fm

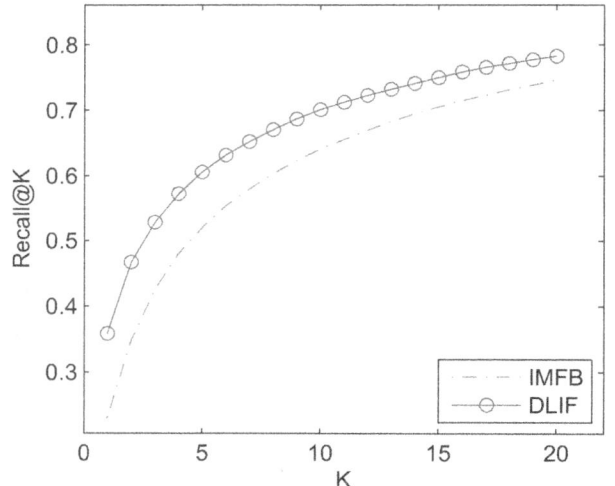

Figure 4: Recall@K of Localized and Baseline Models on Douban

Models,Recall@K	5	10	15	20
IMFB	0.520	0.640	0.705	0.746
DLIF	**0.605**	**0.701**	**0.750**	**0.783**
WLIF	0.602	0.697	0.739	0.780

Table 4: Recall of Localized Models on Douban

baseline, local implicit feedback models give improvement to different extents. Obviously, DLIF, HLIF and WLIF give similar performance. HLIF and WLIF give Recall@20 of 0.373 and 0.380 respectively, and DLIF is a little higher than either of them. Moreover, MLIF gives approximately the performance of IMFB. With a recall@20 of 0.339, it is a bit better than IMFB. The experiments show that our local implicit feedback model can achieve significant improvement over baseline on ranking task.

4.2.4 Recall Performance on Douban

This dataset is handled similarly to Last.fm, and shares the same evaluation methodology with it. Experiments are conducted on three models, IMFB, DLIF and WLIF. We present our results on Douban Music as Table 4, and comparison curves as Figure 4. Figure 4 gives us an abstract impression that a local implicit feedback model outperforms IMFB. As anticipated, a local implicit feedback model gives a consistent improvement in terms of top K precision from Table 4. With a Recall@20 of 0.746 for IMFB, DLIF and WLIF outperform it with 0.783 and 0.780 respectively. Such improvement also can be observed from Recall@5, Recall@10 and Recall@15 in Table 4. Therefore, overall performances show again that local implicit feedback models do work in capturing users' preferences that are neglected by a global model.

4.3 Resolution of Time Granularity

In this part, we evaluate the performance of our local implicit feedback models under different time granularity, and propose an empirical standard to find the most appropriate time granularity for each dataset. For convenience, define

ARP(u) as the set of time-adjacent rating pairs of user u, and ARP as the union of all users' ARP(u). Statistics are done on the percentage of all ARP pairs which are categorized into five groups by their time differences. Time gap groups include: 0(No time difference[6]), $[1M, 1H)$, $[1H, 1D)$, $[1D, 1W)$ and larger than 1 week. M, H, D, W refer to minute, hour, day and week respectively. The results are presented in Figure 5. The statistics is a direct reflection of users' behavior frequency distributions.

From statistics of Yahoo! Music in Figure 5(a), we find that nearly 60% ARP pairs have no timestamp difference. Recalling the results in Figure 2, we find the minute model gives the best performance among all time granularity. This can be explained by our statistics since minute is the *smallest granularity that contains sufficient information*: 60% of ARP pairs have no time difference, indicating that sufficient feedback information can be obtained within a minute. Under the condition of sufficient information, a smaller granularity can lead to a better modeling of users' behaviors.

In Last.fm, it appears that most ARP pairs lie in group $[1M, 1H)$. Using the day as time granularity allows us to cover sufficient information. The results in Table 3 suggest that day is the most proper time granularity, which achieves the best Recall@20 of 0.383. This is consistent with our previous observation. In Douban, we can see that almost 75% of ARP pairs have no difference with each other. The results in Table 4 show that day is the most appropriate. This is also consistent with our previous observation since day is the smallest time granularity with sufficient information.

Through comparison, we discover that the appropriate time granularity depends more on the property of specific dataset, and generally varies in different datasets. From the experiment results, we observe that the most proper time granularity is given by the smallest granularity that contains sufficient information. The most appropriate time granularity are minute, day and day on Yahoo! Music, Last.fm and Douban respectively. Moreover, by analyzing statistics on

---

[6]This means there is no difference in timestamp, due to the limitation of minimal timestamp available on dataset

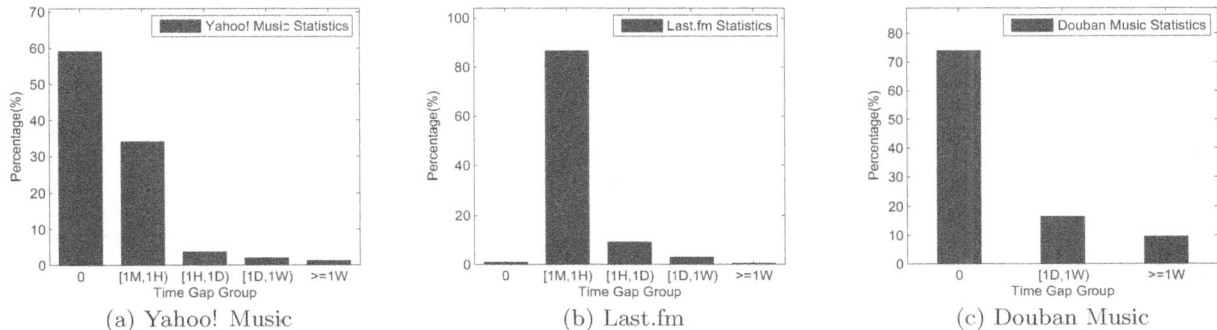

Figure 5: Statistics on Yahoo! Music, Last.fm and Douban Music

Models	RMSE
Integrated.MF	22.346
Integrated.Minute.LIF	**21.879**

Table 5: RMSE Performance on Yahoo! Music

each dataset, we can choose the appropriate time granularity using the empirical standard.

4.4 Integration with Temporal Dynamics

The experimental results presented in Section 4.2 show that local implicit feedback performs well on music datasets. Using a combination of global and local implicit feedback information, our local implicit feedback models capture not only their global interests, but also users' local behaviors that could be indicated through the local rating entries. However, it can be argued that, even if our models captures users' changing interests, it might be largely due to some general influences of changes that existing time-aware models could handle. To answer this question, we conduct a comparison experiment to show that local model captures an unique property of music that differs from properties mined by existing time-aware models on the Yahoo! Music dataset. Here, Integrated.MF stands for the model adopted in [5], the reported best single model on Yahoo! Music, which contains state-of-the-art temporal dynamics modeling approaches. Integrated.Minute.LIF, represents the combination of a local model and Integrated.MF referred to Equation 3. We take the best local implicit feedback model on Yahoo! Music here to integrate with Integrated.MF. That is, Integrated.Minute.LIF is a combination of local implicit feedback (MLIF) and a state-of-the-art temporal approach. Results are shown in Table 5. The combination gives a significant improvement from 22.346 to 21.879, better than all local models on Yahoo! Music dataset. Therefore, it can be concluded that a local implicit feedback model is complementary to time-aware models, and combining it with the best single model reported can provide better performance. In conclusion, our proposed local model does have an unique capability to discover users' potential preferences from a different perspective.

5. RELATED WORK

Collaborative filtering has been an effective technique for recommender systems during the past years. Compared with content-based models [25], CF approaches do not nec-essarily need any attributes of users/items. In general, CF models are often adaptive and flexible since they learn from the dynamically changing user feedback [16]. Neighborhood based models [23, 21] and latent factor models [14] are two canonical approaches to CF. Recently, context-aware models [1] are proposed to capture users' behaviors on specific context. Context-aware models always leverage users' context information such as mood [24], location [19], social relationship [3] etc.. However, these contextual information are often unavailable.

User implicit feedback plays an important role in information retrieval and data mining applications[2, 11]. Due to the limitations on obtaining explicit feedback (e.g. detailed ratings), implicit feedback, has been paid much attention [22] due to availability. For some recommender system applications, users do not always return their explicit feedback due to application limitations. Thus, implicit feedback can be leveraged to improve the recommendation performance. Hu et al. [10] studied CF implicit feedback datasets and proposed to transform users' implicit feedback into training data in a preference-conference format. Koren [12] discovered that incorporating implicit feedback into a neighborhood integrated latent factor model (SVD++) could give significant improvement. However, training with implicit feedback becomes quite expensive in general recommender systems, and is not well studied. G. Takacs proposed a specific unified approach of factor and neighbor based models for large recommender systems[9], differing from our general efficient training one.

Since users' preferences generally change over time, recommender systems should capture both global and local changing interests in order to provide more accurate recommendations. Temporal dynamic models distinguish users/items from different time slots so as to maintain users' changing preferences. Koren [13] proposed a time day bin approach in capturing users' fixed preferences during a time period. Xiang et al.[26] designed a session-based temporal graph and applied a personalized random walk on it. The graph has three types of nodes for users, items, and user sessions, and is used to capture users' long term and short term interests. Chen et al. [5] proposed a multi-resolution temporal CF model that consists of time-dependent user/item bias, latent factors, time-dependent neighborhood, and rating session to cope with Yahoo! Music recommendation. Rendle et al. [20] proposed personalized Markov chains together with MF in capturing both sequential effects and long term user-state, which focuses on influences of sequential actions on the next action. There are also other recommendation approaches involving temporal information in various ways, such as decreasing weights for old data [7], and time-dependent iterative prediction in a growing dataset[15].

6. CONCLUSION

In this paper, we propose a simple and effective local implicit feedback model to mine users' local interests. We also design an efficient training algorithm to speed up the training procedure. Experiments conducted on three datasets show that our local implicit feedback models significantly outperform the global implicit feedback model, and have a capability different from existing time-aware models in capturing users' changing preferences. Meanwhile, we observe that the appropriate resolution changes on different datasets, and also give an empirical standard to determine the most appropriate time granularity. In the future, we plan to take this further to extract personalized local implicit feedback information and other localized properties.

7. ACKNOWLEDGEMENT

Yong Yu is supported by grants from NSFC-RGC joint research project 60931160445.

8. REFERENCES

[1] G. Adomavicius and A. Tuzhilin. Context-aware recommender systems. *Recommender Systems Handbook*, pages 217–253, 2011.

[2] E. Agichtein, E. Brill, and S. Dumais. Improving web search ranking by incorporating user behavior information. In *Proceedings of the 29th annual international ACM SIGIR conference on Research and development in information retrieval*, pages 19–26. ACM, 2006.

[3] R. Andersen, C. Borgs, J. Chayes, U. Feige, A. Flaxman, A. Kalai, V. Mirrokni, and M. Tennenholtz. Trust-based recommendation systems: an axiomatic approach. In *Proceeding of the 17th international conference on World Wide Web*, pages 199–208. ACM, 2008.

[4] O. Celma. *Music Recommendation and Discovery in the Long Tail*. Springer, 2010.

[5] T. Chen, Z. Zheng, Q. Lu, X. Jiang, Y. Chen, W. Zhang, K. Chen, Y. Yu, N. Liu, B. Cao, L. He, and Q. Yang. Informative ensemble of multi-resolution dynamic factorization models. In *KDD-Cup Workshop*, 2011.

[6] P. Cremonesi, Y. Koren, and R. Turrin. Performance of recommender algorithms on top-n recommendation tasks. In *Proceedings of the fourth ACM conference on Recommender systems*, RecSys '10, pages 39–46, New York, NY, USA, 2010. ACM.

[7] Y. Ding and X. Li. Time weight collaborative filtering. In *Proceedings of the 14th ACM international conference on Information and knowledge management*, pages 485–492. ACM, 2005.

[8] A. Ferman, J. Errico, P. Beek, and M. Sezan. Content-based filtering and personalization using structured metadata. In *Proceedings of the 2nd ACM/IEEE-CS joint conference on Digital libraries*, pages 393–393. ACM, 2002.

[9] B. D. G.Takacs, I.Pilaszy. A unified approach of factor models and neighbor based methods for large recommender systems. In *In Proc. of ICADIWT-08, 1st IEEE Workshop on Recommender Systems and Personalized Retrieval*, pages 186–191, 2008.

[10] Y. Hu, Y. Koren, and C. Volinsky. Collaborative filtering for implicit feedback datasets. In *2008 Eighth IEEE International Conference on Data Mining*, pages 263–272. IEEE, 2008.

[11] T. Joachims, L. Granka, B. Pan, H. Hembrooke, and G. Gay. Accurately interpreting clickthrough data as implicit feedback. In *Proceedings of the 28th annual international ACM SIGIR conference on Research and development in information retrieval*, pages 154–161. ACM, 2005.

[12] Y. Koren. Factorization meets the neighborhood: a multifaceted collaborative filtering model. In *Proceeding of the 14th ACM SIGKDD international conference on Knowledge discovery and data mining*, KDD '08, pages 426–434, New York, NY, USA, 2008. ACM.

[13] Y. Koren. Collaborative filtering with temporal dynamics. In *Proceedings of the 15th ACM SIGKDD international conference on Knowledge discovery and data mining*, KDD '09, pages 447–456, 2009.

[14] Y. Koren, R. Bell, and C. Volinsky. Matrix factorization techniques for recommender systems. *Computer*, 42, August 2009.

[15] N. Lathia, S. Hailes, and L. Capra. Temporal collaborative filtering with adaptive neighbourhoods. In *Proceedings of the 32nd international ACM SIGIR conference on Research and development in information retrieval*, pages 796–797. ACM, 2009.

[16] B. Liu. *Web data mining: exploring hyperlinks, contents, and usage data (Second Edition)*. Springer Verlag, 2011.

[17] N. N. Liu and Q. Yang. Eigenrank: a ranking-oriented approach to collaborative filtering. In *Proceedings of the 31st annual international ACM SIGIR conference on Research and development in information retrieval*, SIGIR '08, pages 83–90, New York, NY, USA, 2008. ACM.

[18] J. R. Lorraine McGinty. *On the Evolution of Critiquing Recommenders*. Springer, 2011.

[19] M. Park, J. Hong, and S. Cho. Location-based recommendation system using bayesian user's preference model in mobile devices. *Ubiquitous Intelligence and Computing*, pages 1130–1139, 2007.

[20] S. Rendle, C. Freudenthaler, and L. Schmidt-Thieme. Factorizing personalized markov chains for next-basket recommendation. In *Proceedings of the 19th international conference on World wide web*, WWW '10, pages 811–820, New York, NY, USA, 2010. ACM.

[21] B. Sarwar, G. Karypis, J. Konstan, and J. Reidl. Item-based collaborative filtering recommendation algorithms. In *Proceedings of the 10th international conference on World Wide Web*, WWW '01, pages 285–295, New York, NY, USA, 2001. ACM.

[22] X. Shen, B. Tan, and C. Zhai. Context-sensitive information retrieval using implicit feedback. In *Proceedings of the 28th annual international ACM SIGIR conference on Research and development in information retrieval*, pages 43–50. ACM, 2005.

[23] Y. Shi, M. Larson, and A. Hanjalic. Exploiting user similarity based on rated-item pools for improved user-based collaborative filtering. In *Proceedings of the third ACM conference on Recommender systems*, RecSys '09, pages 125–132, New York, NY, USA, 2009. ACM.

[24] Y. Shi, M. Larson, and A. Hanjalic. Mining mood-specific movie similarity with matrix factorization for context-aware recommendation. In *Proceedings of the Workshop on Context-Aware Movie Recommendation*, pages 34–40. ACM, 2010.

[25] R. Van Meteren and M. Van Someren. Using content-based filtering for recommendation. In *Proceedings of the Machine Learning in the New Information Age: MLnet/ECML2000 Workshop*, 2000.

[26] L. Xiang, Q. Yuan, S. Zhao, L. Chen, X. Zhang, Q. Yang, and J. Sun. Temporal recommendation on graphs via long-and short-term preference fusion. In *Proceedings of the 16th ACM SIGKDD international conference on Knowledge discovery and data mining*, pages 723–732. ACM, 2010.

How Many Bits Per Rating?

Daniel Kluver[1]   Tien T. Nguyen[1]   Michael Ekstrand[1]   Shilad Sen[2]   John Riedl[1]

[1]GroupLens Research
Computer Science and Engineering
University of Minnesota
Minneapolis, MN 55455
{kluver,tien,ekstrand,riedl}@cs.umn.edu

[2]Math, Stats, and Computer Science Dept.
Macalester College
St. Paul, MN 55105
ssen@macalester.edu

ABSTRACT

Most recommender systems assume user ratings accurately represent user preferences. However, prior research shows that user ratings are imperfect and noisy. Moreover, this noise limits the measurable predictive power of any recommender system. We propose an information theoretic framework for quantifying the preference information contained in ratings and predictions. We computationally explore the properties of our model and apply our framework to estimate the efficiency of different rating scales for real world datasets. We then estimate how the amount of information predictions give to users is related to the scale ratings are collected on. Our findings suggest a tradeoff in rating scale granularity: while previous research indicates that coarse scales (such as thumbs up / thumbs down) take less time, we find that ratings with these scales provide less predictive value to users. We introduce a new measure, preference bits per second, to quantitatively reconcile this tradeoff.

Categories and Subject Descriptors

H.1.2 [**Information Systems**]: User/Machine Systems—*Human information processing*; H.5.2 [**Information interfaces and presentation**]: User Interfaces—*Evaluation / methodology*

General Terms

Design, Measurement, theory

Keywords

ratings, recommender systems, information theory, evaluation, metrics

1. INTRODUCTION

Collaborative filtering (CF) recommender systems rely on ratings to understand user preferences and make recommendations. While past research has focused on algorithms that use ratings to accurately predict user preferences [6], more recent work has begun to explore the quality of the ratings themselves [1, 2, 3, 14, 15, 21]. Researchers believe that inconsistencies in user ratings limits CF prediction accuracy. First suggested by Hill et al. [12], this limit has been named the "magic barrier" [11]. CF systems that understand and adapt to noise in user ratings may be able to raise the magic barrier and generate more accurate recommendations.

CF system researchers have explored various ways of understanding and reducing noise in ratings. O'Mahony [15] defines two types of noise that can appear in ratings and affect the quality of predictions: *natural noise* and *malicious noise*. *Natural noise* is noise associated with how a recommender system collects user preferences. *Malicious noise* is biased noise purposely introduced into a recommender system in order to affect predictions. Several researchers have measured natural noise. Both Hill et al. and Cosley et al. measured natural noise by asking users to re-rate items after a period of time has passed [12, 3]. Using this methodology both studies found evidence of noise in user ratings. Amatriain et al. analyzed how characteristics such as rating value, time elapsed, and rating speed affect user noise [1] and found that ratings are a good psychological measure according to classical test theory. Following this work Amatrian et al. proposed an algorithm that collects multiple ratings to reduce noise and increase recommendation quality [2].

Psychology research suggests that a CF system's choice of rating scale may crucially affect the level of rating noise. In work that inspires our approach, Garner uses information theory to measure how much information different rating scales transmit [8]. While he generally finds that increases in scale granularity reduce rating noise, this effect is limited by the "discriminability inherent in the particular rated stimuli." This limitation has been observed in practice on the website YouTube. Prior to 2010, YouTube users could rate movies on a five star scale. However in 2009 YouTube found that almost all ratings were either one or five stars, with the other ratings rarely used. By 2010, YouTube switched to a like/dislike scale, having seen no value in keeping the five-star scale [22].

Although recommender system research has not directly studied how rating scales affect rating noise, they have explored other issues related to rating scales. Cosley et al. [3] found that users preferred using a relatively fine-grained 5-star rating scale with half star increments. They found evidence that the more descriptive rating scale improved prediction quality, but were hesitant to draw strong conclusions. Sparling et al. [21] investigated how cognitive load and rating time varies with rating scale. They found that more granular rating scales require more time, and suggested that system designers may face a fundamental tradeoff between a coarse rating scale that is relatively quick and noisy, or a granular rating scale that is slower with less noise.

We seek to provide system designers with tools to address the rating quality / rating quantity tradeoff raised by Sparling et al. We propose the *preference bits* framework to capture the amount of user preference information in ratings and predictions. We ground the framework in an information theoretic model of human preference, and then explore the model's properties. We apply the preference bit framework to answer three research questions:

RQ1: How much preference information is captured by ratings with different rating scales?

RQ2: How much preference information is provided by predictions with different rating scales?

RQ3: How should a system designer reconcile the tradeoffs between rating time and rating quality?

The reminder of this paper is organized as follows: first we discuss the preference bits framework, followed by an evaluation of how the framework can be used to measure rating and prediction quality. Finally, we discuss limitations and future work.

2. THE PREFERENCE BITS MODEL

In order to address our research questions, we need some method of measuring the quality of information entering and exiting a recommender system. To do this, we introduce the *preference bits* framework for measuring the quantity of information about user preferences. Before we can describe the framework, we must first define preference and introduce the information theoretic concepts it relies on.

2.1 Psychology Theory Background

There is a spectrum of views concerning how preferences are formed and what happens when we measure them [7]. At one end of the spectrum, the philosophy of *articulated values* holds that humans form well reasoned, stable preferences for most things they have experienced. Variance in elicited answers in the philosophy of articulated values is said to come from subtle differences in understanding of questions leading to measuring different preferences. At the other end of the spectrum, the philosophy of *basic values* holds that humans store a small number of opinions or beliefs from which they derive momentary preferences when needed [7]. Variance in elicited answers in the philosophy of basic values is said to come from priming effects, mood, and other psychological phenomena affecting the process used to derive preference.

This paper assumes an intermediate model. We assume that humans can form stable preferences for all items in the item domain. However, we also assume that these preferences are only partially articulated, and that processing is still required when mapping these partial preferences to a rating.

2.2 Information Theory Background

Claude Shannon developed information theory as a way of reasoning about information [20]. One of the fundamental measures of information is called *entropy*, which measures how hard a random variable is to predict. Entropy is defined

as follows where X is a random variable and $P(x)$ denotes the probability mass function of X.

$$H(X) = -\sum_{x \in X} P(x) \log_2 P(x)$$

We measure entropy with *bits* where one bit is the amount of information in a fair coin flip.

The *mutual information* between two random variables is a symmetric measure of how much information the value of one random variable gives us about the value of the other. Mutual information is defined as follows where X and Y are both random variables:

$$I(X;Y) = \sum_{x \in X} \sum_{y \in Y} P(x,y) \log_2 \frac{P(x,y)}{P(x)P(y)}$$

We measure mutual information with *bits* where one bit of mutual information between two variables X and Y means we expect knowing X to reduce the entropy in Y by one bit. Mutual information ranges from zero when the two variables are independent, to the smaller of the two variables' entropy when one variable entirely explains the other.

As an example consider the relationship between ratings (R) on the movie *Titanic* and the gender (G) of the rater. Using information from the MovieLens rating system we can compute $I(G;R) = 0.011$ which would not significantly help us predict the rating given the user's gender, or the user's gender given their rating.

One important result about mutual information is the *data processing inequality* [4]. The data processing inequality applies for any three random variables $X, Y,$ and Z such that X and Z are *conditionally independent* on Y. This means that knowing Y and Z gives us no more information about X than just knowing Y, formally $P(X|Y,Z) = P(X|Y)$. When this property holds, the data processing inequality shows:

$$I(X;Z) \leq I(X;Y)$$

We will use the data processing inequality to derive measures of the amount of information about user preferences that enters and exits the recommender.

2.3 Definition of Preference Bits

To measure preference bits of any factor, we will use information theory. We define the random variable $\Pi(u,i)$ to represent the user u's unknown partial preference for an item i. Because this preference is only partially articulated, we do not assume users will express their preference perfectly. Therefore one $\Pi(u,i)$ may lead to several different ratings.

The preference bits of any variable X is $I(X;\Pi)$, the amount of information X gives us about Π. Of particular interest to us is the *input preference bits* and the *output preference bits* defined as $I(R;\Pi)$ and $I(P;\Pi)$ respectively. Input preference bits measures how much we expect a rating $R(u,i)$ to reduce the recommender's uncertainty about the user's preference. Output preference bits measures how much we expect a prediction $P(u,i)$ to reduce a user's uncertainty about her (potentially unformed) preference.

While the preference bit framework provides logical ways to measure the quality of recommender input and output, we cannot directly measure the metrics it produces due to our assumption that preferences are not fully articulated. To allow measurement of these metrics we will prove the following theorem

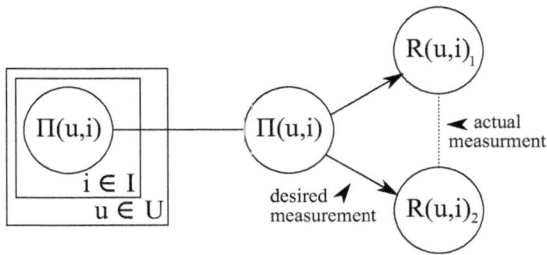

Figure 1: Graphical model of input preference bit measurement.

THEOREM 1. *Let* $X \neq R, \Pi$ *be any random variable:*

$$I(X; R) \leq I(X; \Pi)$$

Theorem 1 follows directly from our assumption that preference $\Pi(u, i)$ is the only measurable variable responsible for $R(u, i)$. This means that X and R must be conditionally independent on Π. Therefore by the data processing inequality, X tells us less about our ratings than it does about preferences. While theorem 1 does not give us an exact measure of the preference bits in X, it can give us a reasonable proxy.

3. INPUT PREFERENCE BITS

Measuring input preference bits: We would like to derive a measure for the input preference bits ($I(R; \Pi)$) of an interface. To overcome the problems with directly measuring preference we will collect multiple ratings on one item. Let R_1 and R_2 represent two ratings for the same item and user drawn independently from $P(R|\Pi)$ for preference Π. We call such pairs of ratings *re-ratings*. Figure 1 shows this model. Because we assume that Π is the only factor determining ratings, any information between R_1 and R_2 must come from information between Π and any single rating. Formally, $I(R_1; R_2) \leq I(R; \Pi)$ because R_1 and R_2 are both samples of R therefore we can apply theorem 1 with R_1 as R and R_2 as X to get

$$I(R_1; R_2) \leq I(R_1; \Pi) = I(R; \Pi)$$

$I(R_1; R_2)$ tells us how much we expect a user's rating for an item will reduce uncertainty about the user's future ratings on that item. We will use this measure as an estimate for input preference bits per rating for the rest of this paper.

Assumptions and limitations: To show the connection between $I(R; \Pi)$ and $I(R_1; R_2)$, we relied on two assumptions that have bearing on how re-ratings should be collected. First, our assumption that two ratings were drawn independently implies that when a user is asked to re-rate an item, they should not remember their previous rating. This can be accomplished by waiting a period of time before asking a user to re-rate. We also assume that preferences are constant and do not drift over time, but this has been shown to be false [1]. To minimize this effect, re-ratings should be collected in a relatively short time span. Any practical experimental design for collecting multiple ratings must balance these two assumptions, inserting an appropriate amount of time between ratings. Because the effects of violating our assumptions will be time dependent, experiments should be designed to place consistent amounts of time between ratings for all conditions.

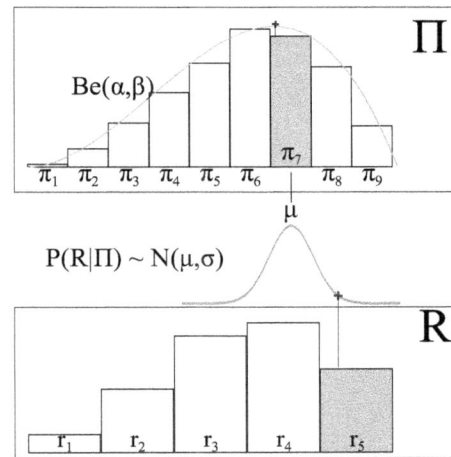

Figure 2: Our rating system model

In practice, we expect that as the time differential between re-ratings increases the amount of measured preference bits per rating will decrease.[1]

3.1 Model Evaluation

To understand how rating scale affects preference bits per rating, we built a probabilistic model of a rating system. Using this model we considered the implications of changing rating scale on true and measured preference bits per rating. We then used this model to assess a strategy for measuring $I(R_1; R_2)$ in practice.

Figure 2 shows a diagram of the model. The true preference Π of a user is modeled as a beta distribution with parameters α and β bucketed into m_T bins. The beta distribution is a natural choice for modeling a random variable over the unit range. By varying α and β many qualitatively different distributions can be achieved. m_T represents the true scale upon which users internally evaluate an item. To pick a user's value for Π, we draw a random value from the beta distribution and take the value of the center of the bucket containing it.

We model the input rating scale as the unit scale divided into m_R buckets. To generate R from Π we calculate Π, add noise from a normal distribution $\mathcal{N}(\mu = \Pi, \sigma)$ and choose the rating bucket (between 1 and m_R) that contains the result. Because a normal distribution is not clamped to the unit range, we treat the smallest and highest rating as extending to negative and positive infinity respectively. This allows us to partially account for the extra stability of extreme ratings noticed in prior research [2]. From $P(\Pi)$ and $P(R|\Pi)$ we can compute $P(R, \Pi)$ and $P(R_1, R_2)$.

We first consider the relationship between the true preference bits per rating (Pbpr) and the rating scale. Figure 3 shows that as we increase the rating scale we find an asymptotic increase in preference bits per rating. We found that the asymptote reached is a function of the preference scale and the amount of noise. As the preference scale increases, we see logarithmic increase in the maximum Pbpr. Likewise,

---

[1] We confirmed this empirically with Xavier Amatriain's dataset [1] where we see a decrease in preference bits per rating from 1.1227 with less than 15 day delay to 0.9757 with a more than 15 day delay.

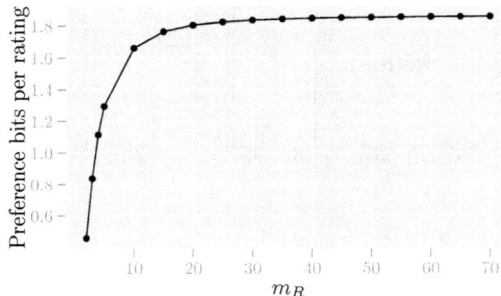

Figure 3: Pbpr rating increases as the size of the rating scale increases ($\alpha = 3$, $\beta = 1$, $m_T = 50$, $\sigma = 0.05$)

when we increase noise the maximum Pbpr decreases. We found this result holds for a wide range of α and β values. This result suggests that there is some number of ratings after which we cannot gain more information by increasing the resolution of our rating interface.

Next we will use our model to assess the accuracy of $I(R_1; R_2)$ as a measure of $I(R; \Pi)$. For $I(R_1; R_2)$ to be a useful measure of $I(R; \Pi)$, we want $I(R_1; R_2)$ to correlate with $I(R; \Pi)$. To evaluate this we compute $I(R_1; R_2)$ and $I(R; \Pi)$ for a range of model parameters for preference distribution ($\alpha, \beta \in \{0.5, 1, 2, \ldots, 6\}$), internal and input scales ($m_T, m_R \in \{2, 3, 4, 5, 10, 15, \ldots, 60\}$) and noise ($\sigma \in \{0.01, 0.1, 0.2\}$). Using this data we found a strong linear dependency between $I(R_1; R_2)$ and $I(R; \Pi)$ ($R^2 = 0.984$). Therefore, we expect $I(R_1; R_2)$ to be a good proxy measure of the true preference bits per rating.

Finally, we used our model to assess our strategy for measuring $I(R_1; R_2)$ on sampled data. We use the Miller-Madow method to estimate mutual information [13]. We estimate the joint probability $P(R_1, R_2)$ as the number of times the two rating value co-occurred divided by the number of ratings. Where m_{R_1} and m_{R_2} are the number of possible values for R_1 and R_2 (m_r in our case), the Miller-Madow estimator is

$$\sum_{r_1} \sum_{r_2} \hat{P}(r_1, r_2) \log_2 \frac{\hat{P}(r_1, r_2)}{\hat{P}(r_1)\hat{P}(r_2)} - \frac{(m_{R_1} - 1)(m_{R_2} - 1)}{2 \log_e 2 * N}$$

The second term corrects for the first order of magnitude in our estimate and substantially improves our estimation. Unfortunately, it has been shown [16] that there is no unbiased estimator for mutual information, therefore it is important that large sample sizes ($N >> m_R^2$) be used to avoid estimator bias. This requirement will make measuring input Pbpr difficult on a per user basis, or for high precision interfaces. While more complex estimators for mutual information may reduce these issues [16], we will focus only on large datasets for this paper.

We used our model to perform a simulation study to evaluate the performance of the Miller-Madow estimator on re-rating data. We discovered that noise was consistently less than bias, and that the correction term can over-correct when noise is small. We found that a sample size of 1000 was sufficient to reduce bias and standard deviation to less than 0.1 bits for rating scales with up to 20 points. This can be used as a guideline for a preferred sample size.

3.2 Empirical Evaluation

Next we use real-world datasets to evaluate the amount of input preference bits in rating inputs. These evaluations then serve as guidelines when we discuss the trade-offs when choosing different rating interfaces.

Datasets: The perfect dataset for this analysis would be a large dataset with re-ratings performed on a variety of rating interfaces. Unfortunately, no such dataset exists. However, we can approximate this dataset using only single ratings on several interfaces with a historical set of ratings to compare against. Therefore we will use the SCALES dataset from Cosley et al. [3]. As can be see in table 1 this dataset is large enough to avoid significant estimator bias but has relatively few users.[2] Therefore we consider these only preliminary results until a more complete dataset can be collected. Ratings in this dataset were entered on three rating interfaces: a 2-point thumbs up / thumbs down, a 6-point scale from -3 to +3 with no zero, and a 10-point 5-star scale with half stars. Re-ratings are made by pairing these user's original ratings (made on a five star interface) with their new ratings. This gives us a way to measure input preference bits. Note, however, that these re-ratings were not entered on the same scale. While our measure is still a lower bound on true input bits for either interface, this does add the confound that some of the three new scales may map to the original interface better than others regardless of actual preference bits.

The 2-point, 6-point, and 10-point scale yielded 0.423, 0.825, and 0.813 input preference bits respectively. As expected the 2-point interface captured much less preference information than the other two interfaces. Surprisingly, the amount of information gained with the 10-point interface was less than the amount gained with the 6-point interface. This contradicts our previous results and suggests that the 6-point interface is more efficient at collecting preference information than the five star interface. This may be because the lack of a zero option actually reduces noise in ratings, but it may also be an artifact of our measure. Because our re-ratings are not taken on the same interface this is at best evidence that the 6-point scale may be better. A follow up study with proper methodology should be performed before a conclusion is reached.

Preference bits per second: One reason why the 6-point interface may have a higher preference bits per rating is because the lack of a zero forces users to think more carefully before rating. While this would lead to more information per rating, it's not clear if it would lead to more overall user information. If the extra information means each rating takes longer this interface could be less efficient overall. Because user attention is a limited resource [10], a general-purpose recommender system should balance the value of information gathered with the value of user attention.

To directly address the quality/effort tradeoff we introduce rating time into our analysis and compute preference bits per second (Pbps). Sparling et al. [21] found that the time required to rate is the best available measure of mental effort. Therefore Pbps is a way to balance the value of information with the value of user effort. We compute Pbps for an interface as Pbpr for that interface divided by the number of seconds per rating on that interface.

---

[2]The original analysis of the scales dataset implies that only 2795 ratings were collected [3]. We contacted the lead author, it appears that the 431 remaining ratings were excluded from the analysis there.

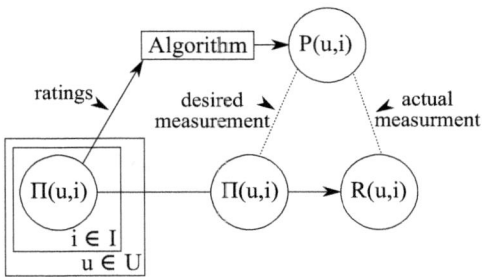

Figure 4: Graphical model of the output preference bit measurement.

Table 2 lists the average time per rating Sparling et al. report [21]. Because they do not asses the 10 or 20 point scales, the listed values are estimated by linear interpolation. We do not estimate a time for the 6-point scale because it is not clear how this interface compares with a traditional six star rating scale. Combining the times per rating from table 2 with our measured Pbpr for the thumbs and five star with half star interfaces we get $\frac{0.423}{3.91} = 0.1082$ and $\frac{0.813}{4.33} = 0.1878$ respectively. While this is a rough estimate, is suggests that, despite the faster ratings, the binary scale gives us information slower than the five star interface. This is evidence that the five star interface with half stars matches how users think about movies better than a thumbs interface.

4. OUTPUT PREFERENCE BITS

Measuring output preference bits: We would like to derive a measure for the output preference bits $I(P; \Pi)$. Ideally, we would compute this value directly, however, as in the last analysis, a user's true preference is hidden. Therefore, we analyze the relationship between a user's rating and the recommender's prediction for that rating. Let $P(u, i)$ represent the recommender's prediction for user u on item i. Figure 4 shows the relationship between preference, prediction, and ratings. If a prediction gives us information about an unknown rating, that prediction should also give us information about the unknown preference that formed the rating. Formally $I(P; R)$ is a lower bound to the output preference bits using theorem 1 with P as X.

$$I(P; R) \leq I(P; \Pi)$$

$I(P; R)$ tells us how much we expect a recommender's prediction to reduce uncertainty about the user's rating. We will use this measure as an estimate of the output preference bits per prediction for the rest of this paper.

4.1 Empirical Evaluation

We apply our methods to compare the efficiency of rating interfaces with different scales.

Datasets: The perfect dataset for this analysis would be large enough to generate meaningful predictions and it would ask the same users to rate the same items on several different rating scales. Unfortunately, no such dataset exists. Re-rating datasets such as Cosley et al's are too small to provide insights into predictive accuracy [3]. Larger datasets with different rating scales for the same items (i.e. Netflix and MovieLens) represent different users and applications, making it difficult to compare results. Therefore, we use two relatively high-precision datasets (MovieLens 10M [18] and

dataset	type	format	scale	users	items	ratings
scales	movies	discrete	varies	77	882	3226
ml	movies	discrete	0.5, 1.0, ..., 5.0	71K	10K	10M
jester	jokes	continuous	−10.0...10.0	59K	140	1.7M

Table 1: The datasets used in our analysis.

Jester [9]) to create synthetic datasets with different scales. Table 1 summarizes both datasets. The Jester dataset is very dense and uses a continuous ratings scale from -10.0 to 10.0. The MovieLens 10M dataset employs a scale from 0.5 to 5.0 with half point precision.

Assumptions and limitations: To show the connection between $I(P; R)$ and $I(P; \Pi)$ we assumed that prediction and ratings are conditionally independent on preference. This means that prediction is not a noise factor, and should have no ability to effect ratings. Unfortunately, this is not true in practice [3]. The Jester interface does not display predictions, and therefore does not suffer from this. Because the MovieLens interface does display predictions, our measured output preference bits may overestimate true output preference bits on the MovieLens dataset.

Choice of scales: In this analysis we must consider three separate rating scales. Users of each system generated ratings using the system's **native scale** (0.5 to 5.0 for MovieLens and -10.0 to 10.0 for Jester). Since we want to analyze the efficiency of different scales, we generate synthetic datasets for different **input scale**s whose precision is less than or equal to the native scale. The procedure for generating the mapping between native scale and input scale is outlined in the next section. Finally, we incorporate a separate prediction **output scale** for two reasons. First, this is common in practice; Netflix only supports whole-star rating input but displays tenths of a star for predicted rating outputs. Second, we found that requiring each input scale to be paired with the same output scale unfairly penalized input scales with less precision.

We considered a variety of input scales. For MovieLens, we chose 10-point (native), 5-point, and 2-point because they are common in practice. The choice for Jester was more complex because the native scale was continuous and our mutual information metric requires a discrete scale. We chose 100-point (for comparison against Sparling et al.'s data [21]), 20-point (buckets of length 1.0) and 10-point, 5-point, and 2-point to enable comparison with the MovieLens dataset. We consider a variety of output scales for both datasets: 2-point, 5-point, 10-point, 20-point, and 100-point.

Methodology: Our analysis uses the LensKit recommender software package [5] with five-fold cross validation. We use an item-item CF recommendation algorithm [17] with recommended parameter settings. Finally, we compute $I(R; P)$ using the Miller-Madow estimator described in section 3.1.

We constructed synthetic datasets that map a native scale to a variety of input scales using a greedy search over possible mappings from native scale to input scale. We initialized the best mapping to a uniform mapping where each input rating represents roughly the same number of native rating values. We evaluated the preference bits captured by predictions with this mapping. We then generated two possible candidate mappings for each threshold between input scale rating bins: one mapping that moves the threshold one native rating

scale points	2	5	10*	20*	100
ML	3.91	4.09	4.33	4.57	5.13
Jester	15.47	16.39	16.55	16.7	17.06

Table 2: Mean rating time per item in seconds for each scale and domain based on Sparling et al. Rating times with a * indicate interpolated values.

to the left and one that moves it to the right. We evaluated the preference bits captured by each of these candidate mappings and recursively repeated this procedure until the best mapping could not be improved. Because this process reduces the number of rating values in our datasets, it also reduces our maximum input preference bits. Therefore, we expect more granular scales to produce higher quality output than coarser scales.

Results: Figure 5 shows the relationship between input scale, output scale, and output preference bits for MovieLens (left) and Jester (right). Each line corresponds to a different input rating scale. The x-axis corresponds to different prediction output scales, and the y-axis shows output preference bits. For example, in Jester, the 5-point input scale (green line) and 10-point output scale ($x = 10$) gives 0.24 output preference bits. In general, increases in input scale granularity up to 10 (orange) lead to increases in output preference bits. For Jester, the 20-point input scale (purple) performs slightly worse, and the 100-point scale (red) performs significantly worse. Since the native scale for the MovieLens dataset is 10-point, we do not know whether performance would decrease for 10-point and 20-point input scales. This suggests that there is a "sweet spot" in rating scale granularity that balances between theoretical maximum input bits and natural rating noise. The prediction output scale (X axis) seems to reach a plateau when it is twice the granularity of the input scale. For example, in the Jester dataset the curve for the 2-point input scale (blue) levels off dramatically at the 5-point output scale. Increases in output rating scale precision never seem to lead to a meaningful decrease in preference bits. Therefore, we consistently use the 100 point output scale in subsequent analysis.

4.2 Reconciling the quantity / quality tradeoff

We have shown that a relatively granular 10-point rating scale seems to capture more preference bits for both rating input and prediction output. This does not, however, tell the whole story. Because user attention is finite [10], and rating on high precision scales takes longer [21], we again face a tradeoff between quality of ratings and the effort needed to obtain them.

This tradeoff is especially relevant with new users. System designers seek to provide new users with valuable predictions as quickly as possible - a challenge often referred to as the *new user problem*. To explore how a rating scale affects this challenge, we break down the previous results according to how many items a test user has rated. More precisely, for each test and train set in the cross fold validation, we build the algorithm's predictive model on the entire training set. We play each test user's ratings forward chronologically and measure the preference output bits in the withheld ratings. This analysis allows us to compare output preference bits per rating (output Pbpr) for different input scales.

To directly address the quantity / quality rating tradeoff raised by Sparling et al. [21] we calculate the output preference bits per second (output Pbps).

Table 2 lists the rating times Sparling et al. report. We use the rating times reported for product reviews as rating times for jokes because in both domains users form an opinion after reading text on the page. Although these assumptions may decrease the accuracy of our analysis, we believe they are close enough to shed light on the relative differences between rating scales.

Figure 6 shows the results for Pbpr and Pbps for MovieLens (left) and Jester (right).[3] Each line corresponds to an input scale. The x-axis shows total rating seconds for a test user calculated by multiplying the number of ratings for a user by the time per rating for that scale. The vertical dotted lines indicate the rating times for a user's n'th rating (n = 1, 2, ...). It is interesting to note that for Jester, while the 20-point scale (purple) clearly beats the 5-point scale (green) in the previous analysis (Figure 5), when we incorporate rating time into the analysis they essentially draw even.

Figure 6 clearly illustrates tradeoffs a designer might consider. For example, consider the times at which the 5-point and 2-point scales reach 0.28 preference bits for the Jester dataset. The 5-point scale gives 0.28 preference bits at approximately 5 ratings or 82 seconds. The binary scale meets the same level of preference bits at approximately 41 ratings or 634 seconds. The Pbps of the 5-point scale ($\frac{0.28}{82} = .0034$) and 2-point scale ($\frac{0.28}{634} = .0004$) clearly indicates that the 5-point scale is more efficient.

The testing set in our analysis was constant size. This corresponds to an assumption that recommender system will eventually capture the same quantity of historical ratings under different scales, despite the varying amounts of time required to generate those ratings. We also tested an assumption of fixed amount of user attention, resulting in smaller training datasets for systems with higher precision scales. However, because rating times only vary slightly across scales, our results were essentially identical to Figure 6.

4.3 Preference bits vs other metrics

We close our discussion of output preference bits by comparing it to other normalized measures of prediction quality. CF researchers most commonly measure prediction accuracy using root mean square error (RMSE) and mean absolute error (MAE) [11]. One attractive aspect of output preference bits is that it is scale free. If everything else is held constant, two scales with rating points associated with different numeric values should yield the same values for output preference bits. Although the standard versions of MAE and RMSE are not scale free, normalized versions of both metrics (NMAE, NRMSE) simply divide the typical measures by the range of the rating scale [19].

We compared output preference bits (for the 100-point output scale), NMAE, and NRMSE across all the analysis we conducted in this section: each rating scale, dataset (Movielens vs Jester), etc. In total, this yields 168 different datasets and observed values for each metric. These 168 observations are not independent because they are based on the same two datasets. However, this is the same general approach

---
[3]The output preference bit values for Jester are generally higher in Figure 6 than 5 because the median number of ratings in Jester is only15 ratings

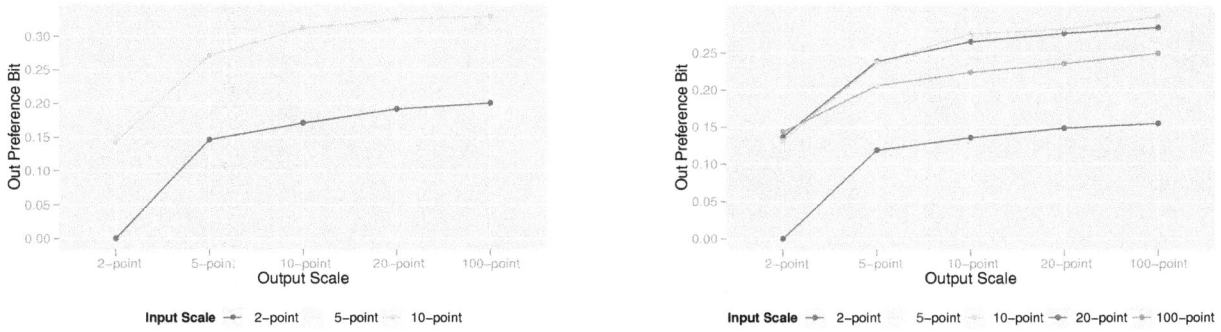

Figure 5: The relationship between input scale, output scale, and output preference bits for MovieLens (left) and Jester (right). Each line corresponds to a different input scale. Output scale granularity increases from left to right.

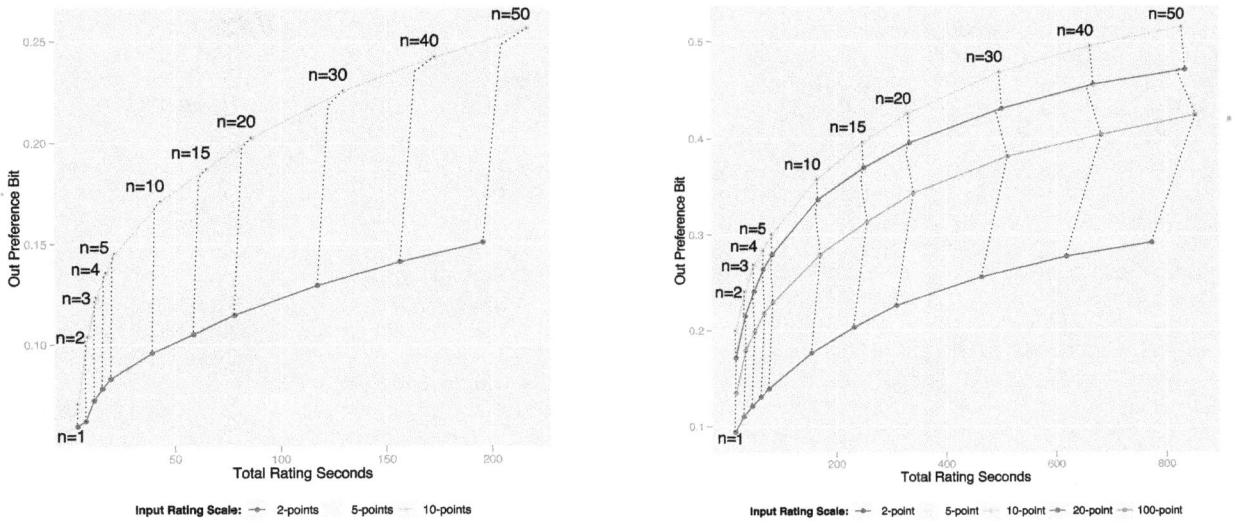

Figure 6: The number of output preference bits for different rating input scales as a function of total time spent rating. MovieLens appears on the left and Jester appears on the right. The vertical dotted lines connect the time it takes for a user to complete a certain number of ratings across all scales.

Herlocker et al. used to evaluate accuracy metrics in [11], and it should provide some intuition.

All three metrics are highly correlated across datasets. However, we found the correlations between NMAE and NRMSE to be much larger (0.98) than the correlations between output preference bits and the other two metrics (-0.73 for each). In addition, the standard version of mutual information and the version with the Miller-Maddow correction exhibited a very high correlation (0.99).

	pref bits	MAE	RMSE
pref bits	1.00	-0.92	-0.90
MAE	-0.92	1.00	0.94
RMSE	-0.90	0.94	1.00

Table 3: Correlation between preference bits, MAE, and RMSE controlling for input scales.

We wondered whether the strong correlations between NMAE and NRMSE were due to their shared normalization procedure. To explore this, we analyzed metric correlations within each input scale and averaged the results across all input scales. Table 3 shows the mean correlation across all input scales for each metric. Note that because the normalizing term is constant given scale, the within-scale correlations for NMAE and NRMSE will be identical to those for MAE and RMSE. The correlation between NMAE and RMSE decreases from 0.98 to 0.94, while the correlations between output preference bits and MAE (-0.92) and RMSE (-0.94) both increase.

These findings suggest that when comparing within an input scale, output preference bits, MAE, and RMSE provide similar evaluations and any one might be preferred. However, across input scales, output preference bits seems to capture different characteristics from NRMSE and NMAE. Because of the theoretical grounding and inherently scale-free nature of output preference bits, we believe output preference bits will make reasoning about differences between rating scales easier, and should therefore be preferred. However, these findings should be verified on other datasets.

5. DISCUSSION AND CONCLUSIONS

Natural noise in user ratings can decrease the accuracy of recommender systems. We develop the preference bits framework to help system designers measure, understand, and reduce noise in recommender systems. We derive measures of the preference bits in ratings and predictions to measure the amount of information users give the a recommender system, and how much information they receive in return. We show how this framework can be used to compare different rating interfaces and find the one that is most efficient.

To look at how noise effects predictions we introduce a technique for reducing the precision of a rating dataset to produce synthetic datasets. This technique enables us to analyze the amount of information we can give to users if we have them rate on different scales. Using these datasets we confirmed ratings on a higher precision rating scale improve predictive accuracy. Future research should verify that these findings with the synthetic datasets holds for natural datasets with different scales. A full user study could reveal insights into the ways users might change their mental rating process for different scales. For example, users may think more carefully about the distinction between an up and down rating than the distinction between 4.5 and 5.0 stars.

To capture the tradeoff between the cost and quality of rating systems we introduced the preference bits per second metric. While we believe this metric is a major step forward, future research can refine it more. For example, the metric should be extended to account for user satisfaction because a good interface is worthless if users do not want to use it. Ultimately, reducing noise should always be balanced against the desire of the users.

We investigated the effect of different rating scales on the preference bits per rating. While we focused on rating scale, many other aspects of a recommender system may affect the amount of noise in ratings. For example, tags displayed alongside a movie ("silent movie", "1920s") may either reduce noise by aiding recall, or, increase noise by biasing ratings. Work is still needed to understand how these types of contextual information reduce noise in ratings.

6. ACKNOWLEDGEMENTS

This work has been supported by the National Science Foundation under grants IIS-0964697, IIS 10-17697, IIS 09-68483.

7. REFERENCES

[1] X. Amatriain, J. Pujol, and N. Oliver. I like it... i like it not: Evaluating user ratings noise in recommender systems. In *UMAP 2009*, pages 247–258. Springer (2009), 2009.

[2] X. Amatriain, J. M. Pujol, N. Tintarev, and N. Oliver. Rate it again: increasing recommendation accuracy by user re-rating. In *RecSys 09*. ACM, 2009.

[3] D. Cosley, S. K. Lam, I. Albert, J. A. Konstan, and J. Riedl. Is seeing believing?: How recommender system interfaces affect users' opinions. In *CHI 03*. ACM, 2003.

[4] T. M. Cover and J. A. Thomas. *Elements of Information Theory (Wiley Series in Telecommunications and Signal Processing)*. Wiley-Interscience, 2006.

[5] M. Ekstrand, M. Ludwig, J. Konstan, and J. Riedl. Rethinking the recommender research ecosystem: Reproducibility, openness, and LensKit. In *RecSys 11*, pages 133–140. ACM, 2011.

[6] M. D. Ekstrand, J. T. Riedl, and J. A. Konstan. Collaborative filtering recommender system. *Foundations and Trends in Human-Computer Interaction*, 4(2):81–173, 2010.

[7] B. Fischhoff. Value elicitation: is there anything in there? *American Psychologist*, 46(8):835, 1991.

[8] W. R. Garner. Rating scales, discriminability, and information transmission. *The Psychological Review*, 67(6):343–352, 1960.

[9] K. Goldberg, T. Roeder, D. Gupta, and C. Perkins. Eigentaste: A constant time collaborative filtering algorithm. *Information Retrieval*, 4(2):133–151, 2001.

[10] F. Harper, X. Li, Y. Chen, and J. Konstan. An economic model of user rating in an online recommender system. *User Modeling*, pages 149–149, 2005.

[11] J. L. Herlocker, J. A. Konstan, L. G. Terveen, and J. T. Riedl. Evaluating collaborative filtering recommender systems. *ACM Trans. Inf. Syst.*, 22(1):5–53, Jan. 2004.

[12] W. Hill, L. Stead, M. Rosenstein, and G. Furnas. Recommending and evaluating choices in a virtual community of use. In *CHI 95*, pages 194–201, 1995.

[13] G. Miller. Note on the bias of information estimates. *Information theory in psychology: Problems and methods*, 2:95–100, 1955.

[14] M. P. O'Mahony, N. J. Hurley, N. Kushmerick, and G. Silvestre. Collaborative recommendation: A robustness analysis. *ACM Transactions on Internet Technology*, 4:344–377, 2004.

[15] M. P. O'Mahony, N. J. Hurley, and G. C. Silvestre. Detecting noise in recommender system databases. In *IUI 06*. ACM, 2006.

[16] L. Paninski. Estimation of entropy and mutual information. *Neural Computation*, 15(6):1191–1253, 2003.

[17] B. Sarwar, G. Karypis, J. Konstan, and J. Riedl. Item-based collaborative filtering recommendation algorithms. In *WWW 2001*, pages 285–295. ACM, 2001.

[18] S. Sen, J. Vig, and J. Riedl. Tagommenders: Connecting users to items through tags. In *WWW 09*, pages 671–680. ACM, 2009.

[19] G. Shani and A. Gunawardana. Evaluating recommendation systems. *Recommender Systems Handbook*, pages 257–297, 2011.

[20] C. Shannon. A mathematical theory of communication. *ACM SIGMOBILE Mobile Computing and Communications Review*, 5(1):3–55, 2001.

[21] E. I. Sparling and S. Sen. Rating: How difficult is it? In *RecSys 11*. ACM, 2011.

[22] Youtube. New video page launches for all users @http://youtube-global.blogspot.com/2010/03/new-video-page-launches-for-all-users.html, Mar. 2010.

High Quality Recommendations for Small Communities: The Case of a Regional Parent Network

Sven Strickroth
Department of Informatics,
Clausthal University of Technology
Clausthal-Zellerfeld, Germany

sven.strickroth@tu-clausthal.de

Niels Pinkwart
Department of Informatics,
Clausthal University of Technology
Clausthal-Zellerfeld, Germany

niels.pinkwart@tu-clausthal.de

ABSTRACT

Traditional recommender systems are well established in scenarios in which "enough" items, users and ratings are available for the algorithms to operate on. However, automatic recommendations are also desirable in smaller online communities which only contain several hundred items and users. Collaborative filters, as one of the most successful technologies for recommender systems, do not perform well here. This paper argues that recommender systems can make use of contextual information and domain specific semantics in order to be able to generate recommendations also for these smaller usage scenarios.

The new hybrid recommendation approach presented in the paper enhances traditional neighborhood-based collaborative filtering techniques through the use of new kinds of data and a combination of different recommendation methods (rule, demographic, and average based). While the algorithmic techniques presented in this paper are suitable (especially) for smaller online communities, they can also be applied to improve the quality of recommendations in larger communities.

The approach was implemented and evaluated in a small regional bound parent education community. A multi-staged evaluation was conducted in order to determine the quality of recommendations: A cross-validation (recall), an expert questionnaire (recommendation quality) and a field study (user satisfaction). The results show that recommenders even for smaller communities are possible and can produce high quality recommendations.

Categories and Subject Descriptors

H.3.3 [**Information Storage and Retrieval**]: Information Retrieval and Search—*Information Filtering*

Keywords

Collaborative Filtering, Recommender System, Small Communities, Sparsity, Implicit Ratings.

1. INTRODUCTION

Recommender systems are very popular both for E-Commerce (e.g. Amazon, Netflix) and the research community [3, 4, 17, 20], as these can calculate potential interesting items for users based on their interests. One of the most successful technologies for this task is Collaborative Filtering (CF) [4, 20]. These traditional

recommender systems are well investigated and well established in scenarios in which a big amount of items, users and ratings are available for the algorithms to operate on (like MovieLens, Netflix, Amazon, …).

However, automatic recommendations are also desirable in smaller online communities which only contain several hundred items and users: Finding interesting items within several hundred alternatives already is not an easy, but time consuming task. It is often assumed that recommendation cannot work or performs badly in such scenarios without special adjustments [20]. Thus, there was little research on this field and there are no optimized approaches available.

We present a new hybrid approach especially suited for smaller online communities. Our approach enhances traditional neighborhood-based collaborative filtering techniques through the use of new kinds of data, contextual information from the community itself and a combination of different recommender methods. We will argue in this paper that recommender systems following this approach are able to generate recommendations also for these smaller usage scenarios.

We designed, implemented and evaluated our approach in a small regional bound parent education community.

In the next section we describe our usage scenario and why recommending interesting articles/items automatically is important in our scenario. The following section illustrates our algorithmic approach and describes the whole recommendation process in detail. The fourth section presents the evaluation of our hybrid recommender system prototype within our usage scenario.

2. USAGE SCENARIO

Our approach was implemented and evaluated in the context of the "Mobile2Learn" community in Germany [21]. This is a small parent community which focuses on education and upbringing of young children (ages up to 6), membership is free. Mobile2Learn is not a plain online community, but a community which combines "real world" events and workshops with new media (a Web 2.0 online platform and mobile phones).

The motivation for creating this community was that appropriate early childhood education is very important for the development of young children. Parents have a major influence on the process of early childhood education processes and development phases of their children. [2] (German context) and [5] suggested that the social origin and education of parents have a bearing on the future and academic achievements of their children. However, it is not easy to reach and motivate especially underprivileged parents – but it is well known that underprivileged parents need more assistance and normally do not take part in (offline) workshops about educational topics cf. [1] for a German report about parental training. Hence, the Mobile2Learn project was started by the community college Goslar, LEB (an institution for rural parent education) and Clausthal University of Technology. The goal of this project was to combine the advantages of regional (offline) workshops/events and new media: parents could access pictures and content of the workshops online (anonymously) and could stay in contact with each other. Also, through the online portal, an intention of the project was to motivate parents to attend events.

In the Mobile2Learn project, the educational activities were structured into six areas (including "learning with all senses", musical education, nature discovery and "speaking and listening"). For each area, six independent thematically related events in different kindergartens were conducted; access to these events was open for everyone (not only parents whose children attend this facility). During an event, all parents were asked to sign on an attendance list. After all events of an area were held, taken photos and articles about the contents of the events were put online. So parents could look up all contents (again) and could find further information. Moreover, also parents who did not attend any events could view the photo galleries and could possibly be motivated to attend events in the future.

In order to reach parents of the target group, a network of partners was established: people and institutions which were in direct contact with parents and where parents go to if they need help (e.g. child doctors, youth welfare office and so on). Partners of this network helped to advertise the project. They were asked to invite and motivate parents to attend a workshop or to register on the online platform.

As already mentioned, the contents of the workshops were available as articles and interactive quizzes on the online platform. A few articles were available without registration on the platform to motivate parents to register. A registration in the system was required, however, to access most articles and all photo galleries. For registration users had to enter an e-mail-address, mobile phone number (optional), nickname, date of birth, gender and residence. Also date of birth and gender for their children could be entered optionally. Most of the articles and photos were provided by pedagogues, but it is possible for parents to submit their own articles or pictures (which were reviewed by pedagogues before publishing). All articles and photo galleries could be rated with one to five stars (dislike to like) and could be commented on by community members. Apart from this, many articles were annotated with ages of children to which they mainly apply to.

A key point of the online platform was that regular personalized messages (campaigns) were sent to parents via SMS and/or e-mail. These campaigns informed parents about upcoming events, new articles or pictures. Also, automatically generated recommendations were included in the e-mail campaigns (recommendations generated by the approach we describe in the next section). This repetitive, direct way of contacting community members was chosen to continuously and actively "push" information about educational opportunities to parents. So, parents were regularly reminded that the project still exists and could easily access all new and "interesting" items.

The Mobile2Learn.de online community was launched in November 2010. As early as this date, users found a small number of 30 articles on that website, which could be commented and rated (with one to five stars). By April 2012 about 250 users were registered; 170 articles and 65 photo galleries were available.

3. ALGORITHMIC APPROACH

In the Mobile2Learn environment it is crucial that users find interesting articles easily – otherwise it is not possible to successfully deliver our educational contents to our users. Users

of the target group have children (i.e. not much time) and might be unwilling to search for interesting articles. Thus, the addition of the website with a recommender system was decided and planned by April 2011. As rating of items was possible since the start of the website and users already read and rated some articles, we did not have to deal with a new system/community.

There is a set of some special properties in our scenario "Mobile2Learn" (which are probably symptomatic for many smaller communities): Small size (few users and items), data sparsity (sparsity level very close to 1; 60 ratings, 151 items, and 175 users at design time), different item types (articles, quizzes, photo galleries) and combination of online/offline items (events).

There seems to be a lack of research focusing on automatic recommendation for small communities. To our knowledge there are no implementations which handle all of these properties. However, there are approaches which handle single aspects: Knowledge-based filtering [3] can easily be used for small datasets (especially if they can be objectively classified on a set of different criteria). They do not rely on ratings or user profiles and allow users to pull recommendations. Filters can also use cold-user data (e.g. demographic data from user profiles) to generate recommendations for users who have not rated (many) items yet [15]. Content-based algorithms [17] can find and calculate similar items based on annotations/tags or content of the items (no ratings required). For recommendations, however, ratings or other (interest-)profiles are required. Collaborative filters [7, 19, 20] are domain independent and can incorporate different item types easily. Instead of using different algorithms, it is possible to (actively) reduce a high sparsity level: A possible approach is aggregating/clustering of articles [10]. Here, several articles are combined/clustered to one (dummy) article with a shared rating (e.g. (un)weighted arithmetic mean). (Content-)Boosting [12] employs a content-based recommender to automatically generate predictions for every single user. Then the generated predictions are used as user ratings and included in a collaborative filtering recommendation process (which generates the final recommendation for the users). [16] and [18] suggested to automatically create ratings for dummy users in order to reduce sparsity and overcome cold-start issues. Apart from relying on (few) explicit ratings, also implicit ratings based on user behavior can be used [9, 13].

Knowledge-based recommenders would work for our structured educational articles (e.g. by article types or prerequisites on age of the children), however, they require active cooperation from users (as well as some content-based approaches). The quality of content-based recommenders heavily depends on consistent tagging and/or good metrics to automatically calculate item similarities. In our scenario, tagging is not an easy task, since within a thematic area all articles require quite similar tags and multiple users are providing content. Furthermore, without good tagging automatic calculation of similarities of different item types (such as photo galleries, articles and quizzes) is impossible. Apart from this, recommendations only contain very similar items and are quite static [18]. Accordingly, those are not applicable in our scenario. As already mentioned, collaborative and demographic filters require special adjustments or "enough" data to work on or in order to generate high quality recommendations. CF based approaches like clustering are not appropriate for smaller scenarios, since these reduce the number of items again – to an even lower number. Also it is not possible to recommend single items, but only whole clusters. A combination of different approaches might be the key for smaller scenarios. Content-boosting, however, is not a viable option since it requires a good content-based recommender. As mentioned above, in the

Mobile2Learn.de scenario, there are few ratings and most articles are quite similar (due to their structure and tags).

We attempted to solve this task by using a hybrid recommender system which incorporates different approaches. The proposed recommender generates a TOP-5 ranking of articles, photo galleries and upcoming events which the user might like most (w/o the presentation of a prediction). To effectively overcome the high sparsity level, implicit ratings were taken into account (this way we gained 1438 ratings; sparsity level could be reduced to 0.94). For this reason, visit times of articles (based on the webserver log) were transformed into ratings. Additionally, our approach makes use of several different sources of data: Demographic information from the user profiles (residence, age, gender) and two types of contextual and sematic information (as [14] has demonstrated that using underlying semantic can improve recommendation quality): (entered and inferred) age of the user's children and location of residence together with information about visited events (based on attendance lists of presence events in kindergartens). Apart from this data, structural information about the articles is also included into the recommendation process to leverage the new item cold-start problem: System-internally, articles are structured as a tree. Whereas the leaves contain the main (educational) content and the inner nodes are (topic) overviews and link pages. It is assumed that the distance of articles in the tree reflects the relatedness of the content. Based on the distance pseudo ratings are generated automatically for all articles. Even if the scenario is a smaller one, we focused on item based CF [19] which performs better than the user-based approach [7]. The entire recommendation process is made up of two phases: The preparation and model building phase as well as the recommendation generation phase (section 3.3 explains our choice of parameters).

3.1 Preparation and model building phase

The first phase consists of three steps: First visit durations are extracted from the webserver log and transformed into the same scale as the explicit ratings. The visit durations were calculated as follows (only for articles, not for other entities): We wrote a small log file parser, which measured the time between two HTTP requests for HTML pages of a user (we ignored images, CSS, JavaScript and AJAX requests). This way we got all visit durations within a session, except for the last article of a session if there is no user logout or other page request (since there is no following request which could be used to calculate the visit duration). We stored the maximum visit duration for each article per user over all sessions which were between three seconds and ten minutes in order to reduce the noise. This is important because we have to remove pages which were just skipped and it is not possible to measure the real reading durations, since we do not know if the user went away from the computer (e.g. to collect a coffee). Apart from these fixed limits, we implemented a dynamic limit to reduce noise more properly: After determining all visit durations and normalizing them by dividing by the number of words of the regarding article, we calculate the standard deviation and remove all visit durations which deviate more than two standard deviations from the average. In order to transform these normalized visit durations into our rating scale, we calculate the standard deviation (σ) again and use the transformation instruction (where m is the mean of all visit durations of a user) shown in Table 1. Using this transformation, we achieve that moderately long visited articles (the majority) get a moderate rating of three. Longer and shorter visited pages get higher resp. lower ratings exploiting the whole rating interval.

Table 1. Instructions for transforming normalized visit durations into ratings

Rating	Interval of normalized visit durations
1	$(-\infty; m - 1.64\sigma]$
2	$(m - 1.64\sigma; m - \sigma]$
3	$(m - \sigma; m + \sigma)$
4	$[m + \sigma; m + 1.64\sigma)$
5	$[m + 1.64\sigma; \infty)$

[9] noted that one should not use negative feedback (e.g. ratings 1 and 2), because by observing the users behavior, one can only infer which items they probably like and thus chose to consume. In our scenario, however, the exclusive use of positive feedback led to excessive positive recommendations for some articles (especially longer ones). We think that in our context it is possible to infer by a small *maximum* visit duration that a user might not like an article, since an uninteresting article never gets read completely/for a long time without being skipped.

Contrary to [13], in our usage scenario it is necessary to normalize the durations with the length of the article before transforming them into ratings. Otherwise longer articles get higher ratings even if the normalized visit duration is the average or below. Too heterogeneous articles might be a reason for this.

The second step includes the generation of the pseudo ratings in order to integrate structural information of the article tree into the CF approach (inspired by [6]). This is done using an adapted filterbot approach [16]: We created filterbots/agents which rate items based on their position and relation in the article tree. On each node (which has at least one child) of the article tree we start a depth-first search which rates the node itself and all child nodes with a rating of 5. Fig. 1 shows an example article tree for which four filterbots are needed. Fig. 1 also shows the resulting ratings: Filterbot no. *1* starts in node *a* and creates ratings for *a* and all child nodes (*b, e, c, d, f, g*) of a. Filterbot no. 2 starts in node *b* and creates ratings for *b, c* and *d*. Filterbot no. 3 and 4 perform accordingly.

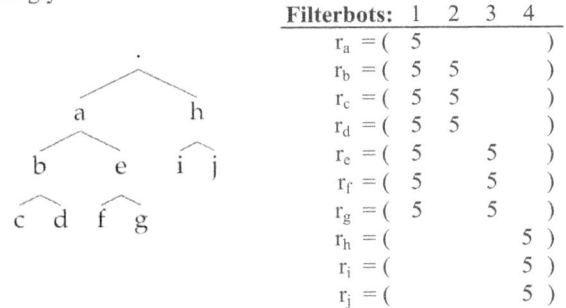

Figure 1. Generalized article tree and related filterbot ratings

By employing this schema (and using a decent metric, like the adjusted cosine similarity metric [19]), we achieve that articles, which are closely related (e.g. *c* and *d*) have a high similarity value ($s(r_c, r_d)=1$) while articles, which are not related (e.g. *f* and *i*) have a low similarity value ($s(r_f, r_i)=s(r_h, r_c)=0$). This schema is employed to be able to recommend new unrated items (new item cold-start problem) and to extend the neighborhood with articles of the same topic.

We also created one filterbot for each thematic area (e.g. "learning with all senses", see section 2) which generates ratings for all articles and galleries of an area. This was done in order to create a link between the articles and the corresponding galleries.

Finally the similarity matrix for the collaborative filtering is calculated. Here the adjusted cosine similarity metric [19] is used. As [20] recommended, we only use positive similarity values (also using negative similarity values led to a worse recommendation quality). Being noisier than explicit ratings, implicit ratings are devalued (multiplied with 0.85) and overridden by explicit ones if available. Photo ratings are aggregated and averaged and stored as a rating for the photo gallery to which they belong. This aggregation is useful here, because it was not possible to generate reliable implicit ratings for photo galleries: Single photos are available through different URLs and change automatically after ten seconds in the slide mode. Also it does not make sense to recommend a single photo in our scenario.

Also, notable for the calculation of the similarity metric in small scenarios, the question how to handle items which have just one rating with exactly one user in common is notable. In smaller or real sparse scenarios those are not rare. If those items are not ignored, the similarity of two items will be 0 (pearson coefficient) or 1 (cosine based metrics) depending on the similarity metric. We evaluated different metrics with such similarities ignored and included (see section 3.3).

The preparation phase does not have to be executed for every recommendation request, but should be recalculated regularly (the system version reported on in this paper executes it once a day).

3.2 Recommendation generation phase

The recommendation generation is based on a multilevel approach with a decreasing degree of personalization and different approaches. If there were less than five candidates generated, the next (less personalized) level is executed until enough recommendations are available or the last level is reached. The criterion for considering an artifact a candidate for recommendation is a predicted rating that exceeds a threshold of 3 (out of 5).

The first level generates rule-based recommendations: Based on location of residence and on visited and upcoming events, recommendations for events are generated. Event recommendations based on collaborative filtering approaches are not possible in this scenario because all events are independent and articles which describe events are published after the events are held. Hence, we constructed simple rules in which contexts (specific) events are recommended: Having a database with personal information, we calculate the distance (in km) between the residence of the user and the location of an upcoming event. If the distance is less than 7 km (average distance + standard deviation of people who already visited events) or the user already attended an event in a location, we insert a recommendation for the nearest event within one topic/area with a "predicted" rating of 5. Such a recommendation stays in the recommendation list until the user clicked on it or the event took place. We assume that each new workshop topic is interesting for all parents. Additionally, recommendations for overview articles and photo galleries of visited events are generated if the user has not accessed these before. These rule-based recommendations are based on generic assumptions about the interest of people (e.g., if you were at a kindergarten workshop, then it is very likely that you'll be interested in the photos). The predicted ratings for recommendations of this type are based on the number of elapsed days since the event took place: We employed the exponential time function $\exp\left(\frac{-t}{T_0}\right)$ for this task, with a value of 810 for the half-time parameter T_0. This value was chosen, so that the items

can show up in the top-5 recommendation list for 413 days (i.e. the predicted rating for an item greater than 3; 413 was arbitrary chosen so that also the first evens could show up in the top-5 recommendation list when we performed the evaluation): $5 \cdot \exp\left(\frac{-t}{810}\right) = t_{\text{min. rating}} = 3 => t \approx 413$. This is based on the assumption that the interest in photo galleries and articles of a past event decays as time moves on and recommendations should not stay in the top-5 recommendation list forever.

The second level consists of a traditional neighborhood-based collaborative filtering algorithm (CF). This is the level which incorporates the ratings created by filterbots. We employed the item-based algorithm as described by [19]. Instead of setting a fixed limit of ratings a user must have done before he gets recommendations, we employed a dynamic limit based on the "significance weighting factor" (also called "devalue") [7]. If an item has less than 5 items in its neighborhood, the similarity/correlation is multiplied by $n/devalue$ (where n is the size of the neighborhood). By employing a $devalue >=$ max. possible rating, it is possible to establish a minimum number of ratings of the current user indirectly. We chose 5 in our scenario. On the one hand this algorithm is able to generate personalized and cross-category recommendations (e.g. items in different categories, the user normally would not have looked at, [18]). On the other hand it has a major drawback. A user must have rated some articles (explicitly or implicitly) in the past in order to get (good) recommendations. However, recommendations should also be generated for new or more passive users. That is why the next two levels were included.

The third level uses alpha-community spaces (AC) combined with the level of agreement method [15]. Here, similar users (according to a property alpha) are grouped together (into an alpha-community) and within each group the level of agreement method is applied: For each item which was rated by at least one member of a group, the average rating within the group is calculated. Afterwards, a threshold is applied and items which are rated by at least by 20 per cent of the members of the group and exceed a minimum average rating are recommended. Instead of using a partition (pairwise disjoint groups) of the space, we relaxed this prerequisite of [15] in order to allow dynamic groups: In our approach, we applied this alpha-community spaces approach considering age of the user (interval of 6 years), age of associated children (interval of age of youngest child to the age of the oldest child), users living close to each other (<= 7 km distance), and visit of same events as alpha properties.

The fourth and last level uses the average rating (AVG) among all users as a prediction of ratings for articles. Of course, thresholds are applied in this case, too: Items have to be rated by at least nine users (called "agreement"). So, recommendations without personalization are possible for "anonymous" users. Recommendations based on average ratings are not optimal, however, "relevant recommendations without personalization are extremely useful for the vast number of anonymous user[s] as there is no user profile" [22, p. 249].

After the last level has been reached or after a level at least five recommendation candidates are available, a final weighting is performed. This is done in order to assign a higher weight to recommendations generated by more personalized approaches. Thus, our hybrid recommendation approach follows the "weighting" approach of the taxonomy of [4]. For an item this was done using a linear combination with weighting factors and the predicted ratings of the different recommenders. We chose a weighting of 1.1 for collaborative prediction, 1.05 for the alpha-community spaces prediction and 1 for all others (based on our experience with test runs). Furthermore, semantics are taken into account to optimize the recommendation candidates: Not only the entered age of the children of the profile is used but also the system tries to infer the child ages the user has major interest in (based on reading behavior of age annotated articles): The idea behind this works as follows: If the users read more than 75 % of articles which are annotated with "ages up to 3", articles for "ages 4 to 6" are devalued (if the user does not have a child in the profile in this age interval) and vice versa. Hence, it is possible to react on varying circumstances in which parents forget or do not enter new children (e.g., the system detects if a parent whose profile only shows a 6-year-old child now also cares for content about new-borns and adapts its recommendations accordingly). The fact if a user attended an event or not did not have an impact on the recommendation of articles, since we wanted to show interesting articles to all users (statistics show that 40 % of the users who attended an event also read articles associated with unattended events and vice versa).

Using this hybrid approach, it is possible to generate recommendations for all types of users (more passive and more active users). The more ratings a user has made, the better and more personalized the recommendations get. Recommendations are presented on the start page, and they are included in e-mail campaigns. Single recommendations (one item) are also sent one-time via SMS text messages if the rating exceeds a predicted rating of 4.

3.3 Parameter optimization

The proposed algorithm is based on several parameters: First of all the threshold – the minimum predicted rating for items to be included into the final top-5 recommendation list –, the similarity metric, the significance weighting, the used alpha communities, the agreement parameter of the alpha communities, the number of minimum ratings for items to be considered in the average based recommendation, and the weight of each recommendation algorithm for the final rating. Apart from the final weighting factor (which is based on practical experience), all parameters have been determined by a parameter optimization based on cross-validations and measuring the recall (see section 4.1, [8]). We tried to optimize one parameter at a time. However, we found for the CF that the devalue parameter, the similarity metric and the inclusion/exclusion of items, which have just one rating of one user in common, were not unrelated. So we performed cross-validations for a set of sensible values. Exemplarily we want to show results of different metrics:

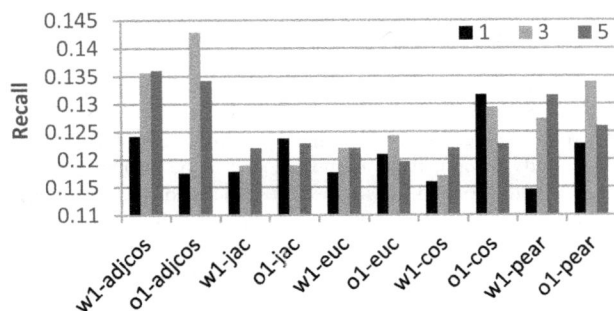

Figure 2. Recall based on different metrics and devalues

As Fig. 2 shows we evaluated different metrics as of: Adjusted cosine (adjcos), Jaccard based (jac, number of similar rated items divided by the number of co-rated items), Euclidean based (euc, 1/(distance + 1)), Cosine and Pearson correlation [19]. We performed cross-validations for each metric with three devalues (1, 3, and 5) and items which have exactly one co-rated item in common included (w1) and ignored (o1), as these are correlated and not independent. The adjusted cosine metric seems to work best, closely followed by the Pearson correlation and pure cosine metric. The peak of "o1-adjcos" with a devalue of 3 seems to be an anomaly (various test runs showed this). We chose the w1-adjcos with 5 as devalue parameter combination. In a manual evaluation we compared the top-5 recommendation lists for 10 % of our users and found that the "w1" variant produced recommendations with higher diversity (compared to the "o1" variant).

Fig. 3 shows the recall in the optimization process after each optimization and integration step. Remarkable is the recall of the plain CF approach (0.02) compared to the random selection of items (0.043). This result confirms the assumption that plain CF (and also alpha-community, Fig. 4) algorithms do not perform well in small scenarios. However, the combination of different algorithms and approaches performs better. We started our optimization by employing a plain CF algorithm and integrated the alpha-communities and average-based approaches (in two steps). Then we optimized the agreement for the average based algorithm, the similarity weight of the CF approach and the parameters of the alpha-communities. At last we integrated the age guessing (domain dependent optimization). With each optimization it was possible to raise the recall. The age guessing does not seem to have a high impact on the recall; however, its effects were clearly visible on the top-5 recommendation lists.

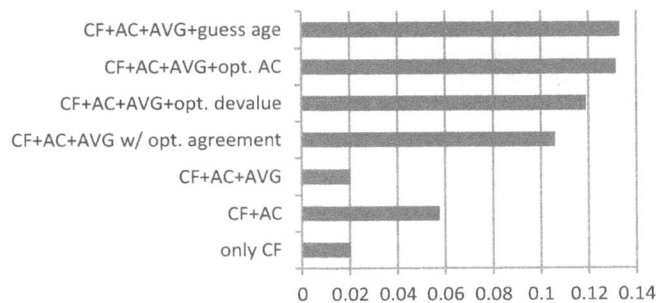

Figure 3. Recall improvements during parameter optimization

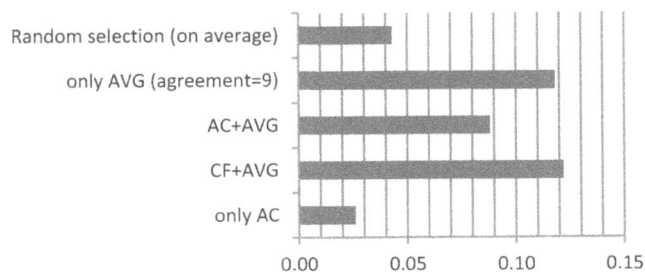

Figure 4. Recall of different recommender combinations

Fig. 4 also depicts the recall of different combinations of the recommendation algorithms. An examination shows that the CF and alpha-community approaches seem to be notably worse than the average-based approach or other combinations. Especially the average-based approach seems to stand out: We suppose the reason for this is our community in which a majority of users read and rated few articles.

4. EVALUATION

Evaluation of recommenders is not an easy task [8] especially for recommenders in smaller communities: A lot of recommenders were only evaluated using an offline evaluation with a metric and cross-validations. This might be a viable option for bigger systems, but smaller systems suffer from a small number of ratings and items [22]. Also, an offline-evaluation cannot really measure user satisfaction [11, 20], but it is often used as a heuristic approximation for it. [11] found that a high prediction accuracy does not always correlate with high user satisfaction.

To face these problems, a multistage evaluation was conducted: An offline evaluation (cross-validation), an expert questionnaire, and a field study (online evaluation).

User satisfaction was central for us, even if it is hard and sophisticated to measure. In addition we also wanted to evaluate the quality of the recommendations as measured by domain experts.

Each stage evaluates different aspects, but also has characteristic weaknesses. There is no single method which combines all these aspects and is applicable to our target group. These three stages complement each other and give an overall view how the recommender performs as a whole. In the following sections we will describe the three evaluations and will present our results.

4.1 Results of our offline evaluation

The offline evaluation was conducted to measure the recommendation quality using a metric. The result is heavily dependent on the employed metric. Common metrics are Mean absolute error (MAE), Rank-based metrics, click-through rate CTR, and Recall/Precision [8, 22]. MAE is not optimal for rank based recommendations, since they do not measure where differences of predicted and actual ratings occur (top or bottom of recommendation list). Rank-based metrics (like Spearmans correlation or Kendalls Tau) compare the order of the recommendation list with an order based on user ratings. Additionally to the position problem, here partial order is also problematic. [22] has shown that CTR and user experience are not consistent. Overall, recall/precision seemed to fit best into our scenario.

For the offline evaluation 2,000 cross-validation runs were carried out. The cross-validations were measured as described in [8]: Rated items by a user were randomly divided into two sets (test set and training set). The test set contained 20 per cent of the rated items (afterwards the test set was cleaned up so that only "relevant" items were in it, i.e. items with rating >= 3). Recall and precision were measured as follows: The recommendation system was initialized and executed on the training set and then recall and precision were measured. Recall is the number of recommended relevant items (i.e. items which are recommended and in the test set) divided by the number of items in the test set and precision is the number of relevant recommended items divided by the number of recommended items (5 here). Recall and precision were not independent on our small data sets (the test-set contained at most 5 items): both are linearly dependent. So it is irrelevant which one or which combination is selected; especially if the number of recommended items is fixed (5 in our scenario). As such, only recall was taken as a measure. The goal of this measurement was to compare the average recall of the proposed approach to the average recall of a random selection of items. We found that the algorithm (recall: 0.13) outperformed a random recommendation (recall: 0.04).

Moreover, we measured the recall for three classes of users, based on their read-behavior: "read < 5 articles" (47 % of user base), "read > 20 articles" (14 % of user base) and users between these two (39 % of user base). Again all recalls (see Fig. 5) were higher than the random selection of items. Fig. 5 shows that the recall for people who read few articles (< 5) is very high (approx. twice as

high as for users who read many articles) and that the recall for users who read many items (0.1) is below the overall recall of our prototype (0.13). A possible explanation for the first is that the test set for a lot of users of this class is empty (hence, these are ignored in the calculation) and a lot of users reflect the majority opinion. For users who read more articles, a reason for this result might be that it is hard or even impossible to measure relevance of recommended articles, which are not in the test set. Based on this initial evidence of system success, we went to the second stage of evaluation.

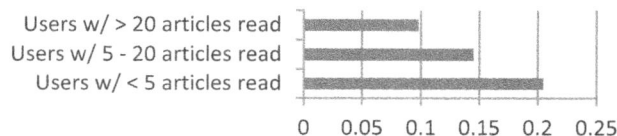

Figure 5. Recall for three different classes of users based on the number of articles read

4.2 Results of our expert questionnaire

In the second stage, an expert questionnaire was conducted to have four pedagogic experts evaluate the quality of the recommendations regarding pedagogic aspects. The experts were the pedagogues of the Mobile2Learn team. These knew the articles and the potential target group and could assess the recommendations best.

Twelve user profiles were taken randomly. However, the selection of profiles should be representative for the Mobile2Learn community. Therefore, we picked user profiles from the three classes mentioned before according to their occurrences in the community. Each profile showed anonymized demographic information about a user (residence, age, gender and children), attended events, all visited pages, visit durations (as "short", < 15 seconds; "mid"; "long", > 60 seconds) and explicit article ratings (if available). Also, for each profile, three different TOP-5 recommendation lists were shown. Two of these were generated by random selection of (unread) items while one was the output of the proposed recommender system. It was the design of the study, that the probability to choose a random selection is as twice as high as the calculated recommendation. The experts were asked to pick of the best of these three recommendation lists based on their own experience and the given user profiles.

The (normalized) results of this study show that three of the experts clearly preferred the recommendations of the algorithm. Overall the random recommendation lists where selected 18 times and the recommendation lists generated by our algorithm 30 times – i.e. each expert selected 7.5 generated recommendation lists on average. Two experts selected the generated recommendations more often than the random ones (11:1 and 9:3), one expert chose the two lists equally often (6:6) and the last expert chose 4 more random recommendations than the generated ones (4:8). Because of our study design, these numbers need to be normalized: This means that the four experts chose the generated recommendation lists to 96 %, 86 %, 67 % and 50 %. The reason for the 50 % of the last expert was that he had different opinions about when galleries should be recommended (only if a user has attended an event or also if the user did not attend any event to advertise them). Overall, for each profile at least one expert chose the recommendation of our algorithm and only for one profile three experts chose the same random generated recommendation list.

Biggest common sense of the experts with the calculated recommendations was for users who did not read many articles. Here, in four of six profiles all experts chose the recommendation of our algorithm. The reason for this might be the small sample of profiles with mid to many read articles (> 5 articles) and the recommendations itself: Especially for users with a detailed profile (many read articles) it is difficult also for humans to interpret the data and imagine/assess which articles/recommendation list might fit best.

Concluding there seems to be a clear tendency for preference of the experts for the recommendations generated by our proposed algorithm.

4.3 Results of our field study

The last part of the evaluation was a field study in order to measure user satisfaction. When the prototype was put productive and a user clicked on a recommendation, we displayed a small feedback box on the recommended page asking if the page is interesting for the user or not. The binary answer, together with the type of recommendation and the rank of the top-5-recommendation list, were stored in a database. Within one month, 43 recommendations were clicked on by 28 unique users. We sent two newsletters to 189 and 199 users and sent 14 recommendations via SMS. We got feedback for 20 recommendations by 17 unique users. All feedback was purely positive. The feedback we got covers all kinds of item types (events, galleries and article) and recommendation methods (rule-, average-, alpha-community- and collaborative filtering-based). Even if the number of 17 users sounds very small, this is approximately 10 per cent of the user base and the same number as the number of "power users" in the community.

18 clicks (42 %) on recommendations came from our newsletters. We cannot definitely say how many users accepted recommendations from SMS messages. However, two users visited the Mobile2Learn website and clicked on the same recommendation which was sent via SMS a few minutes ago.

We cannot know if users who did not give any feedback did not want to give feedback, did not notice the feedback box or if the recommended page was interesting for them or not. Anyway, we analyzed the log files and found out, that even if users did not give any feedback, they often visited a recommended gallery extensively or explored related articles (to the one which was recommended). We suppose that this is a sign for good recommendations, since our users were directed to a right direction which was interesting for them.

Additionally, we observed that the average number of total read articles per user increased from 8.9 (before the prototype was activated) to 9.2 (one month later), despite the fact that ten new users registered in the same time span.

5. DISCUSSION

In this section we present our experiences with explicit ratings and the click behavior of our users. Then we discuss other possible filterbot implementations and the recall evaluation metric we used for our parameter optimization and evaluation. Finally we point out further research tasks for recommendation systems in smaller communities and transferability to other scenarios.

The recommendation algorithm supports explicit ratings, however, this whole recommendation process is mainly based upon implicit ratings (which were generated out of visit times): At the design time, there were only very few explicit ratings (even a lottery for ratings did not encourage people to submit much more ratings). Also in the evaluation period, no noteworthy amount of new ratings was made.

In the evaluation period, we analyzed which ranks the items of the top-5 recommendation had which were clicked on: Clicks on items were not limited to the first recommended items of the top-5 ranking – but we found a clear preference for the first recommended items: 40 % of all clicks were on the first item, 60 % of all clicks were on the first two items and approx. 70 % on the first three items. This means that users did not just click on the first recommended item, but choose on which to click.

We also investigated different implementations of Filterbots in our scenario (see Fig. 6): Three different variants were implemented and tested. Each implementation differs in how similarities are calculated. The first variant is the implementation we described in section 3.1. On closer examination one can see a possible problem: The idea of the design was that the similarity metric reflects the distance in the article tree (see Fig. 1). $s(r_a, r_b) = s(r_a, r_e) = 1/2$ and $s(r_b, r_e) = 1/3$ should hold, but using the adjusted cosine similarity metric (with absence of other ratings) $s(r_a, r_b) = s(r_a, r_e) = s(r_b, r_e) = 1$ holds. Hence, we developed two other filterbot variants, which calculate the similarity of the filterbot ratings and the user ratings separately. Both variants (2 and 3) use the jaccard coefficient to calculate the similarity of the full filterbot rating vectors. For this metric the distance of articles in the tree as described before holds (i.e. $s(r_a, r_b) = s(r_a, r_e) = 1/2$). The difference of variant 2 and 3 lies in the way they combine the filterbot rating similarity with the user ratings similarity: Variant 1 uses the unweighted arithmetic mean. Variant 2 uses a weighted arithmetic mean where the user ratings count twice as high as the filterbot rating similarity. Fig. 6 shows that variant 1 seems to work best in our scenario and that the filterbot rating integration does not corrupt the quality of the recommendations (compared by recall). A possible explanation why variant 2 and 3 are worse than variant 1 might be a weight bias: Variant 2 and variant 3 use a ratio of 1:1 and 1:2 of filterbot rating similarity to user rating similarity – this might be a too high weight for the filterbot ratings. This opens room for further investigations with dynamic weights (based on the number of user ratings).

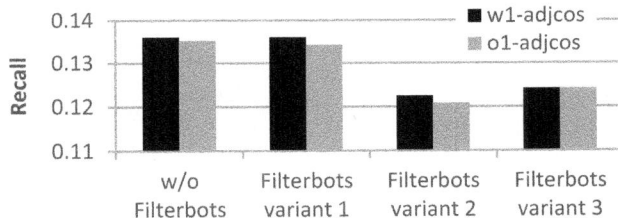

Figure 6. Recall of different Filterbots implementations

The recall metric restricts itself to measure if already good rated items from the test set are recommended based on the items of the training set. So, there is no statement about the quality of items which are recommended, but which are not in the test set. In addition, it is not obvious if a user notices small improvements of the accuracy or if this increases the user satisfaction. Certainly this depends of the concrete situation: When we integrated the age-detection into the final ranking, the recall just improved very slightly, but the impact on the recommended items was clearly visible during test-runs. All in all we increased the quality based on the recall of the recommendations from 0.02 up to 0.13 (factor 6.5) and got positive feedback from both, the domain experts (pedagogues) and users of our system.

Instead of just focusing on tagging the bag-of-words approach combined with TF/IDF weighting could also be considered for classifying articles or determining article similarities. However, this only works for text-based items and cannot be used for articles, which just contain an image, or other item types (like photo galleries or music).

A smaller number of items might lead to decreasing recommendation quality for users who read/rated many items over time. For these users the number of unrated items is even smaller or shrinks and items with higher predictions are recommended first. So, items with a less high prediction are recommended. However, items must have a minimum predicted rating of 3. We have not experienced this so far, but here further research is possible.

One possible problem with CF and also with this approach is diversification: it might happen, that two or more items of the top-5 recommendation are closely related. Further steps might be needed to evaluate and improve diversification.

By applying a multilayered approach, this hybrid recommender algorithm is (or parts of it are) also applicable to bigger or growing systems. If the amount of data grows, the CF layer will generate a full top-5 recommendation without the need to proceed to (possibly worse performing) deeper layers. However, deeper layers can still be reached for new users for which a less amount data is available. Independently, filterbots, alpha-communities, and rule based recommendations can be used to enhance existing recommender systems – these just have to be adapted to the needs of the environment. Of course for scenarios with a bigger amount of data optimizations for speed and performance are becoming more important.

6. SUMMARY

We propose a new recommender approach which was designed for smaller, regionally bound communities. Apart from only using traditional recommendation techniques, it makes use of different kinds of context and semantics: The recommender infers content types which the user is interested in (refers to family context) and also used data about visited "real world" events. This context-awareness and domain specific semantics turned out to be critical for developing the recommendation system in our scenario, a local parent community. We implemented our proposed approach as a prototype and performed a multi-staged evaluation (offline and online). Each stage of the evaluation showed a clear tendency that our approach and recommendation systems can work – even in small scenarios.

Even if the set of special properties sounds special (e.g. different item types, combination of online/offline items), these are quite common for smaller communities. So, the approach presented in the paper is transferable to different smaller communities, but is not bound to that. Aspects of it can also be used to improve the quality of recommendation in larger systems.

7. REFERENCES

[1] Bauer, U. and Bittlingmayer, U. H. 2005. Wer profitiert von Elternbildung? In *Zeitschrift für Soziologie der Erziehung und Sozialisation 25* (3). 263–280.

[2] Büchner, P. 2003. Stichwort: Bildung und soziale Ungleichheit. In *Zeitschrift für Erziehungswissenschaft 6* (1). 5–24.

[3] Burke, R. 2000. Knowledge-based recommender systems. In: Dekker, M. ed. *Encyclopedia of Library and Information Systems 69*. New York, NY, USA. 180–200.

[4] Burke, R. 2007. Hybrid Web Recommender Systems. In Brusilovsky, P., Kobsa, A., Nejdl, W. eds. *The Adaptive Web, LNCS 4321*. Springer Berlin/Heidelberg, 377–408.

[5] Davis-Kean, P. E. 2005. The Influence of Parent Education and Family Income on Child Achievement: The Indirect Role of Parental Expectations and the Home Environment. In *Journal of Family Psychology 19* (2). 294–304.

[6] Ganesan, P., Garcia-Molina, H., and Widom, J. 2003. Exploiting hierarchical domain structure to compute similarity. In: *ACM Trans. Inf. Syst. 21* (1). 64–93.

[7] Herlocker, J. L., Konstan, J. A., Borchers, A., and Riedl, J. 1999. An algorithmic framework for performing collaborative filtering. In *Proc. ACM SIGIR '99.* 230–237.

[8] Herlocker, J. L., Konstan J. A., Terveen, T., and Riedl, J. 2004. Evaluating collaborative filtering recommender systems. In *ACM TOIS '04 22* (1). 5–53.

[9] Hu, Y., Koren, Y., and Volinsky, C. 2008. Collaborative filtering for implicit feedback datasets. In *ICDM'08.* 263–272

[10] Li, Q. and Kim, B. M. 2003. An approach for combining content-based and collaborative filters. In: *Proc. AsianIR'03 11*. Stroudsburg, PA, USA. ACL. 17–24

[11] McNee, S. M., Riedl, J., and Konstan, J. A. 2006. Being accurate is not enough: how accuracy metrics have hurt recommender systems. In *ACM CHI EA '06.* 1097–1101

[12] Melville, P., Mooney, R. J., and Nagarajan, R. 2002. Content-boosted collaborative filtering for improved recommendations. In *Proc. of AAAI'02.* 187–192.

[13] Morita, M. and Shinoda, Y. 1994. Information filtering based on user behavior analysis and best match text retrieval. In *ACM SIGIR '94.* 272–281.

[14] Moshfeghi, Y., Agarwal, D., Piwowarski, B., and Jose, J. M. 2009. Movie Recommender: Semantically Enriched Unified Relevance Model for Rating Prediction in Collaborative Filtering. In *ECIR '09 LNCS 5478.* 54–65.

[15] Nguyen, A.-T., Denos, N., and Berrut, C. 2007. Improving new user recommendations with rule-based induction on cold user data. In *Proc. of ACM RecSys '07.* 121–128.

[16] Park, S.-T., Pennock, D., Madani, O., Good, N., and Decoste, D. 2006. Naïve filterbots for robust cold-start recommendations. In *ACM SIGKDD '06,* 699–705.

[17] Pazzani, M. J. and Billsus, D. 2007. Content-based recommendation systems. In Brusilovsky, P., Kobsa, A., and Nejdl, W. eds. *The adaptive web. LNCS 4321*. Springer-Verlag, Berlin, Heidelberg. 325–341.

[18] Sarwar, B., Konstann, J. A., Borchers, A., Herlocker, J., Miller, B., Riedl, J. 1998. Using filtering agents to improve prediction quality in the GroupLens research collaborative filtering system. In *Proc. of CSCW'98.* 345–354.

[19] Sarwar, B., Karypis, G., Konstan, J. A., and Riedl. J. 2001. Item-based collaborative filtering recommendation algorithms. In *Proc. of ACM WWW'01.* 285–295.

[20] Schafer, J. B., Frankowski, D., Herlocker, J., and Sen, S. 2007. The adaptive web. Springer Berlin/Heidelberg. 291–324

[21] Strickroth, S., Pinkwart, N., and Müller, J. P. 2011. Neue Medien und Präsenzveranstaltungen: Ein didaktisches Modell für die Elternbildung? In Friedrich, S., et al. eds. *DeLFI'11*. Dresden, Germany. TUDpress.

[22] Zheng, H., Wang, D., Zhang, Q., Li, H., and Yang, T. 2010. Do clicks measure recommendation relevancy?: an empirical user study. *ACM RecSys '10*. New York, NY, USA. 249–252.

Finding a Needle in a Haystack of Reviews: Cold Start Context-Based Hotel Recommender System

Asher Levi[*], Osnat (Ossi) Mokryn[†]
School of Computer Science
Tel Aviv Yaffo College
Israel
asherlv2@gmail.com, ossi@mta.ac.il

Christophe Diot, Nina Taft
Technicolor ltd.
Paris, France, Palo Alto, USA
christophe.diot@technicolor.com,
nina.taft@technicolor.com

ABSTRACT

Online hotel searching is a daunting task due to the wealth of online information. Reviews written by other travelers replace the word-of-mouth, yet turn the search into a time consuming task. Users do not rate enough hotels to enable a collaborative filtering based recommendation. Thus, a cold start recommender system is needed.

In this work we design a cold start hotel recommender system, which uses the text of the reviews as its main data. We define context groups based on reviews extracted from TripAdvisor.com and Venere.com. We introduce a novel weighted algorithm for text mining. Our algorithm imitates a user that favors reviews written with the same trip intent and from people of similar background (nationality) and with similar preferences for hotel aspects, which are our defined context groups. Our approach combines numerous elements, including unsupervised clustering to build a vocabulary for hotel aspects, semantic analysis to understand sentiment towards hotel features, and the profiling of intent and nationality groups.

We implemented our system which was used by the public to conduct 150 trip planning experiments. We compare our solution to the top suggestions of the mentioned web services and show that users were, on average, 20% more satisfied with our hotel recommendations. We outperform these web services even more in cities where hotel prices are high.

Categories and Subject Descriptors: H.3.3 [**Information Search and Retrieval**]: Information Search and Retrieval - *Information filtering*

General Terms: Algorithms.

Keywords: Recommender systems, opinion/text mining, context-aware recommender systems, common traits, sentiment analysis.

1. INTRODUCTION

The Internet has overtaken word of mouth as the primary medium for choosing destinations [1]; 63% of consumers plan travel by

---

[*]Part of this work was done while Asher Levi was visiting Technicolor lab in Palo Alto.
[†]Corresponding author

searching the Internet, visiting an average of 22 sites before deciding on a destination.

Producing recommendations for travel is inherently difficult, as an individual rarely rates more than a small number of hotels and thus rich profiles cannot be built. Having limited or no information about the user translates to a user *cold start* recommendation [2, 3, 4]. An intelligent cold start recommender will minimize a new user's effort while still learning enough to recommend the user a product that is likely to be of her interest.

In this paper we design a context-based search recommender system. We show that context-based search can be facilitated for the construction of cold start recommender systems. We further show that contextual information can be mined from review texts, and analyzed for *common traits*[1] per context group. Contextual information has been newly recognized as an important feature when the consumer decides to make a purchase [5, 6, 7]. A lot of research has already been performed in the area of recommender systems and information retrieval. However, most recommender systems focus on recommending the most relevant items to users without taking into account any additional contextual information. Most existing information retrieval systems base their retrieval decisions solely on queries collections, whereas information about search context is often ignored [8].

Users' search patterns are context-based. Among the plethora of reviews, readers opt for recommendations from travelers with comparable needs. A single traveler may share the same needs as other single travelers. A user traveling with her family has different needs from a user traveling on a business trip, i.e. the user context information is an important factor in choosing a hotel. When a user reads reviews she can metaphorically be seen as wearing personalized glasses. Reviews are read through those glasses, and particular words or comments will resonate, positively or negatively, with the reader based upon her needs for her upcoming trip and her personal preferences. Special attention is often given to reviews written with the same intent, or by reviewers from a comparable background. Hence, we define three types of context information. The first is *intent*, or purpose of the trip. We include 5 categories of intent, namely *business trip, single traveler on vacation, family, group, couple.* The second is *nationality.* The third context is user preferences for the different hotel aspects. These were mined from the text using an unsupervised clustering algorithm. We tagged the different clusters found in the text as *location, service, food, room, price-value quality* and the *facilities* (pool, spa, etc). Thus, a user using our system is asked to provide her trip intent, nationality, and preferences for these aspects.

---

[1] In psychology, individuals are often characterized by cardinal traits, while groups (such as nations) can be characterized by common traits.

We obtained data from Venere.com and TripAdvisor.com. The database contains details for each hotel: the hotel's general information, reviews and ratings. In a pre-processing phase, we mined the text and found *common traits* for each context group. These are found in the form of typical words that appear more in text written within that context but are not common for other contexts. A clustering was used to group words that refer to each aspect. Thus, at the end of this pre-processing phase, we have significant words per context, be it intent, nationality, or hotel aspect.

Our recommender system mines the text of the reviews similarly to the user wearing personalized glasses. The user is prompted for her trip intent, nationality and preferences per hotel aspect. We introduce a novel weighted algorithm for context-based text mining. The core idea of the algorithm is to give more importance to reviews of people with the same contexts as the user's. Common traits per the user's context groups and words that describe favorable hotel aspects are given a higher score than other words. We further find the sentiment expressed in the review per context (i.e., positive or negative) and give a corresponding score. Thus, the final score for each review corresponds to that of a user's with comparable needs and preferences and coming from a similar background.

We implemented our system and published it for use, presenting to the users results from our system combined with the top suggestions of the above mentioned web sites. We had over 150 evaluations by friends and colleagues who looked for hotels in four major European cities. Hotels recommended by our system were favored (60.2%) compared to TripAdvisor's and Venere's top suggestions (50.8%). More significant is the fact that our raters said they would not stay in 26.4% of the top hotels recommended by these sites, whereas with our context-based recommender, their dissatisfaction was much lower at 15.9%.

The contributions of our paper are the following: We designed a hotel recommender system that outperforms current leading web sites top suggestions; We define context-based search as a method to overcome the cold start problem for users; Our system is the first we know of that relies mainly on the text of reviews for a cold start recommendation. Hotel ratings and groups' bias are used as tie breakers; We devise a weighted text mining algorithm that leverages common traits found per context group to enable the processing and evaluation of text per users' needs; To find hotel aspects we use a community detection algorithm that leverages the spin glass theory, and changed its distance function to account for the extra clustering overlapping exhibited in the text. This enabled us to find the different hotel aspects in an unsupervised fashion; We held experiments with Mechanical Turk workers and showed that reviews are perceived differently than ratings given by the reviewers, suggesting that sentiment analysis of the text cannot rely on the ratings, although commonly used.

2. RELATED WORK

One of the common and difficult problems for recommender system is the *cold-start* problem, a situation in which the system needs to recommend a product to a new user that has no past information or a new item with very few or no rating [2, 4, 3]. We build a model for domains that by nature don't have a lot of history (or not at all) information about the user, and user cold start recommender system is required.

Despite the abundance of studies targeted at solving the new item problem [4, 9], there has been little work in solving the new user problem. The dominant approach is using a learning phase, in which a user is asked to provide a set of ratings for selected items, in a way that gathers as much information about the user as possible [10]. Another approach presented in [11] exploits the

significance of users' implicit feedback for alleviating the new user problem. In this approach the user has to express interest in items, or organize the items in relative order, without providing explicit ratings for those items. Those approaches use only ratings or relative rankings on items and thus are bounded under a rating recommender system limitations. Moreover, they are missing all the information that can be extracted from the text.

Another approach is the "Metadata" approach; here the metadata of an item is used to create content-based recommender systems. This method relies on systems where the user needs to provide some demographic data. The solution presented in [12] is utilizing the strength of the vector aspect model with user information; they used the demographic information of the user (age, gender and job) as the user's features. A model of relationships between a user's demographic information and an item's metadata was presented by Park et al. [13]. Those solutions use only ratings at their model; They use the user's context information as a feature for building their recommendation; We extend the usage of contextual information. We are not considering only the user general information (e.g, age or gender) and simply profile the user, but rather we are trying to build a more complete behavioral profile that attempts to capture the expectations of the user from our recommendation;

By using a context information that relevant to the current session of search (e.g, for hotel recommendation we use the intent of the trip as one of the context features), we are capturing a more accurate and efficient profile.

3. SYSTEM OVERVIEW

A hotel recommender system typically won't have sufficient historical information to build profiles for individuals. It does, however, have additional data in the form of reviews that is sufficient to enable the characterization of context groups. We give here an overview of our system, which determines *common traits* for groups that share the same context. The core idea of our system is to give more importance to reviews of people with the same context. Our system brings greater importance to the topics those reviewers focus on frequently and also focuses on topics that are associated with the user's stated preferences.

In the hotel arena people can be categorized by their trip intent (such as those who travel as a 'couple', or a 'family', etc.) and nationality, which we refer to as context groups. Using the text reviews from multiple people within a single context group, we can essentially find the common traits of groups such as 'family' travelers (and so on for the other categories). We additionally process the corpus of reviews to identify the vocabulary that is used to describe a particular aspect of a hotel. Once a person using our system specifies her intent, nationality and preferences our system evaluates reviews with accordance with the traits and preferences, and gives a recommendation.

We now give a brief overview of the components and steps of our method, depicted in Figure 1. The top 3 boxes on the left correspond to the pre-processing phase in which we define the common traits of intent and nationality groups, and define the different hotel aspects referenced in reviews, correspondingly. To find common traits for each context group, we extract the nouns and noun phrases (called *features*) from all reviews and find those that are more common for that group. These features are then assigned a weight per each context according to their relative frequency in reviews within that context. The higher the weight, the more important a feature. The common traits of context groups are the higher weight features for that group. Hence, the common traits of Italians consist of a set of features and their weights, while these of Germans may contain largely the same features but with different weights. Common traits of hotel aspects are constructed differently. Here we carried out a clustering task to cluster features based upon co-occurrence in the same sentence. Each feature can only occur in one cluster, and thus each cluster contains the most relevant vocabulary for that aspect. The fourth component of the preprocessing consists of building an opinion lexicon which will allow us to analyze adjectives associated with features, and to give each feature an orientation score depending upon how positive or negative is the sentiment of any associated adjectives.

While the base weight of each feature is one in our system, features that are distinctive of several context groups may have different weight per group. The specific set of weights used in response to a user hotel search will be chosen once the user declares her context and preferences. For example, if a user specifies 'business traveler' as her intent and her nationality, then the set of feature weights used will be those in the 'business traveler' group and the corresponding national group. In our figure, this step corresponds to the "select relevant feature weight for intent" and ".. for nationality" boxes. Similarly, corresponding weights are given to features of important aspects. This implies, for example, that the feature 'air conditioning' will get one weight depending upon its importance for business travelers, a second weight depending upon its importance for the given nationality, and a third weight depending upon its importance per the user preference for the aspect it belongs to. The final weight for each feature is done by combining these three weights (depicted as "build feature score" in the figure).

Next we use our opinion lexicon to give each feature an orientation score. We subsequently combine the features, their weights and orientations to build a score for each sentence. The sentence scores are then combined to give an overall score for each review. This score should reflect the relative importance of the given review for the user. Reviews that are both important and positive are deemed most relevant thereby receiving the highest scores. The final score for each hotel is an average of all of its reviews, each of which is scored from the user's perspective (i.e., based on her context and preferences), and an adjustment bias calculated per the context given.

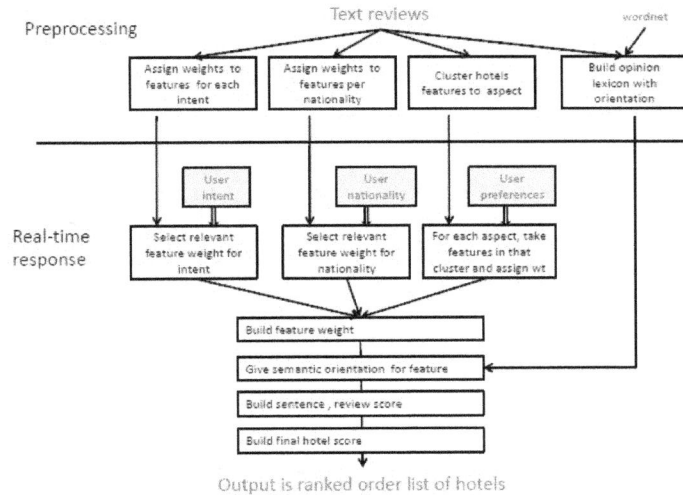

Figure 1: Overview of our approach

4. CONTEXT BASED ANALYSIS

The main idea of our algorithm for context based text analysis is to assign weights to common traits per context. Thus, at the end

of the process, each review is mapped to a score number, based on system's perception of the user's perspective.

4.1 Intent and Nationality Profiling

We find common traits for each context group by mining the text from reviews on a sentence level. Our approach is to extract key features (i.e., words) that are important for each group. It has been shown that a reviewer's vocabulary when commenting on an item was found to converge, in the sense that the most frequently used nouns and noun phrases used correspond to genuine and important features [14]. Similar to [15] we extract features and remove redundant and meaningless items from the candidate features we found.

The basic building block of our algorithm is the trait based weight assigning. For each review written we extract the features and assign each feature a weight that reflects its importance for each context group. Let c denote a general context that can be either an intent (or purpose) p or a nationality n (i.e., $c \in \{p\} \bigcup \{n\}$), and let $freq_f(c)$ denote the frequency of feature f for context c. The frequency of a feature per context is the relative number of occurrences of feature f in sentences appearing in reviews that belong to context c. For example, the frequency of the feature 'WiFi' for Americans is calculated as the ratio of the number of times this feature appeared in sentences written by Americans, divided by the total number of sentences written by Americans. Similarly, avg_f is the average frequency of feature f, $stdv_f$ is its standard deviation, and $dev_f = avg_f - stdv_f$. Using this notation, we define the weight of a feature f for a given context as follows:

$$W_c^f = \begin{cases} 1, & \text{if } |dev_f| < stdv_f \\ Max(0.1, 1 - \frac{dev_f}{stdv_f}), & \text{if } \frac{dev_f}{stdv_f} < -1 \\ 1 + \frac{dev_f}{stdv_f}, & else \end{cases} \quad (1)$$

The majority of features will either be assigned a 1; however those whose frequency is larger than average plus or minus one standard deviation, are assigned values between 1 and 3 or 0.1 and 1 respectively. Hence each feature is assigned a weight in the range [0.1, 3] per context.

4.2 Aspect Profiling

Recall that we ask the user to input their preferences on six aspects. These aspects were not selected at random, but were instead the result of a word clustering analysis we performed on the text. Often in reviews, different words may be used to refer to the same general aspect of a hotel. For example, words like 'area', 'street', and 'metro' may all refer to aspects of a hotel's location. There are many approaches to clustering, hierarchical clustering, partition clustering (e.g k-means) etc. The number of clusters, k, is usually either an input parameter or found by the clustering procedure itself [16]. In our case, clustering would yield the different hotel aspects and therefore should not be supervised but determined by the clustering algorithm over the text itself.

To account for the sparsity and the overlapping characteristics in the network of word features, we build upon an unsupervised community detection[2] [17, 16] technique based on [18]. We build a network graph in which each node corresponds to a feature and each community will correspond to an hotel aspect. Trying to find the maximal modularity is defined as finding a partition that will minimize the energy of the features network graph. The Hamiltonian, denoted in equation 2 is defined in the following way: exist-

---

[2]We use the terms cluster and community interchangeably

ing internal edges and non-existing external links (between formed communities) minimize the Hamiltonian, while existing external links and non-existing internal links increase its value. The algorithm tries to find a partition that minimizes the Hamiltonian, based on the spin glass model for finding a partition that minimizes the energy of the spin glass with the spin states being the community indices.

$$\mathcal{H}(\{\sigma\}) = -\sum_{i \neq j} a_{ij} \underbrace{A_{ij}\delta(\sigma_i, \sigma_j)}_{\text{internal links}} + \sum_{i \neq j} b_{ij} \underbrace{(1 - A_{ij})\delta(\sigma_i, \sigma_j)}_{\text{internal non-links}} \\ + \sum_{i \neq j} a_{ij} \underbrace{A_{ij}(1 - \delta(\sigma_i, \sigma_j))}_{\text{external links}} - \sum_{i \neq j} b_{ij} \underbrace{(1 - A_{ij})\delta(\sigma_i, \sigma_j)}_{\text{external non-links}} \quad (2)$$

Where A_{ij} is a boolean adjacency matrix, $\sigma_i \in 1, 2, \ldots q$ denotes the indices of the communities, with q the number of maximal communities. [19] showed that the division does not depend on q, for large initial q values.

In [19] a_{ij} and b_{ij} where chosen as a function of the probability of two graph nodes to be adjacent under the assumption that when this probability is high the nodes are more likely to belong to same group, or community. In our case, this translates to the probability of two features to appear in a sentence together. However, we found in reviews that very frequent features are often found in sentences together. For example, it is common to find sentences of the following structure:

The location was great and the room was very clean.

Clearly, location and room belong to different hotel aspects, and should therefore belong to different communities. To account for this tendency we instead use the PMI-Pointwise mutual information weight, which measures the information overlapping between two random variables [20], described in (3).

$$PMI_{ij} = \log\left(\frac{p(i \wedge j)}{p(i) \cdot p(j)}\right) \quad (3)$$

Where, $p(i)$ is the probability that the feature i appears in a sentence. Then, $a_{ij} = \gamma \cdot PMI_{ij}$, where γ is a parameter expressing the relative contribution to the energy from existing and missing edges. In our case we chose $\gamma = 1$.

Over the corpus of reviews, our PMI-pointwise improvement of the spinning glass community detection algorithm produced six clusters of different sizes (note that the number of clusters is unsupervised). The identification of these 6 clusters is important as it determined the particular hotel aspects that we chose to ask users their preferences for. Each cluster and the set of features it contains can be thought of, intuitively, as the *common traits* for the aspect associated with this cluster. These clusters are useful as follows. Suppose for example that a user specifies that location is of utmost importance to her. The room cluster identifies a large number of features (or words) that are often used to discuss things inside a hotel room; thus reviews in which these words occur frequently are more important to a user who cares about the room than one who cares about food. After studying the words that ended up in each cluster, we selected the cluster names as indicated in Table 4.2. These clusters can be computed ahead of time as part of the system's preprocessing.

Our weight assigning algorithm for aspect related features relates to the user's preference and is calculated online as follows: Let $u_{pref}(k)$ denote user u's preference for aspect (i.e. cluster) k. If feature f is in cluster k, then we calculate the weight for the feature according to the users preferences as follows:

$$W_{u_{pref}(k)}^f = 1 + \frac{u_{pref}(k)}{5} \quad (4)$$

where, $W_{u_{pref}(k)}^f$ denotes the weight of feature f for user u according to her preference $u_{pref}(k)$. For example, if the user sets their preference for location to 5, and the feature is *train*, then the weight of train for this user is 2. Another user that specifies that location is of importance 1, would have the feature *train* assigned a weight of 1.2. When determining the weight for the feature *train* we only use the user's preference for location (and not for room or food) because the feature 'train' is in the location cluster and cannot be in any other cluster.

4.3 Feature Opinion Orientation

Next we determine the polarity of the opinion expressed in the review on each feature, whether positive or negative, to assign a corresponding sign to a feature's weight. To infer the opinion polarity per feature we use an opinion lexicon. An opinion lexicon is a dictionary of words and word phrases that express positive or negative sentiments. In this work, we consider sentiment words to be adjectives the reviewers use to express opinions on product features, as in [21, 22] . To collect the opinion word list we use a corpus-based approach similar to the approach described in [14, 23]. We extract all the adjectives that appear in the same sentence for each feature.

We then find the semantic orientation of the extracted opinion words. When the reviewer uses a word that expresses a desirable state, then the word is classified as having a positive semantic orientation. Similarly, an undesirable state translates to a negative semantic orientation. We use a bootstrapping lexicon-based approach as in [14]. Manually, we create a set of seed adjectives from the opinion lexicon list with semantic orientation. Then for each adjective in the seed list, we search for a synonym and an antonym in WordNet [24]. Each found adjective in the opinion lexicon is assigned an orientation, and is added to the seed list. The seed list grows in the process. A recent work [25] suggests to consider the influence of aspects on sentiment polarity. However, given that we give the weight per feature and that the aspect counts only for a fraction of the total weight we left the orientation per feature as before.

We used common opinion rules as described in [23]. One is the negation rule, words or phrases like 'no', 'not', etc. take the opposite orientation expressed by the opinion phrase. The other is the *But* clause rules, a sentence containing 'but' also needs special treatment. The opinions before and after a 'but' are usually opposite of each other. First we try to determine the semantic orientation of the feature in 'but' clause. If we cannot get the orientation of the phrase we take the opposite orientation of the clause before the 'but' clause. Phrases such as 'with the exception of', 'except for' etc. behave similarly to 'but' and are handled in the same way. For example, in the sentence "The room was clean except for the bathroom", the opinion about the feature *room* is positive and the feature *bathroom* gets the inverse opinion which is negative. There are also some phrases that contain negation and but words, yet do not change the orientation of the opinion. For example in the phrase "I do not only like the size of the room, but also its style", the 'not', 'but' words do not change the orientation of the opinion words 'like' and 'style'.

Using these rules and our lexicon, we assign an orientation score to each feature f in a given sentence s, denoted $score(f, s)$. It should be clear that the same feature, in two different sentences, could receive different orientations. When many opinion words surround a single feature, they are aggregated as indicated in equation (5).

$$score(f, s) = \sum_{op \in s} \frac{or_{op}}{d(op, f)} \quad (5)$$

Aspect tag	Features per aspect (randomly chosen)
Location	location, area, city, street, ,metro, station, train, distance, bus, airport
Service	staff, service, hotel staff, reception, front desk, luggage lobby, reception, staff, person, wifi
Food	breakfast, morning, food, restaurant, bar, coffee, buffet ,dinner, fruit, terrace, buffet variety, bread, course
Room	bathroom, floor, shower, size, window, door, view, building, tv, water, elevator, balcony, hotel room, lift, bath
General	hotel, night, place, stay, price, experience, trip, value, hotel star, money, rate, money value, deal, quality, cost
Other	pool, spa, gym

Table 1: Spin Glass community detection algorithm results

Here op is an opinion word in sentence s, $d(op, f)$ is the distance (word count) between feature f and opinion word op in sentence s. Also, or_{op} is the orientation $(-1, +1)$ of the opinion word op. Dividing by the distance between the feature and the opinion word is used to give lower weights to opinion words that are farther away from f. When the final score is positive, then the overall opinion of feature f in s is positive, and similarly the reviewer's opinion of the feature is negative when the final feature score is negative.

4.4 Producing a Review Score

We now have a set of weights and their orientation per the user's context for each feature in a review. We combine these elements to produce a single score for a review as follows. Given the user's input on their context, each feature has 3 weights, one for intent, $W_{u_p}^f$, one for nationality $W_{u_n}^f$, and one based on aspect preferences $W_{u_{pref}}^f$. The final weight W_u^f assigned to feature f for user u is the multiplication of these three weights, namely:

$$W_u^f = W_{u_p}^f \cdot W_{u_n}^f \cdot W_{u_{pref}}^f \qquad (6)$$

The weights for each context are multiplied because that allows fine grained differentiation of people within our various groups (such as intent and nationality). Consider a Japanese person who uses our system. Based upon our nationality profiling, we see that the feature 'bath' is important. If that person also marks 'Room' as a hotel aspect that is very important to them (i.e. a preference of 5), then the quality of the bathroom is more important for this user than for a second Japanese person who marks 'room' as low priority and 'food' as high priority. This allows us to differentiate within nationalities by using the intent and preferences (or to differentiate within an intent group by their nationality and preferences).

To produce a score for each sentence, we multiply each feature by its orientation score and sum up the weight scores of all features in a sentence s, namely $\sum_{f \in s} W_u^f \cdot score(f, s)$. Similarly, we sum up the scores of all the sentences in a review to produce a score for a review v, as follows:

$$score(v, u) = \sum_{s \in v} \sum_{f \in s} W_u^f \cdot score(f, s) \qquad (7)$$

Where $score(v, u)$ is the score of review v for user u. The review score captures how important a particular review is for the user based upon their context and preferences.

4.5 Devising a Hotel Score

Next we produce a score for each hotel so that hotels can be ranked and presented to the user in order from highest score to lowest. The major factor in the score of a hotel in our system is the score calculated for reviews based on user context groups and preferences. We term this the hotel orientation score, ho_u, where $ho_u = avg_{v \in R_{(h)}}[score(v, u)]$ and $R_{(h)}$ denotes the set of reviews for hotel h. The second argument is a bias adjustment, denoted

b_{hsn}, which captures the bias of a user with intent p and nationality n, as well as any hotel bias h. (The bias term is explained below.) Thus our final hotel score is given by:

$$S_u^h = ho_u + b_{hpn} \qquad (8)$$

Bias Adjustment to Hotel Score: In our hotel score, the orientation score coming from the text analysis of the reviews is the dominant component of the score, as these values will range from -40 to 80 approximately. Our bias terms range from 0 to 5 and are included primarily to break ties, or to differentiate hotels when their scores are very close. The process of using the star ratings needs to be adjusted for bias because there are systematic tendencies for some traveler groups to rate higher than others. For example, our data analysis shows that reviewers from Spain tend to rate lower in star rating systems than reviewers from the USA.

We compute the bias b_{hpn} for hotel h from traveler with both intent p and nationality n, as follows. Let μ denote the overall average star rating of *all* hotels in the system. The parameter b_h specifies the observed deviations of hotel h from the overall average. We use b_{hp} to denote the observed deviations that travelers with intent p have for hotel h, (and similarly for b_{hn}). These deviations are with respect to the average score of hotel h.

$$b_{hpn} = \mu + b_h + b_{hp} + b_{hn} \qquad (9)$$

The average deviations are shrunk towards zero by using the normalization parameters, $\lambda_1, \lambda_2, \lambda_3$, which are determined by validation on the test set. For each hotel h we set:

$$b_h = \frac{\sum_{r_h \in R_{(h)}} (r_h - \mu)}{\lambda_1 + |R_{(h)}|} \qquad (10)$$

where $R_{(h)}$ is a set of reviews for hotel h, and $\lambda_1 = 30$. The bias of intent group p for hotel h is:

$$b_{hp} = \frac{\sum_{r_{hp} \in R_{(hp)}} (r_{hp} - \mu - b_h)}{\lambda_2 + |R_{(hp)}|} \qquad (11)$$

where $R_{(hp)}$ is a set of reviews for p and h, and where $\lambda_2 = 5$. The bias of a nationality n for hotel h is given by:

$$b_{hn} = \frac{\sum_{r_{hn} \in R_{(hn)}} (r_{hn} - \mu - b_h)}{\lambda_3 + |R_{(hn)}|} \qquad (12)$$

where, $R_{(hn)}$ is the set of reviews from nationality n for h, and $\lambda_3 = 5$.

5. VALIDATION

The dataset used in this study was extracted from two well-known travel search engines, namely, Tripadvisor.com and Venere.com. For each hotel the data contains general information about the hotel (e.g, name, address, average rating, stars, price etc.) and a list of reviews written by hotel guests. The reviews include: travel intent of

119

the reviewer, nationality, rating, review text, and additional metadata. The data was collected for 4 cities in Europe: Munich and Berlin in Germany, and Milan and Rome in Italy. The data includes reviews that were written before January 2011. 84,968 reviews were collected for 1,930 hotels from TripAdvisor, and 52,266 reviews for 1,845 hotels from Venere. For each intent group we obtained thousands of reviews, from 6,541 reviews written by people on business trips to 60,113 reviews written by couples. Overall we collected information for 3,775 hotels corresponding to 137,234 reviews.

Recent researchers have turned to text analysis as there is potentially a great deal more information that could be extracted from reviews then star ratings[26, 27, 28]. In these solutions, the star rating is taken as the overall product rating, and additional information is obtained by analyzing the review text to extract opinions on specific aspects of each item, and thus improve a personalized recommendation. In [29] the authors incorporate a notion of personal scale, that is based on the observation that different users give different values to their describing words to improve sentiment analysis for personal recommendations. Our work differs from these because they use text reviews to build individual profiles and personal scales, thereby not focusing on the cold start problem. In the absence of user profiles, we choose instead to use the text to find the common traits of intent groups and nationalities, and further differentiate hotels by personal preferences.

The underlying assumption of [29] and many others, is that a user's ratings and text correlate. This assumption has not been previously quantified. We further wanted to quantify whether the perception of the reviews differs across context groups.

To this end we designed the following online experiment with Mechanical Turk workers. Each worker was given the text of five different hotel reviews. We used reviews from 50 different hotels' taken from venere.com, where ratings are on a scale of $[1 - 10]$. We then asked the workers, based on the text, to estimate the rate that the reviewer gave. To get meaningful results, we filtered out manually cases in which it was clear the workers did not read the text before estimating the rating[3]. We obtained $50 - 75$ estimates of the star rating for each review, yielding a total of 3715 estimates.

First, we checked whether the workers estimated the ratings were indeed similar to those given by the person writing the text. For each estimate, we computed the difference between the estimate rating and the actual rating, depicted in Table 5 . We averaged all these differences in cases when the estimates where higher than the actual, and computed a second average across all cases when the estimates were lower than the actual ratings. There is a clear skew between the estimated rate and the real rate. The skew is more significant when reviews were perceived as negative by the workers (1.67 more negative vs. 0.94 more positive). Two conclusions are thus possible. The first claims that users give an accurate star ratings while signifying negative aspects when writing down reviews. The other suggests that users are generous with the star ratings while expressing their real opinion in writing. In either case, these result indicate that star ratings do not consistently capture the sentiment in text reviews, and thus the correlation between text and reviews is weak.

To validate the assumption that context matters, we return to our Mechanical Turk experiment and ask whether the perception of the reviews differs across intent groups. For example, we saw that single travelers tend to rank almost the same as others but their text was perceived as much worse by the Mechanical Turk workers.

---

[3]We declined payments when the results seemed random, i.e., the worker did not seem to read the reviews at all. We had one dispute that was ruled in our favor.

The average difference between the perceived rate and the real rate was 1.93 on the average for single travelers. For people traveling in groups, the average difference between perceived and real rate was 1.54. This means that the text reviews of single travelers gives the perception to others of being far more negative than their corresponding rates; whereas for group travelers, the text reviews are only slightly more negative that the ratings would indicate[4].

Type	Average Difference	Reviews
Estimation > Rate	0.94	1474
Estimation < Rate	1.67	2241
Total	1.38	3715

Table 2: Mechanical Turk results, estimating the review's rate.

5.1 Data Analysis

To validate our algorithmic approach that takes into account context we asked ourselves whether different context groups rank hotels differently, whether the intent groups emphasize different things, and whether the tone of the reviews differs across context groups. In our case the different context groups are the different intent groups and different nationalities. The analysis was performed on reviews taken from TripAdvisor, where the hotel ratings are in the range of $[1..5]$. Table 3 shows the average rating for each context group, whether an intent group or nationality. Indeed we see differences across the groups, with the difference between intent groups varying to up to 0.55 (on a five star rating system). Next we computed frequent words used by each of the groups over both datasets, TripAdvisor and Venere. We removed words that are used by all groups. This left us with examples of words that were frequent to one group but not others. Table 3 also shows examples of words that are frequently used by one group, but infrequently or never, appear in the text of other groups. This indicates that the intent of a trip influences the content of reviews that get written, and that the top words (i.e. topics) that interest reviewers also differ by country or culture.

5.2 Tiebreakers Evaluation

The effect of considering the intent of a trip is clear both intuitively and from our results. We wanted to further verify the usefulness of using the nationality, as well as the bias factor used for the final hotel score. The following evaluations were done on our implemented system.

---

[4]In a complementary experiment the workers were asked to estimate reviews' ratings when knowing the intent of the trip. The additional information did not affect the estimations, indicating the above results could not be predicted.

Context	Rate	Typical words
Family	4.15	Air condition, Car, Space, Shuttle, Breakfast
Couple	4.08	Coffee, View, Balcony, Breakfast
Group	4.02	Bar, Money, Bus stop, Shopping, Party
Single	3.8	WiFi, TV, Price, Supermarket
Business	3.6	Internet, Buffet, Park, Bar, Shopping, TV
U.S.A	4.11	Hotel staff, Train station, Lobby, Shuttle
Russia	4.07	Furniture, Style, Bus stop, Air conditioner
Australia	3.97	Food, wifi, Supermarket, Area, Pillow
Netherlands	3.94	Toilet, Hotel front, Coffee, Hotel breakfast
Japan	3.86	Bath, Bed, Room shower, Sightseeing

Table 3: Average ratings given by each intent group/nationality and corresponding distinguishing words

We ran the following experiments with our implemented system. We issued numerous pairs of queries, one with nationality specified and one without. (Similarly for bias.) For each query in the pair, we recorded the top 10 (or top 20) hotels recommended and then compared the two lists. Let S_1 denote the list of top-10 hotels for the query without nationality, and S_2 denote the top-10 list for the same query with a nationality specified. We quantify the difference between these two lists using the Jacard distance:

$$Diff(S_1, S_2) = \frac{(S_1 \bigcup S_2) - (S1 \bigcap S2)}{|S_1| + |S_2|} \qquad (13)$$

Nationality The queries in the experiment are constructed as follows: for each city (4 options), and for each user's travel intent (5 options), we randomly select five different user preference values for each aspect. We calculate the distance $Diff(S_1, S_2)$ for each pair in the formed lists, and compute the average distance across all pairs of queries. This captures the average influence on the search results of including nationality. The total number of queries executed was 2500. We found that the nationality parameter affects 16.6% of the search results, thus, we believe that this piece of context is important to include in our method. We executed the same experiment with 20 hotels in the results sets, and found similar results; in this case we observed that the nationality context affected 15% of hotels recommended.

Bias adjustment. Similarly, we may wonder to what extent the bias adjustment plays a role in affecting the order of hotels presented to a user. Hence, we ran similar experiments. For each city, for each user travel intent, and for each nationality, we randomly select five different user preference values for each aspect. We compare the two obtained sets using our $Diff(S_1.S_2)$ metric. We ran 2500 such experiments. Since the bias parameter is intended to be used as a sort of tiebreaker, to differentiate very closely ranked hotels, we don't expect it to have a large impact; however if it plays no role then it could be eliminated from our method. We found that the bias affected 9% of the search results. Executing the same experiment, but recording the top-20 recommended hotels after each query, we found similar results - the bias influenced 8% of the recommendations. We believe this is a sufficiently influence to warrant retaining the bias parameter in our solution.

6. SYSTEM EVALUATION

Evaluation of a recommender system has to measure whether real people are willing to act based on the recommendations. User satisfaction with a recommender system results is well gauged with an on-line evaluation methodology. We use such a methodology as described in [30]; this methodology doesn't measure absolute user satisfaction but only relative user satisfaction with one system over another.

We implemented our system and made it available on the public web for use. We asked numerous friends and colleagues to evaluate our system and obtained 150 evaluations. Each experiment consisted of the following. The user inserts her search parameters: intent, nationality, aspect preferences, and a price range. Then we present the user a list of six hotels. Some are from our system, and some are the highest star ratings choices from Venere and Tripadvisor. In order to avoid biasing the user, these six hotels are presented in random order and thus the user is unaware of the source of the recommendations. Raters were shown links to the full text reviews to further explore the recommended hotels. For each one of the hotels, raters were asked to express their satisfaction by answering the question "Would you select this hotel?" with three optional answers: Yes, Maybe, No. In addition, the raters were asked to rate all the recommended hotels on the scale of $[1-5]$ to indicate whether

they felt the recommendation had met their search criteria and was to their satisfaction. They were also asked to indicate which aspect was the one that most influenced their decision. Raters were specifically instructed to only select 'intents' that were realistic for them (e.g., if you don't have kids, do not select the 'family' as the intent).

First we look at the overall satisfaction, namely the user's response to the question "would you stay in this hotel?". We averaged the responses over all raters. We see that for 60.2% of the hotels recommended by our system, users stated they would stay there, as compared to 50.8% from the rating systems. Moreover, users stated they would not stay in 26.4% of hotels recommended by the rating systems, compared with a much lower 15.9% dissatisfaction with hotels recommended by our system. To examine this in more detail than just averages, we plot the empirical histogram of the ratings given by our raters in Figure 2. The hotels recommended by our method received more 4 and 5 ratings then those the other method, and similarly our recommended hotels received fewer 1 and 2 ratings than the other method.

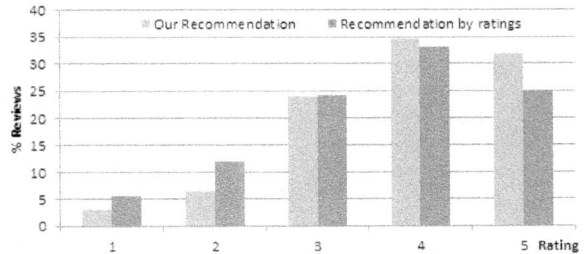

Figure 2: Ratings by recommender method

Interestingly, we observed that the satisfaction/dissatisfaction results varied by country (Germany versus Italy in our data), as can be seen in Table 6. The Italian hotels have higher average price as well as higher price variance than those in Germany. This makes producing a good recommendation in Italy harder, because when the price is reasonable (as in our German hotels) people are more easily satisfied. Satisfaction with the star rating recommendations in Italy was at 47.4%, whereas for our context-based system it rose to 58.8%. Similarly, our system has a more dramatic affect in terms of lowering dissatisfaction for the Italian hotels (dissatisfaction is lowered from 30.1% to 15.4%), than for the German ones.

An important issue in understanding the performance of a cold-start context-based recommender is to assess user consistency. We looked at the reason each rater stated for making their decisions (to stay or not, and their rating of our proposals). We compared that to their preference markings for hotel aspects. In Table 6, we show that our users are indeed very consistent; their decisions were consistent $78\% - 97\%$ of the time (depending upon the case). For the case where users made consistent decisions, we show in $63 - 72\%$ those cases, the user marked that they were satisfied. This indicates that we showed them hotels whose reviews resonated positively for them because the focused on the aspects user care about and make decisions on.

City	System	Price Avg, Stdv	Yes	Maybe	No
Germany	Ours	83, 40	62.2%	21.1%	16.7%
Germany	Stars	83, 40	55.6%	23.0%	21.4%
Italy	Ours	99, 70	58.8%	25.8%	15.4%
Italy	Stars	99, 70	47.4%	22.5%	30.1%

Table 4: User satisfaction by destination country

Type / Reason	Location	Service	Room	Price	Food
Consistent	97.1%	81.6%	88.5%	78.9%	84.4%
Satisfy	65.9%	68.8%	65.7%	72.6%	63.2%

Table 5: Users reason consistency

Next, we checked whether our recommendations resonate well for users with different trip intents. Table 6 details users satisfaction by intent. Satisfaction from our results was higher than for the star rating sites by 13% on average. Single and business travelers were considerably more satisfied with our suggestions than by those of the star ratings systems, and showed a considerable lower dissatisfaction. Interestingly, users who planned to travel in a group were dramatically more satisfied with our system, with 21.4% preferring our systems' suggestions.

Intent	Recommender System	Yes	No
Business	Our System	60.7%	16.0%
	Star Rating	50.0%	25.0%
Single	Our System	65.5%	15.5%
	Star Rating	52.8%	26.4%
Couple	Our System	55.7%	16.3%
	Star Rating	53.7%	22.6%
Family	Our System	52.7%	13.8%
	Star Rating	44.4%	36.1%
Group	Our System	66.6%	17.2%
	Star Rating	45.2%	28.5%

Table 6: Satisfy by intent

7. CONCLUSIONS

We have demonstrated that common traits for groups can be found by preprocessing large samples of text. This is a powerful result, as identifying group traits can later be used for classifying whether unknown individuals belong to the group. Additionally, if common traits of a group are known, text of reviews can be mined to identify the typical crowd of a restaurant or a hotel, for example.

Additionally, an interesting outcome of our Mechanical Turk experiments suggests that there is no strict correlation between how a review is perceived and the corresponding rating given by its author.

8. REFERENCES

[1] N. M. T. Watch, "Online Travel Market," April 2011. [Online]. Available: http://www.newmediatrendwatch.com/world-overview/91-online-travel-market

[2] G. Adomavicius and A. Tuzhilin, "Toward the next generation of recommender systems: A survey of the state-of-the-art and possible extensions," *IEEE transactions on knowledge and data*, 2005.

[3] A. Rashid, I. Albert, D. Cosley, S. Lam, S. McNee, J. Konstan, and J. Riedl, "Getting to know you: learning new user preferences in recommender systems," in *Proceedings of the 7th ACM international conference on Intelligent user interfaces*, 2002.

[4] A. Schein, A. Popescul, L. Ungar, and D. Pennock, "Methods and metrics for cold-start recommendations," in *Proceedings of the 25th annual international ACM SIGIR conference*, 2002.

[5] W. Woerndl and J. Schlichter, "Introducing context into recommender systems," in *Short Paper, Proc. AAAI 2007 Workshop on RecSys in e-Commerce*, 2007.

[6] T. Jiang and A. Tuzhilin, "Improving personalization solutions through optimal segmentation of customer bases," *IEEE transactions on knowledge and data engineering*, 2008.

[7] N. Hariri, Y. Zheng, B. Mobasher, and R. Burke, "Context-aware recommendation based on review mining," *General Co-Chairs*, 2011.

[8] G. Adomavicius and A. Tuzhilin, "Context-aware recommender systems," *Recommender Systems Handbook*, 2011.

[9] Y. Park and A. Tuzhilin, "The long tail of recommender systems and how to leverage it," in *Proceedings of the 2008 ACM conference on RecSys*, 2008.

[10] G. Al Mamunur Rashid and J. Riedl, "Learning preferences of new users in recommender systems: an information theoretic approach," *ACM SIGKDD Explorations Newsletter*, 2008.

[11] L. Zhang, X. Meng, J. Chen, S. Xiong, and K. Duan, "Alleviating cold-start problem by using implicit feedback," *Advanced Data Mining and Applications*, 2009.

[12] X. Lam, T. Vu, T. Le, and A. Duong, "Addressing cold-start problem in recommendation systems," in *Proceedings of the 2nd ACM international conference on Ubiquitous information management and communication*, 2008.

[13] S. Park and W. Chu, "Pairwise preference regression for cold-start recommendation," in *Proceedings of the third ACM conference on RecSys*, 2009.

[14] M. Hu and B. Liu, "Mining and summarizing customer reviews," in *Proceedings of the tenth ACM SIGKDD*. ACM, 2004.

[15] ——, "Mining opinion features in customer reviews," in *Proceedings of the National Conference on AI*. Menlo Park, CA; Cambridge, MA; London; AAAI Press; MIT Press; 1999, 2004.

[16] S. Fortunato, "Community detection in graphs," *Physics Reports*, vol. 486, 2010.

[17] M. Girvan and M. Newman, "Community structure in social and biological networks," *Proceedings of the National Academy of Sciences*, 2002.

[18] J. Reichardt and S. Bornholdt, "Statistical mechanics of community detection," *Physical Review E*, 2006.

[19] ——, "Detecting fuzzy community structures in complex networks with a potts model," *Physical Review Letters*, 2004.

[20] K. Church, W. Gale, P. Hanks, and D. Kindle, "6. using statistics in lexical analysis," *Lexical acquisition: exploiting on-line resources to build a lexicon*, 1991.

[21] B. Liu, "Sentiment analysis and subjectivity," *Handbook of Natural Language Processing,*, 2010.

[22] B. Pang and L. Lee, "Opinion mining and sentiment analysis," *Foundations and Trends in Information Retrieval*, 2008.

[23] X. Ding, B. Liu, and P. Yu, "A holistic lexicon-based approach to opinion mining," in *Proceedings of the international conference on Web search and web data mining*. ACM, 2008.

[24] G. Miller, R. Beckwith, C. Fellbaum, D. Gross, and K. Miller, "Introduction to wordnet: An on-line lexical database*," *International Journal of lexicography*, 1990.

[25] S. Brody and N. Elhadad, "An unsupervised aspect-sentiment model for online reviews," in *Human Language Technologies: The 2010 Annual Conference of the North American Chapter*. Association for Computational Linguistics, 2010.

[26] S. Aciar, D. Zhang, S. Simoff, and J. Debenham, "Informed recommender: Basing recommendations on consumer product reviews," *Intelligent Systems, IEEE*, 2007.

[27] N. Jakob, S. Weber, M. Müller, and I. Gurevych, "Beyond the stars: exploiting free-text user reviews to improve the accuracy of movie recommendations," in *Proceeding of the 1st ACM international CIKM workshop*, 2009.

[28] G. Ganu, N. Elhadad, and A. Marian, "Beyond the stars: Improving rating predictions using review text content," in *12th International Workshop on the Web and Databases*. Citeseer, 2009.

[29] S. Faridani, "Using canonical correlation analysis for generalized sentiment analysis, product recommendation and search," in *Proceedings of the fifth ACM conference on RecSys*, 2011.

[30] C. Hayes and P. Cunningham, "An on-line evaluation framework for recommender systems," 2002.

Review Quality Aware Collaborative Filtering

Sindhu Raghavan
The University of Texas at Austin
sindhu@cs.utexas.edu

Suriya Gunasekar*
The University of Texas at Austin
suriya@utexas.edu

Joydeep Ghosh
The University of Texas at Austin
ghosh@ece.utexas.edu

ABSTRACT

Probabilistic matrix factorization (PMF) and other popular approaches to collaborative filtering assume that the ratings given by users for products are genuine, and hence they give equal importance to all available ratings. However, this is not always true due to several reasons including the presence of opinion spam in product reviews. In this paper, the possibility of performing collaborative filtering while attaching weights or quality scores to the ratings is explored. The quality scores, which are determined from the corresponding review data are used to "up–weight" or "down–weight" the importance given to the individual rating while performing collaborative filtering, thereby improving the accuracy of the predictions. First, the measure used to capture the quality of the ratings is described. Different approaches for estimating the quality score based on the available review information are examined. Subsequently, a mathematical formulation to incorporate quality scores as weights for the ratings in the basic PMF framework is derived. Experimental evaluation on two product categories of a benchmark data set from Amazon.com demonstrates the efficacy of our approach.

Categories and Subject Descriptors

H.2.8 [**Information Systems**]: Database Applications - Data Mining; I.2.7 [**Computing Methodologies**]: Natural Language Processing - Text Analysis

Keywords

Recommender Systems, Probabilistic Matrix Factorization, Collaborative Filtering, Review Quality

1. INTRODUCTION

Collaborative filtering (CF), popularly used in recommendation systems, involves the task of predicting the missing

---

*The first and the second authors have contributed equally to the paper.

scores or ratings in a user-item matrix by collecting preference information from similar users and/or items. The underlying assumption of the CF approach is that those who have agreed in the past, tend to agree again in the future. Such systems are widely deployed in various domains including movie/music recommendation (Pandora[1], Netflix[2]) and product recommendation by several online retailers like Amazon.com[3] and eBay.com[4].

Probabilistic matrix factorization (PMF) [25], which is one of the popular approaches to collaborative filtering infers latent factors of both users and items and estimates ratings based on the interaction of user and item factors. PMF assumes that the ratings given by users for products are genuine, hence it gives equal importance to all available ratings. However, this assumption does not always hold due to several reasons. Often, users unhappy with a seller tend to give a poor rating for the product, which does not necessarily reflect on the quality of the product. Other times, some sellers might deliberately give superior ratings to promote their products, or they might give unjust poor ratings to competitors' products. The presence of such spurious ratings could impact the performance of the underlying collaborative filtering technique [19].

In this paper, the possibility of performing collaborative filtering while attaching quality scores to the ratings is explored. Most online retailers allow users to provide reviews of products in natural language text. Further, some websites also allow users to provide feedback on how useful the reviews have been. We believe that it is possible to assess the quality of a review/rating using this additional information. The quality score can then be used to "up–weight" or "down–weight" the importance given to the individual rating in the user-item matrix. As a result, ratings with a lower quality score will have lower impact on the predicted scores, thereby improving the accuracy of the predictions.

Our approach to collaborative filtering using quality scores consists of two stages. The first stage involves estimating the quality scores for individual ratings in the data set. In order to quantify the quality of ratings, we use the "review helpfulness" score defined by Kim et al. [13], which uses the feedback information provided by the users for individual reviews. This measure provides a reasonable indication of the quality when the amount of feedback is high. However, for more recent reviews that have less feedback, this score might

---

[1]http://www.pandora.com
[2]https://www.netflix.com
[3]http://www.amazon.com
[4]http://www.ebay.com

not necessarily capture the true quality of the rating. For such reviews, the quality score is estimated using a regression model trained on reviews that have sufficient feedback. A variety of features extracted from the review text as well as user and review metadata information are used to train the regression model. In the second stage, the quality scores estimated from the previous stage are used as weights for the ratings in the probabilistic matrix factorization framework.

The novelty of our approach lies in the integration of quality scores based on product reviews with collaborative filtering to improve the performance of recommender systems. On the one hand, there is a large body of work on the analysis of online product reviews, especially in the area of assessing the helpfulness of online reviews [13, 7, 16] as well detecting opinion spam [11, 15, 28, 22]. On the other hand, there is also a fair amount of work in the area of collaborative filtering for recommender systems [14, 30, 27] using various approaches including PMF [25] and its Bayesian variant [26]. To the best of our knowledge, this is the first paper that tries to combine online product review helpfulness with collaborative filtering to improve the overall performance of recommender systems. The efficacy of our approach is demonstrated on two product categories from a benchmark data set from Amazon.com.

The rest of the paper is organized as follows. Related work in the area of opinion spam detection and collaborative filtering is reviewed in Section 2. In Section 3, the mathematical formulation of our model and the two stage approach to collaborative filtering using quality scores are described. The experimental methodology used to evaluate the performance of our approach is described in Section 4. Finally, the results of our experiments are discussed in Section 5.

2. RELATED WORK

Kim et al. [13] proposed a quantitative measure based on the review feedback information to assess the helpfulness of reviews. They trained a regression model using various features extracted from the review text and predict the helpfulness score for new reviews. O'Mahony and Smyth [21] modeled the same problem as a classification task. Rather than predicting a score for helpfulness, they trained a classifier using reputation, content, social, and sentiment based features derived from user and item metadata to classify a review as helpful or unhelpful. Danescu-Niculescu-Mizil et al. [7] studied the correlation of different aspects of review metadata with review helpfulness. The results of their study showed that there is a strong correlation between the signed deviation of the review rating to the average rating of the product. Ghose and Ipeirotis [8] model the helpfulness of reviews as a function of the user subjectivity in the reviews. The user subjectivity is in turn predicted using a classifier trained on reviews that are subjective and objective.

There are several approaches proposed in the literature for opinion spam detection [11, 15, 22, 28]. Jindal and Liu [11] trained a classifier based on user, item, and review metadata to identify different categories of spam in online reviews. Liu et al. [15] proposed a method to detect low quality reviews in order to improve the quality of opinion summarization. They used expensive human annotation for the task of estimating the ground truth. Ott et al. [22] proposed approaches to detect fictitious and imaginative opinions that have been deliberately written to sound authentic. Deceptive spam is not easily noticeable by human readers and hence it cannot be identified by user helpfulness votes, which we consider in our work. Ott et al. acquired 400 samples of spam reviews using Amazon Mechanical Turk[5] and 400 reliable genuine reviews from the Trip Advisor website[6] and trained a classifier using n-gram text features and other linguistic metadata.

There are other approaches in literature that identify spam reviews from the perspective of recommender systems [19, 20, 29]. O'Mahony et al. [20] examined the robustness of various collaborative recommendation techniques in the face of malicious attacks. They derived theoretical results on recommendation accuracy and stability in the presence of malicious agents. Mobasher et al. [19] analyzed various new attack models and their impact on recommendation algorithms through extensive simulation-based evaluation. In a more recent work, Wu et al. [29] proposed a semi–supervised learning algorithm to identify spam reviews/ shillings using user metadata. The spam reviews are then removed from the training set while performing collaborative filtering. However, none of these approaches provide a robust methodology to improve the performance of the recommendation systems in the presence of opinion spam.

The current approaches to recommendation systems are usually classified as content-based [2, 3], collaborative [24, 25, 26], and hybrid recommendations [18, 2, 1]. Content based approaches predict the recommendations from user and item profiles derived from characteristic features of users and items, such us demographic data and product descriptions. An alternative approach to recommendation that is heavily used when rich user and item information is not available is collaborative filtering (CF). CF makes use of past user preferences to make predictions for the future. CF algorithms try to identify similarities between users and items to predict user preferences. There are several memory based and model based approaches to collaborative filtering for recommender systems [27, 14, 25]. The most successful methods for CF are the latent factor models based on probabilistic matrix factorization (PMF) [25, 26]. The PMF model is described in more detail in Section 3. Yifan, Koren, and Volinsky [10] proposed an approach that uses implicit feedback about users likes and dislikes(as produced in signed networks), to assign weights for the raw ratings obtained.

3. MODEL

The two–stage model for performing collaborative filtering with quality scores is proposed here. In *Stage 1*, the quality scores of ratings using the review and user data are estimated. In *Stage 2*, these quality scores are used as weights assigned to ratings and *weighted probabilistic matrix factorization* is performed on the ratings to predict new recommendations.

The following notation is used in the rest of the paper. There are n users and m items in the system. The users are indexed by $i \in \{1, 2, \ldots n\}$ and items by $j \in \{1, 2, \ldots, m\}$. The user-item rating matrix is represented as $Y \in \mathbb{R}^{n \times m}$, where y_{ij} represents the rating given by the ith user to jth item. Given the sparsely populated matrix Y, the task of CF is to estimate the missing entries of Y. To perform CF, the probabilistic matrix factorization approach is used, in which the rating matrix Y, is approximated as a product of

---

[5]https://www.mturk.com
[6]http://www.tripadvisor.com

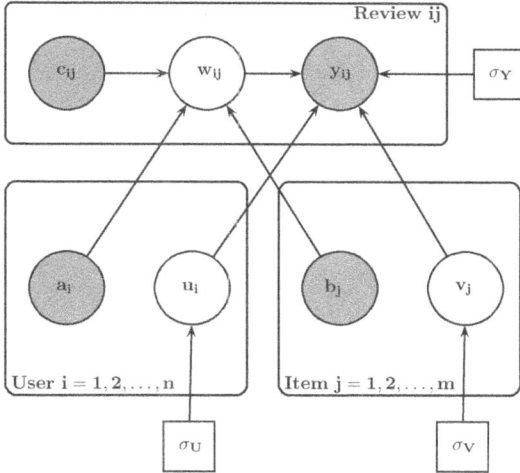

Figure 1: Graphical model for the two stage approach to recommender systems

two low rank matrices, $U \in \mathbb{R}^{n \times d}$ and $V^T \in \mathbb{R}^{m \times d}$, which represent the user and item latent factors respectively. The latent factor vector of user i is denoted as $\mathbf{u_i}$ and that of item j as $\mathbf{v_j}$. Independent Gaussian priors are used for user and item latent factors. With each rating, y_{ij}, a quality score, w_{ij} is associated, which is estimated from the corresponding review. The user and item meta data and review based features, which are used in the estimation of quality scores are represented by $\mathbf{a_i}$, $\mathbf{b_j}$ and $\mathbf{c_{ij}}$ respectively. The standard deviations associated with the user and item latent factors are given by σ_U and σ_V respectively. Finally, The standard deviation associated with the model for ratings is given by σ_Y. The graphical model for our approach is given in Figure 1.

3.1 Stage 1: Quality Score Estimation

Stage 1 involves estimating the quality score for individual ratings. The quality score for a rating is reflective of the authenticity of the rating. Most websites like Amazon and Yelp allow users to indicate if reviews were helpful or not. The amount of positive feedback obtained by a product review is indicative of the authenticity of the corresponding rating. As a result, any measure that uses the feedback information is useful for measuring the quality of the rating. Kim et al. [13] have proposed the measure below to quantify helpfulness for online product reviews based on the feedback information, which we use as the quality score.

$$\text{helpfulness} = \frac{\text{Number of helpful votes}}{\text{Total number of votes}} \quad (1)$$

The quality score computed as described above is a fair indication of quality if the amount of feedback is sufficiently high. However, for more recent reviews that have low feedback, this score might not necessarily capture the true quality of the rating. For such reviews, the quality score is estimated using a regression model trained on those reviews that have sufficient feedback.

First, a regression model is trained using features extracted from those reviews/ratings that have sufficient feedback. The quality score computed using the formula described above is used as the response variable in the regression model. For a review/rating that has low user feedback, the quality score is predicted using the trained regression model. Several features were examined for training the regression model:

- **Text based features**
  In most online websites, ratings are accompanied by reviews written by the users in natural language text. Several different features were extracted from the review text for training the regression model:

  1. *Unigram counts or bag-of-words features*
     Kim et al. [13] have demonstrated that the unigram counts or bag-of-words based features have been very useful for predicting the helpfulness score. In our approach also, bag-of-words features were extracted from the review text to train a regression model.

  2. *Features from topic modeling*
     Each review can be seen as being composed of words from several different latent topics. For instance, words like *amazing*, *awful*, *terrible*, etc. mainly convey a user's subjective opinion about the product. On the other hand, words that describe technical features of the product convey a more objective opinion about the product. Typically, reviews that are less genuine, and hence less useful are dominated by words from the former topic, while the more useful reviews have a larger proportion of words from the latter topic. The overall quality of the review could be influenced by the distribution of such latent topics. One approach to discovering these *latent topics* in natural language text involves using techniques from topic modeling like Latent Dirichlet Allocation (LDA) [5]. In our approach, LDA was applied to discover latent topics in the review text. The latent topic probabilities were then used as features to train the regression model.

- **Metadata based features**
  Along with the natural language text, most reviews are associated with various other information about the user as well as the product, which were used as features to train the regression model. Specifically, *the average rating given to the user*, which indicates how useful his reviews have been, *duration for which the review has been around*, *deviation of the rating from the mean rating of the product*, *length of the title of the review*, and *length of the review text* were used as features in our approach. Note that item based metadata information was not used as features in the regression model since item related features do not necessarily impact the quality of the review.

- **Text and metadata based features**
  In our third model, features extracted from both natural language text and metadata were used to train a regression model.

3.2 Stage 2: Collaborative Filtering

In Stage 2, the quality scores estimated in Stage 1 are used to build a recommendation system based on collaborative filtering. Among the methods used for collaborative filtering, latent factor models have been shown to give the best performance in most scenarios. Thus for our analysis, we adapt the probabilistic matrix factorization framework [25] to incorporate quality scores. This method can be trivially extended to other matrix factorization based models.

The PMF model aims at inferring latent factors of users and items from the available ratings. The missing ratings are estimated based on the interaction of user and item latent factors. These factors represent various hidden dimensions of users' tastes and preferences. The $n \times m$ matrix of ratings, Y, is approximated as $Y = UV^T$, where $U \in \mathbb{R}^{n \times d}$ and $V^T \in \mathbb{R}^{m \times d}$. The priors for U and V are assigned as follows:

$$P(U|\sigma_U^2) = \prod_{i=1}^{n} P(\mathbf{u_i}|\sigma_U^2), \ P(V|\sigma_V^2) = \prod_{j=1}^{m} P(\mathbf{v_j}|\sigma_V^2) \quad (2)$$

where,

$$\begin{aligned} P(\mathbf{u_i}|\sigma_U^2) &= \mathcal{N}(\mathbf{u_i}|0, \sigma_U^2 \mathbf{I}) \\ P(\mathbf{v_j}|\sigma_V^2) &= \mathcal{N}(\mathbf{v_j}|0, \sigma_V^2 \mathbf{I}) \end{aligned} \quad (3)$$

$\mathcal{N}(x|\mu, \sigma^2)$ represents the Gaussian distribution with mean μ and variance σ^2 evaluated at x. For vector valued variables, $\mathcal{N}(\mathbf{x}|\mu, \Sigma)$ represents the multivariate Gaussian distribution with mean μ and variance Σ evaluated at $\mathbf{x}$. Also, σ_U^2 and σ_V^2 are the variances associated with user and item latent factors respectively.

In a traditional PMF setting, the rating matrix Y is modeled as:

$$P(Y|U, V) = \prod_{i=1}^{n} \prod_{j=1}^{m} \left[\mathcal{N}(y_{ij}|\mathbf{u_i}^T \mathbf{v_j}, \sigma_Y^2) \right]^{I_{ij}} \quad (4)$$

where I_{ij} is the indicator variable indicating if the rating y_{ij} is available and σ_Y^2 is the variance associated with the model for the ratings.

We now modify the existing collaborative filtering framework to weight the ratings using the quality scores estimated in Stage 1. The quality score is modeled as a factor which inversely affects the variance of the prediction from the mean of the factor model. The intuition is that higher quality ratings are given a prior with lower deviations from the model and thus their deviations from the model mean are more heavily penalized. On the other hand, low quality scores are allowed larger deviations from the model mean. We keep the same priors for U and V, as in Equation 3. Our new prior on Y is given by:

$$P(Y|U, V) = \prod_{i=1}^{n} \prod_{j=1}^{m} \left[\mathcal{N} \left(y_{ij}|\mathbf{u_i}^T \mathbf{v_j}, \frac{\sigma_Y^2}{w_{ij}} \right) \right]^{I_{ij}} \quad (5)$$

where I_{ij} is the indicator variable indicating if the rating y_{ij} is available.

Maximizing log posterior of observed Y in this model leads to the minimization objective given below, which follows an intuitive interpretation of minimizing the weighted squared error of the observed ratings, regularized appropriately:

$$L(\theta) = \sum_{i,j} I_{ij}[w_{ij}(y_{ij} - \mathbf{u_i}^T \mathbf{v_j})^2] + \lambda_1 \|U\|_F^2 + \lambda_2 \|V\|_F^2 \quad (6)$$

where $\lambda_1 = \sigma_Y^2/\sigma_U^2$ and $\lambda_2 = \sigma_Y^2/\sigma_V^2$ are the regularization parameters and $\|\|_F$ is the Frobenius norm of a matrix. In our experiments we take $\lambda_1 = \lambda_2 = \lambda$. The above objective is a differentiable function in u_i and v_j and the maximization can be performed using the Stochastic Gradient Descent (SGD) algorithm.

	Books	Audio CDs
Total Users	772674	46491
Total Items	493991	29477
No. Ratings in Training	1677892	75518
No. Ratings in Validation	41081	902
No. Ratings in Test	105285	2683

Table 1: Various statistics about the data sets used in experimental evaluation.

4. EXPERIMENTS

4.1 Data set

Amazon.com is an online retailer that sells products like books, movies, furniture etc. Amazon widely employs recommender systems to recommend products to users based on the user's purchase and rating history. In our work, we use the open source data set provided by Jindal and Liu [11]. The categories of *Books* and *Audio CDs* were used for experimental evaluation as they have a reasonable number of users, products, and reviews. The other categories had a very small number of reviews, and hence were not used in our experiments. Each data set consists of three types of information:

1. Review information consisting of user ID, product ID, time stamp, text and title of the review text, rating, and feedback statistics in the form of helpful and unhelpful votes by users.

2. User information consisting of explicit user metadata such as name, location, and a few derived statistics based on the number of reviews written by the user, rank of user, etc.

3. Product information consisting of product specific details like category, brand, price, description, etc.

4.2 Data Preprocessing

Our first preprocessing task was to eliminate multiple reviews for a single user-item pair. This could be due to errors in data transmission on the internet due to which the same review might have appeared multiple times, thereby resulting in duplicate reviews. Other times, it could just be that the user's opinion of a product might have changed over time. In the event that multiple such reviews were present, the latest one was retained and the rest were discarded. Secondly, the entire set of ratings was split into training, validation, and test sets for the recommendation system based on time. The reviews in validation appeared after the ones in training and the reviews in test were posted later than those in validation. Finally, to avoid cold start scenarios, those entries from validation and test set for which either the user or the item was not seen in the training set were removed. Table 1 gives details about the two data sets used in experimental evaluation.

4.3 Regression

Regression models were trained for predicting quality scores using three different sets of features – text, metadata, text and metadata. For the Books data set, all those reviews in the training set that had feedback from more than 50 users were used for training the regression model. Since the

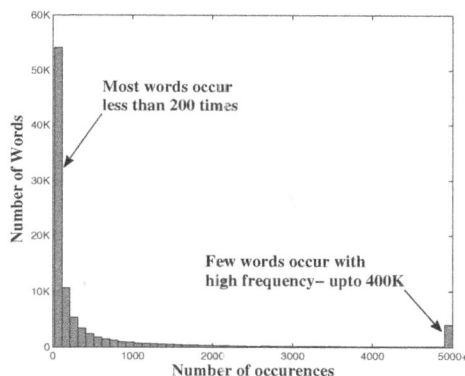

Figure 2: Histogram of words in the reviews in the Books data set.

training set for Audio CDs had reviews with fewer users providing feedback, reviews that had feedback from more than 20 users were used for training the regression model. The methodology used to extract different types of features for training the regression model is described below:

- **Extracting Bag-of-Words features**
  The SRI Language Modeling Toolkit[7] was used to extract unigram counts for bag-of-words features. Commonly occurring stop words were removed while extracting bag-of-words features. Unigram counts were normalized to obtain term frequency values, which were used as features to the regression model. There were around 760,000 unique words in total in the Books data set and around 128,000 unique words in the Audio CDs data set. Following the Zipf's law, a large number of these occurred very rarely in the entire corpus. Figure 2 gives the histogram of words with frequency for the Books data set. The histogram of words from reviews in the Audio CDs data set looked similar. Since a majority of the words occurred very rarely in the corpus and also since quite a few of them were not even valid English words, words that occurred less than 10 times in our corpus were eliminated from the analysis. This reduced the number of features by almost 10 times.

- **Extracting features using LDA**
  David Blei's LDA implementation[8] was used to identify latent topics in the review text. With the reduced set of words, LDA was scalable on the large number of reviews we dealt with. LDA was run with 5, 10, and 50 latent topics and the latent topic probabilities were used as features to train the regression model. The resulting models are called LDA_5, LDA_{10}, and LDA_{50} respectively.

- **Extracting metadata features**
  Metadata-based features were extracted as described in Section 3.1. *All* metadata features were scaled to a value between 0 and 1 by the transformation suggested

in [9], $\tilde{f} = \frac{\log(f+1)}{1+\log(f+1)}$, where f is the original value of the feature.

Regression models were trained using the techniques described below:

- **Logistic Regression**
  Since the quality scores lie in the range 0 to 1, logistic regression was suitable as the predicted values from the logistic regression model lie between 0 and 1. Logistic regression requires the response variable in the training set to be either 0 or 1. An alternate approach to specifying the response variable in logistic regression involves specifying the number of successes and total trials for a given example during training. In our context, this corresponds to specifying the number of users that found the review useful and the total number of users that provided feedback for a given review. Logistic regression models were trained by specifying the response variable in terms of the number of users that provided useful feedbacks and total number of users that provided feedback using features extracted from LDA and metadata.

- **Support Vector Regression**
  Support Vector Regression (SVR) [12] was used for bag-of-words features extracted from natural language text since it can handle any number of features and feature vectors of arbitrary size. One issue with SVR is that it does not guarantee to predict a value between 0 and 1 for the test example. To overcome this, quality scores of the reviews were mapped to real values using the inverse logistic function. The mapped scores were used as response variables to train the SVR model. The predicted value for a test sample was then passed through the logistic function to get the quality score. The linear kernel was used since it performed the best in our preliminary experiments. Note that SVR was used only when feature vectors were not of fixed size, like in bag-of-words, since the implementations of logistic regression and LASSO used in our experiments were inefficient for large feature vectors due to the lack of support for sparse vector representation.

- **LASSO Regression**
  For features extracted from metadata and LDA, a regression model was trained using lasso regression since it helps identify more useful features from the entire set. Like in SVR, quality scores of the reviews were mapped to real values in $\mathbb{R}$ using the inverse logistic function, and the predicted scores were later mapped back to get a value between 0 and 1. The python interface for LASSO in Sklearn package [23] was used in our experimental evaluation.

4.4 Collaborative Filtering

For the baseline estimate, the default implementation of PMF in Graphlab [17], which uses alternating least squares method to perform factorization was used. The number of latent factors was set to $D = 40$. Grid search was performed to tune the regularization parameter λ using the validation data set over the range of 0.01 to 0.8. Note that the weighted PMF approach defined in Section 3.2 has the same objective as the one formulated in [10], though the latter derives the model for a system that has implicit feedback about users

---

[7]http://www.speech.sri.com/projects/srilm
[8]http://www.cs.princeton.edu/ blei/lda-c/

Model	Books	Audio CDs
LR-metadata	0.24	0.25
LR-metadata+LDA_5	0.23	0.24
LR-metadata+LDA_{10}	0.23	0.23
LR-metadata+LDA_{50}	-	0.24
LASSO-metadata	0.30	0.25
LASSO-metadata+LDA_5	0.26	0.25
LASSO-metadata+LDA_{10}	0.26	0.23
LASSO-metadata+LDA_{50}	-	0.26
SVR-bag-of-words	0.27	0.35
SVR-metadata+bag-of-words	0.24	0.27

Table 2: RMSE for 10-fold cross validation on the training set for different regression models in Stage 1. "LR", "LASSO", and "SVR" refer to the models in which quality scores are predicted using logistic regression, LASSO regression, and support vector regression respectively.

likes and dislikes for certain items. The implementation of the latter objective in Graphlab was used to build our new recommendation system. The running time of this implementation is the same as that of the default implementation of PMF in Graphlab.

For the first baseline model, which we refer to as "vanilla PMF", probabilistic matrix factorization was performed by setting weights for all reviews to 1. For the second baseline, which we refer to as "second baseline", quality scores were estimated from the feedback votes for those reviews that had sufficient feedback (50 or more for Books and 20 or more for Audio CDs). For the remaining reviews, the weights were set to the average of the scores that were estimated using feedback information in the previous step and weighted probabilistic matrix factorization was performed. Initial experiments with these two models showed marginal improvement in the performance of latter over former. This observation supported the hypothesis that incorporating quality scores might improve the performance of the recommender system.

5. RESULTS AND DISCUSSION

5.1 Quality Score Estimation

10-fold cross validation was performed on the training set and root mean square error (RMSE) was computed to measure the performance of regression models. Table 2 shows the cross validation RMSE scores for different regression models. Note that due to computational complexity, LDA_{50} did not run till completion on the Books data set. Hence, results for LDA_{50} on the Books data set are not reported. Logistic regression models trained on both LDA and metadata based features perform the best on both data sets. In general, the performance of LASSO and logistic regression models trained only on LDA features is inferior to that of the models trained on both metadata and LDA features. Hence, we do not report results for LASSO and logistic regression models trained only on LDA features. The performance of logistic regression models trained only on metadata features is only slightly inferior to that of the best performing models, thereby indicating that the improvements obtained by adding LDA features are not be substantial. For a given set of features, logistic regression generally performs better than LASSO regression. SVR models trained on bag-of-words features generally perform poorly when compared to the other

models. Here again, the performance of the SVR model trained only on bag-of-words features is substantially inferior to that of the model trained on both metadata and bag-of-words features. These results indicate that there might be a stronger signal in metadata based features for predicting the quality of reviews when compared to the current set of text features considered.

5.2 Collaborative Filtering

Table 3 shows the RMSE for different models on the test set for both Books and Audio CDs. On the Books data set, all models including the second baseline outperform the vanilla PMF, while on Audio CDs data set, a majority of the models outperform the vanilla PMF. Lack of sufficient data in the Audio CDs data set could possibly be the reason for inferior performance of some models. Due to lack of reviews with sufficient feedback, less reliable reviews were used to train the regression models in Stage 1. As a result, the accuracy of the scores predicted by Stage 1 could be inferior, thereby impacting the overall quality of the predictions on the Audio CDs data set.

Logistic regression model trained on metadata features is the best performing model on the Books data set and it results in an improvement of 0.0355 (2.49%) over vanilla PMF and 0.0344 (2.41%) over the second baseline. However, on the Audio CDs data set, SVR trained on metadata and bag-words features is the best performing model, with a performance improvement of 0.0175 (1.27%) over vanilla PMF and 0.0128 (.93%) over the second baseline. In general, models trained only on metadata based features perform better than those trained on both LDA and metadata based features indicating strong signals from the metadata features used – time stamp, length of review text, length of review title, rank of the user and deviation of the rating from the mean rating of the product. Even though models trained with text based features are outperformed by the metadata based models, they still show significant improvement over the baseline models on both data sets, which warrants further investigation into linguistic feature engineering on review text. Overall, our results indicate that incorporating quality scores as weights for ratings in collaborative filtering improves the performance of recommender systems.

In our last experiment, the impact of using ratings with low quality scores in training the PMF was studied. Our hypothesis was that ratings with poor quality scores could possibly affect the predictions adversely, and hence eliminating them during training might further improve the performance of recommender systems. Figures 3 shows the distribution of quality scores from LR-metadata, which is the best performing model on the Books data set. While most of ratings have reasonably high quality scores, a small number of them have fairly poor quality scores. Analysis of the distribution of quality scores from SVR-metadata+bag-of-words, the best performing model on the Audio CDs data set yielded similar results. In our experiment, all ratings with a quality score less than 0.4 were eliminated and PMF was performed with the remaining ratings using the best performing model on both data sets. The results from these experiments, which we call "Best-Model-Low-Quality-Scores-Dropped" are shown in Table 3. Eliminating low quality scores improved the results on the Audio CDs data set considerably, thereby supporting our original hypothesis. However, the performance on the Books data set dropped marginally.

Model	Books	Audio CDs
Vanilla PMF	1.4230 ($\lambda = 0.35$)	1.3739 ($\lambda = 0.30$)
Second Baseline	1.4219 ($\lambda = 0.35$)	1.3692 ($\lambda = 0.20$)
LR-metadata	**1.3875** ($\lambda = 0.25$)	1.3664 ($\lambda = 0.25$)
LR-metadata+LDA_5	1.3972 ($\lambda = 0.20$)	1.3740 ($\lambda = 0.25$)
LR-metadata+LDA_{10}	1.3966 ($\lambda = 0.25$)	1.3779 ($\lambda = 0.25$)
LR-metadata+LDA_{50}	-	1.3731 ($\lambda = 0.25$)
LASSO-metadata	1.3910 ($\lambda = 0.30$)	1.3662 ($\lambda = 0.30$)
LASSO-metadata+LDA_5	1.3952 ($\lambda = 0.30$)	1.3634 ($\lambda = 0.30$)
LASSO-metadata+LDA_{10}	1.3958 ($\lambda = 0.30$)	1.3745 ($\lambda = 0.20$)
LASSO-metadata+LDA_{50}	-	1.3680 ($\lambda = 0.30$)
SVR-metadata+bag-of-words	1.4135 ($\lambda = 0.30$)	**1.3564** ($\lambda = 0.30$)
SVR-bag-of-words	1.4219 ($\lambda = 0.30$)	1.3740 ($\lambda = 0.30$)
Best-Model-Low-Quality-Scores-Dropped	1.3945 ($\lambda = 0.25$)	**1.3389** ($\lambda = 0.30$)

Table 3: Test RMSE for Books and Audio CDs data sets in Stage 2. "LR", "LASSO", and "SVR" refer to the models in which quality scores are predicted using logistic regression, LASSO regression, and support vector regression respectively.

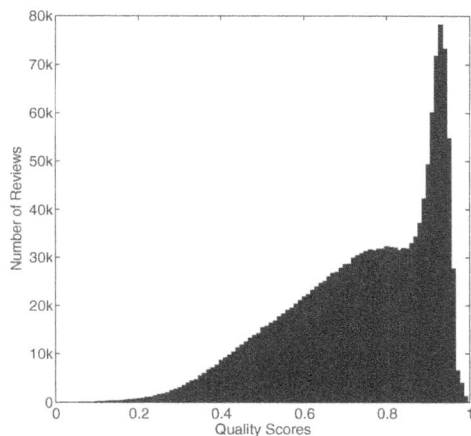

Figure 3: Distribution of quality scores from LR-metadata, the best performing model on the Books data set.

5.3 Quality Indicators

Regression models were analyzed to identify features that impacted the quality of the rating. First, coefficients learned from LASSO and logistic regression on metadata features were examined on the two data sets. Both the regression models had learned similar coefficients for individual metadata features. On both data sets, the review length had the highest positive coefficient, while the deviation of the ratings from the mean rating had the highest negative coefficient from both models. These observations can be intuitively explained as longer reviews are indicators of a thorough analysis of the product by the reviewer and hence the reviewer's rating is highly reliable. On the other hand, a reviewer giving a rating that is highly deviant from the mean rating is likely to be a spammer with a malicious intention of either boosting or degrading a product popularity and hence the negative correlation with the quality. The other features like time stamp and review title length were found to be not very influential in estimating the quality as both regression models assigned low or near-zero coefficients to these features.

Next, regression coefficients learned using LASSO and logistic regression on the topics induced by LDA were examined on the Books data set. Logistic regression assigned more or less the same weights for all topics induced by LDA. However, LASSO regression was able to assign different weights to different topics induced by LDA. There were two topics induced by LDA_5 that had negative coefficients. Some of the words from the former topic included *information, good, great, excellent, guide, books*, while the words from the latter topic included *history, book, war, world, people, american*. The remaining topics had low or near zero coefficients indicating that they did not play a significant role in determining the quality of the rating. While the words in the former topic indicate strong opinions which could be used to mask the real quality of the products, the words in the latter topic mostly describe different categories of books, which might not necessarily describe the quality of the product. In general, we found that LDA was more inclined to clustering thematic topics together rather than topics that were indicative of quality. This inability to distinguish thematic words from quality indicators is possibly one of the reasons for the modest performance of LDA-based features in our experiments. Our analysis of topics induced by LDA_{10} yielded similar results on the Books data set. Further analysis of words induced by LDA on the Audio CDs data set did not yield any interesting observations. Overall, these results emphasize the need for extraction of more sophisticated features from the review text.

In summary, incorporating quality scores or weights to ratings improves the performance of collaborative filtering in recommender systems. Our experiments with different types of features extracted from both review metadata and text indicate that some of the metadata-based features are highly indicative of the quality of the rating. Further, our experiments with text-based features also demonstrate promise, but also indicate the need for extraction of more sophisticated features from review text. Overall, we find that our two stage approach to collaborative filtering is a robust method that is capable of overcoming the negative effects caused by spurious reviews and ratings in recommender systems.

6. FUTURE WORK

Future work includes incorporating several additional features including bigrams and semantic features as described in Kim et al. [13] for learning the regression model to predict quality scores. We would also like to explore measures proposed by Ghose and Ipeirotis [8] for assessing the quality of reviews. The other direction of future work involves exploring feature reduction techniques like PCA to reduce the number of bag-of-words features, which we believe could help improve the performance of the regression model. In our current experiments, LDA could not induce latent word distributions that were reflective of the quality of reviews. To help improve the performance, supervised approaches like supervised LDA [4] can be explored in future. Further, to overcome the lack of sufficient reviews in data sets like the Audio CDs data set, transfer learning approaches [6] can be incorporated for the estimation of quality scores in our framework. Finally, experimental evaluation of our approach on other data sets like the Yelp academic data set and other product categories available in the Amazon data set will also be considered in the future.

7. CONCLUSION

In this paper, a two-stage approach to collaborative filtering that incorporates weights or quality scores for ratings is proposed. Several approaches to estimating quality scores using product reviews associated with the ratings are examined. Experimental evaluation of our approach on two product categories from a large benchmark data set from Amazon.com demonstrates that the proposed two-stage approach performs better than the vanilla PMF that assigns equal importance to the ratings. To the best of our knowledge, this is the first paper that has combined assessing review helpfulness with collaborative filtering to improve the overall performance of recommender systems.

Acknowledgements

We would like to thank Raymond J. Mooney, Sreangsu Acharyya, and Gautam Muralidhar for their valuable feedback on the paper. This research was funded by NSF grants ISS-1016614 and IIS-1116656.

8. REFERENCES

[1] G. Adomavicius and A. Tuzhilin. Toward the next generation of recommender systems: A survey of the state-of-the-art and possible extensions. *IEEE Transactions on Knowledge and Data Engineering*, 2005.

[2] M. Balabanović and Y. Shoham. Fab: content-based, collaborative recommendation. *Communucation ACM*, 1997.

[3] C. Basu, H. Hirsh, and W. Cohen. Recommendation as classification: using social and content-based information in recommendation. *AAAI*, 1998.

[4] D. Blei and J. McAuliffe. Supervised topic models. *NIPS*, 2008.

[5] D. M. Blei, A. Ng, and M. Jordan. Latent dirichlet allocation. *JMLR*, 2003.

[6] J. Blitzer, M. Dredze, and F. Pereira. Biographies, bollywood, boomboxes and blenders: Domain adaptation for sentiment classification. *ACL*, 2007.

[7] C. Danescu-Niculescu-Mizil, G. Kossinets, J. Kleinberg, and L. Lee. How opinions are received by online communities: a case study on amazon.com helpfulness votes. *WWW*, 2009.

[8] A. Ghose and P. G. Ipeirotis. Designing novel review ranking systems: predicting the usefulness and impact of reviews. *ICEC*, 2007.

[9] C.W. Hsu, C.C. Chang, and C.J Lin. A practical guide to svm classification. Technical report, National Taiwan University, 2003.

[10] Y. Hu, Y. Koren, and C. Volinsky. Collaborative filtering for implicit feedback datasets. *ICDM*, 2008.

[11] N. Jindal and B. Liu. Opinion spam and analysis. *WSDM*, 2008.

[12] T. Joachims. Making large-scale support vector machine learning practical. *Advances in kernel methods*, 1999.

[13] S. Kim, P. Pantel, T. Chklovski, and M. Pennacchiotti. Automatically assessing review helpfulness. *EMNLP*, 2006.

[14] Y. Koren, R. Bell, and C. Volinsky. Matrix factorization techniques for recommender systems. *IEEE Computer*, 2009.

[15] J. Liu, Y. Cao, C. Y. Lin, Y. Huang, and M. Zhou. Low-quality product review detection in opinion summarization. *EMNLP-CoNLL*, 2007.

[16] Y. Liu, X. Huang, A. An, and X. Yu. Modeling and predicting the helpfulness of online reviews. *ICDM*, 2008.

[17] Y. Low, J. Gonzalez, A. Kyrola, D. Bickson, C. Guestrin, and J. M. Hellerstein. Graphlab: A new framework for parallel machine learning. *CoRR*, 2010.

[18] P. Melville, R. J. Mooney, and R. Nagarajan. Content-boosted collaborative filtering for improved recommendations. *AAAI*, 2002.

[19] B. Mobasher, R. D. Burke, R. Bhaumik, and C. Williams. Toward trustworthy recommender systems: An analysis of attack models and algorithm robustness. *ACM Transactions on Internet Technologies*, 2007.

[20] M. O'Mahony, N. Hurley, N. Kushmerick, and G. Silvestre. Collaborative recommendation: A robustness analysis. *ACM Transactions on Internet Technologies*, 2004.

[21] M. P. O'Mahony and B. Smyth. Learning to recommend helpful hotel reviews. *RecSys*, 2009.

[22] M. Ott, Y. Choi, C. Cardie, and J. T. Hancock. Finding deceptive opinion spam by any stretch of the imagination. *ACL-HLT*, 2011.

[23] F. Pedregosa, G. Varoquaux, A. Gramfort, V. Michel, B. Thirion, O. Grisel, M. Blondel, P. Prettenhofer, R. Weiss, V. Dubourg, J. Vanderplas, A. Passos, D. Cournapeau, M. Brucher, M. Perrot, and E. Duchesnay. Scikit-learn: Machine Learning in Python . *JMLR*, 2011.

[24] P. Resnick, N. Iacovou, M. Suchak, P. Bergstrom, and J. Riedl. Grouplens: an open architecture for collaborative filtering of netnews. *CSCW*, 1994.

[25] R. Salakhutdinov. Probabilistic matrix factorization. *NIPS*, 2008.

[26] R. Salakhutdinov and A. Mnih. Bayesian probabilistic matrix factorization using markov chain monte carlo. *ICML*, 2008.

[27] X. Su and T. M. Khoshgoftaar. A Survey of Collaborative Filtering Techniques. *Advances in Artificial Intelligence*, 2009.

[28] G. Wu, D. Greene, B. Smyth, and P. Cunningham. Distortion as a validation criterion in the identification of suspicious reviews. *SOMA*, 2010.

[29] Z. Wu, J. Cao, B. Mao, and Y. Wang. Semi-sad: applying semi-supervised learning to shilling attack detection. *RecSys*, 2011.

[30] Y. Zhou, D. Wilkinson, R. Schreiber, and R. Pan. Large-scale parallel collaborative filtering for the netflix prize. *AAIM*, 2008.

Context-Aware Music Recommendation Based on Latent Topic Sequential Patterns

Negar Hariri
DePaul University
School of Computing
Chicago, IL 60604, USA
nhariri@cs.depaul.edu

Bamshad Mobasher
DePaul University
School of Computing
Chicago, IL 60604, USA
mobasher@cs.depaul.edu

Robin Burke
DePaul University
School of Computing
Chicago, IL 60604, USA
burke@cs.depaul.edu

ABSTRACT

Contextual factors can greatly influence the users' preferences in listening to music. Although it is hard to capture these factors directly, it is possible to see their effects on the sequence of songs liked by the user in his/her current interaction with the system. In this paper, we present a context-aware music recommender system which infers contextual information based on the most recent sequence of songs liked by the user. Our approach mines the top frequent tags for songs from social tagging Web sites and uses topic modeling to determine a set of latent topics for each song, representing different contexts. Using a database of human-compiled playlists, each playlist is mapped into a sequence of topics and frequent sequential patterns are discovered among these topics. These patterns represent frequent sequences of transitions between the latent topics representing contexts. Given a sequence of songs in a user's current interaction, the discovered patterns are used to predict the next topic in the playlist. The predicted topics are then used to post-filter the initial ranking produced by a traditional recommendation algorithm. Our experimental evaluation suggests that our system can help produce better recommendations in comparison to a conventional recommender system based on collaborative or content-based filtering. Furthermore, the topic modeling approach proposed here is also useful in providing better insight into the underlying reasons for song selection and in applications such as playlist construction and context prediction.

Categories and Subject Descriptors

H.3.3 [**Information Search and Retrieval**]: Information filtering

Keywords

recommender systems, collaborative filtering, context-aware recommendation

1. INTRODUCTION

Traditional recommender systems have been extensively used in various applications to make recommendations based on users' history of preferences. However, in some applications, failure to consider the users' current situations may result in considerable performance degradation in recommendation effectiveness because users may have different preferences for items in different contexts. To address this problem, the notion of *context-awareness* has been the focus of many research projects.

Based on the classification presented in [1], knowledge of a recommender system about the contextual factors can be of three types: fully observable, partially observable, and unobservable. The contextual information is fully observable if all the relevant contextual factors, their structure and values are explicitly known. On the other hand, if only part of this knowledge is available then it is partially observable to the system. For example, if a restaurant recommender system knows that location, and time are the only important contextual factors and the values of these factors are explicitly given to the system, the contextual knowledge is fully observable in the system. But as it is more common, part of this information might be missing which makes the contextual information partially observable. In unobservable type of knowledge, no explicit information is available about the contextual factors. The way the recommender system will infer and represent context depends on the specific domain and the data available to the system.

In the domain of music recommendation, the setting in which our work is focused, context is usually not fully observable. Moreover, the contextual information may not be captured with a static set of factors, but rather, it is dynamic and should be inferred from users' interactions with the system. More specifically, context is reflected in the sequence of songs liked or played by the user in his/her current interaction with the system, such as a playlist. For example, in Pandora [1], users create different stations by selecting different track seeds or artists. The user can later play each of these stations based on his/her current preferences which can be influenced by different contextual factors such as mood, occasion, social setting, or the task at hand. Given a set of songs in which the user shows interest during an interaction, the recommender system should be able to recommend suitable songs for the current contextual state of the user.

In this paper, we present a music recommender system that captures the changing contextual states of the user

---

[1] http://www.pandora.com

based on the sequence of songs belonging to a playlist or an active interaction session with the system. The recommender system tracks changes in users' preferences and dynamically adapts to these changes. We use a topic modeling approach to map user's interaction sequence to a sequence of latent topics which capture more general trends in user's interests. The latent topics are generated from the top most frequent tags associated with songs, obtained from social tagging Web sites such as last.fm. In order to capture changes in the contextual states over time, we employ sequential pattern mining. Using a training set consisting of human-compiled playlists, sequential patterns are mined over the set of latent topics where each pattern represents a frequent sequence of transitions between topics representing contexts. Given a user's current interaction as the sequence of last w songs, the discovered patterns are used to predict the context for the next song. Additional interaction or ratings by the user during the same session may result in changes to the predicted context. The predicted context is then used to post-filter and re-rank the recommendations produced based on the whole history of the user's preferences.

Mining sequential patterns of topics instead of songs is useful in capturing the general characteristics of songs that are interesting for the user in a given context. Looking at users' interests at a more abstract level makes it easier to track and detect any changes in the users' preferences. Moreover, having topic-based instead of song-based patterns can be specifically useful in handling the cold start problem. A new song which hasn't occurred in the training data may not match sequential patterns obtained from song sequences, but it is likely to match topic-based patterns. Also, it helps to discover patterns at a higher confidence level which is particularly important when the training data for pattern mining is not large enough or too sparse.

In addition to song recommendation, our proposed approach can be used in various applications such as automatic playlist generation where the order of songs and the transition between them is meaningful and is one of the factors affecting the quality of playlists. For example, DJs have special techniques for continuous matching and ordering of songs in a mix. A possible usage scenario for our system would be to ask the user to select an initial sequence of songs for the playlist and produce recommendations based on matching patterns. The user adds one of the recommendations (or a new song outside the recommendation set) to the playlist, and the same process will be repeated again. As we will later discuss, if the user decides to suddenly change the music type in the playlist, our method is able to dynamically adapt to these changes. Similarly, our approach can be used for playlist recommendation, where the user inputs a playlist and looks for similar playlists. In other applications such as music radio, it is still important to track the order of songs liked by the user in order to determine any changes in his/her interests.

The remainder of this paper is structured as follows: Section 2 describes the details of our topic modeling module. In section 3, we introduce the sequential pattern mining component in our system. In section 4, the topic prediction algorithm is described and some evaluations for topic prediction are provided. Our context-aware music recommendation approach is described in section 5.

2. TOPIC MODELING FOR SONG REPRESENTATION

As previously discussed, our system infers the contextual information for each user based on the selection and order of songs in the user's current playlist. To track changes in song characteristics, each song is represented as a set of topics. Instead of manually defining a large number of features for each song, we use social tagging Web sites to automatically extract tags and generate topics.

In our system, the set of top tags for each song are retrieved from last.fm and those with frequency above a minimum threshold are selected. These tags describe various features of the songs including genre, artist name and the era, but they also describe users' attitudes toward the songs, including such feelings as sad, nostalgic, upbeat, and calm. Although people may have different and even contradictory opinions about some songs (particularly those that are related to "mood"), top tags with frequency above a minimum threshold capture the social opinion about each song. While some content-based audio features are not usually contained in tagging data, other characteristics such as cultural norms, references to some events, mood, and theme of the music can be effectively captured. These features can often be very helpful in explaining the commonalities in a set of songs selected by the user. For example, a user who selects "Beauty and the Beast", "The Little Mermaid - Part Of Your World" and "Aladdin - A Whole New World" as seed songs for a station is most likely interested in Disney movie soundtracks. While these songs can be totally different in structural features extracted from music audio, they have been assigned common tags such as "disney", "childhood", "soundtrack".

Instead of using individual tags as features, our system uses Latent topics to represent a contextual state. We use Latent Drichlet Allocation[4] topic modeling approach to reduce the dimensionality and the noise of the feature space, and also to capture the latent relationships between tags and songs. In order to do this, songs are taken as documents and tags as words. After fitting the topic model for K topics, the probability distribution over topics can be inferred for any given song. For each song in our database the set of *dominant topics* can be determined by selecting all topics with probabilities higher than a specific threshold value. This way, for each song a set of one or more topics is selected allowing a mapping of a song sequence into a sequence of latent topics. As we will explain in section 3 this process simplifies tracking the changes in song characteristics and user interests, and it will enable the prediction of future topics.

2.1 Relationships between Music Features and Topics

In this section we illustrate the characteristics of the playlist data set used in our experiments, and explore the potential relationships between some of the songs features and the selected dominant latent topics. Given a song title and the artist name, the set of top tags with frequency of at least 4 were retrieved from last.fm for all the songs in our database. Songs having a minimum of 5 tags were then selected to be used for building the LDA model. We fit a 30-topic LDA model to the set of 48862 selected songs.

Table 1 shows a set of most frequent tags for a sample of ten topics in the fitted model. Based on the frequent tags,

Topic#1	Topic#2	Topic#3	Topic#4	Topic#5	Topic#6	Topic#7	Topic#8	Topic#9	Topic#10	Topic#11	Topic#12
ambient	latin	death	60s	chill	beautiful	electronic	country	reggae	hop	industrial	hard
instrumental	world	thrash	oldies	downtempo	sad	electronica	americana	roots	hip	dark	stoner
soundtrack	streamable	black	roll	chillout	mellow	house	alt-country	rnb	rap	gothic	heavy
classical	spanish	heavy	50s	christmas	melancholy	techno	american	dub	hip-hop	goth	metal
beautiful	para	doom	rockabilly	lounge	acoustic	trance	e french	slow	underground	electronic	nu-metal
age	bossa	brutal	top	electronic	chill	electro	singer-songwriter	ska	old	darkwave	progressive
chillout	fusion	melodic	500	trip-hop	soft	bass	texas	jams	coast	electro	american
experimental	musica	california	radio	electronica	drum	drum	acoustic	chill	hiphop	ebm	angry
movie	que	power	rolling	trip	melancholic	ambient	alt	dancehall	real	wave	crossover
atmospheric	nova	progressive	1960s	ambient	favourite	beat	chanson	jamaica	school	loved	lyrics
world	brazilian	gods	rhythm	hop	chillout	idm	jam	east	electronica	00s
ethereal	african	seixas	time	easy	ballad	experimental	old	jackson	stream	live	hell
chill	party	speed	elvis	cool	singer-songwriter	club	southern	mellow	west	experimental	cool
calm	brasil	swedish	soundtrack	sexy	life	minimal	bluegrass	r&b	beats	post-punk	thrash
electronic	espanol	old	american	radio	easy	party	cash	smooth	york	synth	neo metal

these topics seem to be a mixture of different attributes of songs such as genre and era as well as other features such as mood and theme which represent the social feeling of the music. For example, frequent terms in topic#10 are related to hip-hop music and are representative of some of hip-hop subgenres such as *New York East Coast hip hop*, *West Coast hip hop*, *old school(skool) hip hop*, and *Gangsta rap*. Top terms in Topic # 7 are mostly describing techno/trance music; other terms such as *club* and *Party* describe the social context related to this topic. Top terms of topic#6 are descriptive of songs having a soft mood. Topic#4 is mostly describing rock music in the 50's and 60's.

The genre feature in our database consists of 115 classes. We map these classes to Yahoo! Music Genre Hierarchy which contains 19 categories at the first level. To visualize the relation of genre and LDA topics, each of the genre classes were represented as a vector of topic co-occurrences where the jth entry in the vector of class i is the number of times topic j has been selected as dominant for a song of genre i. The co-occurrence vectors for genre classes have been created based on the set of 9743 songs in our database with known genre categories. The genre vectors were then normalized and projected into the two-dimensional space using Interactive Document Map (IDMAP) method while cosine-similarity was used for distance estimation.

The results of this visualization is shown in Figure 1. Nodes having the same ancestor at the first level are given the same shading or color. Some of the first order categories are marked in the figure. This visualization illustrates that in most cases, different subgenre classes with the same parent at the first level of the hierarchy are placed close to each other and in clusters.

A similar experiment was repeated for different eras of music between 1950 and 2000. The results shown in Figure 2 shows that subcategories of the same era are placed in clusters. Also, it is interesting to see that, the relative distance of different eras in the two-dimensional space is proportional to their time-based distances. For example, subclasses for 50's and 60's music are located close to each other. The same goes for subclasses of 90's and 80's music.

3. SEQUENTIAL PATTERN MINING

The underlying assumption behind our context-aware music recommendation is that a users' current context is reflected in the selected sequence of songs where each song is represented with a set of dominant topics. Based on our training database of playlists, our goal is to find the

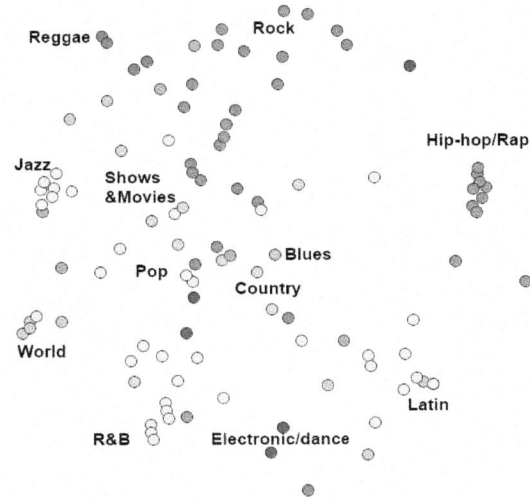

Figure 1: Music Genre Visualization by Topic-based Co-occurrence Analysis

topic-based sequential patterns that occur most frequently. Each frequent pattern is representative of a different context where the user has selected and ordered songs with specific characteristics captured in the topic sets.

Let D be a sequence database where each sequence corresponds to a playlist. In other words, each playlist is a sequence of songs where each song is represented by a set of topics. Let $A = \{t_1, t_2, ..., t_n\}$ be the set of all items which in our problem are LDA topics. A sequence $S = < x_1, x_2, ..., x_n >$ is an ordered list of elements x_i where each element is a subset of A. An item can occur at most once in each element but can occur multiple times in different elements of a sequence. Sequence $X = < x_1, x_2, ..., x_n >$ is a subsequence of $Y = < y_1, y_2, ..., y_m >$ and Y is a supersequence of X (or contains X) if there exists integers $1 < i_1 < i_2 < ... < i_n < m$ such that $x_1 \subset y_{i_1}$ and $x_2 \subset y_{i_2}, ... , x_n \subset y_{i_n}$. Support of X is defined as the number of supersequences of X in the database and is shown as $support(X)$. For a given threshold value $minSupport$, X is called a sequential pattern (SP), if $support(X) >= minSupport$. Contiguous sequential pattern (CSP) are a more restrictive form of sequential patterns which requires each pair of elements x_i and x_{i+1} to appear consecutively in a sequence Y which supports the pattern.

Figure 2: Music Era Visualization by Topic-based Co-occurrence Analysis

Various algorithms have been suggested for efficient sequential pattern mining. GSP [14] is one well-known example that works based on the Apriori principle which states that if a sequence is a sequential pattern then all of its subsequences must also be sequential patterns. Sequential patterns are mined using a candidate generation and pruning approach. Given a support threshold, at the start of the algorithm all frequent items are found. Each item represents a sequential pattern of size one. At each following step of the algorithm, the sequential patterns generated at the previous step will be used to generate new candidate sequences that respect the Apriori principle. The support values are found for each of the candidate sequences and the infrequent sequences are pruned. The remaining sequences are given to the next step and this procedure will continue until no sequential pattern is found in a step or no candidate sequence is generated. In spite of using Apriori pruning, methods similar to GSP are still generating a large number of candidates. Also, they require multiple passes on the database. Inspired with the idea of the FP-growth[6] approach, PrefixSpan[11] can be used for sequential pattern mining without generating candidate sequences at each step. In our experiments, we decided to use PrefixSpan as it has shown to efficient in mining patterns.

4. TOPIC PREDICTION USING SEQUENTIAL PATTERNS

The topic prediction module in our system takes a collection of topic sequential patterns as input and predicts a set of topics by matching the user's active session against the discovered patterns. We define the active session of user u, denoted by $h_u =< s_1, s_2, ..., s_n >$, as the sequence of last n songs that the user has shown interest in. Based on our preliminary experiments, to improve the average recall of topic predictions and to obtain a better balance between recall and precision, we assume that the selected dominant topics for each song are independent. Therefore, the user's activity sequence can be mapped to the corresponding set of single topic subsequences. Table 2 depicts an example user history h_u and the corresponding set of topic sequences h_u^i.

In the next step, each subsequence h_u^i is compared with the set of sequential patterns. A pattern, p, is accepted as a match for topic prediction, if its length is equal to $n+1$, and

Table 2: An example active session and the equivalent set of sequences

$h_u =<< t_1, t_2 >< t_3 >< t_5, t_6 >>$
$h_u^1 =<< t_1 >, < t_3 >, < t_5 >>$
$h_u^2 =<< t_2 >, < t_3 >, < t_5 >>$
$h_u^3 =<< t_1 >, < t_3 >, < t_6 >>$
$h_u^4 =<< t_2 >, < t_3 >, < t_6 >>$

the prefix of size n of p, shown as $prefix_n(p)$, is contained in h_u^i. The set of topics contained in the $(n+1)^{th}$ element of p are selected as candidate topics and the recommendation confidence is calculated as the ratio of support of p to support of $prefix_n(p)$. The recommendation module suggests those topics that have a confidence above a threshold, min_c. Note that if different patterns select a topic with different confidence values, then the maximum of those confidences is selected as the prediction score.

In many situations, a user's preferences in music may change during an active session. If the active session is small, it can result in low precision in predictions because some of the longer patterns that better capture user's behavior are not being used. On the other hand, if the window is large, then the chance that the user's preference changes in this window is high, making it less likely to find a matching sequential pattern. That leads to low recall in predictions made by the system.

In order to overcome this problem, we use the all-k^{th}-order method proposed in [12] for Markov chain models. This idea has been extended in [10] to the context of general sequential patterns. To follow this approach, first, the recommendation engine uses the current active session as an input. If the engine cannot generate any recommendations, the size of active session window is iteratively decreased by removing the oldest song until a recommendation is generated or the window size becomes 0.

To show how the all-k^{th}-order method can be useful in detecting the changes in the user's preferences, consider a playlist of length 9 shown in Table 3. For each song in the playlist, artist name, popular tags from last.fm, and the set of dominant topics are presented. In this example, the user starts by listening to mellow, acoustic rock. The first four songs have tags such as *rock, acoustic, guitar, mellow, singer-songwriter* and topic #6 is common between all these four songs. As shown in Table 1, some of the mentioned tags are appearing as the most frequent terms for this topic.

At the fifth song, there is a sudden change from *acoustic* music to *electronic* and *trip-hop*. Following the all-k^{th}-order method, a match is not found until the sixth iteration (that is when we are only considering the last four songs). The change from *acoustic* to *electronic* did not appear in any sequential pattern seen before, so the system was unable to match any patterns in the first five iterations. For the sequence containing the last four songs, the following patterns are selected:

$$
\begin{aligned}
23 &\rightarrow 6 \rightarrow 30 \rightarrow 20 \rightarrow 6 \\
23 &\rightarrow 6 \rightarrow 30 \rightarrow 6 \rightarrow 6 \\
23 &\rightarrow 6 \rightarrow 18 \rightarrow 20 \rightarrow 6 \\
23 &\rightarrow 6 \rightarrow 18 \rightarrow 6 \rightarrow 23 \\
23 &\rightarrow 6 \rightarrow 18 \rightarrow 6 \rightarrow 6 \\
23 &\rightarrow 25 \rightarrow 18 \rightarrow 6 \rightarrow 6 \\
23 &\rightarrow 25 \rightarrow 18 \rightarrow 20 \rightarrow 23 \\
23 &\rightarrow 25 \rightarrow 18 \rightarrow 6 \rightarrow 23
\end{aligned}
\tag{1}
$$

Table 3: Example of a playlist with corresponding dominant topics and popular tags

Time	Song Title	Artist Name	Popular Tags	Dominant Topics
1	why Georgia	John Mayer	singer-songwriter, mellow, relaxing, chill, male vocalist, easy listening, acoustic, 00's, guitar, rock, happy	6
2	bubble toes	Jack Johhnson	singer-songwriter, chill, acoustic, mellow, rock, summer, surf, male vocalists, pop, relaxing, guitar, happy	6
3	rose parade	Eliot Smith	singer-songwriter, indie rock, folk, acoustic, mellow, chillout, relaxing, bittersweet, lo-fi	6, 20, 23
4	may angels lead you in	Jimmy Eat World	alternative rock, ballads, calm, beautiful, nice, soundtrack, favorites	6, 28
5	sexy boy	Air	electronic, electronica,french, chillout, trip-hop, ambient, downtempo, sexy, 90s, alternative, easy listening, guitar, mellow, relax, female vocal	7, 5
6	what you are	Drill	soundtrack, 90s, alternative, atmospheric, female vocalists, indie, dreamy	23
7	flake	Jack Johhnson	singer-songwriter, acoustic, chill, alternative, rock, male vocalists, easy listening, driving	6, 25
8	what is life	Shaun Mullins	cover, beatles cover, rock, 90s, soundtrack, brass , pop rock, alternative rock, folk, brass	30, 18
9	all mixed up	red house painter	indie, rock, acoustic, 90's, cover, mellow, pop, folk, dreamy, singer-songwriter, sadcore, summery, sweet, alternative rock, female vocalist	6, 20

Based on these patterns, either topic 6 or 23 will have high probability to appear next.

4.1 Topic Prediction Evaluation

In order to evaluate the performance of our topic prediction approach, we compared our approach to several well-known pattern mining approaches. Furthermore, section 4.2 describes how Markov models can be used for topic prediction and compares the results with our sequential pattern mining method.

The database used for this experiment contains 28,963 user-contributed playlists from "Art of the Mix" website[2] in January 2003. This dataset consists of 218,261 distinct songs for 48,169 distinct artists. The average number of songs per playlist is 19.8 and the average number of artists in playlists is 17.1. Top tags were retrieved from the last.fm website for about 71,600 songs in our database. For the rest of the songs either a match was not found in last.fm or there were no top tags available for that song. For the experiment, we built the LDA model for 30 topics and set the threshold for selecting dominant topics to 0.25.

About 7,051 playlists with enough tags (at least 8) for at least 10 songs in the playlist, were used for evaluating the topic predictions. The selected playlists contain 21,783 unique songs. The algorithm was then evaluated using 10-fold cross validation. In each run of the algorithm, sequential patterns were generated using 9 folds of the data and the remaining fold was used as the test set. The support threshold for choosing a sequence as frequent pattern was set to 30.

For each playlist in the evaluation set, the last $w = 7$ songs were selected as the user's active session, the last song was removed and the dominant topics associated with that song were used as target set. Given a confidence threshold, α, the topic prediction module makes recommendations based on the remaining songs in the user's active session. The recommended topics all have recommendation scores of at least α. Figures 3 and 4, compare the precision and recall of the previously mentioned sequential pattern mining approaches for different levels of confidence. According to these figures, contiguous sequential patterns (CSP) produce more accurate results while achieving lower recall levels than general (open) sequential patterns (SP). Also, for $k = 7$, we tested the all-k^{th}-order contiguous sequential patterns, shown as all-k-th order CSP, and all-k^{th}-order general sequential patterns, shown as all-k^{th}-order SP. Similar to the results reported in [10], applying all-k^{th}-order approach re-

[2]http://www.artofthemix.org/

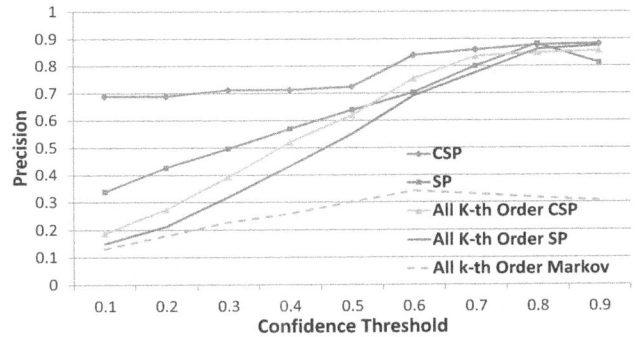

Figure 3: Topic Prediction Precision

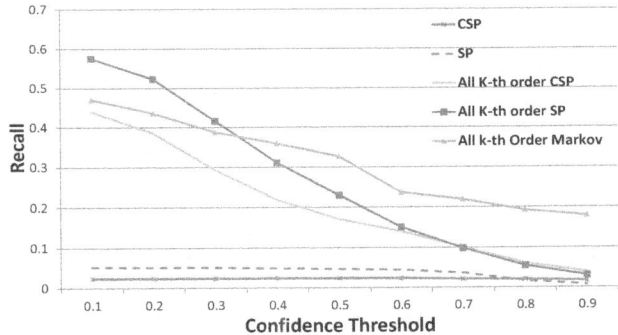

Figure 4: Topic Prediction Recall

duces precision, but significantly improves recall for both SP and CSP methods. Based on precision and recall figures, the best F-score is achieved by all-k^{th}-order SP for confidence level between 0.3 to 0.4. Therefore, in our system we used the all-k-k^{th}-order sequential pattern mining method and accepted the topic predictions with confidence score of at least 0.3.

4.2 Topic prediction using Markov Models

Given a training database of music-listening sessions, a kth-order Markov model can be built where states corresponds to all song subsequences of length k observed in the training data, and actions corresponds to different songs. In this model, the last k actions are used to make predictions about the next action. The transition probability for state s_i and action a_j is computed based on the training sequence database and as the normalized frequency of times a_j has been observed to follow s_j.

For many problems, first-order Markov models have low accuracy in making predictions. Although higher-order modeling can improve the prediction precision, it dramatically increases the model complexity and adds to the running time of the algorithm. Moreover, the algorithm will achieve a much lower recall level in making predictions. Although using the all-k^{th}-order approach can improve coverage, it adds to the state-space complexity even more.

To compare the performance of the Markov model with the sequential pattern mining approaches, the same experiment as in the previous section was set up and the all-k^{th}-order Markov model was used for topic prediction. The order of the model, was similarly set to $k = 7$. The precision and recall of the predictions are compared with sequential pattern mining methods, and are depicted in Figures 3, 4. According to these figures, even without considering the space complexity and long run-time of the all-K^{th}-order Markov model, sequential pattern mining performs better in making a proper balance in recall and precision and achieving a higher F-score.

5. CONTEXT-AWARE MUSIC RECOMMEN-DATION

As previously discussed, given a window of last n songs listened by the user, the topic prediction component makes predictions about the dominant topics of the next song. This is given as the contextual information to the music recommender module.

As described in [2], there are different ways to incorporate the contextual information into the recommendation algorithm. In contextual pre-filtering approaches, the dataset is first filtered, the recommendations are then provided based on the contextualized dataset. On the other hand, a contextual post-filtering method generates recommendations similar to traditional recommender systems. It then filters and re-ranks the recommended items to generate contextual recommendations. In contextual modeling, context is added to the problem as an additional dimension.

In our system, we used a post-filtering approach to integrate the context information. For a given user, u, first an initial ranking is generated using a traditional recommendation algorithm. This initial ordering is generated based on the complete history of user's preferences. In our experiments we used user-based kNN algorithm and computed the similarities between users based on binary cosine similarity metric. Given a user u, the initial recommendation score for a candidate song s is calculated as in equation 2. In this formula, N_u represents the user's neighborhood and $m_{n,s}$ indicates whether neighbor n has listened to song s.

$$CFScore(s) = \sum_{n \in N_u} similarity(u, n) \cdot m_{n,s} \quad (2)$$

In the next step, the recommendations are re-ranked using the contextual information extracted according to the recent preferences of the user (i.e. the last n songs). In order to do that, a contextual score is computed for each song in the recommendation list which shows the suitability of that song for the current context of the user. One heuristic to find the contextual score is to take the average topic probabilities for the candidate song over the set of predicted topics for the next song. Formally, given the active user's session as h_u, the context score of a candidate song s is calculated as follows:

$$contextScore(h_u, s) = \frac{\sum_{t_i \in predictedTopics(h_u)} p(t_i|s)}{|predictedTopics(h_u)|} \quad (3)$$

There are different ways to re-rank the initial ordering of recommendations. We followed equation 4 to compute the final recommendation scores which has shown to work well on our dataset. These scores are used to produce the final ranking of the recommended songs to the user.

$$predictionScore(s) = (contextScore(s) + \alpha_1) \cdot (CFScore(s) + \alpha_2) \quad (4)$$

In this formula, α_1 and α_2 are both smoothing constants and are set to 0.1 in our experiments.

5.1 Evaluation of Recommendations

The goal of this experiment is to determine the performance of the system in making good recommendations for a user's music-listening session (or playlist). We performed a 10-fold leave-one-out cross validation experiment using the same dataset which was used for topic prediction evaluation. For each playlist sequence in the test set, the last song was selected as the target item and was removed from the sequence. The music recommendation module was evaluated with respect to whether it was able to recommend back the removed song. If the song was found by the recommender, its rank in the overall recommendation list is recorded as top recommendations are more valuable for the user. The results for the leave-one-out cross validation was evaluated by computing the *Hit Ratio*, which computes the probability that the removed song is recommended as part of the top N recommendations. Formally, let's denote the top N recommendations for a given playlist, p, as $R_N(p)$. If in this playlist the removed target song, s_p, is part of $R_N(p)$, then it is considered a *hit*. For any given rank N, the hit ratio for the recommendation algorithm is computed as: $h(N) = |p \in testset : s_p \in R_N(p)|/|testset|$.

Figure 5 depicts the hit ratio at different levels of N less than 300 for our algorithm and some other methods including user-based kNN, Matrix Factorization using Bayesian Personalized Ranking (BPRMF) [13], and content-based recommender. Our approach for content-based recommendation and analysis of the results for this algorithm will be discussed in section 5.2. According to Figure 5, the context-aware recommender achieves higher hit ratio at all levels of $N>30$ in comparison to user-based kNN. This improvement is more significant as the number of recommendations increases.

The BPRMF recommender was trained for $m=30$ factors. The results show that both user-kNN and the context-aware recommender achieve higher hit ratio when the number of recommendations is less than 300. The average precision improvement of our method over user-based kNN is presented in Figure 6. As can be seen, our approach has higher average precison than user-based kNN at almost all cut-off levels. Similar to hit ratio results, the improvement in precision is more significant at cut-off values greater than 30.

Similarly, the average precision improvement over BPRM is depicted in Figure 7. The results show that at high ranks,

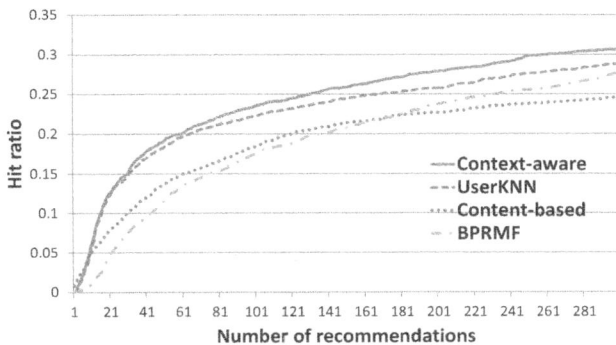

Figure 5: Hit ratio for different number of recommendations

Figure 7: Average precision improvement of our method over BPRMF and content-based recommender

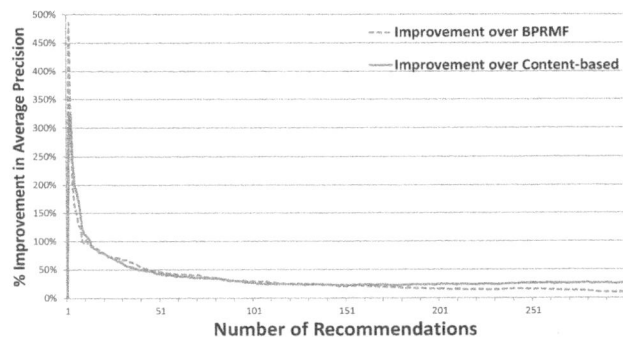

Figure 6: Average precision improvement of our method over user-based kNN

precision of the context-aware recommender can be more than four times larger than BPRMF algorithm. Based on the experiments in this section, our method achieves better performance than user-based kNN and BPRMF in terms of both hit ratio and precision.

5.2 Comparison with Content-based Recommender

This experiment compares the performance of our system against a content-based recommender which produces recommendations based on some of songs attributes including artist, genre, era, and album title. Each song is represented as a binary vector in the attribute space. Given a music listening-session for a user, p, in order to find the prediction score for a song s, first the k nearest neighbors of s are selected from the set of songs in p. The similarity of two songs is calculated as the binary-cosine similarity of their attribute vectors. The recommendation score for song s is then computed as the sum of similarities of s to its neighbors. The songs are then ranked based on the calculated recommendation scores.

As previously explained the dataset which was used in our experiments consists of 7,051 playlists with enough tags (at least 8) for at least 10 songs in the playlist and includes 21,783 unique songs. Beside artist attribute which is known for all the songs, the other three attributes might have missing values. About 9,250 songs have known genre attribute,

album is known for 11,380 songs, and era is known for 2,035 songs.

The content-based recommender was evaluated in a similar manner as described in the previous experiments. The hit ratio results of this experiment are shown in Figure 5. The hit ratio for the content-based recommender is much lower than the other methods. Also, according to the results in Figure 7, the precision of our approach can reach to more than three times that of the content-based algorithm. These results show that using song attributes alone will not result in optimal recommendation performance. In our future work, we are planning to include audio-content data and also investigate if a hybrid of context-aware method and these content-based approach can improve the performance.

6. RELATED WORK

Several researchers have previously investigated the use of contextual information in various applications of recommender systems. An interesting application of context-aware recommender systems is in mobile devices equipped with GPS or other sensors. For example, [7] uses temperature, weather, time and some other contextual factors received from sensors to suggest music to users. [8] uses a user's locations and selects music that fits a place of interest.

There has been some research related to context-aware playlist recommendation. In systems like stereomood [3], users can search for playlists that best fit their feelings or activities. For example, the user can search for happy playlists or playlists suitable for studying or partying. The users can assign mood and activity tags to each song which is played and this information is used by the system to group songs in relevant playlists. In other words, the mood-based recommendation engine uses the social mood given by active users to generate the playlists. In [9] a context-aware playlist recommendation system is proposed which provides users opportunity to browse their music by mood. This system uses a combination of audio and lyric content to classify music by mood. While mood is one of the important factors that can affect the playlist generation, it is not the only factor affecting the users' preferences.

The system described in [3] focuses on social context and proposes Poolcasting as a method to customize musical sequences for groups of listeners. Having a database of human-

[3]http://www.stereomood.com

compiled playlists, they analyzed co-occurrence of songs and artists to measure how much two songs or artists go well together in sequence. Co-occurrence analysis at song level requires much more training data in comparison to our method of finding patterns at topic-level. Also, using this approach, it is hard to find the association of the previously unseen songs/artists while our method is able to make predictions as long as tagging data is available for the songs in the playlist. Similar to our work, [5] defines context in music recommendation as the context of selecting and ordering the songs in the playlist. They generate playlists using both acoustic and social-network data. To extract the social network data, they analyzed a sample of Myspace artist network. For their evaluations, they proposed a method for comparing playlists. To compute the distance between playlists, topic modeling is used to represent songs as a mixture of topics based on their social tags. Each playlist with l songs is then represented as a $l \times d$ matrix where d is the number of topics in the mixture model. A form of cosine-similarity was used to find the similarity of two playlists. Although their representation of songs using topic modeling is similar to our work, our work is different from them in the sense that we are extracting topic-based sequential patterns and use them for context prediction.

7. CONCLUSIONS

This paper has presented a novel approach for context-aware music recommendation which infers the dynamic context of a user from the sequence of songs in his/her active interaction session with the system. Our system mines popular tags for songs from the last.fm Web site. The tag data is used by the topic modeling module which fits an LDA model and infers the topic probability distribution for songs. Each song is then represented as a set of topics which have probabilities above a certain threshold and each playlist is represented as a sequence of topicsets. The topic-based sequential patterns occurring frequently among playlists are then discovered based on a training dataset containing human-compiled playlists.

Given a user's interaction session, all matching topic-based sequential patterns are selected. The selected patterns are used to predict context of the next song. This information is used for contextual post-filtering of the recommendations given by a tradition recommendation algorithm such as the user-based kNN.

Our work differs from prior work in the area of pattern mining for music recommendation which discover patterns for song sequences or focus on analyzing co-occurrences of songs or artists. In contrast, our approach discovers patterns at a more abstract level rather than songs. This generalization makes it easier to track and detect any changes in the users' preferences. Also, it is useful in handling the cold start problem where a new song hasn't occurred in the training data.

The findings reported in this paper introduce numerous additional questions and areas of future work. We plan to extend our work to use other song features such as audio-content features in the recommendation algorithm and also investigate the relationship between LDA topics and audio features. Furthermore, we plan to follow a systematic approach for determining the best number of topics to be used in the topic modeling module.

8. REFERENCES

[1] G. Adomavicius, B. Mobasher, F. Ricci, and A. Tuzhilin. Context-aware recommender systems. *AI Magazine*, 32(3), 2011.

[2] G. Adomavicius and A. Tuzhilin. Context-aware recommender systems. In *Recommender Systems Handbook*, pages 217–253. Springer, 2011.

[3] C. Baccigalupo. *Poolcasting: an intelligent technique to customise music programmes for their audience.* PhD thesis, Universitat Autonoma de Barcelona, 2009.

[4] D. Blei, A. Ng, and M. Jordan. Latent dirichlet allocation. *The Journal of Machine Learning Research*, 3, 2003.

[5] B. Fields. *Contextualize Your Listening:The Playlist as Recommendation Engine.* PhD thesis, University of London, 2011.

[6] J. Han, J. Pei, and Y. Yin. Mining frequent patterns without candidate generation. In *ACM International Conference on Management of Data and Symposium on Principles of Database Systems, SIGMOD/PODS 2000, Dallas, TX, USA, May 15-18, 2000*, pages 1–12. ACM, 2000.

[7] H.Park, J. Yoo, and S. Cho. Context-aware music recommendation system using fuzzy bayesian networks with utility theory. In *Proceedings of the Third international conference on Fuzzy Systems and Knowledge Discovery*. Springer, 2006.

[8] M. Kaminskas and F. Ricci. Location-adapted music recommendation using tags. *User Modeling, Adaption and Personalization*, pages 183–194, 2011.

[9] O. Meyers. *A mood-based music classificationand exploration system*. Master's thesis, Massachusetts Institute of Technology, 2007.

[10] B. Mobasher, H. Dai, T. Luo, and M. Nakagawa. Using sequential and non-sequential patterns for predictive web usage mining tasks. In *Proceedings of the IEEE International Conference on Data Mining (ICDM'2002)*, 2002.

[11] J. Pei, J. Han, B. Mortazavi-Asl, H. Pinto, Q. Chen, U. Dayal, and M. Hsu. Prefixspan: Mining sequential patterns by prefix-projected growth. In *ICDE*, pages 215–224. IEEE Computer Society, 2001.

[12] J. E. Pitkow and P. Pirolli. Mining longest repeating subsequences to predict world wide web surfing. In *USENIX Symposium on Internet Technologies and Systems*, 1999.

[13] S. Rendle, C. Freudenthaler, Z. Gantner, and L. Schmidt-Thieme. Bpr: Bayesian personalized ranking from implicit feedback. In *UAI '09 Proceedings of the Twenty-Fifth Conference on Uncertainty in Artificial Intelligence*.

[14] R. Srikant and R. Agrawal. Mining sequential patterns: Generalizations and performance improvements. In *5th Intl. Conf. Extending Database Technology*, 1996.

CLiMF: Learning to Maximize Reciprocal Rank with Collaborative Less-is-More Filtering

Yue Shi[*]
Delft University of Technology
y.shi@tudelft.nl

Alexandros Karatzoglou
Telefonica Research
alexk@tid.es

Linas Baltrunas
Telefonica Research
linas@tid.es

Martha Larson
Delft University of Technology
m.a.larson@tudelft.nl

Nuria Oliver
Telefonica Research
nuriao@tid.es

Alan Hanjalic
Delft University of Technology
a.hanjalic@tudelft.nl

ABSTRACT

In this paper we tackle the problem of recommendation in the scenarios with binary relevance data, when only a few (k) items are recommended to individual users. Past work on Collaborative Filtering (CF) has either not addressed the ranking problem for binary relevance datasets, or not specifically focused on improving top-k recommendations. To solve the problem we propose a new CF approach, *Collaborative Less-is-More Filtering (CLiMF)*. In *CLiMF* the model parameters are learned by directly maximizing the Mean Reciprocal Rank (MRR), which is a well-known information retrieval metric for measuring the performance of top-k recommendations. We achieve linear computational complexity by introducing a lower bound of the smoothed reciprocal rank metric. Experiments on two social network datasets demonstrate the effectiveness and the scalability of *CLiMF*, and show that *CLiMF* significantly outperforms a naive baseline and two state-of-the-art CF methods.

Categories and Subject Descriptors

H.3.3 [**Information Storage and Retrieval**]: Information Search and Retrieval—*Information Filtering*

Keywords

Collaborative filtering, learning to rank, less is more, matrix factorization, mean reciprocal rank

1. INTRODUCTION

Collaborative Filtering (CF) [1] methods are at the core of most recommendation engines in online web-stores and social networks. The main underlying idea behind CF methods is that users that shared common interests in the past

---

[*]Part of this work was conducted when the first author was an intern at Telefonica Research, Barcelona.

would still prefer similar products/items in the future [22]. While a lot of the CF literature has been devoted to recommendation scenarios where explicit user feedback is present (i.e., typically ratings), CF has also shown to be very valuable in scenarios with only implicit feedback data [12], e.g., the counts of a user watching a TV show, the counts of a user listening to songs of an artist. These counts can be interpreted as a measure of preference and thus a proxy to explicit feedback.

However, in some scenarios even the "count" information is not available, while only binary relevance data exists, e.g., the friendship between users in a Online Social Network, the follow relationship between users (or between a user and an event, etc.) in Twitter[1] or the dating history in online dating sites [20]. Specifically, in these scenarios, we use "1" for a given user-item pair to denote that the user has an interaction (e.g., friendship, follow) with the item, and "0" otherwise. Typically the observed interactions are regarded as positive signals (i.e., indicating relevant items), and although not all items without observed interactions are irrelevant it is safe to assume the vast majority of these items will be irrelevant for the user. In other words, for a given user, the signal "0" indicates an item set containing unobserved items that could be relevant, but are most likely irrelevant. One of the most typical CF methods for those scenarios is item-based CF [9, 15], in which an item-item similarity matrix is first computed, and users are recommended items that are most similar to their past relevant items. However, item-based CF approaches typically require expensive computations in order to construct the similarity matrix. They are thus not a sound solution for large scale scenarios.

Bayesian Personalized Ranking (BPR) [21] has been recently proposed as a state-of-the-art recommendation algorithm for situations with binary relevance data. The optimization criterion of BPR is essentially based on pair-wise comparisons between relevant and a sample of irrelevant items. This criterion leads to the optimization of the Area Under the Curve (AUC). However, the AUC measure does not reflect well the quality of the recommendation lists, since it is not a top-biased measure [34], i.e., the position at which the pairwise comparisons are made is irrelevant to the contribution to the loss: mistakes at the lower ranked positions are penalized equally to mistakes in higher ranked positions, which is not the desired behavior in a ranked list.

---

[1]http://twitter.com/

In view of the drawbacks of previous work, we propose a new CF approach, *Collaborative Less-is More Filtering (CLiMF)*, that is tailored to recommendation domains where only binary relevance data is available. *CLiMF* models the data by means of directly optimizing the Mean Reciprocal Rank (MRR) [29], a well-known evaluation metric in Information Retrieval (IR). Given the analogy between query-document search and user-item recommendation, we can define the Reciprocal Rank (RR) for a given recommendation list of a user, by measuring how early in the list (i.e. how highly ranked) the first relevant recommended item is ranked. The MRR is the average of the RR across all the recommendation lists for individual users. MRR is a particularly important measure of recommendation quality for domains that usually provide users with only few but valuable recommendations (i.e., the *less-is-more* effect [7]), such as friends recommendation in social networks where top-3 or top-5 performance is important.

Taking insights from the area of learning to rank and integrating latent factor models from CF, *CLiMF* directly optimizes a lower bound of the smoothed RR for learning the model parameters, i.e., latent factors of users and items, which are then used to generate item recommendations for individual users.

Our contributions in this paper can be summarized as:

- We present a new CF approach, *CLiMF*, for MRR optimization for scenarios with binary relevance data. We demonstrate that *CLiMF* outperforms other state-of-the-art approaches with respect to making recommendations that are few in number, but relevant.

- We introduce a lower bound of the smoothed RR measure, significantly reducing the computational complexity of RR optimization, and enabling *CLiMF* to scale for large datasets.

The paper is organized as follows. In Section 2 we discuss the related work and position our paper with respect to it. Section 3 presents in detail the proposed *CLiMF* model. Our experimental evaluation is described in Section 4, followed by a summary and conclusions in Section 5.

2. RELATED WORK

The work presented in this paper closely relates to the research on ranking-oriented CF and learning to rank. In the following, we briefly review related work.

2.1 Ranking-oriented CF

A large portion of the Recommender Systems literature has been devoted to the rating prediction problem, as defined in the Netflix prize competition[2]. Latent factor models and in particular Matrix Factorization (MF) techniques, have been shown to be particularly effective [2, 13, 23] for this problem. The main idea underlying MF is to extract latent factor U_i, V_j vectors for each user and item in the dataset so that the inner product of these factors $f_{ij} = \langle U_i, V_j \rangle$ fits the observed ratings.

Several state-of-the-art ranking-oriented CF approaches, that extend upon MF techniques, have been recently proposed. These approaches typically use a ranking oriented objective function to learn the latent factors of users and

---

[2]http://www.netflixprize.com/

items, e.g., CofiRank [31], collaborative competitive filtering (CCF) [33], and OrdRec [14]. The *CLiMF* model presented in this paper can also be regarded as an extension to conventional MF, while it introduces several new characteristics that are presented in Section 3.4, compared to the state-of-the-art.

A ranking-oriented CF that extends memory-based (or similarity-based) approaches has been proposed in Eigen-Rank [16]. Moreover extensions to probabilistic latent semantic analysis [11] that optimize a ranking objective have been proposed in pLPA [17]. However, these methods are all designed for recommendation scenarios with explicit graded relevance scores from users to items.

For the use scenarios with only implicit feedback data, one of the first model-based methods was introduced in [12], where an extension of MF is proposed by weighting each factorization of user-item interaction proportionately to the count of the interactions. A similar approach, one-class collaborative filtering [19], was also proposed to exploit weighting schemes for the factorizations of missing data, which are taken as non-positive examples. However, the computational cost of that work could be inflated due to the large number of non-positive data. In this paper, we study the problem of generating recommendations for the scenarios with only binary relevance data, i.e., where even the count of user-item interaction is not available. In addition, our work directly takes into account an evaluation metric, MRR, when developing the recommendation model, which is also substantially different from the work of [12, 19].

The most similar work to ours is Bayesian personalized ranking (BPR) [21], since it also optimizes a ranking loss (AUC) and deals with binary relevance data. The main benefits of using *CLiMF* lies in its performance in terms of top-k recommendations (i.e., the fraction of relevant items at the top k positions of the list), an issue not addressed by the BPR model. Note that we also leave the detailed discussion of the relationship between *CLiMF* and BPR to Section 3.4, after the presentation of the *CLiMF* model.

2.2 Learning to Rank

Learning to Rank (LTR) has been an active research topic in Machine Learning, Information Retrieval [18] and Recommender Systems [3, 25, 31]. The work in this paper is closely related to one branch of LTR that focuses on direct optimization of IR metrics, for which the main difficulty lies in their non-smoothness with respect to the predicted relevance scores [4]. The approaches proposed in this branch of LTR approximate the optimization of IR measures either by minimizing convex upper bounds of loss functions that are based on the evaluation measures [5, 31, 32], e.g., SVM^{MAP} [34], or by optimizing a smoothed version of an evaluation measure, e.g., *SoftRank* [28] and *generalized SoftRank* [6].

In this paper, we also propose to approximate the Mean Reciprocal Rank (MRR) with a smoothed function. However, our work is different from aforementioned work not only in that we target the application scenario of recommendation rather than query-document search, but also in that we propose an algorithm (CLiMF) that makes the optimization of the smoothed MRR tractable and scalable. We also provide insights about the ability of *CLiMF* to recommend relevant items in the top positions of a recommendation list.

3. CLIMF

In this section, we present the *CLiMF*, Collaborative Less-is-More Filtering, algorithm. We first introduce a smoothed version of Reciprocal Rank by building on insights from the area of learning to rank. Then, we derive a lower bound of the smoothed reciprocal rank, and formulate an objective function for which standard optimization methods can be deployed. Finally, we discuss the characteristics of the proposed *CLiMF* model and its relation to other state-of-the-art recommendation models.

3.1 Smoothing the Reciprocal Rank

The definition of reciprocal rank of a ranked list for user i, as defined in information retrieval [29], can be expressed as:

$$RR_i = \sum_{j=1}^{N} \frac{Y_{ij}}{R_{ij}} \prod_{k=1}^{N} (1 - Y_{ik} \mathbb{I}(R_{ik} < R_{ij})) \tag{1}$$

in which N is the number of items, Y_{ij} denotes the binary relevance score of item j to user i, i.e., $Y_{ij} = 1$ if item j is relevant to user i, 0 otherwise. $\mathbb{I}(x)$ is an indicator function that is equal to 1, if x is true, otherwise 0. R_{ij} denotes the rank of item j in the ranked list of items for user i. Note that the items are ranked in a descending order according to their predicted relevance scores for user i. Clearly, RR_i is dependent on the rankings of relevant items. The rankings of the relevant items change in a non-smooth way as a function of the predicted relevance scores and thus, RR_i is a non-smooth function over the model parameters. The non-smoothness of the RR measure makes it impossible to use standard optimization methods –such as gradient-based methods– to directly optimize RR_i. Inspired by recent developments in the area of learning to rank [6], we derive an approximation of $\mathbb{I}(R_{ik} < R_{ij})$ by using a logistic function:

$$\mathbb{I}(R_{ik} < R_{ij}) \approx g(f_{ik} - f_{ij}) \tag{2}$$

where $g(x) = 1/(1 + e^{-x})$, f_{ij} denotes the predictor function that maps the parameters from user i and item j to a predicted relevance score. The predictor function that we use in our model is the basic and widely-used factor model, expressed as:

$$f_{ij} = \langle U_i, V_j \rangle \tag{3}$$

where U_i denotes a d-dimensional latent factor vector for user i, and V_j a d-dimensional latent factor vector for item j. Even though a sophisticated approximation for the item rank was proposed in [6], it has not been deployed in practice. Notice that in the case of RR_i in Eq. (1), only $1/R_{ij}$ is actually in use. We thus propose to directly approximate $1/R_{ij}$ by another logistic function:

$$\frac{1}{R_{ij}} \approx g(f_{ij}) \tag{4}$$

which makes the basic assumption that the lower the item rank, the higher the predicted relevance score, i.e., $1/R_{ij}$ would approach to 1. Substituting Eq. (2) and (4) into Eq. (1), we obtain a smooth version of RR_i:

$$RR_i \approx \sum_{j=1}^{N} Y_{ij} g(f_{ij}) \prod_{k=1}^{N} (1 - Y_{ik} g(f_{ik} - f_{ij})) \tag{5}$$

Notice that although Eq. (5) is a smooth function with respect to the predicted relevance scores and thus the model parameters U and V, optimizing this function could still be practically intractable, due to its multiplicative nature. For example, the complexity of the gradient of Eq. (5) with respect to V_j (i.e., only for one item) is $O(N^2)$: the computational cost grows quadratically with the number of items N and for most recommender systems N is typically large. In the following, we present a lower bound of an equivalent variant of Eq. (5), for which we derive a computationally tractable optimization procedure.

3.2 Lower Bound of Smooth Reciprocal Rank

Suppose that the number of relevant items for user i in the given data collection is n_i^+. Given the monotonicity of the logarithm function, the model parameters that maximize Eq. (5) are equivalent to the parameters that maximize $\ln(\frac{1}{n_i^+} RR_i)$. Specifically, we have:

$$U_i, V = \underset{U_i, V}{\arg\max} \{RR_i\} = \underset{U_i, V}{\arg\max} \{\ln(\frac{1}{n_i^+} RR_i)\}$$

$$= \underset{U_i, V}{\arg\max} \{\ln \Big(\sum_{j=1}^{N} \frac{Y_{ij}}{n_i^+} g(f_{ij}) \prod_{k=1}^{N} \big(1 - Y_{ik} g(f_{ik} - f_{ij})\big) \Big)\} \tag{6}$$

Based on Jensen's inequality and the concavity of the logarithm function, we derive the lower bound of $\ln(\frac{1}{n_i^+} RR_i)$ as below:

$$\ln(\frac{1}{n_i^+} RR_i)$$

$$= \ln \Big(\sum_{j=1}^{N} \frac{Y_{ij}}{\sum_{l=1}^{N} Y_{il}} g(f_{ij}) \prod_{k=1}^{N} \big(1 - Y_{ik} g(f_{ik} - f_{ij})\big) \Big)$$

$$\geq \frac{1}{n_i^+} \sum_{j=1}^{N} Y_{ij} \ln \Big(g(f_{ij}) \prod_{k=1}^{N} \big(1 - Y_{ik} g(f_{ik} - f_{ij})\big) \Big)$$

$$= \frac{1}{n_i^+} \sum_{j=1}^{N} Y_{ij} \Big(\ln g(f_{ij}) + \sum_{k=1}^{N} \ln \big(1 - Y_{ik} g(f_{ik} - f_{ij})\big) \Big) \tag{7}$$

Note that in the derivation above we make use of the definition of n_i^+, i.e., $n_i^+ = \sum_{l=1}^{N} Y_{il}$. We can neglect the constant $1/n_i^+$ in the lower bound, and obtain a new objective function as:

$$L(U_i, V) = \sum_{j=1}^{N} Y_{ij} \Big[\ln g(f_{ij}) + \sum_{k=1}^{N} \ln \big(1 - Y_{ik} g(f_{ik} - f_{ij})\big) \Big] \tag{8}$$

We can take a close look at the two terms within the first summation. The maximization of the first term contributes to learning latent factors that promote relevant items. However, given one relevant item, e.g., item j, maximizing the second term contributes to learning latent factors of all the other items (e.g., item k) in order to degrade their relevance scores. In sum, the two effects come together to promote and scatter the relevant items at the same time, the main characteristic of the proposed *CLiMF*. In other words, *CLiMF* will lead to a recommendation where some but not all relevant items are at the very top of the recommendation list for a user. We notice that this behavior of *CLiMF* corresponds to the analysis of MRR optimization for a search result list [30], i.e., optimizing MRR results in diversifying ranked documents.

Taking into account the regularization terms that usually serve to control the complexity of the model (i.e. in order to avoid overfitting), and all the M users in the given data collection, we obtain the objective function of $CLiMF$:

$$
\begin{aligned}
F(U,V) = \sum_{i=1}^{M} \sum_{j=1}^{N} Y_{ij} & \Big[\ln g(U_i^T V_j) \\
& + \sum_{k=1}^{N} \ln \big(1 - Y_{ik} g(U_i^T V_k - U_i^T V_j) \big) \Big] \\
& - \frac{\lambda}{2} (\|U\|^2 + \|V\|^2)
\end{aligned} \quad (9)
$$

in which λ denotes the regularization coefficient, and $\|U\|$ denotes the Frobenius norm of U. Note that the lower bound $F(U,V)$ is much less complex than the original objective function in Eq. (5), and standard optimization methods, e.g., gradient ascend, can be used to learn the optimal model parameters U and V.

3.3 Optimization

We use stochastic gradient ascent to maximize the objective function in Eq. (9), i.e., for each user i, we optimize $F(U_i, V)$. The gradients of the objective for user i with respect to U_i and V_j can be computed as below:

$$
\begin{aligned}
\frac{\partial F}{\partial U_i} = \sum_{j=1}^{N} Y_{ij} & \big[g(-f_{ij}) V_j \\
& + \sum_{k=1}^{N} \frac{Y_{ik} g'(f_{ik} - f_{ij})}{1 - Y_{ik} g(f_{ik} - f_{ij})} \big(V_j - V_k \big) \big] - \lambda U_i
\end{aligned} \quad (10)
$$

$$
\begin{aligned}
\frac{\partial F}{\partial V_j} = Y_{ij} & \big[g(-f_{ij}) \\
& + \sum_{k=1}^{N} Y_{ik} g'(f_{ij} - f_{ik}) \big(\frac{1}{1 - Y_{ik} g(f_{ik} - f_{ij})} \\
& - \frac{1}{1 - Y_{ij} g(f_{ij} - f_{ik})} \big) \big] U_i - \lambda V_j
\end{aligned} \quad (11)
$$

where $g'(x)$ denotes the derivative of $g(x)$. Note that we have used a property of $g(x)$, namely, $g(-x) = g'(x)/g(x)$, in the derivation of Eq. (10) and (11) above to simplify the computation.

The learning algorithm for the $CLiMF$ model is outlined in Algorithm 1. We analyze the complexity of the learning process for one iteration. By exploiting the data sparseness in Y, the computational complexity of the gradient in Eq. (10) is $O(d\tilde{n}^2 M + dM)$. Note that $\tilde{n}$ denotes the average number of relevant items across all the users. The complexity of computing the gradient in Eq. (11) is $O(d\tilde{n}^2 M + d\tilde{n}M)$. Hence, the complexity of the learning algorithm in one iteration is in the order of $O(d\tilde{n}^2 M)$. In the case that $\tilde{n}$ is a small number, i.e., $\tilde{n}^2 << M$, the complexity is linear to the number of users in the data collection. Note that we have $\tilde{n}M = S$, in which S denotes the number of non-zeros in the user-item matrix. The complexity of the learning algorithm is then $O(d\tilde{n}S)$. Since we usually have $\tilde{n} << S$, the complexity is $O(dS)$ even in the case that $\tilde{n}$ is large, i.e., being linear to the number of non-zeros (i.e., relevant observations in the data). In sum, our analysis shows that $CLiMF$ is

---

ALGORITHM 1: Learning Algorithm for $CLiMF$

Input: Training set Y, regularization parameter λ, learning rate γ, and the maximal number of iterations $itermax$.
Output: The learned latent factors U, V.
for $i = 1, 2, \ldots, M$ **do**
  % Index relevant items for user i;
  $N_i = \{j | Y_{ij} > 0, 1 \leq j \leq N\}$;
end
Initialize $U^{(0)}$ and $V^{(0)}$ with random values, and $t = 0$;
repeat
  for $i = 1, 2, \ldots, M$ **do**
    % Update U_i;
    $U_i^{(t+1)} = U_i^{(t)} + \gamma \frac{\partial F}{\partial U_i^{(t)}}$ based on Eq. (10);
    for $j \in N_i$ **do**
      % Update V_j;
      $V_j^{(t+1)} = V_j^{(t)} + \gamma \frac{\partial F}{\partial V_j^{(t)}}$ based on Eq. (11);
    end
  end
  $t = t + 1$;
until $t \geq itermax$;
$U = U^{(t)}$, $V = V^{(t)}$

---

suitable for large scale use cases. Note that we also empirically verify the complexity of the learning algorithm in Section 4.4.

3.4 Discussion

We discuss the relationship between the proposed $CLiMF$ and other state-of-the-art recommendation models, and present the insights that highlight the contribution of $CLiMF$ to the area of CF when compared to other models.

Relation to CofiRank: CofiRank [31] was the first work that introduced learning to rank to address CF as a ranking problem. CofiRank makes use of structured estimation of a ranking loss based on NDCG, and learns the recommendation model by minimizing over a convex upper bound of the loss function. The major differences between $CLiMF$ and CofiRank lie in two aspects: First, due to its foundation on the measure of NDCG, CofiRank suits scenarios where graded relevance data, e.g., ratings, are available from users to items, but it might not be appropriate for the scenarios with only binary relevance data, for which $CLiMF$ is tailored. Second, CofiRank and $CLiMF$ root in different classes of methods to achieve learning to rank [18, 32], such as the difference between SVM^{MAP} [34] and $SoftRank$ [28]. CofiRank exploits a convex upper bound of the structured loss function based on the evaluation metric NDCG, and then optimizes the upper bound. However, $CLiMF$ first smooths the evaluation metric RR, and then optimizes the smoothed version of the metric via a lower bound.

Relation to CCF: Collaborative competitive filtering (CCF) [33] was proposed as an algorithm that not only exploits rated items from users, but also the candidate items (or *opportunities*) that were available for the users to choose. The key constraint introduced in CCF is that the utility (or relevance) of a rated item should be higher than any items that are in the candidate set but not rated/selected. $CLiMF$ is similar to CCF in the sense that it also considers the relative pair-wise constraints in learning the latent factors, as shown in the second term with the summation in Eq. (8). However, $CLiMF$ only requires relevant items, while CCF requires all the items in the candidate set, which are not usually available. In practice, CCF needs to include some unrated items together with the rated items to form the candidate set. In addition, CCF is not directly related to any evaluation metrics, while *CLiMF* is designed for MRR optimization.

Relation to OrdRec: OrdRec [14] is an ordinal model that formulates the probability that a rating predictor (a function of the model parameters, such as the latent factors) is equal to a known rating as the probability that the rating predictor falls in the interval of two parameterized scale thresholds corresponding to two adjacent rating values. OrdRec has a point-wise nature, i.e., it does not require any pair-wise computation between any rated/unrated items. Hence, it enjoys the advantage of a computational complexity that is linear to the data size, the same advantage attained by *CLiMF*. However, although OrdRec generally suits to scenarios with implicit feedback data, "count" information is necessary to extract the ordinals, i.e., the ordered preferences of users. For this reason, OrdRec may not be suitable for the scenarios with only binary relevance data. In addition, OrdRec has no direct relation to the ranking-oriented evaluation metrics.

Relation to BPR: BPR [21] models the pair-wise comparisons between positive and negative feedback data (in the scenarios with binary relevance data), and optimizes an objective that corresponds to Area Under Curve (AUC) optimization. BPR is similar to *CLiMF* in the sense that it also directly optimizes a smoothed version of an evaluation metric for binary relevance data, there are though two main differences. First, BPR requires a sampled set of negative feedback data, i.e., a set of unobserved items to be assumed as irrelevant to the users. However, *CLiMF* only requires the relevant items from the users. Second, while BPR aims at promoting all the relevant items, *CLiMF* particularly focuses on recommending items that are few in number, but relevant at top-k positions of the recommendation list, a goal which is attained by promoting and scattering relevant items at the same time, as shown in Eq. (8). Since BPR shares a close relationship with *CLiMF* in terms of modeling and application scenarios, we choose BPR as the main baseline to compare against in the experiments.

4. EXPERIMENTAL EVALUATION

In this section we present a series of experiments to evaluate *CLiMF*. We first describe the datasets used in the experiments and the setup. Then, we compare the recommendation performance of *CLiMF* with two baseline approaches in terms of providing only a few but relevant recommendations at the top positions of the recommendation list. Finally, we analyze the effectiveness and the scalability of the proposed *CLiMF* model.

We designed the experiments in order to address the following research questions:

1. Does the proposed *CLiMF* outperform alternative state-of-the-art algorithms, particularly when recommending just a few but relevant items at top-ranked positions?

2. Is the learning algorithm of *CLiMF* effective for increasing MRR to a local maximum?

3. Is *CLiMF* scalable for large-scale use cases?

Table 1: Statistics of the datasets

Dataset	Epinions	Tuenti
Num. non-zeros	346035	798158
Num. users	4718	11392
Num. friends/trustees	49288	50000
Sparseness	99.85%	99.86%
Avg. friends/trustees per user	73.34	70.06

4.1 Experimental Setup

4.1.1 Datasets

We conduct experiments using two social network datasets from Epinions[3] and Tuenti[4]. The Epinions dataset is publicly available[5], and it contains trust relationships between 49288 users. The Epinions dataset represents a directed social network, i.e., if user i is a trustee of user j, user j is not necessary a trustee of user i. Most microblogging social networks are also directed, such as Twitter. For the purpose of our experiments, we exclude from the dataset the users who have less than 25 trustees. The second dataset collected from Tuenti, one of the largest social networks in Spain, represents an undirected social network, containing friendship between 50K users. Similar to the Epinions dataset, we also exclude the users with less than 25 friends. Note that in these two datasets, friends or trustees are regarded as "items" of users. The task is to generate friend or trustee recommendations for individual users. Statistics on the two datasets used in our experiments are summarized in Table 1.

4.1.2 Experimental Protocol and Evaluation Metrics

We separate each dataset into a training set and a test set under various conditions of user profiles. For example, the condition of "Given 5" denotes that for each user we randomly selected 5 out of her trustees/friends to form the training set, and use the remaining trustees/friends to form the test set. The task is to use the training set to generate recommendation lists for individual users, and the performance is measured according to the holdout data in the test set. We repeat the experiment 5 times for each of the different conditions of each dataset, and the performances reported are averaged across 5 runs. Again, we emphasize that in this work we only consider the observed items as being relevant to the user. Although this setting would underestimate the power of all the recommenders, the comparative results are still useful, since they can be regarded as the approximation of the lower limit of each recommender.

The main evaluation metric that we use in our experiments to measure the recommendation performance is MRR, the measure that is optimized in our model. In addition, we also measure the performance by precision at top-ranked items, such as precision at top-5 (P@5), which reflects the ratio of the number of relevant items in the top-5 recommended items. In order to emphasize the value of "less-is-more" recommendations, we also use the measure of 1-call at top-ranked items [7]. Specifically, 1-call at top-5 recommendations (1-call@5) reflects the ratio of test users who have at least one relevant item in their top-5 recommendation lists.

Finally, as revealed in recent studies from different recommender domains, it is possible that popular items could

---

[3]http://www.epinions.com
[4]http://www.tuenti.com
[5]http://www.trustlet.org/wiki/Downloaded_Epinions_dataset

Table 2: Performance comparison of *CLiMF* and baselines on the Epinions dataset

	Given 5			Given 10			Given 15			Given 20		
	MRR	P@5	1-call@5	MRR	P@5	1-call@5	MRR	P@5	1-call@5	MRR	P@5	1-call@5
PopRec	0.142	0.035	0.166	0.127	0.032	0.134	0.117	0.032	0.136	0.131	0.048	0.210
iMF	0.154	0.059	0.225	0.143	0.059	0.236	0.155	0.063	0.231	0.153	0.059	0.226
BPR-MF	0.241	0.148	0.532	0.167	0.072	0.334	0.177	0.098	0.380	0.216	0.096	0.422
CLiMF	**0.292**	**0.216**	**0.676**	**0.233**	**0.092**	**0.392**	**0.248**	**0.127**	**0.496**	**0.239**	**0.110**	**0.448**

Table 3: Performance comparison of *CLiMF* and baselines on the Tuenti dataset

	Given 5			Given 10			Given 15			Given 20		
	MRR	P@5	1-call@5	MRR	P@5	1-call@5	MRR	P@5	1-call@5	MRR	P@5	1-call@5
PopRec	0.096	0.029	0.138	0.074	0.017	0.080	0.074	0.019	0.088	0.074	0.019	0.086
iMF	0.064	0.020	0.090	0.065	0.017	0.076	0.065	0.021	0.098	0.076	0.023	0.108
BPR-MF	0.096	0.030	0.142	0.075	0.025	0.116	0.075	0.020	0.090	0.076	0.021	0.106
CLiMF	**0.100**	**0.039**	**0.190**	**0.077**	**0.027**	**0.124**	**0.077**	**0.022**	**0.104**	**0.083**	**0.024**	**0.116**

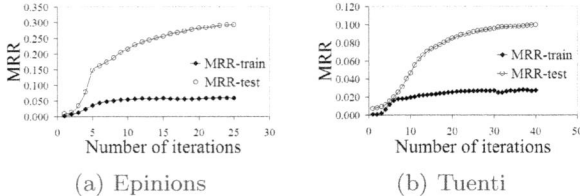

(a) Epinions   (b) Tuenti

Figure 1: Effectiveness of the learning algorithm for *CLiMF* under the "Given 5" condition for both datasets.

heavily dominate the recommendation performance [8, 26, 27]. We also notice this effect in our experiments, namely, recommending the most popular friends or trustees (i.e., those have the most friends or trusters) could already result in a high performance. For this reason, in our experiments we consider the top three most popular items as being irrelevant in order to reduce the influence from the most trivial recommendations [8, 26]. In other words, recommending any of the top three popular friends/ trustees has no contribution to any of the evaluation metrics.

4.1.3 Parameter Setting

We use one fold of randomly generated training-test sets of each dataset under the condition "Given 5" for the purpose of validation, which is used to tune parameters in CLiMF. The values of the parameters that yield the best performance on the validation set are: the regularization parameter $\lambda = 0.001$, the latent dimensionality $d = 10$ and the learning rate $\gamma = 0.0001$.

4.2 Performance Comparison

We compare the performance of *CLiMF* with three baselines, PopRec, iMF and BPR, which are described below:

- **PopRec**. A naive baseline that recommends a user to be a friend or trustee in terms of her popularity, i.e., the number of friends or trusters she has in the given training set. The more friends or trusters the user has, the higher her position in the recommendation list. Note that it is a non-personalized recommendation approach: for any target user, the recommendations are always the same.

- **iMF**: A state-of-the-art matrix factorization technique

for implicit feedback data by Hu et al. [12], as discussed in Section 2. The regularization parameter is tuned to 1, based on the performance on the validation sets.

- **BPR-MF**. Bayesian personalized ranking (BPR) represents the state-of-the-art optimization framework of CF for binary relevance data [21]. BPR-MF represents the choice of using matrix factorization (MF) as the learning model with BPR optimization criterion. Note that the implementation of this baseline is done with the publicly available software MyMediaLite [10]. The relevant parameters, such as the regularization coefficients and the number of iterations, are tuned on the validation sets, which are the same sets that were used for tuning the *CLiMF* model.

The recommendation performances of *CLiMF* and the baseline approaches on the Epinions and the Tuenti datasets are shown in Table 2 and Table 3, respectively.

Three main observations can be drawn from the results: First, the proposed *CLiMF* model *significantly* outperforms the three baselines in terms of MRR across all the conditions and the two datasets. Note that in our experiments, the statistical significance is measured based on the results from individual test users, according to a Wilcoxon signed rank significance test with p<0.01. This result corroborates that *CLiMF* achieves the goal that was designed for and optimizes the value of the reciprocal rank for the recommendations to the individual users. Notice that it is not possible to compare the results in Table 2 and Table 3 across conditions, since different conditions involve a different set of test items, containing different numbers of items. Second, *CLiMF* also achieves a *significant* improvement over the baselines in terms of P@5 and 1-call@5 across all the conditions and the two datasets. The improvement of P@5 indicates that by optimizing MRR, *CLiMF* also improve the quality of recommendations among the top-ranked items. In addition, the improvement of 1-call@5 supports that *CLiMF* particularly contributes to providing valuable recommendations at the top-k positions, i.e., raising the chance that users would receive at least one relevant recommendation among just a few top-ranked items. Compared to BPR, where AUC is optimized, *CLiMF* succeeds in enhancing the top-ranked performance by optimizing MRR, the top-biased metric. As can be also seen from the results, iMF performs worse than both BPR and *CLiMF* in all the conditions of the Epinions dataset and in most of the conditions of the Tuenti dataset.

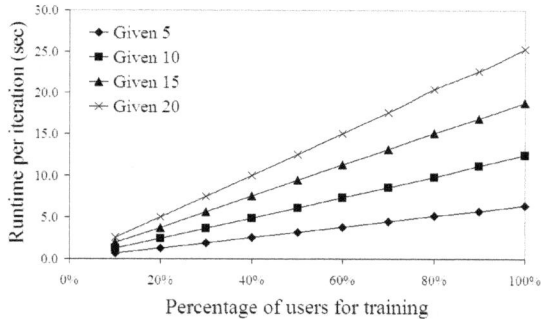

(a) Epinions

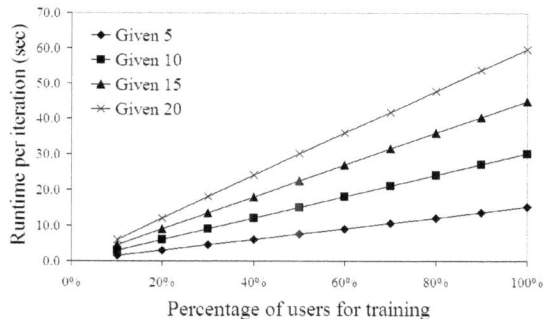

(b) Tuenti

Figure 2: Scalability analysis of *CLiMF* in terms of the number of users in the training set

The reason might be that iMF is particularly designed for implicit feedback datasets with the "count" information as mentioned in Section 2, while it may not be suitable for the scenarios with only binary relevance data. Third, in cases in which users have a lower number of friends/trustees (i.e., the case of "Given 5") the improvement achieved by *CLiMF* over the alternative approaches is relatively larger than the improvement achieved in cases in which users have a higher number of friends/trustees (i.e., the case of "Given 20"). This result suggests that *CLiMF's* key mechanism of scattering relevant items could be particularly beneficial for scenarios under very high data sparseness. Hence, we give a positive answer to our first research question.

4.3 Effectiveness

The second experiment investigates the effectiveness of the proposed learning algorithm for *CLiMF*, as presented in Section 3.3. Figures 1 (a) and (b) show the evolution of MRR with each iteration –as measured in both the training and the test sets– under the "Given 5" condition for the Epinions and Tuenti datasets, respectively. We can see that both MRR measures gradually increase with each iteration and convergence is reached after a few iterations, i.e., nearly after 20 iterations on the Epinions dataset and 30 iterations on the Tuenti dataset. This observation indicates that *CLiMF* effectively learns from the training set latent factors of users and items that optimize reciprocal rank, which consequently also contributes to improving MRR in the test set. With this experimental result, we give a positive answer to our second research question.

4.4 Scalability

The last experiment investigates the scalability of *CLiMF*, by measuring the training time that is required for the training set at different scales. First, as analyzed in Section 3.3, the computational complexity of *CLiMF* is linear in the number of users in the training set when the average number of friends/trustees per user is fixed. To demonstrate the scalability, we use different numbers of users in the training set under each condition: we randomly select from 10% to 100% users in the training set and their known friends/trustees as the training data for learning the latent factors. The results on the Epinions dataset and the Tuenti dataset are shown in Fig. 2(a) and 2(b), respectively. We can observe that for both datasets, the computational time under each condition increases almost linearly to the increase of the number of

users. Second, as also discussed in Section 3.3, the computational complexity of *CLiMF* could be further approximated to be linear to the amount of known data (i.e., non-zero entries in the training user-item matrix). To demonstrate this, we examine the runtime of the learning algorithm against different scales of the training sets under different "Given" conditions. For example, under the "Given 5" condition of the Epinions dataset, there are 5×4718=23590 non-zeros in the training set. The result is shown in Fig. 3, from which we can observe that the average runtime of the learning algorithm per iteration increases almost linearly as the number of non-zeros in the training set increases. The observations from this experiment allow us to answer our last research question positively.

5. CONCLUSIONS AND FUTURE WORK

In this paper we have presented a new CF approach, *CLiMF*, that learns latent factors of users and items by directly maximizing MRR. *CLiMF* is designed to improve the performance of top-k recommendations for usage scenarios with only binary relevance data. We have demonstrated in our experiments that *CLiMF* offers significant improvements over a naive and two state-of-the-art baselines in two social network datasets. We have also experimentally validated that *CLiMF's* learning algorithm is effective for MRR optimization, and has linear computational complexity to the size of the known data, and thus is scalable for large scale use cases.

Future work involves a few interesting directions. First, we would like to extend our *CLiMF* model to suit domains with explicit feedback data, e.g., ratings. Second, it is also interesting to experimentally investigate the impact of *CLiMF* on the recommendation diversity, by exploiting external information resources, such as the categories of items. Third, we are also interested in investigating recommendation models that optimize other evaluation measures, such as mean average precision [24], and in exploring the impact of optimizing different measures on various aspects of recommendation performance [30].

6. ACKNOWLEDGEMENTS

This work is funded, in part, as part of a Marie Curie Intra European Fellowship for Career Development (IEF) award (CARS, PIEF-GA-2010-273739) held by Alexandros Karatzoglou.

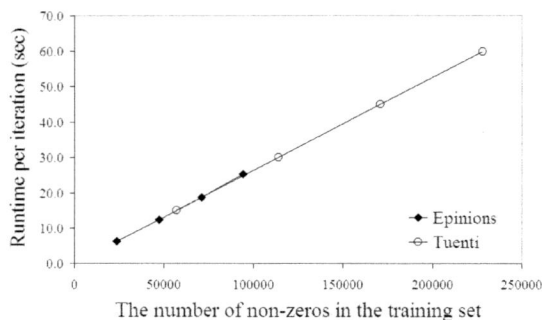

Figure 3: Scalability analysis of *CLiMF* in terms of the scale of the training data

7. REFERENCES

[1] G. Adomavicius and A. Tuzhilin. Toward the next generation of recommender systems: A survey of the state-of-the-art and possible extensions. *IEEE TKDE*, 17(6):734–749, 2005.

[2] D. Agarwal and B.-C. Chen. Regression-based latent factor models. KDD '09, pages 19–28. ACM, 2009.

[3] S. Balakrishnan and S. Chopra. Collaborative ranking. WSDM '12, pages 143–152. ACM, 2012.

[4] C. J. C. Burges, R. Ragno, and Q. V. Le. Learning to Rank with Nonsmooth Cost Functions. In *NIPS*, pages 193–200. MIT Press, 2006.

[5] S. Chakrabarti, R. Khanna, U. Sawant, and C. Bhattacharyya. Structured learning for non-smooth ranking losses. KDD '08, pages 88–96. ACM, 2008.

[6] O. Chapelle and M. Wu. Gradient descent optimization of smoothed information retrieval metrics. *Inf. Retr.*, 13:216–235, June 2010.

[7] H. Chen and D. R. Karger. Less is more: probabilistic models for retrieving fewer relevant documents. SIGIR '06, pages 429–436. ACM, 2006.

[8] P. Cremonesi, Y. Koren, and R. Turrin. Performance of recommender algorithms on top-n recommendation tasks. RecSys '10, pages 39–46. ACM, 2010.

[9] M. Deshpande and G. Karypis. Item-based top-n recommendation algorithms. *ACM Trans. Inf. Syst.*, 22:143–177, January 2004.

[10] Z. Gantner, S. Rendle, C. Freudenthaler, and L. Schmidt-Thieme. Mymedialite: a free recommender system library. In *Proceedings of the fifth ACM conference on Recommender systems*, RecSys '11, pages 305–308, New York, NY, USA, 2011. ACM.

[11] T. Hofmann. Latent semantic models for collaborative filtering. *ACM Trans. Inf. Syst.*, 22:89–115, January 2004.

[12] Y. Hu, Y. Koren, and C. Volinsky. Collaborative filtering for implicit feedback datasets. ICDM '08, pages 263–272, 2008.

[13] Y. Koren, R. Bell, and C. Volinsky. Matrix factorization techniques for recommender systems. *Computer*, 42:30–37, August 2009.

[14] Y. Koren and J. Sill. Ordrec: an ordinal model for predicting personalized item rating distributions. RecSys '11, pages 117–124. ACM, 2011.

[15] G. Linden, B. Smith, and J. York. Amazon.com recommendations: Item-to-item collaborative filtering. *IEEE Internet Computing*, 7:76–80, 2003.

[16] N. N. Liu and Q. Yang. Eigenrank: a ranking-oriented approach to collaborative filtering. SIGIR '08, pages 83–90. ACM, 2008.

[17] N. N. Liu, M. Zhao, and Q. Yang. Probabilistic latent preference analysis for collaborative filtering. CIKM '09, pages 759–766. ACM, 2009.

[18] T.-Y. Liu. Learning to rank for information retrieval. *Foundations and Trends in Information Retrieval*, 3(3):225–331, 2009.

[19] R. Pan, Y. Zhou, B. Cao, N. N. Liu, R. Lukose, M. Scholz, and Q. Yang. One-class collaborative filtering. ICDM '08, pages 502–511, 2008.

[20] L. Pizzato, T. Rej, T. Chung, I. Koprinska, and J. Kay. Recon: a reciprocal recommender for online dating. RecSys '10, pages 207–214. ACM, 2010.

[21] S. Rendle, C. Freudenthaler, Z. Gantner, and S.-T. Lars. Bpr: Bayesian personalized ranking from implicit feedback. UAI '09, pages 452–461. AUAI Press, 2009.

[22] P. Resnick, N. Iacovou, M. Suchak, P. Bergstrom, and J. Riedl. Grouplens: an open architecture for collaborative filtering of netnews. CSCW '94, pages 175–186. ACM, 1994.

[23] R. Salakhutdinov and A. Mnih. Probabilistic matrix factorization. volume 20 of *NIPS '08*, 2008.

[24] Y. Shi, A. Karatzoglou, L. Baltrunas, M. A. Larson, A. Hanjalic, and N. Oliver. TFMAP: Optimizing MAP for top-n context-aware recommendation. SIGIR '12. ACM, 2012.

[25] Y. Shi, M. Larson, and A. Hanjalic. List-wise learning to rank with matrix factorization for collaborative filtering. RecSys '10, pages 269–272. ACM, 2010.

[26] Y. Shi, P. Serdyukov, A. Hanjalic, and M. Larson. Personalized landmark recommendation based on geotags from photo sharing sites. ICWSM '11, pages 622–625. AAAI, 2011.

[27] H. Steck. Item popularity and recommendation accuracy. RecSys '11, pages 125–132. ACM, 2011.

[28] M. Taylor, J. Guiver, S. Robertson, and T. Minka. Softrank: optimizing non-smooth rank metrics. WSDM '08, pages 77–86. ACM, 2008.

[29] E. M. Voorhees. The trec-8 question answering track report. In *TREC-8*, 1999.

[30] J. Wang and J. Zhu. On statistical analysis and optimization of information retrieval effectiveness metrics. SIGIR '10, pages 226–233. ACM, 2010.

[31] M. Weimer, A. Karatzoglou, Q. Le, and A. Smola. Cofirank - maximum margin matrix factorization for collaborative ranking. NIPS '07, pages 1593–1600, 2007.

[32] J. Xu, T.-Y. Liu, M. Lu, H. Li, and W.-Y. Ma. Directly optimizing evaluation measures in learning to rank. SIGIR '08, pages 107–114. ACM, 2008.

[33] S.-H. Yang, B. Long, A. J. Smola, H. Zha, and Z. Zheng. Collaborative competitive filtering: learning recommender using context of user choice. SIGIR '11, pages 295–304. ACM, 2011.

[34] Y. Yue, T. Finley, F. Radlinski, and T. Joachims. A support vector method for optimizing average precision. SIGIR '07, pages 271–278. ACM, 2007.

Ranking With Non-Random Missing Ratings: Influence Of Popularity And Positivity on Evaluation Metrics

Bruno Pradel
LIP6 – Université Paris 6
4 place Jussieu, Paris, France
bruno.pradel@lip6.fr

Nicolas Usunier
LIP6 – Université Paris 6
4 place Jussieu, Paris, France
nicolas.usunier@lip6.fr

Patrick Gallinari
LIP6 – Université Paris 6
4 place Jussieu, Paris France
patrick.gallinari@lip6.fr

ABSTRACT

The evaluation of recommender systems in terms of ranking has recently gained attention, as it seems to better fit the top-k recommendation task than the usual ratings prediction task. In that context, several authors have proposed to consider missing ratings as some form of negative feedback to compensate for the skewed distribution of observed ratings when users choose the items they rate. In this work, we study two major biases of the selection of items: the first one is that some items obtain more ratings than others (popularity effect), and the second one is that positive ratings are observed more frequently than negative ratings (positivity effect). We present a theoretical analysis and experiments on the Yahoo! dataset with randomly selected items, which show that considering missing data as a form of negative feedback during training may improve performances, but also that it can be misleading when testing, favoring models of popularity more than models of user preferences.

Categories and Subject Descriptors

H.4 [**Evaluation metrics and studies**]: Machine learning for recommendation

Keywords

Recommender Systems, Ranking, Evaluation

1. INTRODUCTION

Recommender systems often aim at producing a short list of recommended items. While most prior research has focused on accurately predicting rating values, there is a growing interest in evaluating recommender systems in terms of ranking performance as it seems to better approximate the true task. Recent results [3, 11, 2] tend to show that algorithms having good ranking performance are not the same as those which best predict rating values. These results are somewhat natural because the recent works on ranking involve an important change in the experimental protocol:

unknown ratings are treated as if they were known to be negative, while they were simply ignored in ratings prediction evaluations. Such a shift in the experimental protocol involves a shift in the test data, and it is natural that algorithm should be modified so as to reflect these changes.

In this paper, we propose to analyze two different experimental protocols to evaluate the ranking performance of a recommendation algorithm using the Area Under the ROC Curve (AUC) as performance measure, (1) ignoring missing ratings, and (2) treating them as negative ratings. Our study focuses on the impact of how users choose the items they rate. More specifically, we isolate two specific effects of how the items are selected. The first one, the *popularity effect*, refers to the long tailed distribution of the number of observed ratings per item. The second one, which we call the *positivity effect*, refers to the propensity of users to rate more often the items they like, and this "oversampling" of positive ratings in ratings datasets is *decreasing* with the popularity of the items. While the first effect can easily be observed on any available dataset, we motivate our analysis of the second one using the Yahoo! Music dataset with user-selected and randomly selected songs [14].

We first propose a formal analysis under a simple model of user behavior, in which we show that (1) ignoring missing items leads to a dramatically biased evaluation in the presence of the positivity effect, and (2) considering missing ratings as negative biases the evaluation towards models that favor popular items.

Our analysis suggests that in order to optimize performance when considering missing ratings as negative feedback, one has to model both popularity and user tastes. However, nothing proves that it improves the *real* recommendation performance – it simply proves that we can tune the algorithm to go the same way as the performance measure. Nonetheless, because the popularity effect somewhat compensates the positivity effect, we may expect better performances by considering missing entries as a weaker form of negative ratings. In that sense, we follow [2, 12] and provide an experimental study on the Yahoo! dataset. Our goal is to study an experimental protocol for tuning the hyperparameters of a recommendation algorithm, where the protocol would only use ratings from user-selected items (which are usually available) but with good performances on randomly selected items (which are usually unavailable).

The remainder of the paper is organized as follows. In Section 2, we describe the problem we address. Our formal analysis is then performed in Section 3. The experimental protocol and the results are presented in Sections 4 and 5. We discuss relevant related work in Section 6.

2. TASKS AND MEASURES

The task we address is to predict and evaluate personalized lists of recommended items based on explicit user feedback. A dataset may be decomposed into several subsets (e.g. train/validation/test), and they may not follow the same distribution. Each one of these subsets contains a set of U users and I items which we respectively index by $\{1,...,U\}$ and $\{1,...,I\}$, and, for each user u, we have a subset of size I_u of items $\mathcal{I}_u \subseteq \{1,...,I\}$ together with reference ratings. In our experiments, the ratings are expressed on a 1- to 5-star rating scale, but for ranking evaluations, they will be considered as binary (-1 or $+1$). Items with a 4/5-star ratings are considered *relevant* to the user (and the rating is qualified as *positive*), while other ratings are considered negatives (and the item is then irrelevant)[1].

Offline Evaluation.

We focus on the case where the available dataset contains ratings for *user-selected* items. This is the usual data collection process for ratings data (e.g. Netflix [1] or MovieLens [10] datasets): users chose the items they rated.

However, our underlying goal is to perform well on a more reliable ground truth. We follow Marlin et al. [8] and use their dataset Yahoo! Music with user ratings for *randomly-selected* items [14] for defining this ground truth. The dataset is divided into ratings from the "usual" collection process (i.e. the users choose to rate some items) together with another dataset where some users were asked to rate specific items. These items were chosen randomly to avoid any bias in the choice of the items. We call these datasets Yahoo!^{user} and Yahoo!^{random}. All Yahoo!^{random} users provided ratings in Yahoo!^{user}. Thus, in our experiments, we discuss the development of a recommender system on Yahoo!^{user} as if Yahoo!^{random} were unknown, but the final goal is to perform well on Yahoo!^{random}. Our motivation is that ratings for randomly selected items are usually unavailable, and thus we do not intend to develop a methodology that assumes the existence of such data.

The Area Under the ROC Curve (AUC).

Precision/recall and the AUC are common ways of evaluating recommender systems in terms of ranking (see e.g. [4]). In that paper, we use the AUC, which, for a given user u, counts the number of pairs of (positive, negative) items correctly ordered by the system. In our case, the system produces a score $\hat{R}_{u,i}$ for each (user, item) pair (i.e. taking values in $\hat{R}_{u,i} \in \mathbb{R}$), so that the AUC is computed as[2]:

$$AUC_{\mathcal{I}_u}(\hat{R}_u, R_u) = \frac{1}{N_{\mathcal{I}_u} P_{\mathcal{I}_u}} \sum_{\substack{p \in \mathcal{I}_u : R_{u,p}=+1 \\ n \in \mathcal{I}_u : R_{u,n}=-1}} \mathbf{1}(\hat{R}_{u,p} > \hat{R}_{u,n}), \quad (1)$$

where $\mathbf{1}(predicate)$ is 1 if *predicate* is true (0 otherwise),

---

[1]While many authors consider only 5-star ratings to be $+1$, we consider that both 4- and 5-star ratings are $+1$ because there are very few 5-star ratings in Yahoo!^{random}, and evaluation measures were rather unstable in the usual setting.
[2]Remenber that we binarize the ratings when evaluating. To simplify the notations, we assume here that there are no ties, i.e. all values of $\hat{R}_{u,i}$ are different.

$P_{\mathcal{I}_u} = \sum_{i \in \mathcal{I}_u} \mathbf{1}(R_{u,i}=1)$ is the number of positive ratings, $N_{\mathcal{I}_u} = \sum_{i \in \mathcal{I}_u} \mathbf{1}(R_{u,i}=-1)$ the number of negative ones for u. The AUC on a whole dataset is then computed as the macro-average over users: $AUC(\hat{R}, R) = \frac{1}{U} \sum_{u=1}^{U} AUC_{\mathcal{I}_u}(\hat{R}_u, R_u)$.

The formula given above applies to any set of items $\mathcal{I}_u$. We may restrict $\mathcal{I}_u$ to the set of items with known ratings, ignoring the missing ratings (which is the case when we simply use the notation AUC). Otherwise, we may consider the missing ratings as negative ones, as in [2, 11]. In that case, we consider the transformed observed ratings R^{AMAN} as (AMAN stands for all missing as negatives):

$$R_{u,i}^{AMAN} = \begin{cases} 1 & \text{if } R_{u,i} = +1 \\ -1 & \text{otherwise} \end{cases}$$

and compute AUC^{AMAN} as:

$$AUC_{\mathcal{I}_u}^{AMAN}(\hat{R}_u, R_u) = AUC_{\{1,...,I\}}(\hat{R}_u, R_u^{AMAN}). \quad (2)$$

Notice that AUC^{AMAN} will only be made on datasets with user-selected items, and the true effect of not ignoring missing ratings is to penalize items which are rarely rated.

AUC^{AMAN} was proved to be close to the ATOP measure proposed by [11] when the relevant items represent a small fraction of the items. In [11, 12, 2], precision/recall measures are used instead of AUC/AUC^{AMAN}. All of these works consider missing ratings as negatives. We choose AUC in this work because, even in the presence of missing data, it can be estimated without bias on Yahoo!^{random} and the optimal predictions have a closed form solution (see Section 3.1). To the best of our knowledge, more usual measures in Information Retrieval such as the Normalized Discounted Cumulative Gain do not satisfy any of these two properties.

Ratings Distributions.

Marlin et al. [8] studied the difference between ratings on user-selected items and randomly selected items. In particular, they observed that the marginal distribution of ratings on user-selected items was skewed towards positive ratings. This was in accordance with a poll on users' rating behavior they performed, in which 93.9% of the users claim that they rate an item they love very often, while only 36.5% of them claim that they rate an item for which they are "neutral" with the same frequency. We isolate here two related, yet somewhat more specific "biases" of user-selected datasets.

Let us first consider the *popularity effect*: one can observe on most collaborative filtering datasets that the distribution of the number of ratings per item (whatever the value of the rating) has a long tail distribution: for example on MovieLens, 30% of the movies represent 80% of the number of ratings. Similarly, on Yahoo!^{user}, 4.7% of the songs account for more than 33% of the ratings.

Second, we consider the *positivity effect*, which is a refined version of the skewed marginal distribution of ratings: while positive ratings tend to be oversampled on Yahoo!^{user} in comparison to Yahoo!^{random}, this oversampling is dramatically influenced by the popularity of the items. As a crude approximation, consider Figure 1 which shows the distribution of ratings (on a 1- to 5-star scale) on both the user-selected and the random-selected datasets, depending on the popularity category (Head/Middle/Tail) of the different items (the popularity is computed on Yahoo!^{user}). The density of the ratings on the Yahoo!^{user} is much more stable across the different categories than on the randomly-selected dataset. For instance, 5-star ratings represent about 2% of the Tail ratings on Yahoo!^{random} and about 9% of the ratings on the Head items. On Yahoo!^{user}, they account for more than 20% of the ratings on the Tail items and less than 28% for Head items. More generally, positive ratings tend to be less oversampled for more popular categories. The positivity effect is this inverse relationship between the bias towards positive ratings and popularity.

3. INFLUENCE ON EVALUATION

In this section we present a formal analysis of the effects of popularity and positivity on the evaluation performed by AUC and AUC^{AMAN}. We consider a very simple probabilistic model of ratings, which may exhibit popularity and/or positivity effects and thus allows us to directly analyze these effects on the evaluation measures. Our comparison is between (1) the ranking which optimizes the AUC or the AUC^{AMAN} under the user model, and (2) the ranking which optimizes the AUC under the model of the true underlying user ratings. Our point is to show that these rankings may be very different, and thus the measures we usually compute on validation (user-selected) rating datasets are not optimized by ranking according to the true user tastes. Thus, not only such evaluations are imperfect, but they may lead us to select a suboptimal algorithm or incorrectly choose algorithms' hyperparameters. Before describing our results, we start with a more general analysis of the optimal ranking for the AUC under suitable (binary) rating distributions.

3.1 Optimizing the AUC Under Noise

Fix a given user u, and let us assume that the ratings R_u and the set of rated items $\mathcal{I}_u$ are actually random variables, and we only have observed one realization of these random variable for each user. Then, an optimal prediction $\hat{R}_u^*$ for user u is a vector of scores $\hat{R}_{u,i}^*, i = 1..I$ such that:

$$\hat{R}_u^* \in \operatorname*{argmin}_{\hat{R}_u \in \mathbb{R}^I} \mathbb{E}\left[AUC_{\mathcal{I}_u}(\hat{R}_u, R_u)\right]$$

where $\mathbb{E}[.]$ denotes the expected value, and the expectation is taken over the joint distribution of $(R_u, \mathcal{I}_u)$.

A natural assumption we can make on the distribution of $(R_u, \mathcal{I}_u)$ is that the value of an item's rating does not depend on the other items rated. This assumption between the rating $R_{u,i}$ and $\mathcal{I}_u$ can be written as follows:

$$\exists \alpha_{u,i}, \forall \mathcal{I} \subseteq \{1,..,I\} \text{ such that } i \in \mathcal{I}: \\ \mathbb{P}(R_{u,i} = 1 | \mathcal{I}_u = \mathcal{I}) = \alpha_{u,i} \quad (3)$$

Event though this assumption is quite strong and may not be sufficient to realistically model the bias in how users choose the items they rate, the assumption is still general enough to model ratings data with both positivity and popularity effects, as we shall see in the next Subsection. Now, if we fix $\mathcal{I} \subseteq \{1,...,I\}$ and two items i and j in $\mathcal{I}$, the contribution of the pair (i,j) to $\mathbb{E}\left[AUC_{\mathcal{I}_u}(\hat{R}_u, R_u) | \mathcal{I}_u = \mathcal{I}\right]$ is:

$$\mathbb{E}\left[\frac{p_{i,j}^u(\mathcal{I})\mathbf{1}(\hat{R}_{u,i} > \hat{R}_{u,j}) - p_{j,i}^u(\mathcal{I})\mathbf{1}(\hat{R}_{u,j} > \hat{R}_{u,i})}{N_{\mathcal{I}_u} P_{\mathcal{I}_u}} \Big| \mathcal{I}_u = \mathcal{I}\right]$$

where $p_{i,j}^u(\mathcal{I}) = \mathbb{P}(R_{u,i} = 1, R_{u,j} = -1 | \mathcal{I}_u = \mathcal{I})$. In order to maximize the expected AUC over $\hat{R}_u$, we should thus have

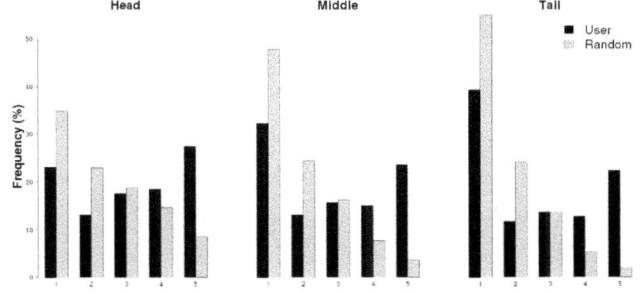

Figure 1: Probability density of ratings on the Yahoo! dataset with *user-selected* items and *randomly selected* items, by category. The *Head* items correspond to the 4.7% most rated items in the user-selected dataset (35.2% of the ratings), the Tail corresponds to the 75.3% less rated items (33.0% of the ratings). The *Middle* items are the remaining ones.

$\hat{R}_{u,i}^* > \hat{R}_{u,j}^*$ whenever $p_{i,j}^u(\mathcal{I}) > p_{j,i}^u(\mathcal{I})$. Since we have[3]:

$$p_{i,j}^u(\mathcal{I}) - p_{j,i}^u(\mathcal{I}) = \mathbb{P}(R_{u,i} = 1 | \mathcal{I}_u = \mathcal{I}) - \mathbb{P}(R_{u,j} = 1 | \mathcal{I}_u = \mathcal{I}),$$

we can claim that maximizing the AUC can be done by ranking i and j according to the sign of $\alpha_{u,i} - \alpha_{u,j}$, whatever the value of $\mathcal{I}$. Thus, when (3) holds, we obtain:

$$\hat{R}_u^* \in \operatorname*{argmin}_{\hat{R}_u \in \mathbb{R}^I} \mathbb{E}\left[AUC_{\mathcal{I}_u}(\hat{R}_u, R_u)\right] \\ \Leftrightarrow \forall i, j \left(\alpha_{u,i} > \alpha_{u,j} \Rightarrow \hat{R}_{u,i}^* > \hat{R}_{u,j}^*\right). \quad (4)$$

3.2 A Single User Profile Scenario

As a simple model of data for user-selected items ratings, suppose that we have a cluster of users which have exactly the same (noisy) tastes, characterized by a random variable R^{true}, so that the optimal ranking we wish to recover is given by sorting the items according to $\mathbb{P}(R^{true} = 1)$. Now suppose that if we pick a random user from this cluster, the rating distribution observed on the items satisfies the following conditions for any i and any $\mathcal{I} \subseteq \{1,...,I\}$:

$$\mathbb{P}(R_{u,i} = 1 | \mathcal{I}_u = \mathcal{I}) = \frac{1}{Z_i} \eta_i \mathbb{P}(R_i^{true} = 1)$$

$$\mathbb{P}(R_{u,i} = -1 | \mathcal{I}_u = \mathcal{I}) = \frac{1}{Z_i} \nu_i \mathbb{P}(R_i^{true} = -1)$$

$$\mathbb{P}(\mathcal{I}_u = \mathcal{I}) = \prod_{j \in \mathcal{I}} Z_j \prod_{j \notin \mathcal{I}} (1 - Z_j)$$

with $Z_i = \eta_i \mathbb{P}(R_i^{true} = 1) + \nu_i \mathbb{P}(R_i^{true} = -1)$.

This corresponds to a simple model where the observed ratings for a user are obtained as follows: each item can have three states (unobserved, observed with rating $+1$, observed with rating -1), and the states of each item are drawn independently. The parameters η_i and ν_i simulate the popularity

---

[3]Since $R_{u,i}$ is binary, the difference $p_{i,j}^u(\mathcal{I}) - p_{j,i}^u(\mathcal{I})$ takes the general form of $\mathbb{P}(A \cap \bar{B}) - \mathbb{P}(\bar{A} \cap B)$ where $\bar{A}$ denotes the complement of event A. This is equal to $\mathbb{P}(A) - \mathbb{P}(B)$: we have (1) $\mathbb{P}(A \cup \bar{B}) = 1 - \mathbb{P}(\bar{A} \cap B)$ (taking the complement) and (2) $\mathbb{P}(A \cup \bar{B}) = \mathbb{P}(A) + \mathbb{P}(\bar{B}) - \mathbb{P}(A \cap \bar{B})$.

and positivity biases: the variation in scale of η_i and ν_i controls the popularity bias, while the ratio between the two controls the positivity bias. Notice that this model satisfies (3), so (4) can be used directly. The motivation behind this model is that it is extremely simple, yet sufficient to exhibit relevant complex phenomena.

Now, if we want to make a prediction for this user cluster, the goal is to find $\hat{R}$ (which does not depend on the user) which maximizes $\mathbb{E}\big[AUC_{\mathcal{I}_u}(\hat{R}, R_u)\big]$.

- When ignoring missing ratings (i.e. using the AUC) an optimal prediction $\hat{R}^*$ satisfies (using (4)):

$$\frac{\eta_i \mathbb{P}\big(R_i^{true}=1\big)}{\eta_j \mathbb{P}\big(R_j^{true}=1\big)} > \frac{\nu_i+(\eta_i-\nu_i)\mathbb{P}\big(R_i^{true}=1\big)}{\nu_j+(\eta_j-\nu_j)\mathbb{P}\big(R_j^{true}=1\big)} \Rightarrow \hat{R}_i^* > \hat{R}_j^*. \quad (5)$$

- For the AUC^{AMAN}, using (2) and (4), we obtain:

$$\eta_i \mathbb{P}\big(R_i^{true}=1\big) > \eta_j \mathbb{P}\big(R_j^{true}=1\big) \Rightarrow \hat{R}_i^* > \hat{R}_j^*. \quad (6)$$

Popularity.

The popularity effect alone (i.e. without any positivity) appears when $\eta_i = \nu_i$ for all i (but the values may depend on i). Then, η_i is the marginal probability of observing the rating for item i for a user, so that items with larger η_i are the most popular ones (Head items), while Tail items are those with smaller η_i.

- When evaluating with the AUC, (5) simplifies to:

$$\frac{\mathbb{P}\big(R_i^{true}=1\big)}{\mathbb{P}\big(R_j^{true}=1\big)} > \frac{\eta_i \eta_j}{\eta_j \eta_i} = 1 \Rightarrow \hat{R}_i^* > \hat{R}_j^*.$$

Thus, the optimal ranking in that case corresponds to the ideal ranking (i.e. matches the true user tastes) despite being evaluated on *user-selected* items.

- When evaluating on observed ratings but considering missing as negatives, Equation 6 directly gives us that an incorrect ranking may actually optimize AUC^{AMAN}, if η_i/η_j is greater than $\mathbb{P}\big(R_j^{true}=1\big)/\mathbb{P}\big(R_i^{true}=1\big)$ for some pair (i, j). That is, considering the missing items in the ranking evaluation process biases the evaluation towards rankings that favor popular items at the expense of highly rated ones.

This simple analysis leads us to the following conclusions: (1) the AUC^{AMAN} has an intrinsic bias towards *popular* (i.e. most rated) items even if these are not the best, and (2) minimizing the AUC on user-selected ratings only, ignoring missing items, leads to the optimal ordering independently of how strong the popularity effect is.

Positivity.

The positivity effect appears when $\eta_i > \nu_i$. First, notice that if η_i and ν_i do not depend on i (i.e. positive ratings are equally over-observed for all items), then both AUC and AUC^{AMAN} are maximized for the ideal ranking (it is clear for the AUC^{AMAN}, and for the AUC, we just have to notice that $x \mapsto \frac{x}{a+b.x}$ is increasing for any $a, b > 0$). Thus, the oversampling of the positive ratings is not, in itself, a problem for ranking (it may be a problem when evaluating with the root mean squared error, though). We now go on to the more general case.

Let us first consider the AUC (thus ignoring missing data for evaluation). From (5), we can observe that only the ratios ν_i/η_i matters, so the *scale* of η_i and ν_i, which controls the popularity, does not. The optimal ranking follows the quantities $f(\nu_i/\eta_i, \mathbb{P}(R_i^{true}=1))$, where $f(a,x) = \frac{x}{a+(1-a)x}$. f increases with x for fixed $a \in (0,1)$ but *decreases* with a. In Figure 1, we can see that good ratings are much more oversampled for Tail items. Equivalently, ν/η is smaller in the Tail than in the Head or in the Middle. To get an order of magnitude, a crude calculation from the histograms assuming that ν_i/η_i is constant in each item category gives $\nu/\eta \approx 0.14$ in the Tail and $\nu/\eta \approx 0.36$ in the Head. These variations can easily lead to items with smaller $\mathbb{P}(R_i^{true}=1)$ have greater $f(\nu_i/\eta_i, \mathbb{P}(R_i^{true}=1))$ (for an order of magnitude, $f(0.14, 0.6) \approx 0.91$ and $f(0.36, 0.75) \approx 0.89$. Thus, evaluating on ratings datasets with user-selected items and ignoring missing data may lead to favor algorithms which incorrectly top-rank Tail items.

For the AUC^{AMAN}, only the value of η_i may lead to inverse the optimal ranking compared to the ideal one. However, what is mostly important is the scale factor, without regard of the ratio ν/η. Even though there may be some local inversions if some similarly popular items have different oversampling characteristics, the difference in scale between the Head and the Tail are huge (on average, there are about 17 times more ratings for Head items than Tail items in our categories). As a good approximation, we can consider that considering missing as negatives renders the evaluation essentially immune to the positivity bias. The evaluation may be no better though, because the reason of that immunity is that it is affected by a much stronger bias.

3.3 Discussion

We presented a simple model under which we can easily analyze popularity and positivity effects. Even if many aspects of practical recommender systems scenarios are not dealt with by our model, we can still show the inherent difficulties and biases of the two evaluation protocols we studied.

An immediate consequence of our analysis is that neither the AUC or the AUC^{AMAN} are accurate evaluations of recommender systems on ratings with user-selected items. These measures do not necessarily favor algorithms which rank according to the true underlying user tastes. More importantly, they may favor algorithms which actually reproduce the underlying bias in the data. Conclusions about the relative performances of the algorithms may thus be incorrect. These results are not specific to the AUC, and actually apply to other measures like precision/recall since the core problem is whether or not one should consider missing ratings as negatives. In general ignoring missing data in the evaluation process may favor algorithms that incorrectly rank high Tail items, but considering missing as negatives may favor models that incorrectly rank high Head items.

Our analysis has focused, until now, on the evaluation measures. They were directed towards the validation or testing of algorithms. However, both the AUC or the AUC^{AMAN} can be considered as training objectives. Even though they cannot be optimized directly, we may consider surrogate losses for efficiency (e.g. the mean squared error). The question is then whether one should ignore missing ratings for training, or imputing them a value corresponding to a negative rating. Considering missing ratings as if they were negative, as in [2], leads to fitting the AUC^{AMAN}, and thus should give good results on this measure (and, more generally, on measures that consider missing as negatives). But the models are not necessarily better on the real task. They may simply fit popularity patterns in the data, and thus improve the AUC^{AMAN}, which is biased in that direction.

Nonetheless, a crucial point is to notice that *popularity compensates for positivity*. Considering missing ratings as if they were negative ratings, we may bias towards popular items because the popularity effect has a greater order of magnitude than the positivity effect (see the previous section). But still, if one consider missing as negatives, but giving them a smaller weight (i.e. a smaller confidence), we may then put the two biases on the same scale, so that one bias (popularity) compensates for the other (positivity). In that sense, we follow the ideas of [12], and we impute a negative rating with a small weight to missing entries. Our experimental section elaborates on the additional degree of freedom given by weighting the rating imputed to missing data, and its relationship to positivity/popularity.

4. EXPERIMENTS

We now describe the experimental protocol we followed to evaluate the impact of considering missing ratings as negatives on (1) the validation errors as measured by the AUC and the AUC^{AMAN}, among other measures, and, (2) on the Yahoo!^{random} dataset, the "true" errors, as measured by the AUC^{rand}. We also study whether we can derive some relevant rule for choosing hyperparameters on a validation set of ratings of user-selected items.

Algorithm.

We use the algorithm proposed by [11]. It belongs to the family of matrix factorization/latent factor models (see [6] for a review on these algorithms in the context of ratings prediction). Such models are considered to be very efficient in the collaborative filtering's literature. The specific algorithm we choose allows to gradually introduce missing data as negative feedback.

The algorithm predicts a ratings matrix $\hat{R} \in \mathbb{R}^{U \times I}$ of the form $\hat{R} = PQ^T$ with $P \in \mathbb{R}^{U \times K}$, $Q \in \mathbb{R}^{I \times K}$, K is a tuning parameter which determines the rank of $\hat{R}$ ($K \ll U, I$), and Q^T denotes the transpose of Q. The objective function regresses directly the observed ratings (not binarized) and missing ratings are considered as 0 entries:

$$L(P,Q) = \sum_{u,i} W_{u,i} \big[(R_{u,i} - \hat{R}_{u,i})^2 + \lambda \sum_{k=1}^{K} (P_{u,k}^2 + Q_{k,i}^2) \big] , \ (7)$$

where $R_{u,i}$ is the value of the (imputed) rating: it equals the true rating value[4] if the rating is observed, and 0 otherwise[5]. The weights $W_{i,u}$ allow us to unequally consider observed and missing ratings. If the rating for (u,i) is observed, then $W_{u,i} = 1$. Otherwise, $W_{u,i} = w_m$. Small values of w_m lead to ignoring missing ratings, while $w_m = 1$ considers missing ratings as if they were observed as the smallest ones. An optimization algorithm for (7) can be found in [11]. Our experiments discuss the influence of hyperparameters λ and

---

[4]For evaluation purposes, we use the binary versions of the ratings, but we use the ratings on a 1 to 5-star scale for training as it only seemed to improve performances.

[5]In [11], the imputed rating is a hyperparameter of the algorithm. We arbitrarily fixed it at 0 as it seemed to be sufficient for our analysis.

w_m on the different metrics considered and to find their optimal values in terms of ranking performance.

Datasets.

We consider the Yahoo!^{user} and Yahoo!^{random} datasets [14] for most of the evaluations, but also use the MovieLens dataset [10] as a sanity-check for some of our claims. Both datasets contain ratings expressed on a 1- to 5-star scale. MovieLens is a well-studied and quite standard ratings data set. The recently released Yahoo! datasets offer the unique opportunity to evaluate performances of an algorithm on users with ratings both on user-selected and random-selected items to study the intrinsic biases of usual datasets.

MovieLens data consist in one million ratings by 6040 users over 3883 items. In our study, only the 3043 items rated at least 20 times are considered (making the minimum number of ratings per user decrease from 20 to 16). Yahoo!^{user} contains 15400 users and 250,000 ratings (users selected the items to rate) over 1000 items (songs) from the Yahoo's LaunchCast Internet Radio Service. Yahoo!^{random} contains a subset of 5400 users of Yahoo!^{user} who were asked to rate 10 randomly selected songs. These users offer the opportunity to train/validate models on Yahoo!^{user}, but evaluate the "true" performance on Yahoo!^{random}. The rank of the factorization (K in (7)), is fixed to 5 on Yahoo! and 10 on MovieLens as these seem to give rather good results.

Evaluation Protocol.

We apply the models in the setting of *strong generalization* [7], to avoid making too many comparisons but still be in the most difficult evaluation setting. On Yahoo!^{user}, all ratings from all users that do not belong to Yahoo!^{random} are used for training the item profiles (the matrix Q in (7)). The remaining $(user, item, rating)$ triplets of Yahoo!^{user} are then used for validation: 20% of the ratings of each user are held out for measuring performance on user-selected items, while the other ratings are used to train the user profiles (the matrix P in (7)) but with Q being held fixed to the value learnt before. The obtained matrix PQ^T is then used for the evaluation on the held-out set of ratings and on Yahoo!^{random}. On MovieLens, 4000 users were randomly selected for training the item profiles, the remaining 2040 are used for the evaluation. In the rest of the paper, the results are averaged over 10 runs.

Performance Measures.

For the performance measures, the term AUC refers to the average AUC on the user-selected validation set (the held-out set of 20% ratings/user with user-selected items, ignoring missing data). AUC^{AMAN} is the AUC when considering missing as negatives on the same validation set. AUC^{rand} is the AUC on Yahoo!^{random} (ignoring missing data).

For a more exhaustive analysis, we also consider the popularity-stratified recall of [12]. For a given item i, define N_i^+ as the number of positive ratings for item i in the data with user-selected ratings. The popularity-stratified recall at k is then defined using a parameter $\beta \in [0, 1]$ as:

$$R@k = \frac{\sum_u \sum_{i \in S_u^{+,k}} 1/(N_i^+)^\beta}{\sum_u \sum_{i : R_{u,i}=1} 1/(N_i^+)^\beta} ,$$

where $\hat{S}_u^{+,k}$ is the number of (known) positive ratings in the top-k items predicted by the system (missing considered as

Measure	$w_m=0$ $\lambda=0.1$	$w_m=0.001$ $\lambda=0.01$	$w_m=0.1$ $\lambda=0.01$	$w_m=0.5$ $\lambda=0.01$
AUC	67.59	67.37	64.60	63.05
AUC^{AMAN}	75.98	83.88	90.83	89.00
$R@10$ ($\beta=$) 0.0	2.83	13.34	33.87	31.80
0.2	2.75	10.83	25.28	22.90
0.5	2.84	7.76	14.21	12.49
1.0	2.44	4.03	4.31	3.31
AUC^{rand}	69.05	74.90	74.43	72.92

Table 1: Illustration of the decorrelation between metrics computed on user-selected and random-selected ratings. None of the measures computed on user-selected items rank the models the same way as the ground truth AUC^{rand}.

$w_m=0.001$	$\lambda=1$	$\lambda=0.1$	$\lambda=0.01$
AUC	65.31	68.84	67.37
AUC^{AMAN}	77.82	86.40	83.88
$R@10$ 0.0	6.06	15.79	13.34
0.2	4.79	12.94	10.83
($\beta=$) 0.5	3.25	9.51	7.76
1.0	0.90	4.16	3.50
AUC^{rand}	69.05	74.90	74.43

Table 2: For all measures, selecting the appropriate λ for fixed $w_m = 0.001$ is not subject to controversy. Same tendencies are noticed for any value of w_m.

negative). For $\beta = 0$, the measure is the usual recall (which is similar to the measure used in [2]). Larger values of β favor top-k items that are relevant to the user but obtained less positive ratings overall. We cannot reliably estimate any form of recall on our ground truth Yahoo!^{random}, nor estimate a value of β which would be interpretable in terms of user behavior. Nonetheless, we can interpret this measure as providing some correction to the bias towards popular items[6] which appears when considering missing as negatives. In that sense, it deserves to be studied in this work. We only exhibit results for recall at 10 due to lack of space. Other values of the cutoff would not change the overall discussion.

5. EXPERIMENTAL RESULTS

In this section, we focus on how the evaluation measures evolve with the parameters w_m and λ, we interpret this evolution in terms of our formal analysis, and we give an application to finding their best values.

No Single Metric is Good.

This first point we verify is that no single performance measure computed on ratings with user-selected items leads to reliable choices of the best algorithm.

Examples of the decorrelation between these metrics and AUC^{rand} are given in Table 1. AUC^{AMAN} and $R@10(\beta = 0, 0.2, 0.5)$ have similar behaviors and both will strongly prefer parameters $w_m = 0.5, \lambda = 0.01$ over $w_m = 0.1, \lambda = 0.01$, while AUC^{rand} ranks the models in reverse order with an absolute difference of 2%. AUC is not doing much better as it highly ranks the model with $w_m = 0, \lambda = 0.1$, which is the worst in terms of AUC^{rand} (69.05%). Finally, $R@10$ with the extreme value $\beta = 1$ estimates that $w_m = 0.1, \lambda = 0.01$ is much better than $w_m = 0.001, \lambda = 0.01$ even though the latter is slightly better on AUC^{rand}. These results indicate that no single measure can reliably be applied to draw conclusions on the relative performances of different models.

Choosing the Optimal λ.

The algorithm we use has two hyperparameters, w_m and λ. w_m controls how much we ignore/consider missing data during training. It should thus affect the relative performances of the models depending on the evaluation protocol. On the other hand, λ only controls the capacity of the

---

[6]Even though the popularity as defined by [12] and by us are different, the global effects that this measure tries to correct are somewhat meaningful for both definitions.

model to avoid overfitting, and is thus an internal tuning parameter. The good news is the following. We just showed that comparing models in general with a single metric on ratings with user-selected items is unreliable for choosing the best one. However, for fixed w_m, models obtained for different values of λ seem to be better or worse than each other on all measures at the same time (or be equivalent on some of them). One can then take *one* optimal lambda, which is the best on all measures. The important point is that the results in terms of AUC^{rand} are perfectly coherent with the other measures. In Table 2, we present and example with various values of λ for $w_m = 0.1$ held fixed. All measures indicate the same optimal λ. We observed the same behavior for all the values of w_m we tried on Yahoo!^{user} and Yahoo!^{random} $(0, 0.001, 0.0025, 0.005, 0.0075, 0.01, 0.02, 0.03, 0.05, 0.1, 0.5, 1$, with λ varying from 10^{-4} to 1 by powers of 10).

We also carried out the same analysis on the MovieLens data (the exact results are omitted) and observed exactly the same behavior. Obviously, one cannot relate this behavior with the true ranking performance on MovieLens because there is no way of computing it. We still obtain an indication that the existence of an optimal λ keeping w_m held fixed is not a matter of a single dataset. This behavior may be due to the specific algorithm we use, but in any case it implies that we can restrict our analysis to the choice of w_m (the extent to which missing data are considered for training). From now on, we only consider the best λ for each w_m.

Influence of w_m.

We now study the influence of gradually taking missing ratings into account by varying w_m from 0 to 1. Figure 2 plots the AUC, AUC^{AMAN}, AUC^{rand} and $R@10(\beta = 0)$ on the Yahoo! dataset. We do not consider other values of β anymore since the usual recall has more widespread usage, and it would be difficult to draw general conclusions with a measure whose parameter depends on the dataset.

First, it is interesting to notice the case where $w_m = 0$ which focuses on observed ratings only. It corresponds to the standard setting of ratings prediction algorithm. Its poor performances in terms of $R@10(\beta = 0)$ or AUC^{AMAN}, noticed in recent studies, have to be contrasted with its competitive $AUC(= 67.23\%)$. AUC, while not being maximized with $w_m = 0$, confirms its supposed sensitivity to positivity and gets higher values for small w_m(68.83% for $w_m = 0.001$). On the other side, AUC^{AMAN} and $R@10$ increase as more importance is given to missing ratings. Considering AUC^{rand}, the optimal value of w_m lies in the interval defined by the optimal value for AUC and the optimal value for AUC^{AMAN}: while models trained by ignor-

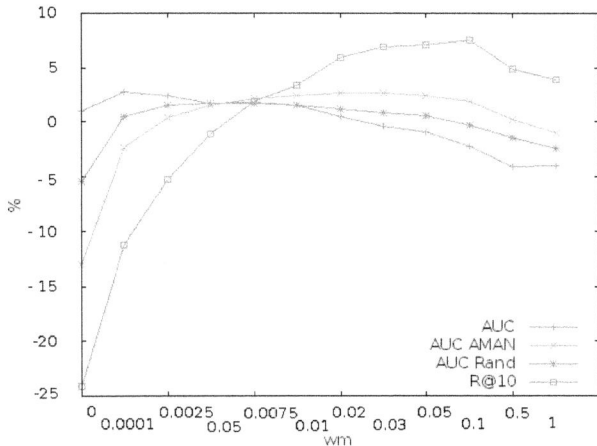

Figure 2: The different performance measures on the Yahoo! dataset, as a function of w_m (with optimal λ for each). The y-axis is normalized as follows: for each measure, we plot the relative difference between the value of a point and the average on all plotted points. Smaller w_m are good in AUC (sensitivity to the positivity effect), while larger values are best for AUC^{AMAN} and recall at 10, which are sensitive to the popularity effect. The optimal value for AUC^{rand} ($w_m = .0075$) lies between the optimal w_m for AUC and AUC^{AMAN} where popularity and positivity compensate each other.

ing ratings are sensitive to the popularity effect (and thus are biased towards the same items than AUC), models that give too much importance to missing data will be biased towards fitting popularity (in the same way as AUC^{AMAN} or $R@10$). Thus, in the interval between optimal AUC and optimal AUC^{AMAN} lie a set of models where popularity and positivity compensate for each other.

The performance of the four optimal models (for each measure considered) are given in Table 3. In that table, we also give the average number of head items (the 4.7% most rated items) in the first top-10, top-20 and top-50 predicted items. The numbers confirm that models trained with higher values of w_m provide more popular items in their recommendation: for the model which is optimal in AUC, 33.74% of top-10 recommended items are head items, against 86% for the model that is optimal for AUC^{AMAN} and more than 96% for the optimal model in terms of $R@10(\beta = 0)$. This confirms the bias towards recommending popular items when giving more importance to missing ratings as negatives, naturally improving performances on measures which have the same bias.

In order to confirm the influence of w_m on AUC, AUC^{AMAN} and recall, we plotted the evolution of these measures with w_m on the MovieLens dataset in Figure 3. We can observe exactly the same behavior as in the Yahoo! dataset. We thus conjecture that on MovieLens as well, the interval of w_m between the optimal AUC and the optimal AUC^{AMAN} contains the optimal model on the "true" underlying distribution, as this interval corresponds to the region where popularity and positivity compensate for each other.

Conclusion.

There is no single way to evaluate a recommender system

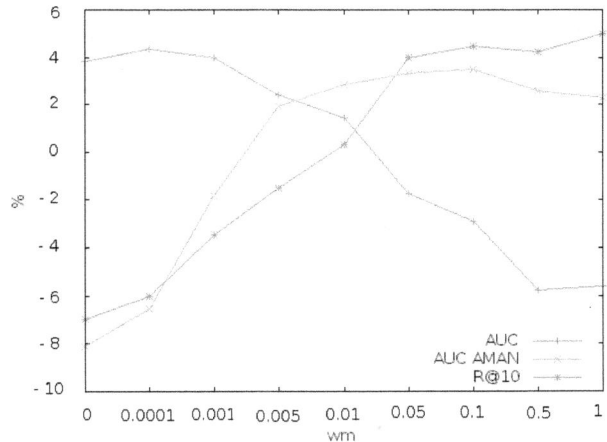

Figure 3: Evolution of performances on the Movie-Lens dataset. The legend is similar to that of Figure 2. We conjecture that the optimal value of the parameter w_m lies between $w_m = 0.0001$ (optimal for AUC) and $w_m = 0.1$ (optimal for AUC^{AMAN}).

	optimal AUC	optimal AUC^{rand}	optimal AUC^{AMAN}	optimal $R@10$
w_m	0.001	0.005	0.03	0.1
AUC	69.18	68.17	66.02	64.17
AUC^{AMAN}	86.38	90.25	91.41	90.66
$R@10$	15.57	25.70	33.74	34.36
AUC^{rand}	74.85	76.08	75.23	74.10
% Head 10	33.74	59.97	85.98	96.07
% Head 20	31.80	52.64	75.85	87.25
% Head 50	25.76	39.22	53.93	60.77

Table 3: Performances of best models for each measure on Yahoo!^{user}. % Head X is the average percentage of the Head items (4.7% most rated) in the top-X predicted items. $R@10$ is $R@10(\beta = 0)$.

in terms of ranking on ratings datasets with user-selected items. If one changes the experimental protocol, it leads to consider different models as being the best. As far as we could see, no single measure provides a satisfactory correlation with the "true" underlying ranking performance.

On the other hand, our experiments suggest that the selection of reasonably good hyperparameters can be performed on ratings with user-selected items by selecting a hyperparameter in the interval between the optimal parameter for the AUC and the optimal parameter for the AUC^{AMAN} (e.g. in the middle of the interval, not close to the bounds since they probably correspond to biased models). The reason is that this interval corresponds to the region where positivity and popularity compensate for each other.

6. RELATED WORK

Missing as Negative for Ranking.

The debate as to whether missing data should be considered as negative is ongoing at least since the beginning of the 2000s (see [4] for a discussion and references). Recently, [13, 9] measured their ranking performance in terms on Normalized Discounted Cumulative Gain (NDCG, see [5]), ignoring missing ratings. The latter measure is widely used in search engine evaluation, and has the advantage of taking into account the value of the rating, without requiring the binarization step. On the other hand, the value that is estimated by taking the NDCG on a sample of observed ratings is not clear (even if these are randomly selected) and this is why we did not choose this measure in our work. [3, 11, 2] use precision/recall measures considering missing as negatives, and show that algorithms with good ranking performances (in such a setting) are not the same as those which accurately predict ratings. In particular, [2] shows that running an SVD on the completed matrix (setting missing values at 0) has good performances compared to algorithms that ignore missing data. This is coherent with both our formal analysis and our experiments. We argue though that one should not conclude that an SVD with missing entries at 0 has better true ranking performances than a similar algorithm which ignores missing ratings, because the former may be severely affected by the popularity bias.

In closer relation to our work, [11] gave the following argument in favor of considering missing as negatives: if a measure ignores missing ratings, it does not penalize a system which ranks unrated items first, even though we may have the prior knowledge that people tend to rate items they like. This argument is similar in spirit to our result that the AUC (ignoring missing ratings) is sensitive to the positivity bias. We somewhat go further though: ignoring missing ratings on real datasets (i.e. exhibiting the positivity bias) should actually *favor* algorithms which have the undesirable property of top-ranking tail items of low true relevance.

Non-Random Missing Data.

The bias involved by users *choosing* the items they rate, and its influence on model fitting has been cast in light in [8] (with, in addition, the release of the Yahoo! dataset we use in this paper), and further studied in [9]. They address the problem of modeling the underlying missing data process from ratings with user-selected items, without any emphasis on popularity. They do not address the problem of evaluating from such data nor comparing algorithms or tuning hyperparameters. In that sense, our work complement theirs and their algorithms should be a viable alternative to the algorithm we chose for generating models with various sensitivities to positivity and popularity.

In [11], the author explicitly addresses the problem of popularity for estimating recall on ratings with user-selected items. Although his practical definition of popularity is different from ours since it considers only the number of observed positive ratings, his proposition of popularity-stratified recall (which we use in our experiments) appears natural from our analysis as well: if one knew the exact effect of popularity, then we could correct this effect from the evaluations in terms of AUC^{AMAN}. However, because the exact effect of popularity/positivity is difficult to obtain at user-level (or user-cluster level), it is unclear if a single parametric correction can be reliable. Moreover, the best parameter (β in $R@10$) of the evaluation measure depends on the data, and it is unclear how one should choose it on a particular dataset. Nonetheless, the popularity-stratified recall provides a viable alternative to the traditional recall which seems severely biased towards popular items.

7. CONCLUSION

We analyzed the influence of the popularity and positivity biases on the training and evaluation of recommender systems, and study the question of whether missing data should be considered as negative feedback or ignored.

Considering missing ratings as negative for evaluation or training may lead to bias models towards popular items. However, it is important to not completely ignore missing ratings, as they give valuable information (at least for training) to compensate for the positivity effect and avoid irrelevant tail items to be predicted as top recommendations.

This property of popularity and positivity effects suggests a methodology for choosing the importance of missing data when training on ratings with user-selected items.

8. ACKNOWLEDGMENTS

This work was supported by Region Ile-de-France and the French Ministry of Industry, project OpenWay III.

9. REFERENCES

[1] J. Bennett, S. Lanning, and N. Netflix. The netflix prize. In *In KDD Cup and Workshop in conjunction with KDD*, 2007.

[2] P. Cremonesi, Y. Koren, and R. Turrin. Performance of recommender algorithms on top-n recommendation tasks. In *RecSys '10*, pages 39–46, 2010.

[3] A. Gunawardana and G. Shani. A survey of accuracy evaluation metrics of recommendation tasks. *J. Mach. Learn. Res.*, 10:2935–2962, 2009.

[4] J. L. Herlocker, J. A. Konstan, L. G. Terveen, and J. T. Riedl. Evaluating Collaborative Filtering Recommender Systems. *ACM Trans. on Inf. Sys.*, 22(1):5–53, 2004.

[5] K. Järvelin and J. Kekäläinen. Cumulated gain-based evaluation of ir techniques. *ACM Trans. Inf. Syst.*, 20(4):422–446, Oct. 2002.

[6] Y. Koren, R. Bell, and C. Volinsky. Matrix Factorization Techniques for Recommender Systems. *Computer*, 42(8):30–37, 2009.

[7] B. Marlin. Collaborative Filtering: A Machine Learning Perspective. Master's thesis, University of Toronto, 2004.

[8] B. Marlin, R. Zemel, S. Roweis, and M. Slaney. Collaborative filtering and the missing at random assumption. In *UAI-07*, pages 267–275, 2007.

[9] B. M. Marlin and R. S. Zemel. Collaborative prediction and ranking with non-random missing data. In *RecSys'09*, pages 5–12, 2009.

[10] MovieLens data. homepage: http://www.grouplens.org/node/73, 2006.

[11] H. Steck. Training and testing of recommender systems on data missing not at random. In *KDD '10*, pages 713–722, 2010.

[12] H. Steck. Item popularity and recommendation accuracy. In *RecSys '11*, pages 125–132, 2011.

[13] M. Weimer, A. Karatzoglou, Q. V. Le, and A. J. Smola. Cofi rank - maximum margin matrix factorization for collaborative ranking. In *NIPS*, 2007.

[14] Yahoo! Webscope Datasets. Yahoo! Music ratings for User Selected and Randomly Selected songs, v1.0, http://research.yahoo.com/Academic_Relations.

Sparse Linear Methods with Side Information for Top-N Recommendations

Xia Ning
Computer Science & Engineering
University of Minnesota, Twin Cities
4-192 EE/CS Building, 200 Union Street SE
Minneapolis, MN 55455
xning@cs.umn.edu

George Karypis
Computer Science & Engineering
University of Minnesota, Twin Cities
4-192 EE/CS Building, 200 Union Street SE
Minneapolis, MN 55455
karypis@cs.umn.edu

ABSTRACT

The increasing amount of side information associated with the items in E-commerce applications has provided a very rich source of information that, once properly exploited and incorporated, can significantly improve the performance of the conventional recommender systems. This paper focuses on developing effective algorithms that utilize item side information for *top-N* recommender systems. A set of sparse linear methods with side information (SSLIM) is proposed, which involve a regularized optimization process to learn a sparse aggregation coefficient matrix based on both user-item purchase profiles and item side information. This aggregation coefficient matrix is used within an item-based recommendation framework to generate recommendations for the users. Our experimental results demonstrate that SSLIM outperforms other methods in effectively utilizing side information and achieving performance improvement.

Categories and Subject Descriptors

H.4.m [**Information Systems**]: Miscellaneous; J.7 [**Computer Applications**]: Computers in other systems—*Consumer products*

Keywords

Recommender system, Sparse Linear Methods, Side information

1. INTRODUCTION

Top-*N* recommender systems have been widely used in E-commerce applications to recommend ranked lists of items so as to help the users in identifying the items that best fit their personal tastes. Over the years, various algorithms for *top-N* recommendation have been developed [12]. The conventional *top-N* recommendation algorithms primarily focus on utilizing user-item purchase profiles to generate recommendations. Such algorithms can be categorized into two classes: collaborative filtering methods and model-based methods. Collaborative filtering methods typically build neighborhood for each user/item by considering the similarities of the purchase patterns among users/items from their profiles, and then recommend new items from the neighborhood. Model-based methods learn to explain the user-item purchase patterns using a specific model. For instance, the most popular matrix factorization (MF) methods present users and items in a common latent space such that the user-item purchase patterns can be explained by the user-item similarities in the space. Recently, a sparse linear method (SLIM) [10] has been developed that leverages the advantages of the above two classes of methods and achieves both better prediction accuracy and run-time performance than the state-of-the-art methods.

With the increased availability of additional information associated with the items (e.g., product reviews, movie plots, etc), referred to as *side information*, there is a greater interest in taking advantage of such information to improve the quality of conventional *top-N* recommender systems. As a result, a number of approaches have been developed from Machine Learning (ML) and Information Retrieval (IR) communities for incorporating side information. Such approaches include hybrid methods [5], matrix/tensor factorization [14, 8], and other regression methods [1].

In this paper, we propose a set of Sparse Linear Methods that utilize the item Side information (SSLIM) for *top-N* recommendation. These methods include collective SLIM (cSLIM), relaxed collective SLIM (rcSLIM), side information induced SLIM (fSLIM) and side information induced double SLIM (f2SLIM). The key idea behind these methods is to learn linear models that are constrained and/or informed by the relations between the item side information and the user-item purchase profiles so as to achieve better recommendation performance. We conduct a comprehensive set of experiments on various datasets from different real applications. The results show that SSLIM produces better recommendations than the state-of-the-art methods.

The rest of this paper is organized as follows. In Section 2, a brief review on related work is presented. In Section 3, the definitions and notations are provided. In Section 4 and Section 5, the methods are described. In Section 6, the materials used for experiments are presented. In Section 7, the results are presented. Finally in Section 8 are the discussions and conclusions.

2. RELATED WORK

Various methods have been developed to incorporate side information in recommender systems. Most of these methods have been developed in the context of the rating prediction problem, whereas the *top-N* recommendation problem has received less attention. In the rest of this section we review some of the best performing schemes for both the rating prediction and *top-N* recommendation problems.

The first category of these methods is based on latent factor models. In [14], Singh *et al* proposed the collective matrix factorization method for both rating prediction and *top-N* recommendation, which collectively factorizes user-item purchase profile matrix and item-feature content matrix into a common latent space such that the two types of information are leveraged via common the item factors. Agarwal *et al* [1] proposed regression-based latent factor models for rating prediction, which use features for factor estimation. In their method, the user and item latent factors are estimated through independent regression on user and item features, and the recommendation is calculated from a multiplicative function on user and item factors. Yang *et al* [16] developed a joint friendship and interest propagation model for *top-N* recommendation, in which the user-item interest network and the user-user friendship network (side information on users) are jointly modeled through latent user and item factors. User factors are shared by the interest network approximation component and the friendship network approximation component so as to enable information propagation. They demonstrated the their model is a generalization of Singh *et al* [14], Koren [9] and Agarwal *et al* [1].

Methods using tensor factorization (TF) have also gained popularity. Karatzoglou *et al* [8] considered the user-item-feature relation as a tensor, and they proposed to use regularized TF to model the relations between the three sets of entities for rating prediction. TF can be considered as a generalization of MF, in which the relations among all the modes (i.e., users, items and features) are jointly learned. Rendle *et al* [11] also treated user-item-feature as a tensor, and they factorized all pairwise interactions in the tensor (i.e, items vs users, items vs context features, users vs context features) rather than the entire tensor for rating prediction.

Another category of algorithms that utilize side information is based on networks. Gunawardana *et al* [6] proposed unified Boltzmann machines for *top-N* prediction, in which user-item profile information and side information are treated as homogeneous features, and interaction weights between such features and user actions are learned in a coherent manner so as to reflect how well such features can predict user actions. Campos *et al* [3] combined content-based and collaborative-filtering based recommendations with Bayesian networks, which are composed of item nodes, user nodes and item feature nodes. During predictions, content information is propagated from purchased items to non-purchased items via feature nodes, and purchase information is propagated from other users to the user of concern via item nodes. Then the two parts are combined to make recommendations.

3. DEFINITIONS AND NOTATIONS

In this paper, the symbols u, t, and $\mathbf{f}$ ($|\mathbf{f}| = l$) will be used to denote the users, items and item side information vectors, respectively. Individual users and items will be denoted using different subscripts (i.e., u_i, t_j). The side information vector for item t_j will be denoted by $\mathbf{f}_j$. The size of user set and item set are denoted by m and n, respectively.

The user-item purchase profile is represented by a binary $m \times n$ matrix M, in which the (i, j) entry (denoted by $m_{i,j}$) is 1 if user u_i has ever purchased item t_j, otherwise it is marked as 0. The i-th row of M, denoted by $\mathbf{m}_i^\mathsf{T}$, represents the purchase profile of user u_i on all items. The j-th column of M, denoted by $\mathbf{m}_j$, represents the purchase profile of item t_j from all users. The side information on all items is represented by an $l \times n$ matrix F. The j-th column of F represents the side information vector of item t_j (i.e., $\mathbf{f}_j$).

All vectors (e.g., $\mathbf{m}_i^\mathsf{T}$ and $\mathbf{f}_j$) are represented by bold lower-case letters and all matrices (e.g., M and F) are represented by upper-case letters. Row vectors are represented by having the transpose supscript$^\mathsf{T}$, otherwise by default they are column vectors. Approximation relation is denoted using $\sim$ and approximation value is denoted by heading a $\sim$ head. The matrix/vector notations are used instead of user/item/side information if no ambiguity is raised.

4. SLIM: SPARSE LINEAR METHODS

In this paper, we focus on incorporating item side information within the Sparse LInear Method (SLIM) that we proposed previously [10]. In SLIM, the recommendation score on an urn-purchased item t_j of a user u_i (denoted by $\tilde{m}_{ij}$) is calculated as a *sparse* aggregation of the items that have been purchased by u_i, that is,

$$\tilde{m}_{ij} = \mathbf{m}_i^\mathsf{T} \mathbf{s}_j, \qquad (1)$$

where $m_{ij} = 0$ and $\mathbf{s}_j$ is a size-n sparse vector of aggregation coefficients. Thus, the model can be presented as

$$M \sim MS, \qquad (2)$$

where S is an $n \times n$ sparse matrix of aggregation coefficients, whose j-th column is $\mathbf{s}_j$ as in Equation 1, and each row $\tilde{\mathbf{m}}_i^\mathsf{T}$ of $\tilde{M} = MS$ represents the recommendation scores on all items for user u_i. The *top-N* recommendations for u_i are obtained by sorting u_i's non-purchased items based on their scores in $\tilde{\mathbf{m}}_i^\mathsf{T}$ in decreasing order and recommending the *top-N* items.

The SLIM method views the purchase activity of user u_i on item t_j in M (i.e., m_{ij}) as the ground-truth item recommendation score. It learns the $n \times n$ sparse matrix S in Equation 2 as the minimizer for the following regularized optimization problem:

$$\begin{aligned} \underset{S}{\text{minimize}} \quad & \frac{1}{2}\|M - MS\|_F^2 + \frac{\beta}{2}\|S\|_F^2 + \lambda\|S\|_1 \\ \text{subject to} \quad & S \geq 0 \\ & \text{diag}(S) = 0, \end{aligned} \qquad (3)$$

where $\|S\|_1 = \sum_{i=1}^n \sum_{j=1}^n |s_{ij}|$ is the entry-wise ℓ_1-norm of S, and $\|\cdot\|_F$ is the matrix Frobenius norm. In Equation 3, MS is the estimated matrix of recommendation scores (i.e., $\tilde{M}$) by the sparse linear method as in Equation 2. The non-negativity constraint is applied on S such that the learned S corresponds to positive aggregations over items. The constraint $\text{diag}(S) = 0$ is also applied so as to avoid trivial solutions (i.e., the optimal S is an identity matrix such that an item always recommends itself). In addition, the constraint $\text{diag}(S) = 0$ ensures that m_{ij} is not used to compute $\tilde{m}_{ij}$. In order to learn a sparse S, SLIM introduces the ℓ_1-norm of S as a regularizer in Equation 3. It is well known that ℓ_1-norm regularization introduces sparsity into the solutions [15]. The matrix S learned by SLIM is referred to as SLIM's aggregation coefficient matrix. Extensive experiments in [10] have shown that SLIM outperforms the state-of-the-art *top-N* recommendation methods.

5. SLIM WITH SIDE INFORMATION

SLIM provides a general framework in which only the aggregation coefficient matrix S is needed for efficient *top-N* recommendations, and this matrix is learned from the user-item purchase profiles. In this section, we present four different extensions of SLIM that are designed to incorporate side information about the items in order to further improve the quality of the recommendations.

5.1 Collective SLIM

The first approach assumes that there exist correlations between users' co-purchase behaviors on two items and the similarity of the two items' intrinsic properties encoded in their side information. In order to enforce such correlations, this approach imposes the additional requirement that both the user-item purchase profile matrix M and the item side information matrix F should be reproduced by the same sparse linear aggregation. That is, in addition to satisfying $M \sim MS$, the coefficient matrix S should also satisfy

$$F \sim FS. \qquad (4)$$

This is achieved by learning the aggregation coefficient matrix S as the minimizer to the following optimization problem:

$$\begin{aligned}
\underset{S}{\text{minimize}} \quad & \frac{1}{2}\|M - MS\|_F^2 + \frac{\alpha}{2}\|F - FS\|_F^2 \\
& + \frac{\beta}{2}\|S\|_F^2 + \lambda\|S\|_1 \\
\text{subject to} \quad & S \geq 0, \\
& \text{diag}(S) = 0,
\end{aligned} \qquad (5)$$

where $\|F - FS\|_F^2$ measures how well the aggregation coefficient matrix S fits the side information. The parameter α is used to control the relative importance of the user-item purchase information M and the item side information F when they are used to learn S. Note that in this method, the side information is actually used to regularize the original SLIM method (i.e., via adding the regularization term $\frac{\alpha}{2}\|F - FS\|_F^2$ into Equation 3). The recommendations are generated in exactly the same way as in SLIM. That is, the recommendation score for user u_i on item t_j is calculated as $\tilde{m}_{ij} = \mathbf{m}_i^\mathsf{T}\mathbf{s}_j$. Since the matrix S is learned from both M and F collectively by using F to regularize the original SLIM method, this approach is referred to as collective SLIM and denoted by cSLIM.

The solution to the optimization problem in Equation 5 is identical to the solution of an optimization problem in the same form as in Equation 3 with M in Equation 3 replaced by $M' = [M, \sqrt{\alpha}F]^\mathsf{T}$.

5.2 Relaxed cSLIM

The second approach also tries to reproduce the item side information using a sparse linear method as in cSLIM, but it uses an alternative approach for achieving this. Specifically, it uses an aggregation coefficient matrix Q to reproduce F as

$$F \sim FQ, \qquad (6)$$

where Q is not necessarily identical to S as in Equation 2. Thus, this method is a relaxation from cSLIM. However, the two aggregation coefficient matrices S and Q are tied by requiring that Q should not be significantly different from S (i.e., $S \sim Q$). The matrix S and the matrix Q in Equation 6 are learned as the minimizers of the following optimization problem:

$$\begin{aligned}
\underset{S,Q}{\text{minimize}} \quad & \frac{1}{2}\|M - MS\|_F^2 + \frac{\alpha}{2}\|F - FQ\|_F^2 \\
& + \frac{\beta_1}{2}\|S - Q\|_F^2 + \frac{\beta_2}{2}(\|S\|_F^2 + \|Q\|_F^2) \\
& + \lambda(\|S\|_1 + \|Q\|_1) \\
\text{subject to} \quad & S \geq 0, Q \geq 0, \\
& \text{diag}(S) = 0, \text{diag}(Q) = 0,
\end{aligned} \qquad (7)$$

where the parameter β_1 controls how much S and Q are allowed to be different from each other. Similar to cSLIM, this method regularizes the original SLIM using item side information by adding the two regularization terms $\frac{\alpha}{2}\|F - FQ\|_F^2$ and $\frac{\beta_1}{2}\|S - Q\|_F^2$ and the recommendations are generated in the same way as in SLIM. Since this method is a relaxation from cSLIM, it is refereed to as relaxed collective SLIM and denoted by rcSLIM.

The optimization problem in Equation 7 can be solved via an approach alternating on solving S and Q. In each iteration, one variable is fixed and the problem becomes a regularized optimization problem with respect to the other variable, and it can be solved using a similar approach of stacking matrices as in Section 5.1. The solution of cSLIM is used as the initial value of S.

5.3 Side Information Induced SLIM

An alternative way to learn the aggregation coefficient matrix S of SLIM is to represent S as a function in the item feature space and thus it captures the feature-based relations of the items. One option of achieving this is to use the item-item similarity matrix calculated as FF^T, that is, the aggregation coefficient from one item to another is calculated as the dot-product of their feature vectors (i.e., item-item feature similarity). However, in this way, the aggregation coefficient matrix is not customized to the user-item purchase profiles M at all, and thus a SLIM with such aggregation coefficient matrix can fit M very poorly. Another way is to learn a weighting matrix W such that the aggregation coefficient value s_{ij} can be represented as a linear combination of item t_i's feature $\mathbf{f}_i$ weighted by item t_j's personalized weighting vector $\mathbf{w}_j$ over individual item features, that is, $s_{ij} = \mathbf{f}_i^\mathsf{T}\mathbf{w}_j$ and $\mathbf{w}_j$ is W's j-th column. In this way, the coefficient matrix S can be represented as a weighted linear combination of the item features F using W, that is,

$$S = F^\mathsf{T}W. \qquad (8)$$

Such weighting matrix W can be learned as the minimizer of the following optimization problem:

$$\begin{aligned}
\underset{W}{\text{minimize}} \quad & \frac{1}{2}\|M - M(F^\mathsf{T}W - D)\|_F^2 \\
& + \frac{\beta}{2}\|W\|_F^2 + \lambda\|F^\mathsf{T}W\|_1 \\
\text{subject to} \quad & W \geq 0 \\
& D = \text{diag}(\text{diag}(F^\mathsf{T}W)),
\end{aligned} \qquad (9)$$

where $D = \text{diag}(\text{diag}(F^{\mathsf{T}}W))$ is a diagonal matrix with the corresponding diagonal values from $F^{\mathsf{T}}W$. D is subtracted from $F^{\mathsf{T}}W$ so as to ensure that m_{ij} is not used to compute $\tilde{m}_{ij}$, and this is equivalent to the constraint $\text{diag}(S) = 0$ in Equation 3. In this method, the recommendation score for user u_i on item t_j is calculated as $\tilde{m}_{ij} = \mathbf{m}_i^{\mathsf{T}}(F^{\mathsf{T}}\mathbf{w}_j - \mathbf{d}_j)$, where $\mathbf{w}_j$ and $\mathbf{d}_j$ is the j-th column of W and D, respectively. Since this method explicitly specifies the aggregation coefficient matrix S as a function of the item side information F, it is referred to as side information induced SLIM and denoted by fSLIM.

The optimal solution to the optimization problem in Equation 9 is $W^* = [\mathbf{w}_1, \mathbf{w}_2, \cdots, \mathbf{w}_j, \cdots, \mathbf{w}_n]$, where $\mathbf{w}_j$ is the optimal solution to the following problem:

$$\underset{\mathbf{w}_j}{\text{minimize}} \quad \frac{1}{2}\|\mathbf{m}_j - MF_{-j}^{\mathsf{T}}\mathbf{w}_j\|_F^2 + \frac{\beta}{2}\|\mathbf{w}_j\|_F^2 + \lambda\|\mathbf{c}\mathbf{w}_j\|_1$$

$$\text{subject to} \quad \mathbf{w}_j \geq 0,$$

where $c_p = \sum_{k=1}^{n} f_{pk}$, and F_{-j} is a matrix with F's j-th column set to 0.

5.4 Side Information Induced Double SLIM

SLIM and fSLIM have their own advantages. SLIM learns the aggregation coefficient matrix S purely from purchase profiles such that it better fits the user-item purchase information. fSLIM forces the aggregation coefficient matrix S to be expressed in the item feature space and therefore it captures useful information from the item features. SLIM and fSLIM can be coupled within one method so as to leverage both their advantages and better learn from purchase profiles and side information concurrently. One way to combine SLIM and fSLIM is to have the user-item purchase profile M reproduced by both SLIM and fSLIM as

$$M \sim MS + M(F^{\mathsf{T}}W - D), \tag{10}$$

where the S and W matrices can be learned as the minimizers of the following optimization problem:

$$\underset{S,W}{\text{minimize}} \quad \frac{1}{2}\|M - MS - M(F^{\mathsf{T}}W - D)\|_F^2$$

$$+ \frac{\beta}{2}(\|S\|_F^2 + \|W\|_F^2) + \lambda(\|S\|_1 + \|F^{\mathsf{T}}W\|_1) \tag{11}$$

$$\text{subject to} \quad S \geq 0, W \geq 0,$$

$$\text{diag}(S) = 0, D = \text{diag}(\text{diag}(F^{\mathsf{T}}W)).$$

In this method, the recommendation score for user u_i on item t_j is calculated as $\tilde{m}_{ij} = \mathbf{m}_i^{\mathsf{T}}\mathbf{s}_j + \mathbf{m}_i^{\mathsf{T}}(F^{\mathsf{T}}\mathbf{w}_j - \mathbf{d}_j)$, where $\mathbf{w}_j$ and $\mathbf{d}_j$ is the j-th column of W and D, respectively. This method is a combination of SLIM and fSLIM and thus it is refereed as side information reduced double SLIM and denoted by f2SLIM.

That the optimal solution of W in the problem in Equation 11 is identical to the first l rows of the optimal solution W' to the problem in Equation 9 with F replaced by $[F, I]^{\mathsf{T}}$ where I is an $n \times n$ identity matrix, and $D' = \text{diag}((F')^{\mathsf{T}}W')$, whereas the optimal S is the last n rows of W'.

6. EXPERIMENTAL METHODOLOGY

6.1 Datasets

We evaluated the performance of different methods on the following real datasets: ML100K, NF, CrossRef, Lib, BBY, and Yelp, whose characteristics are summarized in Table 1.

ML100K The ML100K dataset corresponds to movie ratings and was obtained from the MovieLens research project. The movie plots were fetched from the IMDb database and the words that appear in at least 5 plots are used as the movie side information.

NF The NF dataset is a subset extracted from the Netflix Prize dataset. The item side information was generated as in the ML100K dataset. Only the movies that were rated by 10-30 users were selected.

CrossRef The CrossRef dataset was obtained from cross-ref.org, and contains scientific articles and lists of article citations. All the articles (i.e., references) that have DOI links and are cited by at least 50 other articles were first selected. Then the articles which cite more than 3 of such references were selected. In this way, an article-reference dataset is constructed, in which the articles (the references) are analogous to the users (the items). The words in the reference titles are used as side information. The *top-N* recommendation on CrossRef dataset becomes a task to recommend a reference for a certain article.

Lib The Lib dataset was obtained from the University of Minnesota libraries, and contains the library users and their viewed articles. From the entire library records, the users who viewed at least 5 different articles and the articles that were viewed by at least 10 users were collectively selected to construct an user-article matrix. The words in the article titles are used as the article side information. The *top-N* recommendation on Lib is to recommend an article to a user.

BBY The BBY dataset is a subset of the BestBuy user-product rating and review dataset from BestBuy website (https://developer.bestbuy.com/documentation/archives). The products that were reviewed by at least 5 users and the users who reviewed at least 2 such products were collectively selected so as to construct the dataset. The side information for each item was the text of all the reviews of that item.

Yelp The Yelp dataset is a subset of the academic version of Yelp user-business rating and review dataset downloaded from Yelp (http://www.yelp.com/academic_dataset). The users who reviewed at least 3 businesses and the corresponding businesses were selected to construct the dataset. The side information for each item was constructed from the reviews in a way similar to the BBY dataset.

For the original rating datasets (i.e., ML100K, NF, BBY, Yelp), the multivariate rating values were converted to 1's.

6.2 Evaluation Methodology & Metrics

We applied 5-time Leave-One-Out cross validation to evaluate the performance of different methods. In each run, each of the datasets is split into a training set and a testing set by randomly selecting one of the non-zero entries of each user and placing it into the testing set. The evaluation is conducted by comparing the size-N (by default $N = 10$) recommendation list for each user and the item of that user in the testing set.

The recommendation quality is measured by the Hit Rate (ZR) and the Average Reciprocal Hit-Rank (ARHR) [4]. ZR is defined as follows,

$$\text{HR} = \frac{\#\text{hits}}{\#\text{users}}, \tag{12}$$

where $\#$users is the total number of users, and $\#$hits is the number of users whose item in the testing set is recommended (i.e., hit) in the size-N recommendation list. A

dataset	purchase information						side information					
	#users	#items	#nnzs	rsize	csize	density	desc	#ftr	#nnz	srsize	scsize	sdensity
ML100K	943	1,682	100,000	106.0	59.5	6.30%	plots	2,327	46,915	27.9	20.2	1.20%
NF	3,086	6,909	128,134	41.5	18.6	0.60%	plots	5,941	200,148	29.0	33.7	0.49%
CrossRef	84,260	23,458	466,068	5.5	19.9	0.02%	titles	5,677	149,839	6.4	26.4	0.11%
Lib	13,843	12,123	103,428	7.47	8.53	0.06%	titles	9,991	86,065	7.1	8.6	0.07%
BBY	127,285	7,330	162,451	1.3	22.2	0.02%	reviews	9,686	1,912,444	260.9	197.4	2.7%
Yelp	13,574	6,896	89,608	6.6	13.0	0.10%	reviews	10,305	330,865	48.0	32.1	0.47%

Columns corresponding to purchase information and side information show the dataset statistics for historical profile matrix M and side information matrix F, respectively. Under purchase information, column corresponding to #users, #items and #nnzs show the number of users, items and non-zero values in each dataset, respectively. Column corresponding to rsize, csize and density shows the average row density, the average column density and the matrix density, respectively. Under side information, column corresponding to desc shows the side information types. Column corresponding to #ftr and #nnz show the dimensionality of side information and the number of non-zero values in the side information, respectively. Column corresponding to srsize, scsize and sdensity show the average row density, the average column density and the density of the side information matrix, respectively.

second measure for evaluation is ARHR, which is defined as follows:

$$\text{ARHR} = \frac{1}{\#\text{users}} \sum_{i=1}^{\#\text{hits}} \frac{1}{p_i}, \qquad (13)$$

where if an item of a user is hit, p is the position of the item in the ranked recommendation list. ARHR is a weighted version of HR and it measures how strongly an item is recommended, in which the weight is the reciprocal of the hit position in the recommendation list.

6.3 Side Information Representation

Besides the learning capability of the SSLIM methods, the representation of the side information can impact the overall performance. Since the side information in our datasets is text (e.g., movie plots, product reviews, etc), we investigated different text representations. In all of these schemes, the text of the side information was preprocessed to eliminate stop words and each word was converted to its stem[1].

Binary Representation (F_b) In this scheme, the text of the side information is represented using the bag-of-words model, and the frequency of each word is set to one. The reason for the binarization is that typically the text for an item is short, and there are not many informative words occurring multiple times, and thus a binarized vector is almost same as the original count vector. In addition, since the user-item profile M is binary, intuitively the item-item coefficient matrix Q learned from a binary feature matrix F should be comparable to the aggregation coefficient matrix S learned from M in terms of the values. In this case, the regularization using Q on S (i.e., the $\frac{\beta_1}{2}\|S - Q\|_F^2$ term in rcSLIM) can be more effective.

Normalized TFIDF Representation (F_{tfidf}) For the methods that directly learn from the item text (fSLIM and f2SLIM), it is essential that the text presentation encodes how important a word is in the text. For this purpose, a normalized TFIDF scheme is adopted. The TFIDF scheme [13] is widely used for weighting words in text mining. After the TFIDF scheme is applied on the feature vectors, the feature vectors are normalized to unit length.

Normalized TFIDF Representation with Feature Selection (F_{tfidf_fs}) Another representation scheme is a modification of F_{tfidf} by using feature selection. For each feature vector, the words were sorted in decreasing order according to their weights in the TFIDF representation. Then the highest weighted words were selected until cumulatively they contribute to 90% of the vector length.

---

[1]http://glaros.dtc.umn.edu/gkhome/fetch/sw/cluto/doc2mat-1.0.tar.gz

Normalized TFIDF Binary Representation with Feature Selection ($F_{tfidf\_fs\_b}$) The last side information representation scheme is a compromise of F_b and F_{tfidf_fs}, that is, it converts all the values that are calculated from F_{tfidf_fs} to binary. This scheme tries to use only the words that are considered as important by F_{tfidf_fs} and meanwhile still retain the advantages of the binary representations.

6.4 Comparison Methods

Singh *et al* [14] proposed the collective matrix factorization (CMF) method for relational learning as follow:

$$\underset{U,V,W}{\text{minimize}} \quad \frac{1}{2}\|M - UV\|_F^2 + \frac{\alpha}{2}\|F - WV\|_F^2$$
$$+ \frac{\beta}{2}(\|U\|_F^2 + \|V\|_F^2 + \|W\|_F^2), \qquad (14)$$

where U is an $m \times k$ user factor from M, W is an $l \times k$ feature factor from F, and V is an $k \times n$ item factor which is collectively learned from both M and F. Particularly, $k \ll \min(m, n, l)$. CMF and cSLIM are similar in the sense that a common matrix is learned from both M and F concurrently. However, they are fundamentally different methods. The cSLIM method conforms to linear methods and it models the *top-N* recommendation process as an aggregation on items. On the contrary, CMF models the recommendation process in a low-dimension latent space.

Thu *et al* [7] proposed a weighted regularized matrix factorization (WRMF) method for *top-N* recommendation, which weights the purchase and nonpurchase activities in M differently using a weighting matrix C as follows:

$$\underset{U,V}{\text{minimize}} \quad \frac{1}{2}\|C \circ (M - UV)\|_F^2 + \frac{\beta}{2}(\|U\|_F^2 + \|V\|_F^2). \quad (15)$$

Inspired by this weighting method, we combined WRMF with CMF so as to have a collective weighted regularized matrix factorization method, denoted by CWRMF, as follows:

$$\underset{U,V,W}{\text{minimize}} \quad \frac{1}{2}\|C \circ (M - UV)\|_F^2 + \frac{\alpha}{2}\|F - WV\|_F^2$$
$$+ \frac{\beta_1}{2}\|U\|_F^2 + \frac{\lambda}{2}\|V\|_F^2 + \frac{\beta_2}{2}\|W\|_F^2, \qquad (16)$$

in which M and F are still collectively factorized but errors from M are weighted differently by C. We use WRMF and CWRMF as the comparison algorithms in the experiments. In addition, we use another two collaborative filtering methods for comparison purposes. The itemkNN method is a widely used item-based collaborative filtering method proposed in [4]. The itemSI method is a modification of itemkNN, in which the item similarities are calculated as a linear combination of the similarity values calculated from itemkNN and

Table 3: Performance Improvement over SLIM

feature	itemSI	CWRMF	cSLIM	rcSLIM	fSLIM	f2SLIM
F_b	0.973	0.674	1.095	1.048	0.818	1.008
F_{tfidf}	0.988	0.674	1.090	1.062	0.877	1.026
F_{tfidf_fs}	0.974	0.707	1.090	1.063	0.873	1.027
$F_{tfidf\_fs\_b}$	0.970	0.707	1.113	1.069	0.828	1.012
avg	0.976	0.690	1.097	1.061	0.849	1.018

Each value in the first four rows is calculated as the geometric mean of HR ratios of the corresponding method over SLIM over all the datasets, given the corresponding feature representation scheme used. The values in the last row is calculated as the geometric mean of HR ratios of the corresponding method over SLIM over all the datasets and all the feature presentation schemes.

the cosine similarity values calculated from side information weighted by a parameter α.

7. RESULTS

7.1 Overall Performance

Table 2 presents the detailed results of the SSLIM methods (cSLIM, rcSLIM, fSLIM and f2SLIM), the three methods without side information (itemkNN, WRMF and SLIM) and another two methods that utilize side information (itemSI and CWRMF), with respect to different side information representation schemes (F_b, F_{tfidf}, F_{tfidf_fs} and $F_{tfidf\_fs\_b}$). For the methods itemkNN, WRMF andSLIM, F_{no} is used in Table 2 to denote that side information is not used.

Table 2 shows that SLIM outperforms the other methods that do not utilize side information (i.e., itemkNN and WRMF) on all the datasets except BBY. For the BBY dataset, WRMF performs the best. This conforms to the conclusions as in [10], and thus we use SLIM as the baseline to further evaluate all the methods that utilize side information.

Table 3 summarizes the overall performance of the different methods that utilize side information, with respect to SLIM. Irrespective of the feature representation scheme, cSLIM, rcSLIM and f2SLIM perform better than SLIM with average improvement 9.7%, 6.1% and 1.8%, respectively (the last row in Table 3). This demonstrates that side information contains useful information, and proper incorporation of side information into the recommender systems can bring significant performance improvement.

The methods itemSI, fSLIM and CWRMF perform worse than SLIM. The itemSI method is a trivial extension of itemkNN and it does not involve any learning. fSLIM is a method that learns directly from the side information. The performance of fSLIM indicates that this method may not be able to pick out and highly weight the individual features in the item side information that are most relevant to the recommendations. CWRMF is the worst one and even worse than itemSI. This may be related to the discussion on CMF as in Agarwal *et al* [2], that is, when the side information is sparse, CMF may not work well.

Comparing the gains that can be obtained by utilizing side information across the different datasets, we see that they are not uniform. For the two movie datasets (ML100K and NF), the side information provides minimal benefits, whereas the gains achieved from the other datasets is substantial. We believe that this is due to the fact that the side information used for the movie datasets was quite generic, and does not contain sufficient information.

7.2 Side Information Representation

Table 3 shows that the performance of the side informa-

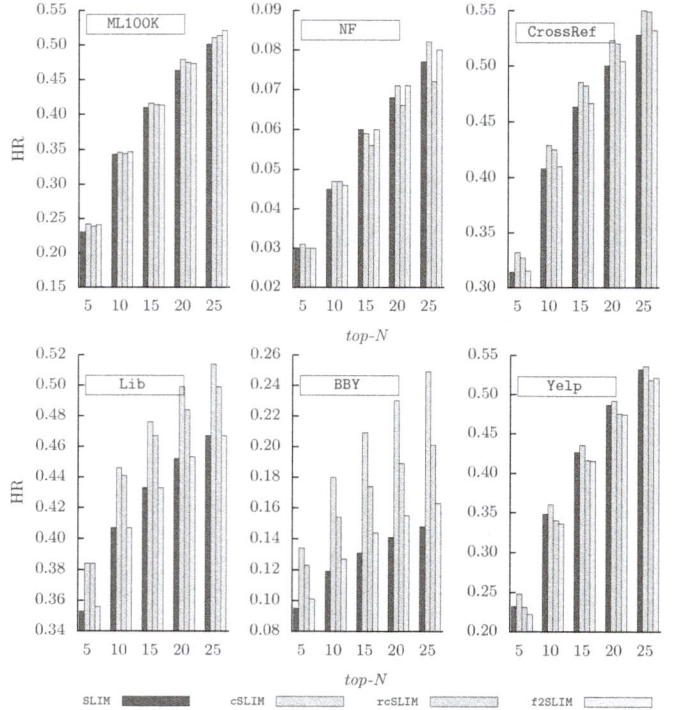

Figure 1: Recommendations for Different N Values

tion representation schemes depends on the recommendation methods. For cSLIM, which uses side information for regularization, the binary feature representations (F_b and $F_{tfidf\_fs\_b}$) lead to better performance than the multivariate feature representations (F_{tfidf} and F_{tfidf_fs}). This may be due to the fact that the binary features are treated homogeneously as the user-item purchase data and thus they can regularize the learning process effectively. However, for the methods fSLIM and f2SLIM, which involve direct learning from side information, F_{tfidf} and F_{tfidf_fs} result in better performance than the binary ones, since they differentiate the importance of features within the representations. In general, fSLIM and f2SLIM prefer the feature representations that encode word importance, so even the binary representation $F_{tfidf\_fs\_b}$, which has feature selection applied, also outperforms F_b, which does not differentiate features at all.

7.3 Recommendation on Different Top-N

Figure 1 presents the performance of SLIM and SSLIM methods (except fSLIM since it performs poorly) for top-N recommendation with different values of N. The $F_{tfidf\_fs\_b}$ side information representation is used for all the methods. For all the datasets, cSLIM consistently outperforms SLIM and other SSLIM methods on all N values (except $N = 25$ for dataset ML100K and $N = 15$ for dataset NF). The rcSLIM method is the second competitive methods over all N values. The f2SLIM method shows performance that is comparable to SLIM and in some cases (i.e., $N = 25$ for ML100K, all the N values for BBY) it outperforms SLIM. It performs consistently worse than SLIM only on Yelp.

7.4 Density Studies on the Purchase Data

To understand how the density of the dataset impacts the gains that can be obtained by utilizing side information, we

Table 2: Performance Comparison on All the Datasets

feature	method	ML100K params				HR	ARHR	NF params				HR	ARHR	CrossRef params				HR	ARHR
F_{no}	itemkNN	10	-	-	-	0.287	0.124	300	-	-	-	0.043	0.015	300	-	-	-	0.404	0.212
	WRMF	5	50	1	-	0.327	0.133	5	400	2	-	0.045	0.017	1	400	10	-	0.352	0.169
	SLIM	2	2	-	-	0.343	0.147	5	0.5	-	-	0.045	0.018	5	0.5	-	-	0.408	0.211
F_b	itemSI	200	0.5	-	-	0.262	0.116	200	0.5	-	-	0.042	0.016	200	0.5	-	-	0.416	0.218
	CWRMF	1	0.5	50	1	0.327	0.139	1	0.5	200	2	0.028	0.009	1	2	200	2	0.168	0.080
	cSLIM	0.5	10	1	-	0.344	0.148	0.1	10	1	-	0.047	0.017	1	5	0.1	-	0.429	0.224
	rcSLIM	0.1	5	1	2	0.344	0.147	5	10	10	1	0.047	0.017	5	10	5	0.5	0.421	0.218
	fSLIM	10	0.5	-	-	0.303	0.130	10	0.1	-	-	0.030	0.011	2	0.1	-	-	0.287	0.139
	f2SLIM	10	1	-	-	0.343	0.146	5	0.1	-	-	0.046	0.017	5	0.1	-	-	0.410	0.211
F_{tfidf}	itemSI	200	0.1	-	-	0.261	0.120	300	0.5	-	-	0.043	0.016	200	0.5	-	-	0.416	0.218
	CWRMF	1	0.5	50	1	0.327	0.139	1	0.5	200	2	0.028	0.009	1	2	200	2	0.168	0.080
	cSLIM	2	5	5	-	0.344	0.147	0.5	10	1	-	0.047	0.017	2	5	0.1	-	0.426	0.220
	rcSLIM	1	2	2	2	0.345	0.147	0.5	10	2	1	0.047	0.018	5	10	5	0.1	0.422	0.217
	fSLIM	2	0.1	-	-	0.317	0.134	5	0.1	-	-	0.035	0.011	2	0.1	-	-	0.292	0.141
	f2SLIM	5	5	-	-	0.341	0.147	10	0.5	-	-	0.046	0.017	5	0.1	-	-	0.413	0.212
F_{tfidf_fs}	itemSI	200	0.5	-	-	0.260	0.118	50	0.5	-	-	0.042	0.015	200	0.5	-	-	0.416	0.218
	CWRMF	1	1	50	1	0.325	0.125	1	2	200	1	0.031	0.010	1	2	200	2	0.174	0.082
	cSLIM	2	5	5	-	0.344	0.147	1	10	1	-	0.047	0.017	2	5	0.1	-	0.425	0.220
	rcSLIM	1	2	2	2	0.345	0.147	1	10	10	0.1	0.047	0.017	5	10	5	0.1	0.422	0.217
	fSLIM	2	0.1	-	-	0.317	0.133	5	0.1	-	-	0.034	0.012	2	0.1	-	-	0.292	0.141
	f2SLIM	5	5	-	-	0.343	0.148	10	5	-	-	0.046	0.017	5	0.1	-	-	0.413	0.212
$F_{tfidf\_fs\_b}$	itemSI	200	0.1	-	-	0.262	0.121	50	0.5	-	-	0.042	0.015	200	0.5	-	-	0.416	0.218
	CWRMF	1	1	50	1	0.325	0.125	1	2	200	1	0.031	0.010	1	2	200	2	0.174	0.082
	cSLIM	0.5	5	2	-	0.346	0.151	0.5	5	0.5	-	0.047	0.018	1	5	0.1	-	0.429	0.223
	rcSLIM	5	10	10	0.1	0.344	0.149	5	10	2	1	0.047	0.019	5	10	5	0.1	0.425	0.220
	fSLIM	20	1	-	-	0.312	0.134	10	0.1	-	-	0.029	0.011	2	0.1	-	-	0.287	0.139
	f2SLIM	10	1	-	-	0.347	0.151	10	1	-	-	0.046	0.017	5	0.1	-	-	0.410	0.211

feature	method	Lib params				HR	ARHR	BBY params				HR	ARHR	Yelp params				HR	ARHR
F_{no}	itemkNN	10	-	-	-	0.385	0.247	200	-	-	-	0.070	0.032	100	-	-	-	0.257	0.092
	WRMF	0.5	200	10	-	0.361	0.221	0.5	100	10	-	0.124	0.062	5	50	10	-	0.327	0.131
	SLIM	5	0.5	-	-	0.407	0.266	10	0.1	-	-	0.119	0.064	20	0.1	-	-	0.348	0.147
F_b	itemSI	300	0.5	-	-	0.407	0.258	100	2	-	-	0.151	0.068	300	1	-	-	0.321	0.127
	CWRMF	1	2	200	2	0.278	0.162	1	0.5	50	2	0.095	0.041	1	0.5	50	2	0.244	0.100
	cSLIM	1	5	0.1	-	0.443	0.283	0.1	20	0.1	-	0.165	0.081	0.1	20	0.5	-	0.361	0.154
	rcSLIM	5	10	2	0.5	0.439	0.288	5	10	10	0.5	0.139	0.079	2	10	10	0.5	0.338	0.144
	fSLIM	2	0.1	-	-	0.358	0.233	10	0.1	-	-	0.107	0.055	10	0.1	-	-	0.319	0.137
	f2SLIM	5	0.5	-	-	0.408	0.267	10	0.1	-	-	0.127	0.068	10	0.5	-	-	0.332	0.138
F_{tfidf}	itemSI	300	0.5	-	-	0.407	0.258	100	2	-	-	0.163	0.072	200	1	-	-	0.318	0.127
	CWRMF	1	2	200	2	0.278	0.162	1	0.5	100	1	0.095	0.041	1	0.5	50	2	0.244	0.100
	cSLIM	2	5	0.1	-	0.436	0.280	2	50	0.1	-	0.169	0.087	2	50	0.1	-	0.351	0.147
	rcSLIM	5	10	2	0.1	0.434	0.286	5	10	10	0.1	0.152	0.081	5	10	10	0.1	0.338	0.143
	fSLIM	2	0.1	-	-	0.367	0.236	2	0.1	-	-	0.115	0.059	10	0.1	-	-	0.353	0.148
	f2SLIM	5	0.5	-	-	0.410	0.266	20	0.1	-	-	0.132	0.069	20	0.1	-	-	0.354	0.148
F_{tfidf_fs}	itemSI	300	0.5	-	-	0.408	0.257	200	2	-	-	0.159	0.069	300	1	-	-	0.308	0.120
	CWRMF	1	2	200	2	0.286	0.167	1	2	100	1	0.102	0.044	1	2	50	2	0.260	0.103
	cSLIM	2	5	0.1	-	0.436	0.280	2	50	0.1	-	0.169	0.087	2	50	0.1	-	0.351	0.147
	rcSLIM	5	10	1	0.5	0.435	0.286	5	10	10	0.1	0.152	0.081	5	10	10	0.1	0.338	0.143
	fSLIM	2	0.1	-	-	0.367	0.236	2	0.1	-	-	0.115	0.059	10	0.1	-	-	0.353	0.148
	f2SLIM	5	0.5	-	-	0.410	0.267	20	0.1	-	-	0.132	0.069	20	0.1	-	-	0.354	0.148
$F_{tfidf\_fs\_b}$	itemSI	300	0.5	-	-	0.407	0.257	100	2	-	-	0.153	0.068	300	1	-	-	0.311	0.120
	CWRMF	1	2	200	2	0.286	0.167	1	2	100	1	0.102	0.044	1	2	50	2	0.260	0.103
	cSLIM	1	5	0.1	-	0.446	0.284	0.1	50	0.1	-	0.180	0.090	0.5	20	0.1	-	0.360	0.156
	rcSLIM	5	10	2	0.1	0.441	0.287	5	10	10	0.1	0.154	0.085	2	10	10	0.1	0.340	0.144
	fSLIM	2	0.1	-	-	0.358	0.234	20	0.1	-	-	0.109	0.057	20	0.1	-	-	0.338	0.141
	f2SLIM	5	0.5	-	-	0.407	0.267	10	0.1	-	-	0.127	0.069	20	0.1	-	-	0.337	0.139

The parameters represented for each method are as follows, respectively: itemkNN and itemSI: the number of neighbors k, and weighting parameter α; SLIM: the ℓ_2-norm regularization parameter β and the ℓ_1-norm regularization parameter λ as in [10]; cSLIM and rcSLIM: the weighting parameter on side information regularizer α, the ℓ_2-norm regularization parameter β (β_1 and β_2) and the ℓ_1-norm regularization parameter λ as in Equation 5 and 7; fSLIM and f2SLIM: the ℓ_2-norm regularization parameter β and the ℓ_1-norm regularization parameter λ as in Equation 9 and 11; WRMF: the regularization parameter β, the number of latent dimensions k and the weight on non-zero values c as in [7]; CWRMF: the weighting parameter on side information α, the regularization parameter λ, the number of latent dimensions k and the weight on non-zero values c as in Equation 15; Columns corresponding to HR and ARHR present the hit rate and average reciprocal hit-rank, respectively. N in this table is 10.

performed a series of experiments in which we removed some of the user-item purchase profile data as follows. For each dataset, we always keep the testing set and side information unchanged, but randomly select a certain percentage (defined as density factor) of non-zero values from each user so as to construct training sets of different information density. Figure 2 presents the results from different recommendation methods on the datasets. The results in these figures are relative to the performance achieved by SLIM at the same density level.

These results show that in general, cSLIM and rcSLIM lead to more significant performance gains when the user-item purchase profiles are sparser. This indicates that when the user-item purchase data is sparse, cSLIM and rcSLIM are more effective in exploiting and incorporating side information and lead to more accurate top-N recommendations. Note that for the dataset ML100K and NF, performance im-

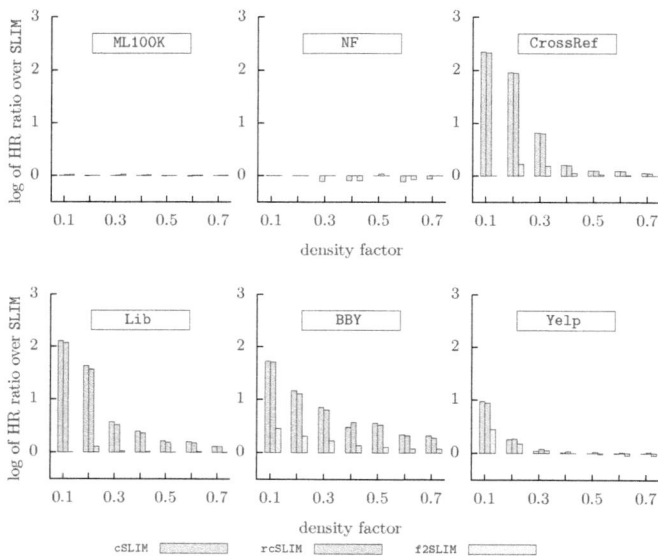

Figure 2: Density Studies

provement from incorporating side information is hardly observed. As discussed in Section 7.1, we believe that this is due to the low quality of the side information.

8. CONCLUSIONS

This paper focused on incorporating side information into the sparse linear methods (SLIM) for *top-N* recommender systems. We developed four different approaches that incorporate side information during the estimation of SLIM's aggregation coefficient matrix. Our experiments showed that the developed methods lead to measurable improvements over the original SLIM methods that relied solely on user-item purchase profiles.

Acknowledgement

This work was supported in part by NSF (IIS-0905220, OCI-1048018, and IOS-0820730), the DOE grant USDOE/DE-SC0005013 and the Digital Technology Center at the University of Minnesota. Access to research and computing facilities was provided by the Digital Technology Center and the Minnesota Supercomputing Institute.

9. REFERENCES

[1] D. Agarwal and B.-C. Chen. Regression-based latent factor models. In *Proceedings of the 15th ACM International Conference on Knowledge Discovery and Data Mining*, pages 19–28, 2009.

[2] D. Agarwal, B.-C. Chen, and B. Long. Localized factor models for multi-context recommendation. In *Proceedings of the 17th ACM SIGKDD international conference on Knowledge discovery and data mining*, KDD '11, pages 609–617, New York, NY, USA, 2011. ACM.

[3] L. M. de Campos, J. M. Fernández-Luna, J. F. Huete, and M. A. Rueda-Morales. Combining content-based and collaborative recommendations: A hybrid approach based on bayesian networks. *International Journal of Approximate Reasoning*, 51(7):785 – 799, 2010.

[4] M. Deshpande and G. Karypis. Item-based top-n recommendation algorithms. *ACM Transactions on Information Systems*, 22:143–177, January 2004.

[5] Z. Gantner, L. Drumond, C. Freudenthaler, S. Rendle, and L. Schmidt-Thieme. Learning attribute-to-feature mappings for cold-start recommendations. *IEEE International Conference on Data Mining*, pages 176–185, 2010.

[6] A. Gunawardana and C. Meek. A unified approach to building hybrid recommender systems. In *Proceedings of the third ACM conference on Recommender systems*, RecSys '09, pages 117–124, New York, NY, USA, 2009. ACM.

[7] Y. Hu, Y. Koren, and C. Volinsky. Collaborative filtering for implicit feedback datasets. In *Proceedings of the 2008 Eighth IEEE International Conference on Data Mining*, pages 263–272, Washington, DC, USA, 2008. IEEE Computer Society.

[8] A. Karatzoglou, X. Amatriain, L. Baltrunas, and N. Oliver. Multiverse recommendation: n-dimensional tensor factorization for context-aware collaborative filtering. In *Proceedings of the fourth ACM conference on Recommender systems*, pages 79–86, 2010.

[9] Y. Koren. Factorization meets the neighborhood: a multifaceted collaborative filtering model. In *Proceeding of the 14th ACM SIGKDD international conference on Knowledge discovery and data mining*, KDD '08, pages 426–434, New York, NY, USA, 2008. ACM.

[10] X. Ning and G. Karypis. Slim: Sparse linear methods for top-n recommender systems. In *Proceedings of 11th IEEE International Conference on Data Mining*, pages 497–506, 2011.

[11] S. Rendle, Z. Gantner, C. Freudenthaler, and L. Schmidt-Thieme. Fast context-aware recommendations with factorization machines. In *Proceedings of the 34th international ACM SIGIR conference on Research and development in Information*, SIGIR '11, pages 635–644, New York, NY, USA, 2011. ACM.

[12] F. Ricci, L. Rokach, B. Shapira, and P. B. Kantor, editors. *Recommender Systems Handbook*. Springer, 2011.

[13] G. Salton and M. J. McGill. *Introduction to Modern Information Retrieval*. McGraw-Hill, Inc., New York, NY, USA, 1986.

[14] A. P. Singh and G. J. Gordon. Relational learning via collective matrix factorization. In *Proceeding of the 14th ACM International Conference on Knowledge Discovery and Data Mining*, pages 650–658, 2008.

[15] R. Tibshirani. Regression shrinkage and selection via the lasso. *Journal of the Royal Statistical Society (Series B)*, 58:267–288, 1996.

[16] S.-H. Yang, B. Long, A. Smola, N. Sadagopan, Z. Zheng, and H. Zha. Like like alike: joint friendship and interest propagation in social networks. In *Proceedings of the 20th international conference on World wide web*, WWW '11, pages 537–546, New York, NY, USA, 2011. ACM.

Scalable Similarity-Based Neighborhood Methods with MapReduce

Sebastian Schelter   Christoph Boden   Volker Markl

Technische Universität Berlin, Germany
firstname.lastname@tu-berlin.de

ABSTRACT

Similarity-based neighborhood methods, a simple and popular approach to collaborative filtering, infer their predictions by finding users with similar taste or items that have been similarly rated. If the number of users grows to millions, the standard approach of sequentially examining each item and looking at all interacting users does not scale. To solve this problem, we develop a MapReduce algorithm for the pairwise item comparison and top-N recommendation problem that scales linearly with respect to a growing number of users. This parallel algorithm is able to work on partitioned data and is general in that it supports a wide range of similarity measures. We evaluate our algorithm on a large dataset consisting of 700 million song ratings from Yahoo! Music.

Categories and Subject Descriptors

H.4.m [**Information Systems Applications**]: Miscellaneous

Keywords

Scalable Collaborative Filtering, MapReduce

1. INTRODUCTION

Today's internet users face an ever increasing amount of data, which makes it constantly harder and more time consuming to pick out the interesting pieces of information from all the noise. This situation has triggered the development of recommender systems: intelligent filters that learn about the users' preferences and figure out the most relevant information for them.

With rapidly growing data sizes, the processing efficiency and scalability of the systems and their underlying computations becomes a major concern. In a production environment, the offline computations necessary for running a recommender system must be periodically executed as part of larger analytical workflows and thereby underly strict time and resource constraints. For economic and operational rea-

sons it is often undesirable to execute these offline computations on a single machine: this machine might fail and with growing data sizes constant hardware upgrades might be necessary to improve the machine's performance to meet the time constraints. Due to these disadvantages, a single machine solution can quickly become expensive and hard to operate.

In order to solve this problem, recent advances in large scale data processing propose to run data-intensive, analytical computations in a parallel and fault-tolerant manner on a large number of commodity machines. Doing so will make the execution independent of single machine failures and will furthermore allow the increase of computational performance by simply adding more machines to the cluster, thereby obviating the need for constant hardware upgrades to a single machine. Another economic advantage of such an approach is that the cluster machines can be temporally rented from a cloud computing infrastructure provider.

When applied to recommender systems, this technical approach requires the rephrasing of existing algorithms to enable them to utilize a parallel processing platform. Such platforms are able to run on a cluster of up to several thousand machines and to store and process amounts of data that were previously considered unmanageable. They typically employ a shared-nothing architecture together with a parallel programming paradigm and store the data in replicated partitions across the cluster. They provide the desired horizontal scalability when the number of machines in the cluster is increased. Furthermore they relieve the programmer from having to cope with the complicated tasks of scheduling computation, transferring intermediate results and dealing with machine failures.

We rephrase and scale out the similarity-based neighborhood methods, a standard approach in academic literature [22]. They have the advantage of being simple and intuitive to understand, as they are directly inspired by recommendation in everyday life, where we tend to check out things we heard about from like-minded friends or things that seem similar to what we already like. They capture local associations in the data which increases serendipity [22] and they are necessary as part of ensembles to reach optimal prediction quality [5]. The item-based variants [23] of the neighborhood methods are highly stable and allow computation of recommendations for new users without the need to rebuild the model. Additionally, they are able to provide instant justifications for their recommendations by presenting the list of neighbor items and the ratings the user already gave to these as explanation. Due to these properties, neighborhood methods are often preferred in industrial use cases [19, 1, 24, 7], although alternative approaches such as the latent factor models are superior in the task of predicting ratings.

We improve the scalability of the similarity-based neighborhood methods by rephrasing the underlying algorithm for pairwise comparisons to MapReduce [8], a popular parallel programming paradigm that has originally been proposed by Google. We demonstrate our approach using Apache Hadoop [2], a widely used, open source platform which implements the MapReduce paradigm.

We already contributed an implementation of the approach presented here to **Apache Mahout** [3], an open-source library of scalable data mining algorithms, where it forms the core of the distributed recommender module.

In this paper, we provide the following contributions:

- We introduce an algorithmic framework that allows scalable neighborhood-based recommendation on a parallel processing platform.

- We describe how to implement a variety of similarity measures in a highly efficient manner in our framework.

- We discuss how to apply selective down-sampling to handle scaling issues introduced by the heavy tailed distribution of user interactions commonly encountered in recommendation mining scenarios.

- We present experiments on various datasets with up to 700 million user interactions.

This paper is organized as follows: After a brief introduction to the MapReduce paradigm for parallel processing, we describe the algorithmic challenges of our approach in Section 2 and related work in Section 3. We describe and in detail derive our algorithm in Section 4. Finally, we evaluate our solution on various datasets in Section 5.

1.1 MapReduce

MapReduce [8], which is inspired by functional programming, has become a popular paradigm for data-intensive parallel processing on shared-nothing clusters.

The data to process is split and stored block-wise across the machines of the cluster in a distributed file system (DFS) and is usually represented as (*key,value*) tuples. In order to efficiently parallelize the computation and offer tolerance against machine failures, data is replicated across the cluster. As the computation tasks should be moved to the data, the runtime system assigns tasks to process data blocks to the machines holding the replicas of these blocks. The computation code is embedded into two functions:

map: $(k_1, v_1) \rightarrow \text{list}(k_2, v_2)$
reduce: $(k_2, \text{list}(v_2)) \rightarrow \text{list}(v_2)$

The data flow in a MapReduce pass is illustrated in Figure 1. At the beginning the **map** function is invoked on the input data in parallel on all the participating machines in the cluster. The output tuples are grouped (partitioned and sorted) by their key and then sent to the reducer machines in the **shuffle** phase. The receiving machines merge the tuples and invoke the **reduce** function on all tuples sharing the same key. The output of that function is written to the distributed file system afterwards.

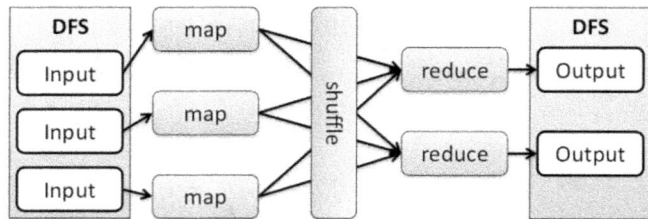

Figure 1: Illustration of the data flow in MapReduce

An optional third function called **combine** can be specified. It is invoked locally after the map phase and can be used to preaggregate the tuples in order to minimize the amount of data that has to be sent over the network, which is usually the most scarce resource in a distributed environment.

combine: $(k_2, \text{list}(v_2)) \rightarrow \text{list}(v_2)$

In addition, Hadoop offers *initialize* functions that are invoked before the map and reduce functions and the system provides a means to broadcast small files to all worker machines in the cluster via a *distributed cache*.

2. PROBLEM STATEMENT

Let A be a $|U| \times |I|$ matrix holding all known interactions between a set of users U and a set of items I. A user u is represented by his item interaction history $a_{u\bullet}$, the u-th row of A. The top-N recommendations for this user correspond to the first N items selected from a ranking r_u of all items according to how strongly they would be preferred by the user. This ranking is inferred from patterns found in A.

2.1 Computational model

Notation hints: $a_{u\bullet}$ denotes the u-th row of the interaction matrix A, $a_{\bullet i}$ denotes the i-th column of A, $|U|$ denotes the number of users which is equal to the number of rows in A. $foreach\ i \in v$ denotes iteration over the indexes of non-zero entries of a vector v, $foreach\ (i, k) \in v$ denotes iteration over the indexes and the corresponding non-zero values of a vector v.

In order to get a clearer picture of the neighborhood approach, it is useful to express the algorithm in terms of linear algebraic operations. Neighborhood-based methods find and rank items that have been preferred by other users who share parts of the interaction history $a_{u\bullet}$. Let A be a binary matrix with $A_{ui} = 1$ if a user u has interacted with an item i and $A_{ui} = 0$ otherwise. For pairwise comparison between users, a dot product of rows of A gives the number of items that the corresponding users have in common. Similarly, a dot product of columns of A gives the number of users who have interacted with both items corresponding to the columns.

When computing recommendations for a particular user with *User-Based Collaborative Filtering* [21], first a search for other users with similar taste is conducted. This translates to multiplying the matrix A by the user's interaction history $a_{u\bullet}$, which results in a ranking of all users. Secondly, the active user's preference for an item is estimated by computing the weighted sum of all other users' preferences for this item and the corresponding ranking. In our simple model this translates to multiplying the ranking of all users with A^{T}. This means the whole approach can be summarized by the following two multiplications:

$$r_u = A^{\mathrm{T}}(A\,a_{u\bullet})$$

To exploit the higher stability of relations between items, another variant of the neighborhood methods called *Item-Based Collaborative Filtering* [23] was developed, which looks at items first and weighs their cooccurrences. This approach computes a matrix of item-to-item similarities and allows for fast recommendations as the model does not have to be recomputed for new users. Expressing the item-based approach translates to simply moving the parentheses in our formula, as $A^{\mathrm{T}}A$ gives exactly the matrix of the item cooccurrences:

$$r_u = (A^{\mathrm{T}}A)\,a_{u\bullet}$$

This shows that both user- and item-based collaborative filtering share the same fundamental computational model. In the rest of the paper, we will focus on the more popular item-based variant.

2.2 Sequential approach

The standard sequential approach [19] for computing the item similarity matrix $S = A^{\mathrm{T}}A$ is shown in Algorithm 1.

Algorithm 1: sequential approach for computing item cooccurrences
---
foreach *item i* **do**
    foreach *user u who interacted with i* **do**
        foreach *item j that u also interacted with* **do**
            $S_{ij} = S_{ij} + 1$

For each item i, we need to look up each user u who interacted with i. Then we iterate over each other item j from u's interaction history and record the cooccurrence of i and j. We can mathematically express the approach using three nested summations:

$$S = A^{\mathrm{T}}A = \sum_{i=1}^{|I|}\sum_{u=1}^{|U|}\sum_{j=1}^{|I|} A^{\mathrm{T}}_{i,u} \cdot A_{u,j}$$

If we wish to distribute the computation across several machines on a shared-nothing cluster, this approach becomes infeasible as it requires random access to both users and items in its inner loops. Its random access pattern cannot be realized efficiently when we have to work on partitioned data.

Furthermore, at a first glance, the complexity of the item-based approach is quadratic in the number of items, as each item has to be compared with every other item. However, the interaction matrix A is usually very sparse. It is common that only a small fraction of all cells are known[1] and that the number of non-zero elements in A is linear in the number of rows of A. This fact severely limits the number of item pairs to be compared, as only pairs that share at least one interacting user have to be taken into consideration. It decreases the complexity of the algorithm to quadratic in the number of non-zeros in the densest row rather than quadratic in the number of columns. The cost of the algorithm is expressed

---
[1] in the datasets we used for our experiments, this ratio varies from 0.1% to 4.5%

as the sum of processing the square of the number of interactions of each single user. Unfortunately, collaborative filtering datasets share a property that is common among datasets generated by human interactions: the number of interactions per user follows a heavy tailed distribution which means that processing a small number of 'power users' dominates the cost of the algorithm.

We will develop a parallelizable formulation of the computation to scale out this approach on a parallel processing platform. As we need to utilize a distributed filesystem, our algorithm must be able to work with partitioned input data. Furthermore, it must scale linearly with respect to a growing number of users. We will enable the usage of a wide range of similarity measures and add means to handle the computational overhead introduced by 'power users'.

3. RELATED WORK

Most closely related to our work is a MapReduce formulation of item-based collaborative filtering presented by Jiang et al. [17]. However they do not show how to use a wide variety of similarity measures, they do not achieve linear scalability as they undertake no means to handle the quadratic complexity introduced by 'power users' and only present experiments on a small dataset. Another distributed implementation of an item-based approach is used by Youtube's recommender system [7], which applies a domain specific way of diversifying the recommendations by interpreting the pairwise item similarities as a graph. Unfortunately this work does not include details that describe how the similarity computation is actually executed other than stating it uses a series of MapReduce computations walking through the user/video graph. Furthermore, a very early implementation of a distributed item-based approach was applied in the recommendation system of the TiVo set-top boxes [1], which suggests upcoming TV shows to its users. In a proprietary architecture, show correlations are computed on the server side and preference estimation is afterwards conducted on the client boxes using the precomputed correlations.

There have also been several works on parallelizing latent factor models: The recommender system of Google News [6] uses a MapReduce based implementation combining Probabilistic Latent Semantic Indexing and a neighborhood approach with a distributed hashtable that tracks item cooccurrences in realtime. This solution is tailored towards the outstanding infrastructure of Google and might not be practical in other scenarios. Another parallelizable implementation of a latent factor model was presented by Zhou et al. [26], where a factorization of the Netflix dataset using Alternating Least Squares is conducted. Mahout [3] contains a MapReduce port of this approach. Similarly, Gemulla et al. [15] propose a stratified version of Stochastic Gradient Descent for matrix factorization on MapReduce. Due to Hadoop's inability to efficiently execute iterative algorithms, these implementations show unsatisfactory performance. Dataflow systems with explicit iteration support such as Stratosphere [12] or specialized systems for machine learning such as GraphLab [20] will pose a solution for the efficient distributed execution of such algorithms in the near future.

4. ALGORITHM

This section discusses the step-by-step development of our algorithmic framework. We start with showing how to conduct distributed item cooccurrence counting for our simple model that uses binary data. After that we generalize the approach to non-binary data and enable the usage of a wide variety of similarity measures. Finally, we discuss means to sparsify the similarity matrix, conduct batch recommendation and apply selective down sampling to achieve linear scalability with a growing number of users.

4.1 Counting item cooccurrences

In order to scale out the similarity computation from Algorithm 1, it needs to be phrased as a parallel algorithm, to make its runtime speedup proportional to the number of machines in the cluster. This is not possible with the standard sequential approach, as it requires random access to the rows and columns of A in the inner loops of Algorithm 1, which cannot be efficiently realized in a distributed, shared-nothing environment where the algorithm has to work on partitioned data.

We need to find a way of executing this multiplication that is better suited to the MapReduce paradigm and has an access pattern that is compatible to partitioned data. The solution is to rearrange the loops of Algorithm 1 to get the row outer product formulation of matrix multiplication. Because the u-th column of A^T is identical to the u-th row of A, we can compute S with only needing access to the rows of A:

$$S = A^\mathrm{T}A = \sum_{u=1}^{|U|} \sum_{i=1}^{|I|} \sum_{j=1}^{|I|} A^\mathrm{T}_{i,u} \cdot A_{u,j} = \sum_{u=1}^{|U|} a_{u\bullet}(a_{u\bullet})^\mathrm{T}$$

Following this finding, we partition A by its rows (the users) and store it in the distributed file system. Each map function reads a single row of A, computes the row's outer product with itself and sends the resulting intermediary matrix row-wise over the network. The reduce function simply has to sum up all partial results, thereby computing a row of S per invocation (Algorithm 2).

This approach allows us to exploit the sparsity of the intermediary outer product matrices by making the map function only return non-zero entries. At the same time we apply a combiner (which is identical to the reducer) on the vectors emitted by the mappers, which makes the system minimize the amount of data that has to be sent over the network. Additionally we only compute the upper triangular half of S, as the resulting similarity matrix is symmetric.

4.2 Generalized similarity computation

Real world datasets contain richer representations of the user interactions than a simple binary encoding. They either consist of *explicit feedback* like numerical ratings that the users chose from a predefined scale or of *implicit feedback* where we count how often a particular behavior such as a click or a page view was observed. We need to be able to choose from a variety of similarity measures for comparing these item interactions, in order to be able to find the one that best captures the relationships inherent in the data. From now on, we drop the assumption that A contains only binary entries and assume that it holds such explicit or implicit feedback data.

Algorithm 2: computing item cooccurrences

```
function map(a_u•):
   foreach i ∈ a_u• do
      c ← sparse_vector()
      foreach j ∈ a_u• with j > i do
         c[j] ← 1
      emit(i, c)
function combine(i, c_1, ..., c_n):
   c ← vector_add(c_1, ..., c_n)
   emit(i, c)
function reduce(i, c_1, ..., c_n):
   s ← vector_add(c_1, ..., c_n)
   emit(i, s)
```

Expressing arbitrary similarity measures: We incorporate a wide range of measures for comparing the interactions of two items i and j by integrating three canonical functions into our algorithm. We first adjust each item rating vector via a function **preprocess()**:

$$\widehat{i} = preprocess(i) \quad \widehat{j} = preprocess(j)$$

Next, the second function **norm()** computes a single number from the preprocessed vector of an item:

$$n_i = norm(\widehat{i}) \quad n_j = norm(\widehat{j})$$

These preprocessing and norm computations are conducted in an additional single pass over the data, which starts with A^T, applies the two functions and transposes A^T to form A.

The next pass over the data is a modification of the approach presented in Section 4.1. Instead of summing up cooccurrence counts, we now compute the dot products of the preprocessed vectors.

$$dot_{ij} = \widehat{i} \cdot \widehat{j}$$

We provide those together with the numbers we computed via the *norm* function to a third function called **similarity()** which will compute a measure-specific similarity value (Algorithm 3).

$$S_{ij} = similarity(dot_{ij}, n_i, n_j)$$

With this approach we are able to incorporate a wide variety of different similarity measures which can be rephrased as a variant of computing a dot product. Note that this technique preserves the ability to apply a combiner in each pass over the data and is therefore highly efficient.

Table 1 describes how to express several common similarity measures through these canonical functions, including cosine, Pearson correlation and a couple of others evaluated by Google for recommending communities in its social network Orkut [24].

Example: The Jaccard coefficient between items i and j (two columns from the interaction matrix A) is computed as the ratio of the number of users interacting with both items to the number of users interacting with at least one of those items. It can easily be expressed by our algorithmic framework, as this example shows:

$$i = \begin{bmatrix} 1 \\ - \\ 3 \end{bmatrix} \quad j = \begin{bmatrix} 2 \\ 1 \\ 5 \end{bmatrix}$$

Pass 1: We start by having *preprocess* binarize the vectors:

$$\widehat{i} = bin(i) = \begin{bmatrix} 1 \\ 0 \\ 1 \end{bmatrix} \quad \widehat{j} = bin(j) = \begin{bmatrix} 1 \\ 1 \\ 1 \end{bmatrix}$$

The second function that is invoked for each of the vectors is *norm*. We let it return the L_1 norm, which gives us the number of non-zero components of each of the binary vectors:

$$n_i = \|\widehat{i}\|_1 = 2 \quad n_j = \|\widehat{j}\|_1 = 3$$

Pass 2: Finally the function *similarity* will be called given the dot product between the preprocessed vectors and their precomputed norms. We have to rearrange the formula of the Jaccard coefficient so that it can be computed from the numbers we have at hand:

$$jaccard(i,j) = \frac{|i \cap j|}{|i \cup j|} = \frac{dot_{ij}}{n_i + n_j - dot_{ij}} = \frac{2}{2+3-2} = \frac{2}{3}$$

Algorithm 3: computing arbitrary item similarities

```
function map(a_u•):
    foreach (i, k_1) ∈ a_u• do
        d ← sparse_vector()
        foreach (j, k_2) ∈ a_u• with j > i do
            d[j] ← k_1k_2
        emit(i, d)

function combine(i, d_1, ..., d_n):
    d ← vector_add(d_1, ..., d_n)
    emit(i, d)

function initialize_reducer():
    n ← load_norms()

function reduce(i, d_1, ..., d_n):
    dots ← vector_add(d_1, ..., d_n)
    s ← sparse_vector()
    foreach j, d ∈ dots do
        s[j] = similarity(d, n[i], n[j])
    emit(i, s)
```

4.3 Sparsification

In order to be able to handle cases with an enormous number of items, we add means to decrease the density of the similarity matrix S to our final implementation.

To get rid of pairs with near-zero similarity, a similarity threshold can be specified, for which we evaluate a size constraint to prune lower scoring item pairs early in the process [4] and eventually remove all entries from S that are smaller than the threshold. Note however that this threshold is data dependent and must be determined experimentally to avoid negative effects on prediction quality.

Furthermore, it has been shown that the prediction quality of the item-based approach is sufficient if only the top fraction of the similar items is used [23], therefore we add another MapReduce step that only retains these top similar items per item in a single pass over the data.

4.4 Batch recommendation

Although the similarity matrix is usually used to compute recommendations online, some use cases such as generating personalized newsletters require batch recommendation for

measure	preprocess	norm	similarity
Cosine	$\frac{v}{\|v\|_2}$	-	dot_{ij}
Pearson correlation	$\frac{(v-\bar{v})}{\|v-\bar{v}\|_2}$	-	dot_{ij}
Euclidean distance	-	$\hat{v}^2$	$\sqrt{n_i - 2 \cdot dot_{ij} + n_j}$
Common neighbors	$bin(v)$	-	dot_{ij}
Jaccard coefficient	$bin(v)$	$\|\hat{v}\|_1$	$\frac{dot_{ij}}{n_i+n_j-dot_{ij}}$
Manhattan distance	$bin(v)$	$\|\hat{v}\|_1$	$n_i + n_j - 2 \cdot dot_{ij}$
Pointwise Mutual Information	$bin(v)$	$\|\hat{v}\|_1$	$\frac{dot_{ij}}{|U|} \log \frac{dot_{ij}}{n_i n_j}$
Salton IDF	$bin(v)$	$\|\hat{v}\|_1$	$\frac{|U| \cdot dot_{ij}}{n_i n_j^2}(-\log \frac{n_i}{|U|})$
Log Odds	$bin(v)$	$\|\hat{v}\|_1$	$\log \frac{\frac{|U| \cdot dot_{ij}}{n_i n_j^2}}{1 - \frac{|U| \cdot dot_{ij}}{n_i n_j^2}}$
Log-likelihood ratio [10]	$bin(v)$	$\|\hat{v}\|_1$	$2 \cdot (H(dot_{ij}, n_j - dot_{ij}, n_i - dot_{ij}, |U| - n_i - n_j + dot_{ij}) - H(n_j, |U| - n_j) - H(n_i, |U| - n_i))$

Table 1: expressing measures with the canonical functions

all users. To achieve that, we need another function called **recommend()** that is invoked with the similarity matrix S and item interaction history $a_{u\bullet}$ of an active user u and returns the top-N items to recommend to that user. A variety of strategies can be applied in the estimation procedure, ranging from simple weighted sum estimation [23], improved by baseline estimates [18], to more advanced techniques that incorporate domain specific knowledge and aim to diversify recommendations [7].

If the similarity matrix fits into the memory of a single mapper instance, the most efficient way of embedding the recommendation computation is by executing a *broadcast join* [9] of the users' item interaction histories and the sparsified similarity matrix. As shown in Algorithm 4, the similarity matrix is broadcasted to all worker machines in the cluster via the distributed cache and the recommendations are computed in a map-only job over the users' interaction histories. Such a job is highly efficient as no reducer is required, which obviates the need for the *shuffle* phase and its associated sorting and network overhead.

In cases with an extreme number of items, the similarity matrix might not fit into the mappers' memory any more. In such a case a less performant *repartition join* [13] has to be used where the items each user has interacted with and their corresponding rows from the similarity matrix are sent over the network to a reducer that joins them after receival.

Algorithm 4: batch recommendation for all users

```
function initialize_mapper():
    S ← load_similarity_matrix()

function map(u, a_u•):
    r ← recommend(S, a_u•)
    emit(u, r)
```

4.5 Linear scalability with a growing user base

Recall that our goal is to develop an algorithmic framework that scales linearly with respect to a growing user base. As described in Section 2.2, the cost of the item-based approach is dominated by the densest rows of A, which correspond to the users with the most interactions. This cost, which we express as the number of item cooccurrences to consider, is the sum of the squares of the number of interactions of each user.

The number of interactions per user usually follows a heavy tailed distribution as illustrated in Figure 3 which plots the ratio of users with more than n interactions to the number of interactions n on a logarithmic scale. Therefore, there exists a small number of 'power users' with an unproportionally high amount of interactions. These drastically increase the runtime, as the cost produced by them is quadratic with the number of their interactions.

If we only look at the fact whether a user interacted with an item or not, then we would intuitively not learn very much from a 'power user': each additional item he interacts with will cooccur with the vast amount of items he already preferred. We would expect to gain more information from users with less interactions but a highly differentiated taste. Furthermore, as the relations between items tend to stabilize quickly [22], we presume that a moderately sized number of observations per item is sufficient to find its most similar items.

Following this rationale, we decided to apply what we call an *interaction-cut*: we selectively down sample the interaction histories of the 'power users'.

We apply this by randomly sampling p interactions from each such user's history, thereby limiting the maximum number of interactions per user in the dataset to p. Note that this sampling is only applied to the small group of 'power users', it does not affect the data contributed by the vast majority of non-'power users' in the long tail. Capping the effort per user in this way limits the overall cost of our approach to $|U|\, p^2$. We will experimentally show that a moderately sized p is sufficient to achieve prediction quality close to that of unsampled data. An optimal value for p is data dependent and must be determined by hold-out tests.

5. EVALUATION

In this section we present the results of a sensitivity analysis for the interaction-cut and conduct an experimental evaluation of our parallel algorithm on a large dataset[1].

We will show that the prediction quality achieved by using an interaction cut quickly converges to the prediction quality achieved with unsampled data. Subsequently, we will analyze the relationship between the size of the interaction-cut, the achieved quality and the runtime for the similarity computation in our large dataset. After that we will study the effects on the runtime speedup if we add more machines to the Hadoop cluster as well as the scaling behavior with a growing user base.

Prediction was conducted with weighted sum estimation enhanced by baseline estimates [18]. In a preprocessing step that has negligible computation cost, we estimate global user and item biases b_u and b_i that describe the tendency to deviate from the average rating μ. This gives us the simple

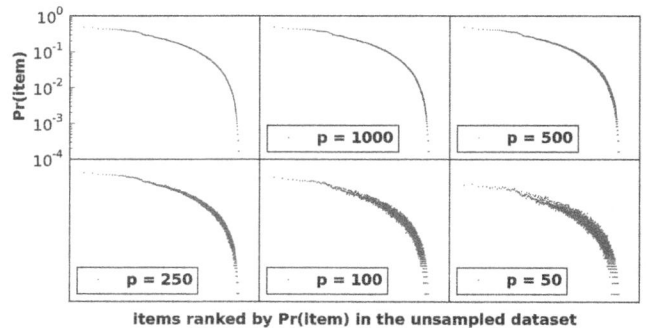

Figure 2: sensitivity of the probability of interaction with an item to an interaction-cut of size p in the Movielens dataset

baseline prediction $b_{ui} = \mu + b_u + b_i$ for the rating of a user u to an item i. To finally predict the rating r_{ui}, we use the normalized weighted sum over the user's ratings to the k most similar items of i, incorporating the baseline predictions:

$$r_{ui} = b_{ui} + \frac{\sum_{j \in S^k(i,u)} S_{ij}(A_{uj} - b_{uj})}{\sum_{j \in S^k(i,u)} S_{ij}}$$

5.1 Effects of the interaction-cut

We conducted a sensitivity analysis of the effects of the interaction-cut. We measured the effect on the probability of interaction with an item and on prediction quality for varying p on the **Movielens**[2] dataset consisting of 1,000,209 ratings that 6,040 users gave to 3,706 movies.

For our first experiment, we ranked the items by their probability of interaction as shown in the plot in the top left corner in Figure 2. Next, we applied the interaction-cut by sampling down the ratings of users whose number of interactions in the training set exceeded p and repeated this for several values of p. In the remaining plots of Figure 2, we retained the order of the items found in the unsampled data and plotted their probabilities after applying the interaction-cut for a particular p. We see that for $p \geq 500$ there is no observable distortion in the ranking of the items, which is a hint that this distribution is independent of the data omitted by sampling down the interactions of the 'power users'.

In our second experiment, we computed the prediction quality (by mean average error) achieved by a particular p by randomly splitting the rating data into 80% training and 20% test set based on the number of users. For this experiment, we additionally used the **Flixster**[3] dataset consisting of 8,196,077 ratings from 147,612 users to 48,794 movies. Again, we applied the interaction-cut by sampling down the ratings of users whose number of interactions in the training set exceeded p and used the 80 most similar items per item for rating prediction. For these small datasets, the tests were conducted on a single machine using a modified version of Mahout's [3] *GenericItemBasedRecommender*.

Figure 4 shows the results of our experiments. Note that the right-most data points are equivalent to the prediction quality of the unsampled dataset. In the Movielens dataset we see that for $p > 400$ the prediction quality converges to the prediction quality of the unsampled data, in the Flixster

---

[1]code to repeat our experiments is available at http://github.com/dima-tuberlin/publications-ssnmm

[2]http://www.grouplens.org/node/73
[3]http://www.cs.sfu.ca/~sja25/personal/datasets/

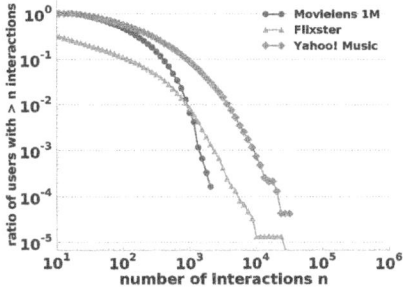

Figure 3: long tailed distribution of the number of interactions per user in various datasets

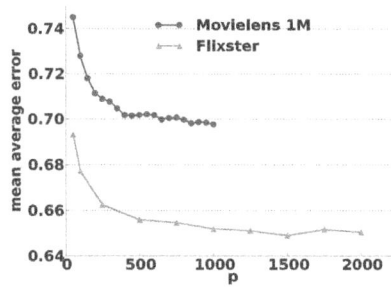

Figure 4: effects of an interaction-cut of size p on prediction quality in various small datasets

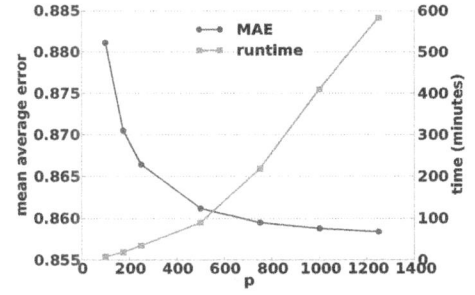

Figure 5: prediction quality and runtime for an interaction-cut of size p in the Yahoo! Music dataset

dataset this happens at $p > 750$. There is no significant decrease in the error for incorporating more interactions from the 'power users' after that. This confirms our expectation that we can compute recommendations based on the user data from the long tail and only samples from the 'power users' without sacrificing prediction quality.

5.2 Parallel computation with MapReduce

The following experiments were conducted on a Hadoop cluster with a MapReduce implementation of our approach. The cluster consisted of six machines running Apache Hadoop 0.20.203 [2] with each machine having two 8-core Opteron CPUs, 32 GB memory and four 1 TB disk drives. The experiments for showing the linear speedup with the number of machines were run on Amazon's computing infrastructure, where we rented *m1.xlarge* instances, 64-bit machines with 15 GB memory and eight virtual cores each.

5.2.1 Prediction quality

To test our approach in a demanding setting, we used a very large ratings dataset[1] which represents a snapshot of the Yahoo! Music community's preferences for various songs that were collected between 2002 and 2006. The data consists of 717,872,016 ratings that 1,823,179 users gave to 136,736 songs.

We used the 699 million training ratings provided in the dataset to compute item similarities and measured the prediction quality for the remaining 18 million held out ratings. We computed the 50 most similar items per item with Pearson correlation as similarity measure with a threshold of 0.01. Figure 5 shows the results we got for differently sized interaction cuts. We see that the prediction quality converges for $p > 600$, similar to what we have observed for the smaller datasets in Section 5.1. We additionally measured the root mean squared error and observed the same behavior, the prediction quality converged to an error of 1.16 here. We see the expected quadratic increase in the runtime for a growing p, which is weakened by the fact there is a quickly shrinking number of users with more than p interactions (from Figure 3 we know for example that only approximately 9% of the users have more than 1000 interactions).

The computational overhead introduced by the 'power users' is best illustrated when we compare the numbers for $p = 750$ and $p = 1000$: we see a decrease of only 0.0007 in the mean average error, yet the higher value of p accounts

for a nearly doubled runtime. We conclude that we are able to achieve the convergence of the prediction quality in this large dataset with a p that is extremely low compared to the overall number of items and thereby results in a computation cost that is easily manageable even by our small Hadoop cluster.

In order to conduct a comparison to a single machine implementation, we tried the *item-based kNN* recommender of MyMediaLite [14], the *GenericItemBasedRecommender* provided by Apache Mahout [3] and the *ItemRecommender* from LensKit [11]. Unfortunately, not one of these was able to complete the computation due to problems with memory handling, although we ran them on a machine which had 48 GB of RAM available.

Based on our findings, we chose to set p to 600 for the following scalability experiments.

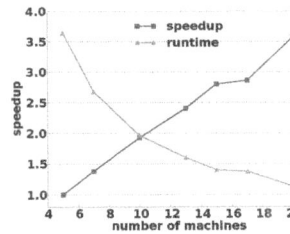

Figure 6: speedup for a growing number of machines in Amazon EC2

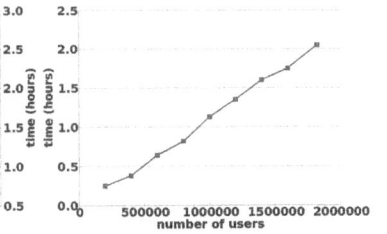

Figure 7: linear runtime increase for a growing user base

5.2.2 Linear speedup with the number of machines

The major promise of parallel processing platforms is seamless horizontal scale out by simply adding more machines to the cluster. This requires the computation speedup to be proportional to the number of machines. To experimentally evaluate this property of our algorithm, we made use of the *ElasticMapReduce* computing infrastructure provided by Amazon, which allows us to run our algorithm on a customly sized Hadoop cluster. We repeatedly ran the similarity computation with an increasing number of cluster machines. Figure 6 shows the linear speedup as expected by us. With 15 machines we were able to reduce the runtime of the similarity computation to less than one hour.

[1]R2 - Yahoo! Music User Ratings of Songs with Artist, Album, and Genre Meta Information, v. 1.0, http://webscope.sandbox.yahoo.com/

5.2.3 Linear scale with a growing number of users

Finally, we evaluate the scaling behaviour of our approach in the case of a rapid growth of the number of users. The Yahoo! Music dataset is already partitioned into several files, with each file containing approximately 77 million ratings given by 200,000 unique users. In order to simulate the growth of the user base, we used an increasing number of these files as input to our parallel algorithm and measured the duration of the similarity computation. Figure 7 shows the algorithm's runtime when scaling from 200,000 to 1,8 million users. We see a perfectly linear increase in the runtime which confirms the applicability of our approach in scenarios with enormously growing user bases.

As the speedup with the number of machines as well as the runtime for a growing number of users scale linearly, we can counter such growth by simply adding more machines to the cluster to keep the computation time constant.

6. CONCLUSIONS

We showed how to build a scalable, neighboorhood-based recommender system based on the MapReduce paradigm. We rephrased the underlying pairwise comparison to run on a parallel processing platform with partitioned data and described how a wide variety of measures for comparing item interactions easily integrate into our method. We introduced a down sampling technique called interaction-cut to handle the computational overhead introduced by 'power users'.

For a variety of datasets, we experimentally showed that the prediction quality quickly converges to that achieved with unsampled data for moderately sized interaction-cuts. We demonstrated a computation speedup that is linear in the number of machines on a huge dataset of 700 million interactions and showed the linear scale of the runtime with a growing number of users on that data.

In future work we intend to explore how our method could be used to scale out recommendation approaches that incorporate similarity computations on large networks [16, 25].

7. ACKNOWLEDGMENTS

A special thanks goes to Ted Dunning, Zeno Gantner and Moritz Kaufmann. The research leading to these results has received funding from the European Union (EU) in the course of the project 'ROBUST' (EU grant no. 257859) and used data provided by 'Yahoo! Academic Relations'.

8. REFERENCES

[1] K. Ali and W. van Stam. Tivo: Making show recommendations using a distributed collaborative filtering architecture. KDD, 2004.

[2] Apache Hadoop, http://hadoop.apache.org.

[3] Apache Mahout, http://mahout.apache.org.

[4] R. J. Bayardo, Y. Ma, and R. Srikant. Scaling up all pairs similarity search. WWW, pp. 131–140, 2007.

[5] R. M. Bell and Y. Koren. Lessons from the netflix prize challenge. *SIGKDD Newsl.*, 9:75–79, 2007.

[6] A. S. Das, M. Datar, A. Garg, and S. Rajaram. Google news personalization: scalable online collaborative filtering. WWW, pp. 271–280, 2007.

[7] J. Davidson, B. Liebald, J. Liu, P. Nandy, T. Van Vleet, U. Gargi, S. Gupta, Y. He, M. Lambert, B. Livingston, and D. Sampath. The youtube video recommendation system. RecSys, pp. 293–296, 2010.

[8] J. Dean and S. Ghemawat. Mapreduce: simplified data processing on large clusters. *Commun. ACM*, 51:107–113, 2008.

[9] D. DeWitt, R. Gerber, G. Graefe, M. Heytens, K. Kumar, and M. Muralikrishna. GAMMA - a high performance dataflow database machine. VLDB, pp. 228–237, 1986.

[10] T. Dunning. Accurate methods for the statistics of surprise and coincidence. ACL, 19:61–74, 1993.

[11] M. D. Ekstrand, M. Ludwig, J. A. Konstan, and J. T. Riedl. Rethinking the recommender research ecosystem: reproducibility, openness, and lenskit. RecSys, pp. 133–140, 2011.

[12] S. Ewen, K. Tzoumas, M. Kaufmann, and V. Markl. Spinning Fast Iterative Data Flows. *PVLDB*, 2012.

[13] S. Fushimi, M. Kitsuregawa, and H. Tanaka. An overview of the system software of a parallel relational database machine GRACE. VLDB, pp. 209–219, 1986.

[14] Z. Gantner, S. Rendle, C. Freudenthaler, and L. Schmidt-Thieme. Mymedialite: a free recommender system library. RecSys, pp. 305–308, 2011.

[15] R. Gemulla, E. Nijkamp, P. Haas, and Y. Sismannis. Large-scale matrix factorization with distributed stochastic gradient descent. KDD, pp. 69–77, 2011.

[16] M. Jamali and M. Ester. Trustwalker: a random walk model for combining trust-based and item-based recommendation. KDD, pp. 397–406, 2009.

[17] J. Jiang, J. Lu, G. Zhang, and G. Long. Scaling-up item-based collaborative filtering recommendation algorithm based on hadoop. *SERVICES*, pp. 490 –497, 2011.

[18] Y. Koren. Factor in the neighbors: Scalable and accurate collaborative filtering. *ACM Trans. KDD*, 4:1:1–1:24, 2010.

[19] G. Linden, B. Smith, and J. York. Amazon.com recommendations: item-to-item collaborative filtering. *Internet Computing, IEEE*, 7(1):76–80, 2003.

[20] Y. Low and J. Gonzalez and A. Kyrola and D. Bickson and C. Guestrin and J. Hellerstein. Distributed GraphLab: A Framework for Machine Learning in the Cloud. *PVLDB*, 2012.

[21] P. Resnick, N. Iacovou, M. Suchak, P. Bergstrom, and J. Riedl. Grouplens: an open architecture for collaborative filtering of netnews. CSCW, pp. 175–186, 1994.

[22] F. Ricci, L. Rokach, B. Shapira, and P. B. Kantor. *Recommender Systems Handbook*. 2011.

[23] B. Sarwar, G. Karypis, J. Konstan, and J. Riedl. Item-based collaborative filtering recommendation algorithms. WWW, pp. 285–295, 2001.

[24] E. Spertus, M. Sahami, and O. Buyukkokten. Evaluating similarity measures: a large-scale study in the orkut social network. KDD, pp. 678–684, 2005.

[25] P. Symeonidis, E. Tiakas, and Y. Manolopoulos. Product recommendation and rating prediction based on multi-modal social networks. RecSys, pp. 61–68, 2011.

[26] Y. Zhou, D. Wilkinson, R. Schreiber, and R. Pan. Large-scale parallel collaborative filtering for the netflix prize. AAIM, pp. 337–348, 2008.

An Approach to Context-Based Recommendation in Software Development

Bruno Antunes
CISUC, University of Coimbra
Coimbra, Portugal
bema@dei.uc.pt

Joel Cordeiro
CISUC, University of Coimbra
Coimbra, Portugal
jfac@student.dei.uc.pt

Paulo Gomes
CISUC, University of Coimbra
Coimbra, Portugal
pgomes@dei.uc.pt

ABSTRACT

A software developer programming in an object-oriented programming language deals with a source code structure that may contain hundreds of source code elements. These elements are commonly related to each other and working on a specific element may require the developer to access other related elements. We propose a recommendation approach that uses the context of the developer to retrieve and rank recommendations of relevant source code elements in the IDE. These recommendations provide a shortcut to reach the desired elements and increase the awareness of the developer in relation to elements that may be of interest in that moment. We have tested our approach with a group of developers and the results show that context has a promising role in predicting and ranking the source code elements needed by a developer at each moment.

Categories and Subject Descriptors

D.2 [**Software Engineering**]: Programming Environments; H.3 [**Information Storage and Retrieval**]: Information Search and Retrieval; I.2 [**Artificial Intelligence**]: Knowledge Representation Formalisms and Methods

Keywords

Recommendation Systems, Context Modeling, Ontologies, Software Development, IDE.

1. INTRODUCTION

The recommendation systems are currently used in a wide variety of domains to help users find relevant information, deal with information overload and provide personalized recommendations of very different kinds of items. The recommendation process is commonly dependent on estimating the utility of a specific item for a particular user [1]. Most of the approaches used in current recommendation systems are mainly focused in estimating how relevant is an item to an user, ignoring any contextual information that could

be used to improve the recommendation process. However, context-based recommendation systems are emerging, taking context into account when providing recommendations to the user [2].

A recommendation system for software engineering has been defined by Robbilard et al. [10] as *"a software application that provides information items estimated to be valuable for a software engineering task in a given context"*. The increasing dimension and complexity of software development projects are fostering the development of such systems, which have been applied to very different tasks such as software reuse, expertise location, code comprehension, guided software changes, debugging, etc.

A software developer programming in an object-oriented programming language, such as Java, deals with a source code structure that may contain hundreds of source code elements. These elements are commonly related to each other and working on a specific element may require the developer to access other related elements. Despite some features that allow the developer to jump between related elements in very specific situations, the primary way to access these elements is by browsing the source code structure, either using the package structure or the outlines provided for individual source code files. As the number of source code elements in the workspace of a developer increase, the need to switch between different elements becomes more frequent and more time is spent in finding the needed elements. We propose a recommendation approach to help developers programming in Java, by recommending source code elements that are considered relevant in the context of the developer at a specific moment. These recommendations provide a shortcut to reach these elements, and a way to increase the awareness of the developer in relation to elements that may be of interest in that moment. The recommendations are provided in a pro-active way and automatically updated as the context of the developer changes.

The recommendation approach uses the context of the developer to retrieve and rank recommendations of source code elements, such as classes, interfaces and methods, in an IDE (Integrated Development Environment). These elements and their relations are represented in a knowledge base using an ontology [15]. The contextual information is implicitly gathered from the interactions of the developer in the IDE, and is used to retrieve and rank relevant source code elements that are then recommended to the developer. We have implemented a prototype, named SDiC[1] (Software Development in Context), that integrates our recommenda-

---

[1]http://sdic.dei.uc.pt

tion approach in an Eclipse[2] plugin. This prototype was submitted to a field experiment with a group of developers. The results show that the use of the context has a promising role in predicting and ranking the source code elements needed by a developer on a specific moment.

The remainder of the paper is organized as follows. In section 2 we make an overview of related work. Then, a description of the knowledge based used is given in section 3. In section 4 we introduce the context model and section 5 explains how this model is used to support the recommendation process. The prototype implemented is presented in section 6, followed by evaluation and result discussion in section 7. Finally, section 8 concludes the work with some final remarks and future work.

2. RELATED WORK

The context of a developer has been modeled in different ways to improve awareness and help locating relevant source code elements through recommendation. Kersten and Murphy [9] propose a model for representing the context associated to a task. The task context is used to help focus the information displayed in the IDE and to automate the retrieval of relevant information for completing the task. Our approach also uses the interactions of the developer to derive the degree of interest of different source code elements to the developer, which is used together with an ontological model of the source code structure to retrieve and rank other relevant source code elements. The context model we use is not attached to tasks and automatically adapts to the changes in the focus of attention of the developer. Warr and Robillard [12] developed Suade, a plugin for the Eclipse IDE that help developers exploring the source code, by suggesting potentially relevant elements for the current context. The context is explicitly provided by the developer and the suggestions are ranked according to their structural dependencies using some heuristics. We do not require an explicit definition of the current context to provide such recommendations, as it is automatically gathered by the system. Heinemann and Hummel [7] propose an approach to recommend methods that may be relevant for the current work of a developer. The recommendations are based on the terms extracted from identifiers preceding method calls. We also explore the lexical dimension of source code elements, which is integrated in our context model and then used to rank the recommendations provided to the developer.

The contextual information was also used to improve the recommendation of reusable software components or useful source code examples. Ye and Fischer [13] propose CodeBroker, a tool that pro-actively suggests reusable components to Java developers using Emacs. The system monitors the JavaDoc comments and signature definitions created by the developer, from which it extracts queries to retrieve matching components. Strathcona, an Eclipse plugin proposed by Holmes and Murphy [8], is used to help developers locating source code examples that are relevant for their current task. When the developer requests for examples related with a class, method or field declaration, the system generates a structural description of these elements to find source code examples that match that description. PARSEWeb was proposed by Thummalapenta and Xie [11] to suggest examples of method call sequences, to help developers going

---
[2]http://eclipse.org

from a specific object type to a desired object type. These approaches focus on recommending external source code elements that can be reused or used as examples to help developers in their current task. We propose an approach to help developers locating relevant source code elements that exist in their workspace.

Some works retrieve recommendations of relevant artifacts based on the contextual information provided by the history associated to a software development project. Hipikat was developed by Cubranic et al. [5] to recommend relevant artifacts from a project's history. The project memory is built from the various kinds artifacts created during a software development project. These artifacts are recommended to the developer when they are considered relevant for a task being performed. eRose [14], presented by Zimmermann et al., provide recommendations of source code elements that were changed together. The changes applied to the source code are mined from a CVS archive and used to predict further changes, reveal hidden couplings and avoid errors produced by incomplete changes. Our approach is focused in the context of the developer at each moment and do not take into account historical information.

3. KNOWLEDGE BASE

The recommendation system is based on a knowledge base that uses an ontology to represent of the source code structure that is stored in the workspace of a developer (see figure 1). This knowledge base is unique for each developer, being built from the source code files with which the developer is working, and maintained as these files are changed. An earlier version of this model has been used to support the search of source code elements in the workspace [3]. The previous model has been extended with additional relations and some statistical information. The source code structure is represented using two top-level elements, one of them representing structural elements and other representing lexical elements.

The structural elements include classes, interfaces and methods. These elements are connected by a set of relations that represent inheritance (*extensionOf* and *implementationOf*), composition (*attributeOf* and *methodOf*) and behavior (*parameterOf*, *returnOf*, *calledBy* and *usedBy*). This way, we create a representation of the source code structure used in the Java programming language, where the static relations between the different elements become explicit.

The lexical elements represent terms that are extracted from the names of the structural elements. The naming convention used in the Java programming language, known as CamelCase, provides a way to distinguish between the various terms used to create a composed name. These terms are extracted from the name of the structural elements represented in the knowledge base and become associated to those elements using an *indexedBy* relation. Each term can be associated to more than one structural element, because it may be used to compose the name of different elements. The number of times a term is used to index a structural element represents its frequency in the knowledge base and is also stored. When two terms are used together to compose the name of a structural element we create an *associatedWith* relation between them. This relation is used to represent the proximity between the terms, the same way co-occurrence is interpreted as an indicator of semantic proximity in linguistics [6]. The number of times the two terms co-occur

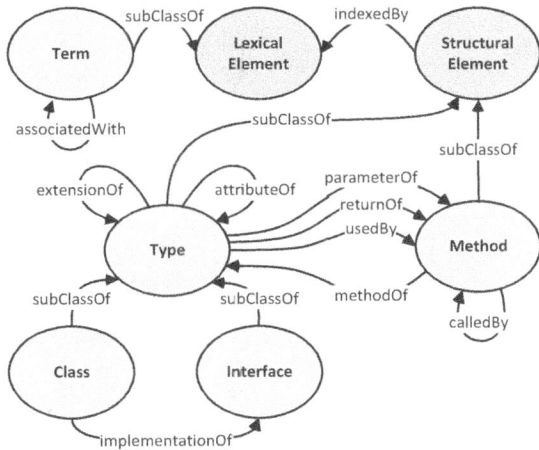

Figure 1: The knowledge base model.

in the names of different structural elements is stored and used as a weight associated to the relation. Terms that co-occur more often have a stronger relation than those that are rarely used together. The lexical elements, and their relations, can be used to find source code elements that are related by the terms that compose their names.

4. CONTEXT MODEL

The context model is based on the source code elements that are more relevant for developers during their work, providing an insight to what is their focus of attention at each moment. The model automatically adapts to the behavior of developers by detecting changes in their focus of attention. This model has been studied before and described in more detail in [4]. The contextual information is implicitly gathered from the interactions of the developer with the source code elements in the workspace. These elements and their relations are represented as the structural context, while the terms used to compose the names of these elements are represented as the lexical context. The relevance of each element is given by a Degree of Interest (DOI), a concept previously introduced by Kersten and Murphy [9].

When the developer interacts with the source code elements in the workspace, these elements are added to the structural context and their DOI is increased according to the type of interaction. When the developer closes or stops interacting with a source code element, its DOI is decreased until it is removed. Based on the source code elements that are more relevant for the developer, we also extract the relevance of the structural relations that are represented in the knowledge base. The relations that exist between the source code elements in the structural context are considered more relevant for the developer. The relevance of these relations is also given by a DOI, which is computed as an average of the DOI of the source code elements that are connected by that relation. The terms extracted from the names of the source code elements that exist in the structural context are added to the lexical context. The relevance of each term is given by its DOI, which is computed as an average of the DOI of the source code elements from which it was extracted.

For instance, imagine a situation where the developer in-

teracts with a class named *ContextModel* and a method of that class named *getContext*. These two elements are added to the structural context and their DOI will reflect the interactions of the developer with that elements over time. Because the two elements are bound by a *methodOf* relation, this relation is also added to the structural context and its DOI is computed as an average of the DOI of the two elements. The terms that comprise the name of the elements, which are *context*, *model* and *get* are added to the lexical context with a DOI computed as an average of the DOI of the elements from where they were extracted.

5. RECOMMENDATION

We have implemented a process that uses the context model described before to retrieve and rank relevant source code elements that are recommended to the developer. Our aim is to help developers reaching the desired source code elements more easily and quickly, decreasing the effort needed to search for that elements in the entire source code structure. This process is based on the fact that most of the source code elements needed by the developer are likely to be structurally and lexically related with the elements that are being manipulated in that moment [4]. The context model plays a central role in this process, providing a mechanism to identify and evaluate the relevance of the source code elements that are being manipulated. We use the structural elements represented in this context model, along with their relations, to retrieve recommendations of elements that are potentially relevant for the developer. These recommendations are then ranked taking into account different components, that represent the retrieval process and the proximity to the context model. Finally, a feedback mechanism allows the system to learn which components are more relevant for the developer and adapt the ranking algorithm in order to favor that components.

5.1 Retrieval

The recommendations are retrieved using the source code elements in the context model with two different approaches, one based on the DOI of the elements and other based on the time elapsed since the elements were last accessed. The first emphasizes the overall importance of the source code artifacts in the context of the developer, while the later privileges the artifacts that were recently accessed by the developer.

The DOI based approach makes use of the relevance of the source code elements that have been manipulated by the developer to identify potentially relevant elements. The recommendations include the top N elements with higher DOI in the context model and all the elements that are structurally related with them. We call N the query size of the recommendation process, and the value of N with which we have achieved the best results is 3 (see section 7), which is the default value used by the system. But this value can also be defined by the developer in specific situations.

The time based approach uses the time, instead of the DOI, to measure the relevance of the source code elements that are being manipulated by the developer. The DOI of an element represented in the context model reflects the relevance of that element during a period of time. But, sometimes, the most relevant elements may not be those with an higher DOI during that period of time, but the ones that have been accessed more recently. The time based approach

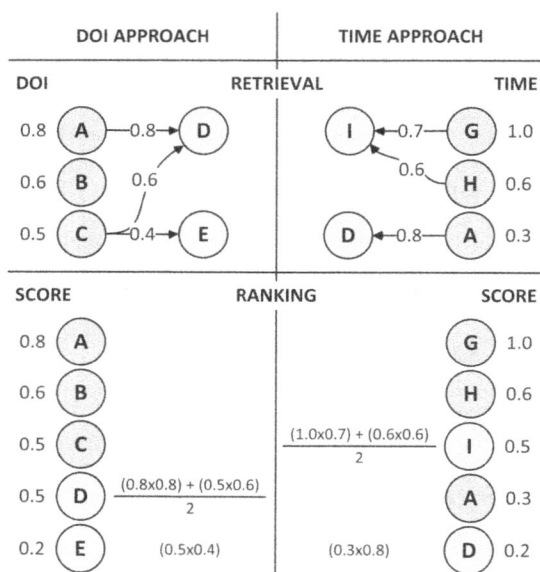

Figure 2: Representation of the retrieval and ranking using the DOI and time based approaches.

favors this aspect, retrieving source code elements that are related with the the elements that have been manipulated more recently. The recommendations retrieved using this approach include the top N elements of the context model that have been accessed more recently and all the elements that are structurally related with them.

A simplified representation of the retrieval process is shown in the upper side of figure 2. The DOI based approach starts by collecting the top 3 source code elements with higher DOI in the context model, shown as elements A, B and C. Then, all the elements that are related with these elements are also retrieved, which are represented by elements D and E. The time based approach first retrieves the 3 elements that were accessed more recently by the developer, represented as elements G, H and A, and then all the elements that are related with them, represented by elements I and D. The values associated with elements A, B and C represent their DOI in the context model. The values associated with elements G, H and A represent the time elapsed since their were accessed, normalized by the time of the most recently accessed element, which is element G. The values associated with the relations represent the DOI of these relations in the context model.

5.2 Ranking

The recommendations retrieved are then ranked taking into account the retrieval process and their proximity to the context model. The final score of each recommendation, which determines its ranking, is computed using four components: DOI, time, structural and lexical. The first two components represent the scores associated with the retrieval process, while the other two represent the scores of the recommendation in relation to the context model. The final score of a recommendation is given by a weighted sum of these scores.

The DOI score represents the score of the elements that were retrieved using the DOI based approach. There are

two types of elements retrieved, those that are in the list of the elements with higher DOI in the context model, and the ones that are structurally related with them. The elements that are retrieved in the list have a score that corresponds to their DOI in the context model. The score of the elements retrieved through a structural relation with the elements in the list is computed using the DOI of the relation and the DOI of the element with which they are related. The DOI of the element in the list is used to normalize the DOI of the structural relation, so that the score of the retrieved element is proportional to the DOI of the element in the list. This way, the score of the retrieved elements take into account the relevance of both the relation and the element that contributed to their retrieval. When an element have a structural relation with more than one of the elements in the list, the score is given by the average of the scores of all the relations. The time score represents the score of the elements that were retrieved using the time based approach. The score is computed in a way that is similar to that used to compute the DOI score. The main difference is that the relevance of each element is computed using the time elapsed since it was last accessed, instead of using its DOI. This time span is normalized by the maximum time span among the elements in the top N list.

As represented in the lower side of figure 2, the elements with higher DOI retrieved in the top N list (A, B and C) have a DOI score correspondent to their DOI in the context model (respectively 0.8, 0.6 and 0.5). Element E was retrieved using a structural relation with element C, thus its DOI score is computed by normalizing the DOI of that relation (0.4) with the DOI of element C (0.5). Finally, element D was retrieved using two structural relations with elements A and C, and its score is computed as an average of the DOI of these relations normalized by the DOI of elements A and C. The same rules apply to the elements retrieved using the time based approach, using the normalized time instead of the DOI of the retrieved elements.

The structural score represents the proximity of the recommendation in relation to the structural context. We define this proximity as a distance between the source code element being recommended and the elements in the structural context. The distance between two source code elements is computed using the structural relations represented in the knowledge base and the relevance of these relations in the structural context. The structural relations allow us to find paths between the source code elements. Instead of using a fixed cost for each relation, the cost of a relation is inversely proportional to the DOI of that relation in the context model. The total cost of a path is given by the sum of the cost of the relations that create the path. This way we take into consideration the current relevance of the structural relations to the developer, assuring that the paths created with relevant relations will have a lower cost. The distance between two source code elements is given by the minimum path cost between those elements. The proximity of the source code element being recommended to the elements in the structural context is computed as an average of the minimum path costs between these elements. We only consider the top 15 elements with higher DOI in the structural context and paths with a maximum of 3 relations, to assure an acceptable performance. The average path cost is normalized, using equation (1), and the structural score is then given by inverting the normalized path cost. The normalization is used to obtain a score value in the interval $[0, 1]$.

The lexical score represents the proximity of the recommendation in relation to the lexical context. We define this proximity as a distance between the terms that comprise the name of the source code element being recommended and the terms in the lexical context. The distance between two terms is computed using the *associatedWith* relations and their weights, which are represented in the knowledge base. These relations allow us to find paths between terms. The cost of these paths is inversely proportional to the weight of the relations that create the path. The weight of each relation represents the frequency of co-occurrence of the two terms and is normalized by the maximum co-occurrence frequency in the knowledge base. By using the weight of the relations between terms to compute the cost of a path, we assure that the paths between terms that co-occur more frequently will have a lower cost. We have also identified a set of terms (such as *get*, *set*, *add*, *to*, *is*, etc) that appear very frequently in the name of the source code elements, especially methods. These very frequent terms co-occur with a variety of other terms and end up connecting almost every term in a distance of a few relations, thus distorting our metric. This problem could be partially solved with a list of very frequent terms that should be ignored, but this would be very limiting, because the top frequent terms vary from workspace to workspace and may include terms that are specific to each workspace. This way, we chose to ignore all the terms that would fall in the top 30% of all term occurrences. The paths that go through one of these terms are discarded. This percentage is based on our observation of a group of knowledge bases from different users, but needs to be further studied. The proximity of the source code element being recommended to the terms in the lexical context is computed as an average of the minimum path costs between terms. Again, we only consider the top 15 terms with higher DOI in the lexical context and paths with a maximum of 3 relations. The average path cost is normalized, using equation (1), and the lexical score is then given by inverting the normalized path cost.

$$1 - \left(\frac{1}{e^x} \right) \qquad (1)$$

5.3 Learning

We have implemented a learning mechanism that uses the recommendations that have been selected by the developer to learn the weights that are associated to the components used in the ranking process. This is not used as a collaborative filtering mechanism, as the learning is limited to the weights used for an individual developer. We use the ranking obtained by the recommendation for each component to find the influence of that component in the final ranking. The final ranking of a recommendation represents its place in the list of all recommendations sorted by their final score, while the ranking of a component represents the place of the recommendation when the list is sorted by the score of that component. The objective is to increase the weights of the components that contribute to promote recommendations and decrease the weights of the components that contribute to demote them.

When the developer selects a recommendation which ranking was influenced by two or more components, the learning

process is initiated. As shown in equation (2), we compute the influence of each component (i_x) using the difference between the final ranking of the recommendation (rf) and the individual ranking obtained for that component (rx). This way, the higher a recommendation is ranked in a specific component, the higher is the influence of that component in the final ranking. The influence obtained for each component is then normalized to the interval $[-1, 1]$, so that the components that had a higher influence get positive values, while those that had a lower influence get negative values. The normalized influence (ni_x) is given by equation (3), where i_{min} and i_{max} represent the minimum and maximum influence among all components.

$$i_x = rf - rx \qquad (2)$$

$$ni_x = 2 \times \left(\frac{i_x - i_{min}}{i_{max} - i_{min}} \right) - 1 \qquad (3)$$

Then, the positive and negative influences must be balanced, so that the weights are increased in the same proportion they are decreased, maintaining their sum as 1. The balanced influence (bi_x) is given by equations (4) and (5), where ni_t represents the sum of the influences in each group and m represents the number of components in each group.

$$bi_x^+ = \frac{ni_x^+}{ni_t^+}; \quad ni_t^+ = \sum_{k=0}^{m^+} ni_k^+ \qquad (4)$$

$$bi_x^- = \frac{|ni_x^-|}{ni_t^-}; \quad ni_t^- = \sum_{k=0}^{m^-} ni_k^- \qquad (5)$$

Because the learning effort needed by the system must depend on how correct is the recommendation, a learning coefficient is applied to the balanced influence. The value of the learning coefficient depends on the final ranking of the recommendation that was selected by the developer. The better the ranking of the recommendation result, the lower the learning effort needed. This way, the learning coefficient (μ) is given by equation (6), where rf is the final ranking of the recommendation and 0.01 is the maximum value for the learning coefficient. Finally, the difference for each weight is obtained by applying the learning coefficient to the balanced influence. The new weight is obtained by adding this difference to the previous weight.

$$\mu = 0.01 \times \left(1 - \left(\frac{1}{e^{0.2 \times rf}} \right) \right) \qquad (6)$$

6. PROTOTYPE

We have developed a prototype that integrates our recommendation approach in Eclipse using a plugin. The recommendations are available through a specific interface, that can be used as an Eclipse View (see 1 in figure 3) or a Window (see 2 in figure 3). The Eclipse View can be integrated in the main window of the IDE and stay visible all the time to keep the recommendations easily accessible. The Window is an alternative that spares space in the IDE, it can be triggered using a combination of keys and is fully functional using only the keyboard. This interface provides recommendations in the form of a list and in what we call a

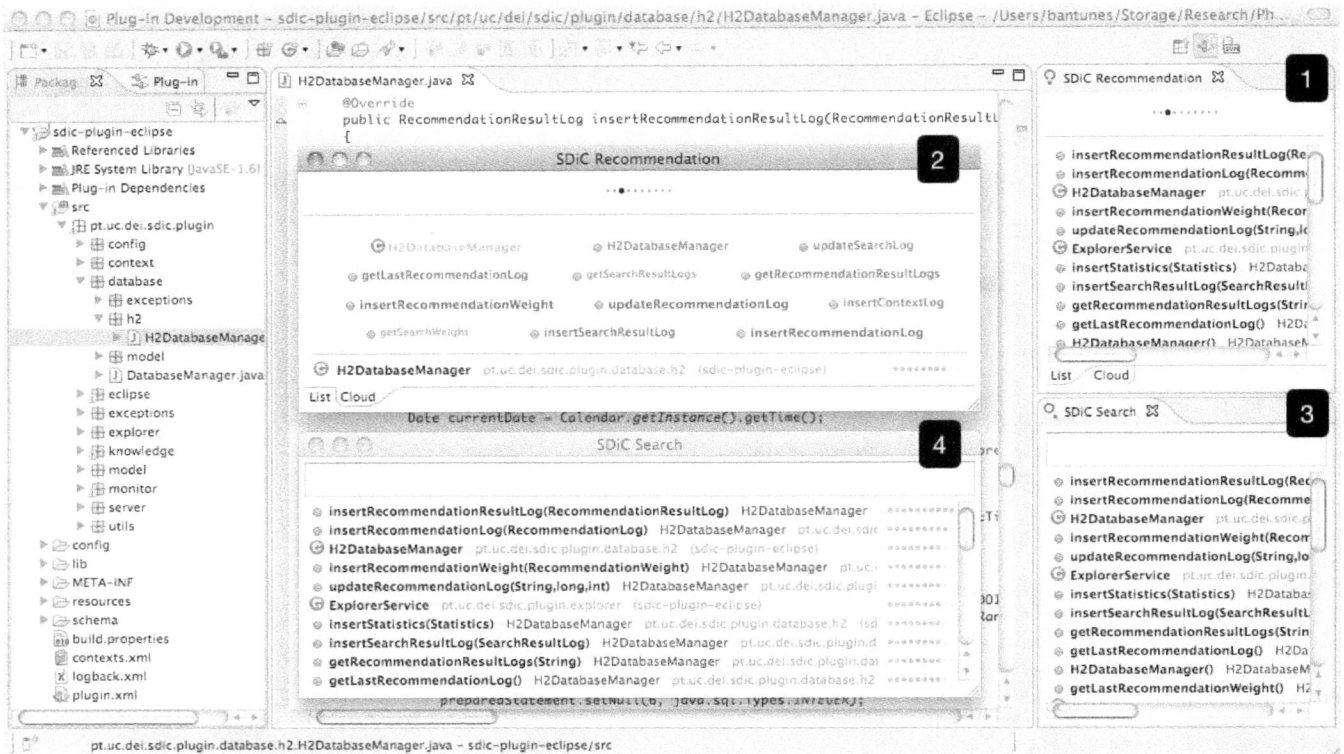

Figure 3: A screenshot of Eclipse running the prototype.

code cloud (see 2 in figure 3). The recommendations are automatically updated as the developer interacts with the IDE, reflecting the changes in the context model. The number of elements of the structural context that are used for the retrieval process, which we call of query size, can be defined by the developer, between 1 and 10, and is shown in the dotted scale in the top of the interface. A maximum of 30 recommendations are presented to the developer each time the recommendations are updated. The prototype also provides a search interface (see 3 and 4 in figure 3), which will not be discussed here, but is important to understand the relevance of our approach. The recommendations were integrated with this interface and are automatically shown to the developer before performing the search. We are assuming that if the recommendations include the source code element desired by the developer, this will avoid the need to perform the search. The query size used for the recommendations shown in the search interface is 3, which is the query size that obtained the best results in our evaluation (see section 7).

7. EVALUATION

We have created an experiment to validate our approach in the field, with developers using the prototype during their work. The experiment was conducted with a group of 5 developers from industry and 10 developers studying computer science, all of them using Eclipse to develop source code in the Java programming language. The application was installed in their work environment and presented as a tool to recommend source code elements inside the IDE.

They have used the application for different periods of time, ranging from 2 weeks to 3 months. The objective of this experiment was to show that our recommendation approach would help developers find the necessary source code elements more quickly and efficiently. More specifically, we wanted to prove that recommendations could be used to avoid the need to perform a search or browse the source code structure to find the needed elements. We also wanted to prove that the context of the developer could be used to identify the most relevant source code elements for the developer and that it would have a positive impact on the ranking of these elements. The results analyzed here refer to the recommendation process only, although being occasionally compared with the search results, for the sake of comprehension.

7.1 Quantitative Study

We wanted to evaluate the capacity of the system in predicting the source code elements that the developers would need in the near future, so that these elements could be proactively recommended to them. Also, we wanted to discover what query size should be used to achieve the better results. This evaluation was performed in the background, by verifying if the source code elements being opened for the first time were already being recommended by the system. This way, we were able to evaluate our approach using the behavior of the developers during their work, without requiring them to use our recommendations. We have implemented a mechanism to store the top 30 recommendations generated by the system with a random query size (between 1 and 10).

Table 1: Percentage of opened source code elements found in recommendations, and average rankings, per query size.

QS	T	P	F	RD	RT	S	L
1	775/1897	40.9%	6.4	7.6	6.4	7.0	5.5
2	998/1867	53.5%	10.0	10.5	10.9	9.7	8.1
3	985/1807	54.5%	12.6	11.7	14.9	11.4	9.7
4	920/1854	49.6%	14.3	13.8	17.5	12.6	10.4
5	863/1838	47.0%	14.6	13.6	18.7	12.1	10.7
6	763/1842	41.4%	15.3	13.9	19.9	12.1	11.5
7	687/1808	38.0%	14.9	13.5	20.2	11.8	11.5
8	630/1770	35.6%	14.8	14.1	20.3	11.8	11.4
9	637/1845	34.5%	14.7	13.6	20.4	11.2	11.2
10	570/1799	31.7%	15.0	14.3	20.1	11.8	11.0
Total	7828/18327	42.7%	13.1	12.5	16.5	11.1	10.0

For each source code element opened for the first time, we have verified if that element was being recommended by the system in that moment. In table 1, we present the results obtained per each query size, where QS represent the query size, T the total number of elements found in recommendations, P the percentage of elements found, F the average final ranking, RD the average retrieval DOI ranking, RT the average retrieval time ranking, S the average structural ranking and L the average lexical ranking. In average, considering all query sizes, 42.7% of the source code elements opened for the first time were already being recommended by the system. The best results were achieved with a query size of 3, with which the system has been able to predict the developer needs in 54.5% of the times. As expected, the results also show that very lower query sizes tend to have worse values, as the number of source code elements used to retrieve the recommendations is not enough to reach the desired element. The higher query sizes also have worse results, which can be explained by the fact that when we increase the number of source code elements in the retrieval process, the recommendations became more dispersed and the probability of finding what the developer needs decreases. We believe that these results are very interesting and show that the context of developers has much to say about their immediate needs. Although the rankings can still be improved, the results also show a tendency for better rankings obtained with the context components, which reveals a positive impact of the context model in the final ranking.

We have also collected information about the recommendations selected by the developers. A total of 348 recommendations were selected. Almost 98% were selected from the recommendations integrated in the search interfaces, which means that the developer had the intention to search for a specific source code element, but the desired element was recommended even before performing the search. When we take into account all the source code elements selected, both from search and recommendation, the recommendations represent about 35% of all selected elements. This means that in 35% of the times in which the developer used our prototype to reach a desired source code element, the need of the developer was satisfied by a recommendation. These numbers are a good indication that our recommendation approach is reducing the effort of the developer on finding

the needed source code elements, avoiding the need to explore the source code structure or perform a search to find these elements. The recommendation exclusive interfaces were almost ignored, especially the *code cloud*, which can be explained by the fact that an interface that provides pro-active recommendations and also allows to perform a search delivers more value to the developer.

The average final ranking of the selected recommendations was 5.4. Although it may still be subject to improvements, we consider that this is a good precision for a recommendation system. The average rankings of the individual components were 9.3 for the DOI component, 9.2 for the time component, 6.6 for the structural component and 5.4 for the lexical component. There is a clear difference between the retrieval and the context components, with the context components obtaining better rankings, showing that the use of context is having a positive impact in the ranking of recommendations. Interestingly, the lexical component achieves better rankings than the structural one, which depicts that the lexical relations between source code elements are contributing to improve the precision of the recommendations. This way, we believe that context should be used not only to retrieve the recommendations, but also to distinguish which can be more relevant to the developer.

7.2 Qualitative Study

The developers were requested to fill an anonymous questionnaire to give their opinion about the recommendations provided by the system. We collected results from 10 out of the 15 developers that used the prototype, which are presented in table 2. The usability of the system was considered good and the relevance of recommendations was rated positively. The utility of the recommendations and the learning mechanism were also rated above the average. The raking of recommendations is something that can be improved and the impact in the productivity could not be verified.

We also asked developers about the things they liked the most and the least, and which suggestions they would give to improve the system. Most of the developers said the recommendations were an easy, fast and useful way to jump into the desired source code elements, avoiding the need to investigate the source code structure, look inside source code files or even perform a search. A few of them said that performing a search was sometimes faster than looking for the desired elements in the list recommendations. Thus, the interface is something that could be improved, in order to find more interesting ways of presenting the recommendations to developers. The capacity of the system in dealing with context transitions is something that could also be improved. Other improvements suggested include using the recommendations in code-completion and to highlight relevant methods directly in the source code.

8. CONCLUSION

We have presented an approach to recommendation in software development that uses a context model of the developer to retrieve and improve the ranking of source code elements that are recommended to the developer. A prototype was experimented by a group of developers. The results obtained are interesting, showing a promising capacity of our recommendation approach in predicting the developer needs, which can be used to reduce the effort of the developer on finding the needed source code elements. The use

Table 2: Questionnaire results per question.

QUESTION	SCALE	AVG	SD
How would you rate the utility of the recommendation functionality?	Very Low (1) - (5) Very High	3.8	0.93
How would you rate the usability of the recommendation functionality?	Very Poor (1) - (5) Very Good	4.3	1.09
How would you rate the impact of the recommendation functionality in your productivity?	Very Low (1) - (5) Very High	3.2	1.26
How would you rate the overall relevance of recommendations?	Very Irrelevant (1) - (5) Very Relevant	4.0	0.45
How often did relevant recommendations appear among all recommendations?	Very Rarely (1) - (5) Very Often	4.0	1.18
How often did relevant recommendations appear well ranked among all recommendations?	Very Rarely (1) - (5) Very Often	3.6	1.16
How would you rate the improvement in ranking of relevant recommendations over time?	Very Low (1) - (5) Very High	3.8	1.71

of the context model is crucial to identify relevant elements and assess their relevance, having a positive impact in the ranking of the recommendations provided to the developer. As future work we plan to improve the recommendation process, especially with regard to accuracy. We need to evaluate the impact of each individual component in the final ranking of recommendations and understand how the retrieval process can be improved to increase the coverage of the algorithm. Because the recommendation of relevant source code elements in the IDE is something the developers are not used to, the interface needs to be improved, seeking innovative ways for providing such recommendations. Also, it is important to evaluate how the time saved using the recommendations compare with other approaches, such as search. Although the results obtained are very promising, there is space for improvements and the approach needs to be validated with a wider range of developers.

9. ACKNOWLEDGMENTS

Bruno Antunes is supported by the FCT scholarship grant SFRH/BD/43336/2008, co-funded by ESF (European Social Fund).

10. REFERENCES

[1] G. Adomavicius and A. Tuzhilin. Toward the next generation of recommender systems: a survey of the state-of-the-art and possible extensions. *Knowledge and Data Engineering, IEEE Transactions on,* 17(6):734 – 749, june 2005.

[2] G. Adomavicius and A. Tuzhilin. Context-aware recommender systems. In F. Ricci, L. Rokach, B. Shapira, and P. B. Kantor, editors, *Recommender Systems Handbook*, pages 217–253. Springer US, 2011.

[3] B. Antunes, J. Cordeiro, and P. Gomes. Context-based search in software development. In *Proc. of the 7th International Conference on the Prestigious Applications of Intelligent Systems (Accepted for publication)*, Montpellier, France, August 2012. IOS Press.

[4] B. Antunes, J. Cordeiro, and P. Gomes. Context modeling and context transition detection in software development. In *Proc. of the 7th International Conference on Software Paradigm Trends (Accepted for publication)*, Rome, Italy, July 2012.

[5] D. Cubranic, G. C. Murphy, J. Singer, and K. S. Booth. Hipikat: A Project Memory for Software Development. *IEEE Transactions on Software Engineering*, 31(6):446–465, 2005.

[6] Z. Harris. Distributional structure. *Word*, 10(23):146–162, 1954.

[7] L. Heinemann and B. Hummel. Recommending api methods based on identifier contexts. In *Proceedings of the 3rd International Workshop on Search-Driven Development: Users, Infrastructure, Tools, and Evaluation*, SUITE '11, pages 1–4, New York, NY, USA, 2011. ACM.

[8] R. Holmes and G. C. Murphy. Using structural context to recommend source code examples. In *Proceedings of the 27th international conference on Software engineering*, ICSE '05, pages 117–125, New York, NY, USA, 2005. ACM.

[9] M. Kersten and G. C. Murphy. Using task context to improve programmer productivity. In *Proceedings of the 14th ACM SIGSOFT International Symposium on Foundations of Software Engineering*, pages 1–11, Portland, Oregon, USA, 2006. ACM.

[10] M. Robillard, R. Walker, and T. Zimmermann. Recommendation systems for software engineering. *Software, IEEE*, 27(4):80 –86, july-aug. 2010.

[11] S. Thummalapenta and T. Xie. Parseweb: a programmer assistant for reusing open source code on the web. In *Proceedings of the twenty-second IEEE/ACM international conference on Automated software engineering*, ASE '07, pages 204–213, New York, NY, USA, 2007. ACM.

[12] F. W. Warr and M. P. Robillard. Suade: Topology-based searches for software investigation. In *Proceedings of the 29th international conference on Software Engineering*, ICSE '07, pages 780–783, Washington, DC, USA, 2007. IEEE Computer Society.

[13] Y. Ye and G. Fischer. Supporting reuse by delivering task-relevant and personalized information. In *Proceedings of the 24th International Conference on Software Engineering*, ICSE '02, pages 513–523, New York, NY, USA, 2002. ACM.

[14] T. Zimmermann, P. Weisgerber, S. Diehl, and A. Zeller. Mining version histories to guide software changes. In *Proceedings of the 26th International Conference on Software Engineering*, ICSE '04, pages 563–572, Washington, DC, USA, 2004. IEEE Computer Society.

[15] G. L. Zuniga. Ontology: Its transformation from philosophy to information systems. In *Proceedings of the International Conference on Formal Ontology in Information Systems*, pages 187–197. ACM Press, 2001.

A Semantic Approach to Recommending Text Advertisements for Images

Weinan Zhang, Li Tian, Xinruo Sun, Haofen Wang, Yong Yu
Dept. of Computer Science and Engineering
Shanghai Jiao Tong University
No. 800, Dongchuan Road, Shanghai, China 200240
{wnzhang, tianli, xrsun, whfcarter, yyu}@apex.sjtu.edu.cn

ABSTRACT

In recent years, more and more images have been uploaded and published on the Web. Along with text Web pages, images have been becoming important media to place relevant advertisements. Visual contextual advertising, a young research area, refers to finding relevant text advertisements for a target image without any textual information (e.g., tags). There are two existing approaches, advertisement search based on image annotation, and more recently, advertisement matching based on feature translation between images and texts. However, the state of the art fails to achieve satisfactory results due to the fact that recommended advertisements are syntactically matched but semantically mismatched. In this paper, we propose a semantic approach to improving the performance of visual contextual advertising. More specifically, we exploit a large high-quality image knowledge base (ImageNet) and a widely-used text knowledge base (Wikipedia) to build a bridge between target images and advertisements. The image-advertisement match is built by mapping images and advertisements into the respective knowledge bases and then finding semantic matches between the two knowledge bases. The experimental results show that semantic match outperforms syntactic match significantly using test images from Flickr. We also show that our approach gives a large improvement of 16.4% on the precision of the top 10 matches over previous work, with more semantically relevant advertisements recommended.

Categories and Subject Descriptors

H.3.3 [**Information Search and Retrieval**]: Search process

General Terms

Algorithms, Experimentation

Keywords

Visual Contextual Advertising, Semantic Matching, Cross-media Mining

1. INTRODUCTION

Nowadays, Web pages no longer contain just textual information. Instead, more and more images have been uploaded and published on the web. For instances, social Web sites like Facebook[1] and Flickr[2] have billions of photo album pages with little text. Compared with the traditional textual Web pages, images become the main contents of these Web pages. Thus, traditional contextual advertising approaches cannot be directly applied to Web pages dominated by images because of the lack of textual information. Therefore, understanding the contents or topics of images and then recommending relevant advertisements based on these images becomes a challenging problem interesting to both academia and industry.

Visual contextual advertising (see Figure 1) refers to finding the most relevant advertisements for a target image without textual information such as tags. It can be regarded as a special case of contextual advertising where images become the context for recommending advertisements. While it is a young branch of contextual advertising, it is more challenging than advertising on textual Web pages because it requires techniques such as computer vision and cross-media transfer learning . In other words, visual contextual advertising aims at semantic matching between two heterogeneous features spaces (i.e., image feature space and text feature space).

Figure 1: An example of visual contextual advertising.

Image annotation [4, 11] is one approach to visual contextual advertising. Intuitively, given a target image, text annotations are extracted based on a model trained by labeled images. Then these annotations are used to search for relevant advertisements, similar to keyword search in traditional contextual advertising. However, since it is time consuming and error prone to obtain high-quality labeled images, the quality of annotations cannot be guaranteed, which leads to poor recommendation performance. On the other hand, since the match process is performed between two heterogeneous feature spaces (i.e., images and text), het-

[1]http://www.facebook.com
[2]http://www.flickr.com

erogeneous transfer learning [30, 9] can be adapted to the image-advertisement match. The state-of-the-art algorithm for visual contextual advertising is ViCAD [8]. It first builds a bridge between the image feature space and text feature space through a feature translation model. Then it uses a method based on a language model to estimate the relevance of each candidate advertisement to the target image. While ViCAD is reported to outperform annotation-based approaches, the advertising precision is still not satisfactory as to be used in real world applications.

With a careful investigation of the performance of previous work, we find that the major weakness of ViCAD as well as the annotation-based approaches comes from mismatches between image tags and text advertisements due to their shortness, ambiguity, and variety. Figure 2 presents some examples which indicate the syntactic mismatch in these approaches. Detailed explanations are as follows.

- **Different term distributions in image tags and text advertisements.** Both ViCAD and annotation-based approaches make use of image-tag co-occurrence data. However, in the image-tag co-occurrence data, if the tag terms (also called text features) translated from target image features are very rarely used or have a different meaning in the advertisement contents (or bid keywords), no advertisement or irrelevant advertisements will be matched.

- **Semantic mismatch between text features and advertisements.** Even if the translated text features are accurate and can syntactically match some advertisements, these advertisements may be semantically irrelevant. This is because current approaches use a syntactic match to retrieve advertisements. Therefore, though the retrieved advertisements contain the image tags, they are irrelevant to the target images.

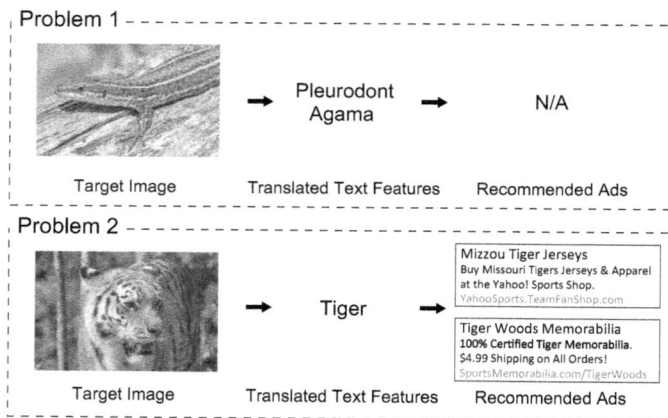

Figure 2: Two main problems in current approaches.

These two problems also occur frequently in traditional Web page contextual advertising. To overcome the syntactic mismatch problem, people use a broad match [12, 15] which finds the semantic relation between different keywords or phrases. In order to optimize the semantic relevance, several semantic approaches to contextual advertising have been proposed [5, 23]. In this paper, we follow this promising direction and propose a semantic approach to tackle the problems of visual contextual advertising. More specifically, we

map the target image to some nodes of interlinked knowledge bases instead of to pure text features. Compared with pure text features, knowledge base nodes have their context and relationships with other relevant nodes, which helps solve the two problems. Using the precision@10 measure on the Flickr test dataset with 230 images, our approach outperforms the syntactic matching approaches by up to 16.4 percent.

To sum up, the contributions of this paper are threefold.

- We identify the problems of syntactic mismatch in existing approaches to visual contextual advertising.

- We propose a knowledge-driven cross-media semantic matching framework to solve these problems. To the best of our knowledge, this is the first work that studies the semantic match between images and text advertisements.

- In the experiment, our approach provides a substantial improvement over the existing approaches, making visual contextual advertising more applicable.

The rest of this paper is organized as follows. In section 2, we discuss related work. In section 3, we present our semantic approach to visual contextual advertising. In section 4, we describe the experiments and analyze the results. Finally, in section 5, we conclude this paper and discuss future work.

2. RELATED WORK

2.1 Contextual Advertising

Contextual advertising refers to placing relevant advertisements on third-party Web pages. The publisher and search engine will share the revenue once any advertisement on their pages is clicked. Studies [28] have shown that the relevance of the advertisements to the content of target pages makes a large difference at the click-through rate. Therefore, the work of matching target Web pages and advertisements is the key point of contextual advertising [23, 6].

Keyword-based approaches [29, 31] are widely used in contextual advertising. They first extract the keywords from a target Web page and then use these keywords to retrieve relevant advertisements just like sponsored search (another kind of Web advertising). However, due to the vagaries of keyword extraction and the lack of content in advertisements, keyword-based approaches always lead to irrelevant advertisements. Besides keyword-based approaches, the authors in [5] imported semantic information to enhance the matching work. They classified both pages and advertisements into a common taxonomy and combined the keyword-based approach with taxonomy matching to rank the advertisements. Moreover, Pak et al. [23] proposed an ESA [14] based approach which makes use of Wikipedia as the knowledge base to improve the performance of contextual advertising. However, they only chose one thousand entities and no link information was used. This work has much room to improve. On the side of efficiency, since analyzing the entire page content is costly and new or dynamically created Web pages cannot be processed to match the advertisements ahead of time, the authors in [2] proposed a summary-based approach to enhance the efficiency of contextual advertising with an ignorable decrease on effectiveness.

2.2 Cross-media Mining

Besides textual content, there are more and more multimedia elements such as images, audio, and video on Web pages. These elements, as pieces of information, are often important and illustrate the topic of a Web page. Mining on these multimedia elements has got considerable attention from both academia and industry. In particular, data mining across different media has become a promising research direction. IJCAI 2009 held a workshop focusing on cross-media information access and mining [1]. Recently, some applications using cross-media mining technologies were developed. Chao et al. [7] proposed TuneSensor, a semantic-driven service to recommend background music for Web photo albums. In contextual video advertising, the systems VideoSense [22] and vADeo [24] have been built based on video content analysis.

Regarding contextual advertising on Web images, image annotation approaches [4, 11] can be leveraged. However, image annotation is not specifically designed for recommending advertisements. The authors of ImageSense [21] first proposed to match advertising with images. But ImageSense mainly used surrounding text for advertisement match while visual relevance acted as a complement to that information. To the best of our knowledge, ViCAD [8] is the only work trying to match advertisements for a target image without any textual information. In ViCAD, the authors built an image-text feature mapping using a graphical model and a language model. Then, the conditional probability of any advertisement for a target image was determined. ViCAD is a very relevant work and will be compared in our experiments.

3. A SEMANTIC APPROACH TO VISUAL CONTEXTUAL ADVERTISING

In the field of contextual advertising, besides the direct syntactic page-ad matching, there are two major frameworks for matching the target Web page and advertisement.

page-keyword-ad
page-taxonomy-ad

In the page-keyword-ad framework [31], advertising keywords are extracted from the target Web page and then advertisements are matched with the keywords. In the page-taxonomy-ad framework [5], pages and advertisements are mapped to the same taxonomic structure and the semantic similarities are calculated using the mapping on the taxonomic hierarchy of pages and advertisements. Besides, the traditional syntactic matching is also combined into this framework.

For visual contextual advertising, the traditional image annotation approach is just like the page-keyword-ad approach and ViCAD corresponds to the syntactic matching. To the best of our knowledge, there is no previous work in visual contextual advertising using any semantic approach. In this section, we propose a semantic approach to visual contextual advertising, with the goal of improving the performance of the advertisement precision.

3.1 Problem Definition

First we formally define the problem of visual contextual advertising. Let $\mathcal{T} = \{t_1, t_2, \ldots, t_m\}$ be the text feature space, where t_i is a text feature and m is the size of the text feature space. Let $\mathcal{A}$ be the advertisement space and each advertisement $a \in \mathcal{A}$ is represented by a text feature vector $(t_a^1, t_a^2, \ldots, t_a^m)$, where t_a^k is the frequency of text feature t_k in a. Similarly, we denote image feature space $\mathcal{V} = \{v_1, v_2, \ldots, v_n\}$, where v_i is an image feature and n is the size of image feature space. The image space is denoted as $\mathcal{I}$. And each image $\iota \in \mathcal{I}$ is represented by an image feature vector $(v_\iota^1, v_\iota^2, \ldots, v_\iota^n)$, where v_ι^k means the frequency of image feature v_k in ι. In addition, the text knowledge base is denoted as $\mathcal{O}_t = \{O_t, E_t\}$, where $O_t = \{ot_1, ot_2, \ldots, ot_\mu\}$ is the node set and $E_t = \{(ot_i, ot_j)\}$ is the edge set. Also the image knowledge base is defined as $\mathcal{O}_v = \{O_v, E_v\}$, where $O_v = \{ov_1, ov_2, \ldots, ov_\nu\}$ and $E_v = \{(ov_i, ov_j)\}$. For a given image $\iota \in \mathcal{I}$, the objective is to find the function $r(\iota, a) : \mathcal{I} \times \mathcal{A} \mapsto \mathbb{R}$ that accurately estimates the relevance of any candidate advertisement a to ι.

3.2 Semantic Visual Contextual Advertising Framework

In this subsection, we discuss the framework of semantic visual contextual advertising. As mentioned in Section 1, we first map the image and advertisement onto some nodes of interlinked knowledge bases. Since the feature spaces of image and text are heterogeneous, the image and text knowledge bases are always different. However, just like the image-text occurrence data, we can still find a way to match the nodes on the two knowledge bases[3]. We propose a framework for semantic matching of images and advertisements by building links between nodes of the image and text knowledge bases.

image-knowledge_{image}-knowledge_{text}-ad

In this framework, first, images and advertisements are mapped to nodes in the image and text knowledge bases, respectively. Then the matching between the nodes of interlinked image and text knowledge bases is processed. With the help of semantic link information in the knowledge bases, syntactic mismatches between the image features and text features can be reduced. Therefore, given a target image ι, the task of finding the best match advertisement can be written as

$$\arg \max_{a \in \mathcal{A}} \mathcal{M}(\psi(\iota), \phi(a)), \tag{1}$$

where

$$\psi(\iota) = \{(ov, \omega_{ov})\}_{ov \in O_v \ and \ \omega_{ov} > 0}, \tag{2}$$

$$\phi(a) = \{(ot, \omega_{ot})\}_{ot \in O_t \ and \ \omega_{ot} > 0}, \tag{3}$$

with

$$\sum_{ov \in O_v} \omega_{ov} = 1 \ and \ \sum_{ot \in O_t} \omega_{ot} = 1.$$

Here ψ and ϕ are the functions mapping text instances to nodes in the text knowledge base and image instances to nodes in the image knowledge base, respectively. Each mapped node is assigned a weight to express its relevance to the image or advertisement. $\mathcal{M}$ is a cross-knowledge base matching function for the two sets of weighted nodes on the combined structure of image and text knowledge bases. To sum up, our framework can be depicted as Figure 3.

---

[3]For ontology engineering, one of the most important processes is to find the match between two ontologies.

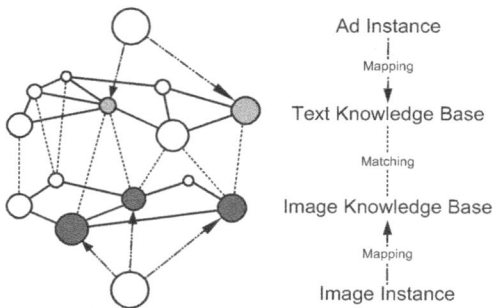

Figure 3: Framework of semantic matching between the image and advertisement.

The specific implementation of the functions in Equation 1 depends on the knowledge used, which will be discussed in detail later.

3.3 Knowledge Bases

In this section, we introduce the specific knowledge bases we use in this framework and the bridging knowledge data between the nodes of two knowledge bases.

Text Knowledge Base: Wikipedia. We use *Wikipedia* as the text knowledge base $\mathcal{O}_t$ in our framework. Wikipedia is a user-contributed online encyclopedia. It contains numerous entities with a formatted article description and interlinks to other relevant entities. Each entity article is written or revised by Web users so as to lead a comprehensive description of the entity. In addition, the interlinks between each two entities serve as auxiliary information and further explanation, which indicates their semantic relatedness. In sum, Wikipedia is a large-scale qualified knowledge base: so far in March 2012, it has more than 3.9 million articles written in English, with 19.67 edits for each article[4].

Image Knowledge Base: ImageNet/WordNet. We choose *ImageNet* [10] as the image knowledge base $\mathcal{O}_v$ in our framework. ImageNet is an image database organized according to *WordNet* [13]. WordNet is composed of synsets, each of which is described by several synonyms. The edges linking two synsets provides the semantic relation between them. The kinds of edges include: `antonym`, `hypernym`, `instance hyponym`, `part meronym`, `derivationally related form`, `member of this domain`, and so on. We regard synsets as concept nodes. The `hypernym` edges are used to construct a node hierarchy. Currently, there are 14.2 million images and 21.8 thousand nodes indexed in ImageNet[5]. Each node is assigned 1000 images on average. Images of each concept are human-annotated and have high quality. Therefore, using ImageNet, each node $ov_i \in \mathcal{O}_v$ is represented as a set of images.

Bridging Knowledge: YAGO. We connect Wikipedia nodes and ImageNet/WordNet nodes using YAGO [27]. Each Wikipedia node is labeled with types in YAGO's taxonomy, which is built on the topology of WordNet. Thus we can obtain a list of WordNet nodes for each Wikipedia node. For example, `Wikipeida::Aristotle` has the type of `WordNet::Person`, `WordNet::Scientist`, etc.

[4]`http://en.wikipedia.org/wiki/Special:Statistics`
[5]`http://image-net.org/about-stats`

3.4 Mapping and Matching Functions

In this section we introduce the definition of the functions in our framework, based on the image and text knowledge bases discussed in Section 3.3.

3.4.1 Text Mapping Function ϕ

Given a candidate advertisement in bag-of-words form, ϕ maps a to the relevant nodes on $\mathcal{O}_t$. Here $\mathcal{O}_t$ represents the set of Wikipedia entities each with an article description. Because advertisement content is in short-text and diversely written, it is usually difficult to directly find Wikipedia entity names in advertisement content. For this reason, we make use of explicit semantic analysis (ESA) [14] to find the most relevant Wikipedia entities for each candidate advertisement.

Here each mapped node weight ω_{ot} in Equation 3 is defined by the ESA association strength. A widely used choice [26] is to select tfidf weighting

$$\omega_{ot} = \sum_{t \in ot} tfidf_{ot}(t) \cdot tf_a(t), \qquad (4)$$

where $tfidf_{ot}(t)$ is the product of the frequency and inverse document frequency of t in the article of ot, $tf_a(t)$ is the frequency of a word or phrase t in the advertisement dataset. Particularly, top-3 weighted nodes are selected.

3.4.2 Image Mapping Function ψ

Given a target image $\iota = \{v_\iota^1, v_\iota^2, \ldots, v_\iota^n\}$, ψ maps ι to the relevant nodes on $\mathcal{O}_v$. Different from mapping advertisement content to nodes of Wikipedia as ϕ, the image mapping function ψ is closer to multi-label classification. Each node in ImageNet has about 1000 image instances; these are used as the training data and the target image is regarded as test data. Specifically, we use a node-level centroid based similarity function $\theta(\iota, ov)$ to obtain the closest k node set C_k to the target image ι. Specifically, θ can be implemented as cosine similarity after the process of principle component analysis (PCA) [18]. Moreover, since ImageNet has a hierarchical structure, we can implement a hierarchical centroid algorithm which leverages the ancestor information in the similarity calculation. Finally, the weight ω_{ov} for the mapped node ov is defined by the (normalized) similarity between ι and ov, calculated as

$$\omega_{ov} = \left(\theta(\iota, ov) \prod_{ov' \in A(ov)} \theta(\iota, ov')^\omega \right)^{\frac{1}{|A(ov)|\omega + 1}}, \qquad (5)$$

where $A(ov)$ denotes the set of ancestors of ov and ω is the weight assigned to each ancestor node; these are combined with a geometric mean. With leveraged ancestor information, ψ is less likely to map ι to irrelevant nodes. Particularly, we set $k = 7$ and $\omega = 0.6$ in our experiment, after preliminary parameter tuning.

3.4.3 Cross-Knowledge Base Matching Function $\mathcal{M}$

Above we have elaborated the text and image mapping functions which map the advertisements and images to Wikipedia and ImageNet/WordNet. In addition, these two knowledge bases could be bridged via YAGO (Section 3.3). Thus we can regard them as a combined knowledge base. Now we introduce the matching function $\mathcal{M}$ between the two disjoint sets of weighted nodes on the combined knowledge base. Here we uniformly use o_i to represent ov_i and ot_i since the two knowledge bases have been combined. We also define the mapped node sets of image and advertisement as O_ι and O_a respectively. We discuss two implementations of the cross-knowledge base matching function $\mathcal{M}$.

LOD Description Overlap (LODDO). This is an approach proposed in [32] for evaluating named entity semantic relatedness on linked open data (LOD). The authors propose to regard the neighborhood of an entity o in LOD as its description $\delta(o)$, defined as the set of entities linked to o. And the similarity between entity o_i and o_j is defined as the description overlap between $\delta(o_i)$ and $\delta(o_j)$.

$$LODDO(o_i, o_j) = \frac{|\delta(o_i) \cap \delta(o_j)|}{\min(|\delta(o_i)|, |\delta(o_j)|)} \qquad (6)$$

Since Wikipedia and ImageNet/WordNet are also members of LOD, this approach can seamlessly be adapted to our matching function. The matching function between the target image ι and a candidate advertisement a can be calculated as the weighted average of the similarity of each image-advertisement entity pair.

$$\mathcal{M}(\psi(\iota), \phi(a)) = \sum_{o_i \in O_\iota} \sum_{o_j \in O_a} \omega_{o_i} \omega_{o_j} LODDO(o_i, o_j) \qquad (7)$$

Hierarchy-based Matching. Taxonomy-based semantic matching has been used in contextual advertising [5]. As has been mentioned in Section 3.3, YAGO does provide a shared taxonomy between Wikipedia and WordNet. Thus we can map the nodes in both knowledge bases to a common taxonomy hierarchy, where we can implement hierarchy-based matching. Matching function $\mathcal{M}$ can be written as

$$\mathcal{M}(\psi(\iota), \phi(a)) = \left(\sum_{o_i \in O_\iota} \sum_{o_j \in O_a} \omega_{o_i} \omega_{o_j} LCA(o_i, o_j) \right)^{-1}, \qquad (8)$$

where $LCA(o_i, o_j)$ means the maximal path length from o_i and o_j to their least common ancestor [5].

In the experiment, we will compare the above two cross-knowledge base matching functions to explore how to provide cross-media semantic matching appropriately.

3.5 Algorithm Chart

So far we have introduced our framework of semantic matching between an image ι and an advertisement a. Now the practical task is to retrieve and rank the relevant advertisements for a given image $\iota,$. Since expansion and matching of graph structures are involved in our matching algorithm, it is very inefficient to traverse the advertisement dataset to perform a match between each advertisement and the target image. Here we propose the algorithm flow to efficiently solve the problem (see Figure 4).

In an offline process, we pre-calculate a set of relevant Wikipedia nodes for each advertisement a using ESA. Thus we can build an inverse advertisement index for each node, like the document index to each keyword in a search engine. For the online process, with a target image ι as input, first we use image mapping function ψ to get k ImageNet nodes $\psi(\iota)$. Then we link mapped ImageNet nodes to Wikipedia nodes via YAGO. With the advertisement index above, we can retrieve the indexed advertisements for each linked Wikipedia node, which lead to the candidate advertisement list

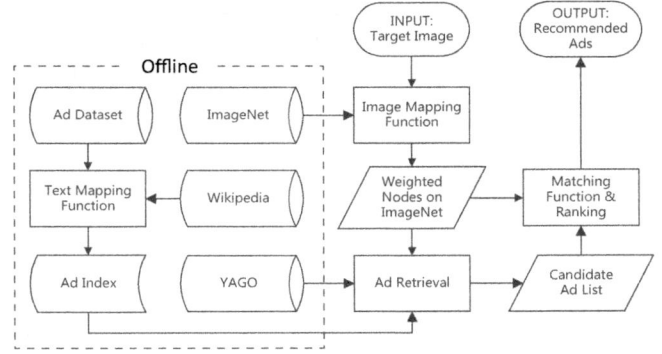

Figure 4: Algorithm Flow Chart.

$\mathcal{L}$. For each advertisement a in $\mathcal{L}$, calculate the similarity with ι via ontology matching function $\mathcal{M}$. Finally, rank the candidate advertisement list in descending order by the similarity score, get the top N advertisements as the output.

3.6 Complexity

For the image mapping process, each image-centroid similarity function θ takes $O(|\mathcal{V}|)$. Thus the image mapping function takes $O(|O_v| \cdot |\mathcal{V}| + |O_v| \log |O_v|)$, where the notation is as in Section 3.1. In practice, the image feature number $|\mathcal{V}|$ is much larger than $\log |O_v|$. Thus the complexity of the image mapping function is $O(|O_v| \cdot |\mathcal{V}|)$.

For the matching process, let n_a be the maximum number of advertisements one Wikipedia node could retrieve, D be the maximum out-degree of WordNet nodes in YAGO. Thus the maximum number of candidate advertisements is $k_\iota \cdot n_a \cdot D$. For the LODDO matching function, the complexity is $O(k_\iota \cdot k_a \cdot N_\delta \log N_\delta)$, where N_δ is the description size. For the hierarchy-based matching function, the complexity is $O(k_\iota \cdot k_a \cdot d)$, where d is the depth of the YAGO taxonomy hierarchy, k_ι and k_a are the maximum number of mapped nodes for images and advertisements respectively. Thus the complexities of the matching processes are $O(n_a \cdot k_\iota^2 \cdot k_a \cdot D \cdot N_\delta \log N_\delta)$ and $O(n_a \cdot k_\iota^2 \cdot k_a \cdot d)$ respectively. In practice, $D \cdot N_\delta \log N_\delta)$ and d are not large numbers ($d < 15$ and $N_\delta < 30$). Uniformly, we use c to denote the upper bound of these two numbers and the matching complexity is $O(n_a \cdot k_\iota^2 \cdot k_a \cdot c)$.

To sum up, the overall complexity of the online algorithm is $O(|O_v| \cdot |\mathcal{V}| + n_a \cdot k_\iota^2 \cdot k_a \cdot c)$.

In our experiment, the average real run time for each test case is 0.751 seconds on a machine with an Intel(R) Core-2(TM) Quad Q8400 CPU with 2 cores at 2.6GHz and 2GB memory. Furthermore, the efficiency can be further improved with the optimization such as parallelization in the image ontology match process and advertisement index pruning.

4. EXPERIMENT

In this section, we introduce the datasets, compare algorithms and evaluation measures, and finally report and discuss our experimental results.

4.1 Datasets

4.1.1 Advertisement Dataset

The textual advertisements can be crawled from a mainstream commercial search engine. Specifically, we use AOL query log [16] as query set and then crawl the delivered advertisements on the search engine result page (SERP) for each query during March 2011. Specifically, there are 9,954,130 queries in the AOL dataset, where 1,118,729 queries attract at least one advertisement. As a result, we collect 1,607,688 unique advertisements.

For each advertisement, we crawl its title, creative, and display URL, as has been shown in Figure 1.

4.1.2 Knowledge Bases

As has been discussed in Section 3.2, there are text and image knowledge bases (Wikipedia and ImageNet/WordNet) and bridging data (YAGO).

Wikipedia - We obtained the Wikipedia dump of Jan.5, 2012. We selected the Wikipedia articles representing concrete concepts using heuristics similar to [14], resulting in a collection of 1,521,080 concept nodes. We use Lucene[6] to build the ESA index from articles describing the concepts.

ImageNet/WordNet - For WordNet structure, we download WordNet 3.0[7] and remove the edges with negative semantics (Antonym). The knowledge base contains 117,659 nodes and 377,592 edges, where 97,666 are Hypernym edges.

For the image data, we take the 1,000 ImageNet synsets released on April 30, 2010 which contain 2,522,812 images. Each image in this dataset has SIFT features extracted and 1000-clustered bag of words. To investigate whether the size of the image knowledge base is large enough to provide relevant advertisements, we will drive an experiment about the performance against the number of ImageNet nodes in Section 4.4.2.

YAGO - To connect Wikipedia and ImageNet/WordNet, we take YAGO dataset of type_star in version yago2core-20120109. On average, each Wikipedia concept is mapped to 25.2 WordNet concepts. In all, 4,564 WordNet concepts have at least one corresponding Wikipedia concept[8].

4.1.3 Target Image Dataset

In our experiment, we use a Flickr image set as our target dataset. This dataset contains 521 thousand images crawled from Flickr during 2010. Considering the large effort of human judgement, we randomly selected 230 images as the target images for testing[9].

The data preprocessing is the same as ImageNet. First we detect the interesting points for each image using SIFT descriptors [19]. Then we cluster 1,000 categories (same as [25]) for all interesting points to obtain a codebook, which turns out to be the image feature space and each image can be represented by image-bag-of-words. These image features are used in the similarity function $\theta(\iota, ov)$.

4.2 Compared Algorithms

Since there are few methods for visual contextual advertising except ViCAD, we compared all the methods that work [8]. The algorithms are listed below.

Annotation + Search (AS). First, the target image is annotated [20]. Then advertisements are retrieved and ranked by a search process using the annotations as query.

---

[6] http://lucene.apache.org/

[7] http://wordnet.princeton.edu/wordnet/download/

[8] Although the ratio of involved WordNet concepts is not high, these concepts are usually the representative category labels, which have links to most of WordNet Concepts.

[9] As a reference, 200 test images were selected in the experiment of previous work [8].

The search engine is built based on Lucene. This work is just like the keyword-based methods used in traditional contextual advertising.

Annotation + Expansion + Search (ASEx). One intuitive approach to adding semantic matching into the traditional AS approach is to expand the extracted annotations using a semantic knowledge base and then search the advertisements with the expanded query set. Specifically, we implement ASEx similar to the work [17].

ViCAD. The heterogeneous transfer learning based ViCAD proposed in [8] has been discussed in Section 2.2.

ImageAdSense. This is our approach and the algorithm has been discussed in Section 3. In order to compare different matching functions, we implement LODDO and the hierarchy-based matching function mentioned in Section 3.4.3, denoted as iAdSense-LODDO and iAdSense-Tree. In order to investigate the impact of a cross-knowledge-base matching function, here we add an algorithm iAdSense-OneLayer, which only has ImageNet/WordNet. The mapping of advertisements to WordNet nodes is based on syntactic match.

4.3 Evaluation Measure

The input of the experiment is a target image ι and the output is k advertisements for ι. As the basis of the evaluation work, we invited six college students to judge the relevance of each image-advertisement pair as below.

- *Relevant.* The advertisement is relevant to the content of the target image, scored as 1.

- *Irrelevant.* The advertisement is not considered relevant to the content of the target image, scored as 0.

Each image-advertisement pair has at least two human judges. Then, we averaged the scores for each image-advertisement pair. Then we evaluated the performance of the algorithms using $P@n$ as the evaluation measure. Precision at position n ($P@n$) is defined to be the fraction of the top-n retrieved advertisements that are relevant [3].

$$P@n = \frac{\sum_{i=1}^{n} \pi_i}{n} \qquad (9)$$

In Equation 9, π_i denotes the average rate score for the pair of the target Web page and the ith recommended advertisements. Since we cannot evaluate every image-advertisement pair, there is no good measure to evaluate the recall of each approach.

4.4 Experimental Results

4.4.1 Overall Performance Analysis

In the first part of the experiment, we judge the overall recommendation performance of the compared algorithms on test dataset. For the 230 test images, each algorithm recommends 10 top ranked advertisements. We use the evaluation measure $P@n$ (see Section 4.3) for the recommendation performance. The result for six algorithms is provided in Figure 5.

From Figure 5 we can have the following observation. (i) Three iAdSense-algorithms provide much better performance than AS, ASEx and ViCAD. The absolute improvement of P@10 of iAdSense-LODDO is 16.4% and 20.7%, compared with ASEx and ViCAD respectively, which verifies the

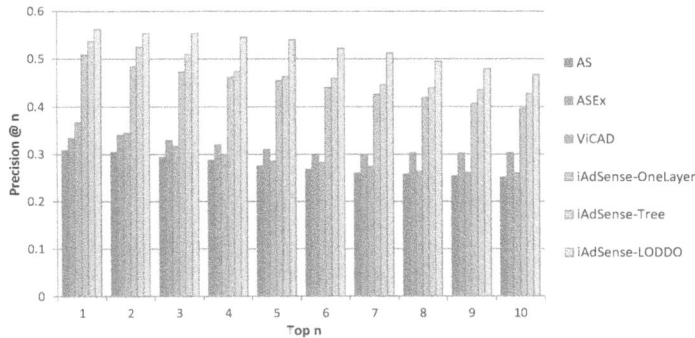

Figure 5: Figure representation of $P@n$ results of all compared algorithms on test dataset.

impact of semantic matching. (ii) In the comparison among these three iAdSense-algorithms, iAdSense-LODDO performs the best. This indicates that the semantic relatedness approach LODDO is well adapted to our framework. iAdSense-Tree has a little lower precision. This is because only hierarchy edges are used in iAdSense-Tree, while iAdSense-LODDO makes use of all edges of each node to provide a more comprehensive semantic description. iAdSense-OneLayer is not as good as others with two layers. This indicates the necessity of semantic text mapping. Syntactically mapping advertisement content to its words in WordNet will import much ambiguity since each word always occurs in several WordNet synsets. (iii) ViCAD outperforms AS but is not as good as iAdSense algorithms. The reason ViCAD is not as good as iAdSense-algorithms is the frequent noise in the tags of training images, which reduces the accuracy of cross-domain feature transferring. In addition, ViCAD is also a syntactic match approach and has the same problems as AS.

To sum up, the above comparison shows that iAdSense is more effective than previous approaches.

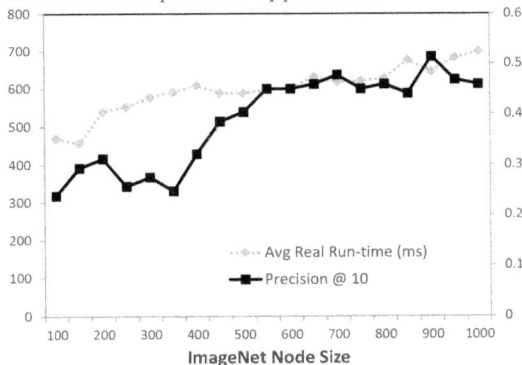

Figure 6: $P@10$ of iAdSense against the number of randomly selected ImageNet nodes.

4.4.2 ImageNet Scale Analysis

As mentioned in Section 4.1.2, we should investigate the recommendation performance of iAdSense against the number of ImageNet nodes and check whether it is enough to take the 1,000 ImageNet synsets released with SIFT features. Specifically, we vary the number of **randomly selected** ImageNet nodes from 50 to 1,000 with a step of 50. Then we evaluate the performance of iAdSense-LODDO in the same way as above[10]. The result is shown in Figure 6.

---

[10]Due to the huge human labeling effort, the test set here is a subset of the test dataset.

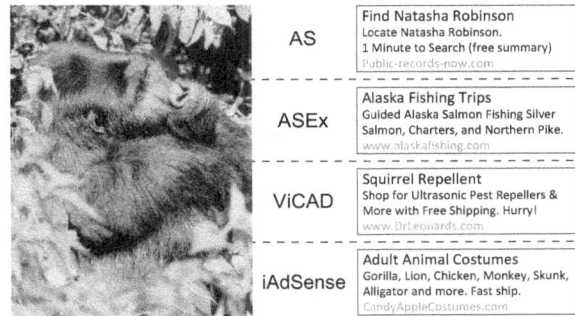

Figure 7: Advertisements recommended by the compared algorithms on one case.

From Figure 6 we can see that (i) as the number of nodes increases, the $P@10$ performance of iAdSense improves and its real run-time increases. (ii) The precision curve has a sigmoid-shaped trend: $P@10$ fluctuates without an obvious increase when number of nodes varies from 50 to 350; in the range of [350, 700], $P@10$ increases rapidly; after 700, $P@10$ fluctuates around 0.475. (iii) The real run-time curve has a stable increase rate against the number of ImageNet nodes. This is because the image mapping process is implemented as a hierarchical centroid algorithm, a memory-based approach, so more ImageNet synsets will surely bring an efficiency decrease. To sum up, 1,000 ImageNet size is a suitable scale for iAdSense considering both effectiveness and efficiency.

4.4.3 Case Study

Here we demonstrate a case that makes a difference among the compared algorithms. Figure 7 provides some advertisements recommended by the four algorithms for a target image about a gorilla. From the results we can find AS recommends an irrelevant advertisement. For ASEx, there is a topic drift between the target image and advertisement, which is caused by annotation expansion. ViCAD recommends a syntactic match advertisement. However, *squirrel* here refers to a brand name instead of a kind of animal, which is a case of semantic mismatch. iAdSense-LODDO recommends a suitable advertisement, where *Gorilla* in the advertisement refers to the animal in the target image.

Finally, we provide more cases of the results of semantic visual contextual advertising with respect to the test dataset. In Figure 8, there are two advertisements listed on the right of each target image. These advertisements are recommended by algorithm iAdSense-LODDO. More demonstrations are presented on the Web site of ApexLab[11].

5. CONCLUSION AND FUTURE WORK

We investigate the current work on visual contextual advertising and point out the problems of semantic mismatch despite a syntactic match between image and advertisement content. In order to solve these problems, we proposed a semantic approach named iAdSense with the help of text and image knowledge bases. In the experiment, iAdSense provides an improvement of 16.4% over the previous approaches, with more semantically relevant advertisements recommended.

In future work, we will explore other knowledge bases to help in this framework. For example, we can use a more com-

---

[11]Online demo. `http://iadsense.apexlab.org`

Figure 8: Case study of iAdSense-LODDO results.

mercially relevant knowledge base to explore a better advertisement mapping. Moreover, we will work on the application of visual contextual advertising to E-commerce such as Taobao[12]. The input will be a product image and some relevant products will be recommended. In this topic, more specific image features will be selected and more information can be obtained from the product pages.

6. REFERENCES

[1] Ijcai-09 workshop on cross-media information access and mining.

[2] A. Anagnostopoulos, A. Broder, E. Gabrilovich, V. Josifovski, and L. Riedel. Just-in-time contextual advertising. *CIKM*, 2007.

[3] Baeza-Yated, R., Ribeiro-Neto, and B. *Modem Information Retrieval*. Addison-Wesley Longman Publishing Co., Inc, Boston, MA, 2008.

[4] K. Barnard, P. Duygulu, D. Forsyth, N. D. Freitas, D. M. Blei, M. I. Jordan, J. K, T. Hofmann, T. Poggio, and J. Shawe-taylor. Matching words and pictures. *Journal of Machine Learning Research*, 3:1107–1135, 2003.

[5] A. Broder, M. Fontoura, V. Josifovski, and L. Riedel. A semantic approach to contextual advertising. In *Proceedings of the 30th annual international ACM SIGIR conference on Research and development in information retrieval*, SIGIR '07, pages 559–566, New York, NY, USA, 2007. ACM.

[6] D. Chakrabarti, D. Agarwal, and V. Josifovski. Contextual advertising by combining relevance with click feedback. In *Proceeding of the 17th international conference on World Wide Web*, pages 417–426. ACM, 2008.

[7] J. Chao, H. Wang, W. Zhou, W. Zhang, and Y. Yu. Tunesensor: A semantic-driven music recommendation service for digital photo albums. In *Proceedings of 16th International Semantic Web Conference*, 2011.

[8] Y. Chen, O. Jin, G. rong Xue, J. Chen, and Q. Yang. Visual contextual advertising: Bringing textual advertisements to images. In *Proceedings of the 24th AAAI Conference*, AAAI'10, 2010.

[9] W. Dai, Y. Chen, G.-R. Xue, Q. Yang, and Y. Yu. Translated learning: Transfer learning across different feature spaces. In *NIPS'08*, pages 353–360, 2008.

[10] J. Deng, W. Dong, R. Socher, L. jia Li, K. Li, and L. Fei-fei. Imagenet: A large-scale hierarchical image database. In *In CVPR*, 2009.

[11] P. Duygulu, K. Barnard, N. de Freitas, P. Duygulu, K. Barnard, and D. Forsyth. Object recognition as machine translation: Learning a lexicon for a fixed image vocabulary. 2002.

[12] E. Even Dar, V. S. Mirrokni, S. Muthukrishnan, Y. Mansour, and U. Nadav. Bid optimization for broad match ad auctions. In *Proceedings of the 18th international conference on World wide web*, WWW '09, pages 231–240, New York, NY, USA, 2009. ACM.

[13] C. Fellbaum. Wordnet: An electronic lexical database. In *Proceedings of 11th Eurographics Workshop on Rendering*. MIT Press, 1998.

[14] E. Gabrilovich and S. Markovitch. Computing semantic relatedness using wikipedia-based explicit semantic analysis. In *Proceedings of the 20th international joint conference on Artifical intelligence*, pages 1606–1611, 2007.

[15] S. Gupta, M. Bilenko, and M. Richardson. Catching the drift: learning broad matches from clickthrough data. In *Proceedings of the 15th international conference on Knowledge discovery and data mining*, 2009.

[16] IAB and PricewaterhouseCoopers. Iab internet advertising revenue report, 2010 full year results, april 2011.

[17] N. James and C. Hudelot. Towards semantic image annotation with keyword disambiguation using semantic and visual knowledge. In *Proceedings of IJCAI-09*.

[18] I. Jolliffe. Principal component analysis. In *Encyclopedia of Statistics in Behavioral Science*, 2005.

[19] D. G. Lowe. Distinctive image features from scale-invariant keypoints. *Int. J. Comput. Vision*, 60:91–110, November 2004.

[20] A. Makadia, V. Pavlovic, and S. Kumar. A new baseline for image annotation. In *Proceedings of the International Conference on Computer Vision*, pages 316–329, 2008.

[21] T. Mei, X. Hua, and S. Li. Contextual in-image advertising. In *Proceeding of the 16th ACM international conference on Multimedia*, pages 439–448. ACM, 2008.

[22] T. Mei, X. Hua, L. Yang, and S. Li. Videosense: towards effective online video advertising. In *Proceedings of the 15th international conference on Multimedia*, pages 1075–1084. ACM, 2007.

[23] A. Pak and C. Chung. A wikipedia matching approach to contextual advertising. *World Wide Web*, 13(3):251–274, 2010.

[24] S. Sengamedu, N. Sawant, and S. Wadhwa. vadeo: video advertising system. In *Proceedings of the 15th international conference on Multimedia*, pages 455–456. ACM, 2007.

[25] J. Sivic, B. C. Russell, A. A. Efros, A. Zisserman, and W. T. Freeman. Discovering object categories in image collections. In *Proceedings of the International Conference on Computer Vision*, 2005.

[26] P. Sorg and P. Cimiano. An experimental comparison of explicit semantic analysis implementations for cross-language retrieval. *Natural Language Processing and Information Systems*, pages 36–48, 2010.

[27] F. Suchanek, G. Kasneci, and G. Weikum. Yago: a core of semantic knowledge. In *Proceedings of the 16th international conference on World Wide Web*, pages 697–706. ACM, 2007.

[28] C. Wang, P. Zhang, R. Choi, and M. Eredita. Understanding consumers attitude toward advertising. In *Eighth Americas Conference on Informatino System*, pages 1143–1148, 2002.

[29] X. Wu and A. Bolivar. Keyword extraction for contextual advertisement. In *Proceedings of the 18th World Wide Web Conference*, pages 1195–1196, 2008.

[30] Q. Yang, Y. Chen, G. rong Xue, W. Dai, and Y. Yu. Heterogeneous transfer learning for image clustering via the social web. In *In Proc. of the 47th Annual Meeting of the ACL and the 4th IJCNLP of the AFNLP*, 2009.

[31] W.-t. Yih, J. Goodman, and V. R. Carvalho. Finding advertising keywords on web pages. In *Proceedings of the 15th international conference on World Wide Web*, WWW '06, pages 213–222, New York, NY, USA, 2006. ACM.

[32] W. Zhou, H. Wang, C. Jiansong, W. Zhang, and Y. Yu. Loddo: Using linked open data description overlap to measure semantic relatedness between named entities. In *Proceedings of Joint International Semantic Technology Conference*, 2011.

[12] http://www.taobao.com

Ads and the City: Considering Geographic Distance Goes a Long Way

Diego Saez-Trumper[*]
Web Research Group
Universitat Pompeu Fabra
Spain
diego.saez@upf.edu

Daniele Quercia
Computer Laboratory
University of Cambridge
United Kingdom
daniele.quercia@cl.cam.ac.uk

Jon Crowcroft
Computer Laboratory
University of Cambridge
United Kingdom
jon.crowcroft@cl.cam.ac.uk

ABSTRACT

Social-networking sites have started to offer tools that suggest "guests" who should be invited to user-defined social events (e.g., birthday parties, networking events). The problem of how to recommend people to events is similar to the more traditional (recommender system) problem of how to recommend events (items) to people (users). Yet, upon Foursquare data of "who visits what" in the city of London, we show that a state-of-the-art recommender system does not perform well - mainly because of data sparsity. To fix this problem, we add domain knowledge to the recommendation process. From the complex system literature in human mobility, we learn two insights: 1) there are special individuals (often called power users) who visit many places; and 2) individuals go to a venue not only because they like it but also because they are close-by. We model these insights into two simple models and learn that: 1) simply recommending power users works better than random but is far from producing the best recommendations; 2) an item-based recommender system produces accurate recommendations; and 3) recommending places that are closest to a user's geographic center of interest produces recommendations that are as accurate as, if not more accurate than, item-based recommender's. This last result has practical implications as it offers guidelines for designing location-based recommender systems and for partly addressing cold-start situations.

Categories and Subject Descriptors

H.4 [**Information Systems Applications**]: Miscellaneous

General Terms

Algorithms

Keywords

Advertisements, Mobile, Social Marketing

---
[*]This work was done while visiting the Computer Laboratory of the University of Cambridge.

1. INTRODUCTION

Social media sites have been recently testing features that return lists of people ("guests") users might want to consider inviting to their events (e.g., law firm parties, birthday parties, *PR*'s club invitations) [5]. Guests are selected based on relevance to the event and to the other fellow guests.

The problem of predicting relevant "guests" for venues or events has thus started to receive attention on the Web but has not been fully explored on mobile-social media platforms such as Foursquare, as discussed in Section 6 on "Related Work". One way of recommending venues to people is to use existing Web-based collaborative filtering algorithms. In Section 2, we show that such algorithms are not effective, mainly because of data sparsity: a venue is visited, on average, by very few users. Therefore, we propose two simple techniques for "recommending guests" that are reasonably accurate and scalable, and whose recommendations are easy to explain. In so doing, we make two main contributions:

- We put forward two proposals - a Bayesian model and a linear regression - that incorporate domain knowledge from the literature of human mobility and that cope with data sparsity (Section 3).

- We evaluate how the models perform against Foursquare data for the whole city of London (Section 4). We find that the simplest model - linear regression - returns the most accurate recommendations for all types of venues.

Before placing this work in the context of relevant literature (Section 6) and concluding (Section 7), we discuss some open questions (Section 5), including that of when our models do *not* work (and, consequently, where future work should go).

2. COLLABORATIVE TARGETING: UNFIT

To begin with, we state our research problem.

> **Problem Statement:** Given a venue (e.g., Italian restaurant), select individuals who are likely to visit it.

This simple problem, if solved, might enable a variety of applications, which include target advertising, commercial property evaluation, and social marketing (as we shall discuss in Section 5).

The problem might be formulated in simple "recommender system" terms - that is, it is the problem of how to recommend venues (items) to people (users). One way of solving it is to run a state-of-the-art matrix factorization algorithm on the inverted *venue-by-people* matrix (whose value m_{ij} is 1, if user j checked-in in venue i; 0 otherwise) and obtain, for each venue, a list of people who

Category	#Venues	#Users
food	1,293	1,566
nightlife	1,075	1,207
travel	850	1,744
home/work/etc.	411	1,037
shops	362	878
arts&entertainment	348	841
parks&outdoors	184	363
education	49	117
Total	4,572	3,110

Table 1: London Foursquare Data. Number of users and venues across venue categories.

Category	Precision@10	Recall@10
food	0.013	0.012
nightlife	0.019	0.018
travel	0.004	0.005
shops	0.003	0.003
home/work/etc.	0.001	0.001
arts&entertainment	0.000	0.000
parks&outdoors	0.000	0.000
education	0.000	0.000

Table 2: Implicit SVD's Precision and Recall across Categories.

might like to visit it. We do just that: we use the state-of-the art *Implicit SVD* method introduced by Hu, Koren and Volinsky [10] and implemented within the Mahout framework. To evaluate its effectiveness, we measure its precision and recall on the following dataset.

Dataset. Foursquare is a mobile social-networking application that allows registered users to share their presence in a venue (e.g., share their "check-in" in a restaurant) with their social contacts. Users can share their check-ins not only on Foursquare but also on Twitter and Facebook. Each venue is associated with a category (e.g., "nightlife", "food") and a sub-category (e.g., "bar", "club", "Italian restaurant"). In 2011, Cheng *et al.* collected 22 million check-ins of 225,098 users [3]. We take the 228,625 check-ins in Greater London, which are generated by 29,044 users across 7,205 venues. To this data in the form *(user,venue)* pairs, with further crawling, we add each venue's category and subcategory. After considering venues and users that disjointly appear at least twice in our *(user,venue)* pairs, we end up with 3,110 users and 4,572 venues in the city of London. Table 1 breaks statistics about users and venues down into the different categories. One can, for example, see that food venues are numerous and attract many users, while educational venues are rare but proportionally attract more users.

Implicit SVD Performance. We arrange this data in a *venue-by-user* matrix and measure the *Implicit SVD*'s precision and recall. For each venue, precision is the probability that a recommended user is relevant ($\frac{relevant \cap recommended}{recommended}$), while recall is the probability that a relevant user will be recommended ($\frac{relevant \cap recommended}{relevant}$). By relevant, we mean users who visited the venue. Also, we consider that the recommendation list for each venue contains the *top-10* recommended users. The results reported in Table 2 shows that precision and recall are extremely low - for some categories, they are even zero. These appalling results have a clear explanation - the data is sparse. There are too few people

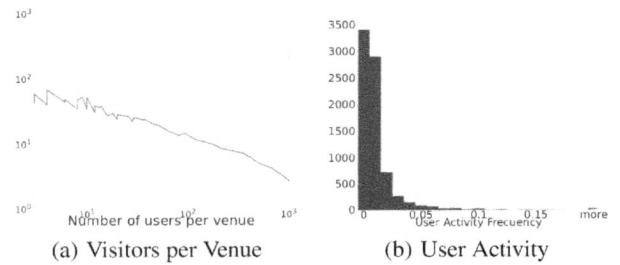

(a) Visitors per Venue   (b) User Activity

Figure 1: *(a)* Number of Visitors per Venue; *(b)* Frequency Distribution of User Activity: this is a user's fraction of visited locations over the total ones.

going to the same venue; indeed, the number visitors per venue is power law (Figure 1(a)).

It thus seems that an alternative mechanism for recommending people is needed. But what sort of mechanism should we use? The widely-used classification algorithm of *SVM* does not work in the presence of data sparsity [18]. Therefore, we need a solution that: 1) is robust to sparsity; and 2) integrates domain knowledge (after all, our goal is to model how people "move" as much as is to model their preferences).

3. DOMAIN-AWARE RECOMMENDATION

We take these two requirements and translate them into a solution that unfolds in three steps:

1. Incorporate domain knowledge from the complex system literature in human mobility (Section 3.1);

2. Deal with data sparsity by using item-based collaborative filtering to model user preferences (Section 3.2);

3. Integrate the previous two steps into a Bayesian model and a linear regression (Section 3.3).

3.1 (Individual) Closeness

For starters, one might go to a venue not only because one likes it but also because one is nearby. Thus, leaving out the users' taste from a moment, one can model the probability of an individual visiting a venue as $p(go|close)$ - i.e., the probability of going to a venue given that it is close - and can do so using Bayes' Law:

$$p(go|close) \propto p_{close} \cdot p_{go} \qquad (1)$$

where p_{close} is the probability of the user being close (being at a certain (log) distance), and p_{go} is the probability of a user going to any venue:

$$p_{go} = \frac{\text{\#venues visited by user } u}{\text{total \#venues}} \qquad (2)$$

This latter probability reflects the general activity of a given user, which is a skewed distribution (Figure 1(b)), as one would expect: the vast majority of users visit few places, while a tiny fraction of (power) users (0.3%) visited roughly 20% of the London venues (within a category).

Literature: How people move. Scientists have long wondered how to measure something as ephemeral as movement. Early studies suggested that humans wander in a random fashion, similar to a so-called "Levy flight" pattern displayed by foraging animals. In 2006, to track human movements, researchers used more than half a million US one-dollar bills as a proxy measure and analyzed their movements as they were passed around over five years [2]. They found many short movements and occasional longer ones. Similar patterns were found by Gonzalez *et al.* who studied the trajectories of 100,000 mobile phone users tracked for six months [7]. These researchers found that people are regular, in that, the vast majority of them move around over a very short distance (from 5 to 10km) and make regular trips to the same few destinations such as work and home on a daily basis (70 percent of the time they were found in their two most frequently visited locations); people occasionally make longer trips when they, for example, go on vacation. More recently, Cheng *et al.* analyzed the movement of Foursquare users across venues and found similar patterns: a mixture of short, random movements with occasional long jumps. As such, the vast majority of users had a small radius of gyration - typically less than 10 miles [3].

Considering Geographic Closeness. To sum up, upon different types of movement (derived from dollar bills, mobile phones, and mobile social-networking applications), researchers in different disciplines have independently concluded that people rarely stray from familiar areas - they travel to a limited number of nearby locations and, consequently, short-range movements are more frequent than long-range ones (i.e., the frequency distribution of distance is exponentially distributed). This is also the case in our London data: Figure 2 plots the probability of one's traveling a certain distance for different venue categories. The distributions (for different categories) are very skewed and all fit the same distribution:

$$p_{close} = k_1 \frac{1}{d_{ui}^\alpha} \qquad (3)$$

where d_{ui} is the distance between the user's (u's) center of geographic interest - which is center of mass or barycenter computed considering the locations where the user has previously checked-in - and the venue i. Interestingly, different venue categories are associated with different α, and the higher α, the less distance matters in one's choice when visiting a venue. Table 3 reports the α's for the different categories. The highest α (2.22) is associated with venues in the category "travel": those include train stations and bus stations, and it makes sense that people travel farther when going to places of limited supply (e.g., not all neighborhoods have a train station). The lowest α's are registered for venues in the categories "nightlife" and "home/work/etc.". That is, one's center of geographic interest revolves around home and work locations, and when going to bars, one goes to nearby ones.

Considering Power Users. Another conclusion from the literature is that not all mobile users are equally mobile. Individuals display significant regularity, yet, when compared to each other, there are few users who travel a lot, while the vast majority have limited travel activity. By framing the problem probabilistically, expression (1) is able to account for those special (power) users. It does so with p_{go} in expression (2), which reflects the extent to which one is a power user or not.

3.2 Likes

The model in expression (1) has only considered whether one user is close or not and whether is a power user or not; but the model has not taken into account personal preferences. To fix that, we need to compute $p(like|go)$ - we need to compute the extent to which a user visits venues that are predictable from his/her past visits/likes. However, to do so, we need a way to measures a user's *likes*. Since our data is sparse (Section 2), we measure likes not

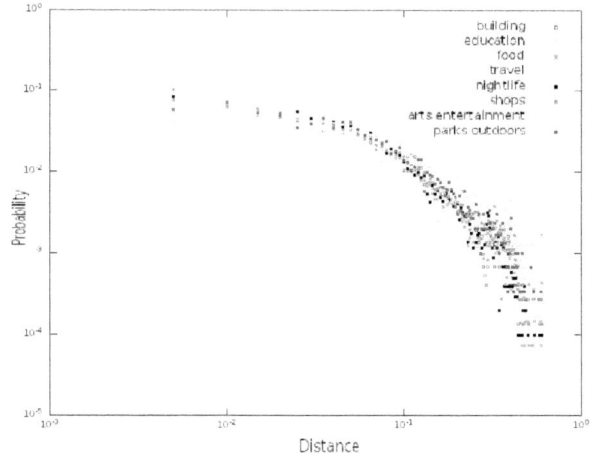

Figure 2: Probability of one's traveling a certain distance across different types of venues (best seen in color).

Category	α
food	1.64
nightlife	1.61
travel	2.22
home/work/etc.	1.62
shops	1.64
arts&entertainment	1.64
parks&outdoors	1.68
education	1.93

Table 3: Why People Visit Different Types of Venues. The higher α, the more one travels farther than usual to reach the venue in that category.

based on similarity among users but among venues. That is, we use an *item*-based collaborative filtering [21], which has been found to work well in such situations: "Unlike traditional collaborative filtering, the algorithm also performs well with limited user data, producing high-quality recommendations based on as few as two or three items." [15]. Rather than matching the user to other similar users, item-to-item collaborative filtering matches each of the user's venues with similar venues. A common way of computing the similarity between two venues is to compute the cosine similarity between two binary vectors: each vector reflects a venue, and a vector's i^{th} position reflects whether the ith user visited the venue or not. Upon a so-constructed venue similarity table, the algorithm finds, for each user, the venues similar to the ones previously visited by the user.

We apply the item-based collaborative filtering algorithm on the *user*-by-*venue* matrix and obtain a rating l_{ui} for each user u and venue i. Figure 3 shows the distribution of the predicted ratings. Upon these ratings, we compute $p(like = l_{ui}|go)$, which is the fraction of venues i visited by u that have predicted ratings l_{ui}:

$$p(like = l_{ui}|go) = \frac{\#\text{venues visited by user } u \text{ with rating } l_{ui}}{\text{total }\#\text{venues visited by user } u}$$

$$(4)$$

3.3 Putting All Together

Having users' whereabouts and preferences at hand, we now

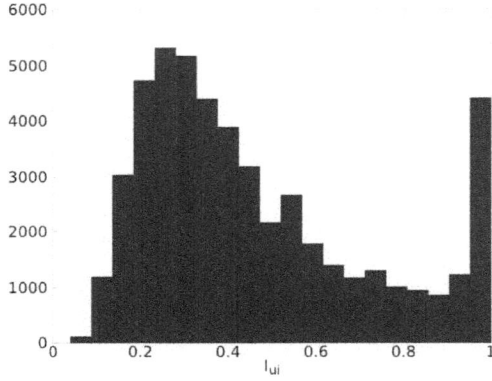

Figure 3: Distribution of Predicted Ratings.

need to predict which users are likely to be at a certain venue. We do so using a Naive Bayesian model, a Bayesian model, and a linear regression.

Naive Bayesian modeling. One simple way of modeling all the three factors together is to compute $p(go|like, close)$ using Bayes' Law:

$$p \quad (go|like, close) \propto$$
$$\propto \quad p_{like} \cdot p_{close} \cdot p_{go}$$
$$\propto \quad p_{like} \cdot p_{go} \cdot k_1 \frac{1}{d_{ui}^{\alpha}}$$

For each pair *(user, venue)*, we compute p_{close} with expression (3) and p_{like} with (4); and for each user, we compute p_{go} with (2). The importance of venue i for user u is then proportional to the above $p(go|like, close)$, and we call it $rank_{u,i}$.

Bayesian modeling. The previous model assumes that whereabouts and preferences are independent. This might well be not the case: those addicted to luxury goods will often be found near Bond Street (a major shopping street in the West End of London with many high price fashion shops). Here preference and whereabout go hand in hand. To go beyond independence, we could model jointly the two attributes:

$$p(go|like, close) = \frac{p_{like|go,close} \cdot p_{go|close}}{p_{like|close}}$$

where:

$$p_{like|go,close} = \frac{\#\text{venues visited by user } u \text{ at distance } d_{ui} \text{with rating } l_{ui}}{\#\text{venues visited by user } u \text{ at distance } d_{ui}}$$

$$p_{go|close} = \frac{\#\text{venues at distance } d_{ui} \text{ visited by user } u}{\#\text{venues at distance } d_{ui}}$$

$$p_{like|close} = \frac{\#\text{venues at distance } d_{ui} \text{with rating } l_{ui}}{\#\text{venues at distance } d_{ui}}$$

Linear Regression. Another approach for combining preferences and whereabouts is to run a linear regression:

$$rank_{u,i} = \alpha + \beta_1 I_{like} + \beta_2 I_{close} + \beta_3 I_{close} \cdot I_{like}$$

where I's are normalized values of whereabouts and preferences: I_{close} is $\frac{1}{log(d_{ui})}$ (the logarithm because the frequency distribu-

tion of distance is very skewed), and I_{like} is l_{ui}. The product $I_{close} \cdot I_{like}$ controls for interaction effects between whereabouts and preferences.

4. EVALUATION

The goal of this work is to predict which users are more likely to visit a given venue. To ascertain the effectiveness of our proposed techniques at meeting this goal, we need to select a desirable metric, measure it, and interpret those measurements. We execute these three steps next.

Metric. We need to find a measure that reflects the extent to which the predicted users for a venue are those who actually visited the venue. One such measure is called percentile-ranking [10]. The percentile-ranking $rank_{u,i}$ of user u for venue i ranges from 0% to 100%: it is 0%, if user u is first in venue i's recommendation list; it is 100%, if the user is last. Percentile-ranks have the advantage over absolute ranks of being independent of the number of users. Our quality measure is then the total average percentile-ranking:

$$\overline{rank} = \frac{\sum_{u,i} gone_{u,i} \cdot rank_{u,i}}{\sum_{u,i} gone_{u,i}} \tag{5}$$

where $gone_{u,i}$ is a flag that reflects whether user u was in venue i: it is 0, if u was not there; otherwise, it is 1. The lower $\overline{rank}$ for a list, the better the list's quality. For random predictions, the expected value for $rank_{u,j}$ is 50% (averaging infinite placements of users for a venue returns the middle position of the list). Therefore, $\overline{rank} < 50\%$ indicates an algorithm better than random. To ease illustration, we covert percentile ranking into ranking accuracy, which is 1, if the percentile ranking is 0% (best); and it is 0, if the percentile ranking is 50% (random):

$$accuracy = \frac{50\% - \overline{rank}}{50\%} \tag{6}$$

Accuracy would be 0 for a random predictor (baseline), and would be 1 for an ideal (oracle) predictor.

Execution. To measure the ranking accuracy, we run a *10-fold* cross validation. That is, we divide the dataset into 10 segments, we take one segment s at a time, consider it to be the testing set, and go through the following steps:

1. For each venue in the *training* set (the venues in all segments other than s), associate it with the users who visited that venue.

2. Train the model using the venues (and corresponding visitors) in the training set.

3. Use the trained model to then infer a rank list of users who are likely to go to each venue in the *testing* set (the venues in s).

We finally compare the users predicted for each venue to those who actually visited it (those who are in the ground truth).

Results. Figure 4 reports the ranking precisions for the individual components of the Bayesian models (first three bars in each venue category) and for the overall models (Naive in the fourth bar, Bayesian in the fifth, and Linear Regression in the sixth). Starting from the first bar in each category (p_{go}), one sees that recommending power users works better than random (accuracy is always well above zero): the more so for shops (.38) than for arts&entertainment venues (.24). Considering only nearby places (second bar in each set) returns more accurate rankings - again, more for shops (.60) than for arts&entertainment venues (.38). However, if one consider only past user preferences (third bar p_{like}), then accuracy is

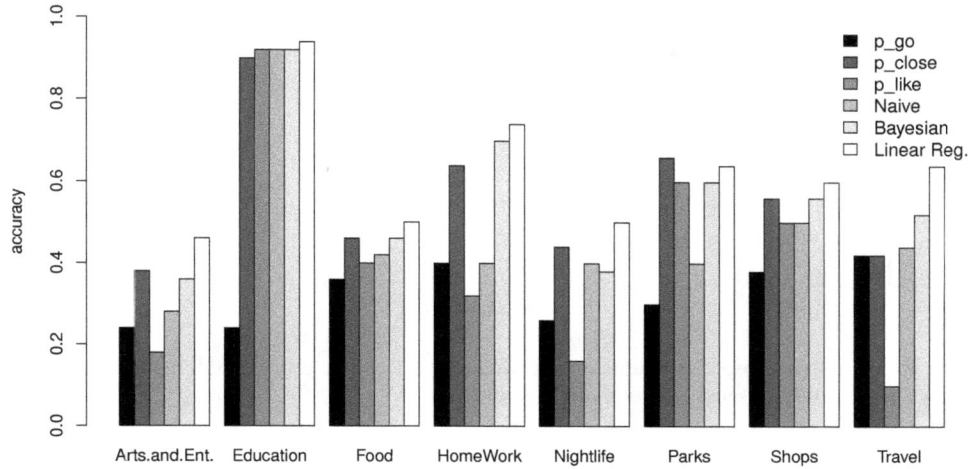

Figure 4: Rank Precision. The rank goes from 0 (random baseline predictions) to 1 (relevant user always ranked first in the recommendation list).

comparable to that of recommendations based on proximity (second and third bars do not differ much). This suggests that the simple concept of geographic distance is as important as that of the user's taste in all venue categories. It also suggests that, by only knowing where a user usually hangs out (without any information on the user's taste), one can produce reasonable recommendations (ideal for cold start situations). If we then combine these previous elements in a Naive Bayesian model, results do not improve; on the contrary, they are worse than those offered by simple geographic proximity for venues in the categories "food" and "arts&entertainment". That might be because the model treats its components as though they were completely independent. However, on average, the Pearson correlation coefficients ρ between each pairs of components are small: $\rho(p_{go}, p_{like}) = .13$, $\rho(p_{go}, p_{close}) = .05$, and $\rho(p_{like}, p_{close}) = .21$. Yet, looking at the fifth bar in each set, one registers improvements with the traditional Bayesian model (in which dependencies are model). Another common reason for which Naive Bayesian does not work well in certain situations is that the addition of redundant components and arbitrary discretization of the random variables skew the learning process, and that seems to be the case here. Indeed, the linear regression (last bar) - which just models taste, whereabouts, and interactions between the two - works best in all categories. As one would expect, for categories characterized by less data sparsity and periodic patterns (e.g., education buildings), the models perform extremely well (accuracy above .90): the performance tend to be comparable to, if not better than, those registered in Web applications.

5. DISCUSSION

Putting Results into Context. For the case of recommending shows on set top boxes, Hu *et al.* had 17K of unique programs (roughly 2× our number of venues) and 32M non-zero ratings (140× ours). In that context of less sparsity, they managed to achieve a ranking accuracy as good as .80 (upon learning from 200 distinct

factors). Thus our results with the linear regression (always above .50 and above .60 for categories such as 'shops' and 'parks' and 'travel') are comparable to those reported in the literature in far more favorable contexts (140× less sparsity). Also, the percentile rankings are expected to slightly improve in more 'realistic' situations. To see why, consider that our data has been collected within a limited time window; by contrast, if one were to crawl the entire Foursquare history, then the resulting data would be still sparse but less so, and, as such, the prediction results would improve, as we have already registered with the category "educational" venues for which the accuracy was above .90.

When It Does not Work. When putting forward new predictive models, one often tends to focus on favorable situations in which predictions are best. Next, we briefly focus on the opposite case - we focus on situations in which prediction are worst. The idea behind this exercise is to find out which aspects future models should consider to increase accuracy. To this end, we run a qualitative study. For each venue i, we compute four predictability and unpredictability measures upon the following quantities: $gone_{ui}$, which reflects whether user u visited venue i; the geographic decay constant α taken from Table 3; the predicted rating l_{ui} for user u and venue i; and the distance d_{ui} between u's geographic center of interest and venue i. More specifically, upon these quantities, for each venue i, we compute:

Geo Predictability. The higher it is, the more the venue's visitors are predictable based on distance. It is higher for venues (e.g., bakery shops) whose visitors travel nearby:

$$P_{geo}^i = \frac{\sum_u \frac{1}{log(d_{ui}^\alpha)} \cdot gone_{ui}}{\sum_u gone_{ui}}$$

It is the average inverse (log) distance for the venue's visitors.

Geo Unpredictability. The higher it is, the less its visitors are predictable based on distance. It is higher for venues (e.g.,

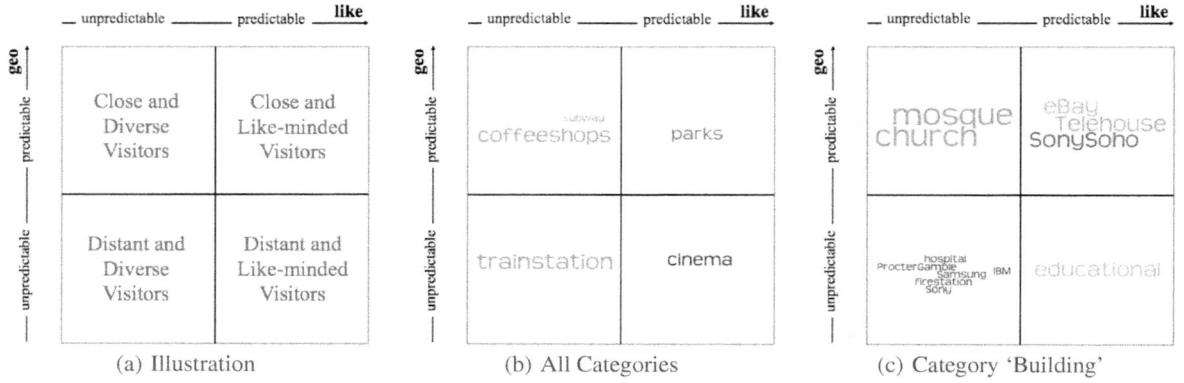

Figure 5: Four-quadrant Predictability Box. Quadrants are defined by the venue's unpredictability and predictability measures, which are based on visitors' geographic closeness (rows) and likes (columns).

airports, high-end restaurants) whose visitors travel farther:

$$U^i_{geo} = \frac{\sum_u log(d_{ui}) \cdot gone_{ui}}{\sum_u gone_{ui}}$$

It is the average (log) distance for the venue's visitors.

Like Predictability. The higher it is, the more its visitors are predictable based on past preferences (past likes). It is higher for venues whose visitors have common preferences:

$$P^i_{like} = \frac{\sum_u l_{ui} \cdot gone_{ui}}{\sum_u gone_{ui}}$$

It is the average predicted ratings for the venue's visitors.

Like Unpredictability. The higher it is, the less its visitors are predictable based on past preferences. It is higher for venues whose visitors have diverse preferences:

$$U^i_{like} = \frac{\sum_u \frac{1}{l_{ui}} \cdot gone_{ui}}{\sum_u gone_{ui}}$$

It is the average inverse predicted ratings for the venue's visitors.

We create four tables that contain the *top*-10 venues ranked by each of those four measures and ask three coders (three Londoners with diverse background - architect, barrister, and medical doctor) to build predictability boxes of the kind in Figure 5(a). For them, that translated into ordering venue categories that are predicted (hard to predict) by geographic distance based on the table ranked by P^i_{geo} (by U^i_{geo}), and categories that are predicted (hard to predict) by user preferences based on the table ranked by P^i_{like} (by U^i_{like}). We consider only the answers for which two out of three coders or all three have independently agreed. In Figures 5(b) and 5(c), word size is proportional to the coders' agreement. For all venue categories (Figure 5(b)), the unpredictable venues (predicted neither by closeness nor by taste) are train stations. That is because train stations are often far from where one hangs out and do not reflect a specific taste in, say, music, bars, clubs, or food. By contrast, local parks and outdoor activities are predictable either by closeness or by taste, suggesting that people prefer their local parks over bigger parks (they stay close), and that residents of the same area tend to be like-minded (a tendency often called "geographic sorting" [1]). Closeness is more informative for predicting visits to coffee shops (one tends to go to local coffee shops);

while user taste is more informative for cinemas in central London areas, where diversified choice of movies motivates visitors to travel farther than usual. For the specific category "buildings" (Figure 5(c)), the unpredictable venues (predicted neither by closeness nor by taste) are companies such as IBM, Procter&Gamble, Samsung whose headquarters are in suburban areas where people with diverse background work but do not hangout, not least because of limited supply of amenities. By contrast, the behavior of employees (mostly interaction designers) of Sony, eBay, Telehouse working in central areas like Soho is predictable either by closeness or by taste. Finally, closeness is more informative for predicting visits to mosques and churches (one tends to go to local religious venues); while user taste is more informative for visitors of university (e.g., UCL's, Birkbeck's) facilities in central areas. From these qualitative results, one can extrapolate two key insights:

1. Predictable situations are those in which people: a) stay close because they have what they need at hand; or b) congregate in places where other like-minded people tend to be (e.g., local parks and cinemas).

2. By contrast, unpredictable situations are those in which people: a) travel because they do not have what they want at hand; or b) go to places that attract individuals of very diverse backgrounds (e.g., coffee shops, train stations).

Future work should go into models that are able to simultaneously account for these (at times) conflicting situations.

Applications. The practical implications of this work go beyond traditional applications of recommender systems:

Target Advertising. The first step when promoting new nightclubs, bars and restaurants is often to identify the target market. Thus, knowing the kind of people who are willing to go to, say, certain restaurants or bars (which is what this work is about) translates into low-cost marketing strategies for bars and restaurants that are willing to attract new crowds.

Commercial Property Evaluation. This is the process of identifying and quantifying the value of commercial properties and is generally carried out by experts who analyze properties similar to the one being valued. A primary factor that affects this assessment is location, yet this factor is generally quantified based on the valuer's expert knowledge of a locality. More recently, well-informed ways of valuing properties have been proposed, and they rely on the *creation and maintenance* of GIS-based property valuation databases. These databases (especially those for commercial properties) might well be enriched by this work - in particular, by knowing how close a venue is to its target audience (the higher the number of potential users who like a type of venue in a neighborhood, the higher the venue's value).

Social Marketing. Social marketing can be defined as a research-driven approach to promote voluntary behavior change in a priority population. A case in point is "Stop the Sores", a social marketing campaign designed to increase syphilis testing in Los Angeles County [17]. Social marketing has its foundation in consumer marketing and consists of three key elements: market research [24], audience segmentation [8], and branding [12]. The second element of segmentation is related to this work and is essential for developing campaign messages that resonate with the target population and helps in identifying the largest or highest-risk subgroup (e.g., swingers, men having sex with men) at minimal cost.

Scalability. The two main parts of this work - which model whereabouts and preferences - are highly scalable:

Whereabout Part. This requires to know a geographic point for each user (where an individual usually hangs out) and one single decay constant α (which is universal in that it equally applies to all users). Learning a point per user and a constant for all is extremely scalable. In addition to being scalable, the models are likely to be generalizable, not least because they have been built upon previous general rules of people's wanderings [2, 3, 7], and, being general, they are also likely to work for any instance of mobility (not only for Foursquare users).

Preference Part. This translates into item-based collaborative filtering. The (computationally) expensive part of this algorithm (venue similarity table) can be computed offline, while what needs to be computed online - matching the user's venues with similar ones - scales independently of the total number of venues and total number of users, in that, it only depends on the number of venues each single user has visited (which is generally extremely low).

6. RELATED WORK: WEB DOCET

The problem of recommending events has been initially tackled on the Web. In this context, researchers have mainly worked on detecting and tracking events [11, 13]. They initially considered how textual content evolves over time and left out network effects. Zhu and Sasha then started to model social interactions and topic evolutions by treating these two elements separately [26]. More recently, Lin *et al.* built a model that considers these two elements simultaneously and showed that it worked upon two very different types of data - Twitter and DBLP [14]. After detecting events, one can then recommend them. That is what Daly and Geyer *et al.* did: they built a system that recommends events in an internal event management service and proposed a new way of recommending events to new users [4]. Before that, Minkov *et al.* had run large user studies in which they evaluated the effectiveness of different strategies for recommending academic talks [16]. They found that, in a situation of limited data sparsity, collaborative filtering approaches work better than content-based ones. The recommendation process generally relies on user ratings but has also been enriched by social

networks at times. A case in point is Golbeck *et al.* who built a recommender system that integrates social networks to offer well-informed movie recommendations [6].

Hence past work on recommending events has mostly gone into Web platforms, while mobile ones have been investigated only recently. Takeuchi and Sugimoto proposed a system that recommends shops based on past visited locations, and found item-based collaborative filtering to work reasonably well [23]. Ricci and Nguyen proposed a system that recommend nearby restaurants using a critique-based model [20]. More recently, for major mobile social-networking services, Scellato *et al.* studied their geographic properties at scale and suggested that these properties could well inform venue recommendation in large cities [22]. Upon mobile phone data in the metropolitan area of Boston, Quercia *et al.* studied strategies for recommending large-scale events (e.g., concerts, baseball matches) and showed how different types of events require different recommendation strategies [19].

Shifting attention from recommending events to recommending people, one sees that most of the work has again gone into Web platforms. Within an enterprise social network, Guy *et al.* proposed ways to recommend people a user is not likely to know but might be interested in [9]. Few months ago, Facebook launched a new feature called "suggested guests" [5]: this returns a list of people (three at the time) a user might want to consider inviting to their event, and the list is compiled based on relevance to the event and to the people who are attending. Since work on recommending people for events has just started on the Web, it comes as no surprise that little work about it has gone into mobile social-networking platforms.

7. CONCLUSION

We have studied different strategies for recommending "guests" for real-world venues and, not surprisingly, found that results are best not only for venues with considerable historical data (e.g., educational institutions) but also for venues that are visited regularly (e.g., work locations). For other types of venues such as restaurants and bars, geographic closeness plays a very important role. Combining user preferences and geographic closeness has the expected result of offering more accurate recommendations, and that result can be achieved by using very simple models - Bayesian or linear regression. Being simple, these models not only are scalable and cost efficient but also produce recommendations that are easy to explain. The main criticism for the new Facebook "suggested guests" feature has been that it "does not offer... any sort of context" [5]. Our recommendations - which depend on whether one has visited similar locations or whether one often hangs out in certain neighborhoods - are likely to be easier to explain than those produced by black-box approaches. In the future, we will work in this direction: on how to recommend "guests" in ways that are easy to explain and that increase serendipitous encounters [25].

Acknowledgments. We thank Tamas Jambor for his help on the *Implicit SDV* algorithm and his valuable insights, and the reviewers for their constructive comments. This work was funded by RCUK through the Horizon Digital Economy Research grant (EP/G065802/1) and by the Spanish Economy and Competitiveness Ministry through the HIPERGRAPH project (TIN2009-14560-C03-01).

8. REFERENCES

[1] B. Bishop. *The Big Sort: Why the Clustering of Like-Minded America Is Tearing Us Apart.* Houghton Mifflin, May 2008.

[2] Brockmann, L. Hufnagel, and T. Geisel. The scaling laws of human travel. *Nature*, 439, January 2006.

[3] Z. Cheng, J. Caverlee, K. Lee, and D. Z. Sui. Exploring Millions of Footprints in Location Sharing Services. In *Proceedings of the 5th International AAAI Conference on Weblogs and Social Media (ICWSM)*, 2011.

[4] E. M. Daly and W. Geyer. Effective event discovery: using location and social information for scoping event recommendations. In *Proceedings of the 5th ACM Conference on Recommender Systems (RecSys)*, 2011.

[5] B. Darwell. Facebook tests 'suggested guests' for events. In *Inside Facebook*, February 2012.

[6] J. Golbeck. Trust and nuanced profile similarity in online social networks. *ACM Transactions on the Web*, 3(4):12:1–12:33, September 2009.

[7] M. C. Gonzalez, C. A. Hidalgo, and A.-L. Barabasi. Understanding individual human mobility patterns. *Nature*, 453(7196), June 2008.

[8] S. Grier and C. A. Bryant. Social marketing in public health. *Annual Review of Public Health*, 26(1), 2005.

[9] I. Guy, S. Ur, I. Ronen, A. Perer, and M. Jacovi. Do you want to know?: recommending strangers in the enterprise. In *Proceedings of the ACM Conference on Computer supported Cooperative Work (CSCW)*, 2011.

[10] Y. Hu, Y. Koren, and C. Volinsky. Collaborative Filtering for Implicit Feedback Datasets. In *Proceedings of the 8th IEEE International Conference on Data Mining (ICDM)*, 2008.

[11] A. Ihler, J. Hutchins, and P. Smyth. Adaptive event detection with time-varying poisson processes. In *Proceedings of the 12th ACM International Conference on Knowledge Discovery and Data Mining (KDD)*, 2006.

[12] K. L. Keller. Branding perspectives on social marketing. *Advances in Consumer Research*, 25(1), 1998.

[13] J. Kleinberg. Bursty and hierarchical structure in streams. In *Proceedings of the 8th ACM International Conference on Knowledge Discovery and Data Mining (KDD)*, 2002.

[14] C. X. Lin, B. Zhao, Q. Mei, and J. Han. PET: a statistical model for popular events tracking in social communities. In *Proceedings of the 16th ACM International Conference on Knowledge Discovery and Data Mining (KDD)*, 2010.

[15] G. Linden, B. Smith, and J. York. Amazon.com recommendations: item-to-item collaborative filtering. *IEEE Internet Computing*, 7(1):76–80, 2003.

[16] E. Minkov, B. Charrow, J. Ledlie, S. Teller, and T. Jaakkola. Collaborative future event recommendation. In *Proceedings of the 19th ACM International Conference on Information and Knowledge Management (CIKM)*, 2010.

[17] A. Plant, J. A. Montoya, H. Rotblatt, P. R. Kerndt, K. L. Mall, L. G. Pappas, C. K. Kent, and D. Klausner. Stop the Sores: The Making and Evaluation of a Successful Social Marketing Campaign. *Health Promotion Practice*, 11(1), 2010.

[18] D. Quercia, H. Askham, and J. Crowcroft. TweetLDA: Supervised Topic Classification and Link Prediction in Twitter. In *Proceedings of the 4th ACM International Conference on Web Science (WebSci)*, 2012.

[19] D. Quercia, N. Lathia, F. Calabrese, G. D. Lorenzo, and J. Crowcroft. Recommending Social Events from Mobile Phone Location Data. In *Proceedings of the 10th IEEE International Conference on Data Mining (ICDM)*, 2010.

[20] F. Ricci and Q. N. Nguyen. Acquiring and Revising Preferences in a Critique-Based Mobile Recommender System. *IEEE Intelligent Systems*, 22(3):22–29, May 2007.

[21] B. Sarwar, G. Karypis, J. Konstan, and J. Riedl. Item-based collaborative filtering recommendation algorithms. In *Proceedings of the 10th ACM Conference on World Wide Web (WWW)*, 2001.

[22] S. Scellato, A. Noulas, R. Lambiotte, and C. Mascolo. Socio-spatial properties of online location-based social networks. In *Proceedings of the 5th International AAAI Conference on Weblogs and Social Media (ICWSM)*, 2011.

[23] Y. Takeuchi and M. Sugimoto. CityVoyager: An Outdoor Recommendation System Based on User Location History. In *Ubiquitous Intelligence and Computing*, Lecture Notes in Computer Science, 2006.

[24] D. C. Walsh, R. E. Rudd, B. A. Moeykens, and T. W. Moloney. Social marketing for public health. *Health Affairs*, 12(2), 1993.

[25] Y. C. Zhang, D. O. Séaghdha, D. Quercia, and T. Jambor. Auralist: introducing serendipity into music recommendation. In *Proceedings of the 5th ACM International Conference on Web Search and Data Mining (WSDM)*, 2012.

[26] D. Zhou, X. Ji, H. Zha, and C. L. Giles. Topic evolution and social interactions: how authors effect research. In *Proceedings of the 15th ACM International Conference on Information and Knowledge Management (CIKM)*, 2006.

BlurMe: Inferring and Obfuscating User Gender Based on Ratings

Udi Weinsberg, Smriti Bhagat, Stratis Ioannidis, Nina Taft
{udi.weinsberg, smriti.bhagat, startis.ioannidis, nina.taft}@technicolor.com
Technicolor
Palo Alto

ABSTRACT

User demographics, such as age, gender and ethnicity, are routinely used for targeting content and advertising products to users. Similarly, recommender systems utilize user demographics for personalizing recommendations and overcoming the cold-start problem. Often, privacy-concerned users do not provide these details in their online profiles. In this work, we show that a recommender system can infer the gender of a user with high accuracy, based solely on the ratings provided by users (without additional metadata), and a relatively small number of users who share their demographics. We design techniques for effectively adding ratings to a user's profile for obfuscating the user's gender, while having an insignificant effect on the recommendations provided to that user.

Categories and Subject Descriptors: H.2.8 Database Applications: Data Mining
Keywords: Recommender Systems, Privacy.

1. INTRODUCTION

Profiling users through demographic information, such as gender, age, or ethnicity, is of great importance in targeted advertising and personalized content delivery. Recommender systems too can benefit from such information to provide personalized recommendations. However, users of recommender systems often do not volunteer this information. This may be intentional – to protect their privacy, or unintentional – out of laziness or disinterest. As such, traditional collaborative filtering methods eschew using such information, relying instead solely on ratings provided by users.

At a first glance, disclosing ratings to a recommender system may appear as a rather innocuous action. There is certainly a utility users accrue from this disclosure – namely, the ability to discover relevant items. Nevertheless, there has been a fair amount of work indicating that user demographics are correlated to, and thus can be inferred from, user activity on social networks [9], blogs [2], and microblogs [12] etc. It is thus natural to ask whether demographic information such as age, gender, ethnicity or even political orientation can also be inferred from information disclosed to recommender systems. Indeed, irrespective of a rating value, the mere fact that a user has interacted with an item (e.g., viewed a specific movie or purchased a product) may be correlated with demographic information.

The potential success of such an inference has several important implications. From the recommender's perspective, profiling users not only improves their own recommendations, but also enables targeted advertising. From the user's perspective, the success of such an inference raises serious privacy concerns. A privacy-conscious user cannot simply withhold all information as this would come at the cost of foregoing the utility gained from using the recommender system in the first place – namely, finding relevant content. Explicitly withholding the user's demographic information does not ensure privacy either, as it may be possible to uncover it through inference. Because the approach of withholding information is often impractical, we believe a more promising approach is that of adding ratings into a user profile with the intent of creating ambiguity. In this paper, we thus explore both the questions of how demographic information can be inferred – from ratings data alone – and that of how to hinder such inference via obfuscating the information disclosed to the recommender system.

In general, any *obfuscation mechanism* employed by the user strikes a tradeoff. This is between (a) the user's privacy, as captured by the recommender's ability to infer her demographic information, and (b) the utility to the user, captured by the accuracy of recommendations she receives. Understanding the nature of such a tradeoff is thus a fundamental question. In this work, we study the above issues in a comprehensive manner, making the following contributions:

- We evaluate several gender inference algorithms on two movie ratings datasets, Movielens and Flixster, and show that a relatively small amount of labeled data (i.e., users who share their gender), is sufficient to predict the gender of users with about 80% accuracy.

- We find that the act of watching a movie, regardless of the rating given, is strongly correlated with one's gender, and we identify movies for which this correlation is high.

- Based on these observations, we propose several obfuscation mechanisms, allowing the users to alter the information they reveal to the recommender service.

- We further evaluate these mechanisms with respect to the trade-offs they achieve between user privacy, as captured by the accuracy of gender-inference, versus user utility, as captured by rating prediction RMSE.

- We establish that quite favorable tradeoffs are feasible; indeed, using 1% additional ratings it is possible to reduce the success of gender inference by 80% while reducing the quality of recommendations only by 1.3%.

To the best of our knowledge, we are the first to study and quantify demographic inference methods that rely solely on rating data. Moreover, we are the first to design and analyze obfuscation mechanisms aiming to preserve gender privacy while maintaining recommendation accuracy.

2. RELATED WORK

Inferring demographics of users has been widely studied in different contexts, and for various types of user-generated data. In the context of interaction networks, the graph structure has been shown to be useful for inferring demographics using link-based information for blog [2] and social network [9] data from Facebook. Other works rely on the textual features derived from writings of users to infer demographics. For instance, Rao et al. [12] use an SVM classifier on Twitter data, and Otterbacher et al. use logistic regression on movie reviews from IMDB [11]. It is useful to note that the prediction accuracy obtained using logistic regression on movie reviews is about 73.7%, lower than that obtained using the same algorithm on movie ratings, albeit for a different dataset. In our setting, the input to the gender inference mechanism is only the movie ratings provided by users, with no metadata about movies or users. While there has been work on collective matrix factorization [13] to take into account attributes of movies and users in addition to ratings for making recommendations, the work does not explore the specific task of inferring user demographics.

Rather than focusing solely on inference, our goal is to be able to use insights gained to design mechanisms that obfuscate users' demographics. Injecting noise for privacy was recently studied in [15, 14] for search privacy, where the goal is to obfuscate search engine queries rather than a user's demographics.

Our work is also related to studies of robustness in recommender systems [10, 3, 1]. The goal of such studies is to evaluate how an attacker can manipulate a recommender system by injecting adversarially selected ratings. In contrast, our study is based on the interaction of a user, whose ratings are not necessarily added in the training set, with the recommender system. Although the user may submit altered ratings, her interest is still in receiving relevant recommendations, albeit without disclosing her gender.

An elegant and formal approach to privacy in recommender systems has been made through differential privacy [8]. Nevertheless, differential privacy guarantees aim at a different goal, which is to ensure that the output of a recommender depends only marginally on the input of any single user. In contrast, we aim at not protecting ratings per se, but the demographic information of each user; this notion cannot be captured within the formalism of differential privacy.

3. PROBLEM DEFINITION

For the sake of concreteness, we assume throughout the paper that the information users wish to protect is their gender; nevertheless, our algorithms are generic, and apply also when different demographic features (age, ethnicity, political orientation, etc.), expressed as a categorical variable, are to be protected.

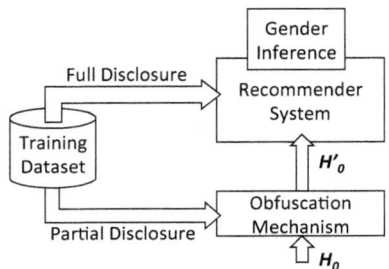

Figure 1: Illustration of the problem setup

3.1 Setup

Our setup is summarized in Figure 1. A user, indexed by 0, views and rates items which, for concreteness, we refer to as movies. We assume that the universe of movies the user can rate comprises a catalog of M movies; the user rates a subset $\mathcal{S}_0$ of the catalog $\mathcal{M} = \{1, 2, \ldots, M\}$. We denote by $r_{0j} \in \mathbb{R}$ the rating of movie $j \in \mathcal{S}_0$ and define the user's *rating profile* as the set of (movie, rating) pairs $\mathcal{H}_0 \equiv \{(j, r_{0j}) : j \in \mathcal{S}_0\}$. The user submits $\mathcal{H}_0$ to an *obfuscation mechanism*, which outputs an altered rating profile $\mathcal{H}'_0 = \{(j, r'_{0j}) : j \in \mathcal{S}'_0\}$, for some $\mathcal{S}'_0 \neq \mathcal{S}_0$. In simple terms, this obfuscation aims at striking a good balance between the following two conflicting goals : (a) $\mathcal{H}'_0$ can be used to provide relevant recommendations to the user, and (b) it is difficult to infer the user's gender from $\mathcal{H}'_0$.

More specifically, we assume that the obfuscated rating profile $\mathcal{H}'_0$ is submitted to a *recommender* mechanism that has a module that implements a *gender inference* mechanism. The recommender mechanism uses $\mathcal{H}'_0$ to *predict* the user's ratings on $\mathcal{M} \setminus \mathcal{S}'_0$, and potentially, recommend movies that might be of interest to the user. The gender inference module is a classification mechanism, that uses the same $\mathcal{H}'_0$ to profile and label the user as either male of female.

Though the implementation of the recommender mechanism might be publicly known, the obfuscation and gender inference mechanisms are not. As a first step in this problem, we take the simple approach that both recommendation and gender inference are oblivious to the fact that any kind of obfuscation is taking place. Both mechanisms take the profile $\mathcal{H}'$ at "face value" and do not reverse-engineer the "true" profile $\mathcal{H}$. (We leave for future work the case when the obfuscation mechanism is known.)

3.2 Training Dataset

We assume that the recommender and inference mechanisms have access to a *training dataset*. This dataset comprises a set of $\mathcal{N} = \{1, \ldots, N\}$ users each of which has given ratings to a subset of the movies in the catalog $\mathcal{M}$. We denote by $\mathcal{S}_i \subseteq \mathcal{M}$ the set of movies for which the rating of a user $i \in \mathcal{N}$ is in the dataset, and by r_{ij}, $j \in \mathcal{S}_i$, the rating given by user $i \in \mathcal{N}$ to movie $j \in \mathcal{M}$. Moreover, for each $i \in \mathcal{N}$ the training set also contains a binary variable $y_i \in \{0, 1\}$ indicating the gender of the user (we map bit 0 to male users). We assume that the training set is unadulterated: neither ratings nor gender labels have been tampered with or obfuscated.

The obfuscation mechanism may also have a partial view of the training set. In the extreme case, the training dataset is public, and the obfuscation mechanism has full access to it. It is interesting however to consider weaker obfuscation mechanisms, that can only access limited statistics (or other queries over the dataset), such as, the average rating of a movie. Though the mechanisms we propose can, *a fortiori*, be implemented when the dataset is public, we will state the training set statistics required in their implementation.

3.3 Matrix Factorization

The main focus of this paper is the design and analysis of mechanisms for gender inference and obfuscation. As such, we fix the recommender mechanism throughout the paper to be matrix factorization [6], since this is commonly used in commercial systems. In short, given the rating profile $\mathcal{H}_0'$, we generate ratings for the set $\mathcal{M} \setminus \mathcal{S}_0$ by appending the provided ratings to the rating matrix of the training set and factorizing it.

More specifically, we associate with each user $i \in \mathcal{N} \cup \{0\}$ a latent feature vector $u_i \in \mathbb{R}^d$. We also associate with each movie $j \in \mathcal{M}$ a latent feature vector $v_j \in \mathbb{R}^d$. We define the regularized mean square error to be

$$\sum_{i \in \mathcal{N} \cup \{0\}, j \in \mathcal{S}_i} (r_{i,j} - \langle u_i, v_j \rangle - \mu)^2 + \lambda \sum_{i \in \mathcal{N} \cup \{0\}} \|u_i\|_2^2 + \lambda \sum_{j \in \mathcal{M}} \|v_j\|_2^2$$

where μ is the average rating of the entire dataset. We construct the vectors u_i, v_j by minimizing the MSE through gradient descent. We use $d = 20$ and $\lambda = 0.3$. Having profiled thusly both users and movies, we predict the rating of user 0 for movie $j \in \mathcal{M} \setminus \mathcal{S}_0'$ through $\langle u_0, v_j \rangle + \mu$.

3.4 Data Description

Flixster. Flixster is an online social network for rating and reviewing movies. Flixster allows users to enter demographic information into their profiles and share their movie ratings and reviews with their friends and the public. The dataset collected by Jamali et al. [5] has 1M users, of which only 34.2K users share their age and gender. We evaluate our techniques on this subset of 34.2K users, who have rated 17K movies and provided 5.8M ratings. The 12.8K males and 21.4K females have provided 2.4M and 3.4M ratings, respectively. Flixster allows users to provide half star ratings, however, to be consistent across the evaluation datasets, we round up the ratings to be integers from 1 to 5.

Movielens. Our second dataset is Movielens from the Grouplens[1] research team. The dataset consists of 3.7K movies and 1M ratings by 6K users. The 4331 males and 1709 females provided 750K and 250K ratings, respectively.

4. GENDER INFERENCE

In this section, we investigate whether inferring a user's gender based on her ratings is indeed possible. We study several different classifiers and evaluate them using the Flixster and Movielens datasets. We use the results of this analysis to inform our design of obfuscation mechanisms (Section 5).

4.1 Classifiers

To train our classifiers, we associate with each user $i \in \mathcal{N}$ in the training set a characteristic vector $x_i \in \mathbb{R}^M$ such that $x_{ij} = r_{ij}$, if $j \in \mathcal{S}_i$ and $x_{ij} = 0$, otherwise. Recall that the binary variable y_i indicates user i's gender, which serves as the dependent variable of our classification. We denote

---

[1] www.grouplens.com/node/73

by $X \in \mathbb{R}^{N \times M}$ the matrix of characteristic vectors, and by $Y \in \{0,1\}^N$ the vector of genders.

We use three different types of classifiers: Bayesian classifiers, support vector machines (SVM) and logistic regression. In the Bayesian setting, we studied several different generative models; for all models, we assume that points (x_i, y_i) are sampled independently from the same joint distribution $P(x, y)$. Given P, the predicted label $\hat{y} \in \{0, 1\}$ attributed to characteristic vector x is the one with maximum likelihood, *i.e.*,

$$\hat{y} = \arg\max_{y \in \{0,1\}} P(y|x) = \arg\max_{y \in \{0,1\}} P(x, y) \quad (1)$$

Class Priors. The class prior classification serves as a base-line method for assessing the performance of the other classifiers. Given a dataset with unevenly distributed gender classes of the population, this basic classification strategy is to classify all users as having the dominant gender. This is equivalent to using (1) under the generative model $P(y|x) = P(y)$, estimated from the training set as:

$$P(y) = |\{i \in \mathcal{N} : y_i = y\}|/N. \quad (2)$$

Bernoulli Naïve Bayes. Bernoulli Naïve Bayes is a simple method that ignores the actual rating value. In particular, it assumes that a user rates movies independently and the decision to rate or not is a Bernoulli random variable. Formally, given a characteristic vector x, we define the rating indicator vector $\tilde{x} \in \mathbb{R}^M$ to be such that $\tilde{x}_j = \mathbb{1}_{x_j > 0}$. This captures the movies for which a rating is provided. Assuming that $\tilde{x}_j$, $j \in \mathcal{M}$, are independent Bernoulli, the generative model is given by $P(x, y) = P(y) \prod_{j \in \mathcal{M}} P(\tilde{x}_j | y)$ where $P(y)$ is the class prior, as in (2), and the conditional $P(\tilde{x}_j | y)$ is computed from the training set as follows:

$$P(\tilde{x}_j | y) = |\{i \in \mathcal{N} : \tilde{x}_{ij} = \tilde{x}_j \wedge y_i = y\}|/|\{i : y_i = y\}| \quad (3)$$

Multinomial Naïve Bayes. A drawback of Bernoulli Naïve Bayes is that it ignores rating values. One way of incorporating them is through Multinomial Naïve Bayes, which is often applied to document classification tasks [7]. Intuitively, this method extends Bernoulli to positive integer values by treating, *e.g.* a five-star rating as 5 independent occurrences of the Bernoulli random variable. Movies that receive high ratings have thus a larger impact on the classification. Formally, the generative model is given by $P(x, y) = P(y) \prod_{j \in \mathcal{M}} P(x_j | y)$ where $P(x_j | y) = P(\tilde{x}_j | y)^{x_j}$, and $P(\tilde{x}_j | y)$ is computed from the training set through (3).

Mixed Naïve Bayes. We propose an alternative to Multinomial, which we refer to as Mixed Naïve Bayes. This model is based on the assumption that, users give normally distributed ratings. More specifically,

$$P(x_j | \tilde{x}_j = 1, y) = (2\pi\sigma_y^2)^{-1/2} e^{-(x_j - \mu_{yj})^2 / 2\sigma_y^2}. \quad (4)$$

For each movie j, we estimate the mean μ_{yj} from the dataset as the average rating of movie j given by users of gender y, and the variance σ_y^2 as the variance of all ratings given by users of gender y. The joint likelihood used in (1) is then given by $P(x, y) = P(y) \prod_{j \in \mathcal{M}} P(\tilde{x}_j | y) P(x_j | \tilde{x}_j, y)$ where $P(y)$, $P(\tilde{x}_j | y)$ are estimated through (2) and (3), respectively. The conditional $P(x_j | \tilde{x}_j, y)$ is given by (4) when a rating is provided (*i.e.*, $\tilde{x}_j = 1$) and, trivially, by $P(x_j = 0 | \tilde{x}_j = 0, y) = 1$, when it is not.

	Flixster		Movielens	
	AUC	P/R	AUC	P/R
Class Prior	0.50	0.39/0.62	0.50	0.51/0.72
Bernoulli	0.72	0.70/0.70	0.81	0.79/0.76
Multinomial	0.75	0.71/0.71	0.84	0.80/0.76
Mixed	0.74	0.71/0.71	0.82	0.79/0.77
SVM	0.82	0.73/0.70	**0.86**	0.78/0.77
SVM ($\tilde{X}$)	0.80	0.72/0.70	0.85	0.78/0.77
Logistic	**0.84**	0.76/0.77	0.85	0.80/0.80
Logistic ($\tilde{X}$)	0.83	0.75/0.76	0.84	0.78/0.79

Table 1: Mean AUC, precision (P) and recall (R)

	Flixster		Movielens	
	Female	Male	Female	Male
Class Prior	0.62/1	0/0	0/0	0.72/1
Bernoulli	0.75/0.80	0.62/0.54	0.57/0.73	0.88/0.78
Multinomial	0.76/0.78	0.63/0.60	0.57/0.73	0.89/0.77
Mixed	0.76/0.81	0.64/0.57	0.57/0.74	0.88/0.78
SVM	0.70/0.95	0.77/0.30	0.80/0.28	0.78/0.97
SVM ($\tilde{X}$)	0.69/0.96	0.77/0.27	0.80/0.28	0.77/0.97
Logistic	0.79/0.85	0.71/0.62	0.69/0.56	0.84/0.90
Logistic ($\tilde{X}$)	0.77/0.87	0.72/0.57	0.73/0.40	0.80/0.94

Table 2: Per-gender precision and recall.

Logistic Regression. A significant drawback of all of the above Bayesian methods is that they assume that movie ratings are independent. To address that, we applied logistic regression. Recall that linear regression yields a set of coefficients $\beta = \{\beta_0, \beta_1, ..., \beta_M\}$. The classification of a user $i \in N$ with characteristic vector x_i is performed by first calculating the probability $p_i = (1 + e^{-(\beta_0 + \beta_1 x_{i1} + ... + \beta_M x_{iM})})^{-1}$. The user is classified as a female if $p_i < 0.5$ and as a male otherwise. The value p_i also serves as a confidence value for the classification of user i. One of great benefits of using logistic regression is that the coefficients β capture the extent of the correlation between each movie and the class. In our case, the large positive β_j indicates that movie j is correlated with class male, whereas small negative β_j indicates that movie j is correlated with class female. We select the regularization parameter so that we have at least 1000 movies correlated with each gender that have a non-zero coefficient.

SVM. Intuitively, SVM finds a hyperplane that separates users belonging to different genders in a way that minimizes the distance of incorrectly classified users from the hyperplane (for a thorough explanation on SVMs see [4]). SVM holds many of the advantages of logistic regression – it does not assume independence in the feature space and produces coefficients. Since our feature space (number of movies) is already quite large, we use linear SVMs in our evaluations. We performed a logarithmic search over the parameter space (C) and found that $C = 1$ gave the best results.

4.2 Evaluation

We evaluate all algorithms on both the Flixster and Movielens datasets. We use 10-fold cross validation and compute the average precision and recall for the two genders across all folds. Additionally, we compute the Area Under the Curve (AUC) using the mean Receiver Operating Characteristic (ROC) curve computed across the folds. For the ROC, the true positive ratio is computed as the ratio of males correctly classified out of the males in the dataset, and the false positive ratio is computed as the ratio incorrectly classified males out of the females in the dataset. The ROC curves are given in Figure 2a and Figure 2b. Table 1 provides a summary of

the classification results for 3 metrics: AUC, precision and recall. Table 2 shows the same results separated per-gender.

We see from the ROC curves that SVM and logistic regression perform better, across both datasets, than any of the Bayesian models since the regression curves for SVM and logistic dominate the others. In particular, logistic regression performed the best for Flixster while SVM performed best for Movielens. The performance of the Bernoulli, mixed and multinomial models do not different significantly from one another. These findings are further confirmed via the AUC values in Table 1. This table also shows the weakness of the simple class prior model that is easily outperformed by all other methods.

In terms of precision and recall, Table 2 shows that logistic regression outperforms all other models for Flixster users and both genders. For the Movielens users, SVM performs better than all other algorithms, while logistic regression is second best. In general, the inference performs better for the gender that is dominant in each dataset (female in Flixster and male in Movielens). This is especially evident for SVM, which exhibits very high recall for the dominate class and low recall for the dominated class. The mixed model improves significantly on the Bernoulli model and results similarly to the multinomial. This indicates that the usage of a Gaussian distribution might not be a sufficiently accurate estimation for the distribution of the ratings.

Impact of user ratings. We assess the importance of the rating value itself (number of stars) versus the simple binary event "watched or not" by applying logistic regression and SVM on a binary matrix, denoted by $\tilde{X}$, in which ratings are replaced by 1. Table 1 shows the performance of these two methods on X and $\tilde{X}$. Interestingly, SVM and logistic regression performed only slightly better when using X rather than $\tilde{X}$ as input, with less than 2% improvement on all measures. In fact, Table 2 indicates that although using X performs better than using $\tilde{X}$ for the dominant class, it is worse for the dominated class. Similarly, the Bernoulli model, which also ignores the rating values, performed relatively close to Multinomial and Mixed. This implies that whether or not a movie is included in one's profile is nearly as impactful as the value of star rating given for the movie. This has important ramifications for obfuscation mechanisms that need to do two things: decide which movies to add to a user profile, and decide which rating to give a movie. This finding suggests that the choice of which movies to add could have a large impact on impeding gender inference. However if the actual ratings do not impact gender inference much, then we could select a rating value that helps maintain the quality of recommendations.

4.3 Analysis of Logistic Regression

We focus on logistic regression to further understand the classification results, since it provides us with coefficients for the movies and confidence in the gender inference. We note that a similar analysis can be done using SVM, which we omit for brevity.

Effect of training set size. Since we use 10-fold cross validation, our training set is large relative to the evaluation set. We use the Flixster data to assess the effect that the number of users in the training set size has on the inference accuracy. In addition to the 10-fold cross validation giving 3000 users in the evaluation set, we performed a 100-fold cross validation using a 300-user evaluation set. Additionally, we

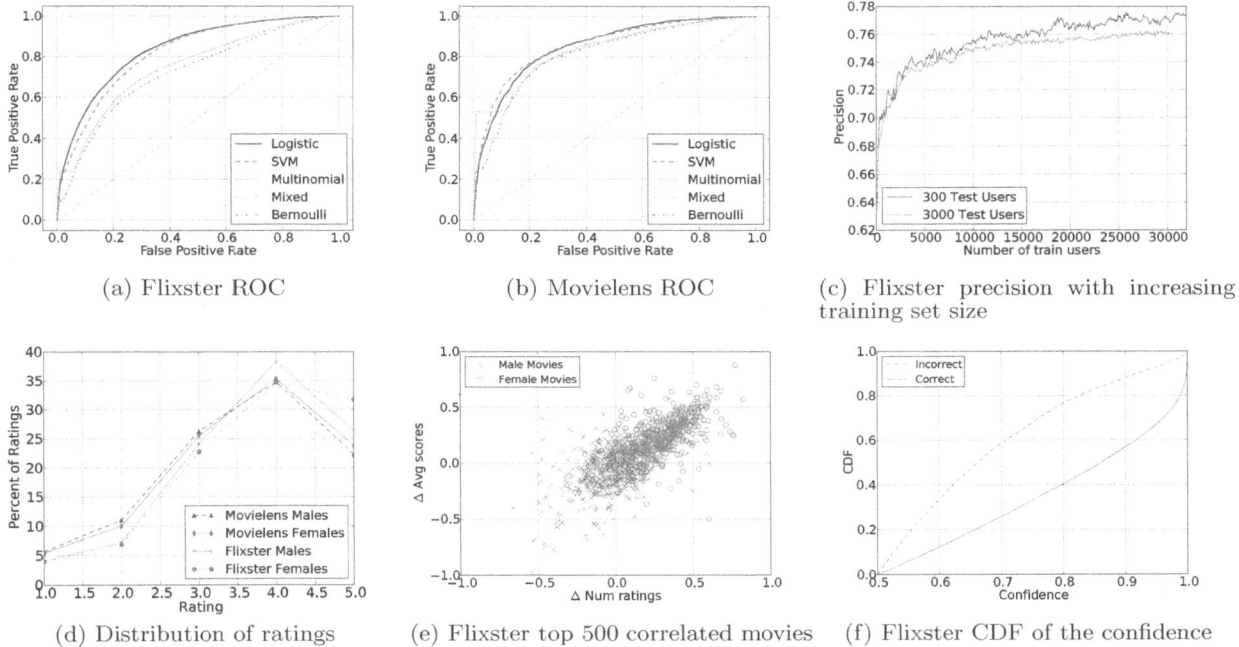

(a) Flixster ROC   (b) Movielens ROC   (c) Flixster precision with increasing training set size

(d) Distribution of ratings   (e) Flixster top 500 correlated movies   (f) Flixster CDF of the confidence

Figure 2: Results of gender classification

incrementally increased the training set, starting from 100 users and adding 100 more users on each iteration.

Figure 2c plots the precision of the logistic regression inference on Flixster for the two evaluation set sizes. The figure shows that for both sizes, roughly 300 users in the training set are sufficient for the algorithm to reach above 70% precision, while 5000 users in the training set reaches a precision above 74%. This indicates that a relatively small number of users are sufficient for training.

Movie-Gender Correlation. The coefficients computed by logistic regression expose movies that are most correlated with males and females. Table 3 lists the top 10 movies correlated with each gender for Flixster; similar observations as the ones below hold for Movielens. The movies are ordered based on their average rank across the 10-folds. We use average rank since the coefficients can vary significantly between folds, but the order of movies does not. The top gender correlated movies are quite different depending on whether X or $\tilde{X}$ is used as input. For example, out of the top 100 most female and male correlated movies, only 35 are the same for males across the two inputs, and 27 are the same for females; the comparison yielded a Jaccard distance of 0.19 an 0.16, respectively. We saw that many of the movies in both datasets align with the stereotype that action and horror movies are more correlated with males, while drama and romance are more correlated with females. However, gender inference is not straightforward because the majority of popular movies are well liked by both genders.

Table 3 shows that in both datasets some of the top male correlated movies have plots that involve gay males, (such as Latter Days, Beautiful Thing, and Eating Out); we observed the same results when using $\tilde{X}$. The main reason for this is that all of these movies have a relatively small number of ratings, ranging from a few tens to a few hundreds. In this case it is sufficient for a small variance in the rating distri-

Female	Male
Broken Bridges	Latter Days
Something the Lord Made	Beautiful Thing
Drunken Master	Birth
Dracula - Dead and Loving It	Eating Out
Young Indiana Jones	Prince of Darkness
Pootie Tang	Mimic
Anne of Green Gables	Show Girls
Another Cinderella Story	Godzilla: Final Wars
The Fox and the Hound 2	Studio 54
Winnie the Pooh	Desperately Seeking Susan

Table 3: Top male and female correlated movies in Flixster

butions between genders with respect to the class priors, to make the movie highly correlated with the class.

Further, we evaluate whether the distribution of movie ratings are different for males and females, and whether movies that are correlated with a gender tend to have more ratings from users of that gender. Figure 2d shows that the rating distribution is similar for females and males in the two datasets. In Figure 2e each dot corresponds to a movie; the x-axis plots the normalized difference in the number of ratings between females and males, and the y-axis (scaled to -1 to 1) shows the difference in the value of the ratings between the sexes. Green circles in the upper-right portion (blue crosses in lower left) of the plot indicate that the pair of features (captured on the two axes) can explain the highly female (male) correlated movies, respectively. While this pair of features explains some of the observed gender correlation, they are not nearly sufficient to explain all of it since more than half of the gender correlated movies lie in the middle of this plot.

Confidence in classification. Finally, the confidence value of the classifier is the obstacle that an obfuscation mechanism needs to overcome when trying to hide the gender from the classifier. The higher the confidence of the classifier in its prediction, the more effort the obfuscation method needs to apply, possibly increasing the impact on the recommendations. Therefore, we evaluate whether the classifier has different confidence values when it outputs a correct or incorrect classification. Figure 2f plots the CDF of the confidence value for correct and incorrect classifications, showing that the confidence is higher when the classification is correct, with a median confidence for incorrect classifications of 0.65, while for correct classification it is 0.85. Moreover, nearly 20% of correct classifications have a confidence of 1.0, which holds for less than 1% of incorrect classifications.

5. GENDER OBFUSCATION

The obfuscation mechanism takes as input a user i's rating profile $\mathcal{H}_i$, a parameter k that represents the number of permitted alterations, and information from the training set to output an altered rating profile $\mathcal{H}'_i$ such that it is hard to infer the gender of the user while minimally impacting the quality of recommendations received. In general, such a mechanism can alter $\mathcal{H}_i$ by adding, deleting or changing movie ratings. We focus on the setting in which the obfuscation mechanism is only allowed to add k movie ratings, since deleting movies is impractical in most services and changing ratings is more suspicious than adding ratings. Because users have different numbers of movies rated in their profiles (and some may have a small number), we do not use a fixed number k but rather we add a number that corresponds to a given percentage of movies in a user's rating profile. In order to add movies into a user's profile, the obfuscation mechanism needs to make two non-trivial decisions:

- Which movies should be added?
- What should be the rating assigned to each movie?

We refer to these added movie ratings as *extra ratings*. We note that the rating values assigned are not "noise" but have some useful value. For example, if this rating corresponds to the average rating over all users, or the predicted rating (using matrix factorization) for a specific user, then the rating value is a reasonable predictor of how the user may have rated had he watched the movie. In this work, we do not aim to provide an exhaustive list of obfuscation mechanisms, instead our goal is to design mechanisms informed by observations from our gender inference study (Section 4).

5.1 Obfuscation mechanisms

To simplify the discussion, we first assume that the obfuscation mechanisms have full access to the training dataset, and can use it to derive information for selecting movies and ratings to add. Later in this section, we amend the assumption of full access to the dataset.

Movie selection. We design three intuitive strategies for selecting movies. Each strategy takes as input a set of movies $\mathcal{S}_i$ rated by the user i, a the number of movies k to be added that corresponds to $p\%$ of i's existing profile, and ordered lists L_M and L_F of male and female correlated movies, respectively, and outputs an altered set of movies $\mathcal{S}'_i$, where $\mathcal{S}_i \subseteq \mathcal{S}'_i$. The lists L_M and L_F are stored in decreasing order of the value of a scoring function $w : L_M \cup L_F \to \mathbb{R}$ where $w(j)$ indicates how strongly correlated a movie $j \in L_M \cup L_F$ is with the associated gender. A concrete example of the scoring function is to set $w(j) = \beta_j$, where β_j is the coefficient of movie j obtained by learning a logistic regression model from the training dataset. We will use this instantia-

tion of the scoring function in our evaluation. Additionally, we assume that $k < \min(|L_M|, |L_F|) - |\mathcal{S}_i|$ and $L_M \cap L_F = \emptyset$.

The movie selection process is as follows. For a given female (or, male) user i, we initialize $\mathcal{S}'_i = \mathcal{S}_i$. Each strategy repeatedly picks a movie j from L_M (or, L_F), and if $j \notin \mathcal{S}'_i$ it adds j to $\mathcal{S}'_i$, until k movies have been added. The set $\mathcal{S}'_i$ is the desired output. The three strategies differ in how a movie is picked from the ordered lists of movies.

1. **Random Strategy.** For a given female (male) user i, pick a movie j uniformly at random from the list corresponding to the opposite gender L_M (L_F), irrespective of the score of the movie.
2. **Sampled Strategy.** Sample a movie based on the distribution of the scores associated with the movies in the list corresponding to the opposite gender. For instance, if there are three movies j_1, j_2, j_3 in L_M with scores $0.5, 0.3, 0.2$, respectively, then j_1 will be picked with probability 0.5 and so on.
3. **Greedy Strategy.** Pick the movie with the highest score in the list corresponding to the opposite gender.

Rating assignment. In Section 4.2, we made a key observation that the binary event of including or excluding a movie in a profile (indicating watched or not) was a signal for gender inference nearly as strong as the ratings. Given that, we aim to assign ratings to the extra movies that have a low impact on the recommendations provided to a user. We propose and evaluate two rating assignments:

1. **Average movie rating.** The obfuscation mechanism uses the available training data to compute the average rating for all movies $j \in \mathcal{S}'_i - \mathcal{S}_i$ and add them to user i's altered rating profile $\mathcal{H}'_i$.
2. **Predicted rating.** The obfuscation mechanism computes the latent factors of movies by performing matrix factorization on the training dataset, and uses those to predict a user's ratings. The predicted ratings for all movies $j \in \mathcal{S}'_i - \mathcal{S}_i$ are added to $\mathcal{H}'_i$.

Access to Dataset. Earlier we assumed the obfuscation mechanism had unrestricted access to the training set. We point out now that our mechanisms described above require access only to the following quantities: (a) for movie selection: ordered lists of male and female correlated movies, and (b) for rating assignment: average movie ratings, and movie latent factors to predict user movie ratings. Note that this information can be found from publicly available datasets, such as the Netflix Prize dataset[2]. Assuming that users in such public datasets are statistically similar overall to those in a particular recommender systems, then we no longer need the assumption of access to the training set.

5.2 Evaluating the obfuscation mechanisms

We evaluate all the permutations of movie selection and rating assignment strategies proposed above. We evaluate values of k corresponding to 1%, 5% and 10% $|\mathcal{S}_i|$ for each user i. The movie scores in lists L_M and L_F are set to the corresponding logistic regression coefficients.

Impact on privacy. We capture the privacy gain that obfuscation brings via the reduced performance in gender inference. Table 4 shows the accuracy of inference for all three movie selection strategies (i.e., random, sampled and

---

[2]http://www.netflixprize.com/index

	Classifier	Strategy	Accuracy with extra ratings			
			0%	1%	5%	10%
Flixster	Logistic Regression	Random	76.5	65.8	46.2	28.5
		Sampled	76.5	60.8	36.6	19.6
		Greedy	76.5	15	1.7	0.1
	Multinomial	Random	71.5	69.3	67	63.5
		Sampled	71.5	68.6	66	61.1
		Greedy	71.5	62	54.3	42.1
Movielens	Logistic Regression	Random	80.2	77.6	71.5	61.1
		Sampled	80.2	75.2	58.6	35.5
		Greedy	80.2	57.7	17.3	2.5
	Multinomial	Random	76.4	75.1	72.9	70.1
		Sampled	76.4	74.9	72.3	68.4
		Greedy	76.4	72.3	66.6	60.4

Table 4: Accuracy of gender inference for different strategies, when rating assignment is average movie rating

	Classifier	Strategy	Accuracy with extra ratings			
			0%	1%	5%	10%
Flixster	Logistic Regression	Random	76.5	65.4	45.5	27.4
		Sampled	76.5	60.5	35.7	18.3
		Greedy	76.5	15.1	1.5	0.1
	Multinomial	Random	71.5	69.5	67.2	63.8
		Sampled	71.5	68.9	66.3	61.5
		Greedy	71.5	63.3	54.9	42.4
Movielens	Logistic Regression	Random	80.2	76.9	68.9	52.7
		Sampled	80.2	73.9	48.9	24.9
		Greedy	80.2	48.4	7.2	0.6
	Multinomial	Random	76.4	74.5	71.8	67.9
		Sampled	76.4	74.3	70.5	65.9
		Greedy	76.4	71.1	64.1	57.3

Table 5: Accuracy of gender inference for different strategies, when rating assignment is users' predicted ratings

greedy) when the rating assigned is the average movie rating. The accuracy is computed using 10 fold cross validation, where the model is trained on unadulterated data, and tested on obfuscated data.

Since the accuracy of inference is the highest for the logistic regression classifier, it would be the natural choice as the inference mechanism for a recommender system. Figures 3a and 3d show the drop in inference accuracy for adding noisy ratings for the two datasets. On adding just 1% extra ratings using the greedy strategy, the accuracy drops to 15% (that is an 80% decrease) and with 10% extra ratings the accuracy is close to zero for the Flixster dataset, as compared with the accuracy of 76.5% on the unadulterated data. Therefore, if the obfuscation mechanism selects movies according to the greedy strategy, adding a small number of movies is sufficient to obfuscate gender. Even when the movies are chosen using the random strategy (which ignores movie scores and hence, the logistic regression coefficients), just 10% additional movies correlated with the opposite gender are sufficient to decrease the accuracy of gender inference by 63% (from 76.5% to 28.5% accuracy). Similar trends are observed for the Movielens dataset.

Our obfuscation mechanism above is using ordered lists that correspond well to the inference mechanism's notion of male or female correlated movies. However, in general, the obfuscation mechanism does not know which inference algorithm is used and thus lists such as L_M and L_F may have a weaker match to such a notion interior to the inference algorithm. We evaluate our obfuscation under such a scenario, with Multinomial Naïve Bayes and SVM classifiers. Our obfuscation still performs well as we see in Table 4, the inference accuracy of the Multinomial classifier drops from 71% to 42.1% for Flixster, and from 76% to 60% for the Movielens dataset (with 10% extra ratings and the greedy strategy). Obfuscation results in a similar decrease in gender inference accuracy of SVMs, results omitted for brevity.

Impact on recommendations. Next, we evaluate the impact on the recommendation quality that the user will observe if she obfuscates her gender. We measure this impact by computing the RMSE of matrix factorization on a held-out test set of 10 ratings for each user. Again, we perform 10 fold cross validation, where the data for users in 9 folds is unadulterated, and one of the folds has users with additional noisy ratings. That is, we use $\mathcal{H}'$ for a tenth of the users, and $\mathcal{H}$ for the rest. This is equivalent to evaluating the change in RMSE for 10% of the users in the system who obfuscate their gender. Figures 3b and 3e show the change in RMSE

due to obfuscation for Flixster and Movielens, respectively, when the ratings added were the same as in Table 4. Overall, we see that obfuscation has negligible impact on RMSE. For Flixster, we see that compared to the case of no extra ratings ("none") the RMSE increases with additional ratings, although negligibly. For Movielens, we observe a slight decrease in RMSE with extra ratings. We conjecture that this may occur because by adding extra ratings we increase the density of the original rating matrix which may improve the performance of matrix factorization solutions. Another explanation could be that the extra ratings are not arbitrary, but somewhat meaningful (i.e., the average across all users). The key observation is that for both datasets, the change in RMSE is not significant, a maximum of 0.015 for Flixster (with random strategy and 10% extra ratings), and 0.058 for Movielens (with sampled strategy and 10% extra ratings).

Analyzing privacy-utility tradeoff. We now take a comprehensive look at the privacy-utility tradeoff of the proposed obfuscation, where the desired high privacy corresponds to a low accuracy of gender inference, and a high utility corresponds to a low RMSE which is often used as a proxy for high quality recommendations. Figures 3c and 3f show the privacy (inference accuracy) on the x-axis and utility (RMSE) on the y-axis. Each point on the curve corresponds to the amount of extra ratings, where the rightmost point corresponds to no additional ratings, and the following points moving left are 1%, 5% and 10% extra ratings. For the Flixster dataset, as we move towards higher privacy the utility decreases. As described above, for Movielens as we move towards higher privacy, the utility increases however only slightly. These plots illustrate the clear trend that our obfuscation mechanism can lead to a substantial reduction in gender inference accuracy yet only incurs very small changes to the quality of the recommendations.

Preserving recommendation quality. We now evaluate the tradeoff when the rating assignment corresponds to the "predicted ratings" approach (Section 5.1). The motivation behind this rating assignment is that, in principle, this obfuscation results in no change in RMSE as compared with the RMSE on unaltered data. In other words, there is no tradeoff to be made on the utility front with this choice of rating assignment. Table 5 shows the accuracy of gender inference when this rating assignment is used. The results are similar those in Table 4 where the rating assignment is the average movie rating. For the Movielens data, the accuracy of gender inference is slightly lower with predicted ratings;

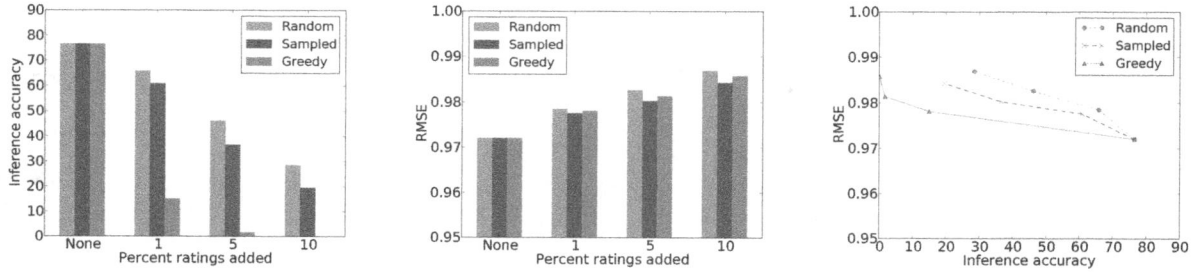

(a) Inference on obfuscation - Flixster (b) RMSE on obfuscation - Flixster (c) RMSE Inference tradeoff - Flixster

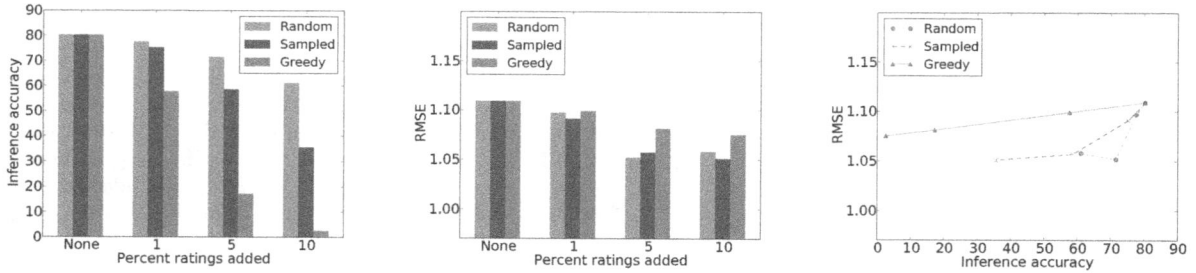

(d) Inference on obfuscation - Movie-lens (e) RMSE on obfuscation - Movielens (f) RMSE Inference tradeoff - Movielens

Figure 3: Effect of obfuscation on inference and recommendations

for example, for the greedy strategy with 1% extra ratings, the accuracy of the logistic regression classifier reduces from 57.7% to 48.4% - and this benefit comes without sacrificing the quality of recommendations.

In conclusion, our experimental evaluation shows that with small amount of additional ratings, it is possible to protect a user's gender by obfuscation, with an insignificant change to the quality of recommendations received by the user.

6. CONCLUSION

In this work we show that a user's rating profile alone can be used to infer her gender with high accuracy. Given a relatively small training set, our inference algorithms correctly predict the gender of users with a precision of 70%-80%. We use the insights from inferring gender to design obfuscation mechanisms that add ratings to a user's profile with the goal of making it hard to infer the user's gender, while posing a minimal impact on recommendation quality. We evaluate the tradeoff in the privacy and utility for different obfuscation mechanisms and show that just 1% additional ratings to a user's profile decreases the inference accuracy by 80%.

Although the focus of this paper is on a relatively simple binary inference of the gender, it raises a red flag regarding the possibility to infer private information about users based on the apparently non-revealing act of rating items for purpose of recommendations, unlike the more explicit actions performed in social networks. We plan to further study the accuracy of more sensitive private information in future work.

Acknowledgments. The authors would like to thank Mohsen Jamali for sharing the Flixster dataset.

7. REFERENCES

[1] G. Adomavicius and J. Zhang. On the stability of recommendation algorithms. In *RecSys*, 2010.

[2] S. Bhagat, I. Rozenbaum, and G. Cormode. Applying link-based classification to label blogs. In *WebKDD/SNA-KDD*, 2007.

[3] Z. Cheng and N. Hurley. Effective diverse and obfuscated attacks on model-based recommender systems. In *RecSys*, 2009.

[4] T. Hastie, R. Tibshirani, and J. H. Friedman. *The elements of statistical learning: data mining, inference, and prediction*. New York: Springer-Verlag, 2001.

[5] M. Jamali and M. Ester. A matrix factorization technique with trust propagation for recommendation in social networks. In *RecSys*, 2010.

[6] Y. Koren, R. Bell, and C. Volinsky. Matrix factorization techniques for recommender systems. *IEEE Computer*, 42(8):30–37, 2009.

[7] A. McCallum and K. Nigam. A comparison of event models for Naive Bayes text classification. In *AAAI*, 1998.

[8] F. McSherry and I. Mironov. Differentially private recommender systems: building privacy into the net. In *KDD*, 2009.

[9] A. Mislove, B. Viswanath, K. P. Gummadi, and P. Druschel. You are who you know: Inferring user profiles in Online Social Networks. In *WSDM*, 2010.

[10] M. P. O'Mahony, N. J. Hurley, and G. C. M. Silvestre. Recommender systems: Attack types and strategies. In *AAAI*, 2005.

[11] J. Otterbacher. Inferring gender of movie reviewers: exploiting writing style, content and metadata. In *CIKM*, 2010.

[12] D. Rao, D. Yarowsky, A. Shreevats, and M. Gupta. Classifying latent user attributes in twitter. In *SMUC*, 2010.

[13] A. P. Singh and G. J. Gordon. Relational learning via collective matrix factorization. In *KDD*, 2008.

[14] V. Toubiana, L. Subramanian, and H. Nissenbaum. Trackmenot: Enhancing the privacy of web search. *CoRR*, abs/1109.4677, 2011.

[15] S. Ye, F. Wu, R. Pandey, and H. Chen. Noise injection for search privacy protection. In *CSE*, 2009.

Distributed, Real-Time Bayesian Learning in Online Services

Ralf Herbrich
Facebook Inc.
1601 Willow Road
Menlo Park, CA 94025, US
ralf@fb.com

ABSTRACT

The last ten years have seen a tremendous growth in Internet-based online services such as search, advertising, gaming and social networking. Today, it is important to analyze large collections of user interaction data as a first step in building predictive models for these services as well as learn these models in real-time.

One of the biggest challenges in this setting is scale: not only does the sheer scale of data necessitate parallel processing but it also necessitates distributed models; with over 900 million active users at Facebook, any user-specific sets of features in a linear or non-linear model yields models of a size bigger than can be stored in a single system.

In this talk, I will give a hands-on introduction to one of the most versatile tools for handling large collections of data with distributed probabilistic models: the sum-product algorithm for approximate message passing in factor graphs. I will discuss the application of this algorithm for the specific case of generalized linear models and outline the challenges of both approximate and distributed message passing including an in-depth discussion of expectation propagation and Map-Reduce.

Categories and Subject Descriptors

I.2.6 [**Computing Methodologies**]: Artificial Intelligence—
Learning

Keywords

Graphical Models, Distributed Machine Learning, Gaming

1. SUMMARY

Over the past decade, the emergence of online services has changed one of the implicit assumptions of machine learning made thus far, namely that the amount of data available for learning can be processed in real-time sequentially by a single computer. Today, rather than having to be concerned about the lack of data and compensating with sophisticated and computationally intensive methods such as structural priors and regularizers, we live in a world where we have more training data available than can be handled by a single computer.

For example, in a single day, an average of 526 million users generate more than 3.2 billion likes and comments on the Facebook platform alone[1]. Such data-streams, which are often the digitization of people interacting with the world, each other or search results, are no longer generated as a single stream but get recorded and analyzed in a fully distributed manner. For example, it is estimated that Google uses around 1 million servers to drive its platform[2].

This shift in data availability necessitates reconsidering machine learning methods from a scaling-out point of view — *distributed machine learning*. In distributed machine learning, we are given a distributed *set of* data streams producing data of varying rate and are required to successively make a prediction as well as incorporating the data item in a distributed way; for conditional or discriminative models, the data is usually divided into a set of features together with a set of target values that *co-occur*. In this setting, key performance criteria for such systems include the ability to be fault-tolerant, i.e., what fraction of machines can fail to still operate successfully, the ability to be state-less, i.e., how many additional bits need to be communicated per request in order to perform prediction and/or training, and the ability to be adaptive to varying rates of data due to distribution shifts in the data.

In this talk, I will present the *sum-product algorithm* [2] and expectation propagation for approximate inference—one of the key building blocks for distributed machine learning. I will discuss how to map the probabilistic messages to network messages and show that the algorithm can be viewed as a real-time generalization of the famous Map-Reduce framework [1] which is a single-pass message exchange in a tree-approximation to a probabilistic model. I will also give an overview of some distributed learning systems already in use in industry applications including the *TrueSkill* ranking and matchmaking system in Xbox Live and collaborative filtering systems used for content recommendation at Facebook.

2. REFERENCES

[1] J. Dean and S. Ghemawat. Mapreduce: Simplified data processing on large clusters. In *In Proceedings of the 6th conference on Symposium on Operating Systems Design & Implementation*, pages 137–150. USENIX Association, 2004.

[2] F. R. Kschischang, B. J. Frey, and H.-A. Loeliger. Factor graphs and the sum-product algorithm. *IEEE Transactions in Information Theory*, 47:498–519, 1998.

---

[1]see http://newsroom.fb.com/.
[2]see http://www.datacenterknowledge.com/.

Recommendation Challenges in Web Media Settings

Ronny Lempel
Yahoo! Labs
Haifa, Israel
rlempel@yahoo-inc.com

ABSTRACT

This paper calls out several research challenges in the art of recommendation technology as applied in Web media sites. One particular characteristic of such recommendation settings is the relative low cost of falsely recommending an irrelevant item, which means that recommendation schemes can be less conservative and more exploratory. This also creates opportunities for better item cold-start handling. Other technical difficulties include analyzing offline data that is heavily biased by the site's appearance, and in a related vein – once a recommendation module's appearance has been designed – defining the correct metrics by which to measure it. Also called out are tradeoffs between personalization and contextualization, as are novel schemes that aim at recommending sets and sequences of items.

Categories and Subject Descriptors

H.3.3 [**Information Search and Retrieval**]: Information filtering

Keywords

Personalization, contextualization, exploration, presentation bias, constrained recommendation, Web media sites

1. INTRODUCTION

The art of recommender systems has evolved tremendously since its early incarnations in the nascent field of eCommerce in the late 1990s. While the field now merits a dedicated conference, many challenges exist - particularly around new applications of recommendation technology in different settings. This extended abstract calls out several challenges arising from the integration of recommendation technology into Web media sites.

2. LOW FALSE POSITIVE COSTS

In many recommendation scenarios, the cost suffered by the user when consuming an irrelevant item – as measured by money or time spent – can be quite high. On the other hand, when browsing a Web media site, the cost of a click leading to a non-engaging story is rather low - no money and very little time is lost by the user. Given the low cost of falsely recommending items, recommender systems can

afford to be less conservative and to take more risks in their output. Two under-explored examples follow.

First, while bootstrapping techniques that proactively address user cold-start problems exist [6], less effort has been devoted to actively overcoming item cold-start issues. Quickly modeling new items is particularly important in Web media sites, where story lifetimes are short. An intriguing direction is to present new items to a carefully selected set of users whose reactions to the items – even if negative – will forecast the response of the rest of the population to the item.

Second, the lower cost of a false positive allows for broader exploration. In many top-(small)-k retrieval settings, items that make it to the top are worthy but also popular, and being presented to users (and receiving generally good feedback) perpetuates their standing [11]. Many other items, that would potentially be appreciated by users, do not get the initial exposure that is necessary for obtaining stature. Meanwhile, exploring the item space for observing rewards is a well known concept in Reinforcement Learning, and has been successfully applied to coarse-grain content filtering problems on the Web [1]. How can one combine exploration with fine-grained personalization in a principled manner? One intriguing direction is the use of latent-space representations of items or users, heavily applied in recommender systems [9], as contexts for certain multi-armed bandit explore-exploit algorithms (e.g. [10]).

3. INFLUENCE OF PRESENTATION BIAS ON CONSUMPTION

Typically, before a system goes into production, offline data is analyzed in order to observe user-item interactions. In many pop-culture domains items have longevity, and users are aware of items regardless of whether the site running the recommendation system actually exposed the items to them. For example, users are aware of old movies (e.g. "Casablanca", released 70 years ago) or non-contemporary bands (e.g. "The Beatles", broken up over 40 years ago) independently of recommender systems. However, in news sites, items are short-lived and users typically are exposed to the latest news while browsing the site rather than arriving to it with an informed agenda of interest. This means that whatever content is showcased by the site has tremendous advantage in terms of exposure to users [3]. Consequently, consumption (browsing) patterns on news sites are heavily biased by the sites' presentation, which limits users' options and obscures their true tastes. Correctly accounting for presentation bias when analyzing offline data is challenging and may require knowing the layout seen by users and considering also the items they chose not to consume [7].

4. RECOMMENDATION MODULES: LAYOUT AND METRICS

Recommendation modules come in many shapes and forms. Some are fixed size, displaying exactly some top-k items. Others allow scrolling through multiple k-sized sets of items. Lately, continuous scrolling modules deliver more items as the user scrolls, leading to an "infinite list" experience.

The differences in how recommendations are shown to the user translate to differences in the metrics that should drive the offline stage of data analysis and modeling. When showing only a few top items, offline learning should probably focus on identifying only the best items available per user per time, whereas discriminating between non top-k items (e.g. the RMSE or AUC metrics) is far less consequential.

Furthermore, online metrics should look beyond just consumption rates, or – as typically manifested in media sites – click-through rates (CTR). It is well known that media sites can gain short-term clicks by promoting sensational stories; however, such policies are detrimental to a site's brand and appeal in the long run. While metrics such as long-term user satisfaction are difficult to quantify, engagement cues other than CTR - dwell time, sharing, rating, commenting – might be better indicators of user satisfaction and can be tapped for learning.

5. CONTEXTUALIZATION VS. PERSONALIZATION

Web media sites often display links to additional stories on each article page. These stories are sometimes tabbed as "related stories" and sometimes as "recommended for you", but the use-case is essentially the same. Given a user visiting a Web page, how should the additional stories be chosen? What should be the relative importance of the story on the page (the context) vs. the person visiting the page? The items that best match the user might be unrelated to the story, and in certain cases displaying them in the story's context may be distasteful and could reflect badly on the site. On the other hand, putting too much emphasis on related context may result in stories that do not match very well the user's taste. Coming up with the right balance of contextualized and personalized factors is a challenging problem, and novel techniques such as Tensor Factorization [8] may prove helpful here.

6. SETS AND SEQUENCES OF ITEMS

In certain media types users often consume several items in rapid succession. Examples include listening to music and browsing online content on a media site. This raises the question of whether items should be recommended independently, based only on their individual projected affinity to the user, or whether the system should aim to recommend a set or sequence of items that are "well consumed together". Recently, two papers attempted to model radio playlists as sets [4] and sequences [5] of items, and clearly there is room for additional investigation. Furthermore, in some domains, constraints might be imposed over the items that are co-consumed together (e.g. playlists may not be allowed to include too many tracks of any individual artists),

further complicating higher order recommendation settings and painting them as *constrained recommendation* problems.

Note that in a different domain, the search engine literature has long considered result set attributes such as diversity to be important [2].

7. REFERENCES

[1] D. Agarwal, B.-C. Chen, and P. Elango. Explore/exploit schemes for web content optimization. In *Proc. Ninth IEEE International Conference on Data Mining (ICDM'2009)*, pages 1–10, 2009.

[2] R. Agrawal, S. Gollapudi, A. Halverson, and S. Leong. Diversifying search results. In *Proc. 2rd ACM Conference on Web Search and Data Mining (WSDM'2009)*, pages 5–14, February 2009.

[3] M. Aharon, A. Kagian, Y. Koren, and R. Lempel. Dynamic personalized recommendation of comment-eliciting stories. In *Proc. 6th ACM Conference on Recommender Systems (RecSys'2012)*, September 2012.

[4] N. Aizenberg, Y. Koren, and O. Somekh. Build your own music recommender by modeling internet radio streams. In *Proc. 21st International World Wide Web Conference (WWW'2012)*, pages 1–10, April 2012.

[5] S. Chen, J. Moore, D. Turnbull, and T. Joachims. Playlist prediction via metric embedding. In *Proc. 18th ACM SIGKDD Conference on Knowledge Discovery and Data Mining (KDD'2012)*, August 2012.

[6] N. Golbandi, Y. Koren, and R. Lempel. Adaptive bootstrapping of recommender systems using decision trees. In *Proc. 4rd ACM Conference on Web Search and Data Mining (WSDM'2011)*, pages 595–604, February 2011.

[7] T. Joachims. Optimizing search engines using clickthrough data. In *Proc. 8th ACM SIGKDD Conference on Knowledge Discovery and Data Mining (KDD'2002)*, pages 133–142, July 2002.

[8] A. Karatzoglou, X. Amatriain, L. Baltrunas, and N. Oliver. Multiverse recommendation: N-dimensional tensor factorization for context-aware collaborative filtering. In *Proc. 4th ACM Conference on Recommender Systems (RecSys'2010)*, September 2010.

[9] Y. Koren, R. M. Bell, and C. Volinsky. Matrix factorization techniques for recommender systems. *IEEE Computer*, 42(8):30–37, 2009.

[10] L. Li, W. Chu, J. Langford, and R. E. Schapire. A contextual-bandit approach to personalized news article recommendation. In *Proc. 19th International World Wide Web Conference (WWW'2010)*, pages 781–792, April 2010.

[11] S. Pandey, S. Roy, C. Olston, J. Cho, and S. Chakrabarti. Shuffling a stacked deck: The case for partially randomized ranking of search engine results. In *Proc. 31st International Conference on Very Large Databases (VLDB'2005)*, pages 781–792, August 2005.

I've got 10 Million Songs in My Pocket. Now What?

[Abstract]

Paul Lamere
The Echo Nest
Somerville Massachusetts
paul@echonest.com

ABSTRACT

The proverbial celestial jukebox has become a reality. With today's online music services a music fan is never more than a few clicks away from being able to listen to nearly any song that has ever been recorded. Recommender systems can play a key role in this new music ecosystem, helping listeners explore, discover, organize and share music. However, in many ways music recommendation is very different than recommendation in other well-studied domains such as books and movies. In this talk we explore how recommender systems can be used in the music space, and the particular challenges that the music domain presents to the designers of recommender systems.

Categories and Subject Descriptors

H.3.3 [**Information Systems Applications**]: Information Filtering

General Terms

Experimentation

Keywords

Recommender Systems, Music

Dynamic Personalized Recommendation of Comment-Eliciting Stories

Michal Aharon, Amit Kagian, Yehuda Koren, Ronny Lempel

Yahoo! Labs, Haifa, Israel

{michala,akagian,yehuda,rlempel}@yahoo-inc.com

ABSTRACT

Media Websites often solicit users' comments on content items such as videos, news stories, blog posts, etc. Commenting activity increases user engagement with the sites, by both comment writers and readers, and so sites are looking for ways to increase the volume of comments. This work develops a recommender system aiming to present users with items – news stories, in our case – on which they are likely to comment. We combine items' content with a collaborative-filtering approach (utilizing users' co-commenting patterns) in a latent factor modeling framework. Building upon previous work, we focus on a continuous, real-time approach to address the problem above.

After an initial training period during which commenting activity of users is observed, the system is tested at each subsequent comment submission event by predicting which story is being commented on by a given user at a given time.

Our results show that we are able to overcome the site's inherent presentation bias and outperform a strong baseline as users' commenting history grows.

Categories and Subject Descriptors

H.3.3 [**Information Search and Retrieval**]: Information filtering

Keywords

Collaborative filtering, user comments, personalization

1. INTRODUCTION

Media Web sites are fiercely competing over online users' attention and engagement. A common feature of many present-day media sites is their attempt to enrich user engagement by soliciting and distributing user generated content (UGC), and in particular users' comments on content. Users are empowered to write comments, are able to rate and share comments, and often may also reply to others' comments, thereby generating discussion threads. Vibrant

Figure 1: Demonstration of a potential application. The application acknowledges a comment submission, and recommends the user other stories to comment on.

discussions around engaging stories may encompass thousands of comments.

This work focuses on news sites, and aims to present to each commenting user a personalized list of stories that may engage the user to the point of commenting on them. Figure 1 depicts such an experience. This is an extreme form of content recommendation, focused on recommending deeply engaging stories that will cause users to actively generate content rather than to merely passively consume it.

In a recently published paper [2], the above problem was tackled in a standard train-test learning framework. The commenting history of users was observed on a training set of stories, and the task was to rank – for each user – all remaining test stories in descending order of the probability of the user commenting on them. Using a combination of story vocabulary and collaborative filtering (CF) signals that utilize co-commenting patterns of users, a latent-factor model was shown to have good predictive signal.

This work follows up and extends upon [2]. Our contributions are the following.

First, we extend the traditional (batch) train-test analysis of [2] with a continuous, dynamic approach that enables personalized recommendation of stories in real time. After an initial training period, the system is tested at each subsequent comment submission event, by predicting which story is being commented on by a given user at a given time. This temporal consideration allows us to manage the available stories in all stages of their (typically brief) life cycle.

Second, we demonstrate the strong effect of a media site's structure and the ephemeral nature of the news articles on the commenting patterns of its users. Media sites often serve thousands of stories, that constantly vary, of which only a few dozens are prominently featured on the site at any given time (e.g. on its main page). The remaining stories are far less discoverable and are only reachable through deliberate effort-consuming search and browse operations. Therefore, at any moment, only a small subset of stories is heavily clicked, read, and eventually commented on. This form of *presentation bias* obscures the ideal commenting preferences of users, who end up commenting on the "better of what's featured" rather than on the "best of what's available". As our evaluation is post-factum, we reveal this bias by demonstrating a non-personalized recommendation baseline that achieves strong results. The baseline follows aggregate commenting trends and predicts, for each commenting instance, stories that have been heavily commented on recently, i.e. the stories that are (presumably) currently featured.

Third, our results show that we are able to overcome the site's inherent presentation bias and outperform the strong baseline above as users' commenting history grows. As our system learns more about its users, the precision of its predictions improves and it better predicts how users' commenting activity will deviate from the overall commenting trends created by the presentation bias.

Finally, as a contribution to evaluation methodology, we develop a framework for prediction that simulates an on-line scenario using off-line data.

A thorough review of prior art on comments analysis and the algorithmic techniques used in this paper appears in [2]. Also related to the current work is a paper attempting to predict the popularity of news articles according to content characteristics such as source, language and context [1].

2. PROBLEM SETTINGS

The goal of this work is to build a dynamic (real-time) personalized story recommendation system for commenting. The system aims to recommend to a user, who is currently active on the site, stories that she is likely to comment on. The notion of the current time is therefore critical - whatever happened in the past is known, as opposed to whatever happens next. This is significant, as comments' dynamics are constantly changing. Stories that are highly trending now may be far less relevant in an hour. Time will play a major role in our experiments, which simulate real-time recommendation using only historical off-line data.

2.1 Data Description

In this work we used data that was extracted by a crawler from the Newsvine website[1] during the period between May 18, 2011 till September 27, 2011 (same as in [2]). The data includes users (u_i), stories (s_j), and comments (c_k). Comments are represented by tuples (user, story, creation time), i.e. $c_k = \left(c_k^u, c_k^s, c_k^t \right)$. Stories are represented by their title, content, and editorial tags (when available). For simplicity, we filtered out users that wrote only one comment during the whole period, together with their corresponding comments. After filtering, our data encompasses $68,649$ stories with $1,036,657$ comments, written by $61,663$ users.

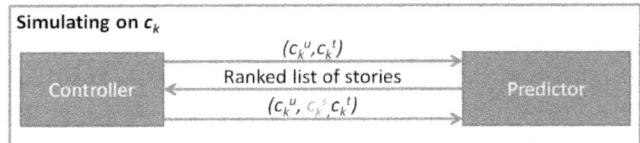

Figure 2: Flow of information between the controller and the predictor in a single step of the simulation.

2.2 Prediction Problem Definition

In order to better illustrate the way we exploit off-line data for simulating an on-line scenario, let us assume the existence of a 'controller' object that is aware of all data in the system, and manages the simulation. Let $\{c_1, c_2, \ldots, c_N\}$ be the sequence of all comments in the data, sorted by their creation time, and let w be the number of training comments. The controller provides the predictor with the sequence $\mathbf{C_{train}} = \{\mathbf{c_1}, \mathbf{c_2}, \ldots, \mathbf{c_w}\}$ for training, together with the story data for the corresponding stories $\left\{ c_j^s \right\}_{j=1}^{w}$ (content, titles, tags). The simulation phase iterates over the remaining comments, $\mathbf{C_{test}} = \{\mathbf{c_{w+1}}, \mathbf{c_{w+2}}, \ldots, \mathbf{c_N}\}$. When simulating on a comment c_k, let us define the available stories $(AS(k))$ as the set of all stories seen so far $\left(\left\{ c_j^s \right\}_{j=1}^{k-1} \right)$. The controller provides the predictor with the user id and the current time $\left(c_k^u, c_k^t \right)$, and the predictor outputs a ranked list of $AS(k)$. The controller then computes the prediction performance by calculating the **rank** $(r(k))$ of the true commented story (c_k^s), defined as the number of available stories ranked higher than c_k^s, which were not commented by c_k^u in the entire data. Finally, the controller reveals c_k^s, with its available data, for possible usage in subsequent predictions. Figure 2 illustrates a single simulation step. All computed ranks are stored for final evaluation of the predictor.

2.3 Evaluation Metrics

Given the ranks computed for the comment set $\mathbf{C_{test}}$, the predictor performance are evaluated using two metrics,

- Mean Reciprocal Rank (mrr) - the average of the inverse ranks $mrr = 1/|\mathbf{C_{test}}| \sum_{c_k \in \mathbf{C}_{test}} 1/r(k)$.

- Recall@K - the ratio of comments from $\mathbf{C_{test}}$ with rank lower or equal K,

$$Recall@K = \frac{1}{|\mathbf{C_{test}}|} \sum_{c_i \in \mathbf{C}_{test}} \mathrm{Ind}(r(i) \leq K),$$

where Ind is the indicator function.

3. BASELINE PREDICTOR

According to the problem setting above, when given a user id and time, we know this user is logged into the system, and available for commenting. It is reasonable, therefore, to inspect recent trends of comments in order to predict the story that the user is about to comment on.

3.1 Description

The baseline predictor we applied is based on recent commenting trends, which we denote as the *Trending Score* (TS). For each available story s in time t, the current rate of its ℓ last comments is computed by

$$TS(s,t) = \frac{(t - t_s(\ell))}{\ell}, \tag{1}$$

---

[1]www.newsvine.com

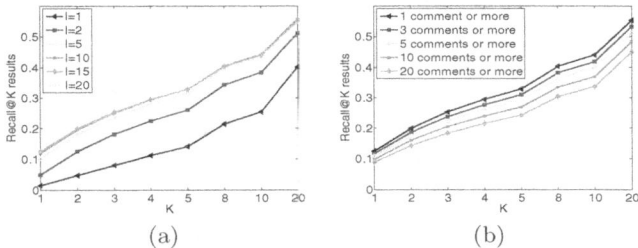

Figure 3: (a) Recall@K results for several values of ℓ, using the baseline predictor. (b) Recall@K results for $\ell = 15$, for users with different history lengths.

where $t_s(\ell)$ is the creation time of the ℓth last comment of s. Interpolation is used for computing TS for stories with less than ℓ comments. The normalized version of the trending score, $0 \le NTS \le 1$ is defined by

$$NTS(s,t) = \max\left\{0, \frac{D - TS(s,t)}{D}\right\}, \quad (2)$$

where D is a fixed large enough comment rate that reflects 'old' stories (e.g. 3 days per comment). In each prediction step (for any user), the baseline predictor ranks the available stories according to a descending order of their NTS values.

3.2 Baseline Results

Figure 3(a) presents recall@K results for several values of ℓ. As expected, for low values of ℓ the results are worse, as every comment written by some user has a strong effect on the story NTS. For higher values of ℓ, recall results are better and more stable. While results for $\ell = \{10, 15, 20\}$ nearly coincide, the value of $\ell = 15$ exhibited the best results, and we adopt it for the rest of this paper. Due to the strong presentation bias of the site, the baseline predictor is quite accurate. It ranks the true commented story in the top 3 places (*Recall@3*) in more than 25% of cases, providing a robust baseline that seems difficult to beat by personalization. Trying to learn some more about the users, we analyzed the recall results for commenters with different amounts of commenting history. In Figure 3(b) we show the recall results for the baseline predictor, for users with different history lengths (note the same user may contribute to all graphs if she commented 20 or more times in our data). We see that 'heavy commenters' are harder to predict using the trending score baseline. Yet, when trying to develop prediction based on personalization, longer history becomes an asset. The following section describes our recommendation scheme, that models deviations from the NTS score based on prior actions of users.

4. PERSONALIZED PREDICTOR

We adopt a latent factor method similar to the one presented in [2]. Our system learns a latent factor of dimension d for each user, tag and co-commenter, where as tags we consider both editorial tags and words from the story's title and content in their base format (after stemming). We deviate from [2] in that we do not expect our model to explain all user-story links, but rather to explain the **deviations** from the dynamic trending score. That is, there is no need to explain the fact that users comment on trending stories, which is to be expected. However, when a user comments

on a non-trending story, our model should have evidence of a strong link between this user and the story. Moreover, if a user does not comment on a trending story, we would expect our algorithm to show a negative link between this user and the story, as the user was likely exposed to this story while commenting, but decided to comment on a different story.

4.1 Prediction Mechanics

Each person in the system is represented by two latent factor vectors, corresponding to the two roles each person plays: $\overline{u_i} \in R^d$ as a user, and $\overline{m_i} \in R^d$ as a co-commenter. Each story s_i is represented by a combination of its commenters and content. Denoting by $\overline{t_i} \in R^d$ a latent factor vector representing a content element (tag) t_i, the story representation for story s_i at time t is defined by

$$\overline{s_i(t)} = \frac{1}{|t_i \in \mathbf{T}_i|} \sum_{t_i \in \mathbf{T}_i} \overline{t_i} + \frac{1}{|m_i \in \mathbf{M}_i(t)|} \sum_{m_i \in \mathbf{M}_i(t)} \overline{m_i}, \quad (3)$$

where $\mathbf{T}_i$ is the set of tags of s_i, and $\mathbf{M}_i(t)$ is the set of all commenters of s_i until time t. Note that the story representation is enriched as more users comment on it. We define a user-story score at time t as the sum of the story's normalized trending score and the inner product between the user's latent factor vector and the story's representation:

$$r_{u_i,s_j,t} = NTS(s_j,t) + \left\langle \overline{u_i}, \overline{s_j(t)} \right\rangle. \quad (4)$$

The left additive in Eq. 4 is responsible for the temporal dynamics. Old or non-popular stories will receive a low score even when they match a user well. This part is constantly changing as time passes. The right additive is the personalization part. It may have both positive and negative values, in different magnitudes, and represents the personalized deviation from the trending score. A large positive value reflects a good match between a user and a story, and therefore even if the story is not very trending, the user is likely to comment on it. Conversely, a negative value represents a bad match between a user and story, so even if the story is trending, this user is unlikely to comment on it.

4.2 Error Function

In the training stage we train all latent factor vectors in order to minimize a cost function. The cost function for minimization is an aggregation of a loss function, computed per all user-story scores over the training period:

$$cost = \sum_{c_k \in \mathbf{C}_{train}} loss(c_k^u, c_k^s, s^-(c_k^u), c_k^t). \quad (5)$$

For each commented story (c_k^s), we also sample one negative story $s^-(c_k^u)$ that indicates a bad match for the user (c_k^u did not comment on it during the whole training period). This loss function, presented in [3], tries to generate a gap between the $r_{u_i,s_j,t}$ values of the positive and negative indications by maximizing their distance. Sampling was done proportionally to $g(NTS)$, where g is a Gaussian function with mean value of 1, and std σ. When sampling stories with high NTS values, we can more comfortably assume these stories were visible to the user prior to commenting, yet the user decided to not comment on them. We train our model by employing stochastic gradient descent with early stopping. Each iteration enumerates all training comments, updating both the trending score for all stories and the various latent factor vectors.

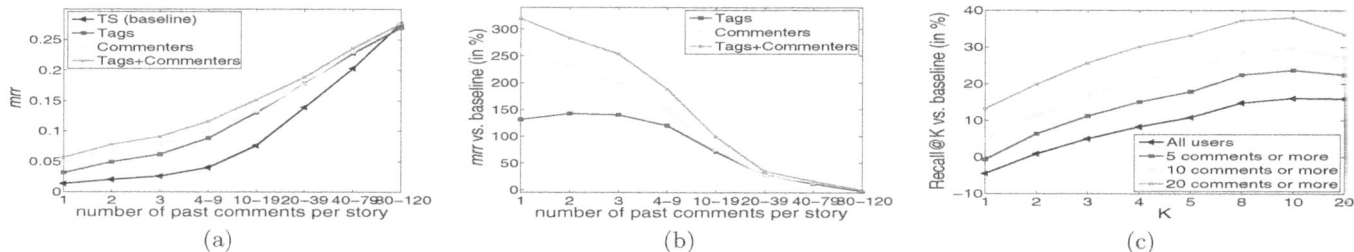

Figure 4: (a) *mrr* results as a function of the number of existing comments per story. Different plots refer to different story representations (tags, commenters or both). (b) percentage of improvement compared to the baseline predictor, for Figure (a). (c) Recall@K improvement (in percentage) of our model's results over the baseline's, for users with different history lengths.

4.3 Continuous learning

Every user-story score changes during simulation because of two reasons. First, the trending score is constantly changing. Second, when comments are added to a story, its representation changes, also affecting the user-story score. However, the learning process described so far does not change the latent factor vectors of the user. Such an update is advocated since as time passes, more user actions are observed and can be exploited for prediction, in particular in cases where user preferences change over time. We therefore retrain the system periodically, as described in Section 5.3.

5. RESULTS

In the following, we set the training set size to $w = 0.5 \cdot N$, the latent space dimension $d = 10$, the standard deviation for sampling $\sigma = 0.02$, and learning rate $\lambda = 0.1$.

5.1 Content Vs. Collaboration

Figure 4(a) shows the *mrr* as a function of the number of past comments of a story. We compare 3 modes of the algorithm - using only tags for story representation, only commenters, and using both. We observe that tags alone improve upon the baseline predictor. However, co-commenting data is even more effective, and results are best when combining the two. Figure 4(b) presents the relative improvement (in %) of each latent factor model compared to the baseline predictor. We see that the additional signals significantly improve over the non-personalized baseline, especially for stories that have yet to accumulate many comments.

5.2 Advantage of prior user information

Figure 4(c) shows *recall@K* results for users with different history lengths. The plots present the percentage improvement of our algorithm (tags + commenters) compared to the baseline results. We see that the baseline predictor deteriorates for users with longer history (see section 3.2). However, our algorithm is able to overcome this deterioration, and achieves greater improvement for these users.

5.3 Continuous learning results

Finally we show how performance changes when applying the continuous learning procedure. Here we initially trained on 20% of our data, and re-trained after 60 days (shorter time periods did not show noticeable improvement in performance). Figure 5 presents *mrr* results per each 4 days of simulation for the three methods: TS (baseline), tags + commenters, and tags+commenters with additinal learning

after 60 days. The latter method coincides with the second one for the first simulation period, and outperforms it afterwards once additional learning is performed.

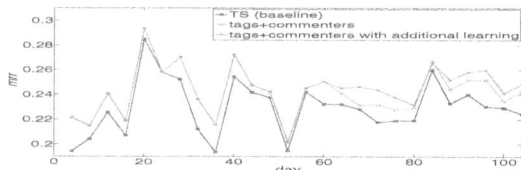

Figure 5: *mrr* per each 4 days of simulation. Note the effect of the additional learning on day 60.

6. CONCLUSIONS AND FUTURE WORK

This work presented a dynamic, real-time latent factor modeling approach for prediction of stories that users are likely to comment on. In addition to story content and user co-commenting patterns, we tap the temporal dynamics of commenting activity that emerge from the presentation bias of media sites. Our method models the personalized deviation of each user from the aggregate commenting behavior, and exhibits good results in a simulation of the real-time scenario that uses off-line data. We examined how personalization improves recommendation for users with varying lengths of commenting history and demonstrated how it can be further improved by re-training of the model over time. We plan to further tune the personalization factor in an online experiment of the proposed user experience.

7. REFERENCES

[1] R. Bandari, S. Asur, and B. A. Huberman. The pulse of news in social media: Forecasting popularity. *CoRR*, abs/1202.0332, 2012.

[2] E. Shmueli, A. Kagian, Y. Koren, and R. Lempel. Care to comment? recommendations for commenting on news stories. In *Proc. 21st International World Wide Web Conference (WWW'2012)*, April 2012.

[3] J. Weston, S. Bengio, and N. Usunier. Large scale image annotation: Learning to rank with joint word-image embeddings. *Machine Learning Journal*, 81:21–35, 2010.

Using Graph Partitioning Techniques for Neighbour Selection in User-Based Collaborative Filtering

Alejandro Bellogín
Information Retrieval Group
Department of Computer Science
Universidad Autónoma de Madrid
alejandro.bellogin@uam.es

Javier Parapar
Information Retrieval Lab
Department of Computer Science
University of A Coruña
javierparapar@udc.es

ABSTRACT

Spectral clustering techniques have become one of the most popular clustering algorithms, mainly because of their simplicity and effectiveness. In this work, we make use of one of these techniques, Normalised Cut, in order to derive a cluster-based collaborative filtering algorithm which outperforms other standard techniques in the state-of-the-art in terms of ranking precision. We frame this technique as a method for neighbour selection, and we show its effectiveness when compared with other cluster-based methods. Furthermore, the performance of our method could be improved if standard similarity metrics – such as Pearson's correlation – are also used when predicting the user's preferences.

Categories and Subject Descriptors

H.3.3 [**Information Search and Retrieval**]: Information Filtering, Clustering

General Terms

Algorithms, Experimentation, Performance

Keywords

Recommender systems, Collaborative Filtering, Clustering, Normalised Cut

1. INTRODUCTION

Collaborative Filtering (CF) is a particularly successful form of personalized Information Retrieval which suggests interesting items to users based on the preferences from similar-minded people [4, 9]. In CF, the most common form of ground user preference evidence consists of ratings, which are explicit relevance values given by users to items of interest. CF algorithms exploit the active user's ratings to make predictions, and thus it has the interesting property that no item descriptions are needed to provide recommendations, since it merely exploits information about past ratings between users and items. Moreover, it has the salient

advantage that a user benefits from others' experience, being exposed to novel recommendations with respect to the latters' personal preferences. Note that this cannot be produced in general by other (content-based) approaches that tend to reproduce the user's past, insofar as they examine the preferences of individual users in isolation [1].

Collaborative filtering approaches can be classified into two main categories: model-based approaches and memory-based approaches. Model-based approaches learn user/item rating patterns to build statistical models that provide rating estimations. Memory-based approaches, on the other hand, compute user/item similarities based on distance and correlation metrics [3], and use these similarities to find similar-minded people of the active user. These people are usually called neighbours, and their preferences are used to predict ratings for the active user.

Memory-based CF algorithms are based on the principle that a particular user's rating records are not equally useful to all other users as input to provide them with item suggestions [4]. Thus, central aspects in these algorithms are how to identify which neighbours form the best basis to generate item recommendations for the active user, and how to properly account for the information provided by them. Typically, neighbourhood identification is based on selecting those users who are most similar to the active user according to a certain similarity metric. In this context, the similarity of two users generally consists of finding a set of items that both users have interacted with, and examining to what degree the users displayed similar behaviours on these items.

Once the active user's neighbours are identified, the more similar a neighbour is to the active user, the more her preferences are taken into account as input to make up recommendations. For instance, a common memory-based approach consists of predicting the relevance of an item for the active user by a linear combination of her neighbours' ratings, which are weighted by the similarity between each neighbour and the user. It is also a common practice to set a similarity threshold (or a maximum number of most similar users) to restrict the set of neighbours, in order to avoid the noisy disruption of long tails of dissimilar users. An instantiation of this algorithm can be formulated as follows [10]:

$$\hat{r}(u,i) = \bar{r}(u) + C \sum_{v \in N_k(u,i)} \text{sim}(u,v)(r(v,i) - \bar{r}(v)) \quad (1)$$

This method establishes that the preference of a user u for a particular unseen item i is given by a numeric rating $r(u,i)$ estimated in the form of $\hat{r}(u,i)$. To provide that estimation, the method takes into account the ratings $r(v, i)$ provided by the k users v who are most *similar* to u, usually called neighbourhood and denoted here as $N_k(u, i)$. The function $sim(u, v)$ measures the similarity between two users u and v, and the constant C is a normalization factor. Thus, the predicted rating of user u for item i is computed over the average rating $\bar{r}(u)$ (this step is not always required [1]), and the sum of u's similarities with her neighbours v, weighted by the deviations of v's ratings for i and average ratings $\bar{r}(v)$. The item-based algorithm can be described analogously [11].

Different ways for building the neighbourhood $N_k(u, i)$ have been proposed: some authors use the concept of *trust* by selecting only the most trustworthy users with respect to some trust metric [9]; other authors split the set of users or items in order to improve the scalability of the recommender systems and their accuracy [8, 15]. Most of the latter approaches use old-fashioned clustering methods such as k-Means or hierarchical clustering. Furthermore, in some situations, external information is used for the data partition, such as the content of the item (genres or tags, in the movie domain).

Cluster algorithms for neighbour selection in CF have not been widely exploited. The few existing approaches [15] produced good results but at the expenses of lower coverage. In this paper, we propose a clustering method which has shown good empirical performance properties in the fields of Information Retrieval and Data Mining and apply it to Recommender Systems. In this process, we provide a general formulation for cluster-based CF methods, framed as neighbour selection methods. We report empirical results confirming where our method outperforms both standard CF methods and other cluster-based algorithms, with the additional advantage that we rely exclusively on data extracted from the rating matrix, that is, no external information is used.

2. GRAPH PARTITIONING BASED CLUSTERING: NORMALISED CUT

Spectral clustering algorithms [14] use graph spectral techniques to tackle the clustering problem transforming it into a graph cut problem. The dataset to be clustered is typically represented as a weighted graph, $G = (V, E, W)$, where V is the set of objects to cluster, E is the set of edges between objects, and W denotes a diagonal matrix whose elements are the weights e_{ij} between vertices $v_i, v_j \in V$. The Normalised Cut ($NCut$) value of a certain cut (a partition of V) of a given graph was introduced by [13]. For a certain cut $w = \{A_1, A_2, ...A_k\}$ of a graph G, $NCut$ is defined as:

$$NCut(A_1, ...A_k) = \sum_{i=1}^{k} \frac{cut(A_i, \bar{A}_i)}{vol(A_i)} \qquad (2)$$

where

$$cut(A, B) = \sum_{i \in A, j \in B} w_{ij}; \quad vol(A) = \sum_{i \in A} \sum_{j=1}^{n} w_{ij} \qquad (3)$$

here, A_1 to A_k are the connected components (ideally, the clusters) in which the graph has been divided and $\bar{A}_i$ are the vertices which are not included in A_i.

A graph cut with a low $NCut$ would represent a cut of the graph in which the weights of the edges which join vertices in different connected components are as low as possible while keeping the volumes of the resulting connected components

as high as possible. This last condition ensures a certain balance between the connected components, trying to avoid trivial solutions. So, a cut of G with a low $NCut$ would correspond to a good clustering of the data.

The minimisation of $NCut$ can be presented as a matrix trace minimisation problem [14]. Let $H = (h_{i,j})$ be a $n \times k$ matrix which will be used to encode the membership of data points to the connected components. The jth column of H contains the membership of connected component A_j (the indicator vector) encoded as follows:

$$h_{ij} = \begin{cases} \frac{1}{\sqrt{vol(A_j)}} & \text{if } v_i \in A_j \\ 0 & \text{else} \end{cases} \qquad (4)$$

Also, let D be a $n \times n$ diagonal matrix such that $d_{i,i} = degree(v_i) = \sum_{j=1}^{n} w_{ij}$ and let L be the Laplacian matrix of graph G (that is, $L = D - W$). Using these matrices, it can be shown that the minimisation of NCut can be written as in Eq. (5)

$$\min_{A_1, ...A_k} \text{Tr}(H^T L H) \text{ s.t. } H^T D H = I \qquad (5)$$

It can be demonstrated that the condition of discreteness on the values of H makes the minimisation problem NP-Hard. If this discreteness of the values of H is relaxed, allowing the indicator columns composing that matrix to have any value in $\mathbb{R}^n$, and the substitution $Y = D^{\frac{1}{2}}H$ is performed, the expression in Eq. (6) is obtained:

$$min_{Y \in \mathbb{R}^{n \times k}} \text{Tr}(Y^T \left[D^{\frac{-1}{2}} L D^{\frac{-1}{2}} \right] Y) \text{ s.t. } Y^T Y = I \qquad (6)$$

The above equation is in the standard form of a trace minimisation problem. Therefore, it can be demonstrated that this equation is minimised by the matrix Y which contains as columns the eigenvectors corresponding to the smallest eigenvalues of $D^{\frac{-1}{2}} L D^{\frac{-1}{2}}$.

Due to the relaxation of the condition of discreteness of the values of H to reduce the complexity of the problem and make it computationally affordable, instead of having an indicator vector for each connected component, we would have a vector in $\mathbb{R}^k$ for each datapoint (the rows of matrix Y). Thus, we have a projection of each data instance in $\mathbb{R}^k$ based on its similarity to the other instances. Hence, some other technique (such as k-Means) should be used to find a discrete segmentation of this space. Once this segmentation has been performed, we can backtrace each projected datapoint to the original one, obtaining the final outcome of the clustering algorithm.

3. NORMALISED CUT FOR RECOMMENDATION

The use of $NCut$ in the recommendation process is proposed as a tool for neighbour selection. Now, we focus on the methods derived when user clusters are used, alternatively, item clusters could be used in a straightforward way and the problem would be casted as an item-based CF method. Hence, in a user-based method, the users will play the role of the nodes of the graph to cut. A good set of neighbours for each user in the collection would be obtained by finding a good $NCut$ of the graph. Besides, in order to perform the complete process of recommendation, we also need to establish a procedure for weighting the edges in the graph (e_{ij}), i.e. the distances among users. We decided to adopt the well-known – and traditionally used – Pearson's correlation similarity. Thus, for each cluster, we obtain a list of users belonging to such cluster. Then, we build a recommender which predicts the rating for user u and item i in the following way:

$$\tilde{r}(u,i) = \frac{\sum_{e \in NC(u)} \text{sim}(u,e)r(e,i)}{\sum_{e \in NC(u)} |\text{sim}(u,e)|} \quad (7)$$

where $NC(u)$ outputs the elements who belong to the same cluster as the target user u, $\text{sim}(u,e)$ represents the similarity between the element e and the current user; finally, $r(e,i)$ is the rating given by user e to item i.

Thus, we summarise our neighbour selection problem for a user-based method as follows. We use the Normalised Cut algorithm to create different user clusterings. Then, we use the information generated by this clustering for the neighbourhood formation, such that for each user u, we find the cluster c_u that user belongs to, then, the rest of the users belonging to c_u, are selected as the potential set of u's neighbours.

4. EXPERIMENT AND RESULTS

In order to empirically compare whether the clusters generated by the Normalised Cut technique are able to enhance standard collaborative recommendations, we have performed the following experiments. Firstly, we have plugged those clusters into a standard user-based CF method and compared their performance against some well-known state-of-the-art recommenders: a user-based CF with Pearson's correlation as similarity measure (UB [10]) and a matrix factorization algorithm with a latent space of dimension 50 (MF [6]). Secondly, in order to show whether the improvement comes from using a cluster-based method or the specific technique that we propose, we compare the performance of our technique against a standard clustering method (k-Means [7, 15]).

These experiments have been carried out using the publicly available dataset called *Movielens 100K*[1]. This dataset contains 943 users, 1,682 items and 100,000 ratings. We performed a 5-fold cross validation using the splits contained in the public package, these splits retain the 80% of the data for training, and the rest for testing. The methodology used in the evaluation corresponds to the *TestItems* approach described in [2], although alternative methodologies (such as the one described by Koren in [6]) have also been evaluated and similar results were obtained. More specifically, the TestItems methodology generates for each user a ranking by predicting a score for every item in the test set. Then, the performance of this ranking is measured using, for instance, the *trec_eval* program[2]. In this way, standard IR metrics such as precision, normalised Discounted Cumulative Gain (nDCG) or Mean Reciprocal Rank could be used.

4.1 Normalised Cut for Neighbour Selection

In order to asses our technique's ability to select good neighbours, we evaluate the *NCut* clustering technique (denoted as NC+P, since Pearson's correlation is used as simi-

Table 1: Results for cluster-based algorithm using Normalised Cut (NC+P). In brackets, the number of eigenvectors used in the Normalised Cut for each k. Statistical significant improvements according to Wilcoxon Test ($p < 0.05$) w.r.t. MF and UB are superscripted with m and u respectively, best values are bolded.

Method	P@5	nDCG@5	Coverage
MF	0.081^u	0.076^u	100%
UB	0.026	0.020	100%
NC+P 50 (100)	0.075^u	0.062^u	100.00%
NC+P 100 (150)	0.101^{mu}	0.085^u	97.82%
NC+P 150 (200)	0.111^{mu}	0.095^{mu}	93.68%
NC+P 200 (250)	$\mathbf{0.112}^{mu}$	$\mathbf{0.097}^{mu}$	79.74%
NC+P 250 (300)	0.111^{mu}	$\mathbf{0.097}^{mu}$	69.06%
NC+P 300 (350)	0.108^{mu}	0.096^{mu}	59.26%
NC+P 350 (400)	0.103^{mu}	0.088^{mu}	54.25%
NC+P 400 (450)	0.094^{mu}	0.083^u	43.36%
NC+P 450 (500)	0.086^u	0.074^u	40.52%
NC+P 500 (550)	0.079^u	0.070^u	35.95%
NC+P 550 (600)	0.079^u	0.068^u	32.03%
NC+P 600 (650)	0.079^u	0.070^u	25.93%

larity) by the Equation 7. We compared it with the standard UB method where the neighbourhood is selected among the set of most similar users. Besides, to put our results in perspective, we also include a well-known method which does not use any neighbour selection and which is typically among the best performing recommenders (MF method).

The results are presented in Table 1, where different values for the size of the cluster (k) have been evaluated against the baselines. In fact, the quality of the results of the spectral methods improve considerably if, in the projection phase, instead of taking only the first k eigenvectors, a higher number of vectors (d) is used [5], for this reason we have taken higher number of eigenvectors (specified between brackets). Due to space constraints, only the results for a specific cut-off ($N = 5$) are reported, but other cut-offs and metrics such as P@50 and nDCG@50 were also tested and a similar trend was observed. Here, we show that cluster-based methods outperform baseline methods for specific values of the cluster size, namely $k = 150$ and 200, and these improvements are statistically significant. In fact only MF and in only one occasion (k=50) achieved statistical significant improvements over our approach. Furthermore, a larger number of clusters does not necessarily mean a better performance, probably because most of the clusters would contain zero or one element when this number is very large.

More importantly, a larger number of clusters tends to produce lower coverage values[3], since the information available to the recommender is not enough, which would produce a well performing method, but a not very interesting one, since it would only be able to suggest items – for some configurations – to less than the 30% of the system's population.

4.2 Sensitivity to the Number of Clusters

In this experiment, we analyse the recommendation performance variations that result when different cluster sizes are used in a cluster-based recommender method. More specifically, we compare our *NCut* method with a state-of-the-art algorithm where the clusters are generated using the k-Means method [15]. Furthermore, since our method has shown good performance once it has been plugged in the standard formulation of user-based CF (previous section), now we also evaluate its performance when no information about the user-neighbour similarity is used.

Figure 1 summarises our results. We denote as *NC* when the cluster-based recommender with $w_1(u, e) = 1$ and the NCut clustering technique are used, as in the previous section, when this technique is used in combination with the standard Pearson's similarity coefficient, we denoted it as *NC+P*. Similarly, *kM* and *kM+P* denote the corresponding k-Means cluster-based methods when a constant similarity $w_1 = 1$ and Pearson values are used, respectively.

Figure 1: Performance obtained (P@5) between the baseline methods (standard CF and cluster-based using k-Means alone or in combination with Pearson's similarity: *kM* and *kM+P*), the NCut method without a similarity function (*NC*), and the combination of a Pearson's similarity function and NCut as the neighbour selection method (*NC+P*). Coverage values are plotted along the secondary axis.

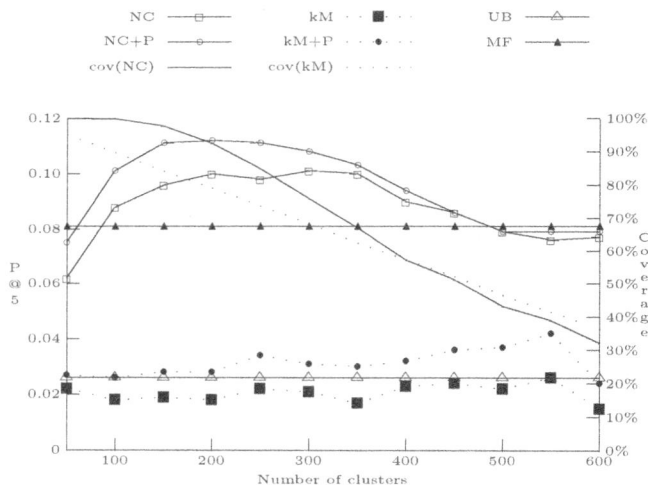

We can observe in the figure, first, that the coverage of both cluster-based methods is very similar, although our method obtains improvements up to 15% when less than 300 clusters are used. Interestingly enough, the performance of our technique reaches a maximum also at that point, whereas the k-Means cluster-based recommender shows a flat performance along the different number of clusters evaluated. It should be noted also that the performance of our method when no similarity information is used (curve *NC*) is comparable to that of *NC+P*, and thus, it becomes apparent that our method alone is improving the way the neighbours are being selected, which leads to a major performance enhancement.

5. CONCLUSIONS AND FUTURE WORK

In this paper, we have presented a new approach for neighbour selection in Recommender Systems. *NCut* is a clustering method based on graph partitioning which is natu-

rally applied to the graph essence of the recommender systems communities. We tested two approaches of exploiting the neighbourhoods determined by the algorithm and both of them greatly improved the performance of standard CF methods. Furthermore, when compared against other cluster-based methods, our method obtained large improvements in performance, while the coverage was comparable to that of those methods.

Although our approach is general enough to work for user- and item-based CF methods, we have decided to focus on user-based methods since they provided better performance in initial experiments, but we aim to further investigate in the future also item clusters along with item-based approaches. We would also explore the behaviour of our proposal when working in larger datasets and the use of other spectral clustering techniques. Besides, we have observed that cluster-based methods tend to reduce the coverage of the system at the expense of its performance, hence, we aim to investigate alternative clustering methods which do not suffer from this behaviour.

6. ACKNOWLEDGMENTS

This work is supported by the Spanish Government (TIN2011-28538-C02-01), and the Government of Madrid (S2009TIC-1542).

7. REFERENCES

[1] ADOMAVICIUS, G., AND TUZHILIN, A. Toward the next generation of recommender systems: A survey of the state-of-the-art and possible extensions. *IEEE Trans. Knowl. Data Eng. 17*, 6 (2005), 734–749.

[2] BELLOGÍN, A., CASTELLS, P., AND CANTADOR, I. Precision-oriented evaluation of recommender systems: an algorithmic comparison. In *RecSys* (2011), pp. 333–336.

[3] DESROSIERS, C., AND KARYPIS, G. A comprehensive survey of neighborhood-based recommendation methods. In *Recommender Systems Handbook*. 2011, pp. 107–144.

[4] HERLOCKER, J. L., KONSTAN, J. A., AND RIEDL, J. An empirical analysis of design choices in neighborhood-based collaborative filtering algorithms. *Inf. Retr. 5*, 4 (2002), 287–310.

[5] JIN, R., DING, C. H. Q., AND KANG, F. A probabilistic approach for optimizing spectral clustering. In *NIPS* (2005).

[6] KOREN, Y. Factorization meets the neighborhood: a multifaceted collaborative filtering model. In *KDD '08* (2008), ACM, pp. 426–434.

[7] MACQUEEN, J. B. Some methods for classification and analysis of multivariate observations. In *Proc. of the fifth Berkeley Symposium on Mathematical Statistics and Probability* (1967), L. M. L. Cam and J. Neyman, Eds., vol. 1, University of California Press, pp. 281–297.

[8] O'CONNOR, M., AND HERLOCKER, J. Clustering items for collaborative filtering. In *ACM SIGIR Workshop on Recommender Systems* (1999).

[9] O'DONOVAN, J., AND SMYTH, B. Trust in recommender systems. In *IUI* (2005), pp. 167–174.

[10] RESNICK, P., IACOVOU, N., SUCHAK, M., BERGSTROM, P., AND RIEDL, J. Grouplens: An open architecture for collaborative filtering of netnews. In *CSCW* (1994), pp. 175–186.

[11] SARWAR, B., KARYPIS, G., KONSTAN, J., AND RIEDL, J. Item-based collaborative filtering recommendation algorithms. In *WWW '01* (2001), ACM, pp. 285–295.

[12] SHANI, G., AND GUNAWARDANA, A. Evaluating recommendation systems. In *Recommender Systems Handbook*. 2011, pp. 257–297.

[13] SHI, J., AND MALIK, J. Normalized cuts and image segmentation. *IEEE Trans. Pattern Anal. Mach. Intell. 22*, 8 (2000), 888–905.

[14] VON LUXBURG, U. A tutorial on spectral clustering. Tech. Rep. TR-149, Max Planck Institute for Biological Cybernetics, 2006.

[15] XUE, G. R., LIN, C., YANG, Q., XI, W., ZENG, H. J., YU, Y., AND CHEN, Z. Scalable collaborative filtering using cluster-based smoothing. In *SIGIR* (2005), ACM, pp. 114–121.

Remembering the Stars? Effect of Time on Preference Retrieval from Memory

Dirk Bollen
Human-Technology Interaction
Eindhoven University of Technology
The Netherlands
d.g.f.m.bollen@tue.nl

Mark Graus
Human-Technology Interaction
Eindhoven University of Technology
The Netherlands
m.p.graus@tue.nl

Martijn C. Willemsen
Human-Technology Interaction
Eindhoven University of Technology
The Netherlands
m.c.willemsen@tue.nl

ABSTRACT

Many recommendation systems rely on explicit ratings provided by their users. Often these ratings are provided long after consuming the item, relying heavily on people's representation of the quality of the item in memory. This paper investigates a psychological process, the "positivity effect", that influences the retrieval of quality judgments from our memory by which pleasant items are being processed and recalled from memory more effectively than unpleasant items. In an offline study on the MovieLens data we used the time between release date and rating date as a proxy for the time between consumption and rating. Ratings for movies tend to increase over time, consistent with the positivity effect. A subsequent online user study used a direct measure of time between rating and consumption, by asking users to rate movies (recently aired on television) and to explicitly report how long ago they watched these movies. In contrast to the offline study we find that ratings tend to decline over time showing reduced accuracy in ratings for items experienced long ago. We discuss the impact these rating dynamics might have on recommender algorithms, especially in cases where a new user has to submit his preferences to a system.

Categories and Subject Descriptors

H.1.2 [Models and principles]: User/Machine Systems-software psychology; H.4.2 [Information Systems Applications]: Types of Systems-decision support; H.5.2 [Information Interfaces and Presentation]:User Interfaces-evaluation/methodology, interaction styles, user centered design

General Terms

Measurement, Design, Experimentation, Human Factors.

Keywords

memory biases, temporal effects, explicit feedback

1. INTRODUCTION

Watching a movie, at home or in a movie theater, provides an experience with all sensory information available and evoking an affective response. After seeing a movie all this information, vividly available, influences our judgments and preference processes. This information will quickly be transferred and encoded into one's memory. The concrete information as well as the emotional and affective responses related to the experience of watching a movie tend to fade over time as the result of a natural forgetting process. This would imply that one's ratings for the same item might differ when given directly after the experience of the item, compared to a few months or years later. For example, the positivity effect that was established in memory research [2,3] shows more effective encoding and recall of positive information, which would suggest that we would rate items more positively if the time between consumption and rating increases.

In the present paper, we investigate the effect of time on preference retrieval in two ways. Using existing offline data from MovieLens[1] we took the difference between release date and rating data as a (less than ideal) proxy for the time between consuming and rating an item. Our data shows that ratings for movies increase over time, showing a positivity effect. In a second online study, we used a more accurate measurement of the time between consumption and rating by asking users after providing ratings to report directly how long ago they had experienced the item. We also asked for explanations for some ratings. This study revealed a negativity effect of time on ratings, but a positivity effect in terms of what items users like to explain.

2. RELATED WORK

2.1 Positivity effects in memory research

In psychological literature the process of memories becoming more positive over time is known as the positivity effect. This effect is part of the Pollyanna principle that describes that pleasant experiences are being processed more efficiently and accurately in memory than less pleasant experiences [2]. This causes retrieval of information from our memories to be biased towards positive aspects. The result of this is that people forget the negative aspects of an experience, while remembering the positive aspects. The positivity effect is a well-established and robust effect in psychology. Walker et al. [3] provide an overview of several studies in what they call the fading affect bias: the process underlying the stronger decrease of the emotional intensity for unpleasant experiences in autobiographical memory. Kwon et al.[1] show that positivity effects hold over different cultures.

2.2 Temporal Effects in Recommender Systems

A number of studies acknowledge the effect of time on user preference. Wang et al. [4] addressed these temporal dynamics, that they call preference evolution, using an Ant Collaborative Filtering algorithm in which users and movies are marked with a

---

[1] Available at: www.grouplens.org

pheromone trail, analogous to how ants communicate. When a user rates an item positively, pheromones are transferred from the user to the item and vice versa. As more ratings are processed, the pheromone patterns for all movies and users become more complex and predictions can be made based on similarity in pheromone patterns between users and movies. Preference evolution is taken into account by letting the pheromone decay over time and processing the ratings in the same order as they were submitted to the system, which results in later ratings contributing more to the final model. While this algorithm takes temporal dynamics into account by putting more weight on recent ratings it provides no understanding in how preferences change as time between consumption and rating increase.

Koren [5] attributed the change in rating behavior to what he calls concept drift. By means of an offline study of the Netflix contest dataset he investigated what common patterns can be found in how ratings evolve over time. An important finding was that as the time between movie release and rating increases, the rating tends to become higher.

3. OFFLINE STUDY: MOVIELENS DATA

Koren's [5] results show that older movies are rated higher on average. However, to provide a good answer to our question, of how the time between viewing and rating influences ratings, the difference between release date and time of rating (as used by Koren [5]) might not be the most accurate proxy.

To more exhaustively investigate memory effects, we study the difference between popular and unpopular movies. The rationale is that popular movies appear more often on television and are discussed more often, this results in the actual experience remaining more freshly in memory and thus being less prone to memory distortions. For less popular movies, the time between release date and rating date is a better proxy for the time that has passed between experience and rating. In other words, any memory distortions should be more prominent in less popular movies, for which we rely on memory more.

For our analysis, we calculate as independent variable Δt, the time difference between the year of release (which is 2000 since we only considered movies released in that year) and the moment of rating by the MovieLens users. For further investigation we divided the movies into popular and unpopular movies, by performing a median split on the number of ratings for each movie.

Figure 1 depicts the results of a multilevel regression with a random intercept for movies and two fixed effects, Δt and popularity. The results show that an increase of time difference (Δt) has a positive significant effect on the mean rating (t=9.12). This effect is qualified by a significant interaction with movie popularity (t=-3.06). Ratings for less popular movies increase about 0.04 stars per year (dotted line), whereas ratings for more popular movies only increase about 0.02 stars per year (solid line).

These results show, like [5], that ratings for movies tend to increase over time. It also shows that this effect is stronger for less popular movies for which we expect that people have to rely on their memory more.

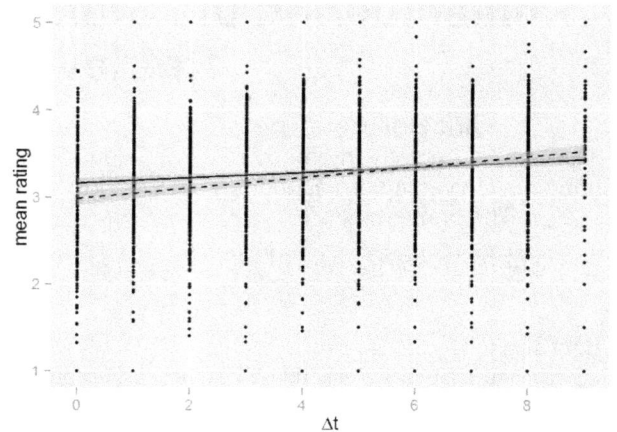

Figure 1: Evolution of the mean rating for movies in the MovieLens dataset, as function of the time between release date and rating (in years). Solid line: more popular movies, dotted line: less popular movies

4. USER STUDY

Our offline MovieLens study provided some initial support for the existence of the positivity effect in which unpopular ratings tend to increase more over time, since users need to rely more on memory than for popular ones. However, the measure Δt used in our offline study is at best a (imprecise) proxy for the time factor we are really interested in: the time in between experiencing (or consuming) and rating.

To the best of our knowledge there are no studies that have taken this time as a measure into account. The goal of our study was to investigate how the time passing between the moment of consumption and the moment of rating influences a user's ratings by asking the users directly for an estimate how long ago they have consumed the item.

4.1 Design

For the experimental setup IMDb was crawled to find information consisting of a movie cover and a synopsis for 156 movies that were aired on public Dutch television in the month prior to the study. These movies were selected in order to maximize the probability that the data would include ratings with a low Δt as it is likely that our participants had at least seen a few of these movies on TV in the month prior. Participants in the study were displayed lists of 10 randomly selected movies at a time and asked to rate the movies they knew. For every movie they rated they were asked to indicate how long ago they watched it via a dropdown menu (six categories: last week, last month, last 6 months, last year, last three years, longer ago). After having provided at least 20 ratings, participants would see a button allowing them to finish the task. They could also continue rating as long as they wanted (one person rated a total of 71 movies out of 156). After the rating task, participants were asked to provide motivations for two ratings they gave. One dropdown menu displayed the titles and ratings of movies that the participant indicated having seen within the last month, another displayed the same information for movies seen a year ago or longer. Below each dropdown menu a text input area allowed the user to enter her explanation.

4.1.1 Participants

In the course of 2 days, 100 participants finished the task (40 male, 60 female, mean age: 29.7 years, SD: 7.9 years). On average they rated 20.41 movies (SD: 9.0 movies).

4.1.2 Results

The first step of the analysis consists of visually investigating the rating distributions. In total 2041 ratings were submitted by the participants. These ratings are visually represented in the mosaic plot in Figure 2, with the width of the columns representing the proportion of ratings in a given timeslot. Only 30% of the provided ratings were of movies seen in the last year.

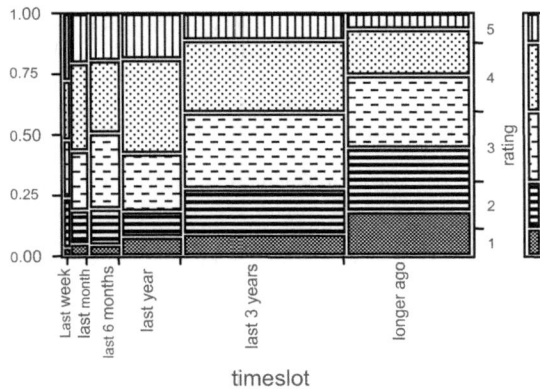

Figure 2: Distribution of Ratings across timeslots

Figure 2 shows that as time progresses the proportion of positive ratings decreases. In the first four timeslots four and five star ratings make up for more than 50% of all ratings, while in the last time slot they only make up for 25%. This seems to run against the hypothesis of a positivity effect.

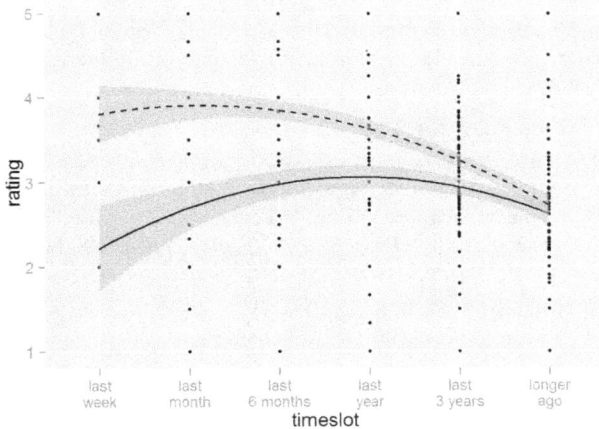

Figure 3: Progression of Average Rating (dotted line for initially high rated items, dashed for initially low rated items)

In order to test for the hypothesized positivity effect a multilevel regression was performed, using the rating value as dependent variable. Random intercepts were included for participant and movie, similar to how biases are handled in biased matrix factorization algorithms. This compensates for differences in ratings by individual users and individual movies that are rated higher or lower on average. As a fixed effect we used Time (a linear effect from 0 to 5 for each timeslot) and its quadratic effect

Time Squared. In addition the movies were divided in initially high and low rated movies to provide more insight into where the pattern observed in Figure 2 originates from. Initially high rated movies (High rated = 1) were movies that in the first three time slots received a high rating (>3 stars).

In Figure 3 we plotted the average rating score as a function of time for each group. The dotted line represents the initially high rated movies, the solid line the initially low rated movies. We see a steady decline in mean rating for high rated movies, and an initial increase followed by a decrease for the mean rating of low rated movies. The multilevel regression, shown in Table 1, confirms the pattern of Figure 3. As far as the influence of time is concerned, the model shows that initially the ratings rise (the linear time coefficient is positive). However, as time progresses the $Time^2$ coefficient weighs heavier on the ratings, and they decrease. Additionally a significant interaction of Time × High Rated shows that for high rated movies the rating decreases faster.

Overall the results seem to suggest a negativity effect, rater than a positivity effect, which is also more pronounced for initially highly rated items.[2]

Table 1: Linear multilevel model

	Coefficient	Std. Err.	t-value
Intercept	2.51	.20	12.708
Time	0.38	.10	3.837
High Rated	1.19	.18	6.445
Time Squared	-.07	.01	-4.979
Time × High Rated	-.19	.04	-4.505

An additional analysis was performed on the explanations people provided for their ratings. After submitting their ratings participants were asked to explain why they gave the rating for a movie they saw in the last month and one they saw longer than one year ago. They could select which rating to explain by selecting it from two drop-down menus, one for each category. Two conclusions can be drawn from this data, displayed in Figure 4. Firstly, given how the ratings are distributed, the choice for explaining movies appears to have a bias toward positive movies. For movies seen more than a year ago four and five star ratings make up for about 30% of ratings, however when asked to explain a rating, participants chose to explain a four or five star rating for almost 75% of the time. For the recent movies (last month), the proportions of explanations and ratings are more similar. For these ratings however participants are more inclined to rate either good or bad movies, as can be seen by the decrease in three star rating explanations. Secondly, the length of the explanation was investigated. Although no significant difference could be found for older versus more recent ratings, people used significantly more characters to explain a positively rated item than a negatively rated item (on average 11 characters more per rating value).

Both these findings concerning the explanation people provided are related to the positivity effect from memory research discussed earlier. As positive experiences are encoded and recalled better, it might have been easier for our participants to describe the positively rated items they had seen longer ago.

---

[2] We also ran an analysis similar to our offline study, using the difference between the time of rating (2012) and the year of release of the movies in this set. This analysis also shows a significant negativity effect (a decrease of 0.02 stars per year).

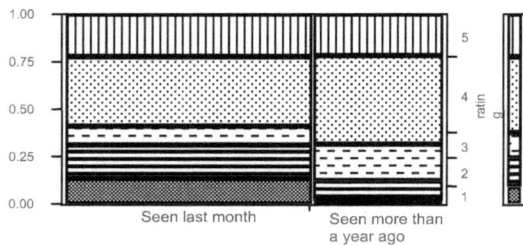

Figure 4: Contingency plot for explanations of ratings for movies seen last month versus more than a year ago

Given that we do find positivity effects in the motivations for the ratings, the question remains why we did not observe a positivity effect on the ratings, as we had expected from our offline study. One interesting observation is that the positivity effect seems to affect the recall of the participant's memory (in their motivations) but not their recognition of poor movies. Many movies that were experienced longer ago seem to be rated rather low, indicating the participants might recognize that some of these movies were not so good, without exactly recalling why (since their memory for bad items is not as accurate, as proposed by the positivity effect). Our setup, using a non-personalized list of movies might contain more of these movies than the usual setups in which data was collected for the Netflix [5] and Movielens datasets, as we will discuss below.

Our results might also suggest a more general memory decay pattern. Both high rated and low rated items seem to regress towards the middle of the scale (see Figure 3). This clearly indicates that our ratings for items we have seen long ago do not resemble well the ratings we give right after experiencing an item.

5. CONCLUSION

This paper investigated the time dynamics underlying ratings in recommender systems. An offline analysis shows a positivity effect in which movies are rated higher as time between release date and rating date increases, especially for movies that are less popular for which we expect people to rely more on their memory. This increase in average rating over time can be mainly attributed to the increase present in low ratings for non-popular movies, which is higher than the increase in high ratings or in popular movies.

However, the real variable of interest was the time difference between the moment of consumption and the moment of rating. In a second experimental study participants were asked to explicitly indicate how long ago they watched a movie. The measure Δt in this study is thus the time difference between moment of consumption and moment of rating as opposed to moment of release and moment of rating. However, in this study a negativity effect instead of a positivity effect was found. Results showed a decline in rating score as movies were rated longer after being watched. Additionally this effect was more pronounced for movies that received high ratings shortly after being watched.

A possible explanation for these contrasting results can be found in the different ways of interaction between actual recommender systems and our experimental setup. In the presented study participants were asked to rate all movies they recognized and indicate how long ago they watched the movie. Recommender systems typically ask new users to rate randomly sampled movies for training, which is similar to our experimental design. However, once enough training ratings have been provided, a recommender system presents movies the user is likely to appreciate. Older movies have more data and are thus more often accurately recommended, which leads to accurate recommendations and thus more high ratings. Because users mainly encounter movies they are likely to appreciate during the interaction, the amount of negative ratings is also to be expected lower than what we found in our online study.

An additional explanation can be found in the result of our user study that over time, ratings tend to regress towards the mean. The best discrimination between good and bad movies can be found when ratings are given no longer than a year after watching a movie.

These dynamics are important to take into account when designing recommendation algorithms. A situation where these dynamics might have a particularly negative influence on recommender accuracy is when a new user enters a system. Often, new users are asked to rate a number of more or less randomly selected movies so that the recommender system has enough information to calculate predictions that can later, through natural interaction, be fine-tuned. In a natural interaction it might be more common that a user rates a movie after having just seen it. With the initial ratings however, the user may rate movies that she has seen long ago, which can lead to suboptimal recommendations. In fact, depending on how data is gathered, current recommender systems may predict how users will like a movie after a certain amount of time, instead of how she will like it right now.

In order to take these effects into account in future recommender systems, two paths can be taken. First of all, users of a recommender system could be asked how long ago they watched a movie when they rate it, similar to how it was done in this study. While this additional data is likely to improve recommendation quality, a negative consequence is that users will have to provide more input. Another way to go about this is investigating the patterns related to the passing of time in more detail, with the goal of understanding and compensating for this influence on rating values.

The authors propose future studies with a more natural interaction with a recommender system, to further investigate the temporal dynamics of preferences.

6. REFERENCES

[1] Kwon, Y., Scheibe, S., Samanez-Larkin, G. R., Tsai, J. L., & Carstensen, L. L. 2009. Replicating the positivity effect in picture memory in Koreans: evidence for cross-cultural generalizability. *Psychology and aging*, 24(3), 748-54. doi:10.1037/a0016054

[2] Matlin, M. W., & Stang, D. J. 1978. *The Pollyanna principle*. Cambridge, MA: Schenkman.

[3] Walker, W. R., Skowronski, J. J., & Thompson, C. P. 2003. Life Is Pleasant — and Memory Helps to Keep It That Way! *Review of General Psychology*, 7(2), 203-210. doi:10.1037/1089-2680.7.2.203

[4] Wang, Y., Liao, X., Wu, H., & Wu, J. 2012. Incremental Collaborative Filtering Considering Temporal Effects. arXiv: 1203.5415v1

[5] Koren, Y. 2009. Collaborative filtering with temporal dynamics. In *Proceedings of the 15th ACM SIGKDD international conference on Knowledge discovery and data mining (KDD '09)*. ACM, New York, NY, USA, 447-456. doi:10.1145/1557019.1557072

Local Learning of Item Dissimilarity Using Content and Link Structure

Abir De, Maunendra Sankar Desarkar, Niloy Ganguly, Pabitra Mitra
Department of CSE, IIT Kharagpur, India
abir.iitkgp@gmail.com, {maunendra,niloy,pabitra}@cse.iitkgp.ernet.in

ABSTRACT

In the *Recommendation Problem*, it is often important to find a set of items *similar* to a particular item or a group of items. This problem of finding similar items for the recommendation task may also be viewed as a link prediction problem in a network, where the items can be treated as the nodes. The strength of the edge connecting two items represents the similarity between the items. In this context, a central challenge is to suitably define an appropriate *dissimilarity function* between the items. For content based recommender systems, the dissimilarity function should take into account the individual attributes of the items. The same attribute may have different importances in different parts of the underlying network. We focus on the problem of *learning a suitable dissimilarity function between items* and address it by formulating it as a constrained optimization problem which captures the local weightages of the attributes in different regions of the graph. The constraints are imposed in such a way that the non-connected nodes show higher value of dissimilarity than the connected nodes. The local tuning of the weights learns the optimal value of weights in various parts of the network: from the portions having rich graph information to the portions having only content information. Detailed experimentation shows the superiority of the proposed algorithm over the Adamic Adar metric as well as logistic regression methodology.

Categories and Subject Descriptors

H.3.3 [**Information Search and Retrieval**]: Information Filtering; H.4 [**Information Systems Applications**]: Miscellaneous

General Terms

Algorithms, Experimentation, Performance

Keywords

Content based recommendation, collaborative filtering, information retrieval.

1. INTRODUCTION

In the recommendation task, the goal is to produce a list of items that a target user might be interested in. A common approach for solving the task is to obtain a set of items that are similar to the items that the user has liked in the past. In this approach, often the items are viewed as the nodes of an underlying network. The strength of the edge connecting two nodes in the network indicates the similarity between the corresponding items. From a social networking perspective, the problem of finding (and recommending) similar items from this network may also be viewed as a link prediction problem.

Many researchers have analyzed the problem of recommending similar items in a pure link-based approach [4]. In [4], the authors suggest different metrics (e.g. Adamic Adar, Jaccard coefficients etc.) to estimate the strength of a candidate edge in the network.

An alternate approach is to make a content based recommendation where each item is associated with a feature vector or attribute profile. The features may hold numeric or nominal values and represent certain aspects of the item (e.g. director, genre, release date for a movie; author, genre, language for a book etc.) In this approach, an important task is to suitably construct a dissimilarity function between the feature vectors for computing the closeness between two items. In a dissimilarity function, the different item attributes are assigned different weights depending on their importances. Many researchers have devised methodologies to derive these weights. [3] uses a poisson regression model to find suitable attribute weights using clickstream data. Some researchers have exploited the knowledge of link information to learn the weights. For example, a hybridization of collaborative filtering and content based recommendation has been presented on a linear regression model based framework in [2]. The method proposed in [1] learns to bias a PageRank-like random walk using supervised approach, so that the walk visits the connected nodes more frequently than other nodes.

In these schemes, the trend is to associate an optimum global weight to each attribute. But importance (or weight) of one attribute may vary widely over items. For example, the role of schooling plays an insignificant role in two celebrities getting connected. However, in case of two common persons, their schooling may possibly play a vital role in recommending each other as a friend. We therefore emphasize that the weights assigned to an attribute over a network should not be constant, instead its importance should be determined taking locality into consideration.

In this technical note, we have addressed the content based recommendation problem from a novel optimization based framework. The framework identifies important attributes and assigns higher weights to those attributes while computing the similarity between the nodes (item). Also, the importance is specific to a region in a network. Finally, depending on the local weights, we develop the sorted list of dissimilarity values of items with a particular item. This is essentially a ranked list of recommendation for that item, or for an user who have liked that item.

2. PROPOSED APPROACH: CONSTRAINED LOCAL LEARNING

Problem Definition: Let x be a particular item. The objective is to make a ranked list of recommendations to x. So, for a given x, the algorithm should generate $A_x(y, S)$, where y's are the recommended items and S's are the corresponding scores. More is the score, higher is the rank of the corresponding item in the list. To find A_x, for a particular item x, we consider all the nodes which have common neighbours with x, as candidates. So, we construct a hypothetical item-item network where two items are connected by an edge if they have been previously accessed/accepted by a certain number of users. The task is to predict whether a presently non existing link may appear in the future. Hence the problem that whether an item y belongs to the list A_x, reduces to the problem of predicting a future edge between nodes x and y in the underlying graph.

In order to understand/predict the dynamics of link formation, we assume that *the dynamics primarily depends on the corresponding locality of the graph*. In other words, the possibility of a node (x) being connected to another node (y) rarely is determined by the nodes that are *far apart* from x and y in the underlying graph. Keeping this in mind, the first step of the algorithm learns the amount of dissimilarity between x and y and their common neighbours. (We term this process as the reference dissimilarity function.) We then use this learning to predict the chance of a link arriving between x and y.

Definitions: We formally define some important terms in Table 1.

Table 1: Important Definitions:

Neighbourhood of a node i	$\Gamma(i)$
Attribute vector of node i	θ_i
Dissimilarity function between nodes i and j	$\Delta_w(i,j) = w^T\|(\theta_i - \theta_j)\|$ [1]

1. Local Weights and Reference Dissimilarity Function: Computation of an appropriate dissimilarity function relies on finding suitable weights associated with each attribute. Weight of one individual attribute may vary widely over the network. Thus the choice of suitable weights specific to a particular region of the graph is crucial. In order to predict an edge between x and y (assuming they are not connected yet), we consider the locality $N = \Gamma(x) \cup \Gamma(y)$ and restrict our discussion to N throughout this section. If there is an edge between two items, we assume that the edge has

[1] $|.|$ denotes the term by term absolute value of a vector (e.g $\|[-2,3]\| = [2,3]$) and w is the weight vector.

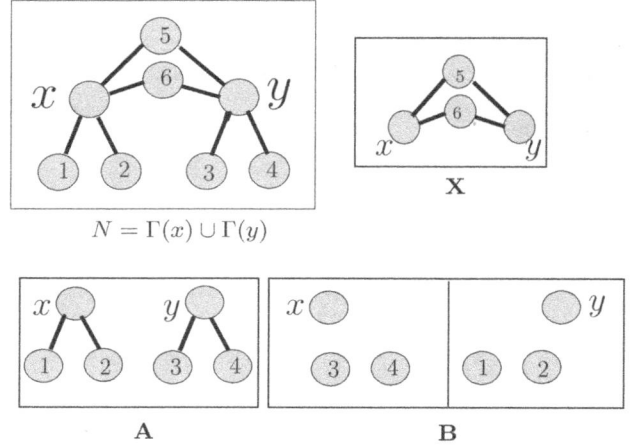

$$N = \Gamma(x) \cup \Gamma(y)$$

Figure 1: Sample graph to determine the possible score between x and y. X : Edges of x and y with their common neighbours give an estimate of minimum value of the reference dissimilarity function Δ_w^{xy}, *under the imposition of* **A**: ($\Delta_w(1,x)$ and $\Delta_w(2,x)$) are not very high (Similarly for the pair $(3,y)$ and $(4,y)$). **B**: For non existing edges, the dissimilarity can be very high w.r.t the reference. So, $\Delta_w(1,y)$, $\Delta_w(2,y)$, $\Delta_w(3,x)$ and $\Delta_w(4,x)$ are relatively high with respect to Δ_w^{xy}.

arrived because the attributes of the two nodes are similar. Hence we assume that the value of the dissimilarity function is low for node pairs which have an edge connecting them and relatively higher for node pairs which are not neighbours. Under these impositions, we wish to minimize the sum of the dissimilarity values of x and y with their common neighbours. This is because this minimization would in turn give the maximum possibility of an edge between x and y appearing in future. It can also be termed as *reference dissimilarity function*. So our objective is to find optimal w, so that, the dissimilarity

$$\Delta_w^{xy} = \sum_{i \in \Gamma(x) \cap \Gamma(y)} \Delta_w(i,x) + \Delta_w(i,y)$$

should be minimum w.r.t w, *assuming that*, the dissimilarity between linking item pairs are low ,*i.e.* $\Delta_w(i,x)$ for $i \in \Gamma(x) \setminus \Gamma(y)$ and $\Delta_w(i,y)$ for $i \in \Gamma(y) \setminus \Gamma(x)$ should be low w.r.t the reference dissimilarity Δ_w^{xy} and for not existing edges, dissimilarity (e.g. $\Delta_w(i,y)$ for $i \in \Gamma(x) \setminus \Gamma(y)$) should be relatively high.

Following the above arguments, the problem of determining the reference dissimilarity function can be cast as the following optimization problem.

$$\Delta^{xy} := \min_w \widetilde{\Delta}_w^{xy}$$

subject to, Ω_w^{xy} :

$$\sum_{i \in \Gamma(x) \setminus \Gamma(y)} \widetilde{\Delta}_w(i,x) \leq \alpha \widetilde{\Delta}_w^{xy} \tag{1}$$

$$\sum_{i \in \Gamma(y) \setminus \Gamma(x)} \widetilde{\Delta}_w(i,y) \leq \alpha \widetilde{\Delta}_w^{xy} \tag{2}$$

$$\sum_{i \in \Gamma(y) \setminus \Gamma(x)} \widetilde{\Delta}_w(i,x) \geq \beta \widetilde{\Delta}_w^{xy} \tag{3}$$

$$\sum_{i \in \Gamma(x) \setminus \Gamma(y)} \widetilde{\Delta}_w(i,y) \geq \beta \widetilde{\Delta}_w^{xy} \tag{4}$$

where $\sum_{i \in G} \widetilde{\Delta}_w(i,x) = \frac{1}{|G|} \sum_{i \in G} \Delta_w(i,x)$,
and α and β are suitable parameters. Symbolically Ω_w^{xy} denotes the set of all constraints. We use standard LP method to solve the optimization problem. The algorithm gives the outputs: *reference dissimilarity value* (Δ^{xy}) and *optimal weights* (w^*).

2. Computation of Actual Dissimilarity Function: Using the optimal weight w^* so derived, we calculate the dissimilarity between x and y,

$$\delta^{xy} = w^{*T}|\theta_x - \theta_y|.$$

3. Computation of Score: The goodness of δ^{xy} needs to be compared w.r.t its locality which is quantified by the value Δ^{xy}. The difference between δ^{xy} and Δ^{xy} would then actually indicate how more similar is x and y than their surroundings and in turn tell us the possibility of formation of new edge. So the score is given as,

$$score(x,y) = (\Delta^{xy} - \delta^{xy}).$$

Choice of α and β: α and β are the parameters that control the inequality constraints. Smaller (larger) value of α (β) allows a lower (higher) dissimilarity between the connected (disconnected) nodes. Here, we have experimentally selected α and β. We have experimented with random items and the optimum values of α and β is found by maximizing the harmonic mean of *true negative and false positive*.

3. EXPERIMENTAL RESULTS

Experimental Setup : We consider part of Movielens, Citeseer, Cora and WebKb datasets for experimentation. As baselines, we have chosen Adamic Adar distance which is a pure link based metric and a logistic regression based recommendation model which is an unconstrained classification model [5].

Datasets used: Movielens [7]: It has 6040 users and 3952 movies. Each user has rated at least one movie. Each movie has features which are a subset of a set of 18 nominal attributes (e.g. animation, drama etc.). We have constructed a hypothetical network where two movies have an edge if they have at least a certain number of common viewers. By choosing the minimum number of common viewers to be 100, we obtain a network with 3952 nodes and 5669 edges.

CiteSeer [6]: The CiteSeer dataset consists of 3312 scientific publications and the citation network consists of 4732 links. Each publication is tagged with a set of keywords. Total number of keywords is 3703.

Cora [6]: The Cora dataset consists of 2708 scientific publications and the citation network consists of 5429 links. Here the total number of keywords is 1433.

WebKb [6]: This dataset consists of 877 scientific publications and the citation network consists of 1608 links. Here the total number of keywords is 1703.

For these four datasets, we generated recommendations with three algorithms, (a) Constrained Local learning (CLL) algorithm proposed in this paper, (b) Adamic Adar metric, (c) Logistic regression. We considered same set of items as queries for all the three algorithms. To build the ground truth, we have removed some edges from the graphs and have noted whether those edges can be predicted back.

Adamic Adar (AA) : Using the method described in [4], we obtain the recommendation scores. **Logistic Regression (LR):** In this method the difference in attributes between x and y is defined by $\theta_{xy} = (0.5\hat{1} - |\theta_x - \theta_y|)$. Here

Table 2: Summary of the datasets , where N is the total number of items, E is the total number of links, $n(a)$ is the number of features, d_{max} is the maximum degree and d_{avg} is the average degree.

Dataset	N	E	$n(a)$	d_{avg}
Movielens	3952	5669	18	2.8689
CiteSeer	3312	4732	3703	2.7391
Cora	2708	5429	1433	3.89
WebKb	877	1608	1703	2.45

$\hat{1}$ is a vector having all elements equal to 1 and dimension same as θ. Since $|\theta_x - \theta_y|$ is a binary vector in all datasets, elements of the term $(0.5\hat{1} - |\theta_x - \theta_y|)$ will be positive or negative depending on whether the features are similar or not. The score between x and y is given by $score(x,y) = 1/(1 + exp(-w^T\theta_{xy}))$. So more is the $score(x,y)$, higher is the rank of y in the list of recommended items to x. The logistic regression algorithm is carried out in two different ways. In the first approach, we randomly choose a part of network as the training data. Using these data, the regression model generates the optimum attribute weight vector w. In the second method, to provide recommendation to an item x, the training data consists of x and it's relation (link or no link) with randomly selected 30 percent nodes. So for each node, the optimum w is different, i.e. w is local to x in this case. Interestingly we observe that, for all the four datasets, the performance of global logistic regression model is very poor. Hence we present the results given by second(local) method of logistic regression for comparison with our method.

For performance comparison we use the following performance metrics.

Metrics 1 and 2: $MeanPrecision(k) = \frac{1}{q}\Sigma_{i=1}^q P_i(k)$; $MeanRecall(k) = \frac{1}{q}\Sigma_{i=1}^q R_i(k)$, where q is the number of queries, $P_i(k)$ is Precision@k for i^{th} query and $R_i(k)$ is Recall@k for i^{th} query. So $MeanPrecision(k)$ is the average of all Precision@k values over the set of queries. Same is with $MeanRecall(k)$.

Metric 3: $AvP(i) = \frac{1}{L}\Sigma_{k=1}^n P_i(k).r_i(k)$, where n is the total number of items, L is the number of retrieved relevant items and $r_i(k)$ is an indicator taking value 1 if the item at rank k is a relevant item or zero otherwise. Thus we obtain $MAP = \frac{1}{q}\Sigma_{i=1}^q AvP(i)$.

Comparison of Results : Figure 2 indicates variations of $MeanPrecision$ against $MeanRecall$ and Table 3 gives a comparative analysis of MAP (Mean Average Precision) values for all datasets. We observe that in all the datasets the overall performance of the proposed approach is much superior to the other two methods. In all these datasets, where the nodes with high degree are connected together, Adamic-Adar metric fails to provide high score. Due to its very poor performance in case of popular items, it produces a poor overall MAP value in all four datasets.

Apart from locality, consideration of constraints is found to be very crucial. Because in the local logistic regression model, although the weights are obtained locally, its performance is substantially poorer than CLL. It is because, even though it is based on local behaviour, no constraint is

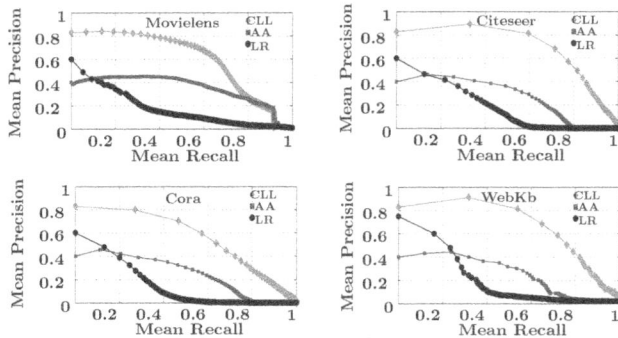

Figure 2: Mean Precision Recall curve

Table 3: Mean Average Precision(MAP), for different algorithms and different datasets

Dataset	CLL	AA	LR
Movielens	0.7740	0.4720	0.2313
CiteSeer	0.6313	0.4161	0.3783
Cora	0.5620	0.3910	0.2923
WebKb	0.6427	0.4029	0.3339

considered. So a suitable formulation of constraint plays a deciding role.

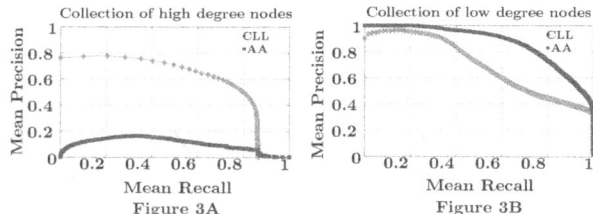

Figure 3: A comparison between CLL and AA algorithm on Movielens dataset

CLL vs Adamic Adar : Figure 3 shows a comparison of Adamic Adar and CLL on Movielens dataset over two parts of the network: on a dense part and on a sparse part. It is clear that in the dense part, high degree nodes being connected together, Adamic Adar shows extremely poor performance [Figure 3A] and CLL performs much better. Interestingly in case of low degree items the results are comparable although Adamic Adar performs a bit better [Figure 3B]. Please note that CLL operates efficiently at a very important zone i.e. the zone where the popular items are being bought along with some other items. In most of the recommendation algorithms this is considered as a superfluous information and given least importance. However there are some definite semantics behind choice of such popular items, which this method (CLL) clearly brings forward. On the other hand, CLL can quickly adjust to regions where link plays a predominant role and performs as good as a pure link based strategy(AA).

Variations on datasets: If we carefully check the MAP values in the Table 3, we observe that MAP value for CLL is much higher in Movielens than the other three datasets. This is because the attributes in Movielens are much more well structured. It actually consists of the genre and it is observed that people usually like movies of similar genre. On the other hand in the citation document space the keywords get diluted through polysemy, synonymy etc. However more important to note that CLL can really take advantage of such attribute structure and increase its MAP much sharply (20%) than Adamic Adar (13%) from the corresponding second best in the list.

4. CONCLUSION AND DISCUSSIONS

A method for local learning of item dissimilarity via a constrained optimization framework has been proposed in this paper. The algorithm assigns more weights to the important features of a particular region of the graph. The important features and the reference score of similarity are estimated in this optimization portfolio. The overall performance of the algorithm is found to be significantly better than two baseline algorithms, namely Adamic Adar and Logistic Regression. An interesting property of our algorithm is that it can also adjust between regions where content is dominating and where link is more important. However, these are initial results, a more detailed theoretical as well as experimental work need to be launched to realize the full potential of the algorithm.

5. REFERENCES

[1] L. Backstrom and J. Leskovec. Supervised random walks: predicting and recommending links in social networks. In *Proceedings of the fourth ACM international conference on Web search and data mining*, WSDM '11, pages 635–644, New York, NY, USA, 2011. ACM.

[2] S. Debnath, N. Ganguly, and P. Mitra. Feature weighting in content based recommendation system using social network analysis. In *Proceedings of the 17th international conference on World Wide Web*, WWW '08, pages 1041–1042, New York, NY, USA, 2008. ACM.

[3] M. Kagie, M. van Wezel, and P. J. Groenen. Choosing attribute weights for item dissimilarity using clikstream data with an application to a product catalog map. In *Proceedings of the 2008 ACM conference on Recommender systems*, RecSys '08, pages 195–202, New York, NY, USA, 2008. ACM.

[4] D. Liben-Nowell and J. Kleinberg. The link prediction problem for social networks. In *Proceedings of the twelfth international conference on Information and knowledge management*, CIKM '03, pages 556–559, New York, NY, USA, 2003. ACM.

[5] A. Popescul, R. Popescul, and L. H. Ungar. Towards structural logistic regression: Combining relational and statistical learning. In *Proceedings of the Workshop on MultiRelational Data Mining*, KDD'02, pages 130–141, 2002.

[6] www.cs.umd.edu/projects/linqs/projects/lbc.

[7] www.grouplens.org.

Design and Evaluation of a Group Recommender System

Toon De Pessemier, Simon Dooms, Luc Martens
Wica, IBBT-Ghent University
G. Crommenlaan 8 box 201
B-9050 Ghent, Belgium
toon.depessemier@ugent.be

ABSTRACT

Though most recommender systems make suggestions for individual users, in many circumstances the selected items (e.g., movies) are not for personal usage but rather for consumption in group. In this paper, we present a recommender system for audio-visual content that generates suggestions for groups of people (such as families or friends) in the home environment. In this context, different group recommendation strategies are evaluated for various algorithms and sizes of the group. An offline evaluation proves the assumption that for randomly composed groups the accuracy of all recommendation algorithms decreases if the group size grows. Besides, the results show that the group recommendation strategy which produces the most accurate results is depending on the algorithm that is used for generating individual recommendations. Consequently, if an existing recommender system for individuals is extended to a recommender system for groups, the group recommendation strategy has to be chosen based on the utilized recommendation algorithm in order to maximize the efficiency of the group recommendations.

Categories and Subject Descriptors

H.5.3 [**Information Interfaces and Presentation**]: Group and Organization Interfaces

General Terms

Algorithms, Experimentation

Keywords

Group recommender, Evaluation, Aggregation strategy

1. INTRODUCTION

Recommender systems can help users to find the most interesting products or content thereby addressing the information overload problem of (online) services. Although

the majority of the currently deployed recommender systems are designed to generate personal suggestions for individual users, in many cases content is selected and consumed by groups of users rather than by individuals. Many researchers have already investigated how the current state-of-the-art recommendation algorithms can be adapted in order to generate group recommendations [6]. In literature, group recommendations have mostly been generated either by aggregating the users' individual recommendations into recommendations for the whole group (aggregating recommendations) or by aggregating the users' individual preference model into a preference model of the group (aggregating preferences) [2].

The first recommendation strategy (aggregating recommendations) generates recommendations for each individual user using a traditional recommendation algorithm. Subsequently the recommendation lists of all group members are aggregated into a group recommendation list, which (hopefully) satisfies all group members. Different methods to aggregate the recommendation lists have been proposed during the last decade. Most of them make a decision based on the algorithm's prediction value, i.e. a prediction of the user's rating score for the recommended item. The second recommendation strategy (aggregating preferences) combines the group members' preferences into a group preference model using a social value function. A social value function describes how the opinions and preferences of individuals affect the group's recommendations. In literature, different social value functions have been proposed to aggregate the members' preferences, but still no consensus exists about the most optimal solution [5]. After aggregating the members' preferences, the group's preference model is treated as a pseudo user to produce group recommendations using a traditional recommendation algorithm.

The goal of this research is to find which of these group recommendation strategies generates the most accurate recommendations for audio-visual content in a home environment. Section 2 provides an overview of related work. Section 3 gives some information about the use case in which the results will be applied, i.e. a group recommender system for audio-visual content in the home environment. Section 4 elaborates on the implemented group recommendation strategies and Section 5 discusses the setup of our experiment. The results are presented in Section 6. Section 7 draws conclusions and points to future work.

2. RELATED WORK

In the domain of movies, Polylens is an extension of Movielens that enables recommendations for groups [6]. Polylens allows users to create and manage their own groups in order to receive group recommendations next to the traditional individual recommendations. Both survey results and observations of user behavior proved that group recommendations are valuable and desirable for the users.

A less obvious use case for group recommendations is a recipe recommender for families [2]. Since all family members typically eat a joint meal at least once a day, recipes and food consumption are good examples of a group activity. In the context of this recipe recommender, the aggregating preferences strategy and the aggregating recommendations strategy were compared. An evaluation with a number of families showed that the aggregating preferences strategy yield slightly better results than the aggregated recommendation lists. This recommender is based on collaborative filtering (CF) and the individual data of group members is aggregated in a weighted manner, such that the weights reflect the observed interaction of group members. As was already remarked by other researchers, this is only one type of recommendation algorithm and one of the possible approaches for aggregating prediction values or recommendation lists [1]. So, an extensive comparison of the two strategies is still missing in literature. Our research compares the aggregated recommendations with the aggregating preferences strategy more thoroughly. In the context of a group recommender for movies, the two strategies are evaluated for different group sizes and different recommendation algorithms.

Research regarding the aggregating recommendations strategy has learned that the influence of the aggregation method is limited [1]. A comparison of the group recommendation lists generated using four commonly-used aggregation methods showed similar results in terms of accuracy for all methods. This study compared the results for groups with a size of 2, 3, 4 or 8 members using the aggregating recommendations strategy and an algorithm based on SVD. Our research investigates the influence of the group size not only for the strategy that aggregates the recommendations, but also for the strategy that aggregates the members' profiles. Moreover, our research also considers other sizes of the group including very large groups, and compares the results for different classes of algorithms.

3. GROUP RECOMMENDER SYSTEM

The recommender system proposed in this paper runs on a home-gateway that aggregates the content of the group members from different sources (local or remote, e.g., external hard drives, recorders, etc.) and provides an overview of their joint collection of content items (songs and videos). For each content item, a list of similar items is provided. Furthermore, personal suggestions are offered based on the preferences of the current users of the system. For scalability reasons, these suggestions are calculated by an external recommendation services and queried by the local client whenever needed. The content items and recommendations can be filtered based on genre and selected for playback on the desired device in the home environment (e.g., the television set). This interaction and viewing behavior is logged as implicit feedback for the recommender system. Besides, explicit feedback can be provided on individual items by the "thumbs up" and "thumbs down" icons or on genres, actors, and directors of the movie by selecting these attributes in

Figure 1: A screenshot of the recommender system

the interface. Figure 1 illustrates this functionality of the recommender system with a screenshot of the user interface.

Groups can be created or changed easily by the users according to the current situation in the home. E.g., a group can be composed for the family members that are going to watch a movie this evening. In addition to adding or removing members of a group, users can assign a personal *importance weight* to each member of the group. These weights can be used to express for example that older people (such as parents) have more influence on the recommendations than younger people (such as children). Three options are possible for these weights: a high, a low, and a neutral importance. The aggregation method or social value function of the grouping strategy (Section 4) takes these importance weights into account during the calculation of the group recommendation list. Changing the group composition or the importance weights has an immediate impact on the group recommendations which are showed in interface. To enable these immediate adjustments to the recommendation list, recommendations are precalculated for every combination of group composition and importance weights. Given the small number of group members in a typical home environment and the limited options for the importance weights (3 possible values), the total number of group combinations remains limited, so that the computation load is still acceptable.

4. RECOMMENDATION STRATEGY

To overcome the cold start problem and evaluate the group recommendation strategies, we used the Movielens (100K) data set in the calculation process of the recommender service. Therefore, the explicit and implicit feedback provided by the users of our system will be converted to the 5-point rating scale of the Movielens system. This way, the combined data set enables the CF to find neighbors for the new users of our system and generate accurate recommendations based on the community knowledge of the Movielens data set.

Before calculating the recommendations, the user's ratings are normalized by subtracting the user's mean rating and dividing this difference by the standard deviation of the user's ratings. Some similarity metrics, such as the Pearson correlation, consider the fact that users are different with respect to how they interpret the rating scale; thereby making the normalization process unnecessary for calculating similarities. However, normalizing the ratings is still meaningful if the ratings of the group members are aggregated into a group rating before the similarities are calculated [5]. After normalization, the ratings can (optionally) be squared to incorporate the quadratic effect of feedback mechanisms. Research has proved that users may not rate items in a linear way; this means: the further away from the middle point of the scale, the larger the differences between subsequent ratings [5]. E.g., the difference between a 5-star and a 4-star rating is more significant than the difference between a 4-star and a 3-star rating.

After transforming the data, group recommendations are calculated based on the preferences of all group members. Three group recommendation strategies are available in the current implementation of the video recommender system: aggregating recommendations, aggregating preferences, or a switching strategy that combines the two previous strategies. The switching strategy generates group recommendations based on the aggregating recommendations strategy if the profile density of the group is below a certain threshold. Above that threshold, the aggregating preferences strategy is used. Switching between a strategy that aggregates the recommendations and a strategy that aggregates the preferences might produce more accurate results than both individual strategies [2].

In case of the aggregating recommendations strategy, the aggregation method calculates for each item the average of the prediction values of each group member's recommendation list. Although several alternative aggregation methods, such as "average without misery" and "least misery", are possible, research has shown that the influence of these aggregation methods on the accuracy of the group recommendations is limited [1]. In case of the aggregating preferences strategy, the members' individual preferences are aggregated into a group preference by calculating the average of the members' rating for each item. By using the same aggregation method (i.e. average) for both aggregating the individual recommendation lists and aggregating the individual preferences, the accuracy of all strategies can be compared (Section 6). If group members have an unequal importance weight, a weighted average is used as aggregation method to take the relative importance of each group member into account. Unfortunately, the influence of the importance weights on the accuracy of the group recommendations could not be evaluated in the experiment of Section 5, since the data set that was used for this research does not contain these weights.

5. EXPERIMENTAL SETUP

A number of state-of-the-art recommendation algorithms are used for comparing different group recommendation strategies. The used implementation of **Collaborative Filtering** (CF) is based on the work of Breese et al [3]. This nearest neighbor CF uses the Pearson correlation metric for discovering similar users in the user-based approach (UBCF) or similar items in the item-based approach (IBCF). As **Content-Based recommender** (CB) the InterestLMS predictor of the open source implementation of the Duine framework [7] is adopted (and extended to consider extra metadata attributes). Based on the actors, directors, and genres of the content items and the user's ratings for these items, the recommender builds a profile model for every user. The used **hybrid recommender** (Hybrid) combines the recommendations with the highest rating prediction of the IBCF and the CB recommender into a new recommendation list. The result is an alternating list of the best recommendations

originating from these two algorithms. A user-centric evaluation comparing various algorithms based on various characteristics showed that this straightforward combination of CF and CB recommendations outperforms both individual algorithms on almost every qualitative metric [4]. As recommender based on matrix factorization, we opted for the open source implementation of the **SVD Recommender** (SVD) of the Apache Mahout project [8]. The recommender is configured to use 19 features, i.e. the number of genres in the Movielens data set, and the number of iterators is set at 50. To compare the results of the various recommenders, the **popular recommender** was introduced as a baseline. This recommender generates for every user always the same list of most popular items, which is based on the number of received ratings and the mean rating of each item.

A major issue in the domain of group recommender systems is the evaluation of the effectiveness. Interviewing groups or performing online evaluations can be partial solutions but are not feasible on a large scale or to extensively test alternative algorithms. Therefore, we are forced to perform an offline evaluation, in which groups are sampled from the users of a traditional single-user data set, as was done by Baltrunas et al [1]. Firstly, all users are randomly assigned to one group of a specific size. Secondly, group recommendations are generated for each of these groups . Thirdly, the recommendations are evaluated individually as in the classical single-user case, by comparing (the rankings of) the recommendations with (the rankings of) the items in the test set of the user. The evaluation of the group recommendations is based on the traditional procedure of dividing the data set chronologically in training set (60%) and test set (40%). For each user, the effectiveness of his group recommendations is evaluated based on his individual ratings in the test set using the Normalized Discounted Cumulative Gain (nDCG) [1]. We opted for a recommendation list of 5 content items, since this is a realistic length for a recommendation list in a TV interface. After calculating the nDCG for each individual user, the average nDCG over all users is calculated as an overall measure of efficiency. The group size is varying from 1 person per group (=individual recommendations) until 10 persons per group. Besides, the results are provided for very large group compositions (group sizes of 15 and 20 persons).

6. EVALUATING STRATEGIES

Figure 2 shows the mean nDCG together with the 95% confidence intervals, according to the recommendation algorithm and the group size. Since groups are randomly created and the accuracy of the recommendations is depending on the composition of the groups, 30 measurements are performed for each combination of group size and algorithm. The average of these 30 measurements is used as an estimation of the effectiveness of the group recommendations and is visualized in Figure 2. The bar series with the prefix "Rec" are using the aggregating recommendations strategy whereas the prefix "Pref" refers to the aggregating preferences strategy. (The switching strategy is not evaluated.) The vertical axis crosses the horizontal axis at the accuracy level of the popular recommender, which is constant for the various group sizes. This way, the bar chart shows the relative improvement of each algorithm with respect to the baseline accuracy of the popular recommender. Since all combinations of group size and algorithm show an accuracy

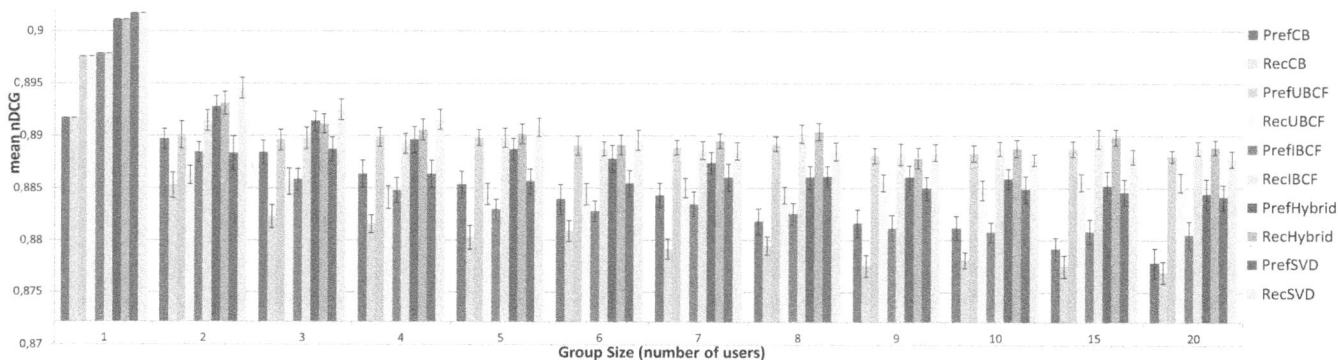

Figure 2: The accuracy of the group recommendation strategies for different algorithms and group sizes.

improvement with respect to the static list of most popular items, this experiment shows that group recommendations are still useful, even for large groups.

As expected, the graph shows a decreasing performance of the group recommendations as the group size increases for all algorithms. The comparison between the aggregating recommendations strategy and the aggregating preferences strategy provides another interesting finding. The grouping strategy that provides the most accurate recommendations depends on the used algorithm. The CB and UBCF algorithm generate the most accurate group recommendations if the group members' preferences are aggregated whereas the results of SVD and IBCF are most optimal if the members' recommendations are aggregated. A possible explanation for these differences in accuracy lies in the way in which the algorithm processes the data. The CB and UBCF algorithm create some kind of user profile to find respectively matching items or similar users. In contrast, the matrix decomposition of SVD and the item-item similarities of IBCF provide less insight into the preferences of the users. So, aggregating the preferences of the group members provides optimal results if the algorithm internally composes some kind of user profile holding his preferences, whereas aggregating the recommendations of the group members is a better option if the users' preferences are less transparent in the data structure of the algorithm. The internal modeling of the user profile can also explain why some combinations of algorithm and strategy (such as PrefSVD) deteriorate faster than others (such as PrefUBCF) as the group size increases. Finally, the results of Figure 2 show that the SVD and hybrid recommender produce the most accurate group recommendations for various group sizes. However these results are only based on the Movielens data; probably the most optimal combination of algorithm and strategy depends on the data and scenario at hand. For the in-home recommender system presented in Section 3, we opted for the hybrid algorithm because of its accuracy and the positive evaluation regarding novelty, usefulness, satisfaction, and trust via previously conducted user tests [4].

7. CONCLUSIONS

We presented a group recommender for audio and video in the home environment and evaluated two commonly-used group recommendation strategies for different algorithms. Neither of these can be designated as the overall winner since the effectiveness of grouping strategies is influenced by

the used recommendation algorithm. If recommender systems for individual users are extended to enable group recommendations, these results can be used to choose the most optimal grouping strategy based on the currently employed algorithm. In the future, we want to investigate the proposed switching strategy that combines the two group recommendation strategies as well as other techniques to combine the strategies. Besides, the accuracy of the grouping strategies will be compared for groups which are composed so that the group members have a high similarity.

8. REFERENCES

[1] L. Baltrunas, T. Makcinskas, and F. Ricci. Group recommendations with rank aggregation and collaborative filtering. In *Proceedings of the fourth ACM conference on Recommender systems*, RecSys '10, pages 119–126, New York, NY, USA, 2010. ACM.

[2] S. Berkovsky and J. Freyne. Group-based recipe recommendations: analysis of data aggregation strategies. In *Proceedings of the fourth ACM conference on Recommender systems*, RecSys '10, pages 111–118, New York, NY, USA, 2010. ACM.

[3] J. S. Breese, D. Heckerman, and C. Kadie. Empirical analysis of predictive algorithms for collaborative filtering. In *Proceedings of the Fourteenth conference on Uncertainty in artificial intelligence*, UAI'98, pages 43–52, San Francisco, CA, USA, 1998.

[4] S. Dooms, T. De Pessemier, and L. Martens. A user-centric evaluation of recommender algorithms for an event recommendation system. In *Proceedings of the workshop on User-Centric Evaluation of Recommender Systems and Their Interfaces at ACM Conference on Recommender Systems (RECSYS)*, 2011.

[5] J. Masthoff. Group modeling: Selecting a sequence of television items to suit a group of viewers. *User Modeling and User-Adapted Interaction*, 14:37–85, 2004.

[6] M. O'Connor, D. Cosley, J. A. Konstan, and J. Riedl. Polylens: a recommender system for groups of users. In *Proceedings of the seventh conference on European Conference on Computer Supported Cooperative Work*, ECSCW'01, pages 199–218, Norwell, MA, USA, 2001.

[7] Telematica Instituut/Novay. Duine Framework, 2009. Available at http://duineframework.org/.

[8] The Apache Software Foundation. Apache Mahout, 2012. Available at http://mahout.apache.org/.

Swarming to Rank for Recommender Systems

Ernesto Diaz-Aviles
diaz@L3S.de

Mihai Georgescu
georgescu@L3S.de

Wolfgang Nejdl
nejdl@L3S.de

L3S Research Center / University of Hannover, Germany

ABSTRACT

Recommender systems make product suggestions that are tailored to the user's individual needs and represent powerful means to combat information overload. In this paper, we focus on the item prediction task of Recommender Systems and present *SwarmRankCF*, a method to automatically optimize the performance quality of recommender systems using a Swarm Intelligence perspective. Our approach, which is well-founded in a Particle Swarm Optimization framework, learns a ranking function by optimizing the combination of unique characteristics (i.e., features) of users, items and their interactions. In particular, we build feature vectors from a factorization of the user-item interaction matrix, and directly optimize Mean Average Precision metric in order to learn a linear ranking model for personalized recommendations. Our experimental evaluation, on a real world online radio dataset, indicates that our approach is able to find ranking functions that significantly improve the performance of the system for the Top-N recommendation task.

Categories and Subject Descriptors: H.3.3 [**Information Storage and Retrieval**]: Information Search and Retrieval–*Information Filtering*

General Terms: Algorithms, Experimentation, Measurement, Performance
Keywords: Collaborative Filtering; Matrix Factorization; PSO; Recommender Systems; Swarm Intelligence

1. INTRODUCTION

The information overload problem facing today's Web users has made the decision-making task really challenging, and in many cases complex to the user, who has to face an overwhelming set of options that outstrips her capability to survey them and reach a decision. How can recommendation systems help users to meet their particular information needs and preferences?

In practice this problem can be cast as a ranking problem, whose goal is to define an ordering among items (e.g., Web sites, news articles, songs, books, or movies) that ranks relevant ones in higher positions of the retrieved list.

In recent years supervised learning-based methods have been proposed to automatically learn an effective ranking model based on training data (e.g., [8, 6]). This task is referred to as "Learning To Rank for Information Retrieval". Learning to rank relies upon

pre-engineered features that characterize the items to be ranked. Such a set of features can be difficult to compute and maintain.

In the absence of high quality *explicit features*, recommender systems can infer *latent factors* about users and items using matrix factorization algorithms for Collaborative Filtering (CF).

This paper presents an approach to learning ranking functions that exploits collaborative latent factors as features. To accomplish this, instead of manually creating an item feature vector, we factorize a matrix of user-item interactions, and use these collaborative latent factors as input to a Swarm Intelligence (SI) ranking method: SwarmRank [5]. We focus in this paper on the item prediction or Top-N recommendation problem in the context of recommender systems, which we consider a harder prediction problem than rating prediction, as only positive feedback is available.

SwarmRank was chosen because it is based on Particle Swarm Optimization (PSO) [11], a global non-linear optimization algorithm inspired in the behavior of biological organisms. PSO does not require, nor approximate, gradients of the cost function, and its resilience to local minima allows it to directly optimize non-smooth Information Retrieval (IR) measures, such as MAP (Mean Average Precision).

Contributions. The main contributions of this paper are as follows: (1) We present SwarmRankCF, a model suitable for the personalized item recommendation task, that seamlessly integrates a SI learning to rank method with a collaborative filtering approach, to derive a personalized ranking function optimized for the ranking task. (2) We present an empirical study on the public dataset *Last.fm Dataset – 1K users* obtained from Last.fm[1], a major social media Internet radio station, demonstrating the superior recommendation performance of SwarmRankCF.

2. BACKGROUND

In this section we introduce the task of learning to rank for IR and provide a background on ranking with swarm intelligence. We also include a brief review of particle swarm optimization and matrix factorization for recommender systems.

2.1 Learning to Rank for IR

Learning to rank for Information Retrieval [10] is a problem formalized as follows. In learning (training), a collection of queries and their corresponding retrieved documents are given. Furthermore, the labels (i.e., relevance judgments) of the document with respect to the queries are also provided. The relevance judgments, provided by human annotators, can represent ranks (e.g., categories in a total order). The objective of learning is to construct a ranking model, e.g., a ranking function, that achieves the best performance on test data.

---

[1]**Last.fm**: http://www.last.fm

In retrieval (test phase), given a query, the learned ranking function is applied, returning a ranked list of documents in descending order of their relevance scores. Suppose that $Q = \{q_1, \cdots, q_{|Q|}\}$ is the set of queries and $D = \{d_1, \cdots, d_{|D|}\}$ the set of documents, the training set is created as a set of query-document pairs, $(q_i, d_j) \in Q \times D$, upon which a relevance judgement (e.g., a label) indicating the relationship between q_i and d_j is assigned by an annotator. Suppose that $Y = \{y_1, \cdots, y_{|Y|}\}$ is the set of labels and $y_{ij} \in Y$ denotes the label of query-document pair (q_i, d_j). A feature vector $\phi(q_i, d_j)$ is created from each query-document pair $(q_i, d_j), i = 1, 2, \cdots, |Q|; j = 1, 2, \cdots, |D|$. The training set is denoted as $T = \{(q_i, d_j), \phi(q_i, d_j), y_{ij}\}$. The ranking model is a real valued function of features:

$$f(q, d) = \vec{p_g} \cdot \phi(q, d) \tag{1}$$

where $\vec{p_g}$ denotes a weight vector. In ranking, for query q_i the model associates a score to each of the documents d_j as their degree of relevance with respect to query q_i using $f(q_i, d_j)$, and sort the documents based on their scores.

2.2 Ranking with Swarm Intelligence

Particle Swarm Optimization

The particle swarm optimization algorithm (PSO) is a population-based probabilistic optimization algorithm inspired by the social behavior of biological organisms, specifically the ability of groups of species of animals to work as a whole in locating desirable positions in a given area. In a PSO algorithm, a population of agents called *particles* move through the solution space of an optimization problem, updating their velocity according to the information collected by the group called *swarm*. PSO algorithms are global non-linear optimization algorithms and do not require nor approximate gradients of the cost function.

In every iteration, each particle is attracted to the best solution that it has found individually, and toward the best solution that any particle in their *neighborhood* has found. In PSO, a neighborhood is defined for each individual particle as the subset of particles which is able to communicate with. The algorithm updates the entire swarm at each time step by updating the velocity and position of each particle in every dimension. An in-depth introduction to PSO is given in [11].

SwarmRank: SI-Based Learning to Rank

SwarmRank [5] is a PSO-based method to handle the task of learning to rank for IR. The goal of SwarmRank is the discovery of good ranking formulas more adapted to the particularities of a specific collection, which are also able to generalize well beyond the training data. The procedure examines important information retrieval features extracted from query-document pairs instances and learns a linear function that combines them optimally.

SwarmRank is an iterative process that learns linear ranking functions of the form of Eq. (1), where $\vec{p_g}$ is the weight vector corresponding to the best global solution found by the *swarm* in PSO. The function represents a linear combination of the query-document pairs feature vectors which associates a real value to each query-document pair as their degree of relevance.

SwarmRank takes as an input the query-document pairs and their corresponding feature vectors. As an instance of PSO, it quantifies the optimality of a solution by means of a fitness function, which serves as a criteria to update the particle and global best solutions. As output, the procedure returns a linear ranking function whose coefficients are given by the dimension values corresponding to the global best solution vector found by the swarm.

Fitness Function. Since the particles represent solutions in terms of weight vectors to be used in a document ranking function, the fitness function measures the quality of the ranking generated by a given particle. SwarmRank takes advantage of the non-linear optimization capabilities of PSO and directly optimizes a non-smooth IR measure, namely Mean Average Precision–MAP.

2.3 Recommender Systems

Let $U = \{u_1, \ldots, u_{|U|}\}$ and $I = \{i_1, \ldots, i_{|I|}\}$ be the sets of all users and all items, respectively. Suppose we are dealing with interactions between these two entities, and for some user $u \in U$ and item $i \in I$ we observe a *relational score* r_{ui}. Thus each instance of the data is a tuple (u, i, r_{ui}), which for example, in the movie recommendation case might correspond to an explicit "rating" given by user u to movie i, or in the case of music recommendation to a "weight" implicitly derived from user u's listening patterns, e.g., how many times the user u has listened to song i. Collaborative Filtering (CF) is a technique used by some recommender systems, whose goal is to make automatic predictions (filtering) about the interests of a user by collecting preferences or taste information from many users (collaborating). Typical CF organizes these tuples into a sparse matrix $\mathbf{R}$ of size $|U| \times |I|$, using (u, i) as index and r_{ui} as entry value. The task of the recommender system is to estimate the score for the missing entries.

Matrix Factorization

Low dimensional linear factor models [9] are one of the most effective approaches for CF. Factor-based algorithms consider that only a small number of *latent* factors can influence the preferences. Their prediction is a real number $\hat{r}_{ui}$ per user item pair (u, i). Some of the most successful realizations of latent factor models are based on *matrix factorization*. In its basic form, matrix factorization estimates a matrix $\hat{\mathbf{R}}$ of size $|U| \times |I|$ by the product of two low-rank matrices $\mathbf{U} : |U| \times k$ and $\mathbf{I} : |I| \times k$:

$$\hat{\mathbf{R}} := \mathbf{U} \, \mathbf{I}^T \,, \tag{2}$$

where k is a parameter corresponding to the rank of the approximation. Each row $\mathbf{U}_u$ in $\mathbf{U}$ and row $\mathbf{I}_i$ in $\mathbf{I}$ can be considered as a feature vector describing a user u and an item i, correspondingly. Thus the final prediction is the linear combination of the factors:

$$\hat{r}_{ui} = \mathbf{U}_u \cdot \mathbf{I}_i^T \tag{3}$$

Singular value decomposition (SVD) provides the best approximation of $\hat{\mathbf{R}}$ to $\mathbf{R}$ with respect to least-square, which is usually extended using regularization to prevent overfitting [9].

3. SI FOR RECOMMENDER SYSTEMS

In this section we present our collaborative ranking approach for learning to rank. The goal of learning to rank is to automatically learn a ranking model from training data, such that the model can sort objects (e.g., documents, songs, movies) according to their degrees of relevance, preference, or importance as defined in a specific application.

In the general learning to rank scenario, a retrieval function is learned from rankings of documents associated to queries. Each document is represented by a vector of predefined features that characterize it, for example, text documents are usually characterized by their term frequency, inverse document frequency, document length and other low-level content characteristics, as well as high-level content features as BM25 scores. In the case of web pages, hyperlink features are also used, such as PageRank or HITS.

In CF, on the other hand, item and user features can be learned based on user-item interactions (e.g, ratings or relevance scores)

Algorithm 1 : SwarmRankCF

Input: (i) Matrix $\mathbf{R}$ of user-item interactions and (ii) The number of factors k

Output: A ranking function $f(u, i) = \vec{p_g} \cdot \phi(u, i)$

1: Extract k collaborative features (i.e., latent factors) $\mathbf{I}$ by factorizing matrix $\mathbf{R}$, e.g, by applying SVD (Section 2.3).

2: Apply SwarmRank to learn a ranking function $f(u, i)$, by optimizing Mean Average Precision (MAP), using U as queries Q, items I as documents D, the item feature vectors $\mathbf{I}$ as $\phi(u, i)$, and $\mathbf{R}$ as the relevance scores Y (Section 2.1).

3: **return** $f(u, i)$

---

using a matrix factorization method, which learns for each item (columns of $\mathbf{R}$), a k-dimensional feature vector (rows of $\mathbf{I}$), where each row of $\mathbf{U}$ is a feature vector that captures the latent factors of the users, and can be considered as a linear predictor, predicting the entries in the corresponding row of $\mathbf{R}$, based on inner products in a k-dimensional space (Section 2.3).

We propose to learn a collaborative ranking model using SwarmRank. To this end, we cast the item recommendation task as a learning to rank problem, where users U can be considered as queries Q, items I as documents D, feature vectors $\phi(u, i)$ as $\mathbf{I}$ (i.e, the latent factors)[2]. Finally, the relevance scores Y are given by $\mathbf{R}$, that corresponds to ratings or implicit feedback information that measures user u's preferences for item i. The approach, named *SwarmRankCF* is summarized in Algorithm (1).

4. EXPERIMENTS AND EVALUATION

We evaluate the recommendation performance of our approach, SwarmRankCF, on a public dataset obtained from *Last.fm*, a major social media Internet radio station. In our evaluation, we empirically compared the recommendation quality of SwarmRankCF to: (i) *PureSVD*: a matrix factorization algorithm based on SVD and proposed for Top-N recommendation tasks [4], (ii) *Most Popular*: a non-personalized method that recommends the most popular artists to every user, and (iii) *Random*: a baseline method, that recommends random artists to users.

Dataset

We conducted experimentation, parametrization and algorithmic fine tuning on the "real-world" public dataset *Last.fm Dataset – 1K users* [3], which represents the whole listening habits, until May, 2009, for nearly 1,000 users.

Specifically, we used in our evaluation a 5-core subset of the dataset, i.e., every user listened to at least 5 songs and each song was listened by at least 5 users. The 5-core statistics are as follows: 242,103 user-item interactions (transactions), 844 unique users and 37,315 unique items (artists).

Evaluation Methodology

We performed a time sensitive split of the dataset S into two sets: a training set S_{train} and a testing set S_{test}. Consider we make the split at time t_{split}, then we put into S_{train} the individual training examples (i.e., user rankings) with timestamps less than t_{split}. Into S_{test}, we put the user rankings with timestamps greater than t_{split}. The recommenders are trained on S_{train} and then their performance is measured on S_{test}.

---

[2] Please note that since we are learning a linear ranking function, the model cannot make use of the user u features, because such features are the same for all items of user u within his rankings.

To evaluate the recommenders we used a variant of the *all-but-1* protocol [1], also known as the leave-one-out holdout method. In particular, we followed a similar schema as the one described in [4].

Our goal is to evaluate the system performance when it suggests Top-N items to a user. For example, recommending the user a few specific artists which are supposed to be the most attractive to him. That is, to find the relative position of these interesting items within the total order of items ranked for a specific user.

For each user u we aggregate his rankings in the test set S_{test}, by accumulating the item frequencies across those rankings in order to produce a single total ranking. The items are again sorted in descending order of their accumulated frequencies. We take one item at random from the the top-10 of the aggregated ranking and hide it. In total, we have $|U|$ hidden items.

Then, for each hidden item i, we randomly select 100 additional items from the test set S_{test}. Notice that most of those items selected are probably not interesting to user u.

We predict the scores for the hidden item i and for the additional 100 items, forming a ranking by ordering the 101 items according to their scores. The best expected result is that the interesting item i to user u will precede the rest 100 random items.

Finally, we generate a Top-N recommendation list by selecting the N items with the highest score. If the test item i is in the Top-N, then we have a *hit*, otherwise we have a *miss*. We measure the quality by looking at the *recall* metric.

More formally, in our recommender systems setting, recall is defined as:

$$\text{recall} := \frac{\sum_{u \in U} hit(u)}{|U|}, \tag{4}$$

where $hit(u)$ is a binary function that returns 1, if the hidden item i is present in u's Top-N list of recommendations, and 0 otherwise. A recall value of 1.0 indicates that the system was able to always recommend the hidden item, whereas a recall of 0.0 indicates that the system was not able to recommend any of the hidden items. Since the precision is forced by taking into account only a restricted number N of recommendations, there is no need to evaluate *precision* or *F1* measures, i.e., for this kind of scenario precision is just the same as recall up to a multiplicative constant.

Experimental Setting

For SwarmRankCF, we used SVD as factorization method to learn the collaborative feature vectors. Furthermore, we set $k = 46$ as the number of latent factors for SwarmRankCF and PureSVD. The rationale behind using the particular value of $k = 46$ is to have an experimental setting with a dimensionality similar to the one used in non-personalized learning to rank tasks, e.g., the number of features in the benchmark dataset LETOR [10]. SwarmRankCF has been implemented in the Java programming language using the optimization framework provided by GenOpt [12]. Specifically, in our experiments we used the PSO with Constriction Coefficient algorithm (PSOCC) with parameter values that are most commonly found in the literature (e.g., [11]). The parameter setting for SwarmRankCF is summarized in Table 1.

Results

Recall was evaluated for different recommendation list sizes: Top-N, where $N \in \{1, 5, 10, 20\}$. The results are presented in Figure 1. SwarmRankCF clearly delivers the best recommendations, achieving significantly better results than the other methods (two-sample t-test, p < 0.025). As expected, SwarmRankCF and PureSVD largely outperform the non-personalized baselines. The reader should note

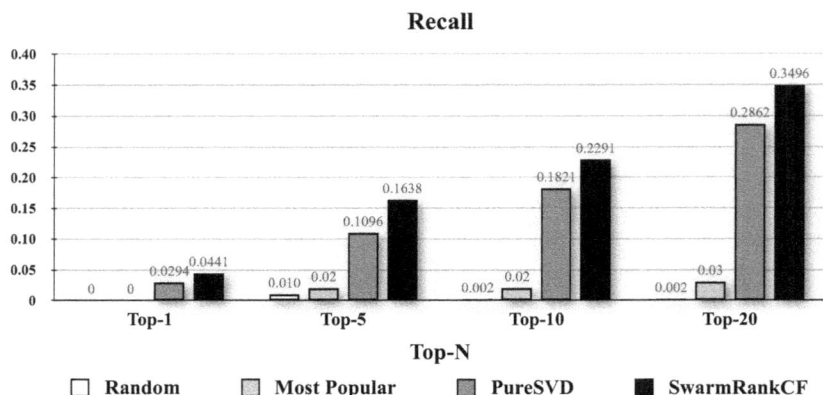

Figure 1: Recommendation performance in terms of recall measure.

Parameter	Value
SI Algorithm	PSO with Constriction Coefficient
Swarm size	50
Neighborhood Topology	Local Best (*lbest*)
Neighborhood Size	10
Cognitive Acceleration	$c_1 = 2.05$
Social Acceleration	$c_2 = 2.05$
Constriction Coefficient	$\chi = 0.72984$
# Dimensions	46 (one per latent feature)

Table 1: SwarmRankCF Parameter Settings.

that both SwarmRankCF and PureSVD use the latent factors learned by factorizing the user-item matrix using SVD, the strong performance of SwarmRankCF may be attributed to its better ability to model ranking semantics based on those features.

5. RELATED WORK

In recent years, the task of learning to rank has drawn a lot of interest in machine learning and several methods have been applied to learn ranking functions and promising results have been obtained. Typical methods include Ranking SVM [8, 7], RankBoost [6] and RankNet [2]. These three methods minimize loss functions that are loosely related to the evaluation measures such as MAP and NDCG. Given the nature of our approach, we are able to directly optimize such measures. All aforementioned algorithms, rely upon a manually selected set of features to build the ranking models. In contrast, our approach uses collaborative latent factors, learned directly from the user-item interactions, as feature vectors to characterize the user and items.

In the field of recommender systems, a lot of attention was directed to the rating prediction task, motivated by the Netflix challenge (e.g., [9]), standard sparse regression and classification methods proved to be successful tackling the challenge, but they cannot be directly applied to the item recommendation problem, as only positive observations are made. State-of-the-art methods for item recommendation are based on matrix factorization models that infer latent factors from the user-item interaction matrix, which we exploit to learn ranking functions for item prediction.

6. CONCLUSION AND FUTURE WORK

In this paper, we proposed SwarmRankCF to address the item recommendation task in the context of recommender systems. SwarmRankCF is a collaborative learning to rank algorithm based on swarm intelligence. While learning to rank algorithms use hand-picked features to represent items, we learn such features based on

user-item interactions, and apply a PSO-based optimization algorithm that directly maximizes Mean Average Precision. Our experimental study demonstrates the recommendation performance gain of SwarmRankCF for the Top-N recommendation task.

For future work, we plan to extend SwarmRankCF to handle stream data. Most learning to rank algorithms work in batch mode, but social media streams, require ranking models to be updated online. Swarm intelligence can help to evolve models when new data enters the system. Another future research direction is to learn the latent factors using swarm intelligence as well, and at the same time optimize the ranking measure.

Acknowledgments This work was funded, in part, by the European Commission FP7 under grant agreements No.247829 and No.287704 for the M-Eco and CUBRIK projects, respectively, and by the NTH School (Niedersächsische Technische Hochschule) for IT Ecosystems.

7. REFERENCES

[1] J. S. Breese, D. Heckerman, and C. M. Kadie. Empirical analysis of predictive algorithms for collaborative filtering. In *Conference on Uncertainty in Artificial Intelligence*, 1998.

[2] C. Burges, T. Shaked, E. Renshaw, A. Lazier, M. Deeds, N. Hamilton, and G. Hullender. Learning to Rank Using Gradient Descent. In *Proceedings of the ICML*, 2005.

[3] O. Celma. *Music Recommendation and Discovery in the Long Tail*. Springer, 2010.

[4] P. Cremonesi, Y. Koren, and R. Turrin. Performance of recommender algorithms on top-n recommendation tasks. In *Proceedings of the ACM RecSys conference*, 2010.

[5] E. Diaz-Aviles, W. Nejdl, and L. Schmidt-Thieme. Swarming to rank for information retrieval. In *Proceedings of the ACM GECCO conference*, 2009.

[6] Y. Freund, R. Iyer, R. E. Schapire, and Y. Singer. An efficient boosting algorithm for combining preferences. *J. Mach. Learn. Res.*, 4:933–969, 2003.

[7] R. Herbrich, T. Graepel, and K. Obermayer. Large Margin Rank Boundaries for Ordinal Regression. In *Advances in Large Margin Classifiers*, 2000.

[8] T. Joachims. Optimizing search engines using clickthrough data. In *Proceedings of the ACM KDD conference*, 2002.

[9] Y. Koren, R. Bell, and C. Volinsky. Matrix factorization techniques for recommender systems. *Computer*, 42:30–37, August 2009.

[10] T.-Y. Liu. *Learning to Rank for Information Retrieval*. Springer, 2011.

[11] R. Poli, J. Kennedy, and T. Blackwell. Particle swarm optimization. *Swarm Intelligence*, 1(1):33–57, 2007.

[12] M. Wetter. Generic Optimization Program – GenOpt. *User Manual, User Manual Version 3.1.0. Lawrence Berkeley National Laboratory.*, 2011.

When Recommenders Fail: Predicting Recommender Failure for Algorithm Selection and Combination*

Michael Ekstrand and John Riedl
GroupLens Research
Dept. of Computer Science, University of Minnesota
{ekstrand,riedl}@cs.umn.edu

ABSTRACT

Hybrid recommender systems — systems using multiple algorithms together to improve recommendation quality — have been well-known for many years and have shown good performance in recent demonstrations such as the NetFlix Prize. Modern hybridization techniques, such as feature-weighted linear stacking, take advantage of the hypothesis that the relative performance of recommenders varies by circumstance and attempt to optimize each item score to maximize the strengths of the component recommenders. Less attention, however, has been paid to understanding what these strengths and failure modes are. Understanding what causes particular recommenders to fail will facilitate better selection of the component recommenders for future hybrid systems and a better understanding of how individual recommender personalities can be harnessed to improve the recommender user experience. We present an analysis of the predictions made by several well-known recommender algorithms on the MovieLens 10M data set, showing that for many cases in which one algorithm fails, there is another that will correctly predict the rating.

Categories and Subject Descriptors

H.3.3 [**Information Storage and Retrieval**]: Information Search and Retrieval—*Information filtering*

Keywords

Recommender systems, evaluation, hybrid recommenders

1. INTRODUCTION

Hybrid recommender systems [1] are a well-known technique for harnessing the strengths of multiple recommenders to produce results that are more accurate or useful than those achieved by individual constituent recommenders. They have

---

*The scripts to re-run the evaluations in this paper are available at http://www-users.cs.umn.edu/~ekstrand/recsys2012/recsys-scripts.tgz.

proven to be powerful means of trimming error; both top-placing systems in the NetFlix prize used hybrids of many algorithms [11].

One key development in the course of the NetFlix prize was *feature-weighted linear stacking* (FWLS) [11], a method of computing a linear combination of individual scoring functions where the blending coefficients are functions of user or item metadata rather than constants.

To win the NetFlix prize, however, contestants created large ensembles of over 100 recommenders. Such large algorithms are impractical to deploy in practice. Therefore, we need to know how to select the algorithms to include in ensembles for production deployments.

Additionally, while FWLS and other hierarchical methods [13] are important advances, they have to date provided little insight into *why* different recommenders are good or bad in particular circumstances. We seek to understand the particular strengths and weaknesses of various recommender algorithms. By understanding when individual algorithms perform well or poorly, we can select algorithms with complementary strengths and combine them in ways that maximize these strengths.

This work also has the potential to provide insight into what "personalities" various algorithms possess, a key prerequisite for selecting and optimizing algorithms for specific user needs. While past work has provided ways of reasoning about user needs in recommendation [7] and using offline evaluations [5] and user studies [12, 2] to evaluate algorithms in the light of user needs, there remains to be gained an understanding of the particular characteristics of the various algorithms in use in a general sense. Developing this understanding will enable future systems to be built more easily and with less costly testing, as algorithms can be selected based on general properties and then validated and tuned with user testing rather than involving users in every step.

To further these aims, we raise and attempt to address the following research questions:

RQ1 Do different recommender algorithms make different errors?

RQ2 Can we identify tractable features of users or items that make them easier or harder for individual algorithms to predict for?

RQ3 Does using these differences in recommendation errors to drive a hybrid recommender improve recommender performance?

The success of hierarchical hybridization strategies such as FWLS suggests that these hypotheses are true, but we seek to demonstrate this more concretely and develop a transparent model for combining recommenders. Our goal is not to immediately produce a more accurate hybrid, but to gain insight into the relative performance of algorithms that can inform future recommender designs and deployments.

To address these questions, we analyzed prediction error of several algorithms using the LensKit recommender toolkit [3] on the MovieLens 10M data set in a cross-validation setup. While prediction accuracy is just one piece of the broader picture of recommender usefulness and suitability, it provides a tractable way to inspect the differing errors made by recommender algoritms.

2. RELATED WORK

Burke [1] provides an overview and taxonomy of hybrid recommender systems, outlining a variety of methods (including *switching* and *weighting*) that can be used to combine individual recommenders into a composite recommender. In this work, we focus primarily on switching hybrids, which pick which recommender to use in each situation and report its result alone; this framing provides the simplest way to study what causes individual algorithms to succeed or fail. More recent developments in hybridization include hierarchical methods like feature-weighted linear stacking [11].

McNee [7] argued for designing and evaluating recommender systems in the context of user needs, picking recommenders that provide specific characteristics that support user information needs. Later work [12] demonstrated that different algorithms that perform similarly on aggregate numeric measures of accuracy exhibit differing user-visible behaviors.

3. METHODOLOGY

We used the MovieLens 10M data set[1] (ML10M) and the LensKit recommender toolkit [3] for our experiments. We used LensKit's evaluation framework to partition the data set in a 5-fold cross-validation configuration. Users were partitioned into 5 sets; for each user in each partition, we randomly selected 20% of their ratings to be the test ratings for that data set, with the remaining ratings plus all ratings from users in the other partitions forming the training set.

We then ran five recommender algorithms on the data, and captured the predictions each algorithm made for each test rating. We used the following algorithms, choosing parameters based on prior results in the research literature and experience tuning LensKit for the MovieLens data sets [3]:

- Item-user mean, the item's average rating plus the user's mean offset with mild Bayesian damping to push means based on few ratings towards the global mean [4]. This algorithm was also the baseline for all others — if they could not make a prediction, the item-user mean was used.

- Item-item collaborative filtering [10] with a neighborhood size of 30 and ratings normalized by subtracting the item-user mean.

- User-user collaborative filtering [9] with a neighborhood size of 30, using cosine similarity over user-mean-normalized ratings [3]. In the predict stage, ratings were normalized by z-score [6, 3].

[1]http://grouplens.org/node/73

- FunkSVD [4, 8] with 30 features and 100 training iterations per feature.

- Lucene as a tag-based recommender. Since the ML10M data set contains tags for movies, we created a document for each movie containing its title, genres, and tags (repeating each tag as many times as it was applied). Recommendation were then computed as in item-item collaborative filtering, with item neighborhoods and scores computed by a Lucene MoreLikeThis query.

After running the recommenders, we processed each test set to discard all users with fewer than 10 test ratings (ultimately using 44,614 of the 69,878 users in ML10M) and splitting their test ratings into two sets: 5 ratings from each user went into a tuning set, and the remaining ratings stayed in the test set.

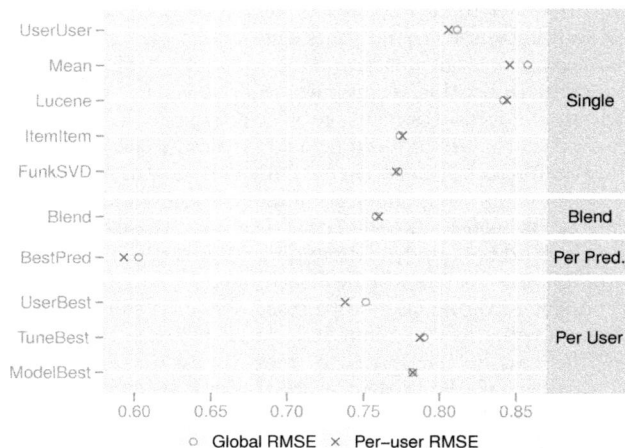

Figure 1: Algorithm accuracy

The "Single" section of Figure 1 shows the overall RMSE achieved by each of the recommender algorithms. "Blend" is the predictions produced by a linear model of the individual algorithms trained against the tuning set.

4. RESULTS

To evaluate and compare algorithm performance, we considered various types of measurements: getting a prediction "right" (within some threshold, such as 1/2 star), being better in absolute error than another algorithm, and providing better accuracy overall for all of a user's predictions. We also tested per-item accuracy and have not yet found any particularly interesting effects; therefore, we only present the per-prediction and per-user results.

4.1 Getting Predictions Right

Algorithm	# Good	% Good	Cum. % Good
ItemItem	1044371	52.23	52.23
UserUser	166008	8.30	60.53
Lucene	90018	4.50	65.03
FunkSVD	53313	2.67	67.70
Mean	21617	1.08	68.78
Unexplained	624291	31.22	100.00

Table 1: Cumulative good predictions (0.5 star)

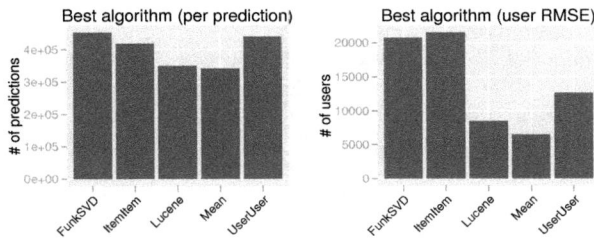

Figure 2: Distribution of best algorithms

Table 1 shows the cumulative "good" predictions for the algorithms we tested. A prediction is considered "good" if it is within 0.5 stars; the table is computed by first picking the algorithm that has the most good predictions. The remaining algorithms are selected and computed by picking the algorithm which has the most good predictions that no prior algorithm has correctly made and adding it to the table. So ItemItem predicts 52% of ratings correctly, UserUser an additional 8%, and so on.

This result provides initial confirmation of H1: algorithms differ in which predictions they get right or wrong. ItemItem gets the most predictions right (52%), but the other algorithms correctly make various predictions until 69% of all test ratings are predicted correctly by at least one algorithm.

This result is robust to higher thresholds; using a threshold of 1.0 stars for good prediction scales the ItemItem hit count up and the other hit counts correspondingly down, but does not change the relative ordering of algorithms.

The existence of differences in the errors made by individual algorithms is further substantiated by the results shown for "BestPred" in Figure 1. This is the RMSE achieved by switching hybrid recommender that uses an oracle to select the best predictor for each individual prediction; the left side of Figure 2 shows how often each algorithm provided the best prediction. This shows that, if we can perfectly predict the best predictor to use, there is room for substantial improvement in error. It therefore provides a lower bound on the error of a switching hybrid comprised of the algorithms in our experiment.

When selecting algorithms to deploy in an ensemble recommender, it is not necessarily desirable just to pick the ones that perform the best. If two algorithms are highly correlated in the errors they make, failing in the same cases, then including both of them will likely not provide much benefit. In selecting algorithms, we look for the following criteria:

- Unique benefit — individual algorithms should contribute unique benefit with respect to the other algorithms in the ensemble.

- Distinguishability — it should be possible to figure out how to blend the algorithms or to select which one to use.

- Tractability — given two algorithms with similar benefit, prefer algorithms that are less expensive to operate.

In general, we found FunkSVD and ItemItem to be highly correlated; the absolute error produced by each has a correlation of $\rho = 0.888$, and ignoring one accrued the greatest benefit to the other. This corroborates their showing in Table 1; if we regenerate the table using FunkSVD first, ItemItem is the last recommender to be picked before Mean. FunkSVD

and ItemItem had the two highest coefficients in the Blend model, however, suggesting that they are contributing unique signal beyond what can be picked up in a threshold analysis.

4.2 Comparing by User

For each user, we determined the algorithm that gave the best RMSE on their test ratings. The right half of Figure 2 shows the distribution of best algorithms. As with optimizing individual predictions, no one algorithm is the winner, suggesting room for intelligent selection of algorithms that will perform better for different users and providing further confirmation that different algorithms do make different mistakes.

To test H2, we tried several regressions to predict relative algorithm performance. Using the log of the user's rating count, their average rating, and the variance of their ratings, we were able to build a logistic model that showed significance and some predictive power for ItemItem being the best algorithm if we ignore FunkSVD (since they have such similar behavior, most users for whom FunkSVD was best ItemItem as their second best algorithm). Table 2 shows the coefficients of this regression; the probability of ItemItem being the best predictor increases with the number of ratings the user has provided and their rating variance, and decreases as their mean rating increases. Relative to the other predictors, therefore, ItemItem does better when more ratings are available (not surprisingly), and is also mildly boosted by low average ratings and high variance in user ratings. Thus H2 is also confirmed — these features correlate with one algorithm beating the others.

	Estimate	Std. Error	z value	Pr(>|z|)
(Intercept)	-2.8966	0.1045	-27.73	0.0000
log10(count)	1.6090	0.0227	70.80	0.0000
mean	-0.1047	0.0224	-4.66	0.0000
var	0.2818	0.0203	13.85	0.0000

Table 2: "Item-item best" prediction model

The ROC curve, computed by holding out 20% of the users as a test set, is shown in Figure 3; its area is 0.68, suggesting that this regression may also be useful from a practical standpoint for identifying when ItemItem is a good choice of recommender.

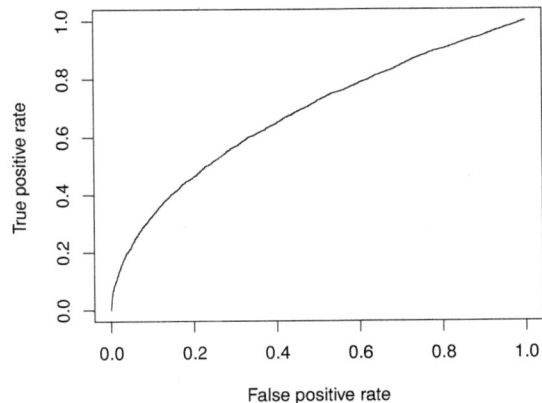

Figure 3: ROC curve for predicting "ItemItem is best"

Building an immediately useful hybrid from these findings has so far been challenging. In Figure 1, the "Per User" section shows the RMSE achieved by three different per-user algorithm switching strategies. *UserBest* uses the algorithm with the best RMSE for that user; it shows a lower bound on the error of per-user switching hybrids optimizing for user RMSE. It shows some improvement over all single algorithms and over the linear blend, particularly when looking at per-user error. It therefore seems that it should be possible to achieve accuracy improvements through intelligent selection of algorithms on a per-user basis.

TuneBest attempts to pick the best algorithm using the predictions for each user's 5 tuning ratings, then apply that algorithm to the remaining test ratings. Using this approach — which does not provide any insight into why the algorithm is selected — achieves an accuracy a bit worse than the best single algorithms (ItemItem and FunkSVD). Its per-user RMSE is slightly better than any single algorithm.

ModelBest uses the 5 probe ratings to train a logistic regression predicting whether either UserUser or Mean will be better than ItemItem for that user based on the log of the number of items the user has rated and an interaction term between that and the variance of their ratings. In turning the modeling from Table 2 around and using it to decide which recommender to use, we found these features to be most useful; variance on its own was not significant and did not make a noticeable contribution in addition to the item count and interaction terms, and mean was similarly unhelpful in improving the regression's power. The regression is then thresholded to decide whether ItemItem or UserUser is used; the threshold was chosen to produce a ratio of UserUser to ItemItem choices that kept in line with the other two models. The resulting predictor beats UserUser alone, but does not improve upon ItemItem. It does, however, beat *TuneBest*; we infer from this that user features are useful for hybridization, but there is still work to do to make this algorithm an actual improvement over the current state of the art.

5. CONCLUSION AND FUTURE WORK

In our experiment, we found that recommenders do indeed fail on different users and items, thus confirming H1. We have also identified user features predicting relative algorithm performance and achieved some success building a hybrid around them, providing preliminary confirmation of H2 and H3, but more work needs to be done to make the algorithm an improvement over the state of the art and to develop a deeper understanding of what makes the various recommenders succeed or fail.

Immediate future work involves continuing to look for features that will help us to select the appropriate recommender to use, as well as investigating blending approaches using hierarchical regressions (similar to FWLS), with the goal of understanding what it is that makes particular algorithms work well, where their individual weaknesses are, and how to combine them into an effective ensemble.

User studies and qualitative investigation of the items and users themselves will likely be helpful in further elucidating the specific behavior of each algorithm. So far, our work has focused only on generic statistics of users and items in the rating set; seeing what actual items are being mispredicted and collecting user feedback on erroneous predictions or bad recommendations will hopefully provide further insight into how the algorithms behave.

Systematic investigation of recommender failures has potential to improve both our understanding of the workings and characteristics of recommender algorithms and our ability to successfully deploy them. By understanding when recommenders fail, we can know better what ones are most likely to work well in particular situations and be more effective at designing and deploying novel recommender systems finely tuned to the particular demands of their domains and users.

Acknowledgements

Our colleagues in GroupLens Research, particularly Tony Lam, Shilad Sen, and Aaron Halfaker, have provided invaluable assistance in this work. We also gratefully acknowledge the support of the National Science Foundation under grants IIS 10-17697 and 08-08692.

6. REFERENCES

[1] R. Burke. Hybrid recommender systems: Survey and experiments. *User Modeling and User-Adapted Interaction*, 12(4):331–370, Nov. 2002.

[2] M. D. Ekstrand, P. Kannan, J. A. Stemper, J. T. Butler, J. A. Konstan, and J. T. Riedl. Automatically building research reading lists. In *RecSys '10*, pages 159–166. ACM, 2010.

[3] M. D. Ekstrand, M. Ludwig, J. A. Konstan, and J. T. Riedl. Rethinking the recommender research ecosystem: reproducibility, openness, and LensKit. In *RecSys '11*, pages 133–140. ACM, 2011.

[4] S. Funk. Netflix update: Try this at home. http://sifter.org/~simon/journal/20061211.html, Dec. 2006.

[5] A. Gunawardana and G. Shani. A survey of accuracy evaluation metrics of recommendation tasks. *J. Mach. Learn. Res.*, 10:2935–2962, 2009.

[6] J. Herlocker, J. A. Konstan, and J. Riedl. An empirical analysis of design choices in neighborhood-based collaborative filtering algorithms. *Inf. Retr.*, 5(4):287–310, 2002.

[7] S. M. McNee, J. Riedl, and J. A. Konstan. Making recommendations better: an analytic model for human-recommender interaction. In *CHI '06 Extended Abstracts*, pages 1103–1108. ACM, 2006.

[8] A. Paterek. Improving regularized singular value decomposition for collaborative filtering. In *KDD Cup and Workshop 2007*, Aug. 2007.

[9] P. Resnick, N. Iacovou, M. Suchak, P. Bergstrom, and J. Riedl. GroupLens: an open architecture for collaborative filtering of netnews. In *ACM CSCW '94*, pages 175–186. ACM, 1994.

[10] B. Sarwar, G. Karypis, J. Konstan, and J. Reidl. Item-based collaborative filtering recommendation algorithms. In *ACM WWW '01*, pages 285–295. ACM, 2001.

[11] J. Sill, G. Takacs, L. Mackey, and D. Lin. Feature-Weighted linear stacking. *arXiv:0911.0460*, Nov. 2009.

[12] R. Torres, S. M. McNee, M. Abel, J. A. Konstan, and J. Riedl. Enhancing digital libraries with TechLens+. In *ACM/IEEE JCDL '04*, pages 228–236. ACM, 2004.

[13] A. Umyarov and A. Tuzhilin. Improving rating estimation in recommender systems using aggregation- and variance-based hierarchical models. In *RecSys '09*, pages 37–44, New York, New York, USA, 2009. ACM.

Constrained Collective Matrix Factorization

Yu-Jia Huang
Department of Computer
Science
Sun Yat-sen University
Guangzhou, China
huangyj9@mail2.sysu.edu.cn

Evan Wei Xiang
Department of Computer
Science and Engineering
The Hong Kong University of
Science and Technology
Hong Kong, China
wxiang@cse.ust.hk

Rong Pan [*]
Department of Computer
Science
Sun Yat-sen University
Guangzhou, China
panr@mail.sysu.edu.cn

ABSTRACT

Transfer learning for collaborative filtering (TLCF) aims to solve the sparsity problem by transferring rating knowledge across multiple domains. Taking domain difference into account, one of the issues in cross-domain collaborative filtering is to selectively transfer knowledge from source/auxiliary domains. In particular, this paper addresses the problem of inconstant users (users with changeable preferences across different domains) when transferring knowledge about users from another auxiliary domain. We first formulate the problem of inconstant users caused by domain difference and then propose a new model that performs constrained collective matrix factorization (CCMF). Our experiments on simulated and real data show that CCMF has superior performance than other methods.

Categories and Subject Descriptors

H.3 [**INFORMATION STORAGE AND RETRIEVAL**]: Information Search and Retrieval

Keywords

Collaborative Filtering, Transfer Learning, Collective Matrix Factorization, Inconstant Users.

1. INTRODUCTION

Collaborative filtering [9] in recommender systems aims to predict users' ratings in the future on a set of items based on a collection of similar users' rating history. In real-life recommender systems, the rating matrix may be extremely sparse. As reported in [11], the density of the available ratings in commercial recommender systems is often less than 1%. To alleviate the sparsity problem in collaborative filtering, one common approach is to pool together the rating data from multiple rating matrices in related domains for knowledge transfer and sharing. Methods like [2, 3] assume that both users and items in an auxiliary domain are related to the target domain, while in practice it is often much easier to find an auxiliary data source with either related users or related items but not both. In this paper, we address the problem of user-sided transfer learning, although the method proposed in this paper can also be applied to item-sided transfer learning. One well-known approach to such one- side-related problem is Collective matrix factorization (CMF) [12]. CMF is proposed for jointly factorizing two matrices with the constraint of sharing one-side (user or item) latent features. CMF improves the prediction in target domain by increasing the rating records for each user/item. When the domain difference is small between source domain and target domain, e.g. transferring knowledge from a movie recommender system to another, this approach can easily improve prediction performance by simply jointing data from two systems.

CMF assumes that all users' features are constant. However, in real life, some users may change their features / preferences across different domains. For example, users having a preference over light music may like horror movies instead of ones about love stories because horror movies entertain audience in much more ways (pictures, stories and so on) than horror music (sounds). Even though the target domain and auxiliary domain are of same type (e.g. movie), users may also behave a little differently when rating different item sets. We refer "inconstant users" to those with changeable preferences across different domains. For these users, it would not hold that the source domain and target domain share a user latent feature matrix. To improve recommendation for inconstant users, we need a new method that takes the change of user feature into account.

To model the change of user feature caused by domain difference, we make two assumptions:

- Bias of latent features are user-dependent. The bias between user features across different domains varies from one user to another. Therefore, to recommend items for both constant and inconstant users, our model should allow user-dependent bias of latent features.

- The change lies in history. What items a user has rated implies what the user was looking for, thus we can infer whether and how much a user has inconstant hobbies through his/her rating history. More specifically, if the items rated by a user are similar to those

*Corresponding author. This work was supported by National Natural Science Foundation of China (61003140, 61033010).

in source/auxiliary domain, we assume that the user tends to have constant preferences and vice versa.

To model the assumptions above, we extend CMF to a new matrix factorization model named Constrained CMF, denoted as CCMF in this paper. Unlike CMF, CCMF does not assume that the users share a same feature matrix in source domain and target domain.

The remainder of this paper is organised as follows. In section 2, we first introduce the problem setting. In section 3, we formulate the model. Then we experimentally validate the effectiveness of the proposed models in section 4. Related work is introduced in section 5. Finally we conclude this paper in section 6.

2. PROBLEM SETTING

In our problem setting, we are given one source domain D^{src}, say music, and one target domain D^{tgt} (movie in our experiments). $u_1, u_2, \ldots, u_n$ denote n users having rating records in both D^{src} and D^{tgt}. In rating matrix R^{src} of D^{src}, users make ratings on m_{src} items $\{v_1^{src}, v_2^{src}, \ldots, v_{m_{src}}^{src}\}$, the rating made by u_i on v_j^{src} is denoted as R_{ij}^{src}. Similarly, in rating matrix R^{tgt} of D^{tgt}, users make ratings on m_{tgt} items $\{v_1^{tgt}, v_2^{tgt}, \ldots, v_{m_{tgt}}^{tgt}\}$, and the rating made by u_i on v_j^{tgt} is denoted as R_{ij}^{tgt}. Both R^{src} and R^{tgt} are observed and another $n \times m_{tgt}$ rating matrix R^{test} is for testing and is unknown while training the model. All the elements from R^{src}, R^{tgt} and R^{test} are at the same scale (e.g. 1- 5 or 0-1 [5]). Our goal is to make use of R^{src} and R^{tgt} to help predict the missing values in R^{test}. Figure 1 shows a toy example of our problem.

Figure 1: A Toy Example

3. CONSTRAINED CMF

A well-known and effective approach to recommender systems is to factorize the user-item rating matrix, and utilize the latent user feature matrix and item feature matrix to make prediction for future ratings [1, 4, 8, 13, 14] .

In order to learn the latent characteristics of the users and items in source and target domain, we employ probabilistic matrix factorization[10] to factorize the user-item matrix. The conditional distributions over the observed ratings in two domains are defined as:

$$p\left(R^{tgt}\left|U^{tgt},V^{tgt},\sigma_R^{tgt2}\right.\right) =$$
$$\prod_{i=1}^{n}\prod_{j=1}^{m_{tgt}}\left[N\left(R_{ij}^{tgt}\left|U_{i.}^{tgt}\left(V_{j.}^{tgt}\right)^T,\sigma_R^{tgt2}\right.\right)\right]^{I_{ij}^{tgt}}, \quad (1)$$

and $p\left(R^{src}\left|U^{src},V^{src},\sigma_R^{src2}\right.\right) =$
$$\prod_{i=1}^{n}\prod_{j=1}^{m_{src}}\left[N\left(R_{ij}^{src}\left|U_{i.}^{src}\left(V_{j.}^{src}\right)^T,\sigma_R^{src2}\right.\right)\right]^{I_{ij}^{src}}, \quad (2)$$

where $N(x\,|\mu,\sigma^2)$ is the probability density function of the Gaussian distribution with mean μ and variance σ^2. I_{ij}^{tgt} and I_{ij}^{src} are the indicator functions that are equal to 1 if user rated item and equal to 0 otherwise.

The zero-mean spherical Gaussian priors are also placed on user and item feature vectors:

$$p\left(U^{src}\left|\sigma_U^{src2}\right.\right) = \prod_{i=1}^{n} N\left(U_{i.}^{src}\left|0,\sigma_U^{src}\mathbf{I}\right.\right), \quad (3)$$

$$p\left(V^{src}\left|\sigma_V^{src2}\right.\right) = \prod_{j=1}^{m_{src}} N\left(V_{j.}^{tgt}\left|0,\sigma_V^{src}\mathbf{I}\right.\right), \quad (4)$$

and $p\left(V^{tgt}\left|\sigma_V^{tgt2}\right.\right) = \prod_{j=1}^{m_{tgt}} N\left(V_{j.}^{tgt}\left|0,\sigma_V^{tgt}\mathbf{I}\right.\right). \quad (5)$

To transfer knowledge about users from auxiliary domain, [12] factorizes two matrices with the constraint of sharing one-side (user or item) latent features. Instead of jointly factorizing rating matrices in source and target domain with a same user feature matrix ($U^{src} = U^{tgt}$), our model factorizes R^{src} with U^{src} and V^{src} and factorizes R^{tgt} with U^{tgt} and V^{tgt}. We transfer the user feature matrix U^{src} learned in source domain to help factorize the rating matrix in target domain by conducting a constraint on U^{src} and U^{tgt} :

$$U_{i.}^{tgt} = U_{i.}^{src} + \frac{\sum_{k=1}^{m_{tgt}} I_{ik}^{tgt}Y_{k.}}{\sum_{k=1}^{m_{tgt}} I_{ik}^{tgt}}, \quad (6)$$

or $U_{i.}^{tgt} = U_{i.}^{src} + \frac{\sum_{k=1}^{m_{tgt}} R_{ik}^{tgt}Y_{k.}}{\sum_{k=1}^{m_{tgt}} R_{ik}^{tgt}}. \quad (7)$

Here $U_{i.}^{src}$ is the prior mean of $U_{i.}^{tgt}$ and the second term is the user feature bias caused by domain difference. $Y_{k.}$ captures the feature bias for a user if the user rated item v_k^{tgt}. Intuitively, the more similar v_k^{tgt} is to the items in source domain, the closer $Y_{k.}$ is to zero. The Gaussian prior for Y is given as follows:

$$p\left(Y\left|\sigma_Y^2\right.\right) = \prod_{j=1}^{m_{tgt}} N\left(Y_{j.}\left|0,\sigma_Y^2\mathbf{I}\right.\right). \quad (8)$$

Hence, through a Bayesian inference, we have the log of the posterior distribution over the user and item feature and Y:

$$\ln p\left((Y,U^{src},V^{src},V^{tgt}\left|R^{src},R^{tgt},\sigma_R,\sigma_Y,\sigma_V^{src},\sigma_V^{tgt},\sigma_V^{src}\right.\right))$$
$$= -\frac{1}{2\sigma_R^2}\sum_{i=1}^{n}\sum_{j=1}^{m_{src}} I_{ij}\left(\left(R_{ij}^{src} - U_{i.}^{src} \cdot V_{j.}^{srcT}\right)^2\right) \quad (9)$$
$$-\frac{1}{2\sigma_R^2}\sum_{i=1}^{n}\sum_{j=1}^{m_{tgt}} I_{ij}\left(\left(R_{ij}^{tgt} - \left(U_{i.}^{src} + \frac{\sum_{k=1}^{M} I_{ik}^{tgt}Y_k}{\sum_{k=1}^{M} I_{ik}^{tgt}}\right)\cdot\right.\right.$$
$$\left.\left. V_{j.}^{tgtT}\right)^2\right) - \frac{1}{2\sigma_Y^2}\sum_{j=1}^{m_{tgt}} Y_{j.} \cdot Y_{j.}{}^T - \frac{1}{2\sigma_V^{tgt2}}\sum_{i=1}^{m_{tgt}} V_{j.}^{tgt}\cdot V_{j.}^{tgtT}$$
$$-\frac{1}{2}\left(\sum_{i=1}^{n}\sum_{j=1}^{m_{tgt}} I_{ij}\left(\ln \sigma_R^{tgt2}\right) + nl\ln \sigma_Y^2 + nl\ln \sigma_V^2\right) + C,$$

where C is a constant that does not depend on the parameters and l denotes the feature number.

Maximizing the log-posterior over latent features with hyperparameters (i.e., the observation noise variance and prior variances) kept fixed is equivalent to minimizing the following sum-of-squared-errors objective functions with quadratic regularization terms:

$$E = \frac{1}{2} \sum_{i=1}^{n} \sum_{j=1}^{m_{src}} I_{ij}^{src} \left(\left(R_{ij}^{src} - U_{i.}^{src} \cdot V_{j.}^{srcT} \right)^2 \right)$$

$$+ \frac{1}{2} \sum_{i=1}^{n} \sum_{j=1}^{m_{tgt}} I_{ij}^{tgt} \left(\left(R_{ij}^{tgt} - \left(U_{i.}^{src} + \frac{\sum_{k=1}^{M} I_{ik}^{tgt} Y_k}{\sum_{k=1}^{M} I_{ik}^{tgt}} \right) \cdot \right. \right.$$

$$\left. \left. V_{j.}^{tgtT} \right)^2 \right) + \frac{\lambda_U}{2} \sum_{i=1}^{n} \|U_{i.}^{src}\|_{Fro}^2 + \frac{\lambda_V^{src}}{2} \sum_{j=1}^{m_{src}} \|V_{j.}^{src}\|_{Fro}^2$$

$$+ \frac{\lambda_V^{tgt}}{2} \sum_{j=1}^{m_{tgt}} \|V_{j.}^{tgt}\|_{Fro}^2 + \frac{\lambda_Y}{2} \sum_{j=1}^{m_{tgt}} \|Y_{j.}\|_{Fro}^2, \qquad (10)$$

where $\lambda_Y = \sigma_R^2 / \sigma_Y^2$, $\lambda_V^{src} = \sigma_R^2 / \sigma_V^{src2}$, $\lambda_V^{tgt} = \sigma_R^2 / \sigma_V^{tgt2}$, and $\|\cdot\|_{Fro}^2$ denotes the Frobenius norm. The graphical model is shown in Figure 2.

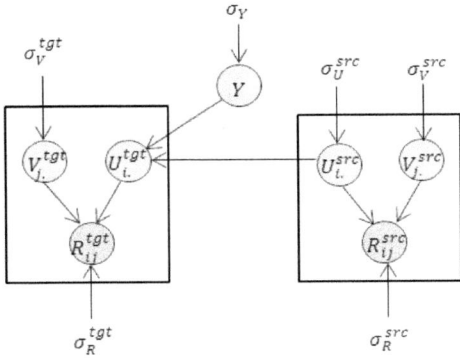

Figure 2: Graphical model of CCMF

To learn CCMF, we first calculate the partial derivatives of the objective function with respect to each variable:

$$\frac{\partial E}{\partial U_{i.}^{src}} = \sum_{j=1}^{m_{src}} I_{ij}^{src} \left(\left(U_{i.}^{src} \cdot V_{j.}^{srcT} - R_{ij}^{src} \right) \cdot V_{j.}^{src} \right)$$

$$+ \sum_{j=1}^{m_{tgt}} I_{ij}^{tgt} \left(\left(\left(U_{i.}^{src} + \frac{\sum_{k=1}^{M} I_{ik}^{tgt} Y_k}{\sum_{k=1}^{M} I_{ik}^{tgt}} \right) V_{j.}^{tgtT} - R_{ij}^{tgt} \right) \cdot \right.$$

$$\left. V_{j.}^{tgt} \right) + \lambda_U \cdot U_{i.}^{src}, \qquad (11)$$

$$\frac{\partial E}{\partial Y_{i.}^{src}} = \sum_{j=1}^{m_{tgt}} I_{ij}^{tgt} \left(\left(\left(U_{i.}^{src} + \frac{\sum_{k=1}^{M} I_{ik}^{tgt} Y_k}{\sum_{k=1}^{M} I_{ik}^{tgt}} \right) V_{j.}^{tgtT} - R_{ij}^{tgt} \right) \cdot \right.$$

$$\left. \frac{I_{ip} x_{ij}^{tgt}}{\sum_{k=1}^{M} I_{ik}} \right) + \lambda_Y \cdot Y_{i.}, \qquad (12)$$

$$\frac{\partial E}{\partial V_{j.}^{src}} = \sum_{i=1}^{N} I_{ij}^{src} \cdot \left(\left(U_{i.}^{src} \cdot V_{j.}^{srcT} - R_{ij}^{src} \right) \cdot U_{i.}^{src} \right)$$

$$+ \lambda_V^{src} \cdot V_{j.}^{src}, \qquad (13)$$

$$\frac{\partial E}{\partial V_{j.}^{src}} = \sum_{i=1}^{N} I_{ij}^{src} \cdot \left(\left(\left(U_{i.}^{src} + \frac{\sum_{k=1}^{M} I_{ik}^{tgt} Y_k}{\sum_{k=1}^{M} I_{ik}^{tgt}} \right) V_{j.}^{tgtT} - R_{ij}^{tgt} \right) \cdot \right.$$

$$\left. \left(U_{i.}^{src} + \frac{\sum_{k=1}^{M} I_{ik}^{tgt} Y_k}{\sum_{k=1}^{M} I_{ik}^{tgt}} \right) \right) + \lambda_V^{src} \cdot V_{j.}^{tgt}. \qquad (14)$$

Then a local minimum of the objective function can be found by performing gradient descent in Y, U^{src}, V^{src} and V^{tgt}.

4. EMPIRICAL ANALYSIS

4.1 Experimental Setting

To check whether CCMF can fit with different settings, we evaluate CCMF on two data sets.

One is a simulated dataset sampled from the Netflix dataset. The Netflix data set used in the experiments is constructed as follows. We first randomly extracted a $10,000 \times 16,000$ dense rating matrix R from the Netflix data, and take the submatrices $R^{tgt} = R_{1 \sim 10000, 1 \sim 8000}$ as the target rating matrix, and $R^{src} = R_{1 \sim 10000, 8001 \sim 16000}$ as the user side source data, so that R^{tgt} and R^{src} share only common users.

The other is a real dataset crawled from Douban[1], which is launched in 2005 and is a Chinese SNS website allowing registered users to record ratings and reviews related to movies, books, and music. We crawled $290,633$ rating records rated by $5,000$ users on $3,000$ musical items and $10,000$ movies. In the experiments, we used the musical rating matrix as auxiliary training data. Then we sampled randomly from the movie rating matrix to generate training data and testing data. The final datasets are summarized in Table 1. We adopt two evaluation metrics: RMSE (Root Mean Square Error) and MAE (Mean Absolute Error).

Table 1: Description of Douban data and Netflix data

dataset		type	sparsity
Netflix	target(training)	movie	<=0.60%
	target(testing)	movie	0.50%
	auxiliary	movie	2.50%
Douban	target(training)	movie	<=0.60%
	target(testing)	movie	0.60%
	auxiliary	music	1.50%

We compare out CCMF methods with two non- transfer learning methods: the UserMean and PMF [10], as well as CMF [12].

4.2 Experimental Results

We denote the CCMF methods that utilize different constraints given in Eqs. (6) and (7) as CCMF1 and CCMF2, respectively. The best results of using different parameters as described in the previous section are reported in Table 2. We can make the following observations:

- UserMean is worse than all other methods.

- Matrix factorization methods (CMF and CCMF) that transfer knowledge from auxiliary domains perform much better than non-transfer method PMF.

- CCMF and CCMF2, which model the inconstant users outperform CMF at all sparsity levels.

- CCMF1 and CCMF2 achieves close results, but as the training data in target domain becomes very sparse (< 0.3%), the performance of CCMF1 deteriorates most slowly.

One challenge of the transfer learning for recommender systems is that it is difficult to recommend items to users who have very few ratings in the target domain. In order to further compare the above methods, we first group all the users based on the number of observed ratings in the target domain, and then evaluate prediction accuracies of different user groups. The experimental results are shown

---
[1]http://www.douban.com

Table 2: Performance Comparisons. Numbers in boldface (i.e. **0.890**) and in Italic (i.e. *0.895*) are the best and second best results among all methods, respectively.

dataset	sparsity of R	without transfer		with transfer		
		UserMean	PMF	CMF	CCMF1	CCMF2
Douban	0.20%	0.876	0.796	0.778	**0.770**	*0.775*
	0.40%	0.863	0.775	0.759	**0.749**	*0.750*
	0.60%	0.860	0.766	0.756	**0.750**	0.750
Netflix	0.20%	1.020	0.969	0.931	**0.914**	0.914
	0.40%	1.007	0.951	0.917	*0.904*	0.902
	0.60%	1.000	0.939	0.905	*0.895*	**0.890**

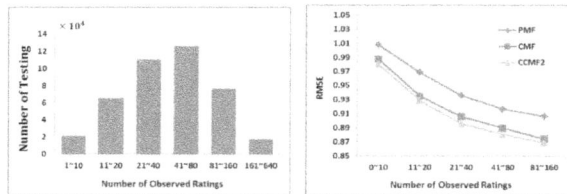

(a) Distribution of Testing Data (b) RMSE Comparison on Different User Rating Scales

Figure 3: Performance Comparison on Different Users

in Figure 3. Users with less than 160 ratings are grouped into 6 classes: "1-10", "11-20", "21-40", "41-80", "81-160", and "161-640", denoting how many ratings users have rated in target domain.

Figure 3a summarizes the distributions of testing data according to groups in the training data (sparsity 0.4%). For example, there are a total 19820 user-item pairs to be predicted in the testing dataset in which the related users in the training dataset have rating numbers from 1 to 10. In Figure 3b, we observe that our CCMF algorithm consistently performs better than other methods when recommending items for all user groups.

5. RELATED WORK

PMF Probabilistic matrix factorization (PMF) [10] is a method for missing value prediction in a single matrix.

CMF Collective matrix factorization (CMF) [12] is proposed for jointly factorizing two matrices with the constraints of sharing one-side (user or item) latent features. However, CMF didn't address the problem that users' may have different interests in different domains.

CST Coordinate System Transfer (CST) was a method proposed by [7] CST addresses the data sparsity problem in a target domain by transferring knowledge about both users and items from auxiliary data sources. In our problem setting, only knowledge about one side (user) from auxiliary domain is available. Hence, CST are not applicable to the problem in this paper.

TCF [6] proposed a framework of Transfer by Collective Factorization that use the binary preference data expressed in the form of like/dislike to help reduce the impact of data sparsity of more expressive numerical ratings.

6. CONCLUSIONS AND FUTURE WORKS

In this paper, we formulated the problem of inconstant users caused by domain difference and presented a new model to address the problem in transfer learning for collaborative filtering. Our method iteratively factorizes the rating matrices in source/auxiliary domain and target domain with a constraint on the user feature matrices for target domain and auxiliary domain. Experimental results on both and

simulated and read data show that CCMF performs better than CMF at various sparsity levels. The study in this paper clearly demonstrates (a) the necessity of taking domain difference and change of user features into account, and (b) the items rated by a user implies to whether and how much the user's feature has changed.

In this paper, we assumed that the rating matrices in auxiliary and target domain are one-sided aligned. But in real life, some users may not have rating records in both domains. Hence, in order to model the domain difference more realistically, for future work, we will extend CCMF so that it can fit partial-users- aligned setting. In the future, we will also extend CCMF in heterogeneous settings, e.g. for transfering like/dislike knowledge from books or music to target domain that involves rating.

7. REFERENCES

[1] Y. Koren. Factorization meets the neighborhood: a multifaceted collaborative filtering model. In *KDD*, pages 426–434, 2008.
[2] B. Li, Q. Yang, and X. Xue. Can movies and books collaborate? cross-domain collaborative filtering for sparsity reduction. In *IJCAI*, pages 2052–2057, 2009.
[3] B. Li, Q. Yang, and X. Xue. Transfer learning for collaborative filtering via a rating-matrix generative model. In *ICML*, page 78, 2009.
[4] I. Murray and R. Salakhutdinov. Evaluating probabilities under high-dimensional latent variable models. In *NIPS*, pages 1137–1144, 2008.
[5] R. Pan, Y. Zhou, B. Cao, N. N. Liu, R. M. Lukose, M. Scholz, and Q. Yang. One-class collaborative filtering. In *ICDM*, pages 502–511, 2008.
[6] W. Pan, N. N. Liu, E. W. Xiang, and Q. Yang. Transfer learning to predict missing ratings via heterogeneous user feedbacks. In *IJCAI*, pages 2318–2323, 2011.
[7] W. Pan, E. W. Xiang, N. N. Liu, and Q. Yang. Transfer learning in collaborative filtering for sparsity reduction. In *AAAI*, 2010.
[8] J. D. M. Rennie and N. Srebro. Fast maximum margin matrix factorization for collaborative prediction. In *ICML*, pages 713–719, 2005.
[9] P. Resnick, N. Iacovou, M. Suchak, P. Bergstrom, and J. Riedl. Grouplens: An open architecture for collaborative filtering of netnews. In *CSCW*, pages 175–186, 1994.
[10] R. Salakhutdinov and A. Mnih. Bayesian probabilistic matrix factorization using markov chain monte carlo. In *ICML*, pages 880–887, 2008.
[11] B. M. Sarwar, G. Karypis, J. A. Konstan, and J. Riedl. Item-based collaborative filtering recommendation algorithms. In *WWW*, pages 285–295, 2001.
[12] A. P. Singh and G. J. Gordon. Relational learning via collective matrix factorization. In *KDD*, pages 650–658, 2008.
[13] K. Yu, S. Zhu, J. D. Lafferty, and Y. Gong. Fast nonparametric matrix factorization for large-scale collaborative filtering. In *SIGIR*, pages 211–218, 2009.
[14] Y. Zhou, D. M. Wilkinson, R. Schreiber, and R. Pan. Large-scale parallel collaborative filtering for the netflix prize. In *AAIM*, pages 337–348, 2008.

Recommending Academic Papers via Users' Reading Purposes

Yichen Jiang
Peking University
Beijing, China
jiangyichen@pku.edu.cn

Aixia Jia
Peking University
Beijing, China
jiaaixia@pku.edu.cn

Yansong Feng
Peking University
Beijing, China
fengyansong@pku.edu.cn

Dongyan Zhao
Peking University
Beijing, China
zhaodongyan@pku.edu.cn

ABSTRACT

The past decades have witnessed the rapid development of academic research, which results in a growing number of scholarly papers. As a result, paper recommender systems have been proposed to help researchers find their interested papers. Most previous studies in paper recommendations mainly concentrate on paper-paper or user-paper similarities without taking users' reading purposes into account. It is common that different users may prefer to different aspects of a paper, e.g., the focused problem/task or the proposed solution. In this paper, we propose to satisfy user-specific reading purposes by recommending the most problem-related papers or solution-related papers to users separately. For a target paper, we use the paper citation graph to generate a set of potential relevant papers. Once getting the candidate set, we calculate the problem-based similarities and solution-based similarities between candidates and the target paper through a concept based topic model, respectively. We evaluate our models on a real academic paper dataset and our experiments show that our approach outperforms a traditional similarity based model and can provide highly relevant paper recommendations according to different reading purposes for researchers.

Categories and Subject Descriptors

H.3.1 [**INFORMATION STORAGE AND RETRIEVAL**]: Content Analysis and Indexing—*Linguistic processing*; H.3.3 [**INFORMATION STORAGE AND RETRIEVAL**]: Information Search and Retrieval—*Information filtering*

General Terms

Algorithms,Experimentation

Keywords

Content-Based Filtering, Paper Recommendation, Topic Model, Concept Terms

1. INTRODUCTION

Researchers are searching academic publications that are related to their interests all the time. However, the rapid development in the volumes of research papers leads to publication overload. As a result, researchers have to spend more and more time on finding related papers. Information Retrieval (IR) technology[11] are used to relieve the case to some extent, but sometimes users, especially junior researchers, may have no idea on how to choose appropriate queries fed into a search engine. On the other hand, Recommender System(RS) is designed to suggest items for users by capturing their interests and needs. This motivates an emerging area, academic paper recommendation, which aims to help users find a set of publications that might be useful for their current research according to their profiles.

Some existing publication recommender systems recommend papers by calculating the relevance between papers. They use content-based features to generate the paper relevance. However, the similarity between academic papers is not just like other similarities such as movie or commodity. In many cases, after reading an academic article, users probably want to find more related papers which solve the same problem or use the same solution. For example, when we are reading the article "Random walk based entity ranking on graph for multidimensional recommendation"[8], we may be curious to know more about the multidimensional recommendation problem or the random walk model. For most existing systems, they recommend similar papers in only one list which means users need to distinguish the problem-related and solution-related papers by themselves.

To address this problem, we satisfy users' specific reading purposes by recommending the most problem-relevant papers and solution-relevant papers to users separately. Due to the copyright issues, most paper recommender systems cannot access full text information. In this case, we only use abstracts to analyze the focused problem and solution aspects discussed before. As we observed, the traditional abstract format includes three parts: a). They first introduce the background of their work and what problems they intend to solve. b). And then describe how they solved this problem and what specific method they finally used. c). Some abstracts also mention the experiments in the end. According to the traditional abstract style, we can find the transition point to split an abstract into a problem part and a solution part. We then build LDA models on each of them, which are in turn utilized to compute the problem-based and solution-based similarities between the target and candidate papers.

In the rest of the paper, we first review related works on recommender system. In the next section, we detail our approach on how to recommend academic papers via researchers' different reading purposes. In Section 4, we describe the experimental setup and discuss our results in detail. We conclude this paper in Section 5 and discuss possible directions for future work.

2. RELATED WORK

Traditional recommender systems are dominated by collaborative filtering based approach and content-based approach.

Collaborative filtering[9] is one of the most successful recommendation approaches which make automatic predictions about a user's interests by collecting preferences or taste information from similar users. This approach has been widely used in e-commerce sites such as Amazon.com[1], eBay[2] and so on. However, it suffers from the *cold-start* problem where a system cannot generate accurate recommendations for new users or new items without enough initial rating history.

Content-based method[4] is also a popular choice in recommender systems, where an item is suggested according to the similarity between user profile and this item. This method has been applied mostly in textual domains such as news recommendation and in hybrid approaches with collaborative filtering[3].

Nitin et al[2] build a recommender system based on subspace clustering approach. The idea is that researchers from the same area tend to be interested in the same article. Hence it is possible to improve search results by recommending previous search queried by other people with similar interests. Since this system is built by collaborative filtering, it cannot avoid the cold-start problem. We are alleviating this problem by recommending papers based on abstract similarity, which can be used to recommend new articles.

Adomavicius et al.[1] find that since content-based systems are designed mostly to recommend text-based items, the content in these systems is usually described with keywords. For example, a content-based component of the Fab system[3], which recommends Web pages to users, represents the page content with its 100 most important words. Basu et al.[5] model the task of assigning technical papers to conference reviewers as a problem of recommending technical papers to the authors based on their interests and background. They show that their content-based retrieval methods can outperform the collaborative filtering methods. This motivates us to adopt the content-based paradigm for paper recommendation, and use concepts to better capture the main content of an academic paper.

Chandrasekan et al.[7] present a concept-based approach which is similar to our work. Their system builds users profiles based on the users' previously published papers and

---

[1]http://www.amazon.com/
[2]http://www.ebay.com/

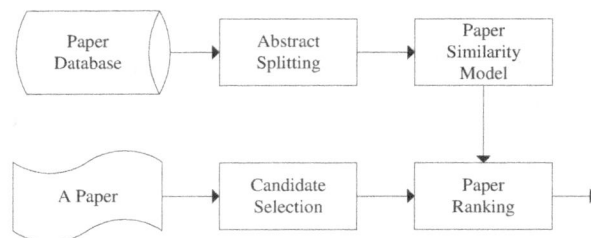

Figure 1: System Overview

compute the similarities between the user profile and the concept profiles of articles in the collection. Papers other than their own are then recommended to the users. This work has the shortcoming that it can only serve the authors who have published papers before. Kodakateri et al. improved this work in[10]. They create user profiles based on the users' previously viewed papers rather than their authored papers, extending the recommendations to CiteSeer users as well as authors. Both Chandrasekan et al.[7] and Kodakateri et al.[10] are building concept profiles based on a predefined set of concepts, which is ACM's Computing Classification System(CCS). However the number of concepts they used is limited, and these concepts are sometimes too general to fully distinguish different ideas. We utilize user tagged academic terms instead of CCS in our model.

3. THE APPROACH

Figure 1 shows the architecture of our system which mainly consists of four components:

1. Abstract Splitting: focuses on splitting abstracts into problem part and solution part. Not all abstract obey the traditional format rules, this component only processes paper abstracts that can be divided into two parts.

2. Paper Similarity Model: extracts features from academic papers to compute the paper similarity. Here we examine three similarity models: TF*IDF model, topic model and concept based topic model.

3. Candidate Selection: selects candidate papers given a target paper based on the paper citation graph.

4. Paper Ranking: ranks the candidate papers given the target by calculating the similarity between the feature vectors.

Both Abstract Splitter and Paper Similarity Model are time-consuming procedures. They need to process every publication in the academic paper dataset, which means we need to run the first two steps offline. The Candidate Selection components narrows down the number of related papers, and the Paper Ranking step only need to compute the paper similarities within a small number of candidate papers. The computing complexity of the last two parts is acceptable. As a result it can meet the requirement of online serving. Each of these modules will be explained in the following sections.

3.1 Abstract Splitting

This module splits an abstract into two parts: problem description and solution description. It is observed that many researchers first introduce the background and problem definition in the abstract, and then describe the solution they used. In order to validate this observation, we conduct a pilot experiment by randomly selecting 200 papers from four international computer science conferences including SIGIR, SIGKDD, RecSys and CIKM, and manually investigate whether these papers contain a problem-solution abstract. The result shows that about 71% of the paper abstracts obey the observed rule and can be spitted into two parts.

According to this convention, we can find a transition point or sentence in the abstract to perform the split. For example, when we see a sentence starting with "In this paper", it probably means that the following content tells us about the proposed solution. Based on these observations, we found that most transition sentences can be matched with certain patterns and summarize them into six regular expression patterns to capture the transitions in an abstract.

3.2 Paper Similarity Model

As we have successfully split an abstract into two parts, we can get a problem content and a solution content for each abstract. We assign the problem content to the problem-based training corpus and solution content to the solution training corpus. We then build similarity models on these two corpora repectively.

3.2.1 TF*IDF Model

The tf*idf weight is a numerical statistic which reflects how important a word is to a document in a corpus. We use $\mathbf{F}_i = (w_1, w_2, ..., w_k)$ to represent the feature vector of paper i where w_j is the tf*idf score of term j in i,

$$w_j = tf * idf(j, d, D) = tf(j, d) \times idf(j, D). \quad (1)$$

Here we uses term importances to represent the paper content.

3.2.2 Topic Model

Topic models are based upon the idea that a document may be viewed as a mixture of various topics, where a topic is a multinomial distribution over words. The latent dirichlet allocation (LDA) model[6] is a generative model that allows sets of observations to be explained by unobserved groups that explain why some parts of the data are similar. LDA estimates the topic-word distribution P(t|z) and the document-topic distribution P(z|d) from an unlabeled corpus using Dirichlet priors for the distributions with a fixed number of topics. As a result, we get P(z|d) for each document and further build the feature vecotr as

$$\mathbf{F} = (P(z_1|d), P(z_2|d), ..., P(z_k|d)). \quad (2)$$

This approach can find similar documents even if they share little common words.

3.2.3 Concept Based Topic Model

Our TF*IDF and LDA models are generated upon the whole vocabulary. However, for a paper abstract, academic topics are often represented by a group of technical terms such as *Latent Dirichlet Allocation*. If we treat the three words individually, we will never know they are actually a

model name. To alleviate this, we gule these academic terms into one word with "_" to form *Latent_Dirichlet_Allocation*. We leverage social tagging resources to define academic terms by extracting users' tagging information from citeulike[3] dataset. In citeulike, users label their interested papers using tags. We believe most tags in this dataset can be used as academic concepts. We count all tags and filter out those appearing less than 50. We match word sequences in the abstracts with the citeulike dataset vocabulary, and merge matched sequence into one academic concept. We then use this processed abstracts to build the LDA similarity model introduced in Section 3.2.2.

3.3 Candidate Selection

Considering the large size of a paper database, it is time consuming to compute similarities between the target paper and all other papers. There are only a small portion of papers that relate to the target paper. We need to narrow down the scope of candidates. Fortunately, the citation graph of academic publications provides meaningful relationship among papers. The fact that researchers cite other papers intending to tell us the referenced papers are similar to their own work in some aspect. For a paper i, we represent its reference set as Ref_i, its citation set as Cit_i, and define i's candidate set as:

$$CandidateSet = Ref_i \cup Cit_i \cup (\bigcup_{j \in Cit_i} Ref_j) \cup (\bigcup_{k \in Ref_i} Cit_k).$$

$$(3)$$

3.4 Paper Ranking

After splitting abstracts and building similarity models on both problem and solution parts, we can obtain weighted vectors for each paper. The Candidate Selection step helps us reduce the number of papers for ranking. We then calculate the cosine similarity between the target paper and each paper in candidate list:

$$sim = \cos\theta = \frac{\mathbf{A} \cdot \mathbf{B}}{\|\mathbf{A}\|\|\mathbf{B}\|} = \frac{\sum_{i=1}^n A_i \times B_i}{\sqrt{\sum_{i=1}^n (A_i)^2} \times \sqrt{\sum_{i=1}^n (B_i)^2}}$$

$$(4)$$

where A and B are feature vectors of two papers. We rank papers by their cosine similarities for both problem- and solution-related candidates. As a result, we get a problem related list and solution related list for each paper.

4. EXPERIMENTAL EVALUATION

We evaluate our models and compare with baselines in this section.

4.1 Experimental Data

We use the dataset built by ArnetMiner[4][12] which contains 1.3 million papers with necessary meta data. But some of the abstracts and references are missing. Our experiments are thus conducted over the subset which contain both abstract and reference information for each paper, counting 349,231 papers in total.

We randomly select 30 papers from the subset and make sure their abstracts can be divided into two parts. For each paper, we recommend both most problem-relevant and most

---

[3]http://www.citeulike.org/
[4]http://arnetminer.org

Table 1: NDCG result

NDCG@5	Problem Relevance		Solution Relevance	
Model Name	Problem	All	Solution	All
TF*IDF	0.8368	0.5928	0.4910	0.7985
LDA	0.8012	0.7068	0.8336	0.7176
LDA+Concept	**0.8403**	0.8320	**0.8867**	0.7324

Table 2: MAP result

MAP@5	Problem Relevance		Solution Relevance	
Model Name	Problem	All	Solution	All
TF*IDF	0.8394	0.5087	0.4483	0.7371
LDA	0.7601	0.8031	0.8313	0.6635
LDA+Concept	**0.8427**	0.7358	**0.8642**	0.7012

solution-relevant papers to users. Besides these two recommended lists, we also build baselines based on the similarity models that are trained by all abstracts instead of problem part or solution part. For problem/solution-relevant recommendations, we ask volunteers to label whether the recommended article describes the same problem/solution with the target paper. The volunteers are required to mark their choice with integer scores ranging from 0 to 3. While 0 represents no similarity and 3 represents highly relevant. And for the baseline, they will make two scores for both problem and solution similarities.

4.2 Evaluation Measures

To properly account for the correctness of top ranked documents and the accuracy of recommendation, we employ the normalized discounted cumulative gain (NDCG) and mean average precision (MAP).

Since users may just notice the top items, we concern mainly about whether the top ranked papers are relevant or not. Therefore, in this work, we use NDCG@N and MAP@N (N = 5) for evaluation where N is the number of top-N papers recommended by our proposed approaches.

4.3 Experimental Results

Table 1 and table 2 are the results on NDCG@5 and MAP@5, respectively. The *Problem/Solution* in the second row means using problem/solution parts only to train models and *All* means using the whole abstract content. We set topic number as 80 to train LDA and LDA+CONCEPT models.

According to the table, we can see that similarity models trained on problem/solution splits outperform those trained on full abstracts. In particular, the highest accuracy is obtained both on NDCG@5 and MAP@5 when we using concept based topic model. For short text, TF*IDF does not perform as well as on long text. In this case, topic models achieves better accuracy. What's more, introducing academic concepts can help LDA understand papers' gists better.

5. CONCLUSION AND FUTURE WORK

In this article, we describe a novel way to recommend academic papers to users from problem-related and solution-related aspects. And our experiments show that the concept based topic model method can achieve a higher accuracy. The key contributions of our work are threefold:

1) we argue that, beyond conventional similarity computations, a better academic paper recommender system should take user's different reading purposes into consideration and provide more focused recommendations accordingly. 2) We propose a topic model based LDA method to flexibly capture the similarity between short texts given different reading purposes, which outperforms a traditional tf*idf method. 3) Our experiments prove that the academic concepts play an important role in conveying the main idea of an abstract, which is potentially crucial for context-based academic recommendations.

Currently, in the Abstract Split step, we analyze the writing patterns for abstracts and further extract template terms by empirical observations. It would be of importance to automatically classify the extracted terms into problem-related or solution-related according to the context information and their positions. It is also necessary to extend our proposed model by further generating user profiles with user click-through history and predicting whether a user prefers to problem-centered papers or method-centered ones.

6. REFERENCES

[1] G. Adomavicius and A. Tuzhilin. Toward the next generation of recommender systems: A survey of the state-of-the-art and possible extensions. *TKDE*, 17(6):734–749, 2005.

[2] N. Agarwal, E. Haque, H. Liu, and L. Parsons. Research paper recommender systems: A subspace clustering approach. *WAIM*, pages 475–491, 2005.

[3] M. Balabanovic and Y. Shoham. Fab: content-based, collaborative recommendation. *Commun. ACM*, 40:66–72, 1997.

[4] C. Basu, H. Hirsh, and W. W. Cohen. Recommendation as classification: Using social and content-based information in recommendation. In *AAAI/IAAI*, pages 714–720. AAAI Press, 1998.

[5] C. Basu, H. Hirsh, W. W. Cohen, and C. G. Nevill-Manning. Technical paper recommendation: A study in combining multiple information sources. *J. Artif. Intell. Res. (JAIR)*, 14:231–252, 2001.

[6] D. M. Blei, A. Y. Ng, and M. I. Jordan. Latent dirichlet allocation. *JMLR*, 3:993–1022, 2003.

[7] K. Chandrasekaran, S. Gauch, P. Lakkaraju, and H. P. Luong. Concept-based document recommendations for citeseer authors. In *AH*, volume 5149 of *Lecture Notes in Computer Science*, pages 83–92. Springer, 2008.

[8] W. Ding and G. Marchionini. A study on video browsing strategies. *Technical Report*, pages 946 – 949, 1998.

[9] D. Goldberg, D. A. Nichols, B. M. Oki, and D. B. Terry. Using collaborative filtering to weave an information tapestry. *Commun. ACM*, 35(12):61–70, 1992.

[10] A. K. Pudhiyaveetil, S. Gauch, H. P. Luong, and J. Eno. Conceptual recommender system for citeseerx. In *RecSys*, pages 241–244. ACM, 2009.

[11] A. Singhal. Modern information retrieval: A brief overview. *IEEE Data Eng. Bull.*, 24(4):35–43, 2001.

[12] J. Tang, J. Zhang, L. Yao, J. Li, L. Zhang, and Z. Su. Arnetminer: extraction and mining of academic social networks. In *KDD*, pages 990–998. ACM, 2008.

Influential Seed Items Recommendation

Qi Liu[1], Biao Xiang[1],Enhong Chen[1], Yong Ge[2], Hui Xiong[2], Tengfei Bao[1], Yi Zheng[1]

[1]School of Computer Science and Technology, University of Science and Technology of China
E-mail: {feiniaol,bxiang,tfbao92,xiaoe}@mail.ustc.edu.cn, cheneh@ustc.edu.cn

[2]Rutgers Business School, Rutgers University
E-mail: yongge@pegasus.rutgers.edu, hxiong@rutgers.edu

ABSTRACT

In this paper, we present a systematic perspective study on choosing and evaluating the initial seed items that will be recommended to the cold start users. We first construct an item consumption correlation network to capture the existing users' general consumption behaviors. Then, we formalize initial items recommendation as the influential seed set selection problem. Along this line, we present several methods, each of which selects seed items according to different rules. Finally, the experimental results on two real-world data sets verify that with different seed items, the users' consumption numbers will be quite different. Meanwhile, the results also provide many deep insights into these selection methods and their recommended seed items.

Categories and Subject Descriptors

H.2.8 [**Database Management**]: Database Applications—
Data Mining

General Terms

Algorithms, Experimentation

Keywords

Seed Items, Item Network, Influential, Popularity

1. INTRODUCTION

Whenever a fresh user comes, there is no or few personal information available, thus how to make recommendations for these cold start users becomes a huge and urgent problem [2]. Actually, many techniques have been proposed to address this kind of problem [2, 3, 6, 7, 8, 10, 11]. For example, one naive method is to recommend the most popular items. Moreover, some methods alleviate the cold start problem by understanding users with their input information [6, 8] or leveraging the meta data of items [11]. Since the user may not want to input her real information for some reasons (e.g., privacy issue) and the meta data is not always available, as an alternative, some other methods try to ask the user to provide a set of ratings for their elaborately selected *seed items* (i.e., the first several items recommended

to a cold start user), so as to collect the user's explicit rating preferences [2, 3, 7, 10] with a few interventions.

However, these related techniques focus on the way of choosing representative items (e.g., with high rating variance or entropy) to predict the preferences of the new user for each specific item. In some real applications, in contrast to this individual perspective strategy, the commercial systems often directly offer some discounted or free items to the new users so as to elicit them to consume as many items as possible. Thus, the profit of the system will increase.

To that end, in the following, we mainly focus on the way of recommending and evaluating seed items from this systematic and marketing perspective. Specifically, we aim to automatically find out the seed items that can bring in more consumptions (i.e., *influential* seed items), without any extra interventions from the users. Along this line, we first construct a weighted and directed item network for capturing the users' general consumption behaviors and the items' consumption correlations. Then, the problem can be formalized as selecting the most influential seed items from this network. Next, we propose several methods (e.g., PageRank based methods), each of which selects influential seed items according to different rules and capturing different information. Finally, we evaluate the presented methods on two real-world data sets Flixster and Douban. The experimental results verify that with different seed items, the users' consumption numbers will be quite different. They also demonstrate that just recommending popular items is not a very effective way to help the system understand and attract users, and better results can be achieved by the algorithms exploiting items' influence or correlations. The main contributions of this paper can be summarized as follows.

- We propose the idea of selecting influential seed items from the systematic and marketing perspective for dealing with the cold start user problem. Along this line, we formalize this problem as the seed items selection for an item network.

- We present many possible methods for seed items selection. The effectiveness of these methods is evaluated on two real-world data sets. The experimental results also provide many deep insights into these selection methods and their recommended seed items.

2. PROBLEM FORMULATION

In the following, we use $U = \{U_1, U_2, ..., U_M\}$ and $I = \{I_1, I_2, ..., I_N\}$ to represent the set of users and the set of items, respectively.

In real applications, when a user registers into a system (e.g., an online movie theater), the system usually offers

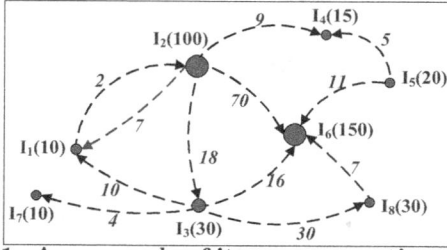

Figure 1: An example of item consumption network.

this cold start user some discounted or free items (e.g., some free movies to watch). There are possibly two kinds of tasks for these free items, the first one is to find out each user's unique preferences (individual perspective), and the other is to attract the user for more consumptions (systematic perspective). Having said, current researches mainly focus on the way of fulfilling the the first task, to our best knowledge, the second aim has so far been overlooked. Since the profit of the system goes directly with the number of users' consumptions, to that end, we focus on the seed item recommendation problem from the systematic perspective.

Along this line, there should be mainly two characteristics of the seed items. At first, they should be *popular* rather than representative. This means we should select the items that are generally enjoyed by the previous users so as to get a potential high acceptance rate. Secondly, they should be *influential*, which means the seed item should have the ability to bring in future consumptions. Please note that, in our scenario, influence/influential is more likely used for measuring the ability of items' consumption correlations, and this is a little bit different from that used in social networks [1].

Then, the problem is how to describe items based on the above two characteristics. Straightforwardly, they can be well represented by item consumption network, as shown by the example in Figure 1. In this network $G = \{I, E\}$, the number in each bracket () is the number of users consumed this specific item, and if there are m users who consumed/enjoyed item I_i before I_j, then we add a directed edge (E_{ij}) and the weight (W_{ij}) should be m. In this way, we can formalize the recommendation problem as selecting the set of seed items from the item consumption network.

3. SELECTION METHODS

In this section, we describe several methods for finding the seed items from the item consumption network.

Popularity. The Popularity method presents items ordered by the number of consumptions that they have been given. It is equivalent to ordering by the probability that a user has enjoyed the item. Popularity is a very straightforward method, and is also easy for calculation.

WDegree. However, the most popular items may not be the most influential. Thus, an alternative approach is to chooses items according to their direct influence abilities, i.e., the number of following consumptions induced by the chosen item. Formally, the WDegree value for each item can be defined as: $WDegree(I_i) = \sum_{j \in [1,N]} W_{ij}$.

Let's take the network in Figure 1 as an example, and suppose we want to select one seed item. According to Popularity, we will choose I_6, which has been consumed 150 times. Instead, according to WDegree, we will select item I_2. We can see that, these two methods directly capture each of the two characteristics of the seed items, respectively.

Intuitively, for better describing each item, the two char-

acteristics should be made use of simultaneously. The idea of PageRank can be well fitted to address this issue. PageRank, introduced by Page et al.[9], was first used for objectively measuring the importance (authority) of web pages.

Though we take advantage of the idea of PageRank, there are still some differences that should be noticed. In PageRank, the *quality* value for each web page is computed recursively by combining the values of the pages that link to this specific page. Thus, a page that is linked to by many pages with high PageRank receives a high rank itself. In contrast, in our situation, the out-edges are used for measuring each item's importance, and if an item links to many items it should receive a high rank value. This can be summarized as: The original PageRank algorithm is used to find out where is the information going, in contrast we want to figure out where is the information from. In the following, based on the idea of PageRank, we illustrate two related methods that can be used for seed set selection.

SPageRank. This is short for Simple PageRank. In this method, each item shares the same initial PageRank value and the same decay factor d. For each item I_i, its PageRank value can be formalized as the following equation:

$$\begin{cases} PR(I_i)^{(0)} = \frac{1}{N} \\ PR(I_i)^{(s+1)} = d \sum_{j, W_{ij} > 0} \frac{PR(I_j)^{(s)}}{InDe(I_j)} + (1-d)PR(I_i)^{(0)} \end{cases}$$

where N is the number of items and $InDe(I_j)$ is the number of edges link to I_j (in-degree). $PR(I_i)^{(s)}$ is the PageRank value for I_i after s (e.g.,30) steps iteration.

WPageRank. In contrast to Simple PageRank, weighted PageRank takes the observed consumptions as weight to learn the PageRank values. For WPageRank, item I_i's PageRank value can be formalized as the following equation:

$$\begin{cases} PR(I_i)^{(0)} = \frac{Popularity(I_i)}{\sum_j Popularity(I_j)} \\ PR(I_i)^{(s+1)} = d_i \sum_{j, W_{ij} > 0} \frac{W_{ij} PR(I_j)^{(s)}}{InWe(I_j)} + (1-d_i)PR(I_i)^{(0)} \end{cases}$$

where $d_i = \frac{max(W_{ij})}{Popularity(I_i)}$, is the observed decay factor of I_i, and $InWe(I_j)$ is the total weight of edges link to I_j.

Circuit. This is a social influence model based on electrical circuit theory to simulate the information propagation process [5] in social networks, which we proposed in [12]. Different from the previous methods which do not consider the possible influence overlaps between seed items and select all the items simultaneously, the Circuit method identifies the independent influence of each item and selects the seeds one by one following a greedy strategy. Specifically, the number of consumptions (popularity) on each item is used for measuring the probability that an event happens on the given item, and the weight of each edge stands for the probability that the information of this event may propagate through this edge. In this way, the seed item set can be selected by solving the social influence maximization problem [5], for more detailed information please refer to [12].

4. EXPERIMENTAL RESULTS

All the experiments were performed on two real-world rating (in the 1-to-5 scale) data sets: Flixster [4] and Douban [1], since the rating records in both of them last for more than 5 years and there are strong correlations between rating orders and the users' consumption (watch) orders [2]. The detailed information is described in Table 1.

---

[1] We collected from the douban.com.
[2] There may be a few exceptions.

Table 1: The description of two data sets.

Data Set	Domain	#User(M)	#Item(N)	#Record	#Edge(|E|)
Flixster	Movie	132,774	28,900	1,928,610	13,540,172
DouBan	Book	25,660	212,647	2,545,054	18,037,755

Since we focus on the positive rating records, we removed the records with rating value lower than 3, and removed the users who gave more than 500 ratings because the correlations between their ratings will be very weak. In Figure 2, we take Flixster data as an example and show the power law distributions of the items' popularity and weighted out-degree. From Figure 2(c) we can see that, though positive correlations between each item's popularity and it's out-degree can be observed, there are still some fluctuations. Thus, most popular items may not be most influential and vice versa.

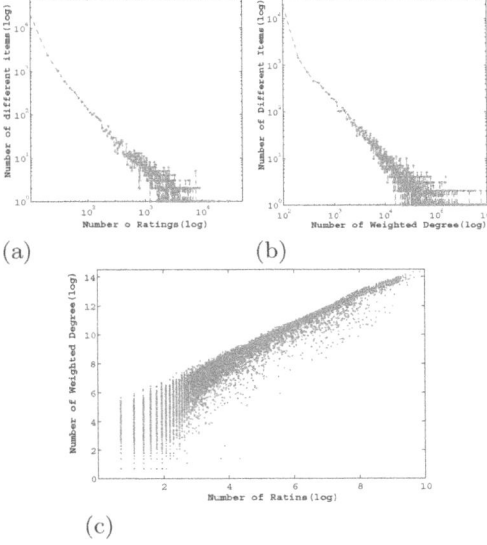

(a) (b)

(c)

Figure 2: The distributions of Flixster items' popularity (a), weighted out-degree (b) and the relationships between popularity and out-degree (c).

4.1 Performance Comparison

In this section, we present a performance comparison of both effectiveness and efficiency between each method. We fix d equal to 0.85 for PageRank based methods [3].

Effectiveness. To evaluate the selected seed items and the effectiveness of each algorithm, we use *Average-Consumption Number-After Given Item* (AveNum) as a metric, which directly measures how many items will be consumed by the test user after the seed item(s). It is formally defined as:

$$AveNum(k) = \frac{1}{|U^k|} \sum_{U_i \in U^k} \frac{\sum_{j \in [1,k]} RN_{U_i S_j}}{k}, \quad k > 0 \quad (1)$$

Where U^k is the set of users in the test set who have consumed k items from the selected item set S. Thus, $RN_{U_i S_j}$ is the number of U_i's consumptions after she consumed seed item S_j. If $k = 0$, then we consider all of U_i's consumptions(this value can be also viewed as baseline). In the experiments, each time 4/5 users and their ratings are used for training, i.e., constructing the consumption network and selecting the seed items S, and then the AveNum value of the remaining 1/5 users are computed for testing. At last, we give the average result of these five testing splits.

We run each method and use their selected Top-10 items as seeds to evaluate the performances. Figure 3 illustrates

---

[3]Different values of d contribute little impact on the results.

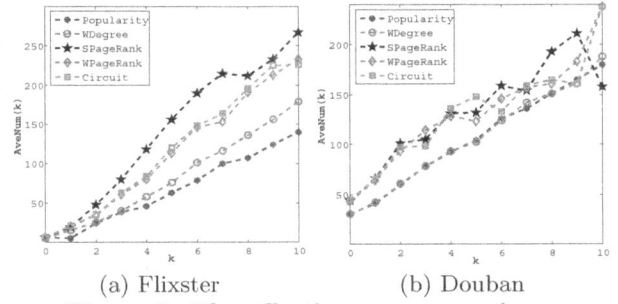

(a) Flixster (b) Douban

Figure 3: The effectiveness comparison.

Table 2: Comparison of the execution time.

Flixster		Douban	
Method	Time(sec.)	Method	Time(sec.)
Popularity	0.015	Popularity	0.58
WDegree	0.075	WDegree	0.93
SPageRank	32.1	SPageRank	56.40
WPageRank	34.8	WPageRank	58.75
Circuit	490	Circuit	470

the AveNum result that we get by the 5 methods with respect to different number of consumed seeds (k). Generally, the more seed items the test users have consumed, the more items they will consume in the future. Since SPageRank, WPageRank and Circuit methods capture both of the two characteristics of seed items, they usually perform better than Popularity and WDegree. However, the WPageRank and Circuit methods, which perform similar with each other, do not perform better than SPageRank, and we think the reasons are the following: Different from SPageRank which only exploits the structure of the item graph, WPageRank and Circuit also rely on the exact observations, and based on these observations, the importance of items' popularity is manually raised or reduced. Thus, these two methods' performance will be impacted if there exists bias in observations, and this manually weight adjustment may break the balance between popularity and influence.

Efficiency. All the methods are performed on the same platform, and Table 2 shows their average execution time [4]. Without a surprise, on both data sets Popularity and WDegree cost the least time due to their simplicity, and since Circuit model has to choose the seed items one by one following a greedy algorithm, it performs the worst.

In summary, Popularity and WDegree are easy for implementation, but this also brings in their poor performance. Considering both effectiveness and efficiency, SPageRank seems to be the most suitable seed items selection method. However, many more deep understandings should be provided before we make this conclusion.

4.2 Understanding Selections

In this section, we give an analysis of the selection results and they provide more insights into each method.

At first, Figure 4 shows the Jaccard similarity coefficient of the seed items (Top-10 for each data split) chosen by the selection methods. We can see that similar coefficient results can be observed from Flixster and Douban. First, the result got by SPageRank is most different from others. Second, since Popularity and WDegree capture different characters of the items, their choosing items are also different from each other. Third, similar to the observations in Figure 1, the output of WPageRank and Circuit have very

---

[4]Note that the time consuming on constructing item network is not included, which are similar for each method.

Table 3: Number of Douban users with different k.

Method	$|U^k|$		
	k=1	k=5	k=10
Popularity	4,605	1,314	59
WDegree	4,442	1,267	68
SPageRank	4,842	526	4
WPageRank	4,540	641	2
Circuit	4,558	638	2

strong correlations. We believe the reason lies in the close relation between Circuit and WPageRank [12].

Then, we compare the shortest distances between the seed items. We set the distance among two neighbor vertexes (e.g., I_i, I_j) as $1 - \frac{W_{ij}}{WDegree(I_i)}$, and after running Dijkstra's algorithm for each seed item set we find that the seed items output by Popularity and WDegree are usually very close to each other(with average distance less than 1). In contrast, the seeds output by SPageRank, WPageRank and Circuit often locate in different subnetworks, and thus no path can be found [5]. As example, Table 3 lists the number of Douban users who consumed k (1, 5 and 10) of the seed items with respect to different methods. From Table 3 we can see that when k equals to 1, the $|U^k|$ are similar for each method, and this also reveal that there is no big difference between the average consumption numbers (popularity) of each single seed item selected by different methods. However, much less users consumed all the 10 seed items that are selected by SPageRank, WPageRank and Circuit, respectively.

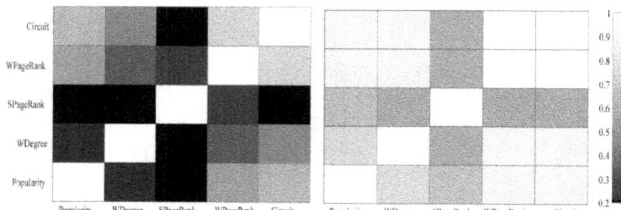

Figure 4: The coefficient of the seed item set selected by each method on Flixster(L) and Douban(R).

Based on the observations from shortest distances and Table 3, we conclude that though the seed items selected by SPageRank, WPageRank and Circuit are also popular items, they are more diverse, e.g., these items may be chosen from different user groups or from different categories. In the future, we plan to find some data sets with available meta data for understanding the selected items more deeply.

5. DISCUSSION

In this section, we analyze the advantages and limitations of current influential seed items recommendation, and show the directions for our future work.

From the experimental results we can see that, by considering the general behavior of users, it is reasonable to find many seed items and recommend them to the cold start users, so as to help the system earn more profit. Furthermore, by exploiting the influence of candidate items, we can get a better recommendation result. However, there are still many limitations of current work. At first, we do not deeply consider the utility of each single user and their personalized acceptance. Second, in this paper, we assume each user's consumption are time ordered and correlated, then make use of the order of users' consumptions to measure

item *influence*, and the consumption correlations produced between different lengths of time are weighted equally. This simplicity also leads to some shortages, for example the long time ago consumption may have little contribution on user's current purchase. Third, none of the existing factors that account for the consumption orders are included in current ideal recommendation strategy, such as the items' life cycles.

According to these limitations, there are many directions for our future work. At first, we plan to evaluate our methods in the real-world applications. Meanwhile, figuring out the way of combining advantages from both representative [2, 7, 10] and general item selections for making more reasonable recommendations. More importantly, we plan to consider domain knowledge and constraints (e.g., some meta data and the time factor) into the seed selection process.

6. CONCLUSION

In this paper, we provide a systematic perspective on choosing and evaluating seed items for cold start user recommendation. We first formalize initial item recommendation as the influential seed set selection problem for the item consumption network. Then, we present several influential seed items selection methods. At last, the performance of these methods are evaluated on two real-world data sets. The experimental study delivers encouraging results, and we hope this work could lead to many future work.

Acknowledgment. This research was partially supported by grants from the Natural Science Foundation of China (61073110, 70890082, 71028002), the National Major Special Science & Technology Projects (2011ZX04016-071), Research Fund for the Doctoral Program of Higher Education of China (20113402110024), the Key Program of National Natural Science Foundation of China (60933013), and the National Science Foundation (NSF, CCF-1018151).

7. REFERENCES

[1] A. Anagnostopoulos, R. Kumar, and M. Mahdian. Influence and correlation in social networks. In *ACM SIGKDD'08.*, pages 7–15, 2008.

[2] M. Crane. The New User Problem in Collaborative Filtering. *University of Otago*, 2011.

[3] N. Golbandi, Y. Koren, and R. Lempel. Adaptive bootstrapping of recommender systems using decision trees. In *WSDM'11*, pages 595–604, 2011.

[4] M. Jamali, and E. Ester. A matrix factorization technique with trust propagation for recommendation in social networks. In *RecSys'10*, pages 135–142, 2010.

[5] D. Kempe, J. Kleinberg, and É. Tardos. Maximizing the spread of influence through a social network. In *ACM SIGKDD'03*, pages 137–146, 2003.

[6] X.N. Lam, T. Vu, T.D. Le, and A.D. Duong. Addressing cold-start problem in recommendation systems. In *ICUIMC'08*, pages 208–211, 2008.

[7] N. Liu, X. Meng, C. Liu, and Q. Yang. Wisdom of the better few: cold start recommendation via representative based rating elicitation. In *RecSys'11*, pages 37–44, 2011.

[8] A. Nguyen, N. Denos, and C. Berrut. Improving new user recommendations with rule-based induction on cold user data. In *RecSys'07*, pages 121–128, 2007.

[9] L. Page, S. Brin, R. Motwani, and T. Winograd. The PageRank citation ranking: Bringing order to the web. *Stanford InfoLab*, 1999.

[10] A. Rashid, I. Albert, D. Cosley, and etc. Getting to know you: learning new user preferences in recommender systems. In *IUI'02*, pages 127–134, 2002.

[11] A.I. Schein, A. Popescul, L.H. Ungar, and D.M. Pennock. Methods and metrics for cold-start recommendations. In *SIGIR'02*, pages 253–260, 2002.

[12] B. Xiang, E. Chen, Q. Liu, H. Xiong, Y. Yang, and J. Xie. A Social Influence Model Based On Circuit Theory. rxiv preprint arXiv:1205.6024,2012.

---

[5] In this situation, the influence of items are almost independent from each other, and no need for finding overlaps.

Discovering Latent Factors from Movies Genres for Enhanced Recommendation

Marcelo Garcia Manzato
Mathematics and Computing Institute – University of São Paulo
Av. Trabalhador Sancarlense, 400, PO Box 668 – 13560-970
São Carlos, SP, Brazil
mmanzato@icmc.usp.br

ABSTRACT

Current approaches on collaborative filtering factorize user-item matrices in order to infer latent factors from ratings previously assigned by users. However, they all have to deal with sparseness, whose workarounds are prone to bias and/or overfitting. This paper proposes a recommender algorithm that is based on a factorized matrix composed of user preferences associated to the movies' genres/categories. The advantage of using such user-genre matrix factorization model is that it requires less computational resources, as the matrix will be less sparse and at lower dimension. We present the experimental results with a dataset composed of real users, comparing the performance of different modules of our algorithm.

Categories and Subject Descriptors

H.3.1 [**Information Storage and Retrieval**]: Content Analysis and Indexing—*Indexing methods*; H.3.3 [**Information Storage and Retrieval**]: Information Search and Retrieval—*Information filtering*

General Terms

Design, Algorithms

Keywords

Collaborative recommender systems, latent semantic analysis, user profiling, users similarity.

1. INTRODUCTION

Recommender systems are an important mechanism for enabling users to deal with the increasing information overload. In its basic form, a recommender system can be designed according to three different strategies: content-based filtering, collaborative and the combination of both. With respect to collaborative filtering, one of the greatest advantages is that it is not limited to the available metadata describing the content, but, on the other hand, a challenge is

how to calculate the similarity of users in order to cluster individuals with similar preferences. In traditional algorithms, the user-item ratings domain is viewed as a vector space and similarity metrics (e.g. Pearson correlation, cosine similarity, etc.) are applied to infer neighbors of users. However, the vectors are of high dimension: for a single user, there will be a vector whose dimension size is the number of items in the system. In addition, the vectors will be redundant, because some users will have similar ratings for the same items [3].

Such limitations in vector space have encouraged researchers to use dimensionality reduction techniques from latent semantic analysis. One well-known technique is Singular Value Decomposition (SVD), which extracts latent semantic relationships between items and users, transforming the vector space into a feature space composed of k topics of interest. Such methodology has been explored by many works [14, 7, 8, 11]; however, they all have to deal with sparseness, which occurs before factorization of the user-item matrix. As many items may not have been rated so far, the methods use an imputation procedure for unknown ratings which can be based on different strategies: user or item averages [14], gradient descent [11], regression based on least-squares [8], among others. Nevertheless, in spite of the recognized efforts, imputation can be very expensive as it significantly increases the amount of data, or it may distort the data considerably [7].

In this sense, this paper proposes a recommender algorithm that is based on a factorized matrix composed of user preferences associated to the movies' genres. Such categories are the unique type of metadata needed from each item, and thus, its extraction is simple and straightforward. The user-genre factorized matrix is used in two directions: i) to infer latent factors of genres, as a specific category may have different concepts, each one with distinct interestingness levels to the user (for instance, two movies classified as science fiction, though the first is about the end of the world and the second is about undead people); and ii) to enrich new users' profiles with predictions of weights for those categories which are still absent in their set of preferences.

This article is organized as follows. Section 2 depicts the related work about collaborative filtering approaches. Section 3 presents the profiling and enrichment mechanisms based on genres' metadata proposed in this paper. Section 4 describes our strategy for predicting unknown ratings based on the model. Section 5 depicts the experimental results of our proposal. Finally, Section 6 presents the final remarks and future work.

2. RELATED WORK

Collaborative filtering algorithms predict unknown ratings based on an aggregate of the ratings of similar users. Computing user similarity in vector space brings some limitations. Because the rating vectors are of high dimension and redundant, the system will have an expensive task to calculate all pairwise similarities [3]. In addition, new users will not have enough ratings to support efficient clustering, and consequently, their chances to be considered at recommendation will be low [1].

Nevertheless, new efforts in collaborative filtering have opened opportunities to deal with the limitations depicted above. Indeed, maybe the most remarkable effort spent on collaborative recommenders was during the Netflix Prize[1], which contributed to the emergence of two new trends of research: the use of matrix factorization models to cluster similar users using a variety of latent semantic factors [7], and the combination of multiple recommender algorithms to improve performance [2, 11].

With respect to latent semantic analysis, some authors [6, 5] propose using matrix factorization techniques to reduce the expensiveness of similarity calculation. Such techniques are also exploited to address the cold start problem, in which the lack of ratings are overcome by a predictor that relies in feature space [12, 10].

The recommender algorithm proposed in this paper also adopts matrix factorization techniques, but is different because our overfitting prevention is accomplished by the creation of a user-genre matrix with lower dimension. In addition, our model is used to enrich user profiles composed of genres and associated weights. With these enriched profiles, the system computes the similarity of users.

The alternative approach of factoring a matrix associated to metadata (e.g. movies' genres) was also considered by Gantner et al. [4], who described a method that maps user or items attributes to the latent features of a matrix factorization model. Our approach differs from theirs because we deal only with genre metadata, while they also explore list of actors, directors and credits for each movie, requiring more efforts to extract such content descriptions.

3. USER-GENRE MODEL

This section presents the profiling mechanism based on genres' metadata adopted in this paper. It consists of: i) a user profile composed of a weight associated to each available genre; ii) a factorization model whose objective is to infer latent factors; iii) an enrichment procedure to support new users profiles; and iv) the users similarity calculation based on the enriched profiles. Next subsections depict such features in details.

3.1 Preferred Genres

A genre cloud corresponds to a set of preferred genres of movies watched so far by the user [9]. Let us denote a given user by $u \in U$, where U is the set of all users, a movie item by $s \in S$, where S is the set of all available items, and a rating value by $r \in \{0.5, 1, 1.5, ..., 5\} \equiv R$. A set of items rated by user u is represented by S_u; and based on this set, we define the rating function for user u as $\delta_u : s \in S_u \mapsto \delta_u(s) \in R$.

We denote by G the global set of genres, G_s the set of genres associated to item s, and N_g the global frequency

[1]http://www.netflixprize.com/

of occurrence of genre g for all items. In this way, a genre cloud is defined as $cloud_g(u, r)$, which contains the set of pairs $(g, n_{g,u,r})$, where $g \in G$ is a genre and $n_{g,u,r} = |\{s \in S_u | g \in G_s \ \& \ \delta_u(s) = r\}|$. Thus, $n_{g,u,r}$ represents the frequency of occurrence of genre g for all items that user u has associated with rating r.

Based on the user genre cloud, it is possible to build a user profile which is composed of weights associated to each available genre. The initial step consists of computing the *tf-idf* weight of each pair $(g, n_{g,u,r})$. Considering that our corpus is the set of all possible ratings, the *tf-idf* value will reflect how important a genre is to a particular rating in the set of all ratings. Thus, if we denote by $R_{g,u}$ the set of different ratings assigned to genre g by user u, we define:

$$tf\text{-}idf(g, u, r) = n_{g,u,r} \log \left(\frac{|R|}{1 + |R_{g,u}|} \right) \quad . \quad (1)$$

After computing the *tf-idf* values for each genre cloud entry, the following step is to calculate a weighted average $w(g, u)$ that represents how much a user u likes genre g. Therefore, we define:

$$w(g, u) = \frac{\sum_{r \in R_{g,u}} tf\text{-}idf(g, u, r) \ r}{\sum_{r \in R_{g,u}} r} \quad . \quad (2)$$

In this way, the user profile will contain a list of all available genres and their associated weights. Considering all users registered in the system, it is possible to create a user-genre matrix, which can be factorized in order to infer latent semantic information. Such task is depicted in the following subsection.

3.2 Factorization

The user-genre matrix factorization is accomplished in this paper with Singular Value Decomposition (SVD). For a user-genre matrix M composed of associated $w(g, u)$ values, its SVD is the factorization of M into three matrices such that $M = V \Sigma T^T$. We truncate those matrices to yield V_k, Σ_k and T_k, in order to decrease the dimensionality of the vector space, and also, to leave only the strongest effects or trends in the model by dropping the smaller singular values [3].

The resulting factorized user-genre matrix can be interpreted as an expression of a topic preference-relevance model [3]. The rows of the $|U| \times k$ matrix V are users' interest in each of the k inferred topics, and the rows of the $|G| \times k$ matrix T are the genres' relevance for each topic. The singular values in Σ are weights for the preferences, representing the influence of a particular topic on user-genre preferences across the system.

In our approach, as we are factoring a user-genre matrix instead of user-item, the number of missing values will be lower because there are less genres/categories in a system than items. Therefore, before user-genre matrix factorization, our model considers a simple strategy of imputation that substitutes the weights $w(g, u)$ which are zero with the user average rating offset. This is defined as [3]:

$$\bar{\delta}_u = \mu + \frac{1}{|S_u|} \sum_{s \in S_u} (\delta_u(s) - \mu) \quad , \quad (3)$$

where μ is the overall average rating. It is worth mentioning that in such cases we also store those zero positions in order to provide an enrichment method to prevent cold start. This procedure is presented in next subsection.

3.3 Enrichment

In spite of the user-genre matrix advantage to provide fewer missing values, there are still cases where new users will not have rated items associated to all available genres. As a consequence, their profiles will lack information about their preferences to those genres, which constitutes the cold start problem. In addition, the weights associated to genres in new user profiles may not be fully reliable, as they depend on the number of past ratings associated by the user (see Equation 2).

In this sense, the profile enrichment procedure has two objectives: i) reducing the effects produced by the lack of information in new user profiles; and ii) adjusting the value of $w(g,u)$ to consider the most relevant topics of interest associated to genre g.

By using the zero positions stored during factorization, we are able to predict new values to those users whose profiles do not have reliable weights associated to particular genres. Such predictions, in contrast to the user average rating offset, reflect the combination of different topics to a genre, and the user's preference to those topics. In this way, to incorporate the topic preference-relevance model, we redefine $w(g,u)$ as:

$$
w'(g,u) = \begin{cases} V_k(u) \cdot \Sigma_k \cdot T_k^T(g) \\ \quad \text{if } w(g,u) = 0 \quad \text{or} \\ (1-\gamma)w(g,u) + \gamma(V_k(u) \cdot \Sigma_k \cdot T_k^T(g)) \\ \quad \text{otherwise} \end{cases} ,
$$
(4)

where vector $V_k(u)$ is the corresponding row of user u in matrix V, vector $T_k^T(g)$ is the corresponding column of genre g in matrix T^T, and γ is a weighting parameter.

As a result, our enrichment algorithm outputs a user profile that contains weight preferences to all available genres in the system. Such values are a combination of user feedback, topic preference and relevance to associated genres. With the set of enriched profiles, our recommender computes the similarities of all users.

3.4 Users Similarity

We compute the adjacency of users for individual u over all remaining enriched profiles using each genre weight $w'(g,u)$ which was defined in last subsection (Equation 4). After being computed using the factorized user-genre matrix, these weights represent the user's preference and relevance combination for different topics associated to the genres. Considering two users u and v, their similarity $sim(u,v)$ is calculated using the Pearson correlation coefficient [1]:

$$
sim(u,v) = \frac{\sum_{g \in G}(w'(g,u) - \bar{w}'(u))(w'(g,v) - \bar{w}'(v))}{\sqrt{\sum_{g \in G}(w'(g,u) - \bar{w}'(u))^2 \sum_{g \in G}(w'(g,v) - \bar{w}'(v))^2}} ,
$$
(5)

where $\bar{w}'(u)$ and $\bar{w}'(v)$ are averages of users u and v for the weights associated to all available genres.

4. RECOMMENDATION

The recommender proposed in this paper is similar to traditional collaborative filtering algorithms [1], but instead of

using the similarity of users calculated from past ratings, it considers the adjacencies of users constructed from the enriched profiles. Thus, considering f_g as a normalizing factor selected as $f_g = \sum_{v \in \hat{U}} sim(u,v)$, where $\hat{U}$ is the set of the most similar users to u, the prediction of unknown rating $\delta_u(s)$ is defined as:

$$
\delta_u(s) = \begin{cases} \bar{\delta}_u + \dfrac{1}{f_g} \sum_{v \in \hat{U}} sim(u,v) \times (\delta_v(s) - \bar{\delta}_v) \\ \quad \text{if } f_g \neq 0 \quad \text{or} \\ \dfrac{1}{|G_s|} \sum_{g \in G_s} w'(g,u) \quad \text{otherwise} \end{cases} .
$$
(6)

Such normalizing factor is used to indicate whether a user does not have any other individual who is similar to her. In such cases, therefore, the prediction is based solely on the simple average of the enriched profile.

5. EXPERIMENTAL RESULTS

This section provides the experimental results of the recommender system proposed in this paper. It consists of comparing the results of different modules that compose the algorithm. In particular, we evaluated the recommender results using the Root Mean Squared Error metric (RMSE) [13], providing the performance for each user considered in the dataset.

The samples that were used as input data for evaluation were gathered from the MovieLens dataset[2]. We randomly selected 500 users, corresponding to 197.211 ratings. This set was split into training and test sets, being the first used to construct the users' profiles, and the second to test our predictions against its actual ratings. This division was made in a way to have exactly 10 ratings for each user in the test set. Consequently, the training and test sets contain 192.211 and 5.000 ratings, respectively. The parameters of our recommender were defined experimentally as follows: in the genre-based collaborative filtering, we select $|\hat{U}| = 100$ most similar users; and $\gamma = 0.5$ for Equation 4.

Figure 1 illustrates the experimental results for each user and corresponding module. Figure 1(a) is presented for comparison purposes only: the prediction for an unknown rating $\delta_u(s)$ is computed as the simple average of $w(g,u)$ values in profiles without enrichment for all genres available in s. As equivalent, Figure 1(b) presents the results for the simple average with enriched profiles module (see second part of Equation 6), where the unknown rating $\delta_u(s)$ is computed based on the redefined $w'(g,u)$ values (Equation 4).

Comparing both strategies presented in Figure 1(a) and 1(b), we can observe a significant improvement of results when the enrichment based on the factorized user-genre matrix is adopted, whose overall RMSE decreased from 1.5350 to 0.9617. This is because the enrichment procedure is reducing the effects of cold start, as unknown preferences for particular genres are predicted from the model, and also, because those weights are adjusted in order to consider the most relevant topics of interest associated to that genres.

Figure 1(c) presents the results for the final recommendation strategy adopted in this paper. The set of enriched profiles are used to compute clusters of similar users, and after that, a collaborative filtering technique is used to predict

---

[2]http://www.grouplens.org/node/12/

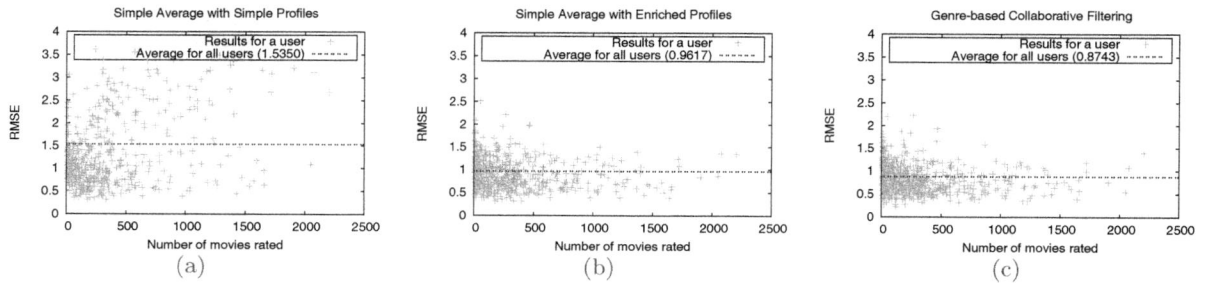

Figure 1: RMSE results for recommenders according to the amount of movies rated.

an unknown rating. The overall RMSE achieved a score of 0.8743, which is better than the simple average illustrated in Figure 1(b). This demonstrates the effectiveness of using the enrichment procedure in association with collaborative methods in order to improve the quality of recommendation.

6. FINAL REMARKS

This paper presented a recommender algorithm based on a factorized user-genre matrix model that is used to discover latent factors from genres in order to enrich users profiles. This enrichment has the main objective to adjust the user preference for a genre by considering the most relevant topics that compose the genre. For instance, an adjusted preference for science fiction will indicate the user's interests for different topics, such as undead people, the end of the world and star wars.

The main advantage of using the proposed model instead of traditional user-item ratings factorization methods is that the matrix will be less sparse because individual weights will be associated to a general information about all the content. Then, by using factorization, it is possible to infer latent semantics without the need of further metadata.

As future work, we plan to investigate more efficient ways to exploit the model in order to address other known problems, such as cold start, overspecialization and sparsity of users.

7. ACKNOWLEDGMENTS

The author would like to thank the financial support from FAPESP, process number 2011/17366-2.

8. REFERENCES

[1] G. Adomavicius and A. Tuzhilin. Toward the Next Generation of Recommender Systems: A Survey of the State-of-the-Art and Possible Extensions. *IEEE Transactions on Knowledge and Data Engineering*, 17(6):734–749, 2005.

[2] R. Bell, Y. Koren, and C. Volinsky. Modeling relationships at multiple scales to improve accuracy of large recommender systems. In *Proceedings of the 13th ACM SIGKDD international conference on Knowledge discovery and data mining*, KDD '07, pages 95–104, New York, NY, USA, 2007. ACM.

[3] M. D. Ekstrand, J. Riedl, and J. A. Konstan. Collaborative filtering recommender systems. *Foundations and Trends in Human-Computer Interaction*, 4(2):175–243, 2011.

[4] Z. Gantner, L. Drumond, C. Freudenthaler, S. Rendle, and L. Schmidt-Thieme. Learning attribute-to-feature mappings for cold-start recommendations. In *2010 IEEE 10th International Conference on Data Mining (ICDM)*, pages 176–185, dec. 2010.

[5] R. Gemulla, E. Nijkamp, P. J. Haas, and Y. Sismanis. Large-scale matrix factorization with distributed stochastic gradient descent. In *Proceedings of the 17th ACM SIGKDD international conference on Knowledge discovery and data mining*, KDD '11, pages 69–77, New York, NY, USA, 2011. ACM.

[6] Y. Koren. Factor in the neighbors: Scalable and accurate collaborative filtering. *ACM Transactions on Knowledge Discovery from Data*, 4(1), 2010.

[7] Y. Koren, R. Bell, and C. Volinsky. Matrix Factorization Techniques for Recommender Systems. *IEEE Computer*, 42(8):30–37, 2009.

[8] M. Kurucz, A. A. Benczúr, and B. Torma. Methods for large scale svd with missing values. In *KDD Cup Workshop 2007*, 2007.

[9] M. G. Manzato and R. Goularte. Peersommender: A Peer-Level Annotation-Based Approach for Multimedia Recommendation. *Journal of Information and Data Management*, 1(2):277–292, 2010.

[10] A. K. Menon and C. Elkan. Dyadic prediction using a latent feature log-linear model. In *2010 IEEE 10th International Conference on Data Mining (ICDM)*, pages 364–373, 2010.

[11] A. Paterek. Improving regularized singular value decomposition for collaborative filtering. In *KDD Cup Workshop 2007*, pages 39–42, 2007.

[12] S. Rendle and S.-T. Lars. Online-updating regularized kernel matrix factorization models for large-scale recommender systems. In *Proceedings of the 2008 ACM conference on Recommender systems*, RecSys '08, pages 251–258, New York, NY, USA, 2008. ACM.

[13] F. Ricci, L. Rokach, B. Shapira, and P. B. Kantor, editors. *Recommender Systems Handbook*. Springer, 2011.

[14] B. M. Sarwar, G. Karypis, J. A. Konstan, and J. T. Riedl. Application of Dimensionality Reduction in Recommender System – A Case Study. In *Proceedings of ACM SIGKDD Conference on Knowledge Discovery in Databases*, Boston, MA, USA, 2000.

Exploiting the Web of Data in Model-based Recommender Systems

Tommaso Di Noia[1], Roberto Mirizzi[1,2], Vito Claudio Ostuni[1], Davide Romito[1]*
[1]Politecnico di Bari – Via Orabona, 4 – 70125 Bari, Italy
[2]HP Laboratories – 1501 Page Mill Road – Palo Alto, CA 94304
t.dinoia@poliba.it, {mirizzi,ostuni,d.romito}@deemail.poliba.it

ABSTRACT

The availability of a huge amount of interconnected data in the so called Web of Data (WoD) paves the way to a new generation of applications able to exploit the information encoded in it. In this paper we present a model-based recommender system leveraging the datasets publicly available in the Linked Open Data (LOD) cloud as DBpedia and Linked-MDB. The proposed approach adapts support vector machine (SVM) to deal with RDF triples. We tested our system and showed its effectiveness by a comparison with different recommender systems techniques – both content-based and collaborative filtering ones.

Categories and Subject Descriptors

H.3.3 [**Information Systems**]: Information Search and Retrieval

Keywords

Model-based RSs, SVM. Linked Data, DBpedia, Linked-MDB, Semantic Web, MovieLens, Precision, Recall

1. INTRODUCTION

The need for a semantic representation of data and user profiles has been identified as one of the next challenges in the field of recommender systems [8]. There are several advantages in the use of semantic data in recommendation tasks spanning from a pure knowledge-based perspective (e.g., we may have a richer representation of data) to more practical points of view such as the easy adaptation of the same approach to different domains. In the recent years, thanks to the Web of Data advance, we are assisting to a flourishing of semantic datasets freely available on the Web encoding machine-understandable RDF triples related to different domains and sometimes representing different points of view on the same domain. All this information

---

*The authors are listed in alphabetical order.

can be exploited to model items and user profiles in an LOD-enabled content-based recommender system where the domain knowledge plays a fundamental role. One of the main components of these systems is represented by the *Content Analyzer* (CA) [6]. This module is responsible for the pre-processing of the information usually coming from textual sources and for extracting keywords used to model both the items and the user profile. The use of the Web of Data can contribute to reduce the effort associated to the definition of the Content Analyzer. In fact, in order to retrieve and use data related to a specific domain of interest, the LOD-based CA has just to formulate SPARQL[1] queries. Moreover, as resources in LOD datasets are identified by unique URIs and are semantically interlinked with each other, the problems related to a keyword-based approach such as synonymy and polysemy [8] are automatically solved.

In this paper we present a model-based approach for a content-based recommender system exploiting exclusively LOD data to represent both the information on the items and on the user profiles. We show how a model-based approach can be easily adapted to cope with the Web of Data. In this paper we use a Support Vector Machine but the overall framework does not rely on any particular classifier. We extensively tested our system to show the effectiveness of the adoption of semantic data for recommendation tasks. The obtained results evidence quality and richness of the information encoded in LOD datasets and we believe they represent a preliminary step towards a new generation of semantic-enabled recommender systems. It is noteworthy that although we performed our experiments in the movie domain, the techniques we proposed are not tied to this particular domain and can be easily adapted to cope with whatever RDF set of triples.

The remainder of the paper is structured as follows: in Section 2 we illustrate the main concepts behind the Web of Data. In Section 3 we detail our model-based approach to recommendation. Section 4 is dedicated to the evaluation. In Section 5 we give a concise overview of related work. Conclusion and future work close the paper.

2. THE WEB OF DATA

The term *Web of Data*, often referred to as *Semantic Web*, *Web 3.0* or *Linked Data*, indicates a new generation of technologies responsible for the evolution of the current Web [2] from a Web of interlinked documents to a Web of interlinked data. The goal is to discover new knowledge and

---

[1]http://www.w3.org/TR/rdf-sparql-query/

value from data, by publishing them using Web standards (primarily `RDF`) and by enabling connections between heterogeneous datasets. As for the traditional Web, the *Web of Data* spans multiple domains – people, movies, music, books, scientific publications, just to cite a few. In particular, the term `Linked Open Data` (LOD) denotes a set of best practices for publishing and linking structured data on the Web. The project includes dozens of `RDF` datasets interlinked with each other to form a giant global graph, the so called `Linked Open Data` cloud. `DBpedia` is a first-class citizen in this cloud since it represents the nucleus of the entire `LOD` initiative [1]. The data are automatically extracted from freely available Wikipedia dumps and each article in Wikipedia is represented by a corresponding resource URI in `DBpedia`. Several `RDF` statements are generated for each resource by extracting information from various parts of the Wikipedia articles (e.g., from the categories at the bottom of the page and from the *infoboxes* at the right side of the page). This allows automatic agents to exploit the extracted structured information by querying the dataset via its `SPARQL` endpoint. Being based on Wikipedia, `DBpedia` is multi-lingual and cross-domain. This feature makes `DBpedia` a hub for `Linked Open Data`: domain-specific datasets can be connected to it to form a single, interconnected data space. Most of the semantic information encoded in `DBpedia` is represented via the properties `dcterms:subject` and `skos:broader`. They are used to represent a relation of hyponymy between resources. In particular, they link respectively a resource (e.g., a movie) to its category (e.g., *American drama films*) and more specific categories to more generic ones (e.g., *American drama films* to *Drama films*). In Figure 1(a) we show an excerpt of the graph containing properties and resources coming both from `DBpedia` and from `LinkedMDB` (i.e., the `RDF` version of `IMDB`).

3. MODEL-BASED RECOMMENDATIONS WITH LINKED DATA

The user profile consists of a model about the user preferences, i.e., a description of the types of items the user is interested in. There are many possible alternative representations of this description, but a common one is a function that for any item predicts the likelihood that the user is interested in that item. The application of Machine Learning techniques is a typical way to achieve the task of learning user profiles in model-based recommender systems. Creating a model of the user preferences from the user history is a form of classification learning wherein each item has to be classified as interesting or not with respect to the user tastes. Model-based recommender systems and in particular content-based ones share some characteristics with text categorization tasks. Machine learning techniques for text categorization/classification has been extensively applied in the field of recommender systems but, to our knowledge, they have not previously been used to build content-based recommender systems that benefits from the usage of `Linked Open Data`. Since in our system the items to be recommended are resources belonging to semantic datasets, we need to build a model able to deal with such data. In reference to Figure 1(a), the items to be recommended are the movies and they are described by the nodes they are connected to. The example is about the movie domain, nevertheless this approach can be extended to any domain covered by the Web

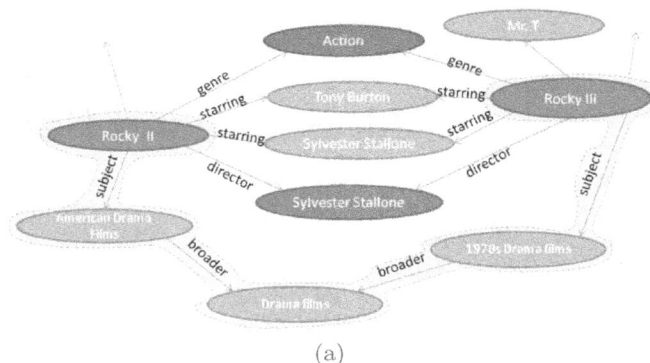

(a)

	STARRING			DIRECTOR	GENRE	SUBJECT+BROADER		
	Tony Burton	Mr. T	Sylvester Stallone	Sylvester Stallone	Action	1970s Drama Films	Drama Films	American Drama Films
Rocky II	0.67	0	0.31	0.66	0.11	0	0.13	0.23
Rocky III	0.67	0.56	0.31	0.66	0.11	0.22	0.13	0

(b)

Figure 1: (a) Sample `RDF` graph extracted from `DBpedia` and `LinkedMDB`; (b) Matrix representation of *property resource-indexes*.

of Data. By exploiting the ontological information encoded via `dcterms:subject` and `skos:broader` properties, we are allowed to perform a semantic expansion of the item description and then to catch implicit relations and hidden information, i.e., information that is not detectable just looking at the nodes directly linked to the item. As an example, if we look at the graph in Figure 1(a), we see that the two movies *Rocky II* and *Rocky III* implicitly have the *Drama films* category in common. The information discovered by exploiting the taxonomic structure of the categories increases the number of common features between two items. In Section 4 we will show how this semantic expansion improves the results of the recommendation.

In our approach we transform the `RDF` graph describing a domain of interest in a feature vector representation that is suitable for the classification task. In a classic *bag of words* model, documents are represented by a set of representative keywords (index terms). We adapt the bag of words model in order to deal with `RDF` triples to obtain a **bag of resources** model. Taken an item from the collection, for each property we extract all the resources that are linked by the current property to the item and we build an index of resources corresponding to that property (i.e., a *property resource-index*). With respect to a given property, each item (i.e., movie) is represented by a vector in a multi-dimensional space, where each dimension corresponds to a resource from the vocabulary. For example, referring to Figure 1(a), the resource-index for the property *starring* is constituted by the resources *Mr. T*, *Tony Burton* and *Sylvester Stallone*. Considering all the properties, each item is represented as a unique vector of weights where each weight indicates the degree of association between the item and the resource with respect to a property. These weights are the TF-IDFs and they are computed distinctly for each property resource-index. Figure 1(b) shows the matrix of TF-IDF weights obtained from the graph in Figure 1(a). We point out that each property resource-index is separated from the others. For example, the resource *Sylvester Stallone* is more frequent in the *starring resource-index* than in the *director resource-index*, because *Sylvester Stallone* starred in more movies than he directed. Another important aspect we want to stress is that if we did not consider the *subject-broader* path (i.e., our semantic expansion), we would lose some matchings between the feature vectors describing the two movies *Rocky II* and *Rocky III*. In fact, as stated before, if we considered only the *subject*, we would not have the category *Drama Films* in the *subject resource-index*.

Support Vector Machine. SVM classifier is based on statistical learning theory developed by Vapnik [14], which uses the principle of Structural Risk Minimization instead of Empirical Risk Minimization, as a supervised machine learning technique. We chose SVM because it is well known that it works well in text classification tasks and our classification problem for learning the user profile has a lot of commonalities with them. Some of these commonalities are the sparse nature of the feature vector and the high dimensionality of the input space. As argued by Joachims [4], SVMs offer two important advantages for text classification task: (1) term selection is often not needed, as SVMs tend to be fairly robust with respect to over-fitting and can scale up to considerable dimensionalities; (2) no human and machine effort in parameter tuning on a validation set is needed. When the decision boundary is not linear we need to transform data into a higher dimensional space using a mathematical transformation known as the kernel trick. We tested three of the main used kernel functions: *Linear*, *Polynomial* and *RBF* kernel. We selected the *RBF* kernel because it proved to be the best performing in our domain. As SVM tool, we used the WEKA[2] SVM(SMO) implementation. The outputs of the SVM are used to build a logistic model able to give posterior probability estimates for the classes. The output of the recommender will be a ranked list of values between 0 and 1 obtained from the logistic model.

4. EVALUATION

The evaluation we present here aims to analyze two different aspects of the proposed approach: (1) inspecting what is the information contained within the datasets of `Linked Open Data` that allows the system to achieve the best results in terms of precision and recall; (2) comparing our system with other relevant approaches, both content-based, collaborative filtering and hybrid ones, in terms of quality of the results. To these purposes, we conducted several experiments using the 1M `MovieLens` dataset. Being our approach based on `LOD` as knowledge base, the first step we had to do was to align the movies in the `MovieLens` dataset with the movies in `DBpedia`. The alignment was mainly done querying `DBpedia` via its `SPARQL` endpoint. A dump of the obtained mapping is available at the url: `http://sisinflab.poliba.it/mapping-movielens-dbpedia-1M.zip`. The test set we extracted from the `MovieLens` dataset has 20 rates per user. For this reason, we were able to compute *Precision@N* and *Recall@N* for values of *N* in the interval [1, 20]. In order to avoid potential bias by some user profiles, we carried out a 5-fold cross-validation. Being our approach based on a binary classifier, we converted the 1-5 Likert scale used by Movielens into a binary one, where ratings above 3 are considered as *like* and the others as *dislike*.

In Figure 2 we plot the precision and recall curves for dif-

[2]`http://www.cs.waikato.ac.nz/ml/weka/`

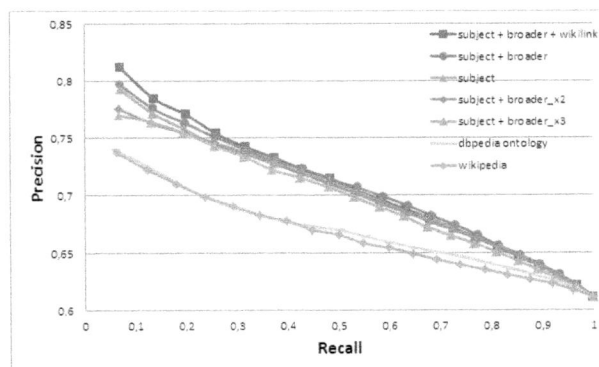

Figure 2: Precision and Recall curves obtained with different groups of properties in LOD.

ferent sets of properties selected from the knowledge base we consider. The intent is to analyze the contribution that ontological information contained within `LOD` datasets may give to the recommendation. In particular, all the curves in Figure 2 refer to our approach and are based on the SVM classifier. We focused on evaluating the importance of the properties `dcterms:subject` and `skos:broader`. Moreover, in this test we evaluated also the importance of the property `dbpedia-owl:wikiPageWikiLink`, that indicates a link between two pages in Wikipedia. Intuitively, if there is a link from one page to another page in Wikipedia, it is reasonable to think the two pages are somehow related. Finally, in this part of the evaluation we also included the properties of the `DBpedia-Ontology` (e.g., in the movie domain some examples of these properties are `dbpedia-owl:starring`, `dbpedia-owl:director`, etc.). The red curve with square markers evidences the best results are achieved for a combination of properties that include: (a) the `DBpedia`-Ontology properties, (b) the `dbpedia-owl:wikilink` property, (c) the `dcterms:subject` property and (d) the first level of the `skos:broader` property (i.e., the categories directly linked to movies and the categories directly linked to them). The next four curves differ from the previous one because they do not include the `dbpedia-owl:wikilink` property and moreover they consider different levels for the `skos:broader` property: the more we use general categories for the recommendation (after the first level), the worse the results are. Finally, in the last two curves we do not consider any taxonomic information at all, but only the standard structured information we can usually find in content-based RS. In this case we obtain comparable results when using only either the `DBpedia`-ontology or Wikipedia. In Figure 3 we compare our approach (indicated by the red curve with square markers) with recent related work, both content-based, collaborative-filtering and hybrid ones. The azure curve with circle markers refers to the hybrid approach presented in [13]. The magenta curve with triangle markers shows the results of precision and recall for a collaborative-filtering approach where the measure of similarity between the ratings of two users is the Pearson correlation coefficient. It is the measure most commonly used in neighborhood-based CF systems [7]. We used the Apache Mahout[3] implementation for CF. The results are fairly comparable to our approach. The orange curve with diamond markers exploits Wikipedia to feed a

[3]`http://mahout.apache.org/`

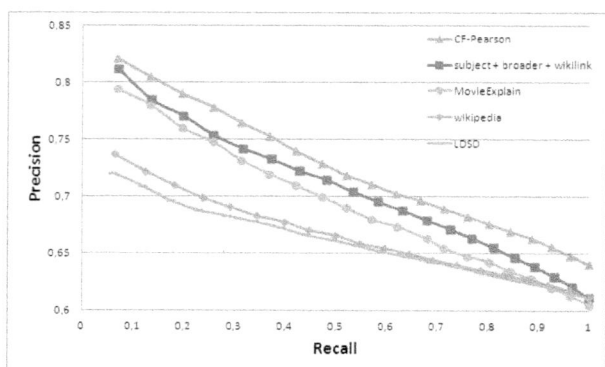

Figure 3: Comparison with several content-based, collaborative filtering and hybrid approaches.

content-based recommender system, as proposed in [5]. Finally, the green curve with dash markers refers to a CB recommender system leveraging DBpedia [9].

5. RELATED WORK

Using the Web of Data as knowledge base for recommender systems is a quite innovative and recent idea. A lot of approaches have been proposed to tackle the well-known issues of recommender systems (both content-based and collaborative filtering), but there are few of them that exploit the huge amount of information encoded in Linked Open Data. Since it would be impossible to give an exhaustive and worth view of all the approaches proposed in state-of-the-art, we refer the interested reader to [7, 12, 10]. In the following we analyze relevant approaches to recommendation exploiting the Web of Data and approaches we compared to in Section 4. One of the approaches we compared to is the *Linked Data Semantic Distance* (*LDSD*) [9]. There DBpedia is used as information source to compute recommendations. Differently than our approach, they do not perform any semantic expansion of the resources. However, such expansion proved to improve the overall quality of the results (cf. Figure 2 and Figure 3). *MoviExplain* [13] is another system we compared to in the evaluation section. The authors present a hybrid approach where they group together users exhibiting highly correlated ratings on set of movies. They leverage only text information sources, while we exploit structured and disambiguated information contained within RDF triples. In [5] the authors use the text content and the hyperlink structure of Wikipedia pages to identify similarities between movies. The aim is to check whether Wikipedia may improve the results of recommendations. However, being the approach based only on unstructured text and on hyperlinks, the approach does not significantly improves the accuracy of the system. In [3] the authors suggest to use Linked Open Data, to alleviate well known issues of CF recommender systems, such as new-user, new-item and sparsity problems. Sen et al. [11] investigate how the use of tags generates high-quality recommendations.

6. CONCLUSION AND FUTURE WORK

As of today, the Web of Data contains a huge amount of structured information publicly available to end-users and service providers. In this paper we have shown how the knowledge encoded in the Linked Open Data cloud can be effectively exploited to model a performing content-based (CB) recommender system. One of the advantages of using the LOD data for CB engines is the mitigation of the *limited content analysis* issue. Indeed, the heterogeneity of topics and contexts represented in the cloud as well as its interlinked nature favors an easy selection and exploitation of new diverse features/properties for a specific domain. The ontological nature of the data we find in LOD datasets has proven to be useful to increase the overall accuracy of the system. The results presented here are an initial step of a comprehensive analysis and investigation on the different uses of LOD data in the field of RSs. We believe the very promising results of our experiments have set a good point in favor of the exploitation of these datasets for recommendation tasks. We are currently working on the modeling of similarity measures different than TF-IDF, such as BM25, and we are investigating on how to have a semantic expansion of the resources used to compute recommendation results by exploiting not only the taxonomy of categories but also other semantic relations occurring within the RDF semantic graph. We are also experimenting the use of other classifiers different from the one used in this paper, such as kNN and Naïve-Bayes and we are performing experiments on different domains.

Acknowledgments. The authors acknowledge partial support of HP IRP 2011. Grant CW267313.

7. REFERENCES

[1] S. Auer et al. Dbpedia: a nucleus for a web of open data. In *Proc. of 6th ISWC and 2nd ASWC*, ISWC'07/ASWC'07, pages 722–735, 2007.

[2] T. Heath and C. Bizer. *Linked Data: Evolving the Web into a Global Data Space*. Synthesis Lectures on the Semantic Web. Morgan & Claypool Publishers, 2011.

[3] B. Heitmann and C. Hayes. Using linked data to build open, collaborative recommender systems. In *AAAI Spring Symposium: Linked Data Meets AI*, 2010.

[4] T. Joachims. Text categorization with support vector machines: learning with many relevant features. In *Proc. of 10th ECML-98*, pages 137–142, 1998.

[5] J. Lees-Miller, F. Anderson, B. Hoehn, and R. Greiner. Does wikipedia information help netflix predictions? In *Proc. of the 7th Int. Conf. on Machine Learning and Applications*, ICMLA '08, pages 337–343, 2008.

[6] P. Lops, M. Gemmis, and G. Semeraro. Content-based recommender systems: State of the art and trends. In *Recommender Systems Handbook*, pages 73–105. 2011.

[7] P. Melville and V. Sindhwani. Recommender systems. In *Encyclopedia of Machine Learning*, pages 829–838. 2010.

[8] S. E. Middleton, D. D. Roure, and N. R. Shadbolt. Ontology-based recommender systems. *Handbook on Ontologies*, 32(6):779–796, 2009.

[9] A. Passant. dbrec: music recommendations using dbpedia. In *Proc. of 9th Int. Sem. Web Conf.*, ISWC'10, pages 209–224, 2010.

[10] F. Ricci, L. Rokach, B. Shapira, and P. B. Kantor, editors. *Recommender Systems Handbook*. Springer, 2011.

[11] S. Sen, J. Vig, and J. Riedl. Tagommenders: connecting users to items through tags. In *Proc. of 18th WWW*, WWW '09, pages 671–680, 2009.

[12] X. Su and T. M. Khoshgoftaar. A survey of collaborative filtering techniques. *Adv. in Artif. Intell.*, 2009, 2009.

[13] P. Symeonidis, A. Nanopoulos, and Y. Manolopoulos. Moviexplain: a recommender system with explanations. In *Proc. of the 3rd ACM Conf. on RSs*, pages 317–320, 2009.

[14] V. N. Vapnik. *The nature of statistical learning theory*. Springer, New York, 1995.

Probabilistic News Recommender Systems with Feedback

Shankar Prawesh and Balaji Padmanabhan
Information Systems and Decision Sciences
College of Business, University of South Florida
4202 E. Fowler Avenue, Tampa, FL 33620

{shankar1, bp}@usf.edu

ABSTRACT

In prior work we addressed a major problem faced by media sites with popularity based recommender systems such as the top-10 list of most liked or most clicked posts. We showed that the hard cutoff used in these systems to generate the "Top N" lists is prone to unduly penalizing good articles that may have just missed the cutoff. A solution to this was to generate recommendations probabilistically, which is an approach that has been shown to be robust against some manipulation techniques as well. The aim of this research is to introduce a class of probabilistic news recommender systems that incorporates widely practiced recommendation techniques as a special case. We establish our results in a special case of two articles using the urn models with feedback mechanism from probability theory.

Categories and Subject Descriptors

H.2.8 [**Database Applications**]: Data mining

General Terms

Algorithms, Top-N

Keywords

News recommender systems, probabilistic sampling, feedback

1. INTRODUCTION

There has been growing evidence of the influence of news recommender systems (NRS) on users. It is considered an important source of news articles for readers, articles which otherwise may get lost due to dynamic environment of news cycles driven by continuous arrival of news articles [12]. It has been noted that once a story appears in most popular list (widely used by media sites), there is an abrupt increase in its popularity and advertising than other stories [1, 7]. As it becomes visible to more readers the number of votes grows at a faster rate.

In prior work we showed that popularity based NRS such as *most emailed* or *most popular*[1] is susceptible to amplifying negligible differences in the initial counts of Nth and $(N + 1)^{th}$ article. Hence, $(N + 1)^{th}$ article which may have "just" missed making the cutoff is often unduly penalized in the most-popular NRS.

---

[1] It is also called Top-N NRS.

These systems are also easily susceptible to manipulation. With some initial effort if a manipulator can get an article into a "most popular" list then the self-reinforcing nature of such lists makes the article remain there with little additional effort. Recently there have been a lot of popular press articles that have highlighted exactly this problem, with manipulation in particular attracting a lot of attention.

Probabilistic NRS has been proposed as a solution to address the aforementioned issues [10]. Briefly, such systems generate recommendations by sampling probabilistically from the pool of articles that could be recommended. An article's probability of being chosen in the recommended list is proportional to its current count/popularity.

This method still generates good recommendations, but permits all articles to have some chance of being recommended. Such a mechanism does not penalize the marginal next articles that might have just missed a hard cutoff in a traditional Top-N list. This mechanism is also more robust against manipulation since it does not suffer as much from the self-reinforcing nature of the hard cutoff lists.

However, there are some limitations of the probabilistic NRS presented in [10]. For one, this approach may select some articles that are not as popular, thereby potentially sacrificing short-term clicks or readership. In an era where page views translate proportionally to advertising revenue this can be a concern in implementing this. Second, giving articles probabilities proportional to their counts is just one method for probabilistically sampling articles and does not provide the media client with any flexibility in implementing such a sampling scheme.

In the present research we propose a novel solution to these problems of the prior probabilistic NRS through a class of probabilistic NRS with *feedback* and discuss various notable properties of it. Feedback models [4] are used in applications where the behavior of the system creates either positive or negative feedback that affects future behavior of the system. For example, the typical "Top N" recommender has a positive feedback mechanism for the articles in the list.

The probabilistic selection mechanism with feedback introduced in this research can be considered as a unified model of selection techniques used for different news recommendation mechanisms based on the count of articles. For example, random selection, probabilistic selection in [10] and Top-N selection can be considered as special cases of a generalized probabilistic NRS introduced in the present research. In general the recommendation probability of an article with count n is proportional to $f(n) = n^{\gamma}, \gamma \in \mathbb{R}$. For a special case with two articles we provide theoretical insights of the proposed recommendation process using results from classical urn models in probability theory [4].

2. RELATED WORK

There has been growing attention towards the study of articles in the most-popular list or articles which are promoted to the *front page*. For example, Berger and Katherine [1] have addressed the issue of virality of most-emailed list at the New York Times; while Lerman and Ghosh [7] have studied the distribution of popularity for articles promoted to the front page of Digg- a popular social new aggregator. In somewhat different approach Prawesh and Padmanabhan [10], have discussed some specific characteristics of Top-N NRS using a thought experiment.

The count of evolution of articles in a probabilistic NRS with feedback exhibits similar behavior as discussed by Khanin and Khanin [6]. They have used a probabilistic model of *positive feedback* to study the pattern of neuron growth. In their context several "neurites" are known to exhibit a pattern of growth and contraction until one of them rapidly grows to become an "axon". The probability that a neurite grows in a time period is modeled to be proportional to its length in the previous period and also depends on the level of competition from other neurites. Our proofs have been adapted from the work of Khanin et al. [6].

Other growth processes where feedback mechanisms have been observed are: the technology dominance of QWERTY and Microsoft's operating systems monopoly [4]. Metcalfe's law- used to value a telecommunication network is also considered a special case of feedback mechanism [11]. In business, positive feedback mechanisms can help a company reach monopoly status by starting with some initial advantage over competitors. Some types of positive externalities particularly contribute to this effect where the values of some systems increase super-linearly with the number of users [4].

In a slightly different context, Pandey et al. [9] have studied the issue of *exploration* and *exploitation* of web-pages using controlled randomness into search result ranking methods. They show that modest amount of randomness leads to improved search results.

3. MODEL

Let us assume that a media site maintains a comprehensive list (CL) of articles. From CL, N articles are selected as "recommendations". The selection of articles is based on probabilistic sampling without replacement. At any given time t the probability that an article-a will be selected in display list (DL) is given by

$$p_a(t) = \frac{c_a{}^\gamma(t)}{\sum_j c_j{}^\gamma(t)} \qquad (1)$$

Where $c_a^\gamma(t)$ represents the count of an article 'a' raised to the exponent γ. While, $\sum_j c_j^\gamma(t)$ represents the sum of counts of articles (those are not yet selected for DL) at time t to the exponent γ. This sampling process is repeated N times to generate the N recommendations in DL.

In the rest of this section we derive analytical results that can illustrate the workings of this selection mechanism in a formal manner. But before we do so, it is worth observing how this selection mechanism can work in practice.

With $\gamma = 1$ we have the probabilistic selection mechanism implemented in [10]. With $\gamma > 1$ we will have a system with positive feedback for the articles with higher counts. These are now going to have an even higher (i.e. more than proportional) of being in a recommended list, which can result in rapidly increasing counts.

With very high γ it is easy to see that the (sampling without replacement) mechanism is similar to the current "Top N" selection. At each stage the article with the highest count is most likely to be selected and the process will end up with a list identical to the Top N.

For values in-between we have varying degrees of positive feedback. It is also interesting here to consider what might happen when $\gamma = 0$. In such a case it is easy to see that all articles have the same probability of being recommended, essentially simulating a random recommender.

From a practical perspective systems like this can permit the site to dynamically manage a recommender list, alternating between exploration and exploitation as needed. Below we present formal theoretical results.

3.1 Analytical Results

To study the count evolution process of articles for the proposed probabilistic NRS, we make following assumptions.

1. Two articles are available for recommendation (article-1 and article-2).
2. Reader upon arrival reads the recommended article with probability p or reads the other with probability $1 - p$.
3. An initial count of articles before the recommender system was implemented is given by $c_1(0)$ and $c_2(0)$; ($c_1(0) > c_2(0)$ *without loss of generality*) respectively.
4. NRS has fairly strong influence on reader's reading behavior (i.e. $p \sim 1$).

The count of two articles at time t has been denoted by $c_1(t)$ and $c_2(t)$ respectively. Let us denote the discrete time points by integer values. At each time, upon arrival of a reader, an article is read and its count is increased by 1. The total count of articles in the system at time t is deterministic and it is given by $c_1(0) + c_2(0) + t$. We focus on the article '1' for subsequent derivation; we also note that theoretical results for article '2' can be obtained in similar way. Let us denote the probability $p[c_1(t + 1) = c_1(t) + 1]$ as $p_{1t}(read)$.

In probabilistic NRS, article-1 can be read in two ways. The article is in the recommended list (with probability $p_1(t)$) and the reader chooses to read the recommended article (with probability p). Or, article-1 can be in the other list $= (CL \setminus DL)$ (with probability $1 - p_1(t)$) and the reader chooses to read the un-recommended article (with probability $1 - p$).

Specifically, the probability that an article '1' is being read at time t is given by

$$p_{1t}(read) = p * p_1(t) + (1 - p) * (1 - p_1(t)) \qquad (2)$$

Substituting the expression for $p_1(t)$ in the above expression from equation (1) we have,

$$p_{1t}(read) = p * \frac{1}{1 + \left(\frac{c_2(t)}{c_1(t)}\right)^\gamma} + (1 - p) * \frac{1}{1 + \left(\frac{c_2(t)}{c_1(t)}\right)^{(-\gamma)}} \qquad (3)$$

So, the process described in generalized probabilistic NRS for the article-1 can be understood as a processes generated through mixture of two processes defined by $\frac{1}{1 + \left(\frac{c_2(t)}{c_1(t)}\right)^\gamma}$ and $\frac{1}{1 + \left(\frac{c_2(t)}{c_1(t)}\right)^{(-\gamma)}}$ respectively. γ can take any real value between

$(-\infty, \infty)$. However, due to symmetric nature of $p_{1t}(read)$ with respect to γ (equation 3), we will discuss the case when $0 < \gamma < \infty$. Similar, analysis can be extended for $-\infty < \gamma < 0$.

3.1.1. $\gamma = 0$

In this case equation (3) is given by $p_{1t}(read) = \frac{1}{2}$. This is equivalent to an article being read randomly at each time step, irrespective of reader's preference (i.e. p). Hence, we do not use any mechanism to incorporate the reader's preference for article recommendation in such a system.

In the following sections for completeness we will discuss the processes generated through the both expressions $\frac{1}{1+\left(\frac{c_2(t)}{c_1(t)}\right)^\gamma}$ and $\frac{1}{1+\left(\frac{c_2(t)}{c_1(t)}\right)^{(-\gamma)}}$. However, it should be noted that with *assumption* 4, equation 3 takes the form of

$$p_{1t}(read) \sim \frac{1}{1+\left(\frac{c_2(t)}{c_1(t)}\right)^\gamma}$$

In this case, count evolution process based on the probabilistic NRS, $p_{1t}(read) = \frac{1}{1+\left(\frac{c_2(t)}{c_1(t)}\right)^\gamma}$ will suffice for the discussion.

3.1.2. $\gamma = 1$

The reading probability of the article '1' is given by

$$prob_{1t}(read) = p * \frac{c_1(t)}{c_1(t) + c_2(t)} + (1-p) * \frac{c_2(t)}{c_1(t) + c_2(t)}$$

$$= p * \{share\ of\ article\ '1'at\ time\ t\} + (1-p) \\ * \{1 - (share\ of\ article\ '1'at\ time\ t)\}$$

The processes generated in this case can be understood as a combination of Pólya urn mechanism and Friedman urn mechanism respectively [5]. Where Pólya urn mechanism corresponds to the probability function $\frac{c_1(t)}{c_1(t)+c_2(t)}$ (i.e. $p = 1$) and a Friedman urn corresponds to the probability function $\frac{c_2(t)}{c_1(t)+c_2(t)}$ (i.e. $p = 0$).

The formulation of Pólya urn process is defined as follows. Let an urn initially contains $c_1(0)$ black balls and $c_2(0)$ white balls. Each time, a ball is drawn randomly from the urn and it is replaced back in the urn with another ball of same color. Using the martingale property of share of the black balls in the urn after time t, (X_t) [5]. It can be shown that

$$\mathbb{E}(X_t) = \frac{c_1(0)}{c_1(0) + c_2(0)}$$

Further, X_t converges with probability 1 to a limiting random variable X_∞ as $t \to \infty$. The distribution of X_∞ depends on the initial share of the black ball. In Friedman urn model each time a ball is drawn randomly from the urn and it is replaced back in the urn with another ball of *different* color. Let us denote the share of black ball in Friedman urn model at time t as Y_t. Then Y_t converges with probability 1 to $\frac{1}{2}$ [5].

One notable property of this probabilistic mechanism is that, it is robust towards maintaining the share of all articles in the system for an influential NRS ($p \sim 1$).

3.1.3. $1 < \gamma < \infty$

From equation 3 we have,

$p_{1t}(read)$

$$= p * \frac{1}{1+\left(\frac{c_2(t)}{c_1(t)}\right)^\gamma} + (1-p) * \frac{1}{1+\left(\frac{c_2(t)}{c_1(t)}\right)^{-\gamma}}$$

In this case the processes generated by probability function $p_{1t}(read)$ can be considered as a mixture distribution of two different processes given by $\frac{1}{1+\left(\frac{c_2(t)}{c_1(t)}\right)^\gamma}$ and $\frac{1}{1+\left(\frac{c_2(t)}{c_1(t)}\right)^{-\gamma}}$. When the probability of article '1' being read is given by, $p_{1t}(read) = \frac{c_1(t)^\gamma}{c_1(t)^\gamma+c_2(t)^\gamma}$, (i.e. $p = 1$). The count evolution processes of articles correspond to a generalized Pólya scheme [6]. This phenomenon is also understood as systems with positive feedback in economics, biology and chemistry [6, 8].

In the context of NRS it leads to a situation where two articles (article-1 and article-2) compete until one article obtains non-negligible advantage in the count share; eventually leading to dominance of the NRS by a single article. The strength of feedback is modeled through the parameter γ. In this case after a random moment of time an article will be always recommended. More precisely, with probability 1 there exists a time t^* such that an article is recommended for all $t > t^*$.

The distribution function $p_{1t}(read) = \frac{c_2(t)^\gamma}{c_1(t)^\gamma+c_2(t)^\gamma}$ (i.e. $p = 0$) corresponds to the case with *negative feedback*. In this case an article with high initial share will become less popular and an article with low initial share will become more popular over time. So, in the present setup the total reading probability (equation 3) is a mixture distribution of positive and negative feedback mechanism.

3.1.4. $0 < \gamma < 1$

The path followed by the probability function $p_{1t}(read) = \frac{c_1(t)^\gamma}{c_1(t)^\gamma+c_2(t)^\gamma}$ generates the subcritical regime in which both articles will have counts of the same order [6]. More precisely, the ratio of counts for articles tends to 1 as $t \to \infty$.

$$\frac{c_1(t)}{c_2(t)} \to 1 \ as \ t \to \infty \qquad (4)$$

It follows from (4) that, $c_i(t) = \frac{t}{2} + o(t); i \in \{1,2\}$ (5)

The behavior of $o(t)$ depends on γ. If $\frac{1}{2} < \gamma < 1$ then there exists nonzero random constant k, such that $\frac{o(t)}{t^\gamma} \to k$ as $t \to \infty$. In this case an article with high count (i.e. article-'1') will maintain higher share for all large enough t.

When $0 < \gamma \leq \frac{1}{2}$, for both articles there exists a sequence $t_n \to \infty$ such that the article will have higher count at time t_n. $o(t)$ in equation (5) are of the order $\sqrt{t}$ if $0 < \gamma < \frac{1}{2}$ and of the order $\sqrt{t \ln t}$ if $\gamma = \frac{1}{2}$.

For the probability function, $p_{1t}(read) = \frac{c_2(t)^\gamma}{c_1(t)^\gamma+c_2(t)^\gamma} = \frac{1}{1+\left(\frac{c_2(t)}{c_1(t)}\right)^{-\gamma}}$, almost surely, $\lim_{t\to\infty} c_i(t) = \frac{t}{2}$. As, when $\gamma < 1$

then counts of both articles grow at the same rate asymptotically [3, 4].

Again, it should be noted that with *assumption* 4, steps discussed in previous sections for the expression $\frac{c_1(t)^\gamma}{c_1(t)^\gamma + c_2(t)^\gamma}$ (i.e. *positive feedback*) will suffice for the derivation of results.

3.1.5. $\gamma \to +\infty$

Finally, we discuss the situation in which proposed NRS will behave like the most-popular NRS. In this case, the total reading probability of the article '1' can be approximated as $p_{1t}(read) \sim p$. This is equal to the reading probability of the recommended article, when recommendation is based on high counts.

In a more general sense this is equivalent to the NRS that uses articles corresponding to the highest count for recommendation (Top-N NRS). The selection process of N articles generated through a processes defined by equation (1) is given by

$$p_a(t) = \frac{c_a^\gamma(t)}{\sum_j c_j^\gamma(t)} = \frac{1}{1 + \sum_{j \neq a}\left(\frac{c_j(t)}{c_a(t)}\right)^\gamma}$$

Without loss of generality we assume that all articles have different count. Further, it can be easily observed that $\forall j$ such that $\frac{c_j(t)}{c_a(t)} < 1, \lim_{\gamma \to \infty} \frac{c_j(t)}{c_a(t)} \to 0$. So, for the article with highest count (among those which are not yet selected for *DL*) the selection probability in DL will be 1, i.e. $\max\{p_a(t)\} \to 1$. Hence, N probabilistic selections correspond to selection of N article with decreasing order of their counts.

4. CONCLUSION

There has been growing awareness towards the various limitations of the "most popular" lists in recommendation systems. For example, in a recent article [2] in New York Times, it has been noted that, being 11th on a top 10 list on the recommendation system is a lot different than being 10th on that list. Nick Bilton in [2] writes *"Being at the top of these lists can generate substantial windfalls. The iTunes App Store, where apps like Angry Birds, Words With Friends and Pages have spent months at the top of the charts, help the app makers collect hundreds of thousands of dollars in revenue, while those who cannot get that visibility founder in obscurity"*.

In a broader context, the widespread use of Top-N based NRS is leading us to less choice of news articles [2]. In several cases it has been also observed or suspected [2] that manipulators artificially inflate the popularity of the items of their interest. A further reason for count amplification in these top lists is the propagation of recommendations over social networks. Once an article (or app) makes such a list they are more likely to be picked up and propagated through social networks.

While some form of increased attention to "good" articles or content is a plus, when this is done so in a disproportional manner, at the expense of other (possibly equally good) articles or apps the mechanism starts creating phenomena that are clearly undesirable. The system influences readership or what succeeds by virtue of an artificial cutoff. It is such weaknesses that attract manipulators to potentially game the system. At the extreme such systems therefore are prone to a high degree of noise.

The use of the proposed probabilistic NRS is one possible solution to deal with these limitations of Top-N NRS. The absence of a hard cutoff eliminates a key component of the self-reinforcing nature of such lists. However, it should be noted that in this new paradigm the term "most popular" will not be appropriate in all cases. Instead it has to be replaced by more appropriate term such as, popular articles.

Possibly more important, probabilistic mechanisms can potentially sacrifice short term revenue for the media site if it selects unpopular articles that are not likely to be read. Our study of feedback functions in this paper makes a novel contribution in the field of NRS by proposing a recommendation technique that elegantly addresses this drawback by allowing the users to control the extent of the feedback in the system.

If a site wants to promote diversity and minimize artificial amplification the model can be parameterized appropriately to select articles. At some point if the goal is to maximize short-term revenue the parameter can be adjusted to mimic the Top-N systems in behavior. The specific feedback function proposed in this paper for NRS therefore permits elegant combination of exploration and exploitation. Dynamically fine-tuning this and experimental results are aspects we are working on in current work.

REFERENCES

[1] Berger, J. and Milkman, K. L. 2011. What Makes Content Viral? *Journal of Marketing Research* (To Appear).

[2] Bilton, N. 2012. Disruptions: Top 10 Lists Lead to Less Choice on the Web. *The New York Times*, April 1, 2012.

[3] Chung, F., Handjani, S. and Jungreis, D. 2003. Generalizations of Pólya's Urn Problem. *Annals of Combinatorics* 7 (2), 141-153.

[4] Drinea, E., Frieze, A. and Mitzenmacher, M. 2002. Balls and Bins Models with Feedback. *In Proceedings of the thirteenth annual ACM-SIAM symposium on Discrete Algorithms (SODA 2002)*.

[5] Freedman, D. A. 1965. Bernard Friedman's Urn. *The Annals of Mathematical Statistics*, vol. 36(3), 956-970.

[6] Khanin, K. and Khanin, R. 2001. A Probabilistic Model for the Establishment of Neuron Polarity. *Journal of Mathematical Biology*, 42, 26-40 (2001).

[7] Lerman, K. and Ghosh, R. 2010. Information Contagion: an Empirical Study of Spread of News on Digg and Twitter Social Networks. *In Proceedings of the 4th International Conference on Weblogs and Social Media (ICWSM'10)*.

[8] Mitzenmacher, M., Oliveira, R. and Spencer, J. 2004. A Scaling Result for Explosive Processes. 11 (2004), *The Electronic Journal of Combinatorics*.

[9] Pandey, S., Roy, S., Olston, C., Cho, J. and Chakrabarti, S. *Shuffling a stacked deck: The case for partially randomized ranking of search engine results*. VLDB Endowment, Norway, 2005.

[10] Prawesh, S. and Padmanabhan, B. 2011. The "top N" News Recommender: Count Distortion and Manipulation Resistance. *In Proceedings of the fifth ACM Conference on Recommender Systems (RecSys'11)*. ACM, New York, NY, USA, 237-244.

[11] Shapiro, C. and Varian, H. 1999. Information Rules, *Harvard Business School Press*. 1999.

[12] Weber, T. E. 2010. Cracking the New York Times Popularity Code. *The Daily Beast*, December 19, 2010.

Collaborative Learning of Preference Rankings

Tim Salimans
Erasmus School of Economics
Rotterdam, The Netherlands
salimans@ese.eur.nl

Ulrich Paquet
Microsoft Research
Cambridge, UK
ulripa@microsoft.com

Thore Graepel
Microsoft Research
Cambridge, UK
thoreg@microsoft.com

ABSTRACT

We propose a model for learning user preference rankings for the purpose of making product recommendations. The model allows us to learn from pairwise preference statements or from (incomplete) rankings over more than two items. We present two algorithms for performing inference in this model, both with excellent scaling in the number of users and items. The superior predictive performance of the new method is demonstrated on the well-known sushi preference data set. In addition, we show how the model can be used effectively in an active learning setting where we select only a small number of informative items for learning.

Categories and Subject Descriptors

I.2.6 [**Artificial Intelligence**]: Learning

Keywords

Recommendation, Collaborative Learning, Preferences, Ranking, Bayes, Active Learning, Approximate Inference

1. INTRODUCTION

Collaborative recommendation has mostly been studied based on explicit feedback in the form of ratings, or based on implicit binary feedback such as observed purchases or clicks. Often real world data sources lie in between these extremes. Explicit ratings of items are rare and hard to obtain, but often the information is richer than a simple binary signal such as click/non-click. For example, users may express relative value judgments in comparing two different products, or they may provide a partial preference ranking over available items. Such rankings can be explicit such as lists of favorite songs, or inferred from implicit information such as play counts for songs. To make efficient use of such information we propose a new bilinear factor model that maps latent user preferences to observed pairwise comparisons or rankings over items. Since feedback is relative to other items, this modeling approach is more robust than models of user

preferences on an absolute scale. Yet it makes more efficient use of available data compared to methods that only allow for binary feedback. Research [2] also shows that people find it easier to formulate their preferences in such a relative way. An additional advantage is that modeling preference rankings directly leads to a ranking of items to be recommended to users, which is the end goal of many recommendation systems.

We present the new user preference model in Section 2, and develop two methods of performing inference in this model in Section 3. In Section 4 we apply the new model to learn correspondents' preferences over sushi items and show that the new method compares favourably to other existing methods. In Section 5 we discuss the potential of the model to guide a more active learning strategy, where we actively and selectively ask the user for relative feedback on different items. Finally, Section 6 concludes.

2. THE MODEL

Each of the N users and M items are represented with low-rank factors: user i with a $K \times 1$ parameter vector $\boldsymbol{u}_i$, and item j with a $K \times 1$ parameter vector $\boldsymbol{v}_j$. As some items are predominantly more popular than others, a univariate bias parameter b_j is added to each. One might also add similar user-specific offsets to the model.

The user and item features are combined into a latent *score* $s_{i,j}$,

$$s_{i,j} \sim N(\boldsymbol{u}_i' \boldsymbol{v}_j + b_j, 1) , \qquad (1)$$

which represents how much user i likes item j. This latent score is generated by a bilinear model similar to the one used in Matchbox [7] and many other papers in the collaborative filtering literature.

The relative ordering of a set of scores determines a user's preference of one item over the next. If user i prefers items $j_1 \succ j_2$ ($\succ$ meaning "is preferred to"), we require that $s_{i,j_1} > s_{i,j_2}$. We then observe a number of pairwise comparisons between the different latent scores $s_{i,j}$. For each user, we denote these by $\boldsymbol{C}^i$, a sparsely filled matrix of dimension $M \times M$, with elements

$$
\begin{aligned}
c^i_{j,j'} &= 1 && \text{if } s_{i,j} > s_{i,j'} \\
c^i_{j,j'} &= -1 && \text{if } s_{i,j} < s_{i,j'} \\
c^i_{j,j'} &= \text{empty} && \text{if unknown .}
\end{aligned}
\qquad (2)
$$

These observed preferences can be explicitly provided by the user in the form of a ranking or a number of pairwise preference statements, or they can be inferred from the behavior of the user, for example by ordering the time spent interacting with different items. Section 4 gives an example application, where we use the model to learn stated preferences over sushi items. Importantly, if the preferences are expressed as a ranking of items, one might always find a set $\{s_{i,j}\}$ that is consistent over the ranked items j, and hence consistent with user i's observations $\boldsymbol{C}^i$. The data likelihood is therefore

$$p(\boldsymbol{C}|\boldsymbol{S}) = \prod_{i=1}^{N} \prod_{(j,j') \in c^i} \mathbb{I}\left[c_{j,j'}^i (s_{i,j} - s_{i,j'}) > 0\right], \quad (3)$$

where $\mathbb{I}[\text{true}] = 1$ and $\mathbb{I}[\text{false}] = 0$. This is similar to the TrueSkill model of [1].

We assign independent normal priors to the user and item vectors, with

$$
\begin{aligned}
(\boldsymbol{u}_i)_k &\sim N(0, \pi), & \text{for } i = 1, \ldots, N, \; k = 1, \ldots, K \\
(\boldsymbol{v}_j)_k &\sim N(0, \pi), & \text{for } j = 1, \ldots, M, \; k = 1, \ldots, K \\
b_j &\sim N(0, \psi), & \text{for } j = 1, \ldots, M.
\end{aligned}
\quad (4)
$$

The two hyperparameters π and ψ are set manually, but can also be inferred from the data, as explained in [6]. For brevity we do not consider this here. The full Bayesian network of our model is given in Figure 1.

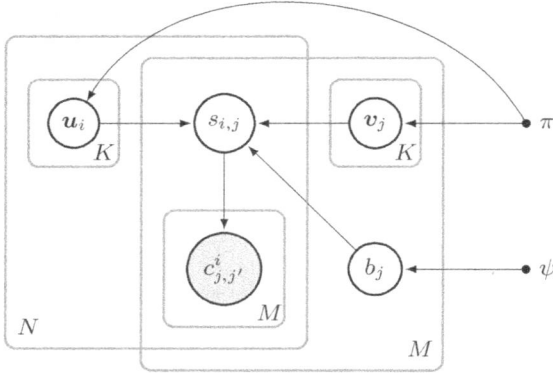

Figure 1: The proposed Bayesian factor model for learning preference rankings

The next section discusses how to infer the posterior distribution of the parameters $\boldsymbol{U}, \boldsymbol{V}, \boldsymbol{b}$ conditional on the observed preferences.

3. BAYESIAN INFERENCE

The model proposed in the last section does not admit a closed form posterior distribution for the parameters $\boldsymbol{U}, \boldsymbol{V}, \boldsymbol{b}$ that we need in order to make recommendations. We therefore propose two strategies for approximating this posterior distribution: a Gibbs sampling algorithm to generate samples from the posterior distribution, and a hybrid message passing algorithm to minimize local divergence measures between the posterior distribution and a factored approximation. The performance and scaling of these two algorithms is evaluated on real world data in Section 4.

3.1 Gibbs Sampling

We can generate correlated samples from the posterior distribution using a Gibbs sampling algorithm that iteratively samples from the conditional distributions $p(\boldsymbol{u}_i|\boldsymbol{V}, \boldsymbol{b}, \boldsymbol{S})$,

$p(\boldsymbol{v}_j, b_j|\boldsymbol{U}, \boldsymbol{S})$ and $p(s_{i,j}|\boldsymbol{s}_{i,/j}, \boldsymbol{u}_i, \boldsymbol{v}_j, b_j, \boldsymbol{C}^i)$, for all i, j, where $\boldsymbol{s}_{i,/j}$ denotes the vector of all scores for user i excluding the j-th. The conditional distributions $p(\boldsymbol{u}_i|\boldsymbol{V}, \boldsymbol{b}, \boldsymbol{S})$ and $p(\boldsymbol{v}_j, b_j|\boldsymbol{U}, \boldsymbol{S})$ are Gaussian and have been used by several authors before. See [6] for their precise form. The full conditional distributions $p(s_{i,j}|\boldsymbol{s}_{i,/j}, \boldsymbol{u}_i, \boldsymbol{v}_j, b_j, \boldsymbol{C}^i)$ are univariate truncated normal, which follows from the Gaussian conditional prior (1) and the truncating likelihood (3).

The $s_{i,j}$ are most efficiently updated by first performing a forward pass over all scores for a given user i, sampling the scores $s_{i,j}$ in the order of the observed preference ranking, followed by a backward pass sampling in the reversed order. (Observe that we do not have to sample those $s_{i,j}$ for which we have no feedback.) We find that this updating schedule does a good job of sampling the relative differences between the scores, but that it is slow in changing the overall level of the scores. To further improve the mixing of the Gibbs sampling algorithm we therefore follow the forward and backward pass by an additional Monte Carlo step that simultaneously shifts all scores for a given user, while leaving the stationary distribution of the Markov chain invariant. The update equation for this step is given as

$$
\begin{aligned}
s_{i,j} &\leftarrow s_{i,j} + d_i, \quad \text{with } j = 1, \ldots, L_i \\
d_i &\sim N(\bar{f}_i - \bar{s}_i, L_i^{-1}),
\end{aligned}
\quad (5)
$$

with L_i the number of items for which user i has provided feedback, $\bar{f}_i$ the mean predicted score for those items, and $\bar{s}_i$ the mean sampled score.

Since sampling the scores using the steps outlined here is relatively quick compared to sampling $\boldsymbol{U}, \boldsymbol{V}$ and $\boldsymbol{b}$, we find that the most efficient implementation of Gibbs sampling resamples $\boldsymbol{S}$ multiple times per iteration.

3.2 Hybrid VB/EP posterior approximation

The Gibbs sampling algorithm outlined in the last section is relatively fast and can be applied at quite a large scale, however for very large data sets a deterministic approximation of the posterior distribution may provide a better trade-off between accuracy and computational cost. An additional advantage of such a deterministic approximation is that it converges to a single mode of the posterior distributions and that it can be represented more compactly than the Gibbs sampling approximation, which reduces the computational cost of generating new recommendations for users given the posterior approximation. We develop a new algorithm to construct such a deterministic approximation, making use of Expectation Propagation (EP) [5] for the ranking likelihood (3) and Variational Bayes for the latent factor model. EP provides an excellent approximation for the uni-modal posterior resulting from the truncated Gaussian in (3), whereas Variational Bayes picks and locally approximates the posterior mode resulting from the product factor in (1).

We approximate the posterior distribution $p(\boldsymbol{U}, \boldsymbol{V}, \boldsymbol{b}|\boldsymbol{C})$ with a fully factorized Gaussian

$$q(\boldsymbol{U}, \boldsymbol{V}, \boldsymbol{b}) = \prod_{i,k} q(u_{i,k}) \prod_{j,k} q(v_{j,k}) \prod_j q(b_j), \quad (6)$$

although our inference algorithm can also be used with a Gaussian approximation that preserves some of these dependencies, e.g. $q(\boldsymbol{U}, \boldsymbol{V}, \boldsymbol{b}) = \prod_j q(\boldsymbol{v}_j, b_j) \prod_i q(\boldsymbol{u}_i)$. In order to optimize this approximate posterior distribution we first approximate the likelihood term $p(\boldsymbol{C}|\boldsymbol{S})$ by a product of univariate Gaussian density functions in $s_{i,j}$, i.e.

$$q(\boldsymbol{C}|\boldsymbol{S}) = \prod_{i,j} \phi(s_{i,j}; \mu_{i,j}, \sigma_{i,j}^2), \qquad (7)$$

with $\phi(\cdot)$ a Gaussian pdf, which we initialize to have infinite variance. The parameters of the likelihood approximation, $\mu_{i,j}$ and $\sigma_{i,j}^2$ are then set using EP. This EP step starts with the construction of a 'pseudo prior' on the $s_{i,j}$,

$$\begin{aligned} q(s_{i,j}) &\propto & \phi(s_{i,j}; \mu_{i,j}^*, \sigma_{i,j}^{2*})/\phi(s_{i,j}; \mu_{i,j}, \sigma_{i,j}^2), \\ \mu_{i,j}^* &=& \mathbb{E}_q s_{i,j} \text{ and } \sigma_{i,j}^{2*} = \mathrm{Var}_q\, s_{i,j}. \end{aligned} \qquad (8)$$

Using this pseudo prior, the algorithm for determining the approximate likelihood terms $\phi(s_{i,j}; \mu_{i,j}, \sigma_{i,j}^2)$ is identical to that used by the TrueSkill rating system [1], with the scores $s_{i,j}$ taking the place of the 'player skills' in that system. We refer the reader to [1] for the specifics on the EP step.

After the EP step, we optimize the posterior approximation using Variational Bayes, i.e., we choose our posterior approximation to solve

$$\max_{q(\boldsymbol{U}, \boldsymbol{V}, \boldsymbol{b})} \mathbb{E}_q \log q(\boldsymbol{C}|\boldsymbol{S}) p(\boldsymbol{U}, \boldsymbol{V}, \boldsymbol{b}) - \log q(\boldsymbol{U}, \boldsymbol{V}, \boldsymbol{b}). \qquad (9)$$

This step can be implemented efficiently using the Variational Bayes Expectation Maximization (VBEM) algorithm. The resulting update equations can be found in [6]. However, note that for our application the expectations with respect to $s_{i,j}$ in (9) follow from $q(\boldsymbol{U}, \boldsymbol{V}, \boldsymbol{b})$ rather than from a separate posterior approximation on $s_{i,j}$, as is more commonly used (e.g. [6]). By avoiding this explicit approximation of $p(\boldsymbol{S}|\boldsymbol{C})$, the posterior approximation $q(\boldsymbol{U}, \boldsymbol{V}, \boldsymbol{b})$ gains in accuracy without increasing computational cost. The VBEM and EP steps are repeated until convergence.

3.3 Parallel Computation

For many real world applications of recommendation algorithms, both the number of users as well as the number of items is very large, necessitating the use of parallel computation to speed up inference. Both algorithms described above can be completely parallelized over users when updating $\boldsymbol{S}$ and $\boldsymbol{U}$, and over items when updating $\boldsymbol{V}$ and $\boldsymbol{b}$, which is an important advantage over the message passing algorithm used in [7]. Since in our application the number of items is quite small compared to the number of users, we found it most efficient to distribute the users and their feedback over multiple threads. Within each thread, $\boldsymbol{S}$ and $\boldsymbol{U}$ can then be updated without requiring any communication across threads. Every update of $\boldsymbol{V}$ and $\boldsymbol{b}$ then requires each thread to submit the sufficient statistics for the update of these variables and to receive the updated values. Since the number of items is relatively low this adds very little overhead and it allows us to speed up inference almost linearly with the number of available computation nodes.

4. LEARNING SUSHI PREFERENCES

In order to compare the new algorithms to existing methods we evaluate them on the sushi preference data of [3]. This data set was generated by asking 5,000 survey correspondents to order a subset of 100 sushi types according to their preferences. Each correspondent provided two such ordered lists containing 10 different sushi types. [3] evaluate their collaborative ranking approach by training on list 'B' and using the model to predict the order of list 'A'.

They measure the performance of their method by the average Spearman correlation between the predicted and realized ranking. We use this measure to compare the performance of the new method to the 'Nantonac' algorithm of [3], and also to compare our two inference algorithms against each other. Using this measure, we found that the maximum predictive accuracy was reached after about 1000 draws of the Gibbs sampler after a burn-in period of 100 draws, or after 50 iterations of the VB/EP algorithm. The corresponding results are shown in Table 1 below.

Table 1: Prediction accuracy of different methods on Sushi preference data

METHOD	SPEARMAN COR. TEST
NEW FACTOR MODEL, GIBBS	0.56
NEW FACTOR MODEL, VB/EP	0.54
NANTONAC [3]	0.49

The results in Table 1 show that the new method compares favorably to that of [3]: The Gibbs sampling version of the new algorithm improves the Spearman correlation of the predictions with the test set by 0.07 in comparison with the Nantonac method, while the deterministic posterior approximation gives an improvement of 0.05. The relatively small performance difference between the Gibbs sampling inference algorithm and the deterministic posterior approximation suggests that the latter is the more practical choice for real world applications, taking into account its benefits discussed in Section 3.2.

5. ACTIVE LEARNING

In order to improve our recommendations we may actively ask users to provide explicit feedback on certain items. This is most commonly done on an absolute rating scale, i.e. by asking the users to rate items. However, some studies indicate that people are better able to formulate their preferences in a relative way, by ranking multiple items, see e.g. [2]. Such relative preference statements can be used directly by the model presented in Section 2.

Asking the user for feedback is costly as it will take time for the user to think about his or her preferences. In addition, users may find it difficult to provide a full ranking of a very large list of items, so the the number of items we can enquire about is limited. When selecting this limited number of items we should take into account that not every item will be equally informative. A popular measure of the amount of information contained in a data point is the entropy reduction in our posterior distribution that we can expect upon conditioning on that data point [4]. By maximizing the expected entropy reduction in our posterior we can select the most informative items to present to the user for feedback.

Since the posterior distribution of the model given in Section 2 is not available in closed form, we cannot maximize the expected entropy distribution exactly. However, we can get an estimate of the amount of information in each possible observation by making use of posterior approximations. In doing so we will focus on the entropy reduction in the posterior distribution of the user parameters $q(\boldsymbol{u}_i)$, which – due to their greater number – are generally much more uncertain than the parameters of the items. To derive an expression for the approximate entropy reduction after obtaining a new observation, we assume a factorized posterior approximation over U, V, d and S, optimized using Variational Bayes. Note that this is not exactly the same as the approximation presented in Section 3.2, where we integrated out S in updating U, V and d. The Variational Bayes EM algorithm then uses the following update equation for the approximate posterior distribution on the user parameters:

$$q(\boldsymbol{u}_i) = N(\mu, \Sigma), \text{ with} \qquad (10)$$

$$\Sigma = \left[\frac{1}{\pi} I_K + \sum_j \mathbb{E}_q \left[\boldsymbol{v}_j \boldsymbol{v}_j' \right] \right]^{-1} \mu = \Sigma \left[\sum_j \mathbb{E}_q \left[\boldsymbol{v}_j \left(s_{i,j} - b_j \right) \right] \right]$$

The entropy of this approximate posterior distribution is given by

$$H(q(\boldsymbol{u}_i)) \propto 0.5 \log |\Sigma| . \qquad (11)$$

After adding a new item l to the ranking of the user we can update the approximate posterior distribution $q(\boldsymbol{u}_i)$ to $q'(\boldsymbol{u}_i) = N(\mu', \Sigma')$, while keeping $q(\boldsymbol{V}, \boldsymbol{d})$ fixed. The new entropy of $q'(\boldsymbol{u}_i)$ is then given by

$$\begin{aligned} H(q'(\boldsymbol{u}_i)) &\propto 0.5 \log |\Sigma'| = -0.5 \log |\Sigma^{-1} + \mathbb{E}_q \boldsymbol{v}_l \boldsymbol{v}_l'| \\ &\propto H(q(\boldsymbol{u}_i)) - 0.5 \log(1 + \mathbb{E}_q \boldsymbol{v}_l' \Sigma \boldsymbol{v}_l) \end{aligned} \qquad (12)$$

In order to maximize the information gain, or entropy reduction, we should thus ask the user to rank that item for which the parameter vector $\boldsymbol{v}_l$ has the highest expected Mahalanobis norm $\|\boldsymbol{v}_l\|_\Sigma$ with respect to the covariance matrix of the current posterior approximation. This has the effect of selecting items that are most informative for exactly those elements of the user vector $\boldsymbol{u}_i$ of which we are most uncertain. Note that for the approximate entropy (12) it does not matter what other item we compare the new item l to, or even whether we have a complete ranking with the new item or just a partial ranking. While this is obviously a very crude approximation, it still gives us a useful rule for actively selecting training examples as shown below.

We evaluate this active selection strategy using the sushi preference data, and we compare the resulting prediction accuracy with that obtained under random selection of the training examples. For each user the data set contains a training set ranking of 10 items of sushi. We actively select a subset of these items for each user by starting out with an empty selection set and subsequently adding that sushi item that minimizes the expected entropy in Equation (12). We then use the resulting selection of training examples to predict the ranking of the test set. For comparison, we do the same while selecting randomly from the remaining sushi items at each iteration. We display the accuracy of the resulting predictions for different numbers of selected items from a minimum of 3 to the maximum of 10. As can be seen from Figure 2 the active selection method leads to faster learning of the correct preferences than random selection of training examples.

Note that the performance measures of the two selection methods in Figure 2 converge as the number of training examples increases because both methods select from the same limited set of 10 potential examples. For small numbers of examples the performance of the active selection method improves much faster than under random selection, indicating the practical value of such an active learning strategy for real life applications, where the user typically only provides feedback on a relatively small fraction of items.

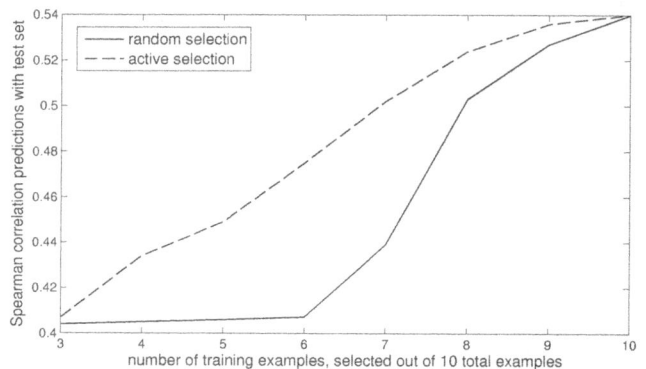

Figure 2: Prediction accuracy obtained using active versus random selection of training examples

6. CONCLUSION

We have proposed a Bayesian factor model to learn users' preference rankings for the purpose of product recommendation. Learning preference rankings with this model can be done quickly and efficiently at large scale, using the two inference algorithms we have developed. The accuracy of our model was demonstrated on a real world data set and was shown to improve upon existing methods. In addition, we have shown that the model can also be used effectively for active preference elicitation. By actively selecting product comparisons to present to the user, we can uncover the user's preferences without requiring large amounts of feedback. This makes the process of preference elicitation much less burdensome on the user, and it can dramatically improve prediction accuracy for real life applications.

7. REFERENCES

[1] P. Dangauthier, R. Herbrich, T. Minka, and T. Graepel. Trueskill through time: Revisiting the history of chess. *Advances in Neural Information Processing Systems*, 20:931–938, 2008.

[2] S. R. Jaeger, A. S. Jørgensen, M. D. Aaslyng, and W. L. Bredie. Best-worst scaling: An introduction and initial comparison with monadic rating for preference elicitation with food products. *Food Quality and Preference*, 19(6):579 – 588, 2008.

[3] T. Kamishima and S. Akaho. Nantonac collaborative filtering. In *Proceedings of The International Workshop on Data-Mining and Statistical Science*, pages 117–124, 2006.

[4] D. J. MacKay. Information-based objective functions for active data selection. *Neural Computation*, 4:590–604, 1992.

[5] T. P. Minka. Expectation propagation for approximate Bayesian inference. In *UAI '01: Proceedings of the 17th Conference in Uncertainty in Artificial Intelligence*, pages 362–369, 2001.

[6] U. Paquet, B. Thomson, and O. Winther. A hierarchical model for ordinal matrix factorization. *Statistics and Computing*, 21(3):1–13, 2011.

[7] D. H. Stern, R. Herbrich, and T. Graepel. Matchbox: large scale online Bayesian recommendations. In *Proceedings of the 18th international conference on World wide web*, pages 111–120, 2009.

Making Recommendations in a Microblog to Improve the Impact of a Focal User

Shanchan Wu
Department of Computer
Science, University of
Maryland
College Park, MD 20742
wsc@cs.umd.edu

Leanna Gong
University of Maryland
College Park, MD 20742
leannagong@yahoo.com

William Rand
Departments of Marketing and
Computer Science, University
of Maryland
College Park, MD 20742
wrand@umd.edu

Louiqa Raschid
University of Maryland
Institute for Advanced
Computer Studies
College Park, MD 20742
louiqa@umiacs.umd.edu

ABSTRACT

We present a microblog recommendation system that can help monitor users, track conversations, and potentially improve diffusion impact. Given a Twitter network of active users and their followers, and historical activity of tweets, retweets and mentions, we build upon a prediction tool to predict the Top K users who will retweet or mention a focal user, in the future [10]. We develop personalized recommendations for each focal user. We identify characteristics of focal users such as the size of the follower network, or the level of sentiment averaged over all tweets; both have an impact on the quality of personalized recommendations. We use (high) betweenness centrality as a proxy of attractive users to target when making recommendations. Our recommendations successfully identify a greater fraction of users with higher betweenness centrality, in comparison to the overall distribution of betweenness centrality of the ground truth users for some focal user.

Categories and Subject Descriptors

H.3.3 [**Information Storage and Retrieval**]: Information Search and Retrieval—*Retrieval models, Information filtering*

Keywords

Recommendation, social media, twitter, microblog

1. INTRODUCTION

The usage of social media has grown considerably in recent years, with microblogging sites being an important area of growth. On a site such as Twitter, one can follow a user and read their tweets. One can initiate a new conversation by tweeting or one can interact by mentioning a user. One can also participate in the diffusion

of a topic by retweeting. Influence in a microblog can be captured in multiple ways. One can generate a lot of content or befriend a lot of users but this may not lead to a large follower network or increase diffusion. Someone who has high betweenness centrality, whose tweets diffuse rapidly or widely outside her immediate follower network, or someone who is mentioned frequently by other users may, is typically considered influential. Other factors such as the level of sentiment or persuasiveness may also play a role in diffusion.

We are interested in analyzing both diffusion and influence in microblogs such as Twitter, from the individual or personalized perspective. We want to understand who will be influenced by a particular focal user. Given a Twitter network of active users and their followers, and their historical activity of tweets, retweets and mentions, we build upon a prediction tool that uses history to predict the Top K users who will retweet or mention a focal user, in the future [10]. Our objective is to make high quality personalized recommendations for each focal user.

Social network and social influence analysis has drawn a lot of research interest. Previous research on social influence had a focus on the measurement of social influence [2] or attempted to maximize user influence [3, 5] at the aggregate level. Our objective is to track those users who will likely be influenced by an individual focal user and to improve the impact of the focal user.

We identify characteristics of focal users such as the size of the follower network and the level of sentiment averaged over all tweets. We demonstrate that these features have an impact on the quality of personalized recommendations, i.e., accuracy of predictions. As the focal user's follower network increases, prediction accuracy decreases. In contrast, we can improve prediction quality for focal users with higher levels of positive sentiment. We note that the focal users with higher levels of positive sentiment appear to have a larger following. Despite a larger following having been shown to decrease prediction accuracy, we are nevertheless able to successfully recommend users who will retweet the more positive focal user (in the future ground truth) with greater accuracy.

We use (high) betweenness centrality as a proxy of attractive and potentially influential users to target when making recommendations. Our recommendations successfully identify a greater fraction of users with higher betweenness centrality, in comparison to the overall distribution of centrality among the ground truth users.

In summary, despite the difficulty of diffusion and influence prediction in evolving and noisy microblog networks, we have been successful in making personalized recommendations with improved accuracy for focal users with high(er) positive sentiment levels. We also are able to successfully recommend users with potentially greater influence (high betweenness centrality).

2. RECOMMENDATIONS IN A MICROBLOG

2.1 Problem Definition

DEFINITION 1. *Future Retweet Prediction: Given a focal microblog user u at a specific time point T and a time interval ΔT, identify K microblog users $S_{T,\Delta T}^u$ who will retweet one (or more) future tweet(s) of user u in the interval $(T, T + \Delta T)$.*

DEFINITION 2. *Future Mention Prediction: Given a microblog user u at a specific time point T and a time interval ΔT, identify K microblog users $S_{T,\Delta T}^u$ who will mention microblog user u one (or more) times in the interval $(T, T + \Delta T)$.*

2.2 Solution Approach

We formalize the retweet prediction and mention prediction problem to be a link prediction problem. Unlike traditional link prediction [7] over a homogeneous network, here we have an evolving hybrid network. In Twitter, there is an explicit Follower network, while Retweets and Mentions reflect a strong(er) relationship between users. These actions can be used to create communication-based networks independent of the follower network. In [10], we proposed and compared several methods to solve the previous two prediction problems. One of the best performing methods was based on a hybrid network and was labeled **WT-COM-BON**. The first step of **WT-COM-BON** is to create a composite weighted hybrid network with the best possible prediction potential as follows:

$$H = r \cdot R + m \cdot M + f \cdot F^*$$

R is the retweet network, M is the mention network, and F^* is a weighted follower network. In F^* a follower relationship is weighted inversely by the number of friends. If a user has a lot of friends, then her attention will be divided among those friends, and hence her weight for each friend should be lower.

The factors r, m, f are decided by a scale factor and a penalty factor. The scale factor scales the matrices so that no elements of any matrix can dominate the others. The penalty factor uses the ground truth from the training data to calibrate the influence of each network R, M, F^* with respect to retweet prediction and mention prediction. We note that the corresponding values of r, m, f, and hence H, are different for retweet prediction and mention prediction.

The next step is to apply the Bonacich centrality [1] metric to H. Bonacich centrality summarizes the total number of paths originating from a node to all other nodes; it uses an attenuation factor α to discount indirect links and β to discount direct links.

$$P = (\beta H + \beta \alpha H \cdot H + \ldots + \beta \alpha^n H^{(n+1)} \ldots)$$
$$= \beta H (1 - \alpha H)^{(-1)}$$

The ranking of the candidate users for future retweet and mention prediction is based on the rank of the values from the matrix P.

2.3 Example Recommendations

We illustrate using examples from a sample Twitter dataset; details of the dataset are provided later. We first consider users who are already widely retweeted. In our sample dataset, this includes

HumanCapLeague (Human Capital League), a blog community of workforce management professionals, Unibul, a credit card merchant company, and MomItForward, a social media site for women. A personalized recommendation for such users, would identify specific users who (1) are more likely to retweet or mention the focal user, and (2) are influential. Targets could include users with high betweenness centrality, or users with a high level of positive sentiment. The recommended action is for the focal user to tailor tweet content, hashtags or links to external pages, to the target users, so that their message is effectively communicated.

Another type of recommendation is based on users who are active and engaged on Twitter but whose tweets have not been widely or effectively diffused as yet. Zaibatsu is a single father of two kids, and he tweets about social media usage, technology, photos, and humor. Users Lizstrauss and TedRubin are both social marketing strategists. Though the target recommendations might be similar in choosing users that have a large follower network, or a high betweenness centrality, the suggested actions would be different. The recommendations would encourage them to mention or retweet the target users, rather than tailor tweets to the target users.

3. EXPERIMENTAL EVALUATION

3.1 Dataset and Metrics

There have been several successful efforts to construct a proxy graph that characterizes the structure of a real network [4, 6]. For this experiment, our objective was different. It was to construct a dataset that reflected a comprehensive history of user interaction and tweet content, over an extended period, for a significant number of active users, given the strict limitations imposed by the Twitter API. We constructed a network of 15,000 users, as well as all their follower (friend) associations within this subnetwork. In choosing these 15,000 users, we focused on active users. Our premise is that the active users generate the most content and have the greatest influence. Thus, following the a large number (15,000) of active users provided us with a dataset that captured a majority of the activity that would have had an influence on these 15,000 users. We note that had we constructed a 15,000 user dataset to reflect the typical distribution of users in the network, we may have been severely limited in our ability to capture a majority of the relevant activity since the average user on Twitter is not very active.

We used the Twitter API to construct the network in the following way: Starting from a seed *active* user, we expanded her follower network and added further active users until we reached 15,000 active users. The test for an active user was as follows based on their most recent 100 tweets: (1) The user should have an average minimum tweet frequency of one tweet per day in this time period. (2) There was at least one retweet in the most recent 100 tweets. the twitter streaming API to collect all tweets published by the 15K active users between April 25, 2011 and June 25, 2011.

For each user, we collected the total number of followers and total number of friends in the entire twitter social network of that user. As we only collected a subset of the users, some of the users may only have a small fraction of her friends or followers in the subset of the users. We used a threshold $X\%$ to filter out those users in the subset by the following way:

- First get a set of the users who has at least $X\%$ of friends and also at least $X\%$ followers from the 15K users. Label this set of users as S.
- Repeat the following loop until the number of users in S is stable, i.e., $|S|$ does not change:
For each user in S, if the number of her friends or the number

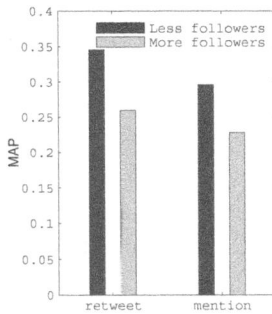

Figure 1: Prediction accuracy for the focal users with the number of followers less than and greater than the average number of followers of all focal users. The left part of the figure is for the focal users of retweet prediction; the right part is for the focal users of mention prediction.

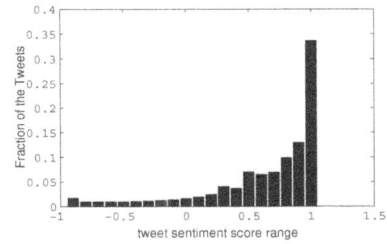

Figure 2: The distribution of the sentiment scores for all of the tweets.

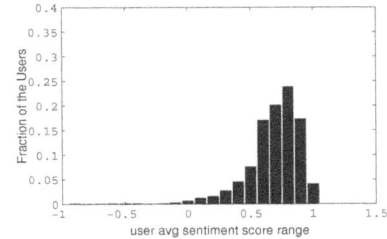

Figure 3: The distribution of the user sentiment scores in the training data for all of the users.

of her followers from S is less than $X\%$ of the total number of his friends or the total number of his followers, remove this user from S.
- Return the set of users S.

We set the threshold $X\% = 2.4\%$. Our crawling statistics shows that 40% of Twitter users were "active users", and we only collected "active users". So with this threshold, we got a subset of users with at least around 6% of their active followers and 6% of their active friends in the subset. We used the first month of our data (from April 25th to May 25th) as a training dataset and we used the second month data (from May 26 to June 25) as a test dataset where we obtained the ground truth. We picked the sets of microblog users who had ground truth in the test dataset for evaluation. 2728 users had retweet ground truth and 4571 users had mention ground truth. The average number of ground truth (retwitterers) for the 2728 users is 4.23, and the average number of ground truth (mentioners) for the 4571 users is 8.64.

The metric that we used for evaluation is MAP (Mean Average Precision). MAP is widely used for evaluating for ranking methods. We set the K value to be 20.

3.2 Impact of User Network

Networking features such as the count of friends and followers, both from the global counts registered on the Twitter profile, as well as the local counts computed in our dataset, were found to be highly significant when creating a model to explain variants of user behavior and the impact of diffusion effectiveness, as reported in [9]. The same holds true for the accuracy of future retweet and mention prediction.

Figure 1 reports on the prediction accuracy for the focal users whose total number of followers is less than, or is greater than, the average number of followers of the focal users. The left part of the figure is for retweet prediction and the right is for mention prediction. The figure demonstrates that it is more difficult to predict for focal users with a larger following. When a user has more followers, more people will potentially read their tweets and retweet or mention her in the future. Some of the future users will be novel users who did not retweet her in the past. Both cases increase the difficulty of prediction.

3.3 Impact of Sentiment

Sentiment has also been widely identified as an important factor of influence and diffusion. We used a dataset and tool [8] trained

for sentiment detection in tweets. In the training dataset, tweets containing positive emoticons like ":)" but not negative emoticons were labeled as positive, and tweets containing negative emoticons like ":(" but not positive emoticons were labeled as negative. A Naive Bayes classifier (NBC) was constructed using the sentiment training dataset of 232K negative tweets and 232K positive tweets. We then used the NBC to classify our training dataset to assign a sentiment score to each tweet, in the range of [-1, +1]. Finally, we averaged the sentiment score over all the tweets of a user to determine a level of sentiment. Figure 2 reports on the distribution of the sentiment scores for each of the tweets of our dataset. Figure 3 reports on the distribution of the user sentiment level computed over all the tweets of each user.

Figure 4 reports on the comparison of focal users with a sentiment level less than, and greater than, the average sentiment level of the focal users, for retweet prediction. Figure 4(a) compares the prediction accuracy while Figure 4(b) presents the number of followers. Figure 4(b) shows that users with a more positive sentiment level are more likely to attract a larger follower network. We have shown in a previous result, that it is more difficult to predict for focal users with more followers. However, for retweet prediction, Figure 4(a) shows that we can predict future users for focal users with more positive sentiment, with higher prediction accuracy. For example, for very positive focal users with user sentiment level > 0.9, the MAP value for retweet prediction is 0.395. In contrast, for very negative focal users with user sentiment score < 0.2, the MAP value for retweet prediction has reduced drastically is 0.253.

3.4 Impact of Centrality

The betweenness centrality of a node v in a network is defined by the expression:

$$g(v) = \sum_{s \neq v \neq t} \frac{\sigma_{st}(v)}{\sigma_{st}}$$

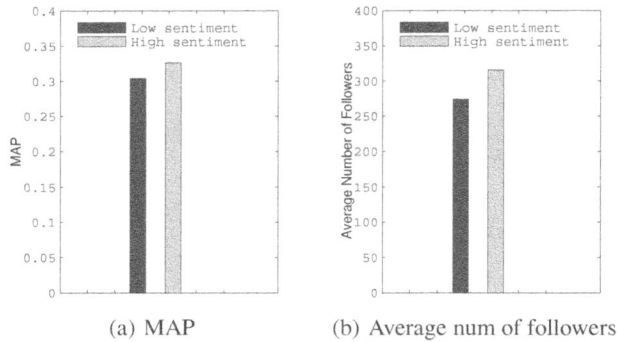

(a) MAP   (b) Average num of followers

Figure 4: Comparison of the focal users with the user sentiment scores less than and greater than the average user sentiment score of the focal users for retweet prediction.

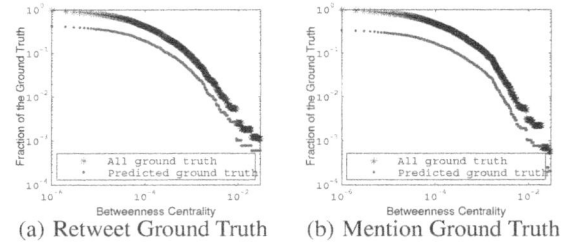

(a) Retweet Ground Truth   (b) Mention Ground Truth

Figure 5: Follower network betweenness centrality distribution of the ground truth users. X axis is the follower network betweenness centrality values; Y axis is the fraction of the users that have betweenness centrality values greater than or equal to the corresponding value on X axis. The upper darker distribution represents all ground truth; the lower lighter distribution is the predicted ground truth.

where σ_{st} is the total number of shortest paths from node s to node t and $\sigma_{st}(v)$ is the number of those paths that pass through node v.

We calculated the betweenness centrality of each user using the follower network in our dataset. We want to evaluate how well our recommended target users are also users with a high betweenness centrality, so that our recommendations are more valuable.

Figure 5 reports on the follower network betweenness centrality distribution of all the ground truth users and the subset that can be predicted by our system. The figures were drawn on the log-log scale and the distribution is somewhat close to a power law distribution. The range of the betweenness centrality values for all users in the network is [0,0.06553]. First, we consider users with low betweenness centrality in the range of [0,0.00002). While 28.9 percent of the retweet ground truth and 24.3 percent of the mention ground truth is in that range, the corresponding values for our target recommendations are 27.7% and 21.7% respectively, i.e., we make a lower fraction of our recommendations in this range.

When the betweenness centrality value increases, the curve of the predicted ground truth is closer to the curve of all ground truth for both retweet prediction and mention prediction. Thus, when we consider users with high betweenness centrality, in the range (0.001,0.06553], we see the opposite effect. A higher fraction of target recommendations is in that range. While 8.2 percent of the retweet ground truth and 12.0 percent of the mention ground truth is in that range, we recommend 8.6% and 12.4% respectively. To summarize, we are successful in recommending users in the retweet and mention ground truth that have a higher betweenness centrality.

4. CONCLUSIONS

We present results on a recommendation system for microblogs. We make recommendations of future retweet and future mention users; we also recommended suggested actions including tailored messages or targeted interactions with users. We show that focal users with higher levels of positive sentiment are associated with a larger follower network and a larger follower network typically reduces the accuracy of our predictions. Nevertheless, we are successful and can make recommendations for focal users with higher levels of positive sentiment with greater prediction accuracy. Our recommendations target future ground truth users with high betweenness centrality values. Those users are potentially more influential. The reason that we are able to identify users with high betweenness centrality values is because our solution is based on a composite network. The users with high centrality values are more

likely to receive all tweets in the system; this increases their likelihood of appearing in both the mention and retweet network. Thus, our prediction method that exploits the composite hybrid network is more likely to identify these more influential users.

Acknowledgements: This research was partially supported by NSF awards CMMI 0753124, IIS 0960963, and IIS 1018361.

5. REFERENCES

[1] P. Bonacich. Power and centrality: A family of measures. *The American Journal of Sociology*, 92(5):1170–1182, 1987.

[2] M. Cha, H. Haddadi, F. Benevenuto, and K. Gummadi. Measuring user influence in twitter: The million follower fallacy. In *ICWSM '10*.

[3] W. Chen, C. Wang, and Y. Wang. Scalable influence maximization for prevalent viral marketing in large-scale social networks. In *Proceedings of the International Conference on Knowledge Discovery and Data Mining (SigKDD)*, 2010.

[4] M. Gjoka, M. Kurant, C. T. Butts, and A. Markopoulou. Walking in facebook: a case study of unbiased sampling of osns. In *INFOCOM'10*, pages 2498–2506, 2010.

[5] D. Kempe, J. Kleinberg, and E. Tardos. Maximizing the spread of influence through a social network. In *KDD '03*, pages 137–146, 2003.

[6] J. Leskovec and C. Faloutsos. Sampling from large graphs. In *Proceedings of the 12th ACM SIGKDD international conference on Knowledge discovery and data mining*, KDD '06, pages 631–636, 2006.

[7] D. Liben-Nowell and J. Kleinberg. The link prediction problem for social networks. In *CIKM '03: Proceedings of the twelfth international conference on Information and knowledge management*, pages 556–559. ACM, 2003.

[8] D. Monner. Tweet sentiment computation. *National Science Foundation SM3 Project Wiki, Smith School of Business*, 2011.

[9] P. Swaroop, Y. Joshi, W. Rand, and L. Raschid. Modeling behavior and its effect on diffusion effectiveness in microblogging. *Technical Report, Smith School of Business*, 2012.

[10] S. Wu and L. Raschid. Prediction with microblogs: Learning in a hybrid network. In *Technical Report, University of Maryland*, 2012.

The Influence of Knowledgeable Explanations on Users' Perception of a Recommender System

Markus Zanker
Alpen-Adria-Universitaet Klagenfurt
Universitaetsstrasse 65-67
Klagenfurt, Austria
mzanker@acm.org

ABSTRACT

Recommender Systems (RS) help online customers in identifying those items from a variety of choices that best match their presumed needs and preferences. In this context explanations summarize the reasons why a specific item is proposed and are capable of increasing the users' trust in the system's results. This paper presents results from an online experiment on a real-world platform indicating that explanations are an essential piece of functionality of a recommendation system, that significantly increases users' perception of the utility of a recommender system, the intention to use it repeatedly as well as the commitment to recommend it to others.

Categories and Subject Descriptors

H.5.m [**INFORMATION INTERFACES AND PRESENTATION**]: Miscellaneous

Keywords

Knowledge-based recommendation, Explanations

1. INTRODUCTION

Recommendation systems support users in managing the information overload in virtual spaces by ranking and filtering items [6, 10]. Explanations provide users additional information *about* the recommendations themselves with the intent of pursuing specific objectives [3]. According to Tintarev and Masthoff [12] these explanation objectives can be manifold such as increasing a system's trustworthiness, leveraging users' confidence in decision making or being persuasive [14].

Since one of the earliest studies [5] on users' preferences for explanations of recommendations considerable interest on this topic has been stirred as for instance summarized in [3]. The purpose of this study is to research and quantify the impact of the explanation feature on the users' overall appreciation of the recommendation system itself. Notably it was of interest if users perceive a recommendation system

differently if it provides explanations compared to no explanations on the result page. In short, we hypothesized that explaining recommendations makes a significant difference in users' perception of the utility of a recommendation system and that it has a positive impact on users' interaction experience, their intention to use the system more often and to recommend it to others.

The object of study was a real tourism recommender system that generated knowledgeable explanations, i.e. explanations that provide natural language arguments that mediate between user characteristics (or the user model) and item properties. For instance the system could state: *"We think you will enjoy this resort with your family due to the available childcare facilities and the very moderate prices"* due to knowledge about the user (family) the item (childcare available, moderate cost level) and some additional domain knowledge (families with children typically have more stringent budget restrictions than for instance double income couples). More details on the mechanism for generating recommendations can be found in [15].

Next we will explain the applied evaluation methodology, detail our research hypotheses and results and finally conclude with a discussion.

2. METHODOLOGY

As the purpose of this evaluation is the confirmation of a causal relationship between the provision of knowledgeable explanations and the users' perception of the system, a *between users online experiment* design was instrumented. The test was conducted on a real-world platform that is Central Europe's most comprehensive information platform about spa resorts [1] that offers its users detailed information and multiple options for searching, browsing and comparing different offers. All users accessing the platform are randomly assigned into group A (no explanations) or group B (knowledgeable explanations) which was automatically measured by a binary variable denoted *explanation*. Over 200 users participated in the experiment and participation was promoted by a banner on the platform that invited users to participate. Students had been encouraged by email to visit the platform, use the RS and participate in the survey. Participants can win prizes of a total worth of several hundred Euros in vouchers for those spa resorts that users nominated as their favorites in the questionnaire. We limited our analysis to those participants who had really used the recommender system before completing the survey, thus

---

[1]See http://www.thermencheck.com

Nr.	Question	Variable
	Perceived Usefulness	
1	The RS increases the quality of my search	Search quality [2]
2	The RS gives me more control over my search	Control [2]
3	The RS saves me time	Time [2]
4	The RS makes it easier for me to find the best matching spas	Convenience [2]
5	The RS provides me better search results I otherwise would find	Result quality [2]
6	The RS addresses my needs	Needs [2]
7	Overall, the RS is useful for my purposes	Utility [2]
8	The RS's search results are personally relevant and interest me beyond this test	Involvement [16]
	Cronbach's alpha for stand. items	**0.907**
	Perceived ease of use	
9	The RS was easy to use	Ease of use [1]
10	I think the RS will also be easy to use for others	Ease of use others [1]
	Cronbach's alpha for stand. items	**0.918**
	Trust	
11	I trust in the information that I receive from the RS	Trust information [16]
12	I do not worry about the use of my input data to the RS	Trust data use [16]
	Cronbach's alpha for stand. items	**0.41**
	Commitment and Experience	
13	It was exciting to use the RS	Experience [16]
14	I would recommend the RS to someone else	Recommendation [16]
15	I think that I would like to use the RS more often	Intention to use [16]
	Cronbach's alpha for stand. items	**0.841**

Table 1: Constructs and their measurement

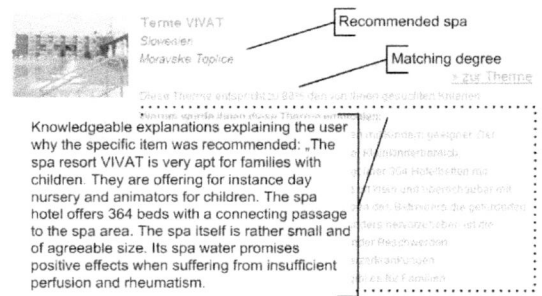

Figure 1: Explanation on result page

number of used cases is 134 out of which 74 had been assigned to group A and 60 to group B.

The platform includes the conversational recommender *Aquarius* that asks users about their preferences and retrieves those spas from a catalogue that matches these requirements according to their descriptions and a knowledge base that mediates between preferences and item descriptions. Recommendations are ranked according to a matching degree from an interval 1 to 100% that indicates the share of weighted preferences that are fulfilled by a specific recommendation. In addition, users who have been assigned to group B will also receive knowledgeable explanations on the result page (see Figure 1) while users in group A receive no explanations at all. However, results and their rankings have been the same in both groups in case of equal user input to the conversational RS. The task of users participating in the experiment was to identify their favorite spa resorts by interacting with the recommender. Then they

completed a post-treatment survey that measured different constructs such as users' perception of the system's usefulness and its ease of use as being the main determinants of the usage intention according to the technology acceptance model (TAM) [2]. Furthermore, the users' trust in the recommender system as well as their intention to use it repeatedly, positive usage experience and their willingness to recommend it to others were measured comparable to the technology acceptance study for recommender websites in tourism by Bauernfeind and Zins [16]. Furthermore, exogenous factors like internet familiarity, web technology experience, experience in visiting thermal spas and frequency of visiting spas, demography (education, gender and age) as well as how they became aware of this survey (invitation by mail, banner on the website) were controlled. None of these variables varied significantly with respect to the explanation variable.

In detail, the measured constructs and their associated variables[2] are enlisted in Table 1. Eight variables measure the construct usefulness as it is perceived by the user, i.e. *the degree to which a person believes that using a particular system would enhance his or her job performance* [2]. The items are based on the scale developed by Davis who showed that the perceived usefulness highly correlates with the actual usage of the system.

The construct *Perceived ease of use* measures *the degree to which a person believes that using a particular system would be free of effort* [2]. For that purpose we used two items from Brooke's system usability scale (SUS) for quick usability evaluations in industry [1]. For both constructs, *Cronbach's alpha* was over 0.90 indicating a high reliability of the items measuring the construct.

The third measured construct is *Trust* in the RS that should measure how the users perceive security aspects of the RS with respect to the information presented (variable *Trust information*) and the processing of the information acquired from the user (variable *Trust data use*). Although, both items have already been used in the context of TAM by [16], they turned out to be not very reliable items here (Cronbach's alpha = 0.41). The reason will be explicated in the context of the dimensionality reduction using principal component analysis, where it becomes obvious that the variable *Trust information* also strongly loads on the factor representing the perceived usefulness of the system that presents "information that can be trusted". Furthermore, we asked users about their *Experience and Commitment* where

---

[2]Some of the measurement items have already been employed in earlier work [7].

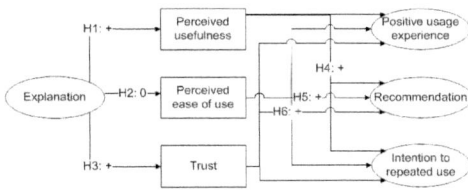

Figure 2: Hypotheses

a positive user experience, the intention to repeatedly use the RS as well as the commitment to recommend it to others is measured. Answers to variables 1-15 are encoded on a 5-point Likert scale ranging from *I strongly disagree (1)*, *I disagree (2)*, *I neither agree nor disagree (3)*, *I agree (4)* to *I strongly agree (5)*. In Figure 2 we depict our hypotheses that we want to confirm. Ellipses represent single variables while rectangles represent constructs. In short, we hypothesize that the explanation feature increases perceived usefulness (H1) and trust (H3) in the recommendation system, while the usability of the system (H2) should not be affected by explanations on the result page. Perceived usefulness (H4), ease of use (H5) and trust (H6) should positively correlate with usage experience as well as intention to repeated use and recommend to others. Next, we will validate the hypotheses and present detailed results.

3. RESULTS

In a first step, we applied factor analysis on variables 1 to 12 to reduce the dimensionality of the data and to assess the factorial validity of constructs in Table 1 using principal component extraction. The resulting three factors (F1 to F3) correspond well to the constructs perceived usefulness (PU), perceived ease of use (PeoU) and trust as depicted in Figure 2.

Variable	F1 PU	F2 PeoU	F3 Trust
Search quality	**.690**		
Control	**.612**		
Time	**.734**		
Convenience	**.676**		
Result quality	**.766**		
Needs	**.797**		
Utility	**.855**		
Involvement	**.602**		
Ease of use	.575	**.717**	
Ease of use others	.570	**.719**	
Trust information	**.569**		.498
Trust data use			.557

Note: Loadings < .4 suppressed.

Table 2: Principle component analysis

As can be seen from Table 2 Factor 1 is very consistent with the items 1-8 of construct perceived usefulness and explains 43.5% of total variance. Factor 2 is also highly correlated with items 9 and 10 thus representing the construct perceived ease of use and additional 12.5% of variance. The third factor is loaded by variables 11 and 12, however 11 is also highly correlated with Factor 1, as question 11 can

also be interpreted to support PU. In total, all three factors extracted 64.5% of the total variance.

Factor	Groups	SS[1]	df	Mean Squares	F	Sign.
Factor 1 (PU)	Between	10.52	1	10.520	11.443	.001*
	Within	108.48	118	.919		
	Total	119	119			
Factor 2 (PeoU)	Between	.061	1	.061	.061	.806
	Within	118.939	118	1.008		
	Total	119	119			
Factor 3 (Trust)	Between	2.425	1	2.425	2.455	.120
	Within	116.575	118	.988		
	Total	119	119			

[1] Sum of Squares

Table 3: ANOVA of explanations on Factors

Next, we apply oneway ANOVA in order to evaluate the impact of the knowledgeable explanations on these three factors as summarized in Table 3. Based on the F-statistics it becomes obvious that only Factor 1 representing the perceived usefulness of the recommendation system is significantly impacted by the provision of knowledgeable explanations ($F_{1,118} = 11.443, p = .001$). Neither the perceived ease of use nor the factor trust vary significantly ($\alpha < 5\%$). The strength of the impact of providing knowledgeable explanations on the result page of the RS on the factor representing the construct perceived usefulness is a moderate one. It is computed by $\eta^2 = \frac{SS_{between}}{SS_{total}} = .088$ that represents the fraction of variance between both groups from total variance. Thus hypotheses H1 and H2 can be confirmed. H3 is not confirmed.

Factor	Experience	Recommendation	Int.[1]
Factor 1 (PU)	.512***	.718***	.619***
Factor 2 (PeoU)	-.008	-.062	-.037
Factor 3 (Trust)	.175*	.158*	.242**

[1] intention to use
* significant with $\alpha < .05$
** significant with $\alpha < .01$
*** significant with $\alpha < .001$

Table 4: Correlation matrix between Factors and Experience and Commitment variables

Furthermore, the impact of the three factors on the users' experience and commitment is measured by giving the Pearson correlation coefficient. As can be seen from Table 4 Factor 1 representing the perceived usefulness strongly correlates with the variables on positive experience as well as commitment to recommend the system to others and to use it repeatedly (accept hypothesis H4) and thus replicates many of the earlier findings on the acceptance of information technology [13]. However, hypothesis H5 has to be rejected as Factor 2 (PeoU) does not show a significant positive correlation with any of the three variables. Finally, H6 has to be accepted, although the correlation between Factor 3 constituting the construct trust is rather low. In the next section we will discuss these findings in more detail.

4. DISCUSSION

The main result of this study is that *knowledgeable explanations significantly increase the perceived usefulness of a recommender system* (H1). This has been shown by an online experiment on a real-world platform involving anonymous users of this website, while up to now almost all evaluation studies about explanations for recommender systems have been conducted on specific research platforms [5] or prototypes [11]. Although explanations on the result page constitute only one of the many technical features and aspects of a recommender system it is an important result that explanations still have a significant impact on the system's perceived usefulness. Furthermore, the study showed that the system's explanation facility is not perceived to make the use of a recommendation system easier (H2). However, if the evaluated system would offer its users tradeoff analysis and critiquing-based navigation functionality like done by [9] then PeoU could be impacted. [16] identified the construct Trust to be a strong influential factor that highly correlates with satisfaction and commitment with recommender websites. However, according to the rejected hypothesis H3 the trust in a recommender system itself - in contrast to a recommender website like *Expedia* or *Active Buyers Guide* evaluated in [16] - is not impacted by the explanation variable.

Furthermore, the perceived usefulness and thus indirectly also the knowledgeable explanations positively impact the users' experience in using the recommendation systems as well as their commitment to repeatedly use it and recommend it to others. This strong interaction between the factor PU and the experience and commitment variables has been shown by strong and highly significant correlation coefficients (H4). Although the perceived ease of use of information technology traditionally influences its adoption rate [13] it is one of the findings in this study that a correlation between factor PeoU and the experience and commitment variables cannot be confirmed, i.e. H5 was rejected. One of the reasons for this could be that recommender systems are focusing on easing information access and are therefore unanimously perceived as very easy to use. This can be substantiated by the observation that the empirical mean of both 'ease of use' variables (i.e. numbers 9 and 10 in Table 1) was > 4.4 and thus the highest among all 15 measured variables on a 5-point Likert scale. Furthermore, hypothesis H6 was only weakly confirmed and the construct trust has only a low correlation with the experience and commitment variables. This contrasts the findings in [16], where trust puts the highest loadings on website satisfaction in their structural model. However, the difference can be explained by the fact that [16] researched recommender websites in general and not in particular recommender systems as a tool. Therefore, Bauernfeind and Zins' study actually researches online experiences with websites in general where trust is an important prerequisite for satisfaction and commitment as also discussed for instance by [4]. However, when focusing specifically on recommender systems as a tool for search and retrieval as done in this study then trust is less important because users do not use it in order to conduct commercial transactions in a narrow sense with it.

5. CONCLUSIONS

This paper neglects accuracy aspects of recommendation systems but evaluated further information systems objectives such as usefulness and trust as postulated by [8]. The findings show that the explanation feature can significantly increase a recommender system's perceived usefulness and thus contributes to increase users' repeated usage intention and their commitment to recommend the service.[3]

6. REFERENCES

[1] J. Brooke. *Usability Evaluation in Industry*, chapter SUS: a quick and dirty usability scale, pages 189–194. Taylor & Francis, 1996.

[2] F. D. Davis. Perceived usefulness, perceive ease of use, and user acceptance of information technology. *MIS Quarterly*, 13(3):319–340, 1989.

[3] G. Friedrich and M. Zanker. A taxonomy for generating explanations in recommender systems. *AI Magazine*, 32(3):90–98, 2011.

[4] S. Grabner-Kräuter and E. A. Kaluscha. Empirical research in on-line trust: a review and critical assessment. *Intl. Jrnl of Hum.-Comp. Studies*, 58(6):783–812, 2003.

[5] J. L. Herlocker, J. A. Konstan, and J. Riedl. Explaining collaborative filtering recommendations. In *ACM CSCW'00*, pages 241–250, Philadelphia, Pennsylvania, USA, 2000. ACM.

[6] D. Jannach, M. Zanker, A. Felfernig, and G. Friedrich. *Recommender Systems - An Introduction*. Cambridge University Press (CUP), 2010.

[7] D. Jannach, M. Zanker, and M. Fuchs. Constraint-based recommendation in tourism: A multi-perspective case study. *Information Technology & Tourism*, 11(2):139–156, 2009.

[8] P. Pu, L. Chen, and R. Hu. A user-centric evaluation framework for recommender systems. In *ACM RecSys'11*, pages 157–164, 2011.

[9] P. Pu and L. Chen. Trust-inspiring explanation interfaces for recommender systems. *Knowl.-Based Syst.*, 20(6):542–556, 2007.

[10] F. Ricci, L. Rokach, B. Shapira, and P. B. Kantor, editors. *Recommender Systems Handbook*. Springer, 2011.

[11] N. Tintarev. Explanations of recommendations. In *ACM RecSys'07*, pages 203–206, Minneapolis, MN, USA, 2007. ACM.

[12] N. Tintarev and J. Masthoff. Designing and evaluating explanations for recommender systems. In Ricci et al. [10], pages 479–510.

[13] V. Venkatesh, , M. G. Morris, G. B. Davis, and F. D. Davis. Perceived usefulness, perceive ease of use, and user acceptance of information technology. *MIS Quarterly*, 27(3):425–478, 2003.

[14] K.-H. Yoo, U. Gretzel, and M. Zanker. *Persuasive Recommender Systems: Conceptual Background and Implications*. Springer, 2012.

[15] M. Zanker and D. Ninaus. Knowledgeable explanations for recommender systems. In *Int. Conf. on Web Intelligence*, pages 657–660, 2010.

[16] A. Zins and U. Bauernfeind. The perception of exploratory browsing and trust with recommender websites. *Information Technology & Tourism*, 8(2):121–136, 2006.

---

[3]We acknowledge the support of New Mediacheck GmbH, Daniel Ninaus and Michael Sperdin for this study.

Social Referral: Leveraging Network Connections to Deliver Recommendations

Mohammad S Amin
LinkedIn, California
mamin@LinkedIn.com

Baoshi Yan
LinkedIn, California
byan@linkedin.com

Sripad Sriram
LinkedIn, California
ssriram@linkedin.com

Anmol Bhasin
LinkedIn, California
abhasin@linkedin.com

Christian Posse
LinkedIn, California
cposse@linkedin.com

ABSTRACT

Much work has been done to study the interplay between recommender systems and social networks. This creates a very powerful coupling in presenting highly relevant recommendations to the users. However, to our knowledge, little attention has been paid to leverage a user's social network to deliver these recommendations. We present a novel approach to aid delivery of recommendations using the recipient's friends or connections. Our contributions with this study are 1) A novel recommendation delivery paradigm called *Social Referral*, which utilizes a user's social network for the delivery of relevant content. 2) An implementation of the paradigm is described in a real industrial production setting of a large online professional network. 3) A study of the interaction between the trifecta of the recommender system, the trusted connections and the end consumer of the recommendation by comparing and contrasting the proposed approach's performance with the direct recommender system.

Our experiments indicate that Social Referral is a promising mechanism for recommendation delivery. The experiments show that a significant portion of users are receptive to passing along relevant recommendations to their social networks, and that recommendations delivered through users' social networks are much more likely to be accepted than those directly delivered to users.

Categories and Subject Descriptors

J.4 [**Computer Applications**]: Social and behavioral sciences

Keywords

Social referral, recommender systems, social network

1. INTRODUCTION

Social Recommender Systems incorporate social ties to either a) Augment relevance of the presented results [4] b) To

explain away the recommendation to the user by presenting social proof [3] or a combination thereof. These ideas are quite powerful in driving relevant engagement from the end user. However, not much attention has been paid in RecSys literature to leverage social ties to *deliver* recommendations. In this scenario a user's connection becomes the channel to route content to the incumbent. The aspects that come into play in this setting are 1) Connection strength or a measure of trust between the users on either side of the connection and 2) The reputation or topical social influence of the user.

Both of these aspects have been studied extensively [5]. We propose leveraging these ideas along with a traditional function of the recommender system i.e. recommending relevant content to the user to generate a *delivery* paradigm via the 'referral' mechanism. Consider how a content provider usually pushes recommendations to target users - They directly deliver relevant products/articles to users via email or online notification streams such as *Facebook Wall*. The interaction of the users' social network with this item is primarily used as a social proof to enhance the credibility of the recommendation. 'Social Referral' is the following alteration to this model : Upon a user's (actor) engagement with a piece of content (say movie, news article or music piece) we surface from their network, other users who will potentially be interested in this content. In this model the content provider does not directly engage with the target user, rather, they nudge the actor to refer this content to their social ties. The selection process of the referral recipients is guided by a function of 1) user-item relevance obtained from the traditional recommender systems and 2) user-user connection strength unveiled from the social network. In this fashion, network ties become effective channels to deliver peer endorsed recommendations i.e. *social referrals*. We describe the properties and efficacy of our system in the context of industrial recommender system for recommending professional groups to end users in a large online professional network. The architecture proposed here is generic and can be applied to any other social settings as well.

2. RELATED WORK

Much work has been done in incorporating Social signals in Recommender Systems. In [4] Kautz et. al. proposed a method to integrate social network with collaborative filtering. It has been shown in [1, 2] that incorporating social network information in traditional recommender improves the outcome of the system. Guy et al.[3] introduced the concept of aggregated familiarity relationships to recommend people to connect to within an enterprise SNS.

From a different perspective, Lerman [7] studied the effect of user recommended stories as opposed to stories recommended by a CF based system and identified that the users are more interested in the former. In a similar study [8], the authors compared movie and book recommendations from friends with online recommender systems and found that incorporating the source of the recommendation introduced a probable bias in favor of the former.

In social science the concept of adoption of ideas and technologies in the social network has been studied [6]. Algorithmic studies to identify influential people in the social network and how they instigate subsequent adoption has also been carried out in [5]. However, any data driven method to deliver recommendation via early adopters and aid them to choose potential followers has not fully been explored. In this paper, we present a system that leverages propagation of relevant group recommendation via group members and aids them to selectively identify a subset of their connections for group recommendation. The selection process takes into account the probable affinity of the potential member to the group and hence, the recipients are more likely to accept the recommendation.

3. PRELIMINARIES

We study Social Referral in the context of professional groups in a large online professional network. A recommendation system has been incorporated on this website that periodically calculates the relevance score of a group for a member. This score is subsequently used to recommend new groups for a member. Members can also share a group that he is a member of to his connections. However, group shares necessitate manual selection of connections which may lead to inclusion of members who may not have the same level of interest for the group or exclusion of potentially interested members. In order to bridge the gap between the group recommendation and group share, and aid members to select connections to share a group with, a seamless integration of these two sources of information is warranted.

Let U $=\{u_1, u_2, \ldots, u_n\}$ be the set of users, G $=\{g_1, g_2, \ldots, g_m\}$ be the set of groups. Let $\eta : u_i \to \hat{U} \subseteq U$ be a function that returns the neighbors of $u_i \in U$ and $\tau : u_i \times g_j \to \mathbb{R}$ be the relevance score of group g_j for user u_i. Then for a given k the objective here is for a 2-tuple (u_i, g_j) to identify $\hat{N} \subseteq U$ such that $\forall u_k \in \hat{N}, \tau(u_k, g_j) \geq \lambda \wedge u_k \in \eta(u_i) \wedge |\hat{N}| \leq k \wedge \forall s(s \in (\eta(u_i) \setminus \hat{N})(\tau(u_k, g_j) > \tau(s, g_j))$, where λ is a predefined threshold. Thus for each member u_i, who we refer to as referrer from here on, and each group g_j, we generate a set of recommendation of top-k target users, $\hat{N}$ from u_i's neighbors, having a minimum relevance score of λ. Figures 1(a), 1(b) delineate the process. After the referrer joins a group, he is prompted to recommend the group to a subset of his network as shown in 1(a). This subset of potential target users are pre-calculated by the system and they are guaranteed to have a minimum level of affinity toward the group. Subsequently, each of the connections receive a personal group recommendation message from the referrer (Figure 1(b)).

4. METHOD

The current group recommendation framework on this website takes into account both content and social aspect

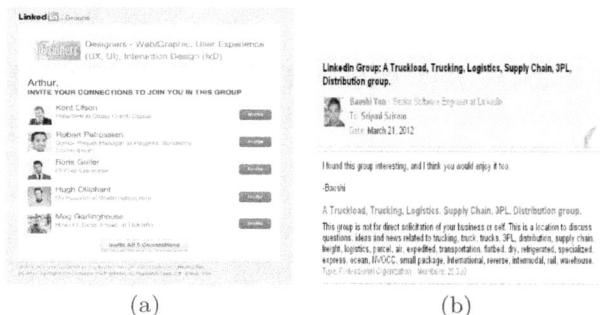

(a)   (b)

Figure 1: (a) Emails sent to referrers, (b)Emails sent to target users

for recommendation purposes. We piggyback off of this framework to obtain user-group affinity scores. This step guarantees that when a group subscriber refers a group to one of his connections, the connection is highly likely to be interested in the group. We also compute the connection strength between members. Since the calculation of connection strength is an orthogonal issue and can be done in a variety of ways, we use an existing in-house algorithm that, given a pair of members, returns the connections strength among them.

From an abstract level, for each (referrer, group) pair, the method searches the neighborhood of the referrer's social network and extracts group affinity scores for those neighbors for the group in question. The potential target users are then reordered based on group relevance score and their connection strength with the referrer.

4.1 Selecting target users

Each (referrer, group) pair can contain multiple target users that can be referred. However, the content of the group may not be equally relevant to all the target users. Analogously, the relationship between the referrer and the target user is also a salient attribute that contributes to the referral operation. We can recommend the target users with strongest connection strength in a members network to refer for a particular group. Alternatively, we could recommend the target users with highest recommendation score, or employ a hybrid approach. Since for each (u_i, g_j) pair we obtain a subset, $\hat{N}$ of member u_i's connections, we can obtain two different ordered list from $\hat{N}$ i.e. $\pi_1 = (\hat{u_{11}}, \hat{u_{21}}, \ldots, \hat{u_{|\hat{N}|1}})$, where $m < n \to \sigma(u_i, u_{m1}) > \sigma(u_i, u_{n1})$ and $\pi_2 = (\hat{u_{12}}, \hat{u_{22}}, \ldots, \hat{u_{|\hat{N}|2}})$, where $m < n \to \tau(\hat{u_{m2}}, g_j) > \tau(\hat{u_{m2}}, g_j)$. Here $\sigma(u_i, \hat{u_{m1}})$ stands for the connection strength between u_i and $\hat{u_{m1}}$. Subsequently, we can target top-k members from either of the two lists.

4.2 Identifying the most relevant group

For each referrer, there might exist multiple groups he/she could refer to his network connections. Instead of flooding the referrer with candidates for all these groups, we have chosen to select one group per referrer to recommend target users for based on different criteria. One criterion is group quality. We generate the quality of a group as a function of recent activity pertaining to that group. We generate the group quality score using the following equation: $Q(g_i) = \frac{ln(1+c+p)}{ln(\alpha+N)}$ Where c, p and N stands for the number of unique commenters, number of unique posters and the total number of members in the group and α represents a user defined parameter. This measure is generated for all the active groups considering their recent activity. The parameter α smooths out the effect of group size with respect to their activity. Alternatively, for each group, we generate a set of potential recommendation target $\hat{N}$ for a specific member , we get: $R(u_i, g_j) = \frac{\sum \sigma(u_k)}{|\hat{N}|}$ where $\forall u_k \in \hat{N}$, $\sigma(u_i, u_k)$ represents the connections strength between u_i and u_k. This measurement depicts the relevance of a group in terms of the average connection strength between the referrer and the target users. We have used both scores to select the group to recommend (named Better_GRP and Stronger_CONN respectively). Again, a combination of these two measures was also employed to compare performance: $RQ(u_i, g_j) = ln(Q(g_j)) + ln(R(u_i, g_j))$. Thus for each referrer, we choose the group with the highest score calculated by any of the three measures defined above and select from the referrer's connections, top-k potential recommendation targets.

5. EXPERIMENTS

Our goal is to test whether social referral is an effective channel for recommendation delivery. In designing experiments, there are two major questions we want to answer: 1) Are referrers receptive to passing along recommendations to their social network connections? 2) For target users, is there a significant difference in the acceptance rate of recommendations coming from social referral, and the acceptance rate of recommendations directly delivered to them? Furthermore, it is also worth investigating what factors affect a referrer's decision on whether to make a referral or not, and what factors affect a target user's decision on whether to accept a recommendation.

5.1 Methodology

The data we used for this experiment came from a professional network, and the kind of recommendation we used was group recommendation for the members of this professional network. We sampled a set of members who joined a group in the last 14 days. These were the referrers. On the other hand, we used a empirically proven in-house recommender system to compute top 100 group recommendations for each member. For each referrer we checked if the group the referrer just joined was also recommended to his/her network connections (aka friends) by the recommender system. Any of such connections is a target user to whom the referrer could make a referral of the group. An email was then sent to each of the referrer (Figure 1(a)). Each email contained links to five target users. A referrer could click on any or all of the five links, which will cause a group join invitation email to be sent to the clicked target user. A total of 85K emails were sent to referrers. As a control group for the experiment, we also sent out another set of recommendation emails directly to target user. The recommendation data and target user list were generated exactly in the same way as the first set of emails that were sent to referrers. The only difference here was that we bypassed the referrer and directly sent those recommendations to target users. A total of 16K emails were directly sent to target users.

5.2 Results

5.2.1 Referrer's Likelihood to Send Out Referral

How likely are referrers to send out a recommendation referral to their network connections? Our experiments showed that about 7.8% of referrers clicked on links in the emails to send out a referral email to their network connections. This is a significant percentage which far outperforms average emails CTR's. Figure 3(a) depicts different referrer CTRs at connection strength (CS) where CTR denotes among all target users with connection strength less than CS, what fraction of them are referred. It showed that the higher the connection strength between the referrer and the target user, the higher the CTR is for the referrers. We also compared overall referrer CTR with referrer CTR's for algorithm Better_GRP, and algorithm Stronger_CONN (Figure 3(a)). The data indicates that for the same connection strength, referrers are more likely to click on a referral link if the quality of the recommended group is higher. This confirms our intuition that referrers are more likely to send referral to their closer friends, and more likely to recommend a group which they think is of a higher quality. Another interesting study we performed was to check if referrers's selections would lead to more relevant recommendations to the target users. It seems intuitive to think that a referrer might remove those recommendations they think are not relevant to the recipients. Figure 3(b) compares the distributions of recommendation relevance scores before and after referrer clicks. Surprisingly, it showed that the distributions of recommendation relevance scores are almost identical before and after referrer selection. The result indicates that for the recommendations produced by an effective recommender system, the referrers are unable to tell which recommendations are more relevant to their network connections.

5.2.2 Target User's Acceptance Rate of Recommendations

We compared the join rate for groups for group recommendations delivered via social referral and those directly delivered to target users (the control group in the experiment). Our experiments showed a strong join rate by target users for group recommendations coming from social referrals. In fact, group recommendations delivered via social referral have more than twice as large acceptance rate (22.5%) than those directly delivered to target users (9.6%) (Figure 2).

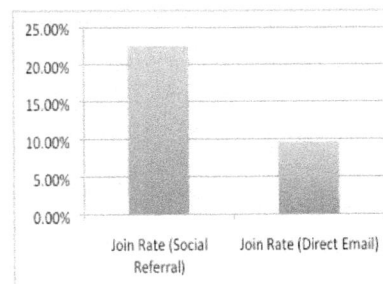

Figure 2: Join rate: Social referral vs. Direct delivery

In addition, as shown in Figure 4(a), there is evident correlation between the join rate and the connection strength CS between referrers and target users. The join rate here denotes among all target users with connection strength less than CS, what fraction of them have joined. The more strongly a target user is connected to a referrer, the more likely that the target user will accept a recommendation referral coming from the referrer.

The target user join rate is also affected by the relevance of recommendations and it is computed in a similar fashion as for Figure4(a). Figure 4(b) showed that more relevant recommendations lead to higher join rates.

5.3 Discussion

One argument against the usage of the social referral is that direct recommendation seems more effective than the overall workflow of social referral. After all, in social referral there are two steps of loss for the recommendations: a recommendation could be ignored by the referrer, or the target user. Using the experiment setup in this paper, a piece of recommendation would have a 9.6% of chance to be accepted via direct recommendation email, but only 2.2% of chance via social referral. However, there are several attractive aspects about social referral. First, the user experience in social referral is better. Although the acceptance rate of 9.6% is significant for direct recommendations, it also means that over 90% of target users find the recommendations not useful. Adding to the fact that these were unsolicited recommendations coming from the system, a significant portion of users will likely be annoyed by them. In a social referral scenario, however, instead of receiving an unsolicited, cold recommendation, the target users receive warm recommendations coming from their network connections directly. Even though about 77% of target users still do not act on the recommendations, it is unlikely to annoy them as these are recommendations coming from their friends who act as filters on behalf of the system. Second, since referrers are already engaging with the recommended objects themselves, it is natural to piggyback the social referral process at various engaging points for free. Although we used emails for the pilot study in this experiment, emails are by no means the only, or the most efficient channel. For example, in the context of a professional social network, the referrals could be brought to a referrer's attention when the referrer joins a group, starts a group discussion, or follows a certain group discussion on the website. Finally, even with emails, an additional 2.2% acceptance rate is not to be easily dismissed for a lot of marketers.

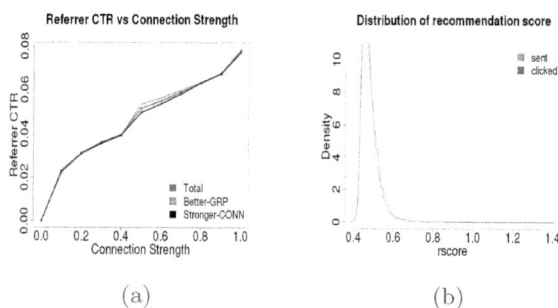

Figure 3: (a) Referrer CTR vs Connection Strength, (b)Distribution of Recommendation Scores Before and After Referral

6. CONCLUSIONS

Our main contributions in the paper are two folds. First, we presented *Social Referral*, a novel recommendation delivery mechanism which leverages a target user's social network for the effective delivery of relevant recommendations

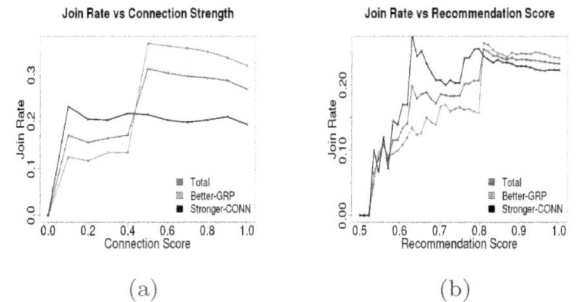

Figure 4: (a) Target User Join Rate vs Connection Strength,(b)Target User Join Rate vs Relevance

to the user. To our knowledge, most previous work has been focused on incorporating social network to improve recommendation relevance or explain recommendations. Little has been explored on leveraging social network for the delivery of recommendations. Second, we carried out a large scale user study with real world data that demonstrated the effectiveness of social referral. Our study showed that referrers are receptive to passing along recommendations to their network connections, and that recommendations delivered via social referrals are more than twice as likely to be accepted by target users than those directly delivered to them.

7. REFERENCES

[1] W. Geyer, C. Dugan, D. R. Millen, M. Muller, and J. Freyne. Recommending topics for self-descriptions in online user profiles. In *RecSys '08: Proceedings of the 2008 ACM conference on Recommender systems*, pages 59–66. ACM, 2008.

[2] G. Groh and C. Ehmig. Recommendations in taste related domains: collaborative filtering vs. social filtering. In *GROUP '07: Proceedings of the 2007 international ACM conference on Supporting group work*, pages 127–136. ACM, 2007.

[3] I. Guy, I. Ronen, and E. Wilcox. Do you know?: recommending people to invite into your social network. In *IUI '09: Proceedings of the 13th international conference on Intelligent user interfaces*, pages 77–86. ACM, 2009.

[4] H. Kautz, B. Selman, and M. Shah. Referral Web: combining social networks and collaborative filtering. *Commun. ACM*, 40(3):63–65, Mar. 1997.

[5] D. Kempe, J. Kleinberg, and E. Tardos. Maximizing the spread of influence through a social network. In *Proceedings of the ninth ACM SIGKDD international conference on Knowledge discovery and data mining*, KDD '03, pages 137–146. ACM, 2003.

[6] J. Kleinberg. *Cascading Behavior in Networks: Algorithmic and Economic Issues*. Cambridge University Press, 2007.

[7] K. Lerman. Social Networks and Social Information Filtering on Digg, Dec. 2006.

[8] R. Sinha and K. Swearingen. Comparing Recommendations Made by Online Systems and Friends. In *In Proceedings of the DELOS-NSF Workshop on Personalization and Recommender Systems in Digital Libraries*, 2001.

Case Study on the Business Value Impact of Personalized Recommendations on a Large Online Retailer

Thiago Belluf
Department of Automation
Engineering
Federal University of Santa Catarina
Florianópolis, Brazil
thiagobelluf@grad.ufsc.br

Leopoldo Xavier
Research Group
Chaordic Systems
Florianópolis, Brazil
xavier@chaordicsystems.com

Ricardo Giglio
Department of Economics
Christian-Albrechts-Universität zu Kiel
Kiel, Germany
r.giglio@economics.uni-kiel.de

ABSTRACT
While much of the recommender systems literature has focused on the off-line evaluation of prediction performance, a few case studies using online controlled experiments that assess the performance of business indicators are available. In this article, we describe the methods and results of an ongoing investigation conducted on the business value impact of personalized recommendations on three different portals of Nova Pontocom, the second largest Latin American online retailer. An on-line controlled experiment (A/B testing), conducted for one month and covering 600,000 distinct users, statistically points out to a general revenue increase in the order of 8-20%. In addition, other consumer behavior metrics such as the number of page views and the more diverse distribution of sales among the products catalog also support the positive impact of personalized recommendations in terms of business value.

Categories and Subject Descriptors
H.3.3 [**Information Search and Retrieval**]; H.4.2 [**Decision Support Systems**]

Keywords
Case study, online controlled experiment, business value, collaborative filtering

1. INTRODUCTION
Recently, several studies have been pointing out some pitfalls with regard to the evaluation of recommender systems that are based purely on off-line prediction performance [9, 11, 12, 13]. Some of these argue that other dimensions, such as novelty and diversity, should be considered in a proper evaluation of the quality of a recommender environment [1, 5, 6, 8]. Moreover, it is also evident in literature that off-line evaluations themselves are of limited scope when one deals with real commercial applications, because there is no guarantee that prediction performance necessarily translates into business value [8, 11]. This is the motivation of some recent works that are trying to identify the impact of recommender systems on real commercial applications. However, some of these are not based on real on-line controlled experiments, such as [3, 7]. Some remarkable exceptions are [15], which tests the conversion performance of

different recommendation algorithms against a human-made control scenario in a mobile internet games context, and [10], which investigates the relevance of the number of clicks as a metric for the online evaluation of recommender systems in a movies website.

This article presents the method and results of a controlled on-line experiment in the three different portals ({pontofrio, casasbahia, extra}.com.br) of Nova Pontocom, the second largest Brazilian online retailer having 17% of market share, 6 million distinct users, and sales totaling more than US$ 2 billion in 2011 figures. The experiment was run for one month, independently in the three portals, summing up to nearly 600,000 different users; 50 million page views, 1 million online orders generating revenues of US$ 230 million. To the best of our knowledge, this is the largest scale controlled experiment aiming to assess the business value impact of personalized recommendations published so far. Users were randomly assigned to a treatment group, in which personalized recommendations were presented, and a control group, with no recommendations. At the end of the experiment, it could be observed that with 95% of statistical significance, there was an overall increase in revenues in the order of 8-20%.

2. METHODS
2.1 Experiment Design
During the test period, from 02/10/2012 to 03/10/2012, the users who visited one of the three websites were randomly assigned to either a treatment group (with recommendations) or a control group (without recommendations) with 95% and 5% of chance, respectively. The test for each website is independent from the others, that is, the persons visiting two such websites are considered in separate tests as being two different users.

As usual in an e-commerce environment, users can navigate and view the recommendations (if in the treatment group) anonymously, but logging in is required for purchasing. In order to keep track of interaction history, we keep a record of the association between anonymous and logged users. In order to guarantee navigation consistency, we choose to keep the user's test group when he logs in. This might be problematic because the user can navigate anonymously (and log in) in more than one machine, what would cause him to participate in different test groups. The workaround applied to this problem is the following: if an anonymous user that is already assigned to a test group logs in and the correspondent logged user was already assigned to a different test group, we reassign anonymous test group to the logged user, maintaining the user experience consistency but making the logged user to participate in different test groups. The same occurs when a logged user logs off: the logged user's test group is assigned to the anonymous user. The test group associations are summarized in Table 1. In the simple case, when it is the first user's interaction with the test, he is randomly assigned to a test group. In subsequent interactions the same test group is used. In addition to these actions taken on-line, when processing the results we also merge the users' anonymous and logged interactions.

Table 1. Rules for anonymous/real users association

	Logged user already has a test group	Logged user does not have a test group
Anonymous user already has a test group	*Use the anonymous user's test group*	*Use the anonymous user's test group*
Anonymous user does not have a test group	*Use logged user's test group*	*Use a random test group*

While observing that no ratings are available for the majority of the products, personalized recommendations presented in the treatment group were generated by seven different collaborative filtering techniques based on the users' interaction history (implicit feedbacks), such as product views, purchases, and shopping cart composition. These techniques are presented to the users in the following different features that attempt to fit specific moments of the purchasing cycle: (a) "Better together", (b) Ultimate Buy: "What other items do customers buy after viewing this item?", (c) Similar Items: "Customers who bought this item also bought", (d) Shopping cart: "What costumers that bought the products in your cart also bought?", (e) View personalized: "You visited this item. You might be interested in these other items", (f) Purchase personalized: "You bought this item. You might be interested in these other items", and (g) "Most popular". Figure 1 shows a screenshot presenting some of the features (a-c) in a product description page (text in Portuguese).

2.2 Evaluation Metrics

Three different tests were performed with regard to the per user numbers of product views, purchases, and revenue generated. Since the main focus in both experiments was to assess the business value of personalized recommendations, we also kept a record of the purchases frequencies per product in each group in order to evaluate sales diversity. In this study, diversity was assessed as the expected information content of the product sales distribution (Shannon entropy) and also by means of Lorenz curves [14]. In short, for each item n ($n = 1,...,N$), we take its share on total sales (p_n) in order to get its information content $p_n*\log(p_n)$, and, thus, diversity is assessed by the normalized expected information content defined as $-\sum \{p_n*\log(p_n)\}/\log(n)$.

A potential problem that is experienced when working with per-user metrics is their accumulation over time, due to returning users. While observing that user averages would always increase, the moments of the distributions are not stable and, thus, no trivial statistical test can be performed. As pointed out by [2], a possible solution for this is the concept of a per-user-day metric, which was implemented in this experiment by considering users who returned to the websites on different days as two separate users. We found no significant serial correlation among daily interactions of a same user, thus exchangeability assumption still holds. Although this workaround might lead to

the loss of relevant information regarding costumer fidelity, we assume here that its impact would at most increase the results that are presented.

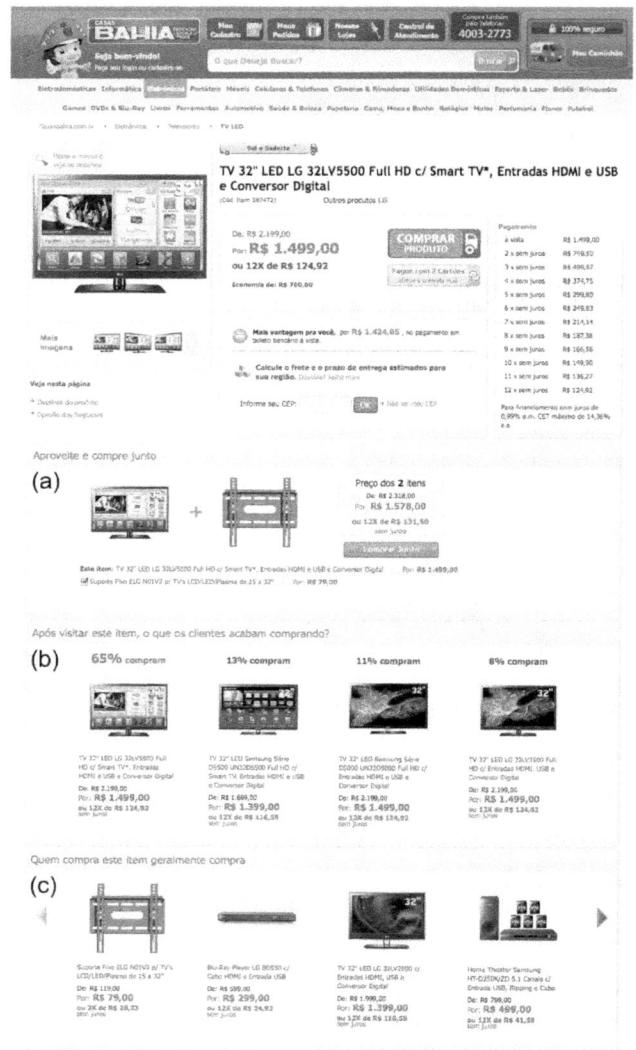

Figure 1. Product description page with recommendations.

Another important remark is related to the diversity of product categories. Some of them, such as electronics and informatics, have longer interaction cycles (in terms of bigger ratio views/purchases) and higher prices. On the other hand, categories such as books and DVDs show bigger order sizes (number of products bought by the same user) and more frequent purchases. Although the aim of this experiment was to evaluate the impact of personalized recommendations on the entire websites, we realized the importance of performing separate testes for each product category. The reason for this is related to confounding problems in statistics when a hidden factor in the observations (here product category affiliation) possesses a strong explanatory power on the user behavior. While observing that the impact of recommendations might not be the same along with a wide diversity of products (i.e. we found it to be stronger in cell phones/accessories than in household appliances), mixing up these categories would lead to misleading results. In a slightly different problem, but still considering on-line controlled experiments, [2] relates this situation to the Simpson's paradox by arguing that mixing data collected on two different days and from different splitting proportions might lead to a negative result, although positive, on both days. We checked for the existence of this situation by considering a simple two-category aggregation problem. It could be seen that merging a category whose control/treatment group difference is positive and statistically significant with another in which the difference is not significant can lead to a negative and statistically significant result. The issue of category aggregation with the aim of building an impact measure of an entire website is further addressed in the next section.

2.3 Statistical Significance

In such a kind of an experiment, it is well known that the pure mean comparison between treatment and control can lead to spurious results due to the play of chance. Hence, we decided to work with confidence intervals for each metric calculated by means of the bootstrapping procedure [4]. In spite of simpler and easy-to-use formulas based on the normality assumption being available, a considerable bias in the confidence intervals was detected when using these methods. These findings are also in accordance to the literature [2]. The existence of a large number of users who had either no or very few interactions along with a few users with lots of interactions (what might be pretty common in online retailing) imprints data with high kurtosis, and, therefore, straightforward, normality-based confidence intervals are inadequate.

While observing that the method of non-overlapping confidence intervals is not a proper hypothesis testing, in order to check whether the two samples are indeed statistically different, we applied a second bootstrapping procedure as follows [4]: Let x_n and y_m ($n = 1,...,N$, $m = 1,...,M$) be two independent samples with respective means $\bar{x}$ and $\bar{y}$. To check the null hypothesis (that is, whether both samples are derived from the same generating process), we first constructed a third sample z_{m+n} by simply concatenating x_n and y_m. Then, we resampled with replacement from z_{m+n} B times ($b=1,...,B$) to get the mean values of its first n elements $\bar{x}_b$, and of its last m elements $\bar{y}_b$. Finally, the test p-value is given by the proportion of resamples, where $\bar{x}-\bar{y} > \bar{x}_b-\bar{y}_b$. We then reject the null hypothesis when $p < 0.05$.

In order to get an estimate of the impact of recommendations on the whole website (and not only by product categories), we compare two aggregated control and treatment scenarios. The control scenario is simply the product between the evaluation metrics mean per user obtained without recommendations (and its correspondent confidence intervals) and the total number of users during the experiment in that particular website and category. The same procedure is applied for the treatment scenario, but we only add the mean per user obtained with recommendations in case there is a positive and statistically significant result in the category-specific hypothesis test. Otherwise, we assume that there is no impact at all and the same value of the treatment for that category is used. Finally, in order to obtain the ratio of difference (and confidence interval) between scenarios, we take an extra caution step by dividing the confidence bounds of the treatment scenario by the upper bound of the control scenario. This is an arbitrary decision that is taken with the aim of reducing the chance of a false-positive result.

3. RESULT ANALYSIS

With regard to the revenue per user, we found statistically significant results in most of the large product categories (by

number of users). In fact, we found no significant result in categories with less than 1,000 users, meaning that for those categories, the experiment period was probably too short. Table 2 depicts particular results for representative categories. The overall figures for the three websites are 8-19%, 9-19%, and 8-22%. Considering page views, the numbers in parenthesis represent the number of users (real and anonymous). For revenue and diversity, they show the total revenue generated and the number of distinct products sold, respectively.

These numbers are higher than those found in [10], for example, at least partially due to the fact that in their experiment the control group contained manually edited recommendations. The analysis of the number of page views per user showed a slightly different picture. While observing that there were many more page views than actual purchases, we found significant differences for all the large categories, as depicted in Table 2. Considering the entire websites, it could be proved that the existence of recommendations increased the number of page views per user in the order of 5-8%, 7-9%, and 5-7%, respectively. Besides the business metrics presented so far, we have also tested for the difference in diversification between the test groups. With regard to the five biggest product categories of each portal (a total of 15 tests), we found statistically significant differences in ten of them in the order (average) of 4% to 15%. Table 2 also depicts these results. The same effect could also be seen by means of Lorenz curves based on the diversity of sales among existing products, as depicted in Figure 2.

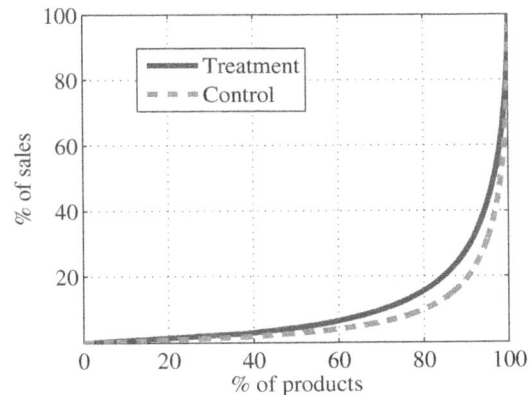

Figure 2. Lorenz curves for the treatment and control groups, product category 'Toys', website casasbahia.

4. CONCLUSION

Recent literature has pointed to the relevance of on-line controlled experiments in assessing the business value impact of recommender systems. In spite of being extremely useful, off-line evaluation metrics present some pitfalls in the evaluation of business indicators, because there is no guarantee, for example, that better prediction performance translates into bigger conversion rates or revenue per user. This article presents the ongoing results from an on-line experiment in the context of three different portals of Nova Pontocom, a large Brazilian online retailer. Statistically significant differences could be observed between the treatment (in which personalized recommendations were presented) and control (no recommendations) groups with regard to key business metrics such as revenue per user (8-20%), sales diversity (4-15%), and page views per user (5-9%). Such results support the use of personalized recommendations in the context of a large multi-item online retailer, even though no product ratings are available. Future research is still needed in order to study how recommendations affect consumer behavior and, thus, business value. We intend to investigate this issue by checking for the difference in performance between the CF algorithms (features), which are presented in different navigational situations through the purchasing cycle.

5. ACKNOWLEDGMENTS

Thanks for the staff of Nova Pontocom for kindly cooperating with the experiment.

Table 2. Confidence intervals for selected product categories.

NS = not significant		pontofrio	casasbahia	extra
Revenue	Cell Phones and Accessories	12 to 20 (10,862,557)	14 to 23 (7,956,723)	10 to 21 (5,586,918)
	Household Appliances	8 to 18 (20,345,591)	9 to 23 (11,497,914)	11 to 25 (10,540,081)
	Electronics and Computers	3 to 14 (14,384,126)	5 to 19 (18,388,180)	2 to 18 (5,703,246)
	Camera, Photo and Video	9 to 26 (3,327,980)	6 to 25 (2,319,508)	6 to 27 (1,557,810)
	Toys	2 to 23 (848,252)	NS (979,049)	NS (682,031)
	TOTAL	8 to 19 (79,930,710)	9 to 19 (64,229,762)	8 to 22 (41,100,389)
Page views	Cell Phones and Accessories	9 to 11 (1,595,856)	11 to 13 (1,371,684)	5 to 7 (1,291,550)
	Household Appliances	4 to 6 (1,077,590)	4 to 7 (1,078,215)	3 to 5 (923,663)
	Electronics and Computers	3 to 6 (892,047)	4 to 6 (686,608)	1 to 4 (575,744)
	Camera, Photo and Video	5 to 8 (319,853)	6 to 9 (286,499)	NS (284,017)
	Toys	5 to 10 (193,048)	9 to 13 (249,422)	5 to 10 (223,822)
	TOTAL	5 to 8 (5,881,020)	7 to 9 (5,498,495)	5 to 7 (5,477,201)
Diversity	Cell Phones and Accessories	-9 to -1 (1,026)	NS (1,020)	NS (951)
	Household Appliances	NS (1,339)	NS (1,252)	2 to 10 (1,237)
	Electronics and Computers	4 to 6 (2,186)	1 to 20 (2,311)	NS (1,908)
	Camera, Photo and Video	NS (856)	1 to 20 (788)	NS (698)
	Toys	2 to 9 (2,006)	5 to 11 (2,235)	1 to 9 (2,179)

6. REFERENCES

[1] Anderson, C. 2006. *The Long Tail.* Hyperion, New York.

[2] Crook, T., Frasca, B., Kohavi, R., and Longbotham, R. 2009. Seven pitfalls to avoid when running controlled experiments on the web. In *Proceedings of the 15th ACM SIGKDD international conference on Knowledge discovery and data mining (KDD '09).* ACM, New York, NY, USA, 1105-1114. DOI=http://doi.acm.org/10.1145/1557019.1557139

[3] Dias, M. B., Locher, D., Li, M., El-Deredy, M., and Lisboa, P. J, G. 2008. The value of personalised recommender systems to e-business: a case study. In *Proceedings of the 2008 ACM conference on Recommender systems (RecSys '08).* ACM, New York, NY, USA, 291-294. DOI= http://doi.acm.org/10.1145/1454008.1454054

[4] Efron, B., and Tibshirani, R. J. 1993. *An Introduction to the Bootstrap.* New York: Chapman & Hall, p. 326.

[5] Fleder, D. M., and Hosanagar, K. 2009. Blockbuster Culture's Next Rise or Fall: The Impact of Recommender Systems on Sales Diversity. *Management Science*, Vol. 55, No. 5, pp. 697-712. DOI=http://dx.doi.org/10.2139/ssrn.955984

[6] Fleder, D. M., and Hosanagar, K. 2007. Recommender systems and their impact on sales diversity. In *Proceedings of the 8th ACM conference on Electronic commerce (EC '07).* ACM, New York, NY, USA, 192-199. DOI= http://doi.acm.org/10.1145/1250910.1250939

[7] Garfinkel, R., Gopal, R. D., Pathak, B. K., Venkatesan, R., and Yin, F. 2006. *Empirical Analysis of the Business Value of Recommender Systems.* Available at SSRN: http://ssrn.com/abstract=958770

[8] Ge, M., Delgado-Battenfeld, C., and Jannach, D. 2010. Beyond accuracy: evaluating recommender systems by coverage and systems. In *Proceedings of the fourth ACM conference on Recommender systems (RecSys '10).* ACM, New York, NY, USA, 257-260. DOI= http://doi.acm.org/10.1145/1864708.1864761

[9] Herlocker, J. L., Konstan, J. A., Terveen, L. G., and Riedl, J. T. 2004. Evaluating collaborative filtering recommender serendipity. *ACM Trans. Inf. Syst.* 22, 1 (January 2004), 5-53. DOI=http://doi.acm.org/10.1145/963770.963772

[10] Jannach, D., and Hegelich, K. 2009. A case study on the effectiveness of recommendations in the mobile internet. In *Proceedings of the third ACM conference on Recommender systems (RecSys '09).* ACM, New York, NY, USA, 205-208. DOI=http://doi.acm.org/10.1145/1639714.1639749

[11] McNee, S. M., Riedl, J., and Konstan, J. A. 2006. Being accurate is not enough: how accuracy metrics have hurt recommender systems. In *CHI '06 extended abstracts on Human factors in computing systems (CHI EA '06).* ACM, New York, NY, USA, 1097-1101. DOI= http://doi.acm.org/10.1145/1125451.1125659

[12] Pu, P., Chen, L., and Hu, R. 2011. A user-centric evaluation framework for recommender systems. In *Proceedings of the fifth ACM conference on Recommender systems (RecSys '11).* ACM, New York, NY, USA, 157-164. DOI= http://doi.acm.org/10.1145/2043932.2043962

[13] Shani, G., and Gunawardana, A. 2011. *Evaluating recommendation systems.* In Recommender Systems Handbook, 257–97. Springer.

[14] Vargas, S., and Castells, P. 2011. Rank and relevance in novelty and diversity metrics for recommender systems. In *Proceedings of the fifth ACM conference on Recommender systems (RecSys '11).* ACM, New York, NY, USA, 109-116. DOI=http://doi.acm.org/10.1145/2043932.2043955

[15] Zheng, H., Wang, D., Zhang, Q. Li, H., and Yang, T. 2010. Do clicks measure recommendation relevancy?: an empirical user study. In *Proceedings of the fourth ACM conference on Recommender systems (RecSys '10).* ACM, New York, NY, USA, 249-252. DOI=http://doi.acm.org/10.1145/1864708.18

The Xbox Recommender System

Noam Koenigstein
Microsoft R&D
Herzliya, Israel

Nir Nice
Microsoft R&D
Herzliya, Israel

Ulrich Paquet
Microsoft Research
Cambridge, UK

Nir Schleyen
Microsoft R&D
Herzliya, Israel

ABSTRACT

A recent addition to Microsoft's Xbox Live Marketplace is a recommender system which allows users to explore both movies and games in a personalized context. The system largely relies on implicit feedback, and runs on a large scale, serving tens of millions of daily users. We describe the system design, and review the core recommendation algorithm.

Categories and Subject Descriptors

H.2.8 [**Database Applications**]: Data Mining

General Terms

Algorithms

Keywords

Recommender Systems, Collaborative Filtering, Xbox

1. INTRODUCTION

The distribution of items bought in a marketplace typically follow a power-law distribution, with popular items exponentially overwhelming less popular ones. As a result, many items in the tail of the distribution, which would have been suited to a particular user, are never exposed. There are a number of reasons for tailoring a user's experience in any marketplace, the most compelling of which are the need to increase user engagement, and the need to give exposure to tail items that would, under normal circumstances, be drowned by a few popular ones.

These can be addressed by a system that models and learns from user engagement. Recommender systems have been particularly effective at this task, to which the systems of MSN News, Amazon, Netflix, and the likes of others can testify. Each of these face unique challenges that are posed by the size and nature of the catalog, type of user feedback, rate of catalog change, scalability, and expected real-time system performance.

1.1 The Xbox dataset

Our dataset comprise of usage information from Xbox consoles. For every user we collect the interaction time with every item on our catalog (either a game or a movie). We have tens of millions of active users, almost all of which are playing Xbox games. Some of our users also enjoy our movies catalog and watch movies purchased using their Xbox consoles. We provide personalized recommendations for both domains using two identical recommendation engines as described in this paper.

The number of Xbox users increase on a daily basis. Similarly, we regularly update our games and movies catalog with the latest products available on the market. With our very large user base, we hardly encounter a "cold start" problem in the items domain, because within a few hours we already have sufficient information on every new item. Similarly, Xbox users generally purchase some games together with their consoles, so we rarely encounter "cold users" in the games domain. Only in the movies domain, we encounter the cold start problem for users that do not watch movies on their consoles. We still hope to address this problem with cross domain learning, however this is out of the scope of this paper. In this paper, we deal only with users that have at least one item in their usage history.

Unlike the well-studied Netflix prize problem, where the user's movie preferences are explicitly given by a five-star rating scale, our data largely comprises of implicit signals, like that of purchasing a game or watching a movie on the Xbox console. The absence of a negative signal poses a challenge to conventional solutions based on the factorization of a "ratings" matrix into two low rank matrices with one representing latent user features, and the other representing latent item features. We therefore augmented our implicit dataset with randomly generated negative signals to prevent trivial solutions as explained in Section 2.

This paper.

In this paper we wish to bring a complete overview of a real world large scale recommender system. We draw a distinction between the tasks of modeling (how hidden parameters are assumed to generate observed data), inference (finding the hidden parameters given the data), and utility (using the model's predictions in order to optimize a reward function). Section 2 is devoted to an overview of the system's architecture and some preprocessing steps to prepare the data. Section 3 explains the modeling and parameter inference, which happens *offline*, and gives an example of an *online* utility function. Finally, in Section 4 we evaluate the model and summarize in Section 5.

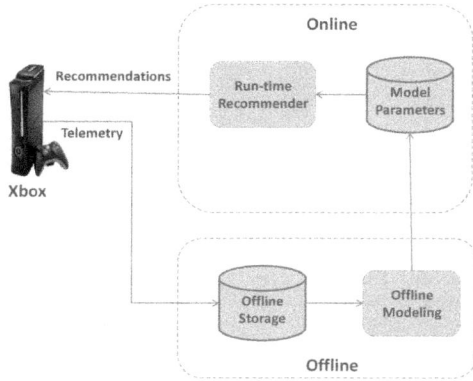

Figure 1: Xbox recommender system architecture

2. SYSTEM DESIGN AND PREPROCESSING

The architecture of the Xbox recommender system consists of an *offline* module and an *online* module (see figure 1). Feedback from the the Xbox consoles ("telemetry") is constantly sent back into the the *offline* module. In the *offline storage* we store information such as users' play time and movies watched by users. This information is then processed into an ownership dataset to be used by the *offline modeling* component for learning the model parameters.

In general, the *offline modeling* component treats the ownership information as an indication that the user liked the item. When processing the ownership information we employ some filters to remove games that were almost never played or movies that were not fully watched, but these are rare events that hardly effect the final recommendations. The ownership dataset comprises of implicitly generated "positive" only signals. Furthermore, it is hard to define credible ordering on those signals (e.g., the user liked a better than b), thus ordering algorithms such as [4, 3] are not useful. We therefore resolve to randomly generating negative signals as we describe below.

On average, an Xbox user has about 18 games and watched about 7 movies. Therefore, we can randomly pick a small number of items from our large catalog and assume that the reason the user do not own them is because she doesn't like them. For every user we pick the same number of "negative" items as the number of items she liked ("positive" items). By doing so, we cancel out the user bias as explained in Section 3. We draw the "negative" items with probabilities proportional to their popularity (training-set frequencies). This approach was highly popular in many of the solutions submitted to the second track in the KDD-Cup'11 [2] competition. Sampling according to popularity penalize popular items, which based on subjective internal evaluations found to increase user satisfaction.

The binary *like-dislike* matrix is then passed on to the *offline modeling*. A detailed description of the inference algorithm implemented in the *offline modeling* component is given in in 3.1.

Every few hours, the learned parameters are uploaded from the *offline modeling* component into the *model parameters* database. The *run-time recommender* in the *online* module is a real-time service that utilize the learned

parameters in order to generate user specific recommendations. Upon a console-query for personalized recommendations, the *run-time recommender* chooses items according to the utility algorithm described in Section 3.2. These recommendations are then sent back to the console and presented to the user under the title "Picks For Me".

3. PREDICTION MODEL

The interaction between users and items lends itself to bilinear models. In them, each user m is represented by $\mathbf{u}_m \in \mathbb{R}^K$, each item n is represented by a similar vector $\mathbf{v}_n$, and the magnitude of their inner product $\mathbf{u}_m^T \mathbf{v}_n$ denotes the user's affinity to the item (personalization). The parameters $\mathbf{U} = \{\mathbf{u}_m\}_{m=1}^M$ and $\mathbf{V} = \{\mathbf{v}_n\}_{n=1}^N$ are unobserved, and should be inferred from data. This core set-up has been widely used in the Netflix prize competition, and we've adopted the Matchbox library of [5] for parameter inference.

The bilinear model has the property that items which often co-occur for a given user, have similar $\mathbf{v}$-vectors. Figure 2 shows $\mathbf{V}$ embedded in $\mathbb{R}^2$ for a number of games. In it, the similarity between vectors of sports games, for example, is clearly visible.

Figure 2: Game feature vectors embedded in $\mathbb{R}^2$, tagged by genre.

In practice some items are more popular than others, for which the personalization term is offset with a bias b_n for each item. We could equally add user-specific biases, but omit it in order to regress a users usage list which has the same number of positive and negative items. The probability that user m is going to like (or dislike) item n is

$$p(l_{mn}|\mathbf{u}_m, \mathbf{v}_n, b_n) = \Phi\left(l_{mn}(\mathbf{u}_m^T \mathbf{v}_n + b_n)\right) \qquad (1)$$

where $l_{mn} = 1$ for the user liking and -1 for the user disliking the item. $\Phi(z) = \int_{-\infty}^z \mathcal{N}(x; 0, 1)\mathrm{d}x$ is the Gaussian cumulative density function, and acts as a link function that maps its argument to a value in the $(0, 1)$ interval.

Our working assumption is that there is sufficient usage data to infer $\mathbf{v}_n$. Alternatively, we can further regress against user or item-specific meta-data, or learn a K-dimensional representation for each meta-data atom if required [5].

Given the data $\mathcal{D} = \{l_{mn}\}$ and parameters $\theta \doteq \{\mathbf{U}, \mathbf{V}, \mathbf{b}\}$, the likelihood for observing the data is

$$p(\mathcal{D}|\theta) = \prod_{(m,n)} p(l_{mn}|\mathbf{u}_m, \mathbf{v}_n, b_n) \ .$$

Our interest lies in how well the data supports the parameters, i.e. the posterior probability of θ given the data. This is given by Bayes' theorem,

$$p(\theta|\mathcal{D}) = \frac{p(\mathcal{D}|\theta)p(\theta)}{p(\mathcal{D})} \ , \qquad (2)$$

which incorporates a prior distribution on the model parameters. In our application, $p(\theta) \doteq p(\mathbf{U})p(\mathbf{V})p(\mathbf{b})$ factorizes as Gaussian distributions over the features and biases.

Bayesian inference comes into its own right when the posterior density $p(\theta|\mathcal{D})$ is used for predictions. Instead of using a single parameter estimate $\hat{\theta}$, which needs to be carefully regularized to avoid overfitting when data is sparse [1], we rather average the likelihood function over all parameters that plausibly explain the data. In other words,

$$p(l_{ij} = 1|\mathcal{D}) = \int p(l_{ij} = 1|\theta) \, p(\theta|\mathcal{D}) \, \mathrm{d}\theta \qquad (3)$$

gives the predictive distribution that user i is going to like item j.

Both the posterior and predictive distributions, given in (2) and (3), require the computation of analytically intractable integrals. The integrals can be evaluated stochastically through Monte Carlo methods, or approximated deterministically through a relaxation into an optimization problem.

3.1 Parameter inference

We approximate the posterior distribution with a one that factorizes with

$$p(\theta|\mathcal{D}) \approx q(\theta) = \prod_m q(\mathbf{u}_m) \prod_n q(\mathbf{v}_n) \, q(b_n) \ .$$

The choice is largely driven by practical convenience, as $q(\theta)$ is "factorized enough" to give rise to computationally tractable optimization problem and algorithm, and it allows us to tractably compute the (approximate) predictive distribution,

$$p(l_{ij}|\mathcal{D}) \approx \int p(l_{ij}|\mathbf{u}_i, \mathbf{v}_j, b_j) \, q(\mathbf{u}_i) \, q(\mathbf{v}_j) \, q(b_j) \, \mathrm{d}\{\mathbf{u}_i, \mathbf{v}_j, b_j\} \ , \qquad (4)$$

where we've replaced $p(\theta|\mathcal{D})$ with $q(\theta)$ in (3). The factorization also ensures that only the matching user and item's approximations play a role, and other random variables are independently averaged away in (4).

We learn $q(\mathbf{v}_n)$ for each item; this is chosen as a K-dimensional factorizing Gaussian distribution $q(\mathbf{v}_n) = \prod_k q(v_{nk})$. As a result, we not only have a mean estimate of each item's (and user's) feature vector, but also encode uncertainty about its location through the variances of $q(\mathbf{v}_n)$. The variance of $q(\mathbf{v}_n)$ is typically small for very popular items, but larger for items that are less frequently played, watched, or bought. As a consequence, popular games (with more "well determined" parameters) will have a stronger effect on determining a user's feature vector, than games for which the model is less certain.

The approximation $q(\theta)$ is found by an algorithm called "Expectation Propagation", which has its roots in old and well-studied approximations[1] in statistical physics; the reader is referred to [5] for further algorithmic details.

---

[1] Expectation Propagation solves for the saddle point of a Bethe free energy with weak consistency constraints.

3.2 Retrieval of Recommendation

After training the model and learning parameter approximations for $\{\mathbf{U}, \mathbf{V}, \mathbf{b}\}$, we are free to choose a utility to optimize. As an example, the default task of retrieving recommendation for a specific user m can be formulated as follows. Let $\mathcal{Q} \doteq \{q(\mathbf{v}_n), q(b_n)\}$ be the set of all item densities, over which we maximize

$$\arg\max_{q_n \in \mathcal{Q}} \int p(l_{mn}|\mathbf{u}_m, \mathbf{v}_n, b_n) \, q(\mathbf{u}_m) \, q(\mathbf{v}_n) \, q(b_n) \, \mathrm{d}\{\mathbf{u}_m, \mathbf{v}_n, b_n\} \ ,$$

to find the items that a user will like with largest predictive probability.

In practice the above expression is analytic if either of the feature distributions is a point mass, or can otherwise be approximated with an analytic function as explained in [5].

4. RESULTS

We evaluate our system using a classification task inspired by the KDD-Cup'11 [2] competition. In Section 4.2 we also give a more traditional mean rank metric.

4.1 Recommendations as a classification task

Results here are based on 10,000 Xbox users with at least 2 games and at most 50 games. Here, we ignored Xbox consoles with more than 50 games because these may belong to larger organizations rather than individuals. Since traditional evaluation (e.g., root mean squared error) do not generalize well to the case of implicit data, we choose a rather new evaluation metric that was recently introduced in the second track of the KDD-Cup'11 [2] competition: For each test user we kept one item in the test set, and trained on the rest of the items. After training, we randomly picked another item for every user which is not present in the training or the test items of that user. The goal is to differentiate the item that the user owned from the item that was randomly picked. The randomly picked items were chosen in two methods: uniformly from the items set, and non-uniformly with probabilities proportional to each item popularity (similar to the process we use to generate implicit negative signals in Section 2). Namely, we define a classification task for every user and measure the overall precision on all users:

$$Precision = \frac{1}{|M|} \sum_{(m,n) \in \mathbf{T}} \mathbf{1}\left[p\left(l_{mn} = 1|\mathcal{D}\right) > p\left(l_{mk} = 1|\mathcal{D}\right)\right], \qquad (5)$$

where $\mathbf{T}$ is the test-set, $p\left(l_{mn} = 1|\mathcal{D}\right)$ is the probability that user m likes item n (3), $p\left(l_{mk} = 1|\mathcal{D}\right)$ is the probability for user m to like the randomly sampled item k, and $\mathbf{1}(\cdot)$ is an indicator function. Note that since each user has just one item in the test-set, there is no real meaning to the recall measurement in this setting.

We had several objectives in choosing this evaluation task: First, it is closely related to the methodology we use in training the model (Section 2). More importantly, we believe this evaluation better relate to real life scenarios: As explained in [2], the proposed metric is related to the common recommendation task of predicting the items that the user is likely to own (rather than predicting a rating value to items the user already owns). In fact, this metric extends the evaluation to the truly missing entries and measures the true generalization power of the algorithm. Finally, there is an advantage to using this metric with negative examples drawn according to their popularity (non-uniformly), because it discourages known trivial solutions where most popular items are always suggested regardless of the user taste. In this way we focus

Domain	Uniform Baseline	Uniform Xbox	Non-Uniform Baseline	Non-Uniform Xbox
Games	84.34%	89.21%	60.21%	75.56%
Movies	65.08%	79.27%	50.88%	69.93%

Table 1: Evaluation results for Xbox personalization recommender for movies and games. We present precision results against a baseline that always chooses the more popular item. We used both uniformly sampled items (over the entire items set) and non-uniformly sampling according to items popularity.

our evaluation on the true personalization power of the algorithm, rather than learning biases.

We evaluate this classification task against a simple baseline that always predicts that the user prefers the more popular item. Table 1 summarize the results. We see that the baseline algorithm performs relatively well when the negative items are uniformly distributed (84.3% and 65.1% in the games and movies domains respectively). However, these results are attributed only to learning popularity of items (biases), without any personalization. When negative signals are drawn according to their popularity, we better observe the precision gained by our personalization recommender over the baseline: an improvement of about 15% and 19% in the games and movies domains respectively. The overall results of the games recommender are better than those of the movies recommender. This is the result of the higher sparsity in the movies dataset (many more items, much less implicit ratings).

Note that in general, if we can sample negative items exactly according to their popularity, we expect that the base precision on the non-uniform test-set will be about 50% (as it is in movies). However, in the games domain some items are so popular that it is impossible to generate negative examples in exactly the same distribution as the positive items[2]. Because we do not assign negative items to users that also has those items in their positive items list, the resulting distribution is still somewhat biased towards the popular items and the baseline algorithm still achieves 60.2% precision on this task.

4.2 Mean Rank

We also present a more "traditional" evaluation using the *mean rank* metric. As before, for each user we keep one game in a test-set, and use the rest of the items to train the model. We then rank the item in the test set against 8 thousand other games from our catalog according to $p(l_{mk} = 1|\mathcal{D})$ - the probability that the user m will like an item k. We measured the mean rank of the test item for every user in the test set.

Figure 3 presents the mean rank (on a log scale) vs. the number of items in users history. Here we used 100,000 users (here we also included users with history longer than 50 items). An interesting observation is that the mean rank values increase with the users history length. Most of our users seem to own several very popular games. As users buy more games, they are more likely to own long tail items that are harder to model, which explains the line trend in Figure 3.

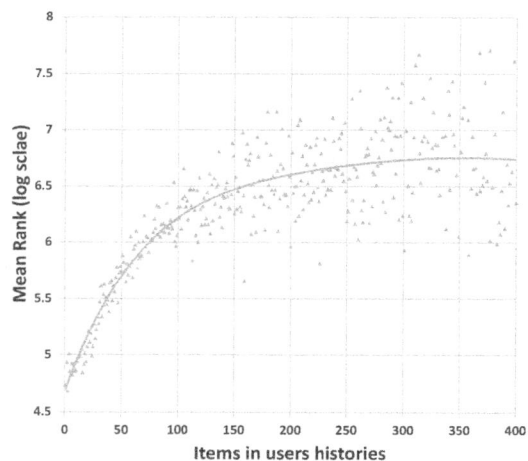

Figure 3: Mean-rank values vs. items in users history.

5. SUMMARY

A recent addition to Microsoft's Xbox Live Marketplace is a recommender system that allows users to explore personalized content tailored individually to their taste. This paper gives a complete overview of Xbox's recommender system in terms of system design, architecture, modeling, inference and utilization. We hope this example of real world large scale recommender system will give insights that would benefit the RecSys community academia and industry alike.

6. ACKNOWLEDGMENTS

The Xbox recommendation system was made possible through the hard work of everyone in the Recommendation Group in Microsoft Entertainment Services. The Matchbox library [5] was developed by the Machine Learning Group in Microsoft Research, Cambridge, UK.

7. REFERENCES

[1] G. Dror, N. Koenigstein, and Y. Koren. Yahoo! music recommendations: Modeling music ratings with temporal dynamics and item taxonomy. In *Proc. 5th ACM Conference on Recommender Systems (RecSys'11)*, 2011.

[2] G. Dror, N. Koenigstein, Y. Koren, and M. Weimer. The Yahoo! music dataset and KDD Cup '11. *Journal Of Machine Learning Research*, 17:1–12, 2011.

[3] U. Paquet, B. Thomson, and O. Winther. A hierarchical model for ordinal matrix factorization. *Statistics and Computing*, 21, 2011.

[4] S. Rendle, C. Freudenthaler, Z. Gantner, and L. Schmidt-Thieme. Bpr: Bayesian personalized ranking from implicit feedback. In *UAI '09: Proceedings of the 25th Conference on Uncertainty in Artificial Intelligence*, 2009.

[5] D. H. Stern, R. Herbrich, and T. Graepel. Matchbox: large scale online Bayesian recommendations. In *WWW*, pages 111–120, 2009.

---

[2]Because the most popular Xbox games are owned by most of our users.

Enlister: Baidu's Recommender System for the Biggest Chinese Q&A Website

Qiwen Liu
Baidu Inc.
Beijing, China
liuqiwen@baidu.com

Tianjian Chen
Baidu Inc.
Beijing, China
chentianjian@baidu.com

Jing Cai
Baidu Inc.
Beijing, China
caijing@baidu.com

Dianhai Yu
Baidu Inc.
Beijing, China
yudianhai@baidu.com

ABSTRACT

In this paper, we describe the concept & design of a real-time question RS (recommender system), the Enlister project, for the biggest Chinese Q&A (Questions and Answers) website and evaluate its performance on massive data from this real-world practice.

We demonstrate how we weigh in among different recommendation algorithms and optimization methods. To enhance recommendation accuracy and handling time-sensitive questions, we propose a large scale real-time RS based on the combination of machine learning algorithms and the stream computing technology. Considering of algorithm flexibility and performance, we use the maximum entropy model as the fundamental model design in the CTR (click-through rate) prediction of recommendation items. In the perspective of the Enlister system architecture, we illustrate how we divide and conquer massive data processing problem with a novel stream computing design which reduces the data process latency down to seconds.

Finally we analyze the online test result and prove our design concept by achieving a series of significant improvements.

Categories and Subject Descriptors

H3.3 [**Information Storage and Retrieval**]: Information Search and Retrieval

General Terms

Algorithms, Performance

Keywords

Machine Learning, Maximum entropy, Stream computing

1. INTRODUCTION

1.1 Background

Baidu Inc. is a leading internet service provider in China. As of January 2012, it ranked 5th in the Alexa global rankings and the 1st in China. One of Baidu's most popular services is the Baidu Knows, which offers a Q&A platform to its users for knowledge and experience sharing.

Today, there are over 400 million questions on the Baidu Knows website, of which more than 170 million questions were answered by users. Around 100 million users search for answers

every day and over 12 new questions are posted online every second.

As the biggest Q&A website in China, the Baidu Knows create an eco-system of knowledge sharing between the website's users.

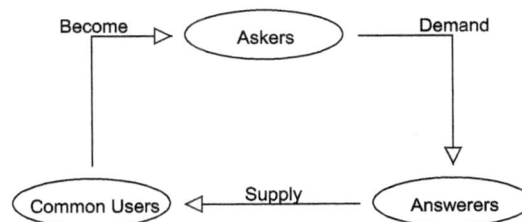

Figure 1: the Baidu Knows eco-system

As shown in Figure 1, typical common users just search in the Baidu Knows for answers with some related keywords. Some of these users will become the Askers to post new questions on the website. Then the users who know the answers to these questions post their answers back and become answerers in this community. As more questions have been answered, the search result quality of common users will be improved. Therefore, the fundamental concept of keeping this community healthy is stimulating the growth of answered questions.

As for the answerers, searching for unsolved questions is often too time-consuming. That's why a question RS is introduced to solve this problem by automatically presenting questions to potential answerers.

1.2 Application Description

We build an intelligent RS to provide the Baidu Knows users with questions that they may be willing to answer. This RS is also capable of tracking down the short-term variation of a user's intention.

After submitting a new answer to an unsolved question, a user will be redirected to a new web page where we put the question recommendation list there. The user could either click through one question to answer it or just leave this page if not interested. The quantitative analysis of the user reactions and recommender performance will be presented in Section 5.

1.3 Contributions

Our Contributions can be summarized as follows:

- We bring a large scale question RS, which is based on the machine learning technology, from fiction to a real internet service. Millions of users have benefitted from this approach.

- We illustrate that the stream computing technology could be very effective in the real-world RSs, which could address the time-sensitive issues explicitly.

- Inspired by the modern search engine ranking technology, we apply a machine learning based CTR prediction methodology to improve the recommendation accuracy.

For your convenience, we organize the remainder of this paper as follows. In Section 2, related work is introduced. The user model and the recommender algorithms are described in Section 3. In Section 4, we will have an overview of the Enlister system architecture. All the evaluations are placed in Section 5. Finally, we have a brief conclusion in Section 6.

2. RELATED WORK

The previous question RS of Baidu Knows is a typical content-based RS [1]. It models the relevance between a user and a question as the cosine similarity degree of the user's preference vector and the term vector extracted from the question. However, this approach has stuck in the mud of blurry optimization objective and wild computational complexity.

As in many other information retrieval [2] and search engine advertising application [3] [4], CTR prediction is introduced to improve the performance of the Q&A RSs [5] [6].

Compared with other RS that had applied CTR prediction, the Enlister has pushed the CTR prediction to the edge of real-time massive-data processing and has proven to be very successful in the real-time RS design.

3. ALGORITHM
3.1 Overview

In our new version of RS, known as the Enlister system, the CTR metric is introduced to characterize the correlation between users and questions. It presents a closer view to the users' experience than the relevance score we used before.

In order to make more accurate predictions, delicate user models are constructed based on the data we collected from the users. On one hand, we expect these models to disclose the nature of our users' choices of answering questions. On the other hand, the models have to be simple enough to accommodate massive calculation and industrial adoption. Based on the user models, the click prediction models are trained according to the user click history and certain features from the user models. With the trained click model, we can easily predict the probability of a specific user clicking on a particular question. After we aggregate probability prediction results and generate a new recommendation list, diversity adjustments are introduced to avoid the potential monotonous problem of the question recommendation and pursue some limited but pragmatic novelty.

3.2 User Model

In the Enlister system, user model is built on both long-term status of user attributes and short-term variation of user interest.

3.2.1 User Attributes

Some user attributes are the basic information of a user, such as age, gender, education, expertise and other tags that one labeled on himself.

3.2.2 User Interest

User interest is essential in the user model data structure. In the Enlister system, three aspects of interest descriptions are generated from three different semantic abstractions:

1) Interest Term Vector

   A vector contains weights of terms, which implicate the correlation between user and the term on sematic level.

2) Related Questions

   Questions are browsed or answered by a user.

3) Abstract Interest Vector

   A vector contains weights of terms, which implicate the correlation between a user and an abstract concept. With the help of PLSA [7] technology, a conceptual topic model is trained from all the questions and answers on the Baidu Knows website which contains millions of question-answer pairs. When a user browse or reply to a question, we can calculate the distribution of conceptual topics to that question by the conceptual topic model mentioned before. Finally, with miner normalization efforts, we could get the abstract interest vector of a user to all questions that he has visited.

3.3 Click Prediction

In the Enlister system, the click model that we created is a kind of probabilistic classification model. It is a binary classification model to calculate the probability of a sample belonging to a class.

We will introduce more details about the click prediction in the following sections, including: 1) Sample Collection; 2) Feature Selection; 3) Classifier Algorithm.

3.3.1 Sample Collection

The samples are collected from the real online log of the original version RS. The positive samples are the questions that the user had examined and clicked. However, the negative samples are not questions that are unvisited by the user, but the questions we randomly choose from the question pool. As besides having no interest, there are many reasons may explain why people do not click on a question in the list. Our offline questionnaire investigation shows that most people did not click on a recommended question because they just did not notice it. Thus the probability that the randomly chosen questions coincide with the user's interest can be negligible, considering the huge size of our question pool.

3.3.2 Feature Selection

From our previous investigations, two factors are essential in the user's decision of following a question. One is the user's attributes. The other is the correlation degree between the user interest and a question.

3.3.2.1 User Attributes

As has been mentioned in Section 3.2.1, user attributes suggest the user's preference in the Q&A community. The basic user attributes contribute to many low level features in our model. For example, a woman who is over 30 with a PhD degree may have a tendency to receive more question recommendations than others who don't have any expertise. Other features from the statistics are also useful in prediction, such as the participation history of a user in the Q&A community. We have observed that the more active a user is, the more likely he will accept the recommended questions.

3.3.2.2 Correlation Degree

The major purpose of the user interest model, which we have described in Section 3.2.2, is to support the calculation of sematic correlation degree between user interest description and an input question. The correlation degree is a combined similarity degree of many natural language aspects consist to the user interest model. For the utilization of the interest term vector, we calculate the number of matched terms, cosine similarity and bm25 between the user interest term vectors and question vectors. Then the matched terms and the semantic similarity are measured between the input question and the related questions in a user's model from both the term angle and the topic angle.

3.3.3 Classification Algorithm

In the Enlister design, two principles are crucial in the classifier selection. First, the classifier should be capable of performing a probabilistic classification, as we need not only the class label, but also the confidence degrees. Second, it needs to be a (generalized) linear classifier, which meets the requirement of online system response latency. Based on these two principles, we choose maximum entropy classifier. The probability P of a user u will click the question q can be calculated as follows:

$$P(c=1|u,q)=\frac{\exp(\sum_{i\in I}\partial_i P_i(c=1|u,q)-\partial_0)}{1+\exp(\sum_{i\in I}\partial_i P_i(c=1|u,q)-\partial_0)} \quad (3.1)$$

With ∂_i $(i\in I)$ are the features.

The global optimization solution is maximizing a logarithmic posteriori against the training set by some optimization methods, such as limited-memory Broyden-Fletcher-Goldfarb-Shanno (L-BFGS) and Stochastic Gradient Descent (SGD). We have tried these two methods in our experiment respectively. The result shows that the L-BFGS optimization is slightly better in this application scenario. After the model training process, we apply 10-fold cross validation to test it. The evaluation criteria include precision, recall etc. More details will be discussed in section 5.

3.4 Diversity Adjustment

Based on previous user research and eye-ball tests, we found the head part and the tail part of a list garnered most attention from the users.

In the Enlister, we use different filter algorithms for the head and the tail. For the head part, we apply a loose filtering algorithm, which only deletes some apparent duplication in the list. While for the tail part, we use a strict filtering algorithm to take out any questions that have noticeable semantic level similarity to each other in the list. In this way, the measurement leads to the novel appearance.

4. SYSTEM SETUP

The most important concept in the Enlister system design is real-time CTR prediction. The major data process can be described as follows.

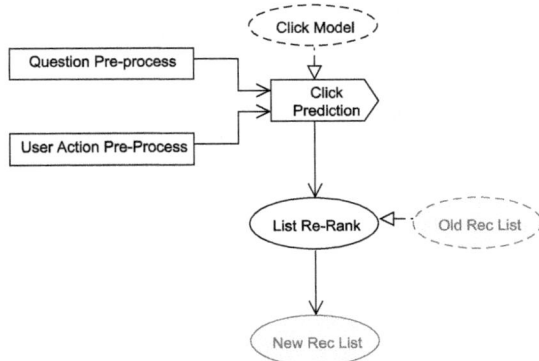

Figure 2: Stream computing model of Enlister

As shown in the Figure 2, the whole process is divided into 3 stages. The first stage includes 2 sections, one of which is used to extract the features of input questions and the other is used to generate user model from user actions in Baidu Knows online service system. The second stage is the CTR prediction where the user-action features, the input question features and the pre-trained click model are arranged together. With all the input parameters, we can calculate the probability of the user being interested in answering the input question. The last stage is the re-rank section. This is where the functionality of the list padding and diversity adjustment algorithm occurs in the whole data process.

For building the data processing flow, we construct multiple logic queues between processing nodes. The processing nodes are grouped into several node groups. Each group represents a simple logic section in the Figure 3, such as the pre-process section and the prediction section. If any group shows sign of lacking processing ability, we could just add nodes to that node group to solve the scalability problem [8].

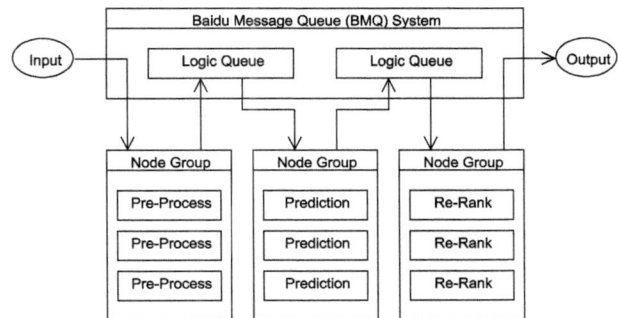

Figure 3: System architecture of Enlister

5. EXPERIMENT & EVALUATION

5.1 Evaluation Metrics

To evaluate the CTR model, we apply standard metrics in our experiments, such as precision, recall and accuracy.

As for a real online RS, users' reaction to the recommendation is a more convincing metric.

5.2 Experiment

5.2.1 Sample Selection

100,000 questions that had been viewed and clicked by users are selected from users' logs as positive sample, which involves 10 thousands users with 10 records per user on average. The negative samples could be built with two different types of data sources, the weak negative or the random negative. The weak negative samples are questions from real recommendation lists that are never clicked by users. The random negative samples are questions that chosen randomly from a large question pool. The data in Table 2 gives us the evaluation of different ways of negative samples selection. It is clear that using the random negative samples in training model is better in this case.

Table 2: Performance on different negative sample sets

	Accuracy	Precision	Recall
Weak negative sample	59.54%	64.16%	55.45%
Random negative sample	82.97%	91.73%	72.46%

5.2.2 Sample Proportion

The ratio of the positive and negative samples proportion is another issue to be settled. To find the proper ratio, we trained the model on datasets with different sample proportion settings. The results in Table 3 indicate that while the proportion ratio of positive and negative samples is 50% to 50%, a better classifier performance is produced.

Table 3: Performance on different sample proportions

Proportion	Accuracy	Precision	Recalls
90%:10%	51.03%	50.52%	99.87%
70%:30%	87.58%	91.14%	84.35%
50%:50%	88.14%	95.07%	80.44%
30%:70%	86.41%	96.84%	75.28%
10%:90%	81.93%	98.37%	65.11%

5.2.3 Optimization Algorithm

As referred in Section 3, SGD and LBFGS algorithms are considered as the optimization algorithm in the maximum entropy model training. The data in Table 4 shows LBFGS is slightly better than SGD in this application. As a result, we choose the LBFGS algorithms for the online evaluation.

Table 4: Performance optimizations on a small dataset

	Accuracy	Precision	Recall
LBFGS	88.08%	95.00%	81.93%
SGD	87.23%	93.42%	78.60%

5.3 Online Evaluation

Enlister was released to the Baidu Knows users and an online evaluation was conducted from Feb. 11th, 2012.

Table 5: Users' reaction to the recommendation

	Previous RS	Enlister	Increase	Promotion
Click	926875	1499940	573065	61.83%
Answer	129229	174622	45393	35.13%

In Table 5, the number of clicks/answers to recommended questions is given. As can be seen from the table, the total number of clicks/answers on the questions from the Enlister RS is increased by a big margin compared with the previous RS.

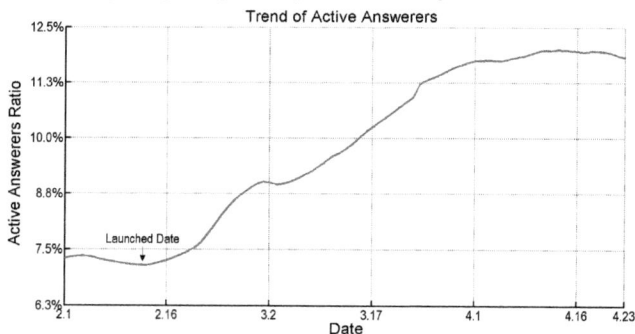

Figure 6: Trend of the active answerers after Enlister was launched

Essentially, as shown in Figure 6, the promotion of the active answerer number suggests a much favorable user experience, where the active answerer is defined as one who answered at least one question in the past 30 days.

6. CONCLUSION

In the Enlister project, we have successfully built a real-time RS that serves millions of users every day. The evaluation data shows that the algorithm and system design fit the recommendation scenario quite well. Great improvement had been made on the accuracy and time-sensitive issues. The number of active users had grown substantially after the system was officially launched.

The result data just illustrates the efficiency and performance of our current solution. Still, there are a lot of other recommendation algorithms and optimization methods to be considered in our future improvement. Two potential aspects are the timing of recommendation and the utilization of relationships between users.

Finally, we hope our work will inspire and encourage more people to build large scale recommender systems for helping users retrieve useful information.

7. ACKNOWLEDGMENTS

The authors would like to thank all the colleagues in Baidu Inc. who contributed to the Enlister project in various ways, especially Hao Tian, Jian Xian, Junyu Cai, Kai Chai, Xin Sun.

The authors are grateful to Dr. Evan Xiang for his comments on the early draft of this paper.

8. REFERENCES

[1] Michael J. Pazzani and Daniel Billsus. 2007. *Collaborative Filtering Recommender System*. Lecture Notes in Computer Science, 2007, volume 4321/2007, 291-324, DOI: 10.1007/978-3-540-72079-9_9

[2] Olivier Chapelle , Ya Zhang, *A dynamic bayesian network click model for web search ranking*, Proceedings of the 18th international conference on World wide web, April 20-24, 2009, Madrid, Spain

[3] M. Regelson and D. Fain. *Predicting click-through rate using keyword clusters*. Proceedings of the Second Workshop on Sponsored Search Auctions, 2006.

[4] Matthew Richardson , Ewa Dominowska , Robert Ragno, *Predicting clicks: estimating the click-through rate for new ads*, Proceedings of the 16th international conference on World Wide Web, May 08-12, 2007, Banff, Alberta, Canada [doi>10.1145/1242572.1242643]

[5] Xin Jin, Yanzan Zhou , Bamshad Mobasher. *A maximum entropy web recommendation system: combining collaborative and content features*. Proceedings of the eleventh ACM SIGKDD international conference on Knowledge discovery in data mining, August 21-24, 2005, Chicago, Illinois, USA [doi>10.1145/1081870.1081945]

[6] Yutaka Kabutoya, Tomoharu Iwata, Hisako Shiohara, Ko Fujimura. *Effective Question Recommendation using Multiple Features for Question Answering Communities*. IPSJ Transaction on Database (TOD), Vol.3, No. 4, 34-47, 2010

[7] Thomas Hofmann. *Probabilistic latent semantic indexing*. Proceedings of the 22nd annual international ACM SIGIR conference on Research and development in information retrieval.50-57, August 15-19, 1999, Berkeley, California, United States [doi>10.1145/312624.312649]

[8] M. Stonebraker, U. Cetintemel, and S. Zdonik: *The 8 requirements of real-time stream processing*, SIGMOD, 2005, Baltimore, Maryland, USA

HeyStaks: A Real-World Deployment of Social Search

Barry Smyth
CLARITY: Centre for Sensor
Web Technologies
University College Dublin
Dublin, Ireland
barry.smyth@ucd.ie

Maurice Coyle
HeyStaks
NovaUCD
University College Dublin
Dublin, Ireland
maurice@heystaks.com

Peter Briggs
HeyStaks
NovaUCD
University College Dublin
Dublin, Ireland
maurice@heystaks.com

ABSTRACT

The purpose of this paper is to provide a deployment update for the HeyStaks social search system which uses recommendation techniques to add collaboration to mainstream search engines such as Google, Bing, and Yahoo. We describe our the results of initial deployments, including an assessment of the quality of HeyStaks' recommendations, and highlight some lessons learned in the marketplace.

Categories and Subject Descriptors

H.3.3 [**Information Search and Retrieval**]: Information Filtering; H.3.5 [**Online Information Services**]: Web-Based Services

Keywords

social search, recommender systems, deployed application

1. INTRODUCTION

Recently researchers have begun to recognise the potential of web search as a platform for a more social and collaborative approach to information discovery. For example, researchers in the area of *collaborative information retrieval* have sought to make web search more collaborative and more social in a variety of ways; see for example [1–3]. The driving insight is that search is an inherently collaborative affair as people frequently search for similar things in similar ways [6, 8, 10, 12]. By embracing collaboration mainstream search deliver improved result relevance especially in the face of ongoing challenges by co called *content farms* and increasingly aggress SEO (search engine optimization) strategies[1]. For instance, novice searchers have much to gain from the search experiences of more expert searchers. But such experiences and recommendations need to be integrated into mainstream search contexts.

---

[1]http://searchengineland.com/google-forecloses-on-content-farms-with-farmer-algorithm-update-66071

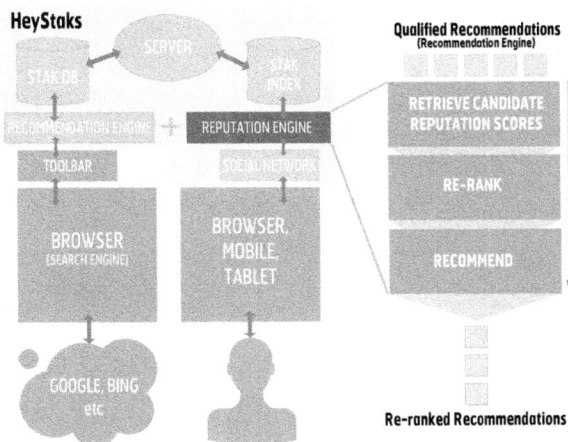

Figure 2: The HeyStaks system architecture and outline recommendation model.

This is a key objective for HeyStaks: an approach to social search that adds a layer of collaboration and recommendation to mainstream search engines like Google and Bing. The recommendation technology that drives HeyStaks has been described in detail elsewhere [5,9–11]. For this purpose of this short paper we review our deployment experiences as HeyStaks has graduated from the research lab to the marketplace. This includes a summary of the lessons learned and key evaluation results based on the first months of usage.

2. A REVIEW OF HEYSTAKS

HeyStaks combines ideas from social networking, content curation, and web search to provide a platform for search collaboration. Users can create named (public or private) staks on topics of interest and they can share these staks with friends and colleagues. As users search, their search experiences are used to populate staks; search histories are stored anonymously in staks. And when a stak member performs a new search, she may benefit from recommendations from her staks, perhaps based on the recent searches of friends. For example a group of vacationing friends might create a stak called *Ski Whistler 2012* to share their searches for their upcoming ski-trip. As they each search the results they find are added to the stak and recommendations are made based on the searches of the group. So, for example, one group member, eager to ensure they stay some-

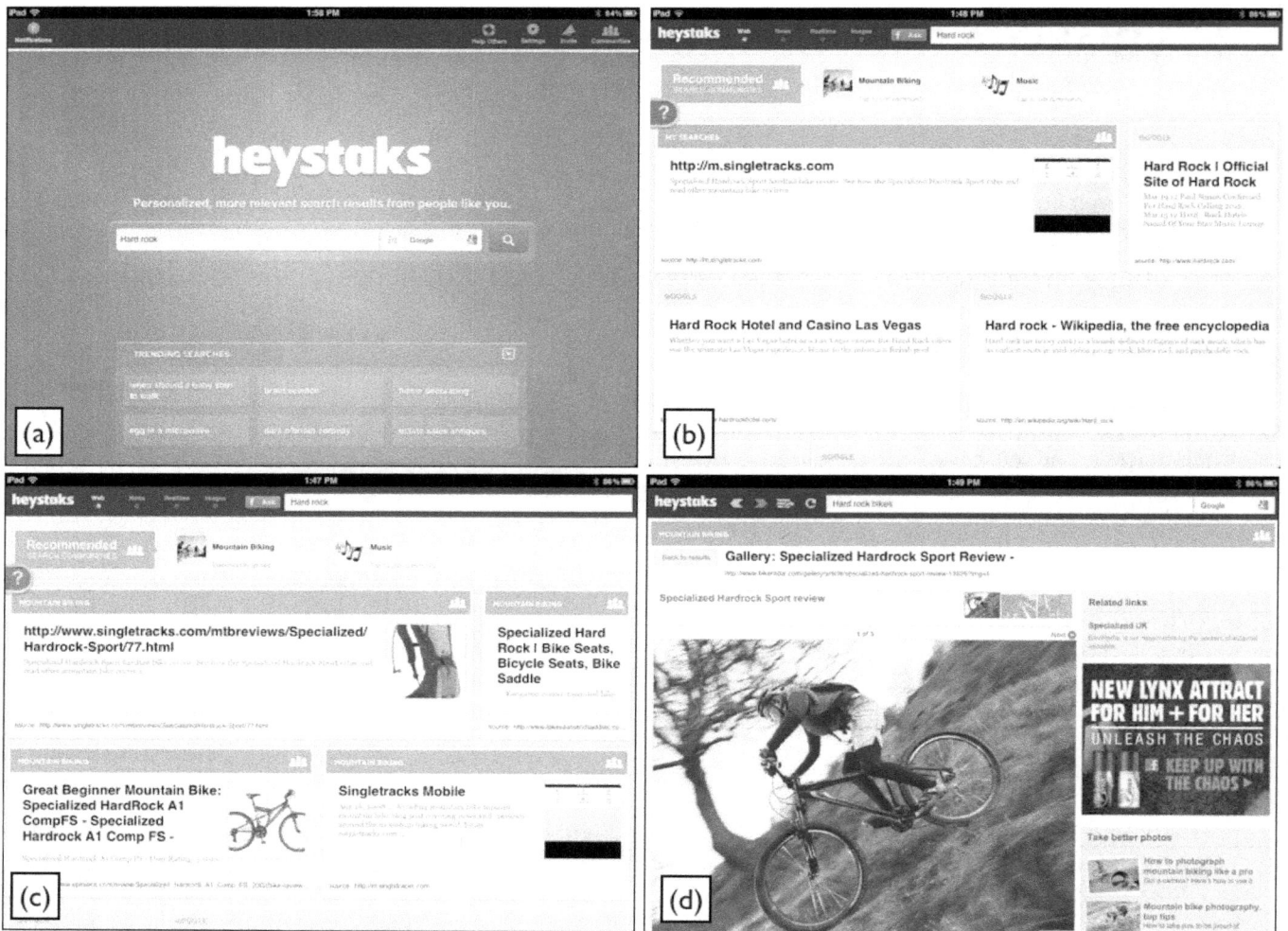

Figure 1: The HeyStaks iPad app: (a) the search homescreen; (b) a search results display with recommended communities/staks; (c) A search results page with recommendations from a joined stak; (d) an individual results page.

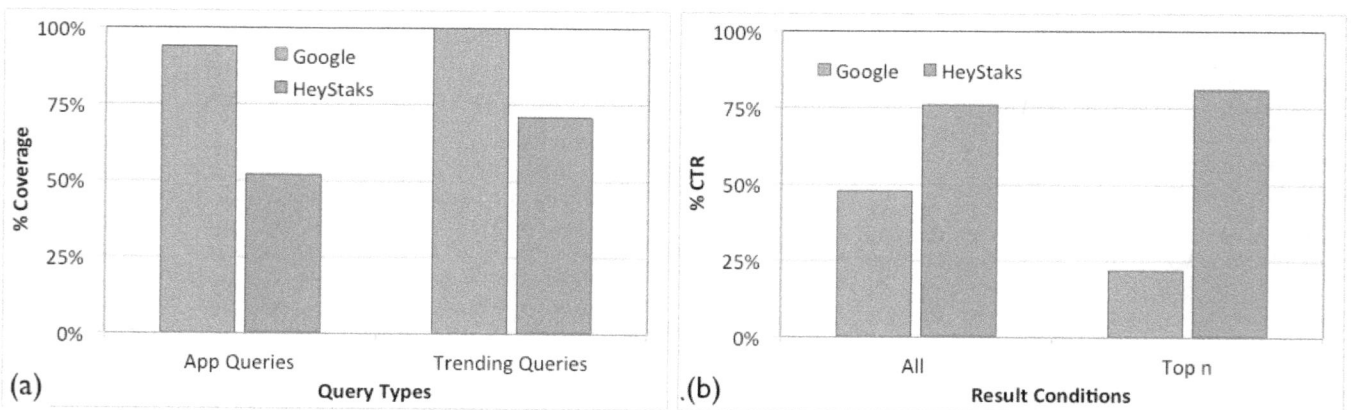

Figure 3: (a) Query coverage for Google and HeyStaks based on in-app and trending queries; (b) CTR for Google and HeyStaks for *All* and *Top n* conditions.

where with free wifi, might search for *"whistler accommodation free wifi"* and benefit from relevant specific results that were found by her friends.

Early versions of HeyStaks delivered social search via a browser plugin, to integrate directly with mainstream search engines at the interface level; examples of this can be seen in [9]. More recently HeyStaks has been deployed through a range of mobile and tablet apps for iOS and Android requiring a different approach to search engine integration via search engine APIs. In the remainder of this section we will provide some examples from the iPad HeyStaks app and summarise the system architecture and basic recommendation components.

2.1 HeyStaks for iPad

Figures 1(a-d) show screenshots from the iPad app. Our user starts with a query for *"hard rock"*; see Figure 1(a). The default results from Google tend towards music and casinos; see Figure 1(b). In this instance our searcher is interested in mountain biking; there is one recent example of this type of result recommended from their personal "My Searches" stak but in addition HeyStaks recommends two staks as possible sources of additional results, one is a "Mountain Biking" stak and the other a "Music" stak; these are presented above the main search results in Figure 1(b). By taping on either of these stak communities the user can join them and benefit from more topically focused recommendations. In this case, the user opts to join the "Mountain Biking" stak and instantly receives a new set of recommendations, this time from the relevant search histories of the most reputable members of this particular stak. The recommendations are inserted at the top of the results list as in Figure 1(c). Finally, the user selects one of the results and views it within the app from where they can chose to tag or share it with their contacts or social networks; see Figure 1(d).

2.2 System Architecture

The HeyStaks system architecture is shown in Figure 2, highlighting two important elements: the client-side apps (browser, mobile, tablet) and the backend HeyStaks server. The client apps provides direct access to key HeyStaks functionality and recommendations and integrate directly with mainstream search engines. The HeyStaks server manages the individual staks (indexing individual pages against query or tag terms and positive or negative votes), the stak database (stak titles, members, descriptions, status, etc.) and the core recommendation and reputation engines.

The recommendation engine at the heart of HeyStaks uses a combination of content-based and social recommendation strategies to identify and rank pages for recommendation to the searcher, based on their target query and stak membership. Its ranking component uses a relevance score to prioritise pages that are similar to the target query and that were frequently selected by stak members for similar search contexts. In addition, this relevance score is combined with the reputation of the stak members who originally sourced the page in question. This reputation score is based on a collaboration model in which reputation propagates between users when results which originated from one user are recommended and selected by another. For the interested reader, these relevance [9] and reputation [4] components have been described and evaluated elsewhere.

3. LIVE-USER EVALUATION

In October 2010 HeyStaks received seed funding to productise its social search technology and bring it to market as a consumer enhancement for traditional search engines. This culminated in key product releases in the second half of 2011. In this section we will summarise recent usage results based on an analysis of approximately 50,000 live users during the first 3 months of 2012, generating an average of about 4 unique searches per user per day. We will focus in particular on the coverage and relevance of HeyStaks' recommendations versus the organic results returned by Google.

3.1 Coverage

By coverage we mean the percentage of search queries for which Google/HeyStaks can return results. Early on coverage was one of the key challenges faced by HeyStaks: in the absence of a critical mass of staks and stak content recommendations were rare and so the potential benefits to new users were limited. This is a classic example of the cold-start problem in conventional recommender systems [7], which we shall return to later.

To evaluate coverage we use two sources of test queries. First there are the queries submitted by the HeyStaks users themselves – we refer to these as *app queries* – and they are important because their coverage tells us how often our current users benefit from HeyStaks recommendations. However, there is the concern that these early-adopter users are probably less representative of future users and so it is also important to understand the likely coverage of HeyStaks as it attracts more mainstream users in the future. To evaluate this coverage we use an alternative set of test queries based on the trending query lists available through Google; we refer to these as *trending queries*.

The results are presented in Figure 3(a). As expected Google offers near perfect coverage for both app and trending queries,; interestingly it fails to hit 100% for the app queries which highlights the somewhat specialised nature of these early-adopter queries. In contrast, HeyStaks does not provide such complete coverage. But it can generate recommendations for more than 50% of app queries and almost 75% of trending queries. In other words, the current users of HeyStaks enjoy at least some recommendations for half of their queries and so the majority of new users will see recommendations during their daily searches. The higher coverage for the trending queries bodes well for future users and reflects well on some of the bootstrapping techniques that have been used to drive coverage during the early stages of deployment (see Section 4).

3.2 Relevance

Query coverage is important but only if the recommendations prove to be relevant. The standard bearer for relevance in web search is the *click-through rate* (CTR) and, all other things being equal, results that attract a higher CTR are likely to be more relevant than those that do not. In this section we calculate two different variations on the CTR measure. First we calculate the CTR across all Google result lists and separately across all HeyStaks recommendation sets. We refer to this as the *All* condition. However, this is imperfect because it is clearly biased towards Google's longer result-lists; for a typical session Google includes 10 results where as HeyStaks includes at most 4 recommendations and usually fewer. As an alternative, we also calculate the CTR for the n HeyStaks recommendations in a session and compare this to the CTR for the top n Google results; we call this the *Top n* condition.

The results shown in Figure 3(b) are clearly very positive. In each condition we can see that HeyStaks enjoys a significantly higher CTR than Google. The difference is particularly striking for the *Top n Results* condition. In this case, HeyStaks enjoys a CTR of more than 80% compared to Google's 22% CTR on a like-for-like basis. This means that HeyStaks results attract selections almost 4 times as frequently as Google, which is surely s strong indicator that HeyStaks' more targeted, social recommendations are proving to be more relevant that the one-size-fits-all organic results from leading mainstream search engines.

4. DISCUSSION

Finally, we would like to briefly outline some of the lessons that have been learned as a result of our deployment experiences over the past 12 months. Many are not unique to the deployment of recommendation apps but we hope that they will nonetheless help others to learn from some of the challenges that we have faced.

Most recommender systems are subject to the cold-start problem and the challenge of making recommendations to users from the start. For HeyStaks this was a considerable challenge since the absence of a critical mass of users and stak content limited recommendations. To partially address this issue we added a number of high-level recommendation triggers or *filters* to staks, basically URL patterns that would trigger the promotion of certain types of organic Google or Bing results. For example, a *Gadgets & Technology* might include filters for *techcrunch.com, theverge.com, or mashable.com* to ensure these results would be promoted when returned by Google or Bing. This provided an opportunity to kick-start recommendations for new users by including reliable filters in newly created staks.

Another important lesson concerned the need to reduce app complexity. The HeyStaks browser apps contain a wide variety of functions, including allowing users to create and maintain their own staks. The motivation here was obviously to use this as an opportunity to crowd-source the early staks and their content. However, this functionality was rarely used and so increased app complexity. The mobile and tablet apps provided a more streamlined service, removing stak creation features, and instead prioritising the recommendation of pre-existing staks for users to join at search time. As well as simplifying the user experience this also helped ensure that new users were recommended high-quality staks on topics that mattered to them.

This simplification limits the opportunity to crowd-source new staks which in turn motivates the development of a semi-automatic stak creation workflow for use in-house. The precise details of this are beyond the scope of this short paper but briefly, by harvesting and clustering trending terms from public search and social network sources it is possible to produce a comprehensive list of stak topics. These topics and their associated seed terms are expanded to produce a comprehensive set of popular queries. And by submitting these queries to a range of search resources (Google, Bing, YouTube, Delicious, etc.) we obtain a large collection of candidate pages. We developed a rating tool to evaluate the quality of these pages, seeding new staks with those pages that passed the ratings test, and extracting recurring filter urls from these pages. Thus it was possible to create a large volume of staks to cover trending topics and ensure that they were bootstrapped with high quality content and thus capable of generating reliable recommendations.

We have provided a brief update on our experiences developing and deploying a recommendation system for social search. Early indications speak to the value of the approach as users benefit from superior search results when compared directly to mainstream search engines like Google. We have also outlined some important deployment lessons that we have learned that may be of interest for other recommender system deployments.

5. ACKNOWLEDGEMENTS

This work is supported by Science Foundation Ireland under grant 07/CE/I1147 and HeyStaks Technologies Ltd

6. REFERENCES

[1] Brynn M. Evans and Ed H. Chi. An elaborated model of social search. *Inf. Process. Manage.*, 46(6):656–678, 2010.

[2] Brynn M. Evans, Sanjay Kairam, and Peter Pirolli. Do your friends make you smarter?: An analysis of social strategies in online information seeking. *Inf. Process. Manage.*, 46(6):679–692, 2010.

[3] Gene Golovchinsky, Pernilla Qvarfordt, and Jeremy Pickens. Collaborative information seeking. *IEEE Computer*, 42(3):47–51, 2009.

[4] Kevin McNally, Michael P. O'Mahony, Maurice Coyle, Peter Briggs, and Barry Smyth. A case study of collaboration and reputation in social web search. *ACM TIST*, 3(1):4, 2011.

[5] Kevin McNally, Michael P. O'Mahony, Barry Smyth, Maurice Coyle, and Peter Briggs. Social and collaborative web search: an evaluation study. In *IUI*, pages 387–390, 2011.

[6] Meredith Ringel Morris, Jaime Teevan, and Katrina Panovich. A comparison of information seeking using search engines and social networks. In *ICWSM*, 2010.

[7] Andrew I. Schein, Alexandrin Popescul, Lyle H. Ungar, and David M. Pennock. Methods and metrics for cold-start recommendations. In *SIGIR*, pages 253–260, 2002.

[8] Barry Smyth, Evelyn Balfe, Jill Freyne, Peter Briggs, Maurice Coyle, and Oisín Boydell. Exploiting query repetition and regularity in an adaptive community-based web search engine. *User Model. User-Adapt. Interact.*, 14(5):383–423, 2004.

[9] Barry Smyth, Peter Briggs, Maurice Coyle, and Michael P. O'Mahony. Google shared. a case-study in social search. In *UMAP*, pages 283–294, 2009.

[10] Barry Smyth, Maurice Coyle, and Peter Briggs. The altruistic searcher. In *CSE (4)*, pages 360–367, 2009.

[11] Barry Smyth, Maurice Coyle, and Peter Briggs. Communities, collaboration, and recommender systems in personalized web search. In *Recommender Systems Handbook*, pages 579–614. 2011.

[12] Barry Smyth, Jill Freyne, Maurice Coyle, and Peter Briggs. Recommendation as collaboration in web search. *AI Magazine*, 32(3):35–45, 2011.

A System for Twitter User List Curation

Igor Brigadir
School of Computer Science and Informatics
University College Dublin, Ireland
igor.brigadir@ucdconnect.ie

Derek Greene, Pádraig Cunningham
School of Computer Science and Informatics
University College Dublin, Ireland
{derek.greene,padraig.cunningham}@ucd.ie

ABSTRACT

With increased adoption of social networking tools, it is becoming more difficult to extract useful information from the mass of data generated daily by users. Curation of content and sources is an important filter in separating the signal from noise. A good set of credible sources often requires painstaking manual curation, which often yields incomplete coverage of a topic. In this demo, we present a recommender system to aid this process, improving the quality and quantity of sources. The system is highly-adaptable to the goals of the curator, enabling some novel uses for curating and monitoring lists of users.

Categories and Subject Descriptors

H.3.3 [**Information Storage and Retrieval**]: Information Search and Retrieval – Information filtering

Keywords

Content curation, Social media monitoring, Network analysis, Social network discovery

1. INTRODUCTION

Storyful[1] is a social media news agency established in 2010 with the aim of filtering newsworthy content from the vast quantities of noisy data on social networks such as Twitter and YouTube. To this end, Storyful invests considerable time into the manual curation of content on these networks. Twitter users can organise the users they follow into *lists*. Storyful maintains user lists as a means of monitoring breaking news. These lists can be constructed manually, but this process is time-consuming, and risks incomplete coverage of all aspects of a news story. Therefore, to support these curation tasks, we have developed and deployed a web-based system for exploring the Twitter network and recommending the important users that form the "community" around a news story (see Fig. 1). Currently the system is being used to monitor over 100 news stories, mining microblogging data for a diverse range of topics, from the United States 2012 presidential election to the political situation in Afghanistan. A video of the system in use is available online[2].

[1] http://www.storyful.com

[2] http://www.youtube.com/watch?v=rMfN59bmEyc

Copyright is held by the author/owner(s).
RecSys'12, September 9–13, 2012, Dublin, Ireland.
ACM 978-1-4503-1270-7/12/09.

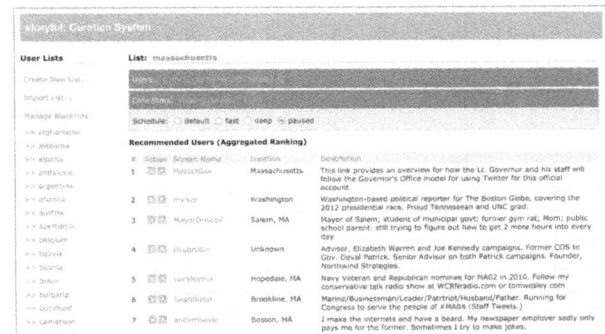

Figure 1: Screenshot of the curation system, showing candidate users recommended for addition to a user list covering the 2012 Republican Party nomination for the state of Massachusetts.

2. CURATION SYSTEM

The input to the system is an initial *seed list* of Twitter users that was manually labelled as being relevant to a particular news story, topic of interest or location. The bootstrap phase retrieves network structure around the egos in the seed list. Information retrieved consists of user profile information, friend and follower links, user list membership information, and tweets. The extent of the exploration process can be easily controlled by pre-set configuration settings – effectively controlling the trade-off between running time and accuracy.

After the initial bootstrap phase, the system maintains two distinct lists of users:

Core list: List of the actual Twitter user accounts used by journalists during content curation for the chosen news story. Initially this will contain the members of the seed list.

Candidate list: User accounts that are not in the core list, but may potentially be relevant for curation. Initially this will consist of the set of non-seed list users identified during the bootstrap phase.

Based on the candidate list, the recommendation engine will then produce a ranked list of potentially relevant users for promotion to the core list. Based on these recommendations, a human curator can select users to add to the core list, or filter incorrect recommendations. Once the core list has been modified, the system updates the network structure around the core list, to reflect (a) changes in membership of the core

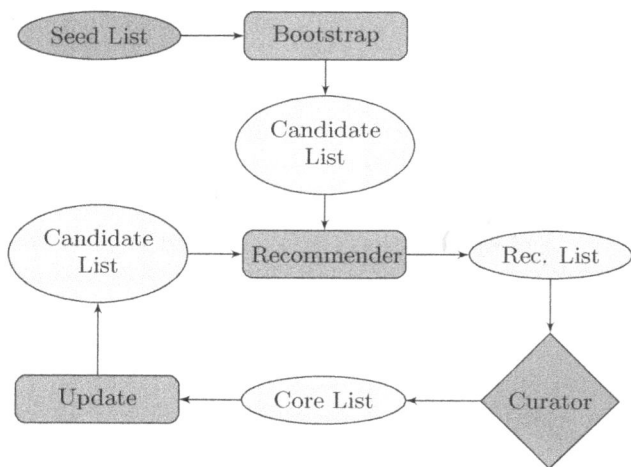

Figure 2: Overview of curation support system, illustrating workflow between phases.

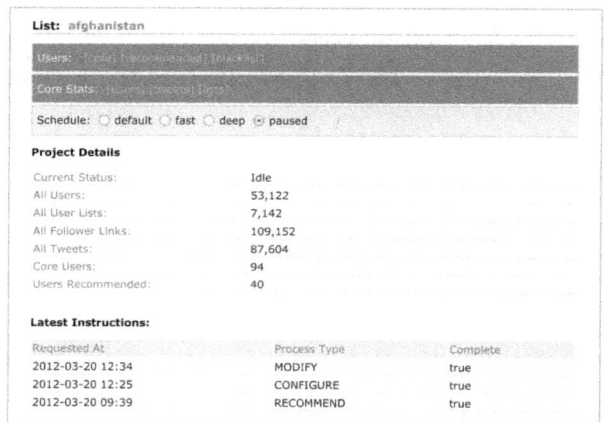

Figure 3: Screenshot presenting data statistics regarding a single user list monitoring the political situation in Afghanistan.

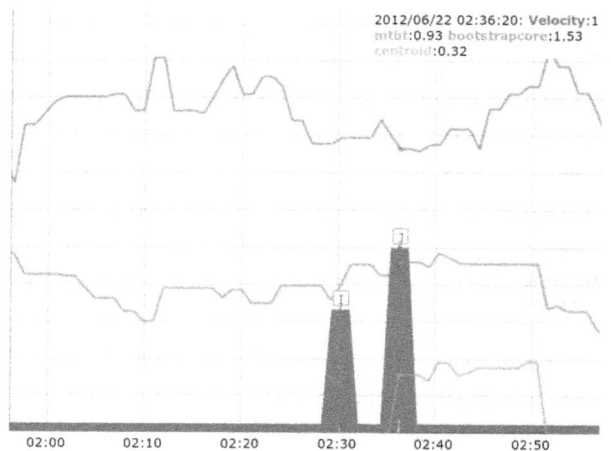

Figure 4: A *velocity chart* showing spikes of tweeting activity corresponding to breaking news events being discussed on the Afghanistan user list.

list, and (b) general changes in the larger Twitter network since the last update. Again the extent of the exploration during the update process can be controlled by user-defined settings. The system then iterates between recommendation and update phases (see Fig. 2).

Rather than using a single view of the network to produce recommendations, we employ a multi-view approach that produces recommendations based on different graph representations of the Twitter network surrounding a given user list, and combines them using an SVD-based aggregation approach [1]. Information from multiple views is also used to control the exploration of the Twitter network. This is an important consideration due to the limitations surrounding Twitter data access. These network views include friend and follower graphs, mention and retweet graphs and views based on how users are sorted into lists by other users – effectively crowd-sourcing the list curation in part.

The system runs on open source software and can be deployed on a single server or an Amazon EC2 instance for example. The specific hardware requirements will depend upon on the total number of lists being monitored. As of April 2012, Storyful are currently monitoring 124 stories, topics, and geopolitical regions using the deployed system.

The curation system is not limited to generating recommendations. When the system is used to monitor lists covering dozens or hundreds of news stories, it will often be important for journalists to focus their limited time on a subset of these lists, where breaking developments are occurring. To facilitate this kind of prioritisation, we monitor the *velocity* of the lists being monitored by the system. The velocity measure is a combination of several indicators including tweet similarity, the level of activity of the core users, and tweet frequency. The velocity measure can detect significant or unusual tweet activity, often indicating a breaking news story (see Fig. 4).

3. CONCLUSIONS

While the curation system presented here is primarily used as a support tool within Storyful for curating lists of sources for online journalism, we are currently investigating its use in other applications that involve social media exploration and

insight. For instance, one current experiment uses the system to identify the presence of extremist groups on Twitter. Another application is the identification of spam accounts or bots that share common links in one or more Twitter network views.

Acknowledgments

This research was supported by SFI Grant 08/SRC/I1407 (Clique: Graph and Network Analysis Cluster). The authors thank Storyful for their participation in the development and testing of the system.

4. REFERENCES

[1] D. Greene, G. Sheridan, B. Smyth, and P. Cunningham. Aggregating Content and Network Information to Curate Twitter User Lists. Under Review, 2012.

CubeThat: News Article Recommender

Sidharth Chhabra
University of Michigan
School of Information
sidc@umich.edu

Paul Resnick
University of Michigan
School of Information
presnick@umich.edu

ABSTRACT

The CubeThat browser extension for Chrome displays recommended additional news stories related to the same topic as the current news story. The recommended stories are organized into clusters, and clusters that the user has already sampled from are grayed out, in order to encourage users to explore multiple aspects of a story. Users can also provide feedback to improve the clustering, by dragging stories from one cluster to another.

Categories and Subject Descriptors

H.5.m Information interfaces & presentation: Miscellaneous

General Terms

Algorithms, Design, Experimentation, Human Factors.

Keywords

Browser Extension, Recommender, Diversity, News, Clustering.

1. INTRODUCTION

News, which was once heavily dependent on print media, is now consumed digitally. The Web provides much more freedom in selection of news sources and types than any other previous media. But Sunstein & others have argued that that freedom will lead to social fragmentation because self-selection of stories will lead people to only expose themselves to like-minded viewpoints [1]. Similarly, Eli Pariser has argued recommender systems, in a race to please the users, create invisible, personal universes of results around them [2]. Among other problems, this could lead people to think that their opinions are more broadly shared than they actually are [3]. However, personalization systems can be designed to deliberately inject diversity into a user's information diet too [7].

Park et. al.[4] found that when people are presented with stories organized into meaningful clusters, in their NewsCube system, it cued people to read more and diverse stories about a single topic. Eveland et. al. [5] found that structuring knowledge so that reading is not a linear activity can be effective in creating dense knowledge structures. In other words, providing affordances while a person is reading an article to traverse other parts of the same topic can lead to denser knowledge structures.

We will demo a browser extension that encourages users to explore multiple aspects of news topics, by recommending other stories related to the current story the user is reading. The recommendations are not personalized, but as in the NewsCube system, the browser keeps track of which clusters each user has already explored and thus cues them to explore additional clusters.

2. ARCHITECTURE

Figure 1 shows the system architecture. A crawler gets stories from popular news sites, and clusters them by topic. The recommendation engine further clusters articles within each topic. Initially, it applies the algorithm used in Newscube [4] to cluster the articles; later it incorporates user feedback to modify the within-topic clustering. Clustered news articles are served through a browser extension (Chrome) which is the interface for the recommender.

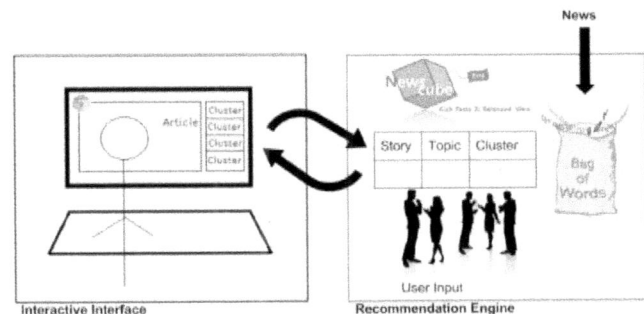

Figure 1: CubeThat Architecture

When a person is reading a news story, he/she can click on the extension icon in the browser header to get recommendations for that article. Figure 2 shows an example recommendation for a news article. For each cluster in a topic, four or less relevant stories are picked. For each story, the source and count of how many Twitter tweets link to the story are shown. A user who disagrees with the clustering of stories can drag a story from one cluster to another and this move affects the recommendations that will be shown to other users as well. If the last story is removed from a cluster, that cluster is deleted, as illustrated in Figure 3, where the sole story from cluster 5 has been moved to cluster 4.

The browser extension tracks which stories users have opened. If a user has previously opened any story in a cluster, that cluster is shaded, as in the original NewsCube interface [4]. Figure 4 shows the recommendation after another story has been read from a different cluster.

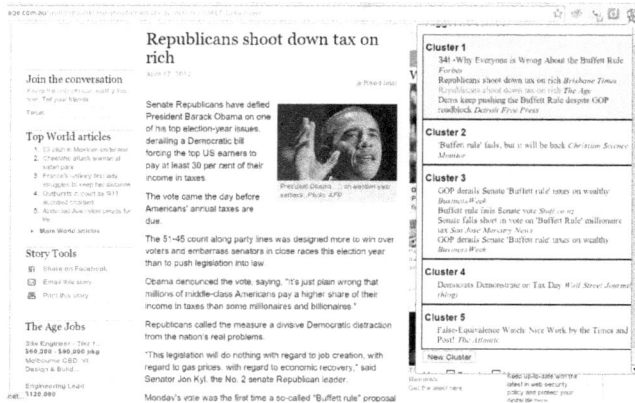

Figure 2. An example recommendation; darkened cluster shows the user has already sampled from it.

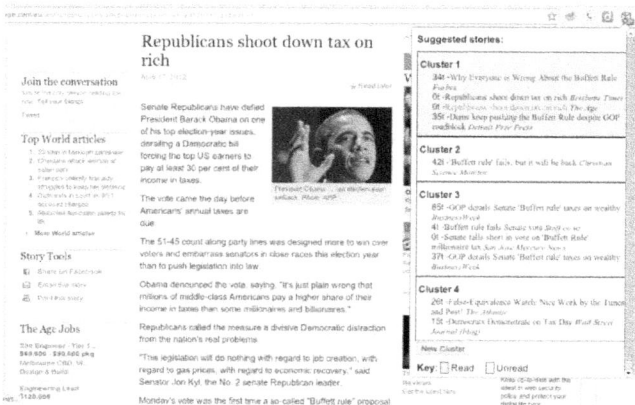

Figure 3. The recommendations after the sole story from cluster 5 is moved to cluster 4.

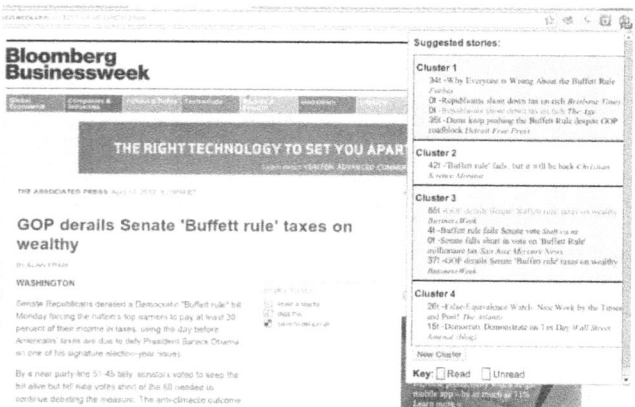

Figure 4. The recommendations after a story from cluster 3 has been read.

2.1 CLUSTER CHANGE

When users reassign stories to different clusters or new stories arrive that are part of an existing topic, we have to find a suitable clustering of new & existing stories that takes into account both the text features and the previous user actions. There has been a lot of research on constrained clustering [6] but here we face multiple constraints together. Again, we propose and use a simple solution to the problem. First, from the story assignment to clusters, as adjusted by the most recent user, we compute the cluster centroid for each cluster in the database. For stories that were not displayed as part of the recommendations, including new stories, we compute the distance from cluster centroids and allocate each to the closest. The simplicity of the algorithm ensures that running time is linear in the number of stories about the topic.

3. CONCLUSION & FUTURE WORK

In this paper, we proposed a novel approach to tackle the concerns regarding polarization due to selective exposure when reading news online. The proposed Chrome extension should cue people to read about more different aspects of a news story. In future work, we plan to incorporate other diversity factors besides the words used in the stories, such as the political ideology of the stories, which we have been able to classify automatically using semi-supervised learning algorithms [8]. The usability of the extension will need to be honed through iterative testing and the efficacy of the extension in encouraging people to consume more diverse news will need to be measured through field trials.

4. REFERENCES

[1] Cass R. Sunstein. 2001. *Republic.Com*. Princeton University Press, Princeton, NJ, USA.

[2] Eli Pariser. 2011. *The Filter Bubble: What the Internet is Hiding from You*. Penguin Group , The

[3] Glenn S. Sanders and Brian Mullen. 1982. Accuracy in perceptions of consensus: Differential tendencies of people with majority and minority positions. European journal of social psychology 13(1).

[4] Souneil Park, Seungwoo Kang, Sangyoung Chung, and Junehwa Song. 2009. NewsCube: delivering multiple aspects of news to mitigate media bias. In *Proceedings of the 27th international conference on Human factors in computing systems* (CHI '09). ACM, New York, NY, USA, 443-452. http://doi.acm.org/10.1145/1518701.1518772

[5] W.P. Eveland and J. Cortese. 2004. How Web Site Organization Influences Free Recall, Factual Knowledge, and Knowledge Structure Density. Human Communication Research, 30: 208–233. doi: 10.1111/j.1468-2958.2004.tb00731.x

[6] Sugato Basu, Ian Davidson, and Kiri Wagstaff. 2008. *Constrained Clustering: Advances in Algorithms, Theory, and Applications* (1 ed.). Chapman & Hall/CRC.

[7] Garrett, R. Kelly, and Paul Resnick. 2011. Resisting Political Fragmentation on the Internet. Daedalus 140(4): 108-120.

[8] Zhou, D. X. and P. Resnick (2011). Classifying the Political Leaning of News Articles and Users from User Votes Fifth International AAAI Conference on Weblogs and Social Media, Barcelona, AAAI.

The Demonstration of the Reviewer's Assistant

Ruihai Dong, Markus Schaal, Michael P. O'Mahony, Kevin McCarthy, and Barry Smyth
CLARITY: Centre for Sensor Web Technologies
School Of Computer Science & Informatics
University College Dublin, Ireland
{firstname.lastname}@ucd.ie

ABSTRACT

User generated reviews are now a familiar and valuable part of most e-commerce sites since high quality reviews are known to influence purchasing decisions. In this demonstration we describe work on the *Reviewer's Assistant* (RA), which is a recommendation system that is designed to help users to write better quality reviews. It does this by suggesting relevant topics that they may wish to discuss based on the product they are reviewing and the content of their review so far.

Categories and Subject Descriptors

H.3.3 [**Information Storage and Retrieval**]: Information Search and Retrieval

Keywords

product reviews, computer support, topic recommendation, recommendation systems

1. INTRODUCTION

It is increasingly important for sites like Amazon and TripAdvisor to help people find and create high-quality reviews since people are increasingly turning to user-generated reviews to support their decision-making. Amazon provides users with the opportunity to rate reviews based on their *helpfulness* and allows prospective customers to rank reviews by their helpfulness score.

In this demo we instead focus on the task of creating new reviews and specifically how recommendation techniques may support users as they write product reviews. This work is inspired by GhostWriter [1], which uses case-based reasoning techniques to make suggestions for the user at review-writing time. Briefly, GhostWriter maintains a case base of review experiences, made up of previously helpful reviews and indexed by the terms that occur in these reviews. As the user writes a new review, the text that they write serves as a query against this case base, GhostWriter retrieves a ranked set of similar review cases, and extracts a set of frequent noun phrases to recommend to the user. Dong et al. [2] adopt a similar approach but compare nouns vs. noun phrases in order to make better suggestions to the user.

In the following sections we describe our approach to re-

view recommendation and how this has been incorporated into a browser plugin and illustrate how it works.

2. THE REVIEWER'S ASSISTANT

The Reviewer's Assistant has been developed as a browser plugin so that it can integrate directly with review systems across a wide variety of web sites, see [2]. Briefly, the Reviewer's Assistant takes the form of an additional recommendation module that appears on review-creation pages. These recommendations are review topics that have been extracted from a database of reviews (on Digital Cameras in this instance) and selected and ranked according to the content of the user's review so far. At any time the user can even select a topic to see an expanded list of relevant review fragments which is a good aid for the review process.

The basic Reviewer's Assistant system architecture has been described in detail elsewhere (see [2]). For the purpose of this demonstration it is sufficient to outline its four main components. The *filtering* module is responsible for extracting and indexing a suitable set of high quality reviews for a given product class. This can be as straightforward as using review quality indicators as a guide (e.g. on Amazon). When a user begins writing a new review, their early content is used as a query against these filtered reviews and the *mapping* component is responsible for identifying a set of k similar reviews ($k = 50$); currently, we use a simple Jaccard similarity metric. Next, from this set of relevant reviews the *extraction* component extract nouns, noun phrases, and ultimately topics from these reviews. The crucial step of identifying and ranking frequent sets of nouns is performed by association rule mining (see [2]).

In order to generate a rich set of suggestions we apply association rule mining both at the review level and at the sentence level to extract a set of association rules as the basis for recommendation. Briefly, our system takes, as input, the set of similar reviews and the current review text by the user and outputs a set of n ($n = 10$) suggestions. If association rules do not lead to a set of n recommended topics then further topics are extracted from reviews based on a simple frequency count as a fallback strategy. Thus as the user continues to type their review, extracted rules can be triggered leading to updated recommendations. Equally, as topics are covered by users in their writing, corresponding recommendations fall away.

In this demonstration we show a novel variation of the Reviewer's Assistant with an extended core recommendation strategy. Instead of noun suggestions that are drawn directly from past reviews we are now presenting topic-based

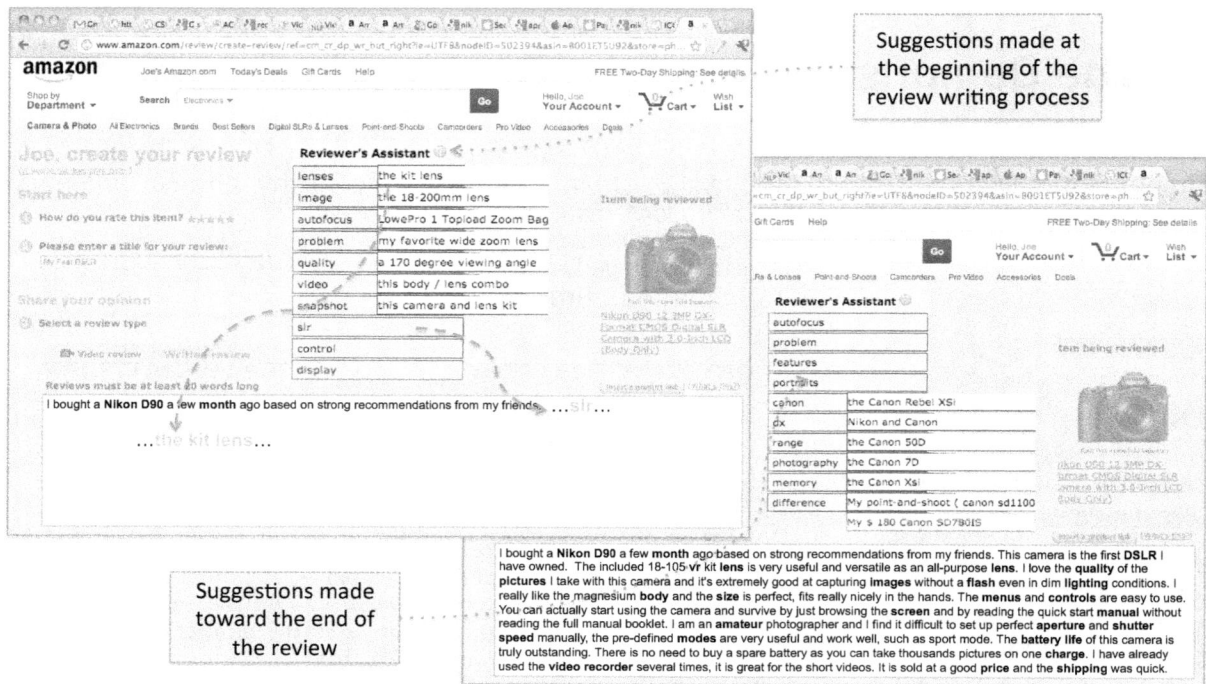

Figure 1: The Reviewer's Assistant in action on Amazon.

suggestions to the user. The benefit of this is two-fold: 1) product topics can be more intuitive for new reviewers than nouns; and 2) by recognizing common topics across suggestions we can improve recommendation ranking and diversity and reduce redundancy. At this moment, we adopt a very simple approach to topic modeling based around a hand-coded set of topics for the target domain, with each topic associated with a synonym set.

3. THE RA IN ACTION

Figure 1 shows the Reviewer's Assistant in action for our user, Joe, who is reviewing a recently purchased Nikon D90 SLR camera on Amazon. Joe is presented with the usual Amazon review creation screen and the figure shows the Reviewer's Assistant overlay; the RA widget can be dragged to any suitable location on screen. The RA presents a dynamic set of updating review suggestions. Figure 1 (left) shows some of the suggestions presented to Joe during the early stages of the review. In this case we see a number of suggestions for some common review topics for this product, including the *lens*, the *image* and *video* capability. As shown, Joe can view review fragments that relate to a particular topic by hovering the mouse over the topic. For example, in this case the fragments *"the kit lens"*, *"the 18-200mm lens"* etc. are displayed for topic *"lens"*. Joe can now directly select suggestions, e.g. *"the kit lens"*, and the selected suggestion will be inserted at the end of his current writing.

In Figure 1 (right) we see a snapshot towards the end of the review writing. This time Joe is presented with additional topics, many of which are more specialised or not uniquely related to the specific product to provide the reviewer with an opportunity to broaden their review. Note also that recommended topics that are already covered by

Joe in this snapshot are highlighted by emphasising the respective topic terms in the sentences of the current writing, including *lens, aperture, shutter, battery life,* etc.

4. CONCLUSIONS

We have evidence that the Reviewer's Assistant supports the writing of better quality and more comprehensive reviews, see [3]. This is of crucial importance for e-commerce sites such as Amazon, TripAdvisor, etc. User trial participants generally provide positive feedback about the quality of suggestions and their overall experience.

5. ACKNOWLEDGMENTS

This work is supported by Science Foundation Ireland under grant 07/CE/I1147.

6. REFERENCES

[1] D. Bridge and P. Healy. The GhostWriter-2.0 Case-Based Reasoning system for making content suggestions to the authors of product reviews. *Knowledge-Based Systems*, 29(0):93 – 103, 2012.

[2] R. Dong, K. McCarthy, M. P. O'Mahony, M. Schaal, and B. Smyth. Towards an Intelligent Reviewer's Assistant: Recommending Topics to Help Users to Write Better Product Reviews. In *Procs. of IUI: 17th International Conference on Intelligent User Interfaces, Lisbon, Portugal, February 14-17, 2012*, pages 159–168, 2012.

[3] R. Dong, M. Schaal, M. P. O'Mahony, K. McCarthy, and B. Smyth. Harnessing the Experience Web to Support User-Generated Product Reviews. In *20th International Conference on Case-Based Reasoning, Lyon, France*, 2012. to appear.

Recommenders for the Enterprise: Event, Contact, and Group[1]

Abigail Gertner
The MITRE Corporation
202 Burlington Road
Bedford, MA 01730
+1 781-271-3130

gertner@mitre.org

Beth Lavender
The MITRE Corporation
202 Burlington Road
Bedford, MA 01730
+1 781-271-6362

lavender@mitre.org

James Winston
The MITRE Corporation
202 Burlington Road
Bedford, MA 01730
+1 781-271-7018

jwinston@mitre.org

ABSTRACT

This extended abstract provides a summary of MITRE's Recommender demonstrations for ACM RecSys 2012. We have three demonstrations: Event Recommender, Contact Recommender, and Group Recommender. MITRE is a technology company with thousands of staff and varied government customers. Help from recommenders is becoming a necessity to find the right resources to bring to bear on large, multidisciplinary problems. We have developed and deployed these recommenders to explore their effectiveness, implementation challenges for an enterprise, and their use models.

Categories and Subject Descriptors

H.4.m [Information Systems Applications]: Miscellaneous

General Terms

Algorithms, Design, Experimentation

Keywords

recommenders, enterprise

1. INTRODUCTION

Recommenders are everywhere in commerce and social networking sites. They benefit from millions of users and are driven by the potential for increased revenue. By contrast, recommenders for business enterprise are scarce. While the enterprise has orders of magnitude fewer users, it has the potential for higher fidelity information about employees to make effective recommendations. The authors are exploring the benefits and use of recommenders at the MITRE Corp, a high tech knowledge-based enterprise.

Currently at MITRE there are three deployed prototype recommender systems to help knowledge workers make connections with pertinent events, people, and groups. The first demo is an 'Event Recommender' available from MITRE's intranet portal. The second demo is a 'Contact Recommender' used in MITRE's external facing social and collaboration system called 'Handshake'. The third demo is a 'Group Recommender' also in MITRE's external Handshake environment. All three leverage the open source recommender framework, Apache Mahout.

2. EVENT RECOMMENDER

MITRE is a diverse technical company with many events for disseminating and sharing technical information. The company's innovation program hosts internal events to bring staff together to form new research topics and proposals. Additionally, there are informal sessions such as lunchtime "brown bags" that are very popular. MITRE has a long standing culture of organizing more formal Technical Exchange meetings (TEMs) to support the networking and sharing customer and technical challenges.

Due to the volume of events, staff may miss an event announcement or not have the time to drill into materials to determine if it's of interest and value. The Event Recommender draws from the corporate feed of events, together with the user's event interests and organizational memberships to make event recommendations.

The initial challenge we had to overcome was seeding the user interest data. Very few events at MITRE required registration so no data from past events attended was available to initialize the staff event interests. We developed an Event Recommender gadget for our intranet portal that provides easy and convenient features for the user to add events to their calendar and remove events from the suggested event list. We are capturing the users "add to calendar" and "remove from list" actions to determine

---

[1] 'Approved for Public Release: 12-2151

their event interest. There has been a pent up demand for the "add to calendar" feature so it is a winning combination.

In addition to collecting user's event interest data, we are addressing the cold start problem by considering the social structure of the corporation. Events may be recommended that are being attended by other members of the user's department, division (collection of departments) or center (collection of divisions). The closer the organizational relationship, the higher the recommendation is weighted. Future versions of the event recommender will also make use of social network connections.

We developed the backend services to persist and feed event and user data to the Event Recommender engine which is using the collaborative filtering algorithm from Apache Mahout [2]. The recommendation is calculated each time the user goes to their portal page or when they select or delete an event from the list.

The demonstration will show the easy to use interface, convenience features and the updated recommendations based upon the user's selection of events.

3. CONTACT AND GROUP RECOMMENDERS

MITRE has an externally facing collaboration and business networking system called Handshake. MITRE staff invite in colleagues to Handshake who are customers, industry or academic partners, or consultants. People can connect or follow each other and join groups for discussion and information sharing. The Contact and Group Recommenders use the Handshake network, profile, tags, and group membership to recommend new groups or colleagues. Recommendations are displayed in widgets that users can add to their Handshake home page.

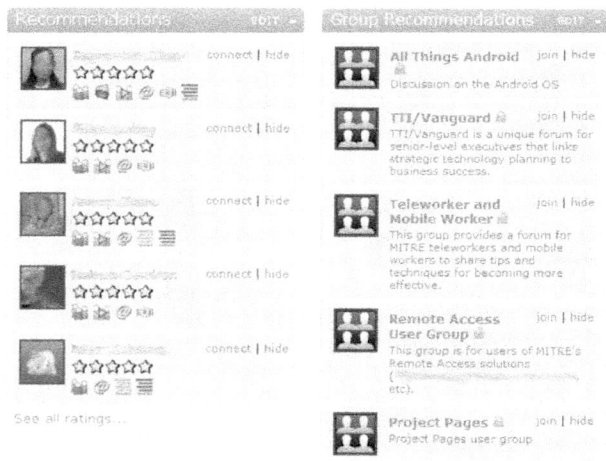

1.1 Contact Recommender

In a large enterprise there are often many people working on similar topics or problems, and there is a significant risk of duplicating work, or missing relevant information, particularly when the organization is geographically dispersed. The Handshake contact recommender [1] is designed to recommend

people to connect with based on common interests or activities. The contact recommender uses information from multiple sources, both within the Handshake environment and on MITRE's intranet, and aggregates that information to determine who would most benefit from being in contact. The sources that are used include MITRE's internal social bookmarking tool, wiki environment, email lists, as well as group memberships and existing connections on Handshake.

The contact recommender presents a list of recommended contacts on the user's Handshake home page, including convenient links to add those people as connections as well as links for retrieving additional explanations regarding the reason for the recommendation. Contact recommendations have been available on Handshake since June of 2011. User studies of the contact recommender have found that the recommendations are relevant and that people are using them to make connections.

1.2 Group Recommender

One of the most frequently requested features since the release of Handshake has been an easier way to find relevant groups to join. Handshake has hundreds of user-created groups and the number of groups is growing all the time. Currently the only ways to find out about groups are by word of mouth, or by manually browsing the list of groups. To address this problem, we have implemented a group recommender for Handshake that uses Apache Mahout's [2] collaborative filtering algorithms to recommend groups. The group recommender is in prototype stage and has not yet been deployed on Handshake.

Group recommendations are subject to a significant cold start problem in that the users who most need recommendations for groups are new Handshake members who belong to very few groups, if any. Therefore, the collaborative filtering approach needs to be augmented by additional recommendation approaches for these new users. In addition to group recommendations derived from collaborative filtering, the group recommender will make use of the enterprise-wide social and organizational structure, similar to the event recommender, in order to recommend groups that may be of interest to new users.

The demonstration will show the recommendation widgets for both contact and group recommendations as they are displayed to users on Handshake.

4. ACKNOWLEDGMENTS

This research was supported by funding from the MITRE Innovation Program. The authors would like to thank Justin Richer and Donna Cuomo for their contributions to this work.

5. REFERENCES

[1] Gertner, A. S., Richer, J., and Bartee, T. Q. 2010. Contact Recommendations from Aggregated On-Line Activity. In *Proceedings of the 8th Workshop on Intelligent Techniques for Web Personalization and Recommender Systems* (Big Island of Hawaii, June 20, 2010). ITWP'10. 44-52.

[2] Owen, S., Anil, R., Dunning, T. and Friedman, E. 2012. *Mahout in Action.* Manning, Shelter Island, New York.

Integrated Content Marketing

[Extended Abstract]

James Griffin
idio Ltd.
Floor 33, Euston Tower
London, UK
james.griffin@idioplatform.com

ABSTRACT

idio is a technology startup that works with major brands in both the UK and the US to drive Customer Centricity through Content Marketing. idio's full-stack platform fuses Natural Language Processing, Data Mining, Web Analytics and Recommendation Engines at scale.

Categories and Subject Descriptors

H.3.5 [**On-line Information Services**]: Web-based Services

Keywords

Personalization, Natural Language Processing, Content Marketing

1. INTRODUCTION

idio are a content marketing technology startup based in London, UK. The core technology is a platform called Integrated Content Marketing (ICM) which provides greater customer understanding through personalized delivery of semantically tagged content through a variety of channels. idio work with a number of major brands in both the UK and USA to help build long-term customer value and brand loyalty.

2. INTEGRATED CONTENT MARKETING

idio's platform fuses a number of technology components: an advanced content aggregation engine that utilises Natural Language Processing techniques to derive semantic meaning from unstructured text; a powerful cross-channel analytics suite capable of identifying and understanding individuals across both brand touch-points and social platforms; and a real-time user modeling and recommendation suite capable of personalizing the customer experience on all channels.

The platform focuses itself on the pure Content Marketing challenges and then provides integrations with various third-party systems to realise content delivery across web, email and social channels. The platform also integrates with the organisation's Customer Relationship Database, and pushes all customer knowledge to this system so that it can be utilised at all points across the organisation.

2.1 Content Ingestion

idio believe the power behind good content strategy lies in breadth and volume of content, and this is facilitated through idio's content aggregation engine. The system is capable of ingesting content of various types from numerous sources, from RSS feeds through Video channels to Social streams. The textual components of this content are passed through Natural Language Processing tehcnologies - some in-house and some third party. These perform Named Entity Recognition, the results of which are disambiguated using against Freebase, a linked-data service. This allows us to build up a feature profile for each item of content:

$$F_c = \{\langle f, w \rangle, ...\}$$

Where w represents the weight of the feature f against content item c. This weight is measured against a corpus of content and gives a measure of relevance.

ICM's management interface gives brand managers and agencies control over the content sources, ingestion process, moderation and content presentation. It allows the user to semantically define content segmentation rules, which not only facilitate realisation of the content strategy but also act as filters for irrelevant content.

2.2 Analytics

idio's analytics suite is capable of gathering implicit interaction data from each customer across all their interaction channels. As the customer interacts with content, the semantic profiles of the content are used to build a semantic profile of the user. These interactions are logged through a combination of Javascript-based web analytics techniques, click tracking and pixel tracking depending upon the channel.

To further augment this, and to partially solve the cold start problem, idio have a social data mining platform that can, at scale, identify users across social networks and apply similar Natural Language Processing techniques to build the user profile from their explicit public profile information.

By studying the feature profiles of content items the user has interacted with and also by analysing their social activity using the same NLP techniques, we can build a feature profile for each user:

$$F_u = \{\langle f, w \rangle, ...\}$$

Where w represents the weight of the feature f against the user u. This weight represents the degree to which the user has shown an interest in that particular feature, and is computed by taking various factors into account, such as how the user interacted with the content (passively or actively) and when it was that the interactions took place.

ICM's management interface presents the user with a "single customer view", which describes the user both semantically and demographically, pulling together data from the organisation's CRM database and idio's customer analytics suite. This data can then be fed into campaign management tools to facilitate segmentation.

2.3 Personalization

The drive towards Customer Centricity in the world of marketing is powered by experiences tailored for the individual. idio acheive this by personalizing all content streams for each individual on all channels. With Guinness, idio power a personalised email campaign which consistently has the highest open and click-through rates across all of Diageo's email campaigns, achieving a 345% increase in open rates and 861% increase in click through.

The personalization engine takes advantage of both the content anlytics and the customer analytics suites to power a Hybrid Recommendation System that fuses two styles of recommendation.

- **Journey analysis.** Using the data from the analytics suite, entire user journeys can be identified and clustered. The current user's journey can then be compared against these clusters to provide a collaborative-style content ranking.

- **Content-based Recommendation.** Using the feature profiles for users and content items, the system performs traditional content-based recommendation. Due to the wide range of domains that the system is deployed under, the similarity measurement algorithms are designed to be plug-and-play, so that simple per-client configuration can tailor the recommendation system.

The engine is architected for scale, and is capable of making content decisions in real-time. The architecture ensures that the user profile is always up-to-date, and allows the engine to use short-term and long-term user profiles to improve recommendation quality.

idio's personalization engine has been honed across many deployments and media types. It began life as a personalized music magazine, and has evolved to handle other media and domains, including applications around business-realted newswire, recipes and healthy living, and lifestyle content. Most recently, it was employed and extended in a TSB-funded project which integrated content from Film, Music and Editorial domains in collaboration with The Guardian, The BFI and Decibel.

2.4 Next-Best-Content

The next generation of ICM will include a technology called Next-Best-Content, which takes its lead from Next-Best-Action, a Predictive Analytics technology that transformed the CRM world in the 90's. Next-Best-Content will build upon idio's customer understanding platform with Adaptive Modeling techniques, allowing the platform to understand the user's journey in terms of their goals, knowledge and context to predict not just what the best item of content is for that user, but also how, on what channel and when

that content should be presented. This technology is in Research & Development phase, set to launch Q1 2013. The project has been part-funded by the Technology Strategy Board in the UK.

2.5 Architecture

The platform is entirely cloud-based and takes advantages of scalable distributed architectures to ensure each component is as fault-tolerant, available and real-time as possible. A blend of technologies is used throughout the stack, with NoSQL & SQL databases working in harmony alongside services written in various languages utilising different serving methodologies. For user modeling, a hybrid real-time and batch processing system ensures that user models are always up to date with the user's latest interactions, and allows the personalization engine to perform recommendations against both a short-term and a long-term user profile.

The outward facing platform interface is primarily a RESTful API, which enables integration with an organisation's encumbant systems.

3. INSTALLATIONS

idio currently have many live projects utilising the platform. The most successful of these include:

- **Guinness 1759 [?]:** This project with Diageo provides a media property for the Guinness brand displaying content around Sport & Lifestyle. Personalized email campaigns are sent weekly, and the system also integrates to deliver content on Twitter and Facebook.

- **Slim.Fast [?]:** This project worked with Unilever to promote the Slim.Fast 1.2.3 plan, and features content around Healthy Living & Weightloss. The site also integrates with a monthly email campaign. In the first 3 months after launch, Slim.Fast received more signups than they had done in the 3 years previous.

- **The Media Briefing [?]:** This project aggregates content around the B2B Media world. It publishes around 200 articles per day, and powers multiple weekly email campaigns.

4. REFERENCES

[1] Guinness 1759. http://1759.guinness.com.
[2] Slim.Fast Magazine. http://magazine.slimfast.co.uk.
[3] The Media Briefing. http://themediabriefing.com.

Using Ratings to Profile Your Health

Neal Lathia
Computer Laboratory, University of Cambridge
15 JJ Thomson Avenue, Cambridge CB3 0FD
neal.lathia, cecilia.mascolo@cl.cam.ac.uk

ABSTRACT

The widespread adoption of mobile technology allows personalised applications to be deployed in an increasing host of contexts; user modelling, profiling, and personalised recommendations are becoming an integral component of mobile information systems. Furthermore, mobile technology enables the recording and collection of facets of daily life, which has given rise to the notion of the *quantified self*; researchers operating at the intersection of computer and social science are now seeking to understand how these mobiles' data can aide the design of health interventions and inform future psychological and social science research. In this work, we describe the design of a personalised mobile application that seeks user feedback and builds a user profile about people's gastrointestinal health using ratings and tags. We describe the application's design and the personalised health insights it provides (and, particularly, why recommendations were not designed as a means for self-diagnosis).

Categories and Subject Descriptors

H.5 [**Information Interfaces and Presentation**]: General; J.3 [**Computer Applications**]: Life and Medical Sciences

General Terms

Design, Human Factors

Keywords

Health, Android Application

1. INTRODUCTION

Mobile phones have become items that we keep within arms length throughout over 75% of the day [1]. Moreover, by ubiquitously connecting us to the web at all times, they are allowing for personalised information systems to be designed and deployed in contexts where they were previously unavailable, such as health, where recommendations can be used to intervene on and change people's behaviour. There is a rising interest in using mobile phones to deliver personalised health-related interventions by, for example, monitoring people's mood throughout the day [2]. Historically,

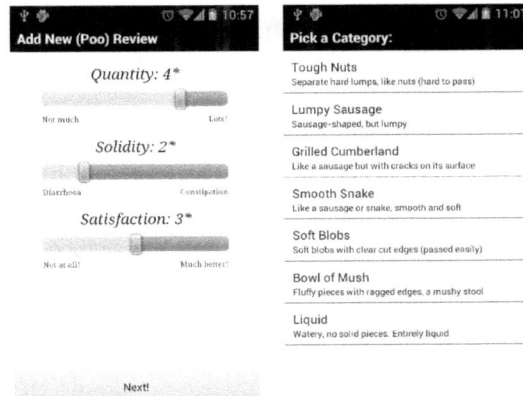

(a) Initial Ratings   (b) Bristol Stool Chart

Figure 1: Adding ratings to *The Poo Review*. The user (a) first gives 3 ratings for *Quantity*, *Solidity*, and *Satisfaction*, and then (b) selects a category from the Bristol Stool Chart.

however, systems that aim to help people manage aspects of their health have been web and survey-based [3] and lack the data granularity and personalisation that recommender systems offer. For example, systems that help people manage minor bowel problems [4] lack a means for users to see if their health is improving, receive personalised information, or check if their goals are being met. We posit that the interaction techniques used in online recommender systems can be used in mobile applications to overcome this deficit. This work describes the design of a mobile application, *The (Poo) Review*[1], that uses ratings and tags to allow users to quickly input data about their gastrointestinal health and receive personalised information about their bowel's habits.

2. APPLICATION DESIGN

An application for the scenario above has two main goals: it should allow users to anonymously *input* data and create a personal profile and convey personalised, *educational* information about the user's health; we designed an application for experience sampling [5] with personalised feedback. The main task of the application is for users to review their current bowel movement via a 4-fold review process: (1) **Numerical Ratings**: users are asked to give a 1-5 star score for

[1]https://play.google.com/store/apps/details?id=com.poo.review

the *Quantity*, *Solidity*, and the *Satisfaction* of passing their current stool (Figure 1(a)). (2) **Categorical Choice**. The second step asks users to tag their review with a category from the 7-level Bristol Stool Chart (Figure 1(b)), which was defined in [6] as a means to monitor stools' intestinal transit time and assess the effectiveness of treatments for bowel diseases. (3) **Current Location**: the user then has the option to add their current location from a pre-defined set of options (e.g., "Restaurant"). (4) **Textual Feedback**. Finally, users have the option of adding a short comment to their review.

Submitting a review results in instant feedback (Figure 2(a)), which comes in two forms. (1) **Personalised Feedback**. The user is given a fact that has been computed from the set of ratings given to date. A range of candidate facts are possible, relating to both time (e.g., the user's 7-day frequency average, the number of reviews in the last 24 hours, the time of the last review), and category (e.g., the last time they input a review with the same category): a personalised fact is chosen randomly from the available candidates. (2) **Generic Fact**. The bottom half of Figure 2(a) is a generic fact under a "Did you know?" heading. These facts have been collected manually from Wikipedia, and include snippets about defecation, the digestive system, minor bowel problems, and healthy eating. The server also returns the Wikipedia link to the source of each fact, which is used to set the target of a "Read More" button that accompanies the snippet. The set of reviews that users input are also used to create a user profile, which is also split into two halves. The first section reports the user's 7-day statistics (number of reviews, average reviews per day). The second section contains the aggregate data. This includes both the total number of reviews as well as distributions of aspects of the reviews: for example, Figure 2(b) shows the aggregate distribution of categories selected by the user.

3. DISCUSSION AND CONCLUSION

This paper has introduced the design of the *The (Poo) Review* Android application: a tool to rate and tag bowel movements in order to generate and deliver personalised health information. Future iterations of the application could extend this in order to visualise correlations between activities, moods, and health (e.g., comparisons with what you eat or other aspects of users' lifestyle).

The current version of the application allows users to review their bowel movements without linking their data to any means for automated diagnosis or comparing them to any notion of "normal" behaviour. Furthermore, personalised facts about reviews are presented as-is, without any qualitative information appended. This was a specific design decision: not only do we assume that the definition of *normal* may be culturally-dependent, but Internet-based interventions for those with bowel problems actively encourage people to *not* over-monitor their bowel movements as this may lead to stress (from questioning whether they are "normal") which may exacerbate any pre-existing problem. Conversely, it would be rather straightforward to encode a means to detect and alarm the user if reviews clearly reflect, for example, early symptoms of bowel cancer [7]: the relation between recommendations and the delivery of health interventions is thus an important issue that future recommender systems in the health domain must tackle.

This application is an initial step into uncovering the po-

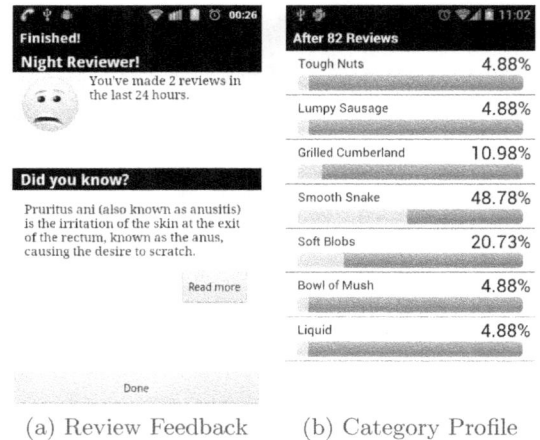

(a) Review Feedback   (b) Category Profile

Figure 2: (a) Example feedback that the user receives after "flushing" a review, including personalised and generic facts, and (b) an example from the user's profile page, showing the distribution of categories generated by all reviews.

tential of merging facets of mobile and recommender system technology in order to tackle the broader context of behavioural intervention via personalised interaction and monitoring techniques. Ratings and tags have become the standard method for web sites to solicit preference data and for users to quickly interact with web content: in the future, similar methods may be used in health-related contexts in order to achieve similar levels of data granularity and analytic insight.

4. REFERENCES

[1] A. K. Dey, K. Wac, D. Ferreira, K. Tassini, J. Hong, and J. Ramos. Getting Closer: An Empirical Investigation of the Proximity of Users to their Smart Phones. In *ACM Ubicomp*, Beijing, China, 2011.

[2] K. Rachuri, M. Musolesi, C. Mascolo, P. J. Rentfrow, C. Longworth, and A. Aucinas. EmotionSense: A Mobile Phone Based Adaptive Platform for Experimental Social Psychology Research. In *ACM UbiComp*, Copenhagen, Denmark, September 2010.

[3] L. Yardley. The Potential of Internet-Delivered Behaviour Change Interventions. *The European Health Psychologist*, 13(3):40–43, September 2011.

[4] L. Morrison. *PhD Thesis*, chapter 6. Development and Qualitative Piloting of "Gut Instincts": A Think-Aloud Study of User Engagement. University of Southampton, 2012.

[5] J. Froehlich, M. Chen, S. Consolvo, B. Harrison, and J. Landay. MyExperience: A System for in Situ Tracing and Capturing of User Feedback on Mobile Phones. In *ACM MobiSys*, San Juan, Puerto Rico, 2007.

[6] S. Lewis and K. Heaton. Stool Form Scale as a Useful Guide to Intestinal Transit Time. *Scand J. Gastroenterol*, 32(9):920–924, September 1997.

[7] Cancer Research UK Blog. Bowel Cancer: 40 Years of Progress But Early Detection is Key. http://scienceblog.cancerresearchuk.org/, April 2012.

Finding a Needle in a Haystack of Reviews: Cold Start Context-Based Hotel Recommender System Demo

Asher Levi[*], Osnat (Ossi) Mokryn
School of Computer Science
Tel Aviv Yaffo College
Israel
asherlv2@gmail.com, ossi@mta.ac.il

Christophe Diot, Nina Taft
Technicolor ltd.
Paris, France, Palo Alto, USA
christophe.diot@technicolor.com,
nina.taft@technicolor.com

ABSTRACT

Online hotel searching is a daunting task due to the wealth of online information. Reviews written by other travelers replace the word-of-mouth, yet turn the search into a time consuming task. Users do not rate enough hotels to enable a collaborative filtering based recommendation. Thus, a cold start recommender system is needed.

This demo describes briefly our cold start hotel recommender system, which uses the text of the reviews as its main data. We define context groups based on reviews extracted from TripAdvisor.com and Venere.com. We introduce a novel weighted algorithm for text mining.

We implemented our system which was used by the public to conduct 150 trip planning experiments. We compare our solution to the top suggestions of the mentioned web services and show that users were, on average, 20% more satisfied with our hotel recommendations. We outperform these web services even more in cities where hotel prices are high.

Categories and Subject Descriptors: H.3.3 [**Information Search and Retrieval**]: Information Search and Retrieval - *Information filtering*

General Terms: Algorithms.

Keywords: Recommender systems, opinion/text mining, context-aware recommender systems, common traits, sentiment analysis.

1. INTRODUCTION

Producing recommendations for travel is inherently difficult, as an individual rarely rates more than a small number of hotels and thus rich profiles cannot be built. Having limited or no information about the user translates to a user *cold start* recommendation [1]. An intelligent cold start recommender will minimize a new user's effort while still learning enough to recommend the user a product that is likely to be of her interest.

A lot of research has already been performed in the area of recommender systems and information retrieval. However, most recommender systems focus on recommending the most relevant items to users without taking into account any additional contextual information. Most existing information retrieval systems base their retrieval decisions solely on queries collections, whereas information about search context is often ignored [2, 3].

Users' search patterns are context-based. Among the plethora of reviews, readers opt for recommendations from travelers with comparable needs. A single traveler may share the same needs as other single travelers. A user traveling with her family has different needs from a user traveling on a business trip, i.e. the user context information is an important factor in choosing a hotel. When a user reads reviews she can metaphorically be seen as wearing personalized glasses. Reviews are read through those glasses, and particular words or comments will resonate, positively or negatively, with the reader based upon her needs for her upcoming trip and her personal preferences. Special attention is often given to reviews written with the same intent, or by reviewers from a comparable background. Hence, we define three types of context information. The first is *intent*, or purpose of the trip. We include 5 categories of intent, namely *business trip, single traveler on vacation, family, group, couple*. The second is *nationality*. The third context is user preferences for the different hotel aspects. These were mined from the text using an unsupervised clustering algorithm. We tagged the different clusters found in the text as *location, service, food, room, price-value quality* and the *facilities* (pool, spa, etc). Thus, a user using our system is asked to provide her trip intent, nationality, and preferences for these aspects.

We obtained data from Venere.com and TripAdvisor.com. The database contains details for each hotel: the hotel's general information, reviews and ratings. In a pre-processing phase, we mined the text and found *common traits* for each context group. These are found in the form of typical words that appear more in text written within that context but are not common for other contexts. A clustering was used to group words that refer to each aspect. Thus, at the end of this pre-processing phase, we have significant words per context, be it intent, nationality, or hotel aspect.

The core idea of the algorithm is to give more importance to reviews of people with the same contexts as the user's. Common traits per the user's context groups and words that describe favorable hotel aspects are given a higher score than other words. We further find the sentiment expressed in the review per context (i.e., positive or negative) and give a corresponding score. Thus, the final score for each review corresponds to that of a user's with comparable needs and preferences and coming from a similar background.

We implemented our system and published it for use, presenting to the users results from our system combined with the top suggestions of the above mentioned web sites. We had over 150 evaluations by friends and colleagues who looked for hotels in four major European cities. Hotels recommended by our system were favored (60.2%) compared to TripAdvisor's and Venere's top suggestions (50.8%). More significant is the fact that our raters said they would not stay in 26.4% of the top hotels recommended by these sites, whereas with our context-based recommender, their dissatisfaction was much lower at 15.9%.

---

[*]Part of this work was done while Asher Levi was visiting Technicolor lab in Palo Alto.

2. SYSTEM OVERVIEW

Figure 1 gives a brief overview of the components and steps of our method. The top 3 boxes on the left correspond to the pre-processing phase in which we define the common traits of intent and nationality groups, and define the different hotel aspects referenced in reviews, correspondingly. To find common traits for each context group, we extract the nouns and noun phrases (called *features*) from all reviews and find those that are more common for that group. These features are then assigned a weight per each context according to their relative frequency in reviews within that context. The higher the weight, the more important a feature. The common traits of context groups are the higher weight features for that group. The fourth component of the preprocessing consists of building an opinion lexicon which will allow us to analyze adjectives associated with features, and to give each feature an orientation score depending upon how positive or negative is the sentiment of any associated adjectives.

While the base weight of each feature is one in our system, features that are distinctive of several context groups may have different weight per group. The final weight for each feature is done by combining these three weights (depicted as "build feature score" in the figure).

Next we use our opinion lexicon to give each feature an orientation score. We subsequently combine the features, their weights and orientations to build a score for each sentence. The sentence scores are then combined to give an overall score for each review. This score should reflect the relative importance of the given review for the user. Reviews that are both important and positive are deemed most relevant thereby receiving the highest scores. The final score for each hotel is an average of all of its reviews, each of which is scored from the user's perspective (i.e., based on her context and preferences), and an adjustment bias calculated per the context given.

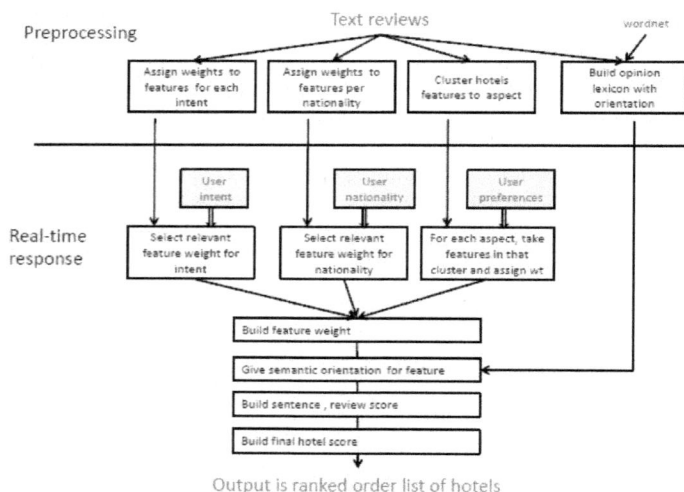

Figure 1: Overview of our approach

3. SYSTEM EVALUATION

Evaluation of a recommender system has to measure whether real people are willing to act based on the recommendations. User satisfaction with a recommender system results is well gauged with an on-line evaluation methodology. We use a methodology as in [4] that measures relative user satisfaction with one system over another. We implemented our system and made it available on the public web for use. We asked numerous friends and colleagues to evaluate our system and obtained 150 evaluations. Each experiment consisted of the following. The user inserts her search parameters: intent, nationality, aspect preferences, and a price range. Then we present the user a list of six hotels. Some are from our system, and some are the highest star ratings choices from Venere and Tripadvisor. In order to avoid biasing the user, these six hotels are presented in random order and thus the user is unaware of the source of the recommendations. Raters were shown links to the full text reviews to further explore the recommended hotels. For each one of the hotels, raters were asked to express their satisfaction by answering the question "Would you select this hotel?" with three optional answers: Yes, Maybe, No. In addition, the raters were asked to rate all the recommended hotels on the scale of $[1-5]$ to indicate whether they felt the recommendation had met their search criteria and was to their satisfaction. They were also asked to indicate which aspect was the one that most influenced their decision. Raters were specifically instructed to only select 'intents' that were realistic for them (e.g., if you don't have kids, do not select the 'family' as the intent).

Table 3 presents the overall satisfactory results. We see that for 60.2% of the hotels recommended by our system, users stated they would stay there, as compared to 50.8% from the rating systems.

Recommender System	Yes	Maybe	No
Our System	60.2%	23.9%	15.9%
Star Rating	50.8%	22.8%	26.4%

Table 1: Would you select this hotel?

To examine this in more detail than just averages, we plot the empirical histogram of the ratings given by our raters in Figure 2. The hotels recommended by our method received more 4 and 5 ratings then those the other method, and similarly our recommended hotels received fewer 1 and 2 ratings than the other method.

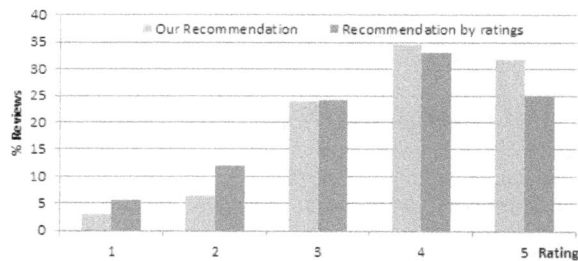

Figure 2: Ratings by recommender method

4. REFERENCES

[1] A. Schein, A. Popescul, L. Ungar, and D. Pennock, "Methods and metrics for cold-start recommendations," in *Proceedings of the 25th annual international ACM SIGIR conference on research and development in IR*. ACM, 2002.

[2] G. Akrivas, M. Wallace, G. Andreou, G. Stamou, and S. Kollias, "Context-sensitive semantic query expansion," in *AIS, 2002.(ICAIS 2002)*. IEEE, 2002.

[3] G. Adomavicius and A. Tuzhilin, "Context-aware recommender systems," *Recommender Systems Handbook*, 2011.

[4] C. Hayes and P. Cunningham, "An on-line evaluation framework for recommender systems," 2002.

Yokie - Explorations in Curated Real-time Search & Discovery using Twitter

Owen Phelan, Kevin McCarthy, and Barry Smyth
CLARITY Centre for Sensor Web Technologies
School of Computer Science and Informatics
University College Dublin, Ireland
firstname.lastname@ucd.ie

ABSTRACT

Our research involves developing technology and techniques that apply the vast sea of real-time web data to interesting problems and topics. In this demo, we will present the ongoing development of a novel real-time search and discovery service named Yokie[1] (http://yok.ie, early technology description originally published in [1]). Yokie uses the large volume of hyperlink-laden messages on social networks like Twitter as the basis of its content and ranking systems. Curated sets of users (or "Search Parties") form the basis of sourcing the content from the networks, and the metadata of the containing messages form the basis of ranking and contextual retrieval of the hyperlinks. Each hyperlink is indexed with a compound set of terms from multiple tweets (should the given hyperlink be shared more than once). This indexing step is a novel example of collaborative tagging of resources. The application is live with more than 100 users, who have performed approximately 1000 queries. We will demonstrate the main techniques and novel ranking and retrieval techniques and user features.

Categories and Subject Descriptors

H.4 [**Information Systems Applications**]: Miscellaneous

General Terms

Applications, Algorithms, Experimentation, Theory

Keywords

Search, Discovery, Information Retrieval, Relevance, Reputation, Twitter

1. DEMONSTRATION DESCRIPTION

The Yokie system can be broken down into several main components, all of which are briefly discussed in this section.

Curated "Search Parties"

A key aspect of the system is the curated list of content sources, what we've termed a *Search Party*. Users can curate dedicated search engines for personal and community use. These parties can be simple user lists, keywords, geographical metadata, or based on algorithmically generated lists. In the current prototype we have curated several seed Search Parties: *Technology*, *Irish Interest* and *World News* Twitter users (each containing 1000 users).

Content Acquisition

The system uses Twitter's API to acquire content as defined by the search parties. These messages are then stored and indexed using the service described in the following subsection. This component also carries out real-time language classification and finds other messages that contain the same URL so the system can calculate item popularity, and extracts tags so as to allow collaborative tagging.

Storage & Indexing

Once content is acquired, it is pushed to the Storage and Indexing subsystem. This is responsible for extracting metadata regarding the tweets, for instance timestamp data, hashtags (#linsanity, etc.), user profile information, location, etc. as well as the message content itself. The main content, the urlID of the URL mentioned in the message, and the timestamp is pushed to an indexer for storage and querying. In our current implementation we use Apache Solr[2] for this. We also store the remaining extracted metadata in a MongoDB[3] NoSQL database for easy retrieval and fast MapReduce functionality.

---

[1] Yokie - http://yok.ie

---

[2] Apache Solr - http://lucene.apache.org/solr/
[3] MongoDB - http://mongodb.org

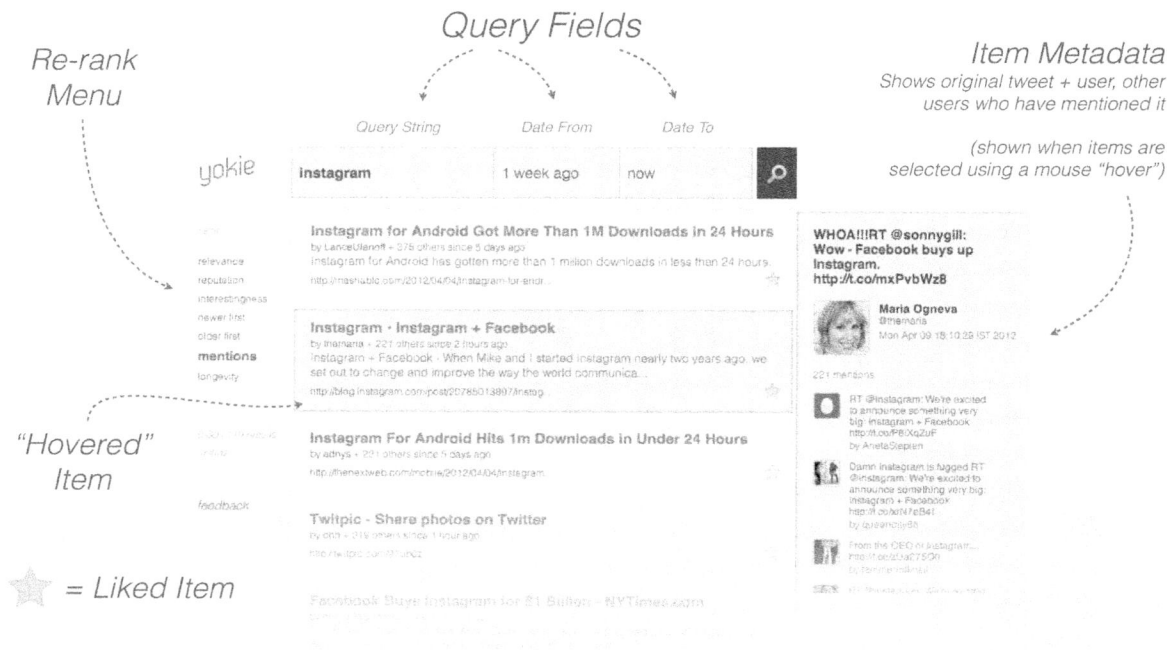

Figure 1: Yokie takes a traditional approach to Search-system layout, with query, ranking and results-list views. It also includes a UI pane for viewing extra metadata related to the item.

UI & Querying

The system is presented with a query interface, currently comprising of a query string field, and two temporal fields, *date from* and *date to*. The system takes as input a query string with an associated time window, which can be either a natural language query (e.g.. "1 day ago", "now", "last week", etc) or a fixed date ("12 June 2012"). These natural-language date strings are then parsed into UNIX timestamps. The UI also allows users to drill-down on results to explore related content such as the original tweet that the URL was shared with, the time and date it was shared, and the related Tweet mentions (if any). Each item contains a rating *star*, which users select when they believe an item is relevant to the query. The re-ranking menu allows users to re-rank the results. The querying subsystem parses user queries, based on the notation {QueryString, SearchParty, T_{max}, T_{min}}.

User & Item-based Result Ranking

Users can rank using typical relevance, which is vanilla Term Frequency Inverse Document Frequency scoring (TFxIDF) [2]. However, a set of ranking strategies beyond standard relevance have been devised using the added contextual metadata of the microblogs. These include ranking using temporal features such as item age and longevity (which is the size of the time window between the first and last mentions of the given hyperlink). Popular items can be ranked as we capture sample tweets that mention the hyperlink. Reputation is a user-based sum of follower-counts for each search-party member who have mentioned the link.

The premise of this strategy is high-volume in-links for users can be a sign of high user reputation. It is easily possible to derive a range of Reputation scores as we are specifically "listening" to people for content, as opposed to relying on a document graph that ignores content publishers. The novel Interestingness ranking algorithm is a function of popularity, level of user interactions of the items, and binary ratings.

2. CONCLUSION & FUTURE POTENTIAL

Yokie allows users to discover and rerank topical and niche webpages before they appear in old-style search engines like Microsoft's Bing. It has a considerable future in research of novel UI and ranking, collaborative search party curation, collaborative tagging, indexing and retrieval, user reputation, to name but a few.

3. ACKNOWLEDGEMENTS

This work is generously supported by Science Foundation Ireland under Grant No. 07/CE/I1147 CLARITY CSET.

4. REFERENCES

[1] Owen Phelan, Kevin McCarthy, and Barry Smyth. Yokie - a curated, real-time search and discovery system using twitter. In *2nd Workshop of Real-time and Social Web. At Recommender Systems 2011*, RSWEB'11, 2011.

[2] Fabrizio Sebastiani. Machine learning in automated text categorization. *ACM Comput. Surv.*, 34:1–47, March 2002.

pGPA: A Personalized Grade Prediction Tool to Aid Student Success

Mark Sheehan
Dept. of Computer Science & Information Systems
Bradley University
Peoria, IL, U.S.A.

mmsheehan@mail.bradley.edu

Young Park
Dept. of Computer Science & Information Systems
Bradley University
Peoria, IL, U.S.A.

young@bradley.edu

ABSTRACT

Many educational institutions are starting to make use of their scholastic data to improve the academic experience for their students. To aid in this endeavor we have developed a research prototype implementation of a collaborative filtering-based tool called the personalized Grade Prediction Advisor (pGPA). The goal of this prototype tool is to demonstrate the potential of recommender technology by providing grade predictions for upcoming courses in a student's academic career to support decision-making for administrators, students, educators, and academic advisors. In this demonstration we briefly describe the underlying technology and potential applications of pGPA. We then present how a user can interact with pGPA to produce and interpret personalized grade predictions for an individual student or group of students.

Categories and Subject Descriptors

K.3.1 [**Computers and Education**]: Computer Uses in Education – *collaborative learning.*

General Terms

Algorithms, Design, Experimentation

Keywords

Collaborative filtering, Grade prediction, Student success, Software tool

1. INTRODUCTION

Recently, recommender technology has seen application in many domains including the academic realm [1]. One major problem in the educational sphere that has been the focus of such research is improving the educational experience and success rate of students in higher education. Researchers have used recommender technology to predict the likelihood of student dropout [2], help students select appropriate majors, minors, and classes [3], and identify learning patterns and performance [4].

We set out to design and build our own recommender technology-based tool to help individuals at all echelons of the academic system make appropriate and well-informed decisions. The result of our efforts is a collaborative-filtered recommender system that utilizes past student performance, temporal elements, and contextual information to predict a student's grades for the upcoming semester. We have developed this concept into a tool we call the personalized Grade Prediction Advisor (pGPA).

The prototype of pGPA is designed to showcase the effectiveness of our approach towards reliably predicting student grades. This prototype allows users to select a subset of students from the dataset we have, determine the semester to predict grades for, configure the prediction properties, and then displays the results to the user. In addition, this prototype provides a means of examining how these predictions could be used by the proposed stakeholders of pGPA to help guide their decision-making.

2. THE pGPA PROTOTYPE

Underlying Technology: pGPA is a collaborative-filtered recommender system that uses an individual's academic history in the form of grades earned and important temporal and contextual information such as when these grades were earned and how many hours the students carried during each semester to predict a student's grades. We ultimately chose to implement both user-based and item-based collaborative filtering using the Pearson Correlation Coefficient as the similarity metric in the pGPA prototype. We leave decisions such as the number of items and minimum similarity threshold used for the prediction calculation to the user in order to highlight the impact that different criteria have on the prediction capability of this system and allow for experimentation and tweaking in the future.

Potential Applications: The final version of pGPA is designed to be a software tool that can be adopted and used by academic institutions to help improve decision-making at all levels. We examine some of the potential uses of pGPA for each of our stakeholders to highlight how pGPA's grade predictions can provide value to Students, Instructors, Academic Advisors, and Administrators.

Students will be able to better manage their time and schedules to account for the amount of work they will likely need to do in order to succeed in the upcoming semester. Additionally, the knowledge of what grade a student may receive in a course could provide motivation to work harder and achieve more in order to push a borderline grade to the next level. Students may also use their grade predictions to add or drop courses in a semester to improve their chances for success.

Instructors can use grade predictions to better understand students' needs and individual deficiencies to help create personalized assignments for students. They may also be able to identify and recommend additional groups or resources for students that may be able to help the student improve his or her grade.

New academic advisors can benefit from grade predictions for students because these predictions may represent vital information about a student's chance for success that the advisor may not be aware of without extensive experience. These predictions may also help the advisor to convince a student to re-examine his or her schedule in order to improve the chances for success.

Armed with predictions for all students in a program or curriculum, administrators can identify likely dropouts or low performers and intervene before the semester starts. Administrators can also use the predictions made for a set of students in an intervention or study program to identify the effectiveness of such programs and allocate funding accordingly. Lastly, administrators can utilize the fact that they employ pGPA to improve student success rates in their school for marketing purposes.

3. pGPA DEMONSTRATION

To demonstrate pGPA we utilize a dataset that encompasses 10,000 grades earned by 600 Computer Science students in a university over a 20 year time frame. The steps below outline how a user can utilize this data in pGPA to produce and evaluate grade predictions and what potential advice pGPA may administer for its target stakeholders.

Step 1: Select student(s) for prediction

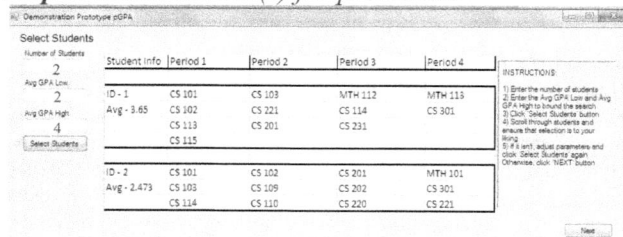

The user selects a set of students to make predictions for. This can be done by selecting a range of student performance and inputting the number of students to make predictions for or by selecting individual students from a list. The user can then examine the selected data to determine if it is acceptable.

Step 2: Identify period for prediction

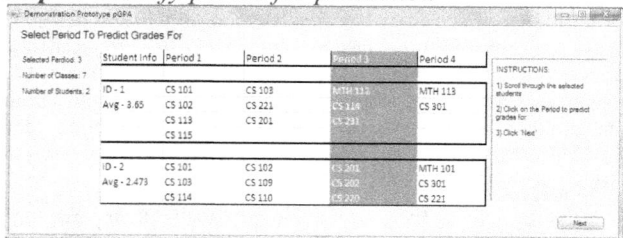

The user then identifies the period to make predictions for. The pGPA prototype then displays general information like the number of students who took courses during that period and the number of courses that this prediction will include.

Step 3: Configure parameters for prediction

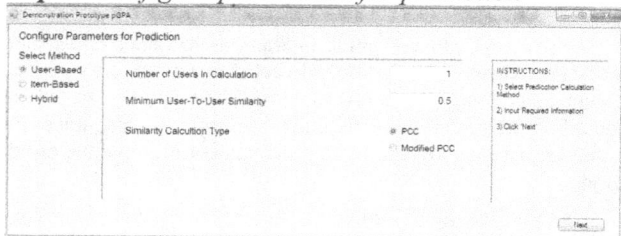

The user then has the option of changing the parameters of the predictions mechanism chosen. This will include settings like the number of nearest users (user-based) or items (item-based)

or the minimum similarity threshold used in the prediction calculation

Step 4: View results and explanation

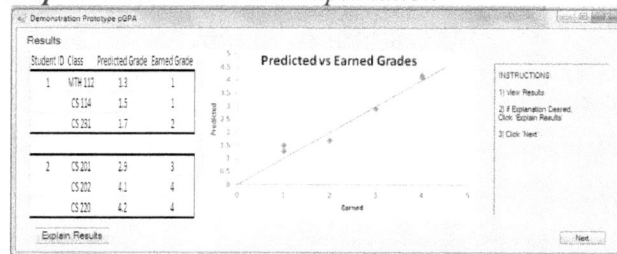

The user will be presented with several graphics that characterize the predictions. If desired, the user can also view a summarization of the factors that influenced the prediction.

Step 5: Show recommendations & advice

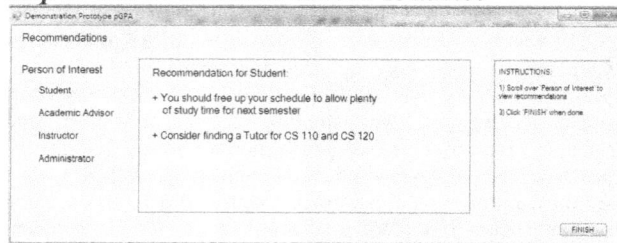

The pGPA prototype displays possible advice that could be recommended to the stakeholders such as (in the case of a student) "You should either remove a course from your proposed schedule or make sure you have a lot of study time scheduled so that you can improve your chances for success".

4. CONCLUSION & FUTURE WORK

We have demonstrated the research prototype tool called pGPA using a real-world academic dataset. This tool can be used to investigate various prediction strategies and it can be developed into a software package that higher education institutions can utilize to improve student success. We plan to test this tool with other datasets, add more features, and refine the prediction mechanism. We will also explore the commercialization and adoption of this tool.

5. ACKNOWLEDGEMENTS

We sincerely thank Dr. K. Sohn for the dataset.

6. REFERENCES

[1] Jannach D., Zanker M., Felfernig A., Friedrich G., *Recommender Systems: An Introduction*. Cambridge University Press, 2010.

[2] Tanner T., Toivonen H., Predicting and Preventing Student Failure Using the k-Nearest Neighbour method to predict student performance in an online course environment, *IJLT, 5*, (4), pgs 356 – 377, 2010.

[3] Vialardi C., Bravo J., Shafti L., Ortigosa A., Recommendation in Higher Education Using Data Mining Techniques, *Educational Data Mining* (Cordoba, Spain, 2009), pp. 190 – 199.

[4] Thai-Nghe N., Horvath T., Schmidt-Thieme, L., Factorization Models for Forecasting Student Performance, *Educational Data Mining* (Eindhoven, Netherlands, 2011), pp. 11-20.

Recommending Interesting Events in Real-time with Foursquare Check-ins

Max Sklar
Foursquare
568 Broadway
New York, NY
max@foursquare.com

Blake Shaw
Foursquare
568 Broadway
New York, NY
blake@foursquare.com

Andrew Hogue
Foursquare
568 Broadway
New York, NY
ahogue@foursquare.com

ABSTRACT

Foursquare is a location-based social application that helps users explore the world around them and share their experiences with friends. When foursquare users visit places, they "check in" using their mobile phones, indicating they are at that place. People check in for a variety of reasons: to keep up with friends, get tips about places, redeem rewards, and keep track of their personal history [2]. In aggregate, billions of these check-ins reveal distinct patterns about when places are popular [5] and allow us to build a unique place recommendation engine which can identify and recommend interesting events in real-time based on statistical deviations from past historical trends.

Categories and Subject Descriptors

H.2.8 [**Information Systems Applications**]: Data mining, Spatial databases and GIS; H.3.3 [**Information Systems Applications**]: Information filtering; H.3.5 [**Information Systems Applications**]: Web-based services; G.3 [**Probability and Statistics**]: Time series analysis; I.5.1 [**Pattern Recognition**]: Models—*Statistical*

Keywords

real-time event identification, foursquare, spatiotemporal data, machine learning

1. INTRODUCTION

One of the oldest components of foursquare is a system called "trending" which ranks nearby places by the number of people currently checked in [1] . We use this system to detect interesting "events" where many people check in at the same time. While this is useful, it tends to surface expected events at large venues, such as transportation hubs and office buildings. In order to identify truly interesting events, we have created a system that detects events that are both large and anomalous by leveraging machine learning and past historical data. We call the measure that this system produces "off-trending."

Consider Figure 1, which shows the number of check-ins per hour for two venues over a period of 20 weeks in 2011. Penn Station, shown in Figure 1(a), is one of the most popular places in New York City, and often has more people checked in than any other place in the city. The number of

check-ins per hour fluctuates from 0 to well over 200 with a very predictable pattern. There is a strong weekly periodicity with distinct peaks located at commuting times on weekdays. Although these periodic events are quite large in magnitude, they aren't very interesting to users since they appear regularly.

In Figure 1(b), we see a popular New York bar called The Scratcher. This venue has far fewer check-ins per week, and cannot compare to Penn Station in terms of raw popularity. However, there are occasional events at The Scratcher, as shown in week 5 on Friday night, that constitute a large deviation from historical behavior. These are the kinds of events that we aim to identify and surface to our users. Instead of surfacing common popular places like airports and train stations, we aim to surface anomalous events like street fairs, parties, and gallery openings. The key to identifying these events in real-time is modeling the expected number of check-ins each place is likely to receive.

2. TEMPORAL MODELS OF PLACES

We define the "off-trending" score of a venue, Ω_v, to be an estimate of how *unusually* busy the place is (that is, how busy the venue is relative to historical trends). To calculate the off-trending score, we first learn a probabilistic model for each place that yields a negative binomial distribution [4] over the number of people we expect to check in at any given time:

$$P(k|t, \alpha_w, \beta) = \frac{\Gamma(\alpha_w + k)}{\Gamma(\alpha_w)\Gamma(k+1)} \frac{\beta^{\alpha_w} t^k}{(\beta + t)^{\alpha_w + k}}$$

where k is the number of people who have checked in, t is the elapsed time period, and w is the current hour of the week. The parameters of the probabilistic model, α_w, are parameterized by the hour of the week w (a number between 1 and 168), and a tunable parameter β which controls the variance of the distribution:

$$\alpha_w = \beta C_{(10-\text{day})} P(w).$$

$C_{(10-\text{day})}$ is an estimate of the number of check-ins per week computed using an exponential moving average with a 10-day half life, and $P(w)$ is the probability of a check-in at the place occurring at weekhour w. The term $C_{(10-\text{day})} P(w)$ can be thought of as the expected number of check-ins for a particular hour of the week.

2.1 Learning the Parameters

For each popular place on foursquare we estimate the 168 parameters of $P(w)$ by maximum likelihood from all of the

(a) Penn Station

(b) The Scratcher

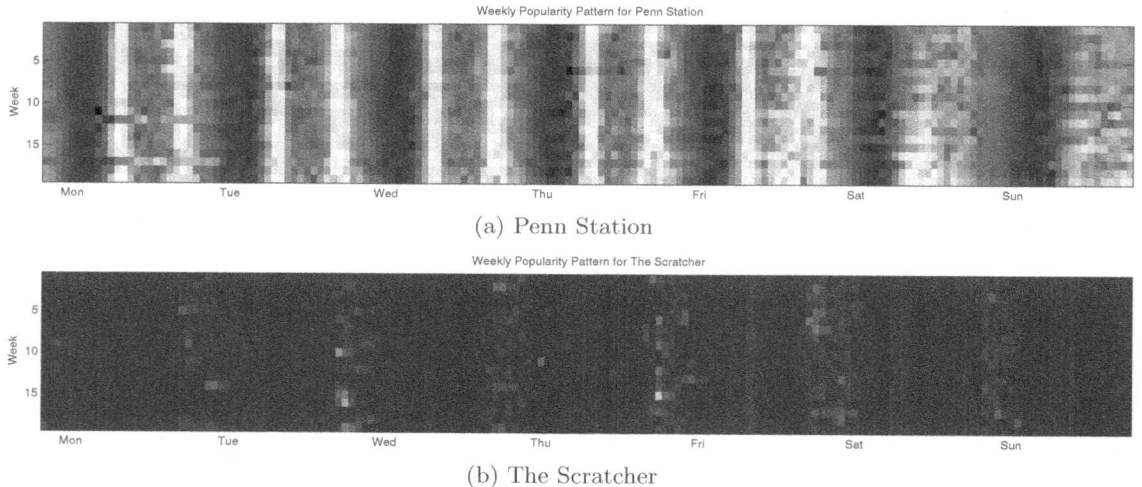

Figure 1: Weekly check-in patterns for two places in New York City: Penn Station (top) and The Scratcher (bottom). Each colored cell represents a single hour in a given week. Red entries indicate hours with many check-ins, and blue entries correspond to hours with very few check-ins. Note there are strong periodic events that happen at Penn Station which need to be accounted for in order to not overwhelm more anomalous events at less popular places (such as the event at The Scratcher on Friday of Week 5).

check-ins that have occurred at that place. We use a Dirchlet prior [3] to smooth the estimates and avoid over-fitting:

$$P(w) = \frac{C_w + \theta_w}{\sum_{i=1}^{168}(C_i + \theta_i)}.$$

C_w represents the number of check-ins that occur at the place at week hour w. The hyper-parameters of the Dirchlet distribution, θ_w, are optimized over a large subset of places via gradient descent using the log-likelihood of held-out data as an objective function. Every time a user checks in on foursquare, we increment both the weekhour-based counts C_w, as well as a global popularity term $C_{(10-\text{day})}$, which estimates the number of check-ins a venue receives per hour by averaging check-in counts from many previous days.

2.2 Ranking Places

Given $P_v(k|t, \alpha_w, \beta)$ for each place v in a candidate set with a current value of k people checked in, we can compute $\Omega_v = \sum_{j=0}^{k} P_v(j|t, \alpha_w, \beta)$ based on the cumulative density function of the negative binomial distribution. We call this quantity the "off-trending" score for a venue, which indicates how unlikely it is for us to see at least k people checked in to a place.

2.3 Summary

The off-trending score is a critical component of foursquare's recommendation engine, allowing us to identify and surface interesting nearby events in real-time. When a user opens

foursquare's Explore recommendation engine on their mobile device, the off-trending score is combined with signals about the user's past history, the preferences of the user and their friends, and other global signals in order to deliver a ranked list of recommendations for interesting places nearby the user might want to discover. Unlike other recommendation systems such as those for movies or books, understanding timeliness is critical to building a great place recommendation engine that can suggest activities to users in real-time.

3. REFERENCES

[1] foursquare for developers. https://developer.foursquare.com/, 2012.

[2] J. Lindqvist, J. Cranshaw, J. Wiese, J. Hong, and J. Zimmerman. I'm the mayor of my house: examining why people use foursquare - a social-driven location sharing application. In *Proceedings of the 2011 annual conference on Human factors in computing systems*, CHI '11, pages 2409–2418, New York, NY, USA, 2011. ACM.

[3] T. P. Minka. Estimating a dirchlet distribution. 2000.

[4] T. P. Minka. Estimating a gamma distribution. 2002.

[5] A. Noulas, S. Scellato, C. Mascolo, and M. Pontil. An empirical study of geographic user activity patterns in foursquare. In *Proceedings of the 5th Int'l AAAI Conference on Weblogs and Social Media (ICWSM)*, pages 570–573, July 2011.

An Open Framework for Multi-source, Cross-domain Personalisation with Semantic Interest Graphs[*]

Benjamin Heitmann
Digital Enterprise Research Institute (DERI)
National University of Ireland, Galway, Ireland
benjamin.heitmann@deri.org

ABSTRACT

Cross-domain recommendations are currently available in closed, proprietary social networking ecosystems such as Facebook, Twitter and Google+. I propose an open framework as an alternative, which enables cross-domain recommendations with domain-agnostic user profiles modelled as semantic interest graphs. This novel framework covers all parts of a recommender system. It includes an architecture for privacy-enabled profile exchange, a distributed and domain-agnostic user model and a cross-domain recommendation algorithm. This enables users to receive recommendations for a target domain (e.g. food) based on any kind of previous interests.

Categories and Subject Descriptors

H.3.3 [**Information Search and Retrieval**]: Information filtering

General Terms

Architecture, Algorithm, User model

Keywords

Personalisation, Cross-domain, Multi-source, Domain-agnostic

1. INTRODUCTION

Personalised recommendations have proven themselves to greatly enhance the user experience of searching, exploring and prioritising new and interesting content on social networking sites, like Facebook, Twitter and Google+. However all of these social networking sites are also each the centre of an ecosystem which allows external services and 3rd

[*]The work presented in this paper has been funded in part by Science Foundation Ireland under Grant No. SFI/08/CE/I1380 (Líon-2) and by the ADVANSSE project funded by Cisco Systems. In addition, I would like to thank my PhD supervisor Dr. Conor Hayes as well as Dr. Maciej Dabrowksi for their support and feedback.

party sites to enhance the user experience. Music streaming services, news papers or image sharing sites can post to a users activity stream on behalf of the user, thus contributing interests to his profile.

While this increases the value of the social network for the user and results in stronger user attachment, this also introduces completely new challenges for building recommender systems. The input data consists of user profiles which are (a) distributed across different services (*multi-source profiles*), and (b) not associated with a specific domain or inventory (*domain-agnostic profiles*) [1]. A source of background knowledge which provides connections between interests is required. Finally, the recommendation algorithm needs to provide *cross-domain recommendations* [8] targeted to a specific domain and inventory, in order to e.g. recommend posts from a food blog to users without food interests, or select advertisements from a pool of ads.

In order to provide users with the benefit of multi-source, cross-domain recommendations, users currently need to accept a trade-off [3] regarding their *privacy, trust, data ownership and control*: Privacy and personalisation are currently at odds [11]. Enabling 3rd party services requires sharing of user data with external services, thus introducing the potential for data leaks. In addition, the social networking operator, e.g. Facebook, owns all of the user data. The user has no control to whom his data is sold.

The goal of my research is to provide the same recommendation capabilities, namely multi-source and cross-domain recommendations, outside of closed social networking ecosystems. Towards this goal I am developing an open framework, as illustrated in figure 1. It addresses fundamental issues in the architecture and methodology of making recommendations on heterogeneous, distributed data. It consists of (i) an architecture for exchanging and aggregating of user profile data, (ii) user model, (iii) background knowledge and (iv) recommendation algorithm. In my PhD thesis, I plan to make contributions to the following three research areas:

Architecture for privacy-enabled profile exchange: Can user profile data and user profile fragments be exchanged in a decentralised way, while protecting the privacy of the users?

Distributed and domain-agnostic user model: What kind of data structure is required in order to allow merging profile fragments from multiple sites? What background knowledge can be used to build domain-agnostic user profiles? How can domains and genres and their constituent entities be defined in a flexible and universal way?

Cross-domain recommendation algorithm: Which class of recommendation algorithm can process domain-agnostic user profiles in order to provide recommendations for a specific target domain and inventory? Which data sets and metrics can be used for evaluating the performance of cross-domain recommendations?

2. BACKGROUND

I propose a framework, which addresses shortcomings in the way that current social networking sites provide cross-domain recommendations. Namely, I address the closed nature of current social network ecosystems, the missing portability of user profiles and interests, and the proprietary and secret recommendation algorithm.

In social networking ecosystems one site typically has the role of the *hub site*, which provides the main entry point for the whole ecosystem and stores the user profile data. Prominent hub sites are the social networking sites Facebook, Twitter and Google+. *Third party services* can provide value-added and personalised services for the user of an ecosystem.

In order to give the user a sense of privacy, while aggregating his profile data from external services, current social networking ecosystems are fundamentally *closed* [6]: The user model and all aggregated data about a user belong to the social network operator, e.g. Facebook. This is meant to protect the privacy of the user when his profile data is exchanged between the hub site and an external 3rd party service. However, it also creates user lock-in and data silos [3]. Users can not control to whom their data is sold, as the data generated by the users is the actual product. As an alternative, I propose an open architecture for exchange of user profile fragments in section 3. It uses open standards to enable a decentralised ecosystem in which users are not dependent on social network operators for recommendations.

All major social networking sites cater to general purpose sharing, as they are not limited to a particular genre or domain (like music or sports) or to a specific inventory (such as books in stock at a physical warehouse). This results in domain-agnostic user profiles. However, these user profiles are *not portable* between social networks [1], as they use identifiers for the interests which are specific to each social network operator. In addition, it is difficult to merge user profiles from different sources. Lastly, different social networks have different ways of classifying interests, so that it is difficult to specify a "food" recommendation in a universal way. In section 4, I present a distributed and domain-agnostic user model based on semantic interest graphs. It enables portability and merging of user profiles, as well as the definition of domains to categorise interests.

Facebook provides a so-called "social plugin" for cross-domain recommendations, which allows an external site to recommend content or items from its inventory to users without prior interest in the domain of the external site. E.g. a weblog about rugby, can use this plugin to recommend its articles to users without prior preferences about "sports". However, Facebook's cross-domain recommendations use *a proprietary algorithm*, so that users without a Facebook account can not benefit from this new kind of personalisation. In section 5, I introduce a graph-based algorithm which can use semantic interest graphs and domain definitions from section 4 to provide recommendations from a target domain. In addition, I introduce the challenges for

evaluating the user utility and performance of cross-domain recommendation in section 6.

3. ARCHITECTURE FOR PRIVACY-ENABLED PROFILE EXCHANGE

We propose a novel, open architecture as an alternative to closed architectures for exchanging and aggregating user profile data. An architecture prescribes (a) the standards as well as (b) the roles and (c) the communication pattern between the different participants. This makes it possible to align the different interests of all stakeholders. By implementing an architecture, all *individual* participants agree on the same technical principles, which in turn allows the architecture to guarantee the identified requirements on a *global* level.

Requirements: The architecture must support a federated and scalable ecosystem with any number of hub sites, 3rd party services and users. The architecture must enable interoperability of user profile data between 3rd party services and hub sites. At the same time, user data must not be allowed to leak to any unauthorised parties. These could be other web sites, adversaries or just other users. In other words, the architecture must be privacy-enabling and interoperable *at the same time*.

Our architecture prescribes the usage of the following standards: Linked Data [2], and the Friend-of-a-Friend (FOAF) and Semantically Interlinked Online Communities (SIOC) vocabularies allow the description of domain-agnostic user profiles. WebIDs securely connect a user identity to the information in a user profile and can be used for authenticating a user. The WAC vocabulary allows the user to authorise third party services for accessing different parts of his profile information.

The interplay between FOAF, WebIDs and the WAC vocabulary requires the participants to perform one of three roles: profile storage services, data consumers and user agents. We specify a communication pattern, which prescribes how the participants need to interact with each other according to their role. This communication pattern is described in full in [6], as well as a qualitative evaluation of the presented architecture based on the evaluation framework for privacy-enhanced personalisation suggested by Wang and Kobsa [11].

We summarise the qualitative evaluation as follows: The architecture provides a universal ecosystem, as all participants will support the same standards and implement the same communication pattern. The architecture is scalable, as there are no bottlenecks or central points of failure, due to the decentralised nature of the used standards. WebIDs allow anybody to host any number of identities on any server, and the WAC vocabulary allows authorising the access of a resource on the same server on which it is stored. For profile storage and data consumption existing standards and infrastructure from the World Wide Web and the Web of Data, such as HTTP and RDF are reused, thus making future adoption by service providers easy.

4. DISTRIBUTED AND DOMAIN-AGNOSTIC USER MODEL

The presented architecture enables aggregating domain-agnostic user profiles from multiple sources. In order to use

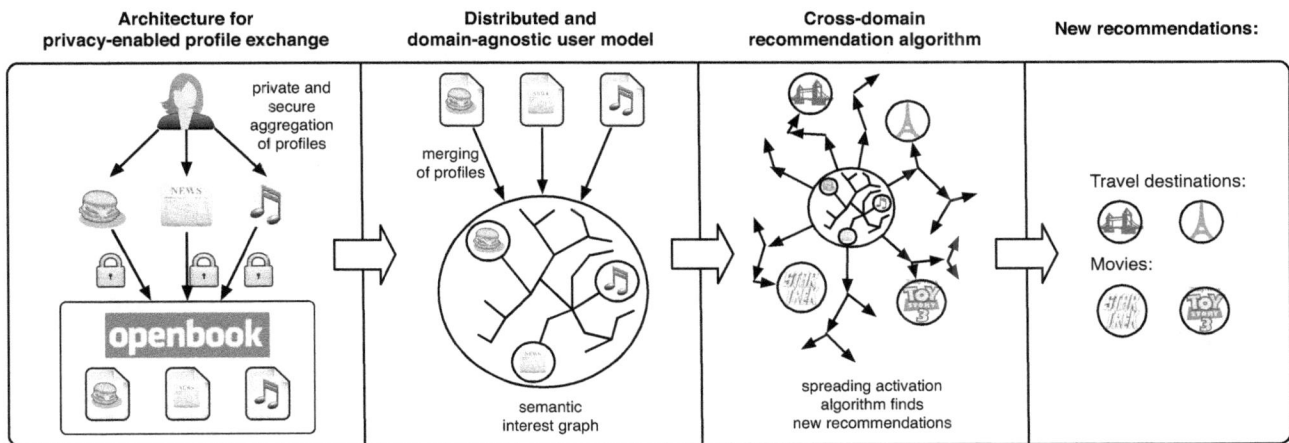

Figure 1: Overview of the proposed open framework for multi-source, cross-domain personalisation

these profiles for personalisation, a novel user model which fulfils the following requirements is needed.

Requirements: The data structure of the user model needs to allow merging of aggregated profiles from different sources. It must be able to model interests from many diverse domains, as well as the connections between interests. The representation of the interests must not be dependent on a specific inventory of items, i.e. it must not use identifiers which only exist at one data source. In order to utilise the user model for recommendations, a source of background knowledge which provides connections between interests is required. Finally, in order to provide cross-domain recommendations, it must be possible to define the target domain and its constituent items. These domain definitions are required in order to differentiate domains, or to use preferences in one domain for recommendations in another domain.

We propose using semantic interest graphs to represent the user profiles. Traditional user profile representations are not suitable to the requirements. *Vectors of item ratings* can not be exchanged between sites with different inventories, and they can not represent interests outside of the inventory. *Lists of plain text items*, which can represent tags or frequent words, can represent any kind of interest independent of the inventory. But they provide only limited support for finding relationships between interests.

In a *semantic interest graph*, each interest is represented by an RDF entity. RDF uses a graph data model [2], which allows RDF entities to be linked to each other. By leveraging existing knowledge engineering from the Linking Open Data community project, vocabularies and ontologies which already define any kind of interests can be reused. We propose to use concepts from the DBpedia knowledge base as *background data*, which is an RDF version of Wikipedia, and provides RDF concepts for any page or category from Wikipedia. In addition, each DBpedia concept is linked to other DBpedia concepts, which allows discovering relationships between concepts.

Using DBpedia concepts to represent user profiles as semantic graphs, enables exchanging and merging of profile data and independence from specific domains and inventories. The concept identifiers stay the same, independent of the site on which the profile was created. Profiles can be

easily merged when all sites use identifiers from DBpedia. Any concept from Wikipedia can be used as interest, independent of any domain or inventory restrictions.

In order to define domains and genres in a flexible and universal way, we propose to use the reasoning capabilities of RDF Schema and SKOS. RDF Schema allows defining hierarchies of classes and class instances. The Simple Knowledge Organisation System (SKOS) vocabulary allows defining relationships between categories and topics, such as broader or narrower categories. Together, RDF Schema and SKOS allow defining of reasoning rules to determine which entities belong to a domain. E.g. Pizza is categorised as WorldCuisine which has a super-category of Food, so it belongs to the Food domain.

Open issues: What process is required for specifying domains and entity types, and can this be automated based on e.g. the inventory of a site? How should the rule based reasoning process and the recommendation process be interleaved in order to support cross-domain personalisation?

5. CROSS-DOMAIN RECOMMENDATION ALGORITHM

After exchanging user profile data via the proposed architecture, and merging them to an interest graph with the proposed methodology, an appropriate recommendation algorithm is required. Cross-domain recommendation is a novel personalisation approach, which can utilise user preferences in one domain (e.g. music) to suggest recommendations in another domain (e.g. movies) [8].

Requirements: The algorithm must be graph-based, as both the input data and the background data for the algorithm are both graph-based. The algorithm must be able to take the semantics of the graph into account, e.g. by differentiating between different types of connections or different entity types. In addition, it must be able to differentiate between items from a target domain and from other domains.

We propose using a spreading activation (SA) algorithm as a recommendation algorithm. Spreading activation is inspired by the fact that human memory retrieves memories by association. Crestani [4] describes the algorithm as follows: An activation value spreads to all direct neighbours of

315

the starting nodes. As soon as the activation threshold of a node is reached, it counts as activated. Activated nodes are used as the starting points for the next activation phase. *Unconstrained* SA will quickly cover the whole graph, however the activation can be constrained. Possible *constraints* include the distance to the start nodes, the number of out-links of a node (fan-out constraint), the type of the link (path constraint), or the type of activated nodes (activation constraint).

In order to use spreading activation to provide recommendations in a target domain, the path and activation constraints can be used. Based on the target domain definition, the weights of all link types which belong to entities from the domain are set higher then other link types. In addition, the spreading step is repeated until enough nodes from the target domain are activated, using the activation constraint. This allows e.g. recommending book authors to a user who has only travel interests and destinations in his user profile.

Open issues: The algorithm can be parameterised in different ways, however the effect of e.g. the fan-out penalty on the results need to be measured using the performance metrics of the overall evaluation framework. An additional possibility is performing a user based evaluation.

6. EVALUATION

We are planning to evaluate the presented framework for multi-source, cross-domain personalisation as part of the ADVANSSE [5] collaboration project with Cisco Ireland. ADVANSSE has the goal of providing personalisation embedded in the IT landscape of current enterprises, which are characterised by large and heterogeneous information systems and social platforms. This results in user profiles being spread across multiple systems and user interests from many diverse domains.

Requirements for the ADVANSSE prototype: The implementation must be very scalable, in order to provide sufficient performance on large datasets from enterprise IT systems. In order to achieve this, the implementation should distribute the processing among multiple servers.

In order to provide a very scalable implementation, we are implementing our spreading activation algorithm with the Apache Giraph Java library for large-scale graph processing. Giraph runs on top of Apache Hadoop, which is an implementation of the MapReduce framework. As background data we use a subset of DBpedia which consists of 11 million entities and 40 million edges. For evaluation purposes we use user profiles from the question answering sites of the Stack-Exchange network. The StackExchange network consists of sites from very different domains, such as StackOverflow for computing questions, cooking, photography and bicycle sites.

Viewing the algorithm as a link predication task provides a general evaluation framework [9], in which area under the curve (AUC) and precision can be used as performance metrics. Comparable baseline approaches are provided by Linked Data Semantic Distance (LDSD) [10] and Random Walk with Restart (RWR) [7].

Open issues: The execution speed of the algorithm can be improved by pre-selecting a smaller subset of the data. However what is the trade-off in user utility of the personalisation results?

7. CONCLUSIONS AND FUTURE WORK

In this extended abstract I have presented an overview of my PhD research. My goal is to enable cross-domain recommendations from multi-source user profiles outside of closed social networking ecosystems. Towards this goal I am developing an open framework, which covers all parts of a recommender system. I propose an *open architecture* for profile data exchange as an alternative to current, closed architectures. It empowers users by using open standards to protect their privacy. I present a *distributed and domain-agnostic user model* to represent the user interests based on Linked Open Data standards and DBpedia concepts. It enables merging of profile fragments and the definition of domains to categorise interests. *Spreading activation* is a graph-based and content-based recommendation algorithm, which can provide cross-domain recommendations.

The goal of my final year research will be a systematic evaluation of the performance and user utility of cross-domain recommendations, as both the execution speed and the user utility depend on the parameterisation of the algorithm.

8. REFERENCES

[1] F. Abel, E. Herder, G. Houben, N. Henze, and D. Krause. Cross-system user modeling and personalization on the social web. *User Modeling and User-Adapted Interaction (UMUAI)*, 22(3):1–42, 2011.

[2] C. Bizer, T. Heath, and T. Berners-Lee. Linked Data - The Story So Far. *International Journal on Semantic Web and Information Systems*, 5(3):1–22, 2009.

[3] R. Chellappa and R. Sin. Personalization versus Privacy: An Empirical Examination of the Online Consumers Dilemma. *Information Technology and Management*, 6(2):181–202, 2005.

[4] F. Crestani. Application of spreading activation techniques in information retrieval. *Artificial Intelligence Review*, 11(6):453–482, 1997.

[5] B. Heitmann, M. Dabrowski, A. Passant, C. Hayes, and K. Griffin. Personalisation of Social Web Services in the Enterprise Using Spreading Activation for Multi-Source, Cross-Domain Recommendations. In *AAAI Spring Symposium on Intelligent Web Services Meet Social Computing*, 2012.

[6] B. Heitmann, J. G. Kim, A. Passant, C. Hayes, and H.-G. Kim. An architecture for privacy-enabled user profile portability on the Web of Data. In *Int. Workshop on Information Heterogeneity and Fusion in Recommender Systems (HetRec 2010)*, 2010.

[7] G. Jeh and J. Widom. Scaling personalized web search. In *World Wide Web Conference*, 2003.

[8] A. Loizou. *How to recommend music to film buffs: enabling the provision of recommendations from multiple domains*. PhD thesis, University of Southampton, 2009.

[9] L. Lü and T. Zhou. Link prediction in complex networks: A survey. *Physica A: Statistical Mechanics and its Applications*, 390(6):1150–1170, 2011.

[10] A. Passant. dbrec – music recommendations using dbpedia. *ISWC*, 2010.

[11] Y. Wang and A. Kobsa. Technical Solutions for Privacy-Enhanced Personalization. *Intelligent User Interfaces: Adaptation and Personalization Systems and Technologies*, 2009.

Exploiting the Characteristics of Matrix Factorization for Active Learning in Recommender Systems

Rasoul Karimi, Christoph Freudenthaler, Alexandros Nanopoulos,
Lars Schmidt-Thieme
Information System and Machine Learning Laboratory
Samelsonplatz 1 University of Hildesheim 31141 Hildesheim Germany
[karimi, freudenthaler, nanopoulos, schmidt-thieme]@ismll.uni-hildesheim.de

ABSTRACT

Recommender systems help web users to address information overload. However their performance depends on the number of provided ratings by users. This problem is amplified for a new user because he/she has not provided any rating. To address this problem, active learning methods have been proposed to acquire those ratings from users, that will help most in determining their interests. However, different from the classic active learning, users (the "oracle") are not always able to provide an answer for queries. The easiest way to solve this problem is to ask most popular items, i.e items which have received many ratings from training users. But it is static and presents the same items to all users regardless of the ratings they have provided so far. In this paper we propose a method that improves the most popular selection strategy using the characteristics of matrix factorization. It finds similar users to the new user in the latent space and then selects item which is most popular *among the similar users*. The experimental results show the proposed method outperforms the most popular method both in terms of error and the number of received ratings.

Categories and Subject Descriptors

I.2.6 [**Artificial Intelligence**]: Learning Parameter learning

Keywords

Recommender System, Active Learning, New User Problem

1. INTRODUCTION

Recommender systems help web users to address information overload in a large space of possible options [3]. In many applications, such as in e-commerce, users have too many choices and too little time to explore them all. Moreover, the exploding availability of information makes this problem even tougher.

There are several techniques for recommendation. Collaborative filtering is a traditional technique that is widely applied [1, 13, 15]. It makes automatic predictions about the interests of a user by reusing taste information from other users. The underlying assumption of the collaborative filtering approach is that those who agreed in the past tend to agree again in the future.

Collaborative filtering methods fall into two main categories:
memory-based algorithms and model-based algorithms. In memory-based techniques, the value of the unknown rating is computed as an aggregate of the ratings of some other (usually, the N most similar) users for the same item [13]. Model-based collaborative techniques provide recommendations by estimating parameters of statistical models for user ratings. Nevertheless, recent research (especially as has been demonstrated during the Netflix challenge[1]) indicates that Matrix Factorization(MF) is a superior prediction model compared to other model-based collaborative methods [14].

Evidently, the performance of collaborative filtering depends on the amount of information that users provide regarding items, most often in the form of ratings. However, a well identified problem is that users are reluctant to provide information for a large amount of items. This fact impacts negatively the quality of generated recommendations. A simple and effective way to overcome this problem, is by posing queries to new users in order that they express their preferences about selected items, e.g., by rating them. Nevertheless, the selection of items must take into consideration that users are not willing to answer a lot of such queries. To address this problem, *active learning* methods have been proposed to acquire those ratings from users, that will help most in determining their interests [5, 8].

However, different from the classic active learning, users (the "oracle") may not know the answer of queries. For example in the movie recommendation scenario, if the new user has not watched the queried movie, he/she is not able to provide a rating for that movie. The easiest way to solve this problem is to ask most popular items, i.e items which have received many ratings from training users. Hopefully, the item which is known for many training users is also known for the new user and he/she can rate it. But the most-popular selection strategy is static and presents the same items to all users regardless of the ratings they have provided so far. An advanced active learning algorithm adapts to the earlier answers given by the new user to improve the preference of elicitation process. In this paper we propose

---

[1] www.netflixprize.com

a method which improves the most popular selection strategy using the characteristics of matrix factorization. It finds similar users to the new user in the latent space and then selects an item which is most popular *among the similar users*. As the similar users change during the active learning process based on the new user's answers to the queries, the active learning algorithm is also automatically adapted to these answers.

The rest of this paper is organized as follows: In Section 2, the related work is reviewed. As the proposed method relies on matrix factorization, a short introduction to matrix factorization is introduces in section 3. The proposed method is explained in section 4 followed by the experimental results in Section 5. Finally, the conclusions are stated in Section 6.

2. RELATED WORK

Active learning, in the context of the new-user problem, was introduced by Kohrs and Merialdo [12]. This work suggested a method based on nearest-neighbor collaborative filtering, which uses entropy and variance as the loss function to identify the queried items. Mamunur et al. [15] expanded this work, by considering the popularity of items and also personalizing the item selection for each individual user. Boutilier et al. [2] apply the metric of expected value of utility to find the most informative item to query, which is to find the item that leads to the most significant change in the highest expected ratings.

Jin and Si [8] developed a new active learning algorithm based on Aspect Model(AM) [7, 6] which is similar to applying active learning for parameter estimation in Bayesian networks [17]. It uses the entropy of the model as the loss function. However, it does not directly minimize the entropy loss function, because the current model may be far from the true model and relying only on the current model can become misleading. To overcome this problem, that work proposes to use a Bayesian network to take into account the reliability of the current model. This Bayesian approach is, however, complex and intractable for real applications (demands excessive execution time). Karimi et. al. [9] applied the simple most popular item selection to AM. The results show that it competes in accuracy with the Bayesian approach while its execution time is in the order of magnitude faster than the Bayesian method.

Karimi et. al [10] developed a non-myopic active learning which capitalizes explicitly on the update procedure of MF model. First, this method queries items that updating the new user features with the provided rating will change the features as much as possible. Its goal is to explore the latent space to get closer to the optimal features. Then, it exploits the learned features and slightly adjusts them. Karimi et. al [11] by being inspired from existing optimal active learning for the regression task, exploits the characteristics of matrix factorization and develops a method which approximates the optimal solution for recommender systems.

All aforementioned methods assume that new users are always able to provide ratings to the queried items. Some methods relax this assumption and develop algorithms for the realistic scenario in which new users might not rate the queried item. [18, 4] leverage a decision tree to address this issue. But training the decision tree is costly in terms of time and memory specially when the size of the dataset is large.

3. MATRIX FACTORIZATION IN RECOMMENDER SYSTEMS

Matrix Factorization (MF) is the task of approximating the true, unobserved ratings-matrix R by $\hat{R} : \mathbb{R}^{|U| \times |I|}$. It maps both users and items to a latent space of dimensionality k. In this space, user-item interactions are modeled as inner products. In the latent space, each item i is represented with a vector $h_i \in R^k$. The elements of h_i indicate the importance of factors in rating item i by users. Some factors might have higher effect and vice versa. In the same way, each user u is represented with a vector $w_u \in R^k$ in the latent space. For a given user the element of w_u measure the influence of the factors on user preferences. Different applications of MF differ in the constraints that are sometimes imposed on the factorization. The most common form of MF is finding a low-rank approximation (unconstrained factorization) to a fully observed data matrix minimizing the sum-squared difference to it.

The resulting dot product, $h_i^T w_u$, captures the interaction between user u and item i. However, the full rating value is not just explained by this interaction and the user and item bias should also be taken into account. It is because part of the rating values is due to effects associated with either users or items,i.e biases, independent of any interactions.

By considering the user and item bias, the predicted rating is computed as follows [14]:

$$\hat{r}_{ui} = \mu + b_i + b_u + h_i^T w_u \qquad (1)$$

in which μ is the global average, b_i is the item bias and b_u is the user bias. The major challenge is computing the mapping of each item and user to factor vectors $h_i, w_u \in R^k$. The mapping is done by minimizing the following squared error [14]:

$$Opt(S, W, H) = \sum_{(u,i) \in S} (r_{ui} - \mu - b_u - b_i - h_i^T w_u)^2 + \qquad (2)$$

$$\lambda(\|h_i\|^2 + \|w_u\|^2 + b_u^2 + b_i^2)$$

in witch λ is the regularization factor, and S is the set of the (u, i) pairs for which r_{ui} is known, i.e the training set (S).

When a new user enters the recommender system, the prediction model should be updated to learn the new user latent features. As there are already a lot of users in the recommender system, training the whole of the model from scratch needs a lot of time. Therefore, we switch to online updating and develop the active learning algorithm based on that. Online updating means after a first training with all users, further retraining is only done for new users. For the online updating, we use the method introduced in [16]. In this method after getting a new rating from the new user, the user's latent features are initialized to a random setting and then are learned using all ratings of the new user.

4. PROPOSED METHOD

There is a trade-off between the information-theoretic value of a new rating and the likelihood of a user being able to provide a rating. A comprehensive method should take both criteria into account. However, in this paper we stick with the second criterion and leave the first one as the future work. Therefore, the aim of the proposed active learning is to choose a query which is likely to be answered by the new user. As the process of active learning starts without any rating from the new user and ends up just with a few queries (e.g 10 queries), relying solely on the new user's answers does not provide *enough* information to find suitable queries specially when the number of items is large. An appropriate solution to address this problem is to link the new user to the training users and exploit available information from them. In this paper we use the characteristics of matrix factorization, as the prediction model of the recommender system, to make such connection.

When the training of the matrix factorization is done users and items are mapped into the latent space. Users (or items) with similar rating behaviors are mapped into the same region. We exploit this characteristic of matrix factorization to find similar users to the new user. Then the item which is most popular *among the similar users* is selected for the next query. Hopefully, the new user is able to provide a rating for an item which has already been rated by users with similar rating behavior. In this way, the drawback of the global most popular, which asks *static* queries regardless of the answers of the new user, is fixed while the new user is connected to the training users to exploit their information. The number of the similar users k is a hyper parameter. It should not be too large because it affects the scalability of this method in real time. Algorithm 1 describes the proposed active learning.

Algorithm 1 Proposed Active Learning

1: **loop** {repeats until N queries}
2:    find k-similar users
3:    find most popular item among the k-similar users
4:    ask the query
5:    **if** user provides a rating **then**
6:       update the new user features according to [16]
7:    **end if**
8: **end loop**

The accuracy of the proposed method heavily depends on how good the similar users are found. If the computed similar users are accurate, then approximating the rating behavior of the new user using the similar users is also accurate. Otherwise the performance is affected. In order to find similar users cosine similarity is used which is reasonable because in matrix factorization the ratings are predicted by the inner product of user and item features.

5. EXPERIMENTAL RESULT

In this section, we examine experimentally the performance of the proposed method.

5.1 Experimental Setup

The main challenge in applying active learning for recommender systems is that users are not willing to answer many queries in order to rate the queried items. For this reason, we report the performance of all examined methods in terms of prediction error (RMSE) versus the number of queried items, which is simply denoted as the *number of queries*. For the new user i, the RMSE is computed as follows:

$$RMSE = \sqrt{\frac{1}{N} \sum_{u,i \in M_u} (r_{ui} - \hat{r}_{ui})^2} \qquad (3)$$

where M_u is the set of the test items of user u, $\hat{r}_{ui}$ is the predicted rating (by MF) of user u for item i, and r_{ui} is the true (actual) rating. Thus, we examine the problem of selecting at each step, the item for which each new user u will be queried to provide a rating. The item has to be selected in order to minimize the $RMSE$ based on the MF model.

Non-myopic active learning [10] and most popular item selection,i.e item which has received most ratings from training users, are used as the baseline. We use 50 percent of the Netflix dataset in our experiments. The dataset was randomly split into training and test sets, containing 70% and 30% respectively. Each test user is considered as a new user. 30% of the ratings of each test user is separated to compute the error (test data) and the queries are selected from the remaining items (pool data). In our experiment, 10 queries are asked from each new user.

5.2 Results

Figure 1 illustrates the comparison between the proposed method, non-myopic active learning [10], and most popular selection in terms of RMSE as a function of the number of queried items.

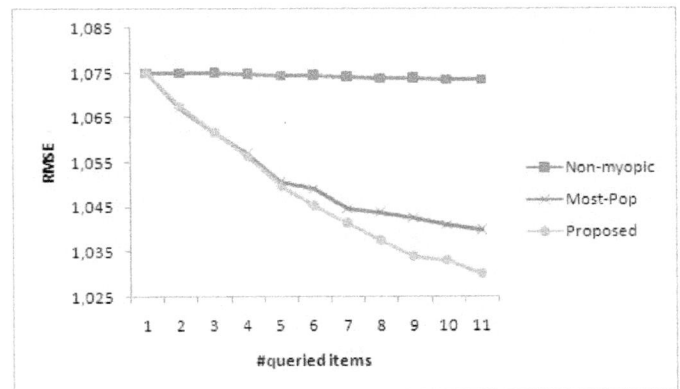

Figure 1: RMSE results of the proposed active learning, non-myopic and most popular versus the number of queries

In the first queries, the most popular method competes with the proposed method. This happens because during the first queries, a little knowledge about the rating behavior of the new user has been revealed. Therefore, it is difficult to find *real* similar users. But as more ratings are provided by the new user, the similar users become closer to the real similar users which consequently increases the probability that the new user provides an answer to the queried item. Another interesting observation of figure 1 is that the non-myopic method does not work at all. This was expected to happen because this method only aims to choose a query which reduces the test error and supposes that the new user always is able to provide a rating to the queried item. However, in this paper we have relaxed this assumption, meaning that the queried item might not get a ratings. Obviously, this assumption has a significant effect on the performance of active learning methods.

Table 1: The average number of ratings received from the new user after 10 queries

Proposed	2.6468
Most Pop	2.6374
Non-myopic	0.0007

The performance of the active learning methods depends on the number of provided ratings by the new user. The more the ratings, the better the performance. Table 1 shows the average number ratings provided by a new user in all methods. The proposed method receives more ratings compared to most popular while the non-myopic method receives almost no ratings.

6. CONCLUSIONS

In this paper we proposed a new active learning method for the new user problem in recommender systems. It improves the most popular selection strategy using the characteristics of matrix factorization. First it finds similar users to the new user in the latent space and then selects item which is most popular *among the similar users*. Hopefully, the new user is able to provide a rating for an item which has already been rated by users with similar rating behavior. In this way, the drawback of the global most popular, which asks *static* queries regardless of the answers of the new user, is fixed while the new user is connected to the training users to exploit their information.

As the future work, we plan develop a criterion for informative queries and then combine it with the proposed active learning in the way that the trade-off between both criteria is met.

7. ACKNOWLEDGMENTS

This work is co-funded by the European Regional Development Fund project REMIX under the grant agreement no. 80115106.

8. REFERENCES

[1] G. Adomavicius and A. Tuzhilin. Toward the next generation of recommender systems: A survey of the state-of-the-art and possible extensions. *IEEE Transactions on Knowledge and Data Engineering*, 17(6):734–749, 2005.

[2] C. Boutilier, R. S. Zemel, and B. Marlin. Active collaborative filtering. In *Conference on Uncertainty in Artificial Intelligence(UAI)*, 2003.

[3] R. Burke. Hybrid recommender systems. *User Modeling and User Adapted Interaction*, 12(4):331–370, 2002.

[4] N. Golbandi, Y. Koren, and R. Lempel. Adaptive bootstrapping of recommender systems using decision trees. In *WSDM*, pages 595–604. ACM, 2011.

[5] A. S. Harpale and Y. Yang. Personalized active learning for collaborative filtering. In *Proceedings of the 31st annual international ACM SIGIR conference on Research and development in information retrieval*, pages 91–98. ACM, 2008.

[6] T. Hofmann. Collaborative filtering via gaussian probabilistic latent semantic analysis. In *Proceedings of the 26th annual international ACM SIGIR conference on Research and development in informaion retrievall*, pages 259–266. ACM, 2003.

[7] T. Hofmann and J. Puzicha. Latent class models for collaborative filtering. In *International Joint Conference on Artificial Intelligence*, pages 688–693. Morgan Kaufmann Publishers Inc., 1999.

[8] R. Jin and L. Si. A bayesian approach toward active learning for collaborative filtering. In *Proceedings of the 20th conference on Uncertainty in artificial intelligence*, 2004.

[9] R. Karimi, C. Freudenthaler, A. Nanopoulos, and L. Schmidt-Thiem. Active learning for aspect model in recommender systems. In *IEEE Symposium on Computational Intelligence and Data Mining (CIDM)*. IEEE, 2011.

[10] R. Karimi, C. Freudenthaler, A. Nanopoulos, and L. Schmidt-Thiemee. Non-myopic active learning for recommender systems based on matrix factorization. In *IEEE Information Reuse and Integration (IRI)*. IEEE, 2011.

[11] R. Karimi, C. Freudenthaler, A. Nanopoulos, and L. Schmidt-Thiemee. Towards optimal active learning for matrix factorization in recommender systems. In *23th IEEE International Conference on Tools With Artificial Intelligence (ICTAI)*. IEEE, 2011.

[12] A. Kohrs and B. Merialdo. Improving collaborative filtering for new users by smart object selection. In *International Conference on Media Features (ICMF)*, 2001.

[13] J. A. Konstan, B. N. Miller, D. Maltz, J. L. Herlocker, L. R. Gordon, and J. Riedl. GroupLens: Applying collaborative filtering to usenet news. *Communications of the ACM*, 40(3):77–87, 1997.

[14] Y. Koren, R. Bell, and C. Volinsky. Matrix factorization techniques for recommender systems. *Computer*, 42:30–37, 2009.

[15] A. M. Rashid, I. Albert, D. Cosley, S. K. Lam, S. M. McNee, J. A. Konstan, and J. Riedl. Getting to know you: Learning new user preferences in recommender systems. In *International Conference on Intelligent User Interfaces (IUI)*, pages 127–134. ACM Press, 2002.

[16] S. Rendle and L. Schmidt-Thieme. Online-updating regularized kernel matrix factorization models for large-scale recommender systems. In *ACM Conference on Recommender Systems (RecSys)*, pages 251–258. ACM, 2008.

[17] S. Tong and D. Koller. Active learning for parameter estimation in bayesian networks. In *Advances in Neural Information Processing Systems(NIPS)*, 2000.

[18] K. Zhou, S.-H. Yang, and H. Zha. Functional matrix factorizations for cold-start recommendation. In *Proceedings of the 34th international ACM SIGIR conference on Research and development in Information Retrieval*, SIGIR '11, pages 315–324. ACM, 2011.

Dynamically Selecting an Appropriate Context Type for Personalisation

Tomáš Kramár
Faculty of Informatics and Information
Technologies
Slovak University of Technology
Bratislava, Slovakia
kramar@fiit.stuba.sk

Mária Bieliková
Faculty of Informatics and Information
Technologies
Slovak University of Technology
Bratislava, Slovakia
bielik@fiit.stuba.sk

ABSTRACT

Narrowing down the context in the ranking phase of information retrieval has been shown to produce results that are more relevant to searcher's need. We have identified two types of contexts that could be used in the process of personalisation. We research these contexts in the domain of personalised search, but show that our approach can be used for any kind of personalisation or recommendation. We focus on two aspects of the context: temporal context and activity-based context and describe a more general personalisation framework based on lightweight semantics, that can leverage any type of context.

Categories and Subject Descriptors

H.3.3 [**Information Search and Retrieval**]: Information filtering

Keywords

search context, personalised search, search intent

1. INTRODUCTION

Recommendation systems and search engines play a crucial role in accessing the amount of content that can be found of the Web nowadays. Users are usually able to filter the information by issuing short keyword queries, but this model has several known disadvantages. The number of keywords is usually low, typically 1-3 keywords [8] and many of the words are ambiguous. The queries are almost never accurate [5], they are either too generic or too specific, but almost never exactly aligned with the specific intent the user has in mind. When we combine the impact of each of the described problem, we come to a conclusion, that finding the relevant document is indeed a difficult task, both for the user and the search engine.

A number of approaches have been researched, each with the specific goal of helping people find relevant content,

preferably without changing the established paradigm of searching by keywords. The approaches range from modifying the query in-place to better capture the specific intent the user has in mind, i.e., query reformulation [3], learning to rank [9], or relevance feedback [12].

Each of the existing approaches leverages some sort of context to infer user's search intent and personalize the search results accordingly. Most of the time, the search context is long-term, built from all available data, e.g. learning to rank approaches use clickthrough data to learn user's preference based on her past search result clicks. The other extreme, short-term context is used in relevance feedback, where search results are modified according to user's last action, with a basic assumption, that a click represents a form of implicit feedback on the relevance of the clicked search result.

In all cases, the time by which the context is bound is determined statically and does not adapt to what is appropriate in the given situation. Long-term search context has the advantage that much data about the user is available, so the personalisation system can make reliable model of user's long-term interests. However, when a new, previously unseen goal appears, the information in the long-term model can shadow it, leading to bad personalization. E.g., an aquarium guide looking to buy a new *Barracuda* hard drive - the term *Barracuda* is associated with a sea fish in the model of his interests and the search results are personalized towards results that deal with sea and fish, while the guide is in fact looking for information on the specific model of the hard drive. On the other hand, the short-context can contain so little information, that no personalisation is possible, because no search goal can be inferred. We believe that the solution to this problem is to select the size of context dynamically at the time of search, and that the search context should be as specific as possible.

We have identified two types of search context that may be helpful in personalizing the search:

- *Activity-based context*, which is based on user's actions; what is she doing, how is she interacting with the search engine in terms of clicking on the search results. The activity-based search context is basically bounded by the changes in information-seeking goals. This context starts, and ends with each change in information-seeking goal.

- *Recurrent temporal context*, which is based on user's preferences in time. We assume that users form some kinds of recurrent habits that correlate with time, e.g., someone may be searching for work-related goals most of the time during work-week, while on weekends, the search goals are very different. The temporal habits may happen on a larger scale, e.g. searching for asparagus recipes in spring, during the asparagus season.

2. CONTEXT MODELING AND PERSONALISATION FRAMEWORK

We model user's context using lightweight metadata extracted from accessed documents. Normally, a personalisation system would have access to this type of information in a direct way, but for the purposes of our research, we are collecting this information using a logging proxy server [1]. Usage of the proxy server is completely transparent for the users and does not affect system performance. Users have to explicitly opt-in for the proxy and they are aware that their activity is logged and we provide tools to selectively remove logged data or opt-out.

For each accessed document we extract document metadata, along with other user interaction indicators (mouse and keyboard activity and time spent on page), which are used to calculate implicit feedback on the page.

The context model is a hypergraph $H :=< V, E >$, with a set of vertices $V = A \cup P \cup T$, where A represents a set of users accessing the pages: $A = (a_1, a_2, \cdots, a_k)$, P represents a set of pages $P = (p_1, p_2, \cdots p_l)$ and T represents a set of terms $T = (t_1, t_2, \cdots, t_m)$; and a set of edges $E = (a, p, t)|a \in A, p \in P, t \in T$ and $P \cap T = \emptyset, A \cap P = \emptyset, A \cap T = \emptyset$. Using this representation is advantageous, as it allows for good denormalization and allows us to track each of the vertices independently. It may seem intuitive to merge accesses and pages, but this models allows us to abstract page from access, and if the document represented by the URL (page) changes, we can create new vertex in the graph and connect it respectively.

We use following series of steps to produce an interest model that can be used for recommendation or personalisation.

- First, search context can be selected by any arbitrary method. Context selection is based on selection of actions, e.g., we can use a long-term context and select all documents ever viewed by the user, or we can leverage activity based search context and select only the documents that she viewed for the purposes of her current information goal.

- Next, we extract a subgraph H_c from the original hypergraph H, by selecting only vertices $A_c \in A$, that match the context restriction established in previous step. The subgraph also contains vertices adjacent to A_c, i.e., $P_c \in P$ and $T_c \in C$ and their connecting edges $E_c \in E$.

- The subgraph H_c now represents the interest model for current context. We further enrich it by finding its nearest neighbors in terms of similarity in access patterns and similarity of the acquired document metadata T. This interest model and models of similar interests can now be used for personalisation or recommendation.

To validate this context acquisition approach we ran synthetic experiments [10] and experiments in a live system [11].

In both cases we used a simple form of long-term context, by using $H_c = H$. In both cases we evaluated our approach in the domain of personalised search and used two simple query reformulation approaches:

- search the query history in the interest models (extracted from URLs of accesses to search engine, find reformulation patterns of the current query and select the one with highest satisfaction score, calculated by considering implicit feedback on clicked results, assuming, that unsatisfactory search results lead to clicks with low implicit feedback scores and to a further query reformulation until the user is satisfied.

- search the metadata in the interest models and find terms that are often co-occurring with the terms from the current query. Select the most frequent co-occurrences and extend the original query.

Using this experimental setup, we were able to substantially increase the relevance of search results and thus validate our approach. The proposed framework has two important properties: both context, and the personalisation method can be selected independently and we believe that using a different form of context than the long-term context, one could improve the relevance of recommendations even more.

3. ACTIVITY-BASED CONTEXT

Basic idea of an activity-based context is that information from previously accessed documents can help to understand the current information goal. E.g., imagine a series of queries: *watermelon; used cars; bentley; jaguar* – it is easy to see that the current query *jaguar* is connected with two previous queries, *used cars* and *bentley* and that the search goal in this case is to buy a car. With this knowledge, the personalisation can be moved towards search results that deal with used jaguar cars.

The problem of finding activity-based search context is stated as follows. The user is issuing queries and clicking on search results. The goal of activity-based search context detection is, for a current query, select the previous queries (and their respective clicked search results), that are part of the same search goal as the current query. This problem is equivalent to the problem of detecting the search session, when we define the search session as a set of queries following the same goal. Note that the queries that are part of the same search session may not be issued sequentially, but can be interleaved.

The problem of detecting search sessions is well researched, although the definition of a search session usually varies. The approaches could be categorized into three classes. The methods that are based on time, e.g. [6], are not appropriate for this definition of search session. They assume that a search session is bounded by a higher inactivity time, i.e. if the query is issued after a specified time from the last query elapsed, the query is considered to start a new session. The pause in the interaction with search engine can be caused by external factors (e.g., phone call, lunch) and user can return to the original task afterwards. Similarly, the search goal can change rapidly, without a pause in the interaction. The other class of approaches is based on lexical similarity between queries [7]. Two queries are considered part of the same session if they share common terms. However, this class of approaches does not deal with semantic similarity, e.g., queries *information retrieval* and *IR*. The third class of approaches uses signals from the search results page, such as snippets from the search results page [4].

The existing methods have two downsides: firstly, most of them do not consider multitasking in search, and restrict the search session to sequential queries. Once the search session is terminated, in cannot be reopened to add new queries. Secondly, existing methods that are using search results consider all search results equal.

We designed a method that considers multitasking in web search and incorporates implicit feedback on the search results. We build a query model, where we leverage the same lightweight semantic document metadata as in our personalisation framework. Each query is associated with the respective clicked search results. We leverage a negative implicit feedback and assume that a search result that was viewed for less than a specified number of seconds was not relevant to the search goal and therefore closed. We remove these documents from the query model.

To put the query into a session, we look at the clicked search results and their respective metadata. For each query, a metadata vector is built and compared to metadata vectors of currently active search session. When the vectors are similar (in terms of shared metadata), the query is considered part of the session and its metadata is merged into session model. At any given time, a stack of previous search sessions is being maintained and if the query model does not match the current session model, the stack of sessions is searched, until a match is found or the stack is exhausted. When a match is found, the matching search session is moved to the top of the stack and becomes active search session. Old search sessions are removed from the stack automatically, so that they cannot become active.

We have experimentally evaluated this approach against a set of 245 manually segmented queries of 3 different users. We have compared the results with other approaches (temporal with a window of 5 and 30 minutes and lexical). The results are described in Table 1. *Precision* indicates the internal coherence of the session. First, queries from automatically detected session are linked to the manually detected sessions and precision is calculated as the ratio of the cardinality of the best match to the cardinality of the whole session. *Recall* indicates the completeness of the session. It is calculated as the ratio of cardinality of the best match against the manually created sessions to the cardinality of the best-matched session. Best results were obtained by combining the lexical approach with our session matching method (referenced in the table as *metadata*) as a fallback, i.e., the first heuristics of a session match is the lexical similarity of queries and if it is negative, the document metadata are compared.

The preliminary experiments seem promising and we believe that applying activity based context to a personalisation process will yield better results that using a long-term context.

4. RECURRENT TEMPORAL CONTEXT

We postulate that there are patterns in user behavior changes that are related to time. For example searching for summer resorts before summer holidays, or searching for which movies to watch on Sunday night. We believe that

Table 1: Results of search session segmentation using various approaches.

approach	precision	recall	f-score
lexical	0.985 ± 0.078	0.499 ± 0.410	0.662
temporal30m	0.932 ± 0.153	0.489 ± 0.380	0.641
temporal5m	0.974 ± 0.101	0.346 ± 0.373	0.510
metadata	0.992 ± 0.055	0.276 ± 0.312	0.432
lexical+meta	0.969 ± 0.108	0.540 ± 0.411	0.694

these patterns are recurrent and appear more or less regularly during the same periods of day, week, month or year.

To the best of our knowledge, there are no studies that would support or disprove our hypotheses. There are many works that study the temporal dynamics of search, but all of them analyze it only globally, and do not focus on individual patterns. For example, in [13] authors analyze transactional logs from the Excite search engine using Markov matrices and Poisson sampling and explore variations in aggregated users' searches related to changes of time in day. Or similarly, Beitzel et al. [2] analyzed a partially manually topically labeled large-scale transactional log of a search engine and analyzed topical patterns in the queries and results, but again, only in an aggregated form.

In our work, we analyze changes in user interest for the top-100 users from the publicly available AOL search engine. We focused on analyzing the patterns of interest changes during work hours and free time. For each user in the top-100 log, we have separated the clicked search results into two distinct clusters using two setups: work-week vs. weekend and work-hours (9:00 - 17:00) vs. free-time (the rest). Next, we compared how well were the two clusters separated from each other in each of the two setups.

For this analysis, we used HTML code of the search results and extracted keywords, description and title of each page (provided by the author) and it's respective ODP categories. For each user, these vectors of textual metadata were converted to binary vectors, 1 indicating presence of the metadata in the vector, 0 otherwise. These vectors were then treated as vectors in Euclidian space and we calculated standard Davies-Bouldin score for each user's clusters in both setups. Davies-Bouldin score represents how well are the clusters separated from each other and how coherent are they internally. The results for both setups are displayed in Figure 1.

The Davies-Bouldin scores suggest, that our hypotheses are not valid in general, but there exist some users, for which they are true. In both setups, there are some users for whom the score is low, which means that they indeed search for semantically different information during weekend and during work-week or during work-time and free-time hours. What is interesting, is that the calculated scores cover the whole range of scores nearly linearly, without any peaks. This suggests that the probability of a user changing interests on the two temporal conditions exhibits uniform distribution.

5. CONCLUSIONS AND FUTURE WORK

In our work, we try to design a framework for personalisation based on context constraints, with the aim of selecting the proper type of context dynamically. So far, we have

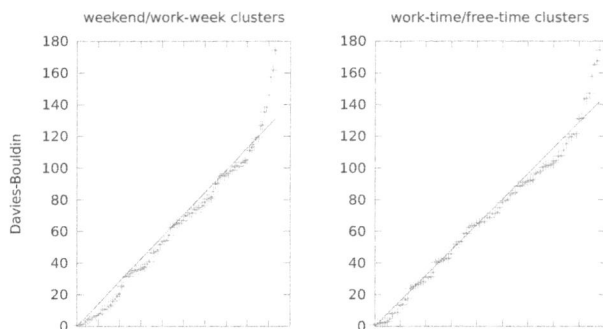

Figure 1: Davies-Bouldin cluster scores calculated for the two setups: work-week/weekend clusters and work-time/free-time hours assigned to each of the top 100 active users, scored ascendingly by the value of the score. Lower values of the score denote more tight and separated clusters. Each plot contains a best linear fit.

designed and evaluated a framework based on lightweight semantics and nearest neighbors. We have identified two types of context that might be helpful in constraining the personalisation process and have done some initial work in order to understand them better. We have designed and evaluated a novel approach for search session detection that is based on implicit feedback and designed with multitasking in mind. We have also done an initial analysis of temporal recurrent patterns in user searching behavior and are now working on extending our work to automatically detect different types of recurrent patterns in clickthrough data automatically. Eventually, we plan to design a method for selecting an appropriate type of context (e.g., long-term, activity-based, or temporal) based on the analysis of the contexts found by each of the methods.

Our approach is not limited to search personalisation only, it can be used for any kind of personalisation or recommendation. We believe that the two identified context types can be found in other domains, e.g. a user researching a certain topic in a news portal is apt for an activity-based context recommendation [14], or a user searching for information on Olympic games is apt for a temporal context based recommendation.

6. ACKNOWLEDGMENTS

This work was partially supported by the Scientific Grant Agency of the Ministry of Education of Slovak Republic, grants VG1/0675/11, VG1/0971/11 and by the Slovak Research and Development Agency under the contract No. APVV-0208-10.

7. REFERENCES

[1] M. Barla. Towards social-based user modeling and personalization. *Information Sciences and Technologies Bulletin of the ACM Slovakia*, 3:52–60, 2011.

[2] S. M. Beitzel, E. C. Jensen, A. Chowdhury, D. Grossman, and O. Frieder. Hourly analysis of a very large topically categorized web query log. In *Proceedings of the 27th annual international ACM SIGIR conference on Research and development in information retrieval*, SIGIR '04, pages 321–328, New York, NY, USA, 2004. ACM.

[3] C. Biancalana, A. Micarelli, and C. Squarcella. Nereau: a social approach to query expansion. In *Proceeding of the 10th ACM Workshop on Web Information and Data Management*, WIDM '08, pages 95–102, New York, NY, USA, 2008. ACM.

[4] M. Daoud, M. Boughanem, and L. Tamine-Lechani. Detecting session boundaries to personalize search using a conceptual user context. In S.-I. Ao and L. Gelman, editors, *Advances in Electrical Engineering and Computational Science*, volume 39 of *Lecture Notes in Electrical Engineering*, pages 471–482. Springer Netherlands, 2009.

[5] D. Downey, S. Dumais, D. Liebling, and E. Horvitz. Understanding the relationship between searchers' queries and information goals. In *Proc. of the 17th ACM conference on Information and knowledge management*, CIKM '08, pages 449–458, New York, NY, USA, 2008. ACM.

[6] E. Herder. *Forward, back and home again. Analyzing user behavior on the Web.* (phd) dissertation, University of Twente, 2006.

[7] B. J. Jansen, A. Spink, C. Blakely, and S. Koshman. Defining a session on web search engines: Research articles. *J. Am. Soc. Inf. Sci. Technol.*, 58:862–871, April 2007.

[8] B. J. Jansen, A. Spink, and T. Saracevic. Real life, real users, and real needs: a study and analysis of user queries on the web. *Inf. Process. Manage.*, 36:207–227, January 2000.

[9] T. Joachims. Optimizing search engines using clickthrough data. In *Proceedings of the eighth ACM SIGKDD international conference on Knowledge discovery and data mining*, KDD '02, pages 133–142, New York, NY, USA, 2002. ACM.

[10] T. Kramár, M. Barla, and M. Bieliková. Disambiguating search by leveraging the social network context based on the stream of user's activity. In *UMAP '10: Proc. of the 18th Int. Conf. on User Modeling, Adaptation, and Personalization*, pages 387–392, Hawaii, USA, 2010. Springer.

[11] T. Kramár, M. Barla, and M. Bieliková. Personalizing search using metadata based, socially enhanced interest model built from the stream of user's activity. *Journal of Web Engineering*, 2012. (submitted).

[12] Y. Lv and C. Zhai. Adaptive relevance feedback in information retrieval. In *Proceedings of the 18th ACM conference on Information and knowledge management*, CIKM '09, pages 255–264, New York, NY, USA, 2009. ACM.

[13] S. Ozmutlu, A. Spink, and H. C. Ozmutlu. A day in the life of web searching: an exploratory study. *Inf. Process. Manage.*, 40:319–345, March 2004.

[14] D. Zeleník and M. Bieliková. News recommending based on text similarity and user behaviour. In *Proceedings of the 7th International Conference on Web Information Systems and Technologies*, WEBIST, pages 302–307, Noordwijkerhout, The Netherlands, 2011. SciTePress.

Utilising Document Content for Tag Recommendation in Folksonomies

Nikolas Landia
University of Warwick
Coventry CV4 7AL
UK
N.Landia@warwick.ac.uk

ABSTRACT

Real-world tagging datasets have a large proportion of new/ unseen documents. Few approaches for recommending tags to a user for a document address this new item problem, concentrating instead on artificially created post-core datasets where it is guaranteed that the user as well as the document of each test post is known to the system and already has some tags assigned to it. In order to recommend tags for unseen documents, approaches are required which model documents not only based on the tags assigned to it in the past (if any), but also the content.

The focus of my research is on utilising the content of documents in order to address the new item problem in tag recommendation. I apply this methodology first to simple baseline tag recommenders and then the more advanced tag recommendation algorithm FolkRank [3][4].

One of my main contributions is a novel adaptation to the FolkRank graph model to use multiple word nodes instead of a single document node to represent each document. This enables FolkRank to recommend tags for unseen documents and makes it applicable to full real-world tagging datasets, addressing the new item problem in tag recommendation.

Categories and Subject Descriptors

H.3.3 [**Information Storage and Retrieval**]: Information Search and Retrieval—*Information Filtering*

General Terms

Algorithms

Keywords

tag recommendation, social bookmarking, FolkRank

1. INTRODUCTION

With the advancement of Web 2.0, tagging is a popular methodology for many user-driven document organisation

applications, such as social bookmarking websites. The automatic generation of tag recommendations aid the social tagging process and lead to a more valuable document organisation overall. Tags are an unstructured organization method where each user has the liberty of choosing and/or making up any string of characters to be used as a tag for a document. The task of tag recommendation is to automatically suggest a set of tags to a user for a document that he is in the process of tagging.

The data contained in social tagging systems is often described as a folksonomy. A folksonomy is a tuple (U, D, T, A) where U is the set of users, D is the set of documents, T is the set of tags and $A \subseteq U \times D \times T$ is the set of tag assignments. A tag assignment $a \in A$ is a triplet (u, d, t) and means that user u has assigned tag t to document d [4]. Thus a folksonomy can be modelled as a hyper-graph with the adjacency tensor given by a 3-dimensional binary matrix $F = [f_{i,j,k}]_{|U| \times |D| \times |T|}$ where each entry $f_{i,j,k} \in \{0, 1\}$ specifies whether or not user u_i tagged document d_j with tag t_k. A post in the folksonomy consists of a **set** of tags T_{ij} assigned by a user u_i to a document d_j. The set of posts is given as $P \subseteq U \times D \times 2^T$ where each post $p \in P$ is a triplet (u_i, d_j, T_{ij}). We use the notation $(u_i, d_j, \emptyset)$ for query posts where the set of tags T_{ij} is unknown and to be predicted.

2. RELATED WORK

Existing tag recommendation solutions can be categorised into approaches which model and analyse the folksonomy data in order to come up with recommendations; and content-based approaches where the textual content and/or metadata of documents is considered.

Methodologies relying on the folksonomy data include Hypergraph [12][10], Graph [4][9][5] and Collaborative [1][7][13] approaches. While hypergraph approaches try to capture and analyse all characteristics of the data in their models, graph-based and collaborative approaches can be described as reductionist methods since they reduce the 3-dimensional folksonomy data to one or several 2-dimensional projections. In content-based approaches, the textual content of the documents is used for either Tag/Keyword Extraction [7] or with document classification techniques [2][11].

Hybrid approaches which combine several tag recommenders also exist and have been explored in [8]. The challenge with hybrid approaches is how to combine the recommendations given by the individual components of the system to achieve the best effect.

The effectiveness of tag recommendation algorithms is usually evaluated against simple baseline recommenders. These include recommending the most popular tags in the system, the most used tags of the user and the most popular tags of the document.

3. PERSONALISING CONTENT-BASED TAG RECOMMENDATION

My first contribution was a hybrid tag recommender which generates predictions based on the content of documents and personalises the resulting tag list to reflect the preference of the user which is learnt from the user profile [6]. The existing documents in the system are represented by their fulltext content using a bag-of-words representation with each word having an importance score with regard to the document (Tf-Idf score). The known training documents are organised into groups based on their content by a hierarchical clustering technique. This allows the system to assign the query document to a cluster in the hierarchy and thus identify a content-based neighbourhood of existing documents. The initial set of candidate tags then consists of past tags not only related to the query document but also to documents in the neighbourhood. This gives the system the potential to recommend tags for new documents that have not been tagged yet. The hierarchical clustering also provides a set of content words which are deemed most important for each cluster of documents. If the set of candidate tags for a test post does not contain a sufficient number of tags with scores above a preset threshold, these cluster-related words can be used to generate new tags for the test post.

4. EXTENDING FOLKRANK WITH CONTENT DATA

While utilising the immediate neighbourhood of the query document and user is a simple and effective way to generate and rank a small set of candidate tags, there is a limit to how many of the real tags can be found this way with a reasonable accuracy. If the user decides to use a tag that he has not used before and that also has not been used by his peers for the query document (or its immediate neighbourhood), then a more advanced ranking technique is required to filter out the most probable tags from the large number of total tags in the system.

FolkRank is a graph-based tag recommendation algorithm [3][4] which is modelled on Google's PageRank. Similarly to PageRank, the key idea of FolkRank is that a document which is tagged by important users with important tags becomes important itself. The same holds symmetrically for users and tags. Users, documents and tags are represented as nodes in an undirected multi-edge tri-partite graph with uniform edge weights (all edges have weight 1). All co-occurrences of users and documents; users and tags; and documents and tags are edges between the corresponding nodes.

The importance or rank of each node is calculated by an iterative weight-spreading algorithm, in a similar fashion to PageRank. To find a set of tag recommendations for a new post $(u_i, d_j, \emptyset)$, where the set of tags T_{ij} is not known (and is denoted by the empty set), the nodes in the graph representing u_i and d_j are given a high preference weight and the iterative weight-spreading algorithm is executed until the weights in the graph stabilise. The nodes which represent tags are then ranked in descending order of weight,

and the top K ranked tags are selected as tag recommendations, where K is a predefined number usually set to a value between 1 and 10.

One drawback of the FolkRank algorithm is its complexity and long runtime. However, more importantly, FolkRank can only generate successful recommendations for posts where the user and document are already known to the system. This makes FolkRank only applicable to artificially created post-core datasets of level 2 or higher with any success. When trying to apply FolkRank to a query post with a new user and/or new document, it defaults to recommending either the most popular tags of the user, the most popular tags of the document or the most popular tags in the system, depending on whether the document, the user or both are new. While the new user problem is not as prominent since each user is only new during his first post, the new document problem is present in the vast majority of posts. Post-cores of level 2 or higher thus only capture a small fraction of the real-world tag recommendation problem. In the Bibsonomy dataset, only about 15.2% of all posts are included in the post-core level 2 subset.

In the full Bibsonomy dataset, the number of posts where the user is new (first encountered) is equal to the number of users; this is the case in roughly 1% of all posts. The number of posts where the document is new (first encountered) is equal to the number of documents; which comprises 90% of all posts. A further consideration is that FolkRank cannot generate and recommend new tags which do not exist in the training data. In Bibsonomy the number of tag assignments where the tag is new (first encountered) is equal to the number of tags and is the case in 7% of all tag assignments.

4.1 Extension of Folkrank with Content Data

In order to overcome the new item problem and make FolkRank applicable to full real-world datasets, we include the content of documents in the tag recommendation process. For test posts where the query user is new (as well), we have to default to the most popular tags found to be related to the query document and cannot personalise these to the user, which is acceptable since the user does not have a tagging profile yet.

4.1.1 ContentFolkRank

Our approach for including content data into FolkRank is to include the word content of documents directly into the FolkRank graph. We adapt the graph to use triplets $(user, word, tag)$ instead of $(user, document, tag)$. Each tag assignment in the training data (u, d, t) is converted to a set of tag assignments with words instead of documents $\{(u, w_1, t), (u, w_2, t), ..., (u, w_k, t)\}$ where each of the words $w_l \in d$. The test vector for each test post (u_q, r_q) is then given by $(u, w_1, w_2, ..., w_k)$ where each $w_q \in d_q$.

We first change the FolkRank graph from having multiple edges with uniform edge weights to a single-edge graph with weighted edges. If we set the weight of the edge between two nodes n_1 and n_2 to the number of nodes N which have an edge to both n_1 and n_2, our graph configuration is equivalent to the original FolkRank multi-edge graph. However, using single weighted edges reduces the runtime of the recommender considerably and allows for easier manipulation of edge weights.

We then create custom rules for setting the weights of edges connecting different types of nodes, namely user - word edges, word - tag edges and user - tag edges. We want the sum of the weights of edges connecting any user u to words nodes from any one document d to be either zero or equal to a pre-defined constant. This would mean that regardless of the number of words that a document is represented by, the sum of weights of edges connecting u to the word nodes representing d will always be the same. To achieve this and additionally include the varying importance of different content words to the document, we use the Tf-Idf scores of the words in the document, where the Tf-Idf scores are normalised to sum 1. Since several documents, for example d_a and d_b, tagged by the same user u_a can contain the same word w_a, the weight of the edge between u_a and w_a is set to the sum of the Tf-Idf scores of w_a in d_a and d_b. The same holds for edges between word and tag nodes. The weight of the edges between user and tag nodes, u_a and t_a, is set to the number of posts (documents) in which u_a has used t_a.

The preference vector for each test post is given by $(u, w_1, w_2, ..., w_k)$ where each $w_q \in d_q$. The preference weight of the user u is set to a predefined constant PW, while the preference weight for each word w_q is set proportional to its Tf-Idf score in r_q, and is given by $pw(w_q) = PW * tfidf(w_q, r_q)$. Since the Tf-Idf weights are normalised to sum 1 per document, the sum of the preference weights of all words $w \in d_q$ is equal to the preference weight attributed to the user u (and is equal to PW).

5. POSTRANK

An issue which exists in plain FolkRank (as well as our adaptation ContentFolkRank) is the problem of balancing edge weights. Due to the fact that a post can have a variable number of tags assigned to it, the user and document nodes of the post can be connected to a variable number of tag nodes each. The difficulty is then deciding the weight to give to the single edge between the user and document node. One can either keep the weight at 1 regardless of the number of tags in the post or set the user-document weight relative to the number of tags in the post. Both are incomplete solutions only addressing one part of the problem. The problem is to choose edge-weights so that from the perspective of any one node in the graph, an equal total edge weight goes to different types of nodes (ie all document nodes and all tag nodes) while preserving an equal (or explicit) distribution of edge weights between nodes of the same type (ie each document node should get the same weight from a user regardless of the number of tags in their respective posts). In order to address the edge-weight balancing problem, I am currently working on a solution which includes an additional type of node, representing posts themselves, in the graph. This allows to balance edge weights in the context of each post without affecting the weights in other posts which contain the same user, document or tag.

6. EXPERIMENTAL EVALUATION

6.1 Bibsonomy Dataset

The dataset consists of tagging data from the social bookmarking website Bibsonomy[1] [3]. The system and data is divided into web site bookmark tagging data and publication BibTex tagging data. We concentrate on website bookmarks and evaluate our recommenders on this subset of Bibsonomy.

[1]http://www.bibsonomy.org/

6.2 Evaluation Methodology

6.2.1 Train-Test Split

We split each dataset by date to produce a training and test set, similarly to [7], where the last 3 months of tagging activity is the test set. Table 1 shows the percentage of new users, documents and tags in our test set for the Bibsonomy Bookmark Full dataset.

	% New Items	% Posts with New Items
Users	68%	17.5%
Documents	92.5%	91.6%
Tags	39%	17% *

* This is the average percentage of new tags per test posts, since one post can have multiple tags

Table 1: Bibsonomy Bookmark Test Set New Items

6.2.2 Evaluation Measures

We use Recall@N to evaluate the success rate of the recommenders where N is the predefined number of tags to recommend. Recall is calculated per test post and then averaged over posts to give the overall recall on the test set.

6.3 Results

Figure 6.3 shows the recall on the Bibsonomy Bookmark Full dataset when recommending tag sets of different sizes. In these initial results, the source of content words which are used in the ContentFolkRank recommender is the title of the documents. As expected, FolkRank performs better than the baseline recommenders at all levels since it is able to aggregate much wider-reaching connections in the data than the immediate neighbourhoods from which the baseline recommendations are constructed. By including content in the FolkRank graph, the ContentFolkRank recommender shows a considerable improvement over plain FolkRank.

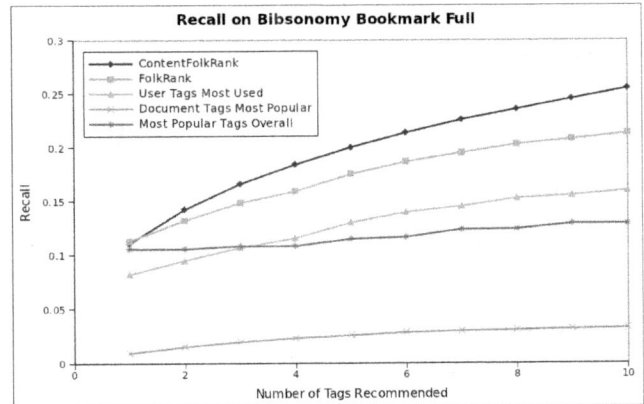

Figure 1: Recall on Bibsonomy Bookmark Full

7. CONCLUSION

In order to utilise tag recommendation algorithms in real-world tagging systems, it is important that the algorithms can recommend tags for previously untagged documents since the vast majority of posts in these systems contains new documents. The new item problem for documents can be addressed by building recommenders which consider the content of the documents in their models. My research is focused on this task and I have applied this methodology first to baseline recommenders [6] and then to the more potent graph-based recommender FolkRank. The inclusion of document content data in tag recommendation systems shows an improvement over purely folksonomy-based recommenders and makes the content-aware recommenders viable for use in real-world tag recommendation systems.

8. FUTURE WORK

In the future, I plan to fully evaluate the suggested ContentFolkRank recommender on further datasets such as CiteULike[2] and Delicious[3] as well as complete the development of the PostRank recommender. I also plan to examine the usefulness of topic models with regard to FolkRank. After learning a set of topics from the content of the training documents, the FolkRank graph could be created with nodes which represent topics instead of documents or words. A query document would then be mapped onto a set of topics and the document-related part of the preference vector would be a set of weights which represent the membership of the document to each of the topics. With ContentFolkRank, especially when using content information from metadata and fulltext content, the need arises to reduce the dimensionality of the document representation and size of the graph. While topic models would be one way to achieve this, I also plan to explore traditional dimensionality reduction alternatives.

9. REFERENCES

[1] J. Gemmell, T. Schimoler, M. Ramezani, and B. Mobasher. Adapting k-nearest neighbor for tag recommendation in folksonomies. In *Proceedngs of the 7th Workshop on Intelligent Techniques for Web Personalization and Recommender Systems*, 2009.

[2] P. Heymann, D. Ramage, and H. Garcia-Molina. Social tag prediction. In *SIGIR '08: Proceedings of the 31st annual international ACM SIGIR conference on Research and development in information retrieval*, pages 531–538, New York, NY, USA, 2008. ACM.

[3] A. Hotho, R. Jäschke, C. Schmitz, and G. Stumme. Information retrieval in folksonomies: Search and ranking. In Y. Sure and J. Domingue, editors, *The Semantic Web: Research and Applications*, volume 4011 of *Lecture Notes in Computer Science*, pages 411–426. Springer, 2006.

[4] R. Jaeschke, L. B. Marinho, A. Hotho, L. Schmidt-Thieme, and G. Stumme. Tag recommendations in folksonomies. In *Knowledge Discovery in Databases: PKDD 2007, 11th European Conference on Principles and Practice of Knowledge Discovery in Databases*, 2007.

[5] H.-N. Kim and A. El Saddik. Personalized pagerank vectors for tag recommendations: inside folkrank. In *Proceedings of the fifth ACM conference on Recommender systems*, RecSys '11, pages 45–52, New York, NY, USA, 2011. ACM.

[6] N. Landia and S. Anand. Personalised tag recommendation. In *Proceeding of the Recommender Systems and the Social Web Workshop, held in conjunction with the ACM conference on Recommender Systems*, 2009.

[7] M. Lipczak, Y. Hu, Y. Kollet, and E. Milios. Tag sources for recommendation in collaborative tagging systems. In *Proceedings of the ECML/PKDD 2009 Discovery Challenge Workshop, part of the European Conference on Machine Learning and Principles and Practice of Knowledge Discovery in Databases*, 2009.

[8] M. Lipczak and E. Milios. Learning in efficient tag recommendation. In *RecSys '10: Proc. the 4th ACM Conference on Recommender Systems*, pages 167–174. ACM, 2010.

[9] M. Ramezani, J. Gemmell, T. Schimoler, and B. Mobasher. Improving link analysis for tag recommendation in folksonomies. In *Proceedings of the 2nd Recommender Systems and the Social Web held in conjunction with the 4th ACM conference on Recommender systems.*, 2010.

[10] S. Rendle, L. Balby Marinho, A. Nanopoulos, and L. Schmidt-Thieme. Learning optimal ranking with tensor factorization for tag recommendation. In *Proceedings of the 15th ACM SIGKDD international conference on Knowledge discovery and data mining*, KDD '09, pages 727–736, New York, NY, USA, 2009. ACM.

[11] Y. Song, Z. Zhuang, H. Li, Q. Zhao, J. Li, W.-C. Lee, and C. L. Giles. Real-time automatic tag recommendation. In *SIGIR '08: Proceedings of the 31st annual international ACM SIGIR conference on Research and development in information retrieval*, pages 515–522, 2008.

[12] P. Symeonidis, A. Nanopoulos, and Y. Manolopoulos. Tag recommendations based on tensor dimensionality reduction. In *Proceedings of the 2008 ACM conference on Recommender systems*, RecSys '08, pages 43–50, New York, NY, USA, 2008. ACM.

[13] Z. Xu, Y. Fu, J. Mao, and D. Su. Towards the semantic web: Collaborative tag suggestions. In *Proceedings of the Collaborative Web Tagging Workshop at the WWW 2006*, Edinburgh, Scotland, 2006.

---

[2]http://www.citeulike.org/
[3]http://delicious.com/

Using Group Recommendation Heuristics for the Prioritization of Requirements

Gerald Ninaus
Institute for Software Technology
Graz University of Technology
Inffeldgasse 16b
8010 Graz, Austria
gerald.ninaus@ist.tugraz.at

ABSTRACT

Group recommendation heuristics have been successfully applied in different domains such as Interactive Television and e-Tourism. Our work focuses on the improvement of requirements prioritization techniques on the basis of group recommendation technologies. First, we analyse the impact of preference visibility on the outcome of the requirements prioritization process. In this context we evaluate the resulting software quality and the satisfaction of stakeholders with the requirements prioritization. Second, we analyse standard group recommendation heuristics on a dataset originated from a study conducted at our university. Finally, we propose new heuristics to improve the prediction quality and conclude with an evaluation of the different recommendation heuristics.

Categories and Subject Descriptors

D.2 [**Software Engineering**]: Requirements Engineering; D.2.1 [**Requirements/Specifications**]: Requirements Negotiation; H.5 [**Information Interfaces and Presentation**]: Modelling Environments

General Terms

Requirements Engineering, Human Factors, Experimentation

Keywords

Requirements Prioritization, Recommender Systems, Group Recommendation, Group Decision Making

1. INTRODUCTION

Limited resources in software projects make it necessary to prioritize requirements [7] [13], which eases the planning of subsequent software releases. Furthermore, requirements prioritization supports the conflict resolution between stakeholder preferences. Because of the complexity of the requirements prioritization task, this process requires collaborative work contributed by multiple stakeholders. Systematic requirements prioritization is crucial for the successful completion of a software project since then the available resources are used only for the most relevant requirements. However, establishing consensus about the prioritization of requirements is a challenging task. There are many effects which must be considered, such as cognitive dissonance reduction [5], group polarization [14], and the primacy effect [3]. To simplify this task, tools can recommend prioritizations of requirements as a function of the team preferences. However, if such a prediction of a group decision is only based on the initial preferences of the stakeholders, the prediction quality will be low [11]. A challenge in this context is changing group member preferences during the decision-making process [11]. Another challenge is the primacy effect in preference elicitation [3]: the outcome of this phase will depend on the sequence in which preferences have been inserted. The psychological literature shows that consensus about topics formed early in discussions is cognitive resistant to changes. Additional information added later will be assimilated to already chosen consensus and it is very unlikely that another option is chosen [8]. This phenomenon can be explained by the *assimilating effect* based on the dissonance theory [5] which states *that individuals are motivated to reduce psychological incongruity or discrepancy* that may arise by adding new information to a present perception [2]. The result is that stakeholders will perceive already selected options more attractive than new options [2] and this leads to a bias of group preferences depending on the order of the incoming preferences. Unfortunately, this effect is increased if there is a high group identity, because the fear of exclusion from the group is higher [8]. To reduce this effect, a brainstorming phase in which stakeholders become aware of their own preferences should be established. To raise the willingness of stakeholders to report their honest concerns, this brainstorming phase should be implemented in an anonymous fashion [6]. In a software environment, for example, this can be done if requirements can be added without authentication. First related study results will be presented in the following Sections.

An important issue in the group prioritization process is the factor of *fairness*. The degree of perceived fairness influences the willingness of group members to accept compromises in the resolution of disputes and their trust in other stakeholders [8]. Especially in environments with a high amount of requirements it is necessary to provide tool support in the prioritization process to achieve a maximum degree of perceived fairness. One way to provide such a tool support is to present recommendations of reasonable requirements prioritizations to decision makers.

In our work we investigate different *recommendation heuristics* with respect to the achieved *prediction quality*. We also show how anonymity (preferences are not connected to stakeholder names)

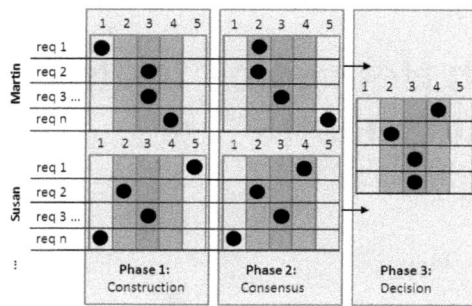

Figure 1: The three decision making phases in IntelliReq.

in group decision processes can help to improve the quality of requirements prioritizations. The major contributions reported in this paper are to show that anonymity in decision making influences the quality of the prioritization and to present new heuristics for group recommendations with better prediction quality.

The remainder of the paper is organized as follows. In Section 2 we provide an overview of the IntelliReq decision support environment developed at the Graz University of Technology. In Section 3 we show the impact of the factors *anonymity*, *consensus*, and *decision diversity* on the *quality of the prioritization*, the *stakeholder satisfaction* with the prioritization, and the *quality* of software artefacts. Section 4 covers standard recommendation heuristics including our new proposed heuristics. Section 5 gives an overview of the empirical study conducted at our university. In Section 6 we conclude the paper with an outlook on future work.

2. INTELLIREQ DECISION SUPPORT

IntelliReq is a group decision support tool developed at our university. One feature is the group based requirements negotiation support which will be discussed in detail in the remainder of this paper. The requirements prioritization process consists of three different phases (see Figure 1). The first phase is called *construction* in which the individual stakeholder preferences are collected. The second phase is denoted as *consensus* and is used to discuss and adapt stakeholder preferences. In the third phase (*decision*), the project manager finalizes the group prioritization and can attach explanations for decisions. This explanation can be seen by all group members. To evaluate the impact of anonymized preference elicitation (it is hidden which stakeholder defined which preference), IntelliReq supports two different modes of interaction. One mode uses anonymised preference elicitation, in the other mode user preferences are visible.

3. ANONYMITY, DIVERSITY, AND OUTPUT QUALITY

Within the scope of our study we evaluated the impact of the factors *anonymity*, *consensus*, and *decision diversity* (the standard deviation between the final prioritization results of different requirements) on the *output (software) quality* and the *satisfaction* of the stakeholders with the prioritization. The empirical study has been conducted within the course *Object-oriented Analysis and Design* (N=39 software teams of size 5-6; 15.45% female, 84.55% male) and showed that there are impacts as can be seen in Figure 2. These impacts can be summarized as follows:

- *Anonymous preference elicitation increases consensus of the group:* The hypothesis is that anonymous preference elic-

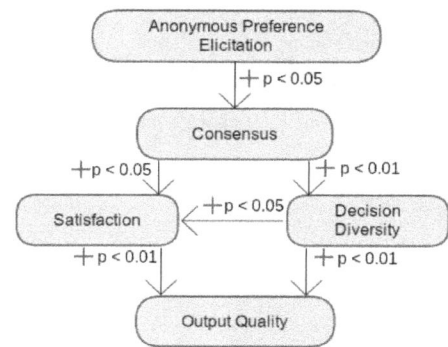

Figure 2: The result of the IntelliReq study (N=39 software teams of size 5-6) as an impact graph. Starting with Anonymous Preference Elicitation each following phase is influenced according to the arrows.

itation helps to decrease the commitment [1] which is the degree of a person's unwillingness to change an already articulated opinion. Our study showed that the anonymous preference elicitation leads to a significantly higher degree of consensus between group members.

- *Decision diversity influences the satisfaction of stakeholders:* Our study revealed a significant correlation between a high standard deviation in the prioritization of different requirements and the satisfaction of the stakeholders with the prioritization of the requirements.

- *Consensus increases decision diversity:* Our study results show that a high consensus leads to a high decision diversity. We hypothesise that a dissent makes it necessary to find a compromise in the group. This compromise will often be oriented towards the average of the initial group member preferences. If there are many compromises necessary, the final group prioritization will be located in the middle of the scale. This leads to a small standard deviation in the prioritization of the requirements.

- *Consensus increases satisfaction of stakeholders:* Our study showed that groups with a high consensus regarding the requirements are more often satisfied with the prioritization.

- *Satisfaction and decision diversity increase the output quality:* When evaluating the quality of the software developed by the different groups (teams), we could detect a significant increase of the output quality of software artefacts in situations where there was a high decision diversity and satisfaction with the requirements prioritization.

4. GROUP DECISION HEURISTICS

We hypothesise that the effects described in the previous section could be enhanced by the use of group recommendation heuristics. Consequently, we want to evaluate the best heuristics to predict stakeholder voting behaviour in the requirements prioritization process. In the afore mentioned study conducted at our university we collected a dataset of approximately 10.000 decisions including the preferences of stakeholder and the resulting prioritizations of the software teams. With this dataset we evaluated different group recommendation heuristics [9] and compared predicted with real decisions taken by the participants of our study. For the evaluation we used the precision metric according to Formula 1. The following subsections describe the different applied heuristics that are based on following parameters: **median** m**, standard deviation** sd**, heuristic** h **and requirement** r.

$$precision(h) = \frac{\# \ correctly \ predicted \ group \ preferences(h)}{\# \ predicted \ group \ preferences(h)} \quad (1)$$

Least Distance

This heuristic determines (recommends) for each requirement r_i the priority value p with the lowest distance to the other elements in the set of distinct user preferences (PREF) where $p \in PREF$. This criteria is formalized in a corresponding evaluation function (see Formula 2 and Table 1).

$$Priority(r) = selectmin(p, \sum_{pref \in PREF} |pref - p|) \quad (2)$$

	Martin	Susan	Peter	Pauline	recom. priority
r_1	1	3	1	2	2
r_2	5	4	4	3	4
r_3	2	1	3	1	2

Table 1: Application of *Least Distance* heuristic.

Majority Voting

This heuristic recommends a priority value which represents the majority of stakeholder votes related to a specific requirement. An example for the recommendation result of the majority heuristic is given in Table 2.

	Martin	Susan	Peter	Pauline	recom. priority
r_1	1	3	1	2	1
r_2	5	4	4	3	4
r_3	2	1	3	1	1

Table 2: Application of *Majority Voting* heuristic.

Average Value

This heuristic determines the average value (see Formula 3) and round the result (see Formula 4) of the declared stakeholder preferences for each requirement. This value is taken as a recommendation of the priority value for the corresponding requirement.

$$AVG(r) = \frac{\sum_{i=1}^{\#user} pref(i,r)}{\#user} \quad (3)$$

$$Priority(r) = \begin{cases} rounddown(AVG(r)) & AVG(r) < 0.5 \\ roundup(AVG(r)) & AVG(r) \geq 0.5 \end{cases} \quad (4)$$

Random Priority Selection

The random heuristic has been integrated only for evaluation purposes. This heuristic has - as expected - the weakest performance, which can be seen in the *Study Result* Section.

Median Based

Inspired by a survey of the voting behaviour of six-person juries conducted by the University of Chicago [12], we implemented a new heuristic called *Median Based Heuristic* (see Formula 5). In the original study [12], jury members were asked about their initial preferences for the penalty. These initial punishment preferences were compared with the final decision after group deliberating. The finding was that *there was a severity shift for the high-punishment cases, and a leniency shift for the low-punishment cases* [12]. Such choice shifts are explained by the *Group Polarization Theory* [14].

In a similar fashion, our algorithm calculates the recommendations depending on the median of the initial preferences. In our study, the preferences are distributed on a six point scale. In this context, the algorithm has three possible states (see also Formula 5):

- The median is one or two: Calculate the average (see Formula 3) of the preferences and round **down** the result
- The median is three or four: Use the *Least Distance* heuristic (see Formula 2).
- The median is five or six: Calculate the average (see Formula 3) of the preferences and round **up** the result

$$Priority(r, m) = \begin{cases} rounddown(AVG(r)) & m = 1, 2 \\ LDIS(r) & m = 3, 4 \\ roundup(AVG(r)) & m = 5, 6 \end{cases} \quad (5)$$

Ensemble Based

For a further improvement of the prediction quality we introduced a combination of different heuristics. For each recommendation task the algorithm calculates the *Majority* (MAJ), the *Least Distance* (LDIS), and the *Median Based* (MDB) heuristic. If two heuristics recommend the same result, this result is the final recommendation. If there are three different recommendations, the *Median Based* heuristic is the final recommendation. The heuristic is shown in Formula 6.

$$Priority(r, m) = \begin{cases} MDB(r, m) & MDB(r, m) = MAJ(r) \\ MDB(r, m) & MDB(m, r) = LDIS(r) \\ LDIS(r) & LDIS(r) = MAJ(r) \\ MDB(r, m) & else \end{cases} \quad (6)$$

Standard Deviation Based

The hypothesis for this heuristic is that the best heuristic depends on the degree of conformity of the initial preferences. For example, if the group has a high degree of consensus, the *Least Distance* heuristic will perform best. If there is a high dissent, the *Average* heuristic will perform better. The highest standard deviation in our dataset is 2.50. We divided the dataset into three subsets. Each subset has the same standard deviation range ($\frac{2.50}{3} \approx 0.84$). Next we tested the heuristics mentioned in this paper on the dataset to find the heuristic which performs best on each subset (see Table 3). For this evaluation we used the complete dataset of the *Consensus* and *Decision* phases (see Figure 1) which included the anonymous and the non-anonymous setting. The combination of the different heuristics can be found in Formula 7.

$$Priority(sd, r, m) = \begin{cases} LDIS(r) & sd < 0.84 \\ MDB(r, m) & 0.84 \geq sd < 1.67 \\ AVG(r) & 1.67 \geq sd \leq 2.50 \end{cases} \quad (7)$$

Thereafter we generated a function which uses the heuristic depending on the standard deviation and the results of Table 3.

Group	SD From	SD To	Best Heuristic
1	0	0.84	LDIS
2	0.85	1.67	MDB
3	1.68	2.50	AVG

Table 3: Groups based on *Standard Deviation*.

5. EMPIRICAL STUDY

An overview of the study results can be found in Table 4.[1] Although the *Median Based* heuristic is outperformed by the *Least Distance* (see Table 4), the *Median Based* performs better in an environment with visible preferences. In the study conducted at the University of Chicago [12], on which this heuristic is based on, the same observation was made in decision making processes where group member preferences are visible. The *Standard Deviation* heuristic is out of competition as this heuristic is defined for this specific dataset. Future studies with new datasets will show whether this combination of different heuristics will have an improved prediction quality. When comparing the prediction quality of the remaining heuristics, the *Ensemble* heuristic performs best.

Heuristic	Consensus	Decision
LDIS (least distance)	0.619	0.733
MAJ (majority voting)	0.576	0.719
AVG (average value)	0.617	0.702
RAN (random selection)	0.167	0.188
MDB (median based)	0.618	0.732
ESB (ensemble)	0.629	**0.739**
SDB (sd based)	**0.636**	0.722

Table 4: Comparison of heuristics of the *Consensus* and the *Decision* phase (best results are marked bold).

6. CONCLUSION AND FUTURE WORK

The heuristics discussed in this paper are basic group recommendation heuristics [9]. Our initial analysis shows the applicability of these heuristics in terms of prediction quality. As part of our future work we want to investigate the positive effect of recommendations on the dimensions *decision diversity, satisfaction with requirements prioritization*, and the *software quality* resulting from the requirements prioritization process. In addition, we want to conduct an in-depth evaluation of the user acceptance of the determined group recommendations - up to now only the majority voting based recommendations has been analysed [4].

Furthermore, we want to analyse whether if the *Standard Deviation* heuristic (as described in Formula 7) has the same prediction quality on other data sets . We also want to use this heuristic for group recommendation in our upcoming IntelliReq user studies. An important task for future work is the analysis of preference reversal [2] in requirements engineering and the impact on the prioritization quality and the stakeholders' satisfaction with the prioritization. In this context we want to develop new methods to improve stakeholder satisfaction.

Another topic of interest is the different interpretation of preferences. Are the group members arguing of the same topic? Does everybody has the same understanding of a given term? It has been

shown that team members often enter a decision process from different viewpoints. Therefore it is necessary to find a consensus on the interpretation of shared information. This is especially important as the interpretation of issues has an massive impact on the decision making and is therefore considered as crucial [10].

7. REFERENCES

[1] R. B. Cialdini. The science of persuasion. *Scientific American*, pages 76–81, Feb. 2001.

[2] J. R. Curhan, M. A. Neale, and L. Ross. Dynamic valuation: Preference changes in the context of face-to-face negotiation. *Journal of Experimental Social Psychology*, 40(2):142 – 151, 2004.

[3] A. Felfernig, G. Friedrich, B. Gula, M. Hitz, T. Kruggel, G. Leitner, R. Melcher, D. Riepan, S. Strauss, E. Teppan, and O. Vitouch. Persuasive recommendation: serial position effects in knowledge-based recommender systems. In *Proceedings of the 2nd international conference on Persuasive technology*, PERSUASIVE'07, pages 283–294, Berlin, Heidelberg, 2007. Springer-Verlag.

[4] A. Felfernig, C. Zehentner, G. Ninaus, H. Grabner, W. Maalej, D. Pagano, L. Weninger, and F. Reinfrank. Group decision support for requirements negotiation. In L. Ardissono and T. Kuflik, editors, *Advances in User Modeling*, volume 7138 of *Lecture Notes in Computer Science*, pages 105–116. Springer Berlin / Heidelberg, 2012.

[5] L. Festinger. *A Theory Of Cognitive Dissonance*. Stanford University Press, 1957.

[6] P. Grünbacher. Collaborative requirements negotiation with easywinwin. In *Proceedings of the 11th International Workshop on Database and Expert Systems Applications*, DEXA '00, pages 954–960, Washington, DC, USA, 2000. IEEE Computer Society.

[7] H. Hofmann and F. Lehner. Requirements engineering as a success factor in software projects. *IEEE Software*, 18(4):58–66, Jul/Aug 2001.

[8] E. A. Lind, L. Kray, and L. Thompson. Primacy effects in justice judgments: Testing predictions from fairness heuristic theory. *Organizational Behavior and Human Decision Processes*, 85(2):189–210, July 2001.

[9] J. Masthoff. Group recommender systems: Combining individual models. In *Recommender Systems Handbook*, pages 677–702. 2011.

[10] S. Mohammed and E. Ringseis. Cognitive diversity and consensus in group decision making: The role of inputs, processes, and outcomes. *Organizational Behavior and Human Decision Processes*, 85(2):310 – 335, 2001.

[11] E. Molin, H. Oppewal, and H. Timmermans. Modeling group preferences using a decompositional preference approach. *Group Decision and Negotiation*, 6:339–350, 1997. 10.1023/A:1008612828114.

[12] D. Schkade, C. R. Sunstein, and D. Kahneman. Deliberating about Dollars: The Severity Shift. *Ssrn Electronic Journal*, 2000.

[13] K. Wiegers. First Things First: Prioritizing Requirements. *Software Development Online*, 7:48–53, Sept. 1999.

[14] J. A. Zuber, H. W. Crott, and J. Werner. Choice shift and group polarization: An analysis of the status of arguments and social decision schemes. *Journal of Personality and Social Psychology*, 62(1):50 – 61, 1992.

---

[1]Due to the limited space only the results without differentiation between visible and non-visible preferences are shown.

Beyond Lists: Studying the Effect of Different Recommendation Visualizations

Denis Parra
School of Information Sciences
University of Pittsburgh
135 North Bellefield Avenue,
Pittsburgh, PA, USA
+1 412 6249437

dap89@pitt.edu

ABSTRACT

Recommendation Systems have been studied from several perspectives over the last twenty years –prediction accuracy, algorithmic scalability, knowledge sources, types of recommended items and tasks, evaluation methods, etc. - but one area that has not been deeply investigated is the effect of different visualizations and their interaction with personal traits on users' evaluation of the recommended items. In this paper, I survey visual approaches that go beyond presenting the recommended items as a textual list or as annotations in context. I also review related literature from recommendations' explanations. In this thesis, I aim to understand how different visualizations and some personal traits might influence users' assessment of recommended items, particularly in domains where multidimensional data or contextual constraints are involved. I present the prototype of 2 recommendation visualizations and then briefly propose the research approach of this investigation.

Categories and Subject Descriptors

H.3.3 [**Information Search and Retrieval**]: Information Filtering. H.5.2 [**User Interfaces**]: Interaction Styles, User-centered design.

General Terms

Design, Experimentation, Human Factors.

Keywords

Recommender Systems, Adaptive Interfaces, Visualization of Recommendations.

1. INTRODUCTION

Recommendation Systems (RS) aim to provide users relevant items from a crowded information space. RS have become popular commercially and as a research field in the latest 10 years as evidenced by online contests such as the Netflix Prize, and by the interest in related conferences like ACM RecSys.

On the research field, among the several issues that has been investigated over the years –prediction accuracy, algorithmic scalability, knowledge sources, types of recommended items and tasks, evaluation methods, etc. - the effect of different visualizations and their interaction with personal traits over users' preference on recommended items has been studied but not at the same extent as the aforementioned issues. One can identify three main reasons that make this an important area of research: a) *Recommendation Transparency*: Explaining how the recommendations were generated to RS users has shown to have a positive effect in users' trust in the RS [1], and visualizations of recommendations can provide better ways than textual lists to comprehend how recommendations were generated. b) *Multivariate Data*: most current state-of-the-art recommender systems rely not only on one dimension of user feedback –such as user ratings-- but also on implicit feedback, time, location, and many other forms of contextual information. In addition, some domains such as Event Recommendations pose additional constraints, such as the limited life time of the items. Visualizing recommendations beyond textual lists might facilitate incorporating several dimensions [2] when presented to users to make sense of the recommendations. c) *The Effect of Personal Traits*: Considering human traits as factors affecting how users evaluate recommendations has been studied [3, 4], but not their interaction with rich visualizations as those implemented in [5-7].

In this research proposal, I aim to bridge the lack of investigation on the effects of rich visualizations, interface interactions and some traits of users upon how users' evaluate recommendations: users' perceived relevance, novelty and serendipity, users' satisfaction, and users' trust on the recommendation process. In section 2, I survey related work about visualizing and explaining recommendations. Then, in section 3, I introduce two systems that will be used to test the influence of different visualizations. In section 4 I briefly describe my research plan, and in section 5 I summarize the challenges and expectations of this research proposal.

2. RELATED WORK

Research on visualizing recommendations beyond a textual list or as *annotation in context* [8] is not abundant. Some of these studies have shown the positive effect on user satisfaction of visualizing recommendations and allowing richer interaction [5-7, 9]. PeerChooser [5] was introduced by O'Donovan et al. to show movie recommendations interactively. The active user was depicted as the center node in an ego-network, and PeerChooser's users could explore their nearest neighbors, movie genres, and check their recommendions in different ways.

a)   b)

Figure 1. Screenshots of a) Conference Navigator 3, and b) CoMeT.

The recommended movies were presented as top-N, clusters and rating predictions on user-selected items. A study with 25 users that compared four recommendation methods showed that adding user interaction significantly improved the system's rating predition and users' satisfaction. On a different domain, Gotz and Wen [9] developed an approach called Behavior-Driven Visualization Recommendation (BDVR), by which they detect patterns of user behavior in a data analysis system called HARVEST. Based on the detected behavioral patterns, HARVEST recommend other visualizations to analyze the data. A user study with 20 users, that considered two different tasks per user on three possible scenarios, showed that using the BDVR on HARVEST decreased the average time needed to complete the tasks and also the error rate, compared to those users that were assigned to the group without BDVR support. Despite its novel approach, BDVR innovates in the way that recommendations are identified based on user behavior, rather than the way in which the recommendations are presented to the user –a link blinking that suggest another way to analize the data to the user. Another interesting work using visualizations to show user recommendations is SmallWorlds [6], a visual recommender system implemented as a Facebook application that utilizes the profiles of the active user and her connections to generate recommendations. A user study allows the authors of SmallWorlds sustaining that system's transparency and interface interaction increases user satisfaction, and even under the constraints of the Facebook API – that allows to get information only from the user's direct connections– the recommendations enhanced by pre-existing friends' information boost the users' satisfaction on recommendation predictions. The main limitation of this study is the assessment based only on user preferences of movies, although Facebook profiles provides various other types of items for recommendation. More recently, in the music domain, Gou et al. present SFViz [7], a sunburst visualization to allow users exploring and finding friends interactively under a context of interest. Their visualization and iteraction method is novel, and although they introiduce a case study using the social network and tags of a last.fm dataset, no user study is presented to empirically assess their design.

The second related area of research is explaining recommendations to users and how visualizations help increasing users' trust in the system and overall user satisfaction. Zhao et al. [10] present Pharos, a content-centric system able to recommend items, people and communities. They try to tackle the cold-start problem and also explain the recommendations by visualizing a social map with terms organized in latent communities. A within subjects study with ten

users shows that Pharos helped the subjects to complete exploratory tasks faster and better than BlogCentral, an existent tool. Although user knowledge and tasks were considered and they didn't affect the significant differences between Pharos and BlogCentral, the small amount of subjects calls for a larger user study to generalize these results. Another drawback of this system is the lack of personalization. The social map displayed the same communities and terms to every user, and users' feedback suggests adding this feature in an future version. Zhang et al. [11] go beyond textual explanation by presenting a visual interface for a critiquing-based RS. In an e-commerce system, they present various critiques by a set of meaningful icons, and their results show how the visual presentation and the aided interaction improve the shopping experience of the users. However, the visualizations are not rich visualizations as those presented in [5-7].

The main limitation of the aforementioned studies, excepting Pharos [10], is the absence of users' traits in the analysis of factors interacting with visualizations on users' perceived relevance of the recommendations. The importance of users' traits on recommendation peformance is shown in [3, 4]. Knijnenburg et al. [3], show that the users' domain knowledge is a factor influencing what kind of interaction method they prefer in an energy-saving RS. Tkalcic et al. [4] show how affective parameters, based on users' emotive responses, improve the performance of a recommender system. These two studies do not neglect user traits but they do not consider rich visualizations (such as ego-network graphs, sunburst plots, circle packs, etc.) in their evaluations.

Building on the results of these articles, I will consider the users' domain knowledge and the type of visualization as the main independent factors of my investigation.

3. WORK UP-TO-DATE

I plan to test the effect of rich recommendation visualizations and their possible interaction with user's domain knowledge in two systems developed in the PAWS lab[1] at the University of Pittsburgh: Conference Navigator and CoMeT (Collaborative Management of Talks).

---

[1] http://adapt2.sis.pitt.edu/wiki/

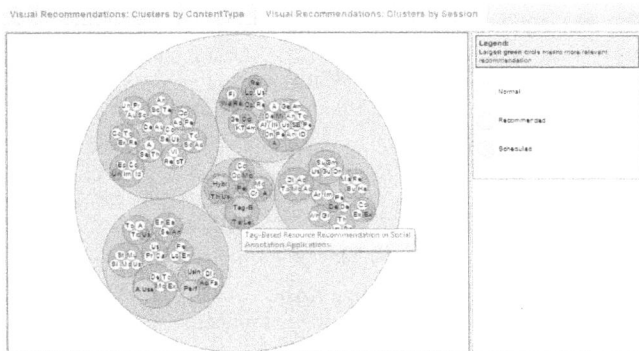

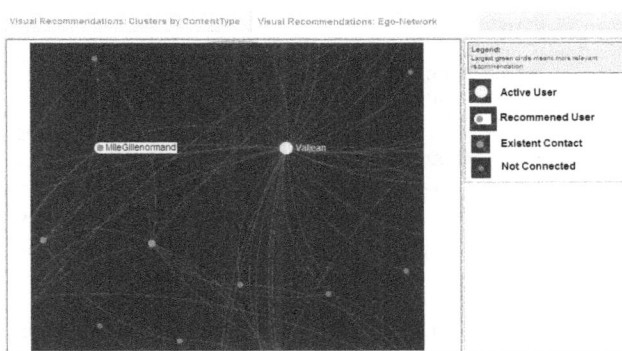

a)   b)

Figure 2. Screenshots of a) recommendations as a circle pack using d3.js b) ego-network visualization based on sigma.js.

3.1 Conference Navigator

Conference Navigator[2], shown in Figure 1.a), is a system which supports attendees at academic conferences. It provides usual information about a conference such as the list of papers with authors and abstract, the schedule of the conference, details of each talk and in some occasions the list of attendees.

Conference Navigator allows users scheduling the talks they are interested to attend, and in addition, it allows users connecting each other as followed/follower, and also as reciprocal connections. Furthermore, Conference Navigator has useful features that present aggregated users' behavior, such as the ranking of the most scheduled papers, the most active contributors, the institutions of the most active contributors, a cloud of the tags which users have attached to papers, and a feature similar to Amazon.com about papers: "people who scheduled paper X also scheduled paper Y". Conference Navigator also provides a personalized textual list of recommended talk constructed upon content and tag-based algorithms. Since the talks are part of a larger event, the conference, they can be classified inside days or sessions. Moreover, talks present characteristics such as popularity or whether they are recommended that make them prone to be presented in a richer visualization than a list.

3.2 CoMeT

CoMeT[3], shown in Figure 1.b), stands for Collaborative Management of Talks. It is a system that aggregates information about open academic lectures and talks in the Pittsburgh area, mainly at the University of Pittsburgh and at Carnegie Mellon University. Some of the talks are manually entered to the system by users and the rest are automatically collected by web crawlers. CoMeT provides socially aggregated data about the talks: how many people has viewed, bookmarked and e-mailed each one. CoMeT features the most popular talks of each day in its homepage, but it also provides different ways to navigate and search throughout all the talks available. Although some talks are entered as isolated events, they usually belong to series of talks or to special department events, what makes them good candidates to be grouped and classified under several categories. The social activity associated to talks, and their temporal restrictions, makes

them good candidates to be recommended and presented in visualization richer than lists. Currently, CoMeT provides users a textual list of recommended talks by e-mail once per week.

4. RESEARCH PLAN

4.1 Prototypes of Recommendation Visualizations

Choosing a set of layouts or visual structures to represent multivariate data that convey recommendations is a difficult task. However, I first decided choosing only two types of visualizations given the number of users that are usually active in our systems and the statistical power that we might need to find a significant effect in case that one exists, which is around 30 users. Under this circumstance, the ego-network seems a good model to pick since it has been already studied but not under the effects that I have considered, what would make our results more comparable with previous research. On the other hand, I decided choosing a circle pack layout (also called nested circle layout) since it has been shown to be an efficient way to represent a tree structure [12], and in this case I want to represent multivariate hierarchical data. A conference program where days, sessions and papers form a tree is a good example of this kind of structure. Finally, it is important noticing that I am undertaking an incremental empirical approach, where I plan testing these visualizations in two conferences during the summer of 2012. Depending on the results, I might test other visualizations in conferences at the beginning of 2013.

4.1.1 Circle Pack Layout

Considering a layout that shows concentric circles helps visualizing a tree, which is one of the possible ways to structure the contents of a conference consecutively as days, sessions and talks. Figure 2.a) shows a prototype of talk recommendation for Conference Navigator, where the largest circle represents the conference, the largest circles inside represent days, and inside them, sessions and talks. The talks which the user has scheduled to attend are presented with orange color, the ones not scheduled with white, and the recommended talks with green. The different size of the recommended talks (the green circles) represent their recommendations score: the larger the recommendation score, the larger the circle. Planned interaction will allow the users clicking on a circle and obtaining additional information of the day, session or talk clicked on. Furthermore, they will be able to rate the recommended talks and we will track user interactions with the system.

---

[2] http://halley.exp.sis.pitt.edu/cn3/

[3] http://halley.exp.sis.pitt.edu/comet/

4.1.2 Ego-Network Layout

This model has been already explored in [5, 6] and a prototype implemented in Conference Navigator is shown in Figure 2.b). The active user will be represented as the central node in a network, and every node in the network will be at most 3 hops away from the central user. The other nodes will represent neighbors, articles, and in some cases users' tags –depending on the method used to provide the recommendations: content-based or tag-based.

4.2 User Study

In the user study, I plan comparing the 2 aforementioned layouts (Circle Pack and Ego-Network) with a textual list of recommendation, and a recommendation list enhanced with facets in both Conference Navigator and CoMeT. The reason for comparing with a faceted list comes for the advice given by Hearst in [11], where faceted lists provide support to present multivariate data beyond lists, that is difficult to overcome by other richer visualizations in terms of user performance to complete search tasks.

5. DISCUSSION AND FUTURE WORK

The proposal presented in this paper is ongoing work, in an early stage of development, but it highlights its importance and introduces the research and initial evaluation plan.

I have outlined the reasons, surveyed the related work, and presented an initial description of the approach to investigate the influence of visualizations in users' perception of recommendations. First, regarding the reasons, there is a lack of research that integrates the influence of different visualizations, interactions and user traits in recommender systems. Moreover, the need of transparency in recommendations, the multivariate nature of the data, and the lack of studies incorporating the effect of personal traits makes this proposal and important research goal. The related work highlights the positive effect of different interaction on user satisfaction, but it presents limitations. They are restricted to recommend movies [5, 6]–an area already deeply investigated whose results cannot be generalized to other fields, such as event recommendations--, their recommendation visualizations are not rich in interaction [9], their user studies have few users [10] or they completely lack a user study [7].

The expected number of subjects for the study in Conference Navigator and CoMeT is around 20 and 30. However, a power analysis is necessary to establish the minimum number of users needed to detect an effect in case it exists. This analysis will be conducted in the next stage of this research.

6. ACKNOWLEDGEMENTS

I appreciate the feedback provided by Roman Bednarik to clarify some terms that might lead to confusion, and some ideas to describe better the purpose of this research.

7. REFERENCES

[1] Tintarev, N. and Masthoff, J. 2011. Designing and Evaluating Explanations for Recommender Systems. Recommender Systems Handbook. Springer US.

[2] Knijnenburg, B. P., Reijmer, N. J. M. and Willemsen, M. C. 2011. Each to his own: how different users call for different interaction methods in recommender systems. In Proceedings of the Proceedings of the fifth ACM conference on Recommender systems, Chicago, Illinois, USA.

[3] Tkalčič, M., Burnik, U. and Košir, A. 2010. Using affective parameters in a content-based recommender system for images. User Modeling and User-Adapted Interaction, 20, 4, 279-311.

[4] O'Donovan, J., Smyth, B., Gretarsson, B., Bostandjiev, S. and Höllerer, T. 2008. PeerChooser: visual interactive recommendation. In Proceedings of the twenty-sixth annual SIGCHI conference on Human factors in computing systems (CHI '08). ACM, New York, NY, USA.

[5] Gretarsson, B., O'Donovan, J., Bostandjiev, S., Hall, C. and Höllerer, T. 2010. SmallWorlds: Visualizing Social Recommendations. Computer Graphics Forum, 29, 3, 833-842.

[6] Gou, L., You, F., Guo, J., Wu, L. and Zhang, X. L. 2011. SFViz: interest-based friends exploration and recommendation in social networks. In Proceedings of the 2011 Visual Information Communication - International Symposium (VINCI '11). ACM, New York, NY, USA, article 15, 10 pages.

[7] Gotz, D. and Wen, Z. 2009. Behavior-driven visualization recommendation. In Proceedings of the 14th international conference on Intelligent user interfaces (IUI '09). ACM, New York, NY, USA, 315-324.

[8] Herlocker, J. L., Konstan, J. A., Terveen, L. G. and Riedl, J. T. 2004. Evaluating collaborative filtering recommender systems. ACM Trans. Inf. Syst., 22, 1, 5-53.

[9] Zhao, S., Zhou, M. X., Yuan, Q., Zhang, X., Zheng, W. and Fu, R. 2010. Who is talking about what: social map-based recommendation for content-centric social websites. In Proceedings of the fourth ACM conference on Recommender systems (RecSys '10). ACM, New York, NY, USA, 143-150.

[10] Zhang, J., Jones, N. and Pu, P. 2008. A visual interface for critiquing-based recommender systems. In Proceedings of the the 9th ACM conference on Electronic commerce. Chicago, IL, USA.

[11] Hearst, M. 2009. Search User Interfaces. Cambridge University Press.

[12] Guilmaine, D., Viau, C., McGuffin, M. J.2012. Hierarchically Animated Transitions in Visualizations of Tree Structures. In Proceedings of the 11th International working Conference in Advanced Visual Interfaces (AVI 2012), Capri, Italy.

The User-Centered Design of a Recommender System for a Universal Library Catalogue

Simon Wakeling
iSchool, University of Sheffield
Regent Court, 211 Portobello
Sheffield, UK
+44 (0)114 222 2630

s.wakeling@sheffield.ac.uk

ABSTRACT

This paper describes a collaborative project between the University of Sheffield's iSchool and OCLC (an international library cooperative), the aim of which is to develop a prototype recommender system for WorldCat.org, the aggregated catalogue of OCLC's member libraries. This paper describes a user-centered approach, utilizing both qualitative and quantitative methods, which aims to establish how and why users engage with library catalogues and WorldCat.org in particular, whether there is a need for recommendations in the library domain, and if so what type of recommendations best support the information-seeking needs of users. An outline of the proposed methodology is provided, along with a report on work completed to date. An analysis of UK library catalogues shows the prevalence of recommender systems to be very low, while initial results from focus group interviews and a pop-up survey show a significant demand for recommendations from two key user-groups (students and academics).

Categories and Subject Descriptors

H.3.3 [**Information Storage and Retrieval**]: Information Search and Retrieval – Information filtering, Relevance feedback, Retrieval models, Search process, Selection process.

General Terms

Design, Experimentation, Human Factors, Theory.

Keywords

Library, User-Centered Design, Recommender Systems.

1. INTRODUCTION

The problem of information overload, identified as a key issue for information seekers, is one that particularly applies to the library community. The huge growth in available resources has led to a wealth of accessible information with the potential to inform scholarship and facilitate teaching and research, while the provision of that information over the web has dramatically

increased the range and scope of services available to users [18]. Despite these undoubted benefits, however, it has become increasingly clear that traditional models of information retrieval

frequently fail to best connect users with potentially relevant material. As Baez et al note, the sheer volume of available material often demands that searches are narrowed to identify core resources rather than broadened to take in potentially useful but perhaps ill-defined items [3]. This is exacerbated by the increasing preference of users – born out of practice and familiarity - for self guided web-based searches, bypassing the traditional role of the librarian or information professional [14]. Furthermore in an increasingly competitive information market-place, institutions are becoming ever more conscious of the need for their services to match the expectations of users in both appearance and functionality.

The application of recommender systems to diverse library and research systems is potentially a key step towards ensuring that all types of library users are properly serviced, as a growing body of evidence suggests that recommender systems can help users find and use information more efficiently [14, 18]. This raises several key questions for researchers seeking to apply the principles of recommendation to the library domain. Since such applications are required to serve a diverse community of users, previous implementations have tended to orientate themselves towards a generic user model [2] – a problem given the increasingly diverse user-base [13]. That this diversity is manifest not only in a demographic sense, but also from a task-orientated perspective, further complicates the development of systems that can truly be described as "useful" – in other words closely attuned to the context of the search [14].

This AHRC funded research project represents a collaborative venture between the University of Sheffield's iSchool, and OCLC (the International library consortium), and forms the basis of the author's doctoral thesis. We aim to explore the role of recommendations in a library context, with the goal of producing a prototype system for WorldCat.org.

The principle research questions for the project are defined as follows:

1. What do library catalogue users want from recommender systems?
2. What criteria influence and shape their recommendations?
3. What data are required to make recommendations relevant, reliable and effective?
4. How should recommendations be presented to users to facilitate discovery?

The objectives for the research are:

1. Establish user needs and expectations from library-based recommender systems;

2. Perform an analysis of existing systems for library book and media recommendations;

3. Establish 'important' criteria for making and assessing library-based recommendations;

4. Analyze the data resources held by OCLC, including bibliographic records, transaction logs and library holdings data;

5. Design and build an interactive recommender prototype for the WorldCat 'universal' library catalogue; and

6. Evaluate the prototype in a lab-based setting and operational environment.

2. OCLC AND WORLDCAT.ORG

OCLC is an international not-for-profit library cooperative. Founded in 1967 by a consortium of Ohio libraries, its purpose was to "create a nonprofit membership organization dedicated to the public purposes of furthering access to the world's information and reducing information costs" [9]. Most importantly this meant the development of a networked platform on which to build shared resources, thereby increasing access to information while lowering operational costs among member libraries. Today OCLC has more than 72,000 member libraries in 170 countries. OCLC's operations cover a broad range of activities, including the provision of library management systems, library advocacy, research and product development, and cataloguing services. Aside from its ownership of the Dewey Decimal Classification system, it is perhaps best known for WorldCat.org[1], the aggregated catalogue of the holdings of its member libraries. Representing the largest bibliographic database in the world, WorldCat contains more than 250 million bibliographic records and more than 1.8 billion holdings. This catalogue is publically accessible via a web interface (see Figure 1). Users of this service can search the catalogue and access a range of supplementary information about items, including reviews and ratings (both user generated and imported from third parties), links to online retailers selling the item, and related content based on classification number.

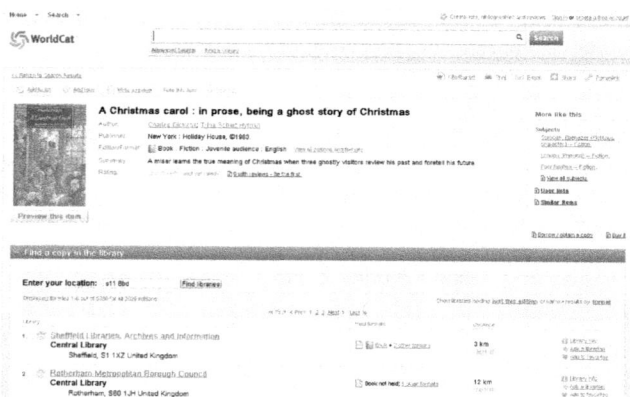

Figure 1 - www.worldcat.org

Users are also able to create customized lists of items. While these features are visible to all users, creation of a WorldCat.org account is required in order to add tags and reviews, and create lists. At present the number of accounts created as a proportion of the total user-base is low, with a significant majority of users accessing the system without logging in.

---

[1] http://www.worldcat.org

3. RELATED WORK

There are a number of existing recommender systems within the library domain. Significant work has been done with regards to the recommendation of journal articles, with two principle methods being employed. Systems such as *TechLens* [18] and *Synthese* [15] supply the user with recommendations through traditional collaborative filtering algorithms utilizing citation data in lieu of user generated ratings. Taking a different approach, Ex Libris' *bX* service mines link resolver logs to identify articles downloaded by individual users during the same session, and uses this data to infer relationships between items [16]. It has been suggested that recommender systems powered by access data generate more diverse and interesting recommendations – particularly for newly published papers [11]. Interestingly it has also been shown that the two methods – citation based and usage based – demonstrate a high level of complimentarity; not only does testing show that they rarely provide the same recommendations, but they can rarely *both* provide recommendations related to a given specific article [16].

Several relatively recent projects have also attempted to implement OPAC (Online Public Access Catalogue) recommendation functions. The University of Huddersfield in the UK has developed a system that mines circulation logs to present users viewing an item on the catalogue with suggested alternatives [10], while the BibTip project utilizes three software agents to build recommendations based on implicitly inferred data [7]. An *Observation Agent* observes all the titles selected within each discrete user session, passing the data to an *Aggregation Agent* which collates co-occurrences and builds links between items. Finally a *Recommendation Agent* mines these links to provide recommendations to users in the form of hypertext links in the browser. A third model for providing OPAC recommendations is provided by *Library Thing For Libraries* (LTFL). *Library Thing* is an online service allowing members to catalogue their book collections, and supplement this catalogue with a wide array of ratings, reviews and tags. LTFL is sold to libraries as an OPAC overlay, using a small piece of JavaScript to query *Library Thing*'s database for the ISBN of the item being viewed in the OPAC. Relevant tags, reviews and recommendations are then exported to the OPAC interface and can be viewed by the user.

4. METHODOLOGY

As the title of this project indicates, a key feature of the proposed system development process is that it should be "user-centered". User Centered Design (UCD) describes a design process "in which users have a deep impact on the design by being involved as partners with designers throughout the design process" [1]. Seen as particularly relevant to the development of technological innovations, and as originally proposed by Norman and Draper [8], the primary purpose of UCD is to ensure that finished products combine optimal functionality with intuitive usability. At the heart of all UCD processes is a focus on engaging with users to understand their needs and requirements.

In accordance with these principles our initial work has been focused on better understanding the users of WorldCat.org, establishing whether there a demand for recommendations, and if so identifying the type of tasks that recommendations can support. To this end a series of Focus Groups with WorldCat.org users have been conducted, and a pop-up survey attached to WorldCat.org (see section 5). Furthermore OCLC have provided access to WorldCat.org transaction logs, which will be the subject of analysis on two counts; as a means of further investigating user behaviour within the system, and as a potential data source for developing recommendations (see section 6). It is hoped that the first stage of analysis can provide greater clarity about the types of tasks undertaken on the system, and provide more data about the prevalence of known item vs subject searching.

The results of this research will feed into an iterative system design process that will evaluate potential data sources and produce a prototype system for testing and evaluation.

5. PRELIMINARY RESULTS

5.1 Comparison of UK OPACS

In order to ascertain the extent to which recommendation feature in current library catalogues, a comparison of UK OPACS was undertaken [17]. 118 University and 211 Local Council library OPACS were analyzed, and the incidence of recommendations noted. For the purposes of this study, a recommendation was characterized as:

The presentation to the user of one or more specific resources not directly related to the ranking of the search results, and not requiring the use of a faceted browsing function (e.g. "find more by this author").

Results showed that only 11% (n=13) of University OPACS and 2% (n=4) of public library OPACS offer recommendations, indicating that the the incorporation of recommendations into the OPAC is extremely limited. While these results can partly be explained by the prevalence of legacy Library Management Systems unable to support recommender functionality, it also appears that librarians have yet to persuaded that the value of and demand for recommender systems justifies their implementation becoming a priority.

5.2 Focus Group Interviews

Focus group research offers "a way of collecting qualitative data, which – essentially – involves engaging a small number of people in an informal group discussion (or discussion), "focused" around a particular topic or set of issues"[19]. Focus groups constitute an established methodology within Library and Information Science [12], and a number of previous studies have used the methodology to investigate the use of online catalogues [4, 6].

The research was conducted in three stages, based on geographical location: Australia and New Zealand (21 March – 8 April 2011), the UK (9 – 17 May 2011) and the US (25 – 27 October 2011). Participants were drawn from four of WorldCat.org's main user-groups; Librarians, Students, Academics and Booksellers. 21 focus group interviews were conducted, with a total of 118 participants. The questions posed were deliberately broad, and aimed to explore how and why users engage with WorldCat.org, what they perceived as its strengths and weaknesses, and how they believed the system could be improved. Recommender systems were not explicitly mentioned by the moderator. Results indicate a strong desire for recommender functionality to support specific information-seeking tasks (specifically relating to subject searches) among several key user-groups, particularly students and academics. The full results of the research are currently being prepared for publication.

5.3 Pop-up Survey

To determine the make-up of the WorldCat.org user-base, and to obtain quantitative data about the types of task the site is used for, a pop-up survey was developed for WorldCat.org, and offered to users between 4 – 19 April 2012. In addition to gathering basic demographic information, the survey attempted to identify the purpose and context for the user's engagement with the system, and explored their views on the utility of recommendations. A total of 2,917 users completed the survey. Initial findings suggest a significant demand for recommendations, with 65% of respondents who accessed the survey from the WorldCat.org homepage stating that some form of recommendations would be useful. Further analysis is underway, and it is hoped that the data will reveal the types of users and tasks that can be best supported by recommendations.

6. FUTURE WORK

The following work will be undertaken in the remaining eighteen months of this project:

6.1 Users and Library Recommendations

Based on the findings of the user-centred research, one or more experiments will be developed that investigate how users of online library catalogues respond to recommendations in a variety of contexts. These experiments will consist of giving participants a number of information seeking scenarios, and asking them to rate or rank various recommendation lists, each generated by different existing book recommender systems. In order to control variables, a basic web interface will be developed to display the recommendations to participants, thereby negating the potential influence of interface design. The development of the scenarios and the selection of participants will be heavily influenced by the data collected in the first stage of the research. It is also possible that the raw rating or ranking data generated during the experiment can be supplemented with the participants' reflections on the exercises – perhaps during a post-experiment interview or questionnaire.

6.2 System Design and Evaluation

The low number of users with WorldCat.org accounts is likely to represent a significant challenge for the prototype design, since the system will have only very limited data about the user for most sessions. Given these limitations, a key aspect of the design process will be the extent to which a variety of information seeking contexts can be supported. A number of potential data sources for recommendation generation have been identified, and will be evaluated according to the user needs revealed in earlier phases of the project. These data sources include selected library circulation records, WorldCat.org transaction logs, explicit user ratings, OCLC member library holdings, and a variety of metadata fields that supplement item records within the catalogue. The results of this evaluation will inform not only system design, but also allow for the selection of a subset of WorldCat.org holdings for the prototype system to work on. It is anticipated that the design will be an iterative process, with additional user testing being conducted to refine the recommendation algorithm and interface design. OCLC have committed the time of one software engineer to assist with system development. Once complete, it is hoped that the prototype system will be tested and evaluated in a usability laboratory setting, either at the University of Sheffield or the OCLC headquarters in Dublin, Ohio.

7. CONCLUSION

Although a variety of approaches to recommendation in the library domain have been attempted, the prevalence of systems in public access catalogues remains low. This suggests that library professionals remain to be convinced that such systems offer enough value to users to justify their implementation. In taking a highly user-centered approach to the development of a recommender system for a library catalogue, it is anticipated that this project will go some way to establishing recommender functionality as an important facet of library systems. It is also hoped that the methodology used can be employed in other domains, particularly those in which recommender systems have not previously been widely applied, and for systems where recommendations must be made with only minimal data about individual users.

8. ACKNOWLEDGMENTS

The author thanks his PhD supervisors, Paul Clough and Barbara Sen (both University of Sheffield), and Lynn Silipigni Connaway (OCLC).

9. REFERENCES

[1] C. Abras, D. Maloney-Krichmar, and J. Preece. User-Centered Design. In W. Bainbridge (Ed.) *Encyclopedia of Human-Computer Interaction*. Sage Publications, Thousand Oaks, CA, 2004.

[2] H. Avancini, L. Candela, and U. Straccia. Recommenders in a personalized, collaborative digital library environment. *Journal of Intelligent Information Systems*, 28(3): 253-283, 2007.

[3] M. Baez, A. Birukou, F. Casati, and M. Marchese, M. Addressing Information Overload in the Scientific Community. *Internet Computing*, 14(6): 31 – 38, 2010

[4] W.K. Berger and R.W. Hines. What Does the User Really Want? The Library User Survey Project at Duke University. *Journal of Academic Librarianship*, 20(5-6): 306-309, 1984

[5] J. Bollen, and H. Van de Sompel. An Architecture for the Aggregation and Analysis of Scholarly Usage Data. *In Proceedings of the 6th ACM/IEEE-CS Joint Conference on Digital Libraries* (Chapel Hill, NC, USA, June 11-15, 2006). http://public.lanl.gov/herbertv/papers/jcdl06_accepted_version.pdf

[6] L.S. Connaway, D.W. Johnson, and S.E. Searing. Online Catalogs from the Users' Perspective: The Use of Focus Group Interviews. *College & Research Libraries*, 58(5):403-420, 1997.

[7] M. Mönnich, and M. Spiering. Adding Value to the Library Catalog by Implementing a Recommendation System. *D-Lib Magazine*, 14(5/6), 2008 http://www.dlib.org/dlib/may08/monnich/05monnich.html

[8] D. Norman, and S. Draper (Eds). *User Centered System Design: New Perspectives on Human-Computer Interaction*, L. Erlbaum Associates Inc., Hillsdale, NJ,1986.

[9] OCLC. Public Purpose. OCLC, Dublin, OH 2012 http://www.oclc.org/ca/en/about/purpose/default.htm

[10] D. Pattern. Using "circ_tran" to show borrowing suggestions in HIP, blog, 17 November 2005 http://www.daveyp.com/blog/archives/49.

[11] S. Pohl, F. Radlinski, and T. Joachims. Recommending Related Papers Based on Digital Library Access Records. In *Proceedings of the 7th ACM/IEEE-CS joint conference on Digital libraries* (Vancouver, Canada, June 17-22, 2007) *JCDL '07* http://radlinski.org/papers/Recommending_PohlJCDL.pdf

[12] M. Von Seggern, and N. Young, N. The focus group method in libraries: issues relating to process and data analysis. *Reference Services Review*, 31(3):272-284, 2003.

[13] A. F. Smeaton, and J. Callan. Personalisation and Recommender Systems in Digital Libraries. *International Journal on Digital Libraries*, 57(4):299-308, 2005.

[14] R. Torres, S.M. McNee, M. Abel, J.A. Konstan, and J. Riedl. Enhancing Digital Libraries with TechLens+. In *Proceedings of the Joint Conference on Digital Libraries* (Tuscon, Arizona, USA, June 7-9, 2004. *JCDL 2004* http://www-users.cs.umn.edu/~mcnee/torres-jcdl2004.pdf.

[15] A. Vellino, and D. Zeber, D. A Hybrid, Multi-Dimensional Recommender for Journal Articles in a Scientific Digital Library. In *Proceedings of the 2007 IEEE/WIC/ACM International Conferences on Web Intelligence and Intelligent Agent Technology* (Silicon Valley, US, November 2-5, 2007). *WI '07* http://delivery.acm.org/10.1145/1340000/1339671/3028a111.pdf?ip=143.167.102.38&acc=ACTIVE%20SERVICE&CFID=89751095&CFTOKEN=65824140&__acm__=1331653818_f04a29e122216ee73b288c88e15ac9df

[16] A. Vellino. A Comparison Between Usage-based and Citation-based Methods for Recommending Scholarly Research Articles. In *Proceedings of ASIS&T 2010 Annual Meeting* (Pittsburgh, PA, USA, October 22–27, 2010). *ASIS&T 2010* http://www.asis.org/asist2010/proceedings/proceedings/ASIST_AM10/submissions/330_Final_Submission.pdf

[17] S. Wakeling, P. Clough, B. Sen, and L.S. Connaway. "Readers who borrowed this also borrowed...": Recommender Systems in UK libraries, *Library Hi Tech*, 30(1): 134 – 150, 2012.

[18] J. Webster, S. Jung, and J. Herlocker. Collaborative Filtering: a new approach to searching digital libraries, *New Review of Information Networking*, 10(2): 177 – 191, 2004.

[19] S. Wilkinson, (2004). "Focus group research." In D. Silverman (ed.), *Qualitative research: Theory method, and practice* pages 177–199. Sage, Thousand Oaks, CA, 2004

Reducing the Sparsity of Contextual Information for Recommender Systems

Dusan Zelenik
Faculty of Informatics and Information
Technologies
Slovak University of Technology
Bratislava, Slovakia
zelenik@fiit.stuba.sk

Maria Bielikova
Faculty of Informatics and Information
Technologies
Slovak University of Technology
Bratislava, Slovakia
bielik@fiit.stuba.sk

ABSTRACT

Our work focuses on the improvement of the accuracy of context-aware recommender systems. Contextual information showed to be promising factor in recommender systems. However, pure context-based recommender systems can not outperform other approaches mainly due to high sparsity of contextual information. We propose an idea to improve accuracy of context based recommender systems by context inference. Context inference is based on effect discovered by analyses of the context as a factor influencing user needs. Analyses of the news readers reveals existence of behavioural correlation which is the main pillar of proposed context inference. Method for context inference is based on collaborative filtering and clustering of web usage (as a non-discretizing alternative to association rules mining).

Categories and Subject Descriptors

H.3 [**Information Storage and Retrieval**]: Clustering Information Filtering; H.2 [**Database Applications**]: Data mining

General Terms

Algorithms

Keywords

context, recommender system, clustering, user behaviour

1. MOTIVATION

Context-aware recommender systems have become very popular since variety of contextual information could be acquired. With an increase of the smart-phone popularity and available features which they provide, we are able to associate user needs with contextual information. From the high level context types such as location, time, weather, to the low level context types such as humidity, noise, movement,

we study the impact of the context on the user behaviour and needs. However context itself has shown to be insufficient when it comes to accuracy of context-aware recommender systems. Context is therefore used as a secondary aspect for generating recommendation.

One reason for low accuracy is high sparsity of contextual information. High sparsity is caused by various natures of users and their preferences [5]. Some users do not want to share their personal information such as location, thus causing missing contextual information. Poor context information leads to low accuracy in prediction. On the other hand, some users are willing to expose even personal contextual information such as emotions. They are willing to answer question and explicitly express contextual information, which is then useful in context-aware recommendation.

Our idea is to propagate contextual information from one user to another in order to reduce the sparsity of data. We propose the propagation of the context by exploiting a correlation in users' behaviour. We assume that users' behaviour is not random, it is based on context of the user. For instance, Perse [14] discovered association between negative mood and tendency to watch competition-style programs as a result of the need to experience happiness. Action-style programs are selected when viewers are in a positive emotional state. However, even if some associations are valid for majority of users, we expect that there are associations which could be discovered only for a subset of users. This leads to clustering of users by their behaviour. Identifying clusters of similar users helps to identify how to propagate the context between users.

We discover associations between the need and context using alternative to standard association rules mining. The difference is in non-discrete values which we use. For example, wrong discretization causes noise, as we lose the ability to compare them and thus sort them. Therefore we expect to achieve higher accuracy when we use proposed value based associations discovery instead of item based. To accomplish value based associations discovery we combine standard techniques from machine learning such as x-means [6] and vector distance computation (Euclidean distance).

2. RELATED WORK

Context inference has received relatively little attention in the literature when it comes to implicit inference from the logs of user activity. It is caused by the selective approaches to context incorporation. Specific solutions work with specific contextual information. Kahng et al. [8] demonstrate the predefined context as one of the factors for document ranking in information retrieval process. As an example they introduce weather and its impact on the user's interest in song listening. This empiric context selection emerges from the observation made by Baltranus et al. [2]. They research the relevance of the context in the system explicitly by asking the user. They showed that supposed context has positive impact on the success of their method. On the other hand, research by Asoh et al. [1] proves that there is a significant difference between the real and supposed reaction to the context. One way or another, this could be understood as an explicit form of the context acquisition. And unfortunately, we are still unable to persuade and engage everyone into explicit feedback.

Therefore we work with the acquired context and users' behaviour to infer missing contextual information. To stress the unavailability of contextual information we pick the work of Bermingham et al. [3]. They propose a solution to discover the sentiment from microblogs. The sentiment is a derivation of the emotional context. Microblogs are perfect source for discovering this type of the context. However it is domain specific and could not be used as a generic solution.

Riboni et al. [16] announced a hybrid of statistical analyses and ontological reasoning in order to acquire the context. Utilization in the COSAR project shows better results by combining both of these approaches. We have decided to use statistical approach boosted by empirically observed effect of users behaviour correlation. We understand the correlation of the behaviour as the correlation of the contextual information. Konomi et al. [10] present connections between people formed by co-presence at places. These connections are based on geo-location but correlate with social connections.

User behaviour is often represented by a set of actions performed by the user. Kramar [11] observed the effect of changing the behaviour with the change of current context. He identified that multiple personas are present in the behaviour of individual. The same effect was exploited by Park et al. [13]. They clustered user's behaviour by actions to improve query suggestions. They have actually used client side logs to cluster the behaviour, which outperformed the state-of-art approaches. From multiple personas of an individual, we expanded to multiple personas of all users in the system. Our intention is to supplement missing contextual information using multiple personas.

Combining multiple personas of more users will improve results in context inference. Research made by Cadiz et al. [4] or Rahnama et al. [15] enables us to work with more users and in various systems. By using standardized frameworks and unifying context-aware systems, we are able to gather usage logs. Contextual information on activities from various systems improves our abilities to infer missing context. The only drawback of such framework is the redundancy of some information and higher complexity. Including information on the past, present and future context [12] increases the complexity even more. Several approaches have been presented to address this problem. Komninos et al. [9] work with vector representation of action and propose solution to reduce complexity, even the complexity caused by vector weighting issues. Reduction of the complexity is important even for mobile devices where computational resources are constrained. Dargie et al. [7] discuss the need to reduce time to recognise the context and its essence in real-time systems.

3. CORRELATION IN SIMILAR USERS' BEHAVIOUR

We have studied the effect of correlation in users' behaviour to propose an exploitation which would help to reduce the sparsity of contextual information. Our idea is to propagate contextual information only to users whose behaviour highly correlates.

3.1 Contextual Information

To prove our concept we have decided to work with database of web usage recorded by news portal SME.sk[1]. This news portal is the biggest local news portal with more than 20 thousands active readers at the peak. Every click recorded includes time, IP address, user identifier and article identifier. Further information such as category, section, author, publishing time for article are also provided. We used this database before for content-based recommendation [17] which enables us to compare results achieved by our previous work.

To prepare database for further research, we add the context which is not in database. We use services such as `wunderground`[2] and `ip2location`[3] to add information on weather and location. We also process timestamp to store time derivatives (such as day of week, part of day, etc.). Location which is extracted from IP address is very rough and for dynamic block of IP addresses, the location is almost untraceable. We have also applied a simple rule based algorithm to extract information on location (home, work, outside) using time and IP address. It is based on repeating IP addresses during work days (from 8 AM to 5 PM), during night and weekends. If the IP address was used by user during work hours many times, we add the context of location respectively (at work).

Dataset prepared in this way contains contextual information which is acquired with both high and low confidence. Low confidence causes sparsity what negatively affects further recommendation process.

3.2 User habits

We have analysed news reading with focus on various context types. We presume that user has habits which are affected by context. We also presume that some users have similar habits thus their behaviour is affected by context similarly. We have mostly analysed the time as the most popular context (see Fig. 1). The figure shows that majority of users are influenced by forthcoming events. The figure proves that majority of users have similar habits. For instance, they read about cooking, when they are going to cook for Christmas.

We also recognized same habits in smaller groups of users. For example, local football games are commented on this site, which attracts some users with interest in football. These users have same interests. It could also mean that these people are also similarly influenced by the same context. Influence of context means correlation in their behaviour. We use this effect of correlation in behaviour to form clusters of similar users. Knowing similar users enables us to propagate contextual information correctly.

---

[1] http://www.sme.sk
[2] http://www.wunderground.com
[3] http://www.ip2location.com

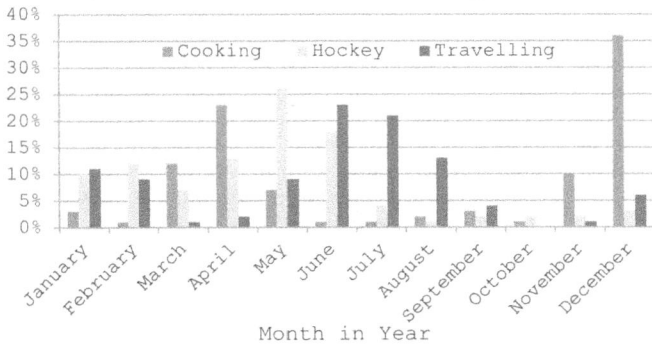

Figure 1: **Readers are influenced by forthcoming events. We can observe the impact as an increase of displayed articles on specific topics (monitoring real users at SME.sk).**

Table 1: **Example of context inference. Actions 1, 2 are similar to actions 6, 7. Unknown context of location could be propagated to user B using known location from user A.**

action	user	hour in day	day in week	section	location
1	A	22	2	sport	work
2	A	21	3	sport	work
3	A	11	7	cooking	home
4	A	11	7	cooking	home
5	A	12	7	cooking	home
6	B	19	3	sport	N/A
7	B	18	5	sport	N/A
8	B	12	7	cooking	N/A
9	B	12	7	cooking	N/A
10	B	13	7	cooking	N/A

4. USER MODEL ENRICHMENT

Our user model represent the measure of user interest in item. Similarly to association rules we work with patterns. Every pattern consist of a condition and filter. In our case of news recommending, the filter is a combination of the section and the category in news portal. There are around 420 combinations which could be used.

Condition expresses the context of the user which has to be valid when the rule is applied to recommendation process. We understand condition as a set of contexts which form a condition together. Conditions are used to find situation of the user. Condition and current situation of the user must be matching when we want to apply the filter.

User model contains only the most frequent patterns. But even if condition does not match current situation of the user, we are still able to find the best matching condition. Every context has its value which represents the importance with a condition. Calculation of vector distances between conditions and situation of the user results in best matching rule which is then applied.

4.1 Context Inference

We have build the user model by processing user activity which has been already recorded. Some contextual information could be acquired directly using services and processing attributes. However, some contextual information could be missing or the confidence of the information is very low. We propose the context inference which leads to the reduction of the sparsity. Table (see Tab. 1) demonstrates how could the contextual information be propagated to another. These actions could be represented as vectors and clustered using all attributes except missing location. User A has complete contextual information in this example. User B is considered to be similar to User A since their behaviour is similar. In result, contextual information for missing values is inferred using known values and similar actions.

Context inference is basically executed in following steps.

1. **Cluster users' actions.** User identifier is ignored at this moment. We use x-means clustering [6].

2. **Compute users' similarity.** Co-presence of actions made by different users reveals their behavioural similarity.

$$similarity(u,v) = \frac{\sum_{i=0}^{n}(|U \cap C_i| + |V \cap C_i|)}{|U| + |V|} \quad (1)$$

where U and V are sets of actions for users u and v and C is the set of clusters.

3. **Find the best matching actions.** Only actions of similar users are used.

4. **Propagate contextual information.** Only in case it is available.

Context inference is followed by enriching the user model. It is done by finding the most frequent actions and extraction of conditions and associated items. In order to keep values not discretized we use clustering to find clusters of similar actions per user. Complexity of such computation is however very high, thus we process only the most recent actions. This also keeps the user model up-to-date.

4.2 Recommendation

There are more options how to incorporate context into recommender systems. Our user model enables us to recommend items by filtering them using rules stored in the model. Current situation of the user is used to find the best matching items. Items are used as filters on the dataset of potential recommendations. Here we can work with content-based approaches, collaborative filtering or other.

Content-based recommendation is generated using items which are fetched from the model using current user conditions. In this alternative we are searching for the items in the dataset which are similar to items from user model. Item in the user model is not necessarily one of items in dataset. It could be only the set of keywords. We propose to use category and section in our dataset of news.

Collaborative filtering is another very popular approach to generate recommendations. To use our model with this approaches we change the pair of condition and item to condition and user. This enables us to reveal the most similar users whose situations were very similar to the current situation of the user.

5. EVALUATION STRATEGY

As we already mentioned we work with database where both low and high confident contextual information is present. We propose to evaluate our method for sparsity reduction by using only records with high confident contextual information. We randomly select contextual information and simulate its confidence to be very low. Then we apply context inference and compare inferred results with original values.

We also propose evaluation for context-aware recommendation. We want to compare results achieved by recommending news both with and without inferred contextual information. Experiment is conducted with real people who are separated into two groups. One group receives recommendations generated only with original contextual information. Another group receives recommendations generated with inferred context (reduced contextual sparsity). In this case we also incorporate test for statistical significance.

Another approach to evaluate our approach to sparsity reduction is to compare our recommender system to others to show expected improvement.

6. CONCLUSIONS AND FUTURE WORK

In our work we face significant drawback of common sparsity in contextual information. We presented our proposal for solving sparsity in contextual information using context inference. We showed analyses of the web usage for news portal and revealed the effect of behavioural correlation. Clustering users by the web usage splits users with similar preferences into groups where the association between context and need is also similar. In such group we can propagate missing contextual information or context value which is lacking in confidence. Our approach solves problems which are often present in frameworks which are gathering contextual information from more sources.

In our future work we plan to generate news recommendations using inferred context and compare results to our previous work where we used pure content-based recommendations [17]. We also plan to apply this method for smartphones where we encounter higher variety of context types. Context which could be acquired on smart-phone is more complex than on the web what is big challenge for us.

Our main contribution is in reducing the sparsity of contextual information thus improving accuracy of context-aware recommender systems. We have also designed an alternative to association rules mining which respects numeric values.

7. ACKNOWLEDGMENTS

This work was partially supported by the Scientific Grant Agency of the Ministry of Education of Slovak Republic, grant VG1/0675/11 and by the Slovak Research and Development Agency under the contract No. APVV-0208-10.

8. REFERENCES

[1] H. Asoh. An Analysis of Differences between Preferences in Real and Supposed Contexts. *Knowledge Creation Diffusion Utilization*, 2010.

[2] L. Baltrunas, B. Ludwig, and F. Ricci. Context relevance assessment for recommender systems. In *Proc. of the 15th int. conf. on Intelligent user interfaces*, pages 287–290. ACM, 2011.

[3] A. Bermingham and A. Smeaton. Classifying sentiment in microblogs: is brevity an advantage? pages 1833–1836. ACM, 2010.

[4] A. Cádiz, S. González, and K. Mens. Orchestrating context-aware systems: a design perspective. In *Proc. of the first int. workshop on Context-aware software technology and applications*, pages 5–8. ACM, 2009.

[5] D. Carlson and A. Schrader. A wide-area context-awareness approach for Android. *Proceedings of the 13th Int. Conf. on Information Integration and Web-based Applications and Services*, page 383, 2011.

[6] A. M. Dan Pelleg. X-means: Extending k-means with efficient estimation of the number of clusters. In *Proc. of the 17th Int. Conf. on Machine Learning*, pages 727–734, San Francisco, 2000. Morgan Kaufmann.

[7] W. Dargie. Why is context-aware computing less successful? *Proc. of the 5th ACM International Workshop on Context-Awareness for Self-Managing Systems*, pages 1–3, 2011.

[8] M. Kahng and S.-g. Lee. Ranking in context-aware recommender systems. In *Proc. of the 20th int. conf. companion on World wide web*, pages 65–66, New York, New York, USA, 2011. ACM.

[9] A. Komninos, A. Plessas, V. Stefanis, and J. Garofalakis. Application of dimensionality reduction techniques for mobile social context. In *Proc. of the 13th int. conf. on Ubiquitous computing*, pages 583–584. ACM, 2011.

[10] S. Konomi. Colocation networks: exploring the use of social and geographical patterns in context-aware services. In *Proc. of the 13th int. conf. on Ubiquitous computing*, pages 565–566. ACM, 2011.

[11] T. Kramar. Towards contextual search: social networks, short contexts and multiple personas. In *User Modeling, Adaption and Personalization*, pages 434–437. Springer, 2011.

[12] K. Oku, S. Nakajima, J. Miyazaki, S. Uemura, H. Kato, and F. Hattori. A Recommendation System Considering Users' Past/Current/Future Contexts. *ids.csom.umn.edu*, pages 3–7, 2010.

[13] H. Park, I. Cho, and S. Ji. An empirical study on web search behavior through the investigation of a user-clustered query session. *Modeling and Using Context*, pages 233–245, 2011.

[14] E. M. Perse. Implications of cognitive and affective involvement for channel changing. *Journal of Communication*, pages 49–68, 1998.

[15] H. Rahnama, S. Jamshidi, S. Johns, and A. Shepard. CAMPUS: context aware mobile platform for uniformed security. In *Proc. of the 13th int. conf. on Ubiquitous computing*, number Figure 1, pages 489–490. ACM, 2011.

[16] D. Riboni and C. Bettini. COSAR: hybrid reasoning for context-aware activity recognition. *Personal and Ubiquitous Computing*, 15(3):271–289, Aug. 2010.

[17] D. Zelenik and M. Bielikova. News Recommending Based on Text Similarity and User Behaviour. In *Proc. of the 7th Int. Conf. on Web Information Systems and Technologies*, pages 302–307, Noordwijkerhout, The Netherland, 2011. ACM.

4th ACM RecSys Workshop on Recommender Systems and the Social Web

Bamshad Mobasher
DePaul University, Chicago
mobasher@cs.depaul.edu

Dietmar Jannach
TU Dortmund, Germany
dietmar.jannach@tu-dortmund.de

Werner Geyer,
IBM Research, Cambridge MA
werner.geyer@us.ibm.com

Andreas Hotho
University of Würzburg
Germany
hotho@informatik.uni-wuerzburg.de

ABSTRACT

The new opportunities for applying recommendation techniques within Social Web platforms and applications as well as the various new sources of information which have become available in the Web 2.0 and can be incorporated in future recommender applications are a strong driving factor in current RS research for various reasons:

(1) Social systems by their definition encourage interaction between users and both online content and other users, thus generating new sources of knowledge for recommender systems. Web 2.0 users explicitly provide personal information and implicitly express preferences through their interactions with others and the system (e.g. commenting, friending, rating, etc.). These various new sources of knowledge can be leveraged to improve recommendation techniques and develop new strategies which focus on social recommendation.

(2) New application areas for recommender systems emerge with the popularity of the Social Web. Recommenders can not only be used to sort and filter Web 2.0 and social network information, they can also support users in the information sharing process, e.g., by recommending suitable tags during folksonomy development.

(3) Recommender technology can assist Social Web systems through increasing adoption and participation and sustaining membership. Through targeted and timely intervention which stimulates traffic and interaction, recommender technology can play its role in sustaining the success of the Social Web.

(4) The Social Web also presents new challenges for recommender systems, such as the complicated nature of human-to-human interaction which comes into play when recommending people and can require more interactive and richer recommender systems user interfaces.

The goal of this workshop, which continuous a series of successful RSWeb-Workshops since ACM RecSys 2009, was to bring together researchers and practitioners to explore, discuss, and understand challenges and new opportunities for recommender systems and the Social Web.

The technical program of the one-day workshop comprised a set of technical papers on recent and ongoing research, which were selected for presentation and discussion at the workshop in a formal review process. In addition, the workshop featured informal breakout sessions on recent topics.

Overall, we received 13 paper submissions from 12 different countries, out of which 8 long and 1 short paper were selected for presentation and inclusion in the proceedings.

The submitted papers addressed a variety of topics related to Social Web recommender systems including:

- The usage of microblogging data for personalization
- Personalization of news feeds
- Trust and reputation in the Social Web
- Generating recommendations for groups
- Recommendation of interesting users to connect with
- Exploiting the Social Graph for improved recommendations
- Resource recommendation in Social Web applications
- Tag-based recommendation and folksonomies
- Cross-domain personalization
- Fusion of various Web 2.0 data sources
- Scalability of recommender systems

The list of papers, the workshop schedule and pointers to downloadable versions of the papers can be found at the workshop's homepage given below. The electronic proceedings appear in the ACM Digital Library.

http://ls13-www.cs.uni-dortmund.de/homepage/rsweb2012

Categories and Subject Descriptors

H.3.3 [**Information Search and Retrieval**]; H.4.2 [**Decision Support Systems**]

Keywords

Recommender Systems, Personalization, Social Web.

RecSys'12 Workshop on Human Decision Making in Recommender Systems

Marco de Gemmis
University of Bari Aldo Moro
Via E. Orabona, 4, Bari, Italy
degemmis@di.uniba.it

Alexander Felfernig
Graz University of Technology
Inffeldgasse 16b/2
Graz, Austria
alexander.felfernig@ist.tugraz.at

Pasquale Lops
University of Bari Aldo Moro
Via E. Orabona, 4, Bari, Italy
lops@di.uniba.it

Francesco Ricci
University of Bolzano
Piazza Domenicani 3,
Bozen-Bolzano, Italy
fricci@unibz.it

Giovanni Semeraro
University of Bari Aldo Moro
Via E. Orabona, 4, Bari, Italy
semeraro@di.uniba.it

Martijn C. Willemsen
Eindhoven University of Technology
P.O. Box 513
5600MB Eindhoven
M.C.Willemsen@tue.nl

ABSTRACT

Interacting with a recommender system means to take different decisions such as selecting an item from a recommendation list, selecting a specific item feature value (e.g., camera's size, zoom) as a search criteria, selecting feedback features to be critiqued in a critiquing based recommendation session, or selecting a repair proposal for inconsistent user preferences when interacting with a knowledge-based recommender. In all these situations, users face a decision task. This workshop (Decisions@RecSys) focuses on approaches for supporting effective and efficient human decision making in different types of recommendation scenarios.

Categories and Subject Descriptors

H.3.3 [**Information Search and Retrieval**]; H.4.2 [**Decision Support Systems**]

General Terms

Algorithms, Experimentation, Human Factors, Theory.

Keywords

Recommender Systems, Decision Making, Decision Psychology, Recommender Algorithms, Decision Biases.

1. OVERVIEW

The complexity of decision tasks, limited cognitive resources of users, and the tendency to keep the overall decision effort as low as possible lead to the phenomenon of bounded rationality [5], i.e., users employ decision heuristics rather than trying to take an optimal decision. Furthermore, preferences of users are usually not stable entities easily retrieved from memory when requested. More often, preferences are constructed [1] while in a

specific decision environment and are therefore likely to change throughout a recommendation session or between sessions, due to contextual effects. Decision making under bounded rationality is a door opener for different types of non-conscious influences on the decision behavior of a user. Theories from decision psychology and cognitive psychology are trying to explain these influences, for example, decoy effects [2] and defaults [3] can trigger significant shifts in item selection probabilities; in group decision scenarios [4], the visibility of the preferences of other group members can have a significant impact on the final group decision.

The major goal of this workshop was to establish a platform for industry and academia to present and discuss new ideas and research results that are related to the topic of human decision making in recommender systems. The workshop consisted of technical sessions in which results of ongoing research were presented, informal group discussions on focused topics, and a keynote talk.

The topics of papers submitted to the workshop can be summarized as follows:

- Avoidance of decision biases: decision biases can lead to suboptimal user decisions – the detection of potential biases can significantly improve the perceived quality of recommender systems.

- Intelligent preference elicitation: the understanding of user preferences is a major precondition for the determination of relevant recommendations. Research areas covered by this year's submissions are opinion formation, preference learning, and preference relaxation.

- Emotions in recommender systems: emotions play a major role in the context of human decision making since the emotional state has a major impact on the decision outcome. Related submissions are tackling the challenge of integrating knowledge about emotions into the underlying recommendation processes.

- New application domains: the variety of application domains for recommendation technologies is increasing. An example domain is software engineering where recommender systems are applied to support processes such as requirements engineering, software quality assurance, and software reuse.

- Empirical studies: the design of recommender user interfaces can have a major impact on the outcome of a decision processes. The results of related empirical studies have been reported in a couple of submissions to this year's *Decisions@RecSys* workshop.

2. FURTHER INFORMATION
The workshop material (list of accepted papers, invited talk, and the workshop schedule) can be found at the *Decisions@RecSys 2012 workshop webpage*: http://recex.ist.tugraz.at/RecSysWorkshop2012.

3. REFERENCES
[1] R. Bettman, M. Luce, and J. Payne. Constructive Consumer Choice Processes, Journal of Consumer Research, 25:187-217, 1998.

[2] J. Huber, W. Payne, and C. Puto. Adding Asymmetrically Dominated Alternatives: Violations of Regularity and the Similarity Hypothesis, Journal of Consumer Research, 9:90-98, 1982.

[3] M. Mandl, A. Felfernig, J. Tiihonen, and K. Isak. Status Quo Bias in Configuration Systems, 24th International Conference on Industrial, Engineering & Other Applications of Applied Intelligent Systems (IEA/AIE 2010), Syracuse, NY, 105-114, 2011.

[4] J. Masthoff. Group recommender systems: Combining individual models. In F. Ricci, L. Rokach, B. Shapira, and P. Kantor, editors, Recommender Systems Handbook, pages 677-702. Springer, 2011.

[5] H. Simon. A Behavioral Model of Choice. Quarterly Journal of Economics, 69(1): 99-118, 1955.

4th Workshop on Context-Aware Recommender Systems (CARS 2012)

Gediminas Adomavicius
University of Minnesota, USA
gedas@umn.edu

Linas Baltrunas
Telefonica Research, Spain
linas@tid.es

Ernesto William de Luca
University of Applied Sciences Potsdam, Germany
deluca@fh-potsdam.de

Tim Hussein
University of Duisburg-Essen, Germany
tim.hussein@uni-due.de

Alexander Tuzhilin
New York University, USA
atuzhili@stern.nyu.edu

ABSTRACT

CARS 2012 builds upon the success of the three previous editions held in conjunction with the 3rd to 5th ACM Conferences on Recommender Systems from 2009 to 2011. The 1st CARS Workshop was held in New York, NY, USA, whereas Barcelona, Spain, was home of the 2nd CARS Workshop in 2010. In 2011, the 3rd CARS workshop was held in Chicago, IL, USA.

Categories and Subject Descriptors

H.3.3 [**Information Storage and Retrieval**]: Information Search and Retrieval – *information filtering, relevance feedback, retrieval models, search process, selection process.*

General Terms

Algorithms, Design, Experimentation, Human Factors, Measumerment, Performance.

Keywords

Recommender systems, context-aware systems, contextual information, context modeling.

1. WORKSHOP GOALS

The importance of contextual information has been recognized by researchers and practitioners in many disciplines, including e-commerce personalization, information retrieval, ubiquitous and mobile computing, data mining, marketing, and management. While a substantial amount of research has already been performed in the area of recommender systems, the vast majority of existing approaches focuses on recommending the most relevant items to users and does not take into account any additional contextual information, such as time, location, weather, or the company of other people. Therefore, this workshop aims to bring together researchers with wide-ranging backgrounds to identify important research questions, to exchange ideas from different research disciplines, and, more generally, to facilitate discussion

and innovation in the area of context-aware recommender systems (CARS). In particular, the workshop covers the following topics:

- Context modeling techniques for recommender systems;
- Context-aware user modeling for recommender systems;
- Data sets for context-dependent recommendations;
- Algorithms for detecting the relevance of contextual data;
- Algorithms for incorporating contextual information into recommendation process;
- Algorithms for building explicit dependencies between contextual features and ratings;
- Interacting with context-aware recommender systems;
- Novel applications for context-aware recommender systems;
- Mobile context-aware recommender systems;
- Context-aware group recommendations;
- Large-scale context-aware recommender systems;
- Evaluation of context-aware recommender systems.

Additional information about the workshop and its program can be found on the workshop website.[1]

2. WORKSHOP FORMAT

CARS will be organized as a full day event, in which authors of selected papers will present their work to the workshop audience.

Depending on the number of participants, interactive elements such as plenary discussions are planned.

---

[1] http://cars-workshop.org

Workshop on Recommendation Utility Evaluation: Beyond RMSE – RUE 2012

Xavier Amatriain
Netflix, Inc.
100 Winchester Cir
Los Gatos, CA 95032, USA

xamatriain@netflix.com

Pablo Castells
Univ. Autónoma de Madrid
Fco. Tomás y Valiente 11
Madrid, 28049 Spain

pablo.castells@uam.es

Arjen de Vries
Cent. Wiskunde & Informatica
Science Park 123, 1098 XG
Amsterdam, Netherlands

arjen@acm.org

Christian Posse
LinkedIn, Corp.
2029 Stierlin CT, Mountain
View, CA 94043, USA

cposse@linkedin.com

ABSTRACT

Measuring the error in rating prediction has been by far the dominant evaluation methodology in the Recommender Systems literature. Yet there seems to be a general consensus that this criterion alone is far from being enough to assess the practical effectiveness of a recommender system. Information Retrieval metrics have started to be used to evaluate item selection and ranking rather than rating prediction, but considerable divergence remains in the adoption of such metrics by different authors. On the other hand, recommendation utility includes other key dimensions and concerns beyond accuracy, such as novelty and diversity, user engagement, and business performance. While the need for further extension, formalization, clarification and standardization of evaluation methodologies is recognized in the community, this need is still unmet for a large extent. The RUE 2012 workshop sought to identify and better understand the current gaps in recommender system evaluation methodologies, help lay directions for progress in addressing them, and contribute to the consolidation and convergence of experimental methods and practice.

Categories and Subject Descriptors

H.3.3 [**Information Search and Retrieval**]: *information filtering.*

General Terms

Algorithms, Measurement, Performance, Experimentation, Standardization, Theory.

Keywords

Utility, evaluation, methodology, metrics, recommender systems.

1. INTRODUCTION

Measuring the error in predicting held-out user rating values has been by far the dominant offline evaluation methodology in the Recommender Systems (RS) literature [6,9]. Yet there seems to be a general consensus in the community that this criterion alone is far from being enough or even adequate to assess the practical effectiveness of a recommender system in matching user needs [12]. The end users of recommendations receive lists of items rather than rating values, whereby recommendation accuracy metrics –as surrogates of the evaluated task– should target the quality of the item selection, rather than the numeric system scores that determine this selection. Furthermore, as far as the order of recommended items determines the set of elements that the user will actually consider for consumption, effectiveness assessment methodologies should target item rankings. For this

reason, metrics and methodologies from the Information Retrieval (IR) field –where ranking evaluation has been studied and standardized for decades [11]– have started to be adopted by the RS community [3,4,6,9]. Gaps remain between the methodological formalization of tasks in both fields though, which result in divergences in the adoption of IR methodologies for RS, hindering the interpretation and comparability of empirical observations by different authors.

On the other hand, there is a growing realization that accuracy is only one among several relevant dimensions of recommendation effectiveness. The value of novelty, for instance, has been recognized as a key dimension of recommendation utility for users in real scenarios, in-as-much as the purpose of recommendation is inherently linked to discovery in many application domains [2,7,10]. Closely related to novelty, diversity is also a desirable quality to enrich the user's experience and enhance his array of relevant choices [1,8]. Novelty and diversity are generally positive for businesses as well, by favoring the diversity of sales and helping leverage revenues from market niches [5]. As a matter of business performance enhancement, the value added by recommendation can be measured more directly in terms of on-line clickthrough rate, conversion rate, sales order size increase, returning customers, increased revenue, etc. On the other hand, web portals and social networks commonly face multiple objective optimization problems related to user engagement, requiring appropriate evaluation methodologies for optimizing along the entire recommendation funnel, from the initial click to the real user engagement in subsequent downstream utilities. Other potentially relevant dimensions of effective recommendations for consumers and providers may include confidence, coverage, risk, cost, robustness, etc.

While the need for further extension, formalization, clarification and standardization of evaluation methodologies is recognized in the community, this need is still unmet for a large extent. When engaging in evaluation work, researchers and practitioners are still often faced with experimental design questions for which there are currently not always precise and consensual answers. Room remains for further methodological development and convergence, which motivated this workshop.

2. SCOPE AND GOALS

RUE 2012 gathered researchers and practitioners interested in developing better, clearer, and/or more complete evaluation methodologies for recommender systems –or just seeking clear guidelines for their experimental needs. The workshop provided an informal setting for exchanging and discussing ideas, sharing experiences and viewpoints. RUE sought to identify and better understand the current gaps in recommender system evaluation methodologies, help lay directions for progress in addressing them, and contribute to the consolidation and convergence of experimental methods and practice. As a particular focus of interest, the workshop sought a better understanding of the methodological connections from confluent disciplines (IR, Machine Learning, HCI, etc.), as well as the scientific vs. industrial perspectives.

Specific questions raised and addressed by the workshop included, among others, the following:

- What are the unmet needs and challenges for evaluation in the RS field? What changes would we like to see? How could we speed up progress?

- What relevant recommendation utility and quality dimensions should be cared for? How can they be captured and measured?

- How can metrics be more clearly and/or formally related to the task, contexts and goals for which a recommender application is deployed?

- How should IR metrics be applied to recommendation tasks? What aspects require adjustment or further clarification? What further methodologies should we draw from other disciplines?

- What biases and noise should experimental design typically watch for?

- Can we predict the success of a recommendation algorithm with our offline experiments? What offline metrics correlate better and under which conditions?

- What are the outreach and limitations of offline evaluation? How can online and offline experiments complement each other?

- What type of public datasets and benchmarks would we want to have available, and how can they be built?

- How can the recommendation effect be traced on business outcomes?

- How should the academic evaluation methodologies improve their relevance and usefulness for industrial settings?

- How do we envision the evaluation of recommender systems in the future?

3. COVERED TOPICS

The accepted papers and the discussions held at the workshop addressed, among others, the following topics:

- Recommendation quality dimensions.
  - Effective accuracy, ranking quality.
  - Novelty, diversity, unexpectedness, serendipity.
  - Utility, gain, cost, risk, benefit.
  - Robustness, confidence, coverage, usability, etc.
- Matching metrics to tasks, needs, and goals.
  - User satisfaction, user perception, human factors.
  - Business-oriented evaluation.
  - Multiple objective optimization, user engagement.
  - Quality of service, quality of experience.
- Evaluation methodology and experimental design.
  - Definition and evaluation of new metrics, studies of existing ones.
  - Adaptation of methodologies from related fields: IR, Machine Learning, HCI, etc.
  - Evaluation theory.

- Practical aspects of evaluation.
  - Offline and online experimental approaches.
  - Simulation-based evaluation.
  - Datasets and benchmarks.
  - Validation of metrics.

The workshop opened with a keynote talk, followed by the presentation of accepted papers and open discussions. The accepted papers and a summary of discussions are available in the workshop proceedings, which can be reached from the workshop website at http://ir.ii.uam.es/rue2012.

4. REFERENCES

[1] Castells, P., Wang, J., Lara, R., Zhang, D. Workshop on novelty and diversity in recommender systems – DiveRS 2011. *8th ACM International Conference on Recommender Systems (RecSys 2011)*. Chicago, IL, 393-394.

[2] Celma, O. and Herrera, P. A New Approach to Evaluating Novel Recommendations. *5th ACM International Conference on Recommender Systems (RecSys 2008)*. Lausanne, Switzerland, 179-186.

[3] Cremonesi, P., Garzotto, F., Negro, S., Papadopoulos, A. V., and Turrin, R. Comparative evaluation of recommender system quality. *ACM Conference on Human Factors in Computing Systems (CHI 2011)*. Vancouver, Canada, 1927-1932.

[4] Cremonesi, P., Koren, Y., and Turrin, R. Performance of Recommender Algorithms on Top-N Recommendation Tasks. *4th ACM International Conference on Recommender Systems (RecSys 2010)*. Barcelona, Spain, 39-46.

[5] Fleder, D. M. and Hosanagar, K. Blockbuster Culture's Next Rise or Fall: The Impact of Recommender Systems on Sales Diversity. *Management Science* 35, 5, 2009, 697-712.

[6] Herlocker, J. L., Konstan, J. A., Terveen, L. G., and Riedl, J. T. Evaluating collaborative filtering recommender systems. *ACM Trans. on Information Systems* 22, 1, 2004, 5-53.

[7] Lathia, N., Hailes, S., Capra, L., and Amatriain, X. Temporal Diversity in Recommender Systems. *33rd Annual International ACM SIGIR Conference on Research and Development in Information Retrieval (SIGIR 2010)*. Geneva, Switzerland, 210-217.

[8] McNee, S. M., Riedl, J., and Konstan, J. A. Being Accurate is Not Enough: How Accuracy Metrics have hurt Recommender Systems. *ACM Conf. on Human Factors in Computing Systems (CHI 2006)*. Montréal, Canada, 1097-1101.

[9] Shani, G. and Gunawardana, A. Evaluating Recommendation Systems. In Ricci, F. et al (Eds.), *Recommender Systems Handbook*. Springer, 2011, 257-297.

[10] Vargas, S. and Castells, P. Rank and Relevance in Novelty and Diversity Metrics for Recommender Systems. *8th ACM International Conference on Recommender Systems (RecSys 2011)*. Chicago, IL, 109-116.

[11] Voorhees, E. M. and Harman, D. K. *TREC: Experiment and Evaluation in Information Retrieval*. MIT Press, Cambridge, MA, 2006.

[12] Willemsen, M. C., Bollen, D. G. F. M., Ekstrand, M. D. UCERSTI: 2nd workshop on user-centric evaluation of recommender systems and their interfaces. *8th ACM International Conference on Recommender Systems (RecSys 2011)*. Chicago, IL, 395-396

Recommender Systems Challenge 2012

Nikos Manouselis
Agro-Know Technologies 17
Grammou Str., 15235 Vrilissia,
Athens, Greece
nikosm@ieee.org

Alan Said
Technische Universität Berlin
Ernst-Reuter-Platz 7, TEL 14,
DE-10587 Berlin, Germany
alan@dai-lab.de

Domonkos Tikk
Gravity R&D
Expo tér 5-7, 1011
Budapest, Hungary
domonkos.tikk@
gravityrd.com

Jannis Hermanns
moviepilot
Mehringdamm 33, DE-10961
Berlin, Germany
jannis@moviepilot.de

Benjamin Kille
Technische Universität Berlin
Ernst-Reuter-Platz 7, TEL 14,
DE-10587 Berlin, Germany
kille@dai-lab.de

Hendrik Drachsler
Open University of the
Netherlands Centre for
Learning Sciences and
Technologies (CELSTEC)
PO-Box 2960, 6401 DL
Heerlen,The Netherlands
hendrik.drachsler@ou.nl

Katrien Verbert
Katholieke Universiteit Leuven
Computerwetenschappen
Dept. Celestijnenlaan 200A,
B-3001 Leuven, Belgium
katrien.verbert@
cs.kuleuven.be

Kris Jack
Mendeley Ltd. 144a
Clerkenwell Road London,
EC1R 5DF, United Kingdom
kris.jack@mendeley.com

ABSTRACT

The Recommender System Challenge 2012 invited participants to work on two tracks with real-world datasets and to submit their contributions that would be related to specific problem contexts. First of all, it asked participants to develop new algorithms and to compare them to other algorithms in given settings; in addition, it asked participants to explore with new recommendation methods, services, as well as added-value services related to recommendation.

Categories and Subject Descriptors

D.2.8 [**Software Engineering**]: Metrics - complexity measures, performance measures; H.3.3 [**Information Storage and Retrieval**]: Information search and retrieval - information filtering, relevance feedback; H.3.4 [**Information Technology and Systems Applications**]: Decision support; H.3.5 [**Online Information Services**]: Data Sharing; H.5.1 [**Multimedia Information Systems**]: Evaluation/methodology

General Terms

Algorithms, Design, Experimentation, Human Factors, Measurement

Keywords

Recommender Systems, dataset, challenge, competition, context-aware, scientific paper recommendation

1. INTRODUCTION

As in other research areas, the availability of datasets in recommender systems can be considered as key for research and application purposes. These datasets serve as benchmarks to develop new algorithms and to compare them to other algorithms in given settings. Furthermore, they can be used for experimenting with new recommendation methods, services, as well as added-value services related to recommendation (such as supporting visualization and argumentation). The *2012 Recommender System Challenge* was organized in conjunction with the 6th ACM Conference on Recommender Systems (RecSys'12). The challenge focused on two tracks, each with a real-world dataset: the Benchmarking Track on *Context-Aware Movie Recommendation* (CAMRa), and the Exploratory Track on *Scientific Paper Recommendation* (ScienceRec).

The 2012 Challenge builds upon the tradition of a number of contests, challenges and workshops for recommender systems that have taken place in the past, e.g. the series of CAMRa challenges that where organized in conjunction with RecSys in 2010 [1, 4] and 2011 [5,6] as well as a series of workshops and challenges on Technology Enhanced Learning, e.g. the DataTEL challenge [2] and RecSysTEL workshop [3].

2. CAMRA TRACK

Following CAMRa-2010 and CAMRa-2011, the CAMRa-2012 Track focused on context-aware recommendation of movie-related news from moviepilot.com. The track addressed context-aware recommendation and context-aware evaluation, as well as live evaluation of recommendations. For this, a dataset from moviepilot has been made available, with information related to concepts from the world of cinema, e.g. single movies, movie universes (such as the world of Harry Potter movies), upcoming details (trailers, teasers, news, etc).

At the end of the challenge, a live evaluation session took place during which participants demonstrated how algorithms trained on offline data where evaluated online, on real users. The aim of this track was to find the right audience for a given movie. This movie has not necessarily been released already (as it might be in production), so the overall goal is to generate a large impact on the recommended item in terms of interaction in the social networks of those users to whom the movies are recommended. The aim of the hands on session was to expose participants into the types of information that they would be expecting to handle when working on such systems, as well as bring them in direct contact with the technical team of a deployed service where recommendation is core.

3. SCIENCEREC TRACK

This track focused on recommendations to users about scientific papers that they might be interested in, using a data set that comes from the Mendeley system[1]. The aim was to share recommendation approaches and discuss issues like:

- the types of scientific recommendation services that social research platforms like Mendeley can implement

- the types of data sets that could help advance research around scientific paper recommendation

Submissions used the already published Mendeley dataset [2] which came out after the 1st DataTEL Challenge of the 2010 Workshop on Recommender Systems in Technology Enhanced Learning (RecSysTEL) [3]. The Track asked participants to use and evaluate their approaches in an off-line manner, as well as invited them to propose their approaches for relevant services, navigational interfaces, visualizations of recommendations etc. Thus it welcomed submissions that combined the data set with the Mendeley API[2].

During the interactive part of the workshop, participants had the opportunity to discuss with the Mendeley technical team about ways in which their ideas and proposals could be incorporated into a large-scale real-world system like Mendeley, and get insight into the way that such systems are being maintained and extended in the industry.

4. OVERVIEW

The Recommender Systems Challenge 2012 took place on September 13th, 2012, as a full day workshop that included sessions both focusing on algorithmic evaluations as well as hands-on experience with real datasets, APIs, recommender systems software and systems.

5. ACKNOWLEDGMENTS

The involvement of Nikos Manouselis has been carried out with European Commission funding support and, more specifically, the FP7 agINFRA project[3].

---

[1] http://www.mendeley.com
[2] http://dev.mendeley.com
[3] http://www.aginfra.eu

Katrien Verbert is a postdoctoral fellow of the research foundation - Flanders (FWO).

6. REFERENCES

[1] Gediminas Adomavicius, Alexander Tuzhilin, Shlomo Berkovsky, Ernesto W. De Luca, and Alan Said, 'Context-awareness in recommender systems: research workshop and movie recommendation challenge', in *Proceedings of the fourth ACM conference on Recommender systems*, RecSys '10, pp. 385–386, New York, NY, USA, (2010). ACM.

[2] Kris Jack, James Hammerton, Dan Harvey, Jason J Hoyt, Jan Reichelt, and Victor Henning, 'MendeleyÊijs reply to the datatel challenge', *Procedia Computer Science*, **1**(2), 1–3, (2010).

[3] Nikos Manouselis, Hendrik Drachsler, Katrien Verbert, and Olga C. Santos, 'Recsystel preface 2010', *Procedia Computer Science*, **1**(2), 2773 – 2774, (2010). Proceedings of the 1st Workshop on Recommender Systems for Technology Enhanced Learning (RecSysTEL 2010).

[4] Alan Said, Shlomo Berkovsky, and Ernesto W. De Luca, 'Putting things in context: Challenge on context-aware movie recommendation', in *Proceedings of the Workshop on Context-Aware Movie Recommendation*, CAMRa '10, pp. 2–6, New York, NY, USA, (2010). ACM.

[5] Alan Said, Shlomo Berkovsky, and Ernesto W. De Luca, 'Group recommendation in context', in *Proceedings of the 2nd Challenge on Context-Aware Movie Recommendation*, CAMRa '11, pp. 2–4, New York, NY, USA, (2011). ACM.

[6] Alan Said, Shlomo Berkovsky, Ernesto William De Luca, and Jannis Hermanns, 'Challenge on context-aware movie recommendation: Camra2011', in *Proceedings of the fifth ACM conference on Recommender systems*, RecSys '11, pp. 385–386, New York, NY, USA, (2011). ACM.

RecSys'12 Workshop on Interfaces for Recommender Systems (InterfaceRS)

Nava Tintarev
Dept. Computing Science
University of Aberdeen
Aberdeen, AB24 5UA, UK
n.tintarev@abdn.ac.uk

Rong Hu
Human Computer Interaction Group
EPFL
CH-1015, Lausanne, Switzerland
rong.hu@epfl.ch

Pearl Pu
Human Computer Interaction Group
EPFL
CH-1015, Lausanne, Switzerland
pearl.pu@epfl.ch

Categories and Subject Descriptors

H.1.2 [**Models and principles**]: User/Machine Systems – *human factors*, *software psychology*; H.5.2 [**Information Interfaces and Presentation**]: User Interfaces – *evaluation/methodology, interaction styles, user centered design*

General Terms

Human Factors, Design, Experimentation, Performance.

Keywords

Recommender systems, human-computer interaction, user experience, user interface, user studies, evaluation

1. MOTIVATION AND GOALS

Since the emergence of recommender systems, a large majority of research has focused on objective accuracy criteria and less attention has been paid to how users interact with the system and the efficacy of interface designs from users' perspective. Well-designed user interfaces have the capability of enhancing user interaction experience and overall satisfaction. For example, explanation interfaces can increase user confidence in their decision choices and inspire user trust and loyalty to the used system. Nowadays, a variety of novel recommendation technologies have been developed to meet different needs (e.g., group and social recommenders). Recommender systems have also extended to new application platforms (e.g., mobile devices). In addition, heterogeneous information resources have been incorporated into recommender systems (e.g., psychological factors, social media). This brings forward new challenges in designing effective and efficient interfaces for these new recommender applications.

As a half-day workshop, ACM RecSys'12 first workshop on Interfaces for Recommender Systems (InterfaceRS) aimed to bring together researchers and practitioners around the topics of designing and evaluating novel intelligent interfaces for recommender systems in order to: (1) share research and techniques, including new design technologies and evaluation methodologies (2) identify next key challenges in the area, and (3) identify emerging topics. This workshop aims at creating an interdisciplinary community with a focus on the interface design issues for recommender systems and promoting the collaboration opportunities between researchers and practitioners.

2. TOPICS OF INTEREST

Topics of interests included, but were not limited to:

- Explanation interfaces
- Context-aware interfaces
- Social interfaces
- Ubiquitous and mobile interfaces
- Affective Interfaces
- Spoken and natural language interfaces
- Collaborative multi-user interfaces
- Conversational interfaces
- Adaptive interfaces
- Argumentation and persuasion interfaces
- Example- and demonstration-based interfaces
- New approaches to designing interfaces
- Empirical studies and evaluations of new interfaces
- Evaluation methods and metrics (e.g., evaluation questionnaire design)
- User studies
- Human factors (e.g., privacy, trust, confidence)
- Comparative studies across domains
- Design guidelines

3. WORKSHOP FORMAT

The workshop combined the presentations of accepted technical papers and was concluded by a panel discussion. Additional information about the workshop, including the list of accepted papers, schedule, proceedings, and program committee, is provided on the workshop website (https://www.abdn.ac.uk/~csc284/InterfaceRS/index.html).

The accepted papers will also be published online in CEUR-WS proceedings (http://CEUR-WS.org).

1st Workshop on Recommendation Technologies for Lifestyle Change 2012

Bernd Ludwig
University Regensburg,
Germany
bernd.ludwig@ur.de

Francesco Ricci
Free University of
Bozen-Bolzano, Italy
fricci@unibz.it

Zerrin Yumak
Ecole Polytechnique Fédérale
de Lausanne, Switzerland
zerrin.yumak@epfl.ch

ABSTRACT

The workshop on Recommendation Technologies for Lifestyle Change will be an opportunity for discussing open issues, and propose technical solutions for the designing of intelligent information systems that can support and promote lifestyle change. The objective of these systems is to provide users with up-to-date information, and help them to make choices in every day life activities establishing a sustainable compromise between quality of life, individuality, and fun.

Categories and Subject Descriptors

H.3.5 [**Information Storage and Retrieval**]: Online Information Services; H.4 [**Information Systems Applications**]: Miscellaneous; H.5.1 [**Information Interfaces and Presentation**]: Multimedia Information Systems; J.3 [**Computer Applications**]: Life and Medical Sciences

General Terms

Recommender Systems, Persuasive Technologies, Medical Applications

Keywords

User Models, User Preferences, Behaviour Change, Pervasive Computing

1. GOALS OF THE WORKSHOP

In today's society, particularly in the affluent society, lifestyle is influenced by technology, and the abundance of financial resources. For instance, a large variety of computer games are excessively used, and people often travels by individualized transportation means, such as car, just for fun. Moreover, the idea that technique and money can buy anything spreads also to health management: people believe that medical knowledge can be immediately applicable in case of illness, as technical knowledge can be used for repairing a broken car.

This results in lifestyles that do not care about the negative long-terms effects on the environment, but also about well being of individual persons. The most prominent example of this is represented by various types of chronic illnesses in developed countries that result from poor lifestyle choices.

In this context, the aim of this workshop is to explore possibilities for recommender systems to support users in taking decisions related to various aspects of their lifestyle; we call them Lifestyle Change Recommender Systems (LSCRS). There are three main challenges for LSCRSs: firstly, such systems have to assess the user's context for delivering such recommendations. Secondly, in order to promote any change in user's lifestyle, they have to recommend a tailored sequence of items, mostly actions, taking into account the dependencies between the recommended items and the effects of each item recommendation. Thirdly, LSCRS have to be designed to favor the user's continuous attention, to enable the explanation of the reasons for the suggested changes in the user's future behavior, and to recall the changes already effectuated.

Hence, in order to provide an effective support to lifestyle change, recommender systems need to provide communicative capabilities, e.g, with multi-modal dialogue systems. Recommendation technologies have to initiate a feedback-change-loop that could contribute to lowering the risks of severe illnesses for many individual users and improving the overall environmental situation.

In order to discuss recent developments and advances in this area, the workshop focusses on the following topics:

- Surveys of lifestyle related activities and technological approaches to monitoring them;
- Context modeling for activity recommendations;
- Formal models of sensor data for monitoring every day activities;
- User models for every day life recommendations that provide user-tailored content;
- Motivational models for lifestyle, every day activities, and environmental responsibility;
- Recommendations of sequences of items (e.g. physical exercises for a whole week, planning meals for a month);
- Measures of the effectiveness for lifestyle change recommender systems;
- Approaches to combine sensor data and interactive user input in LSCRS;
- Strategies to cement behavioral change;
- Strategies for situation- and user-aware presentation of recommendations;
- Persuasive technologies for interaction with and among users on their personal situation, their habits, and

their options to change their lifestyle Recommendation of activities for leisure time and lifestyle;

- Recommendation of information sources (e.g. forum entries, blogs) for LSCRS.

2. PROGRAM OF THE WORKSHOP

During the workshop, participants will present their papers and discuss contributions to the field addressing a variety of issues:

- As recommendations in this area are more dependent on the personal history of individual users rather than on the collective behavior and attitudes of many users as in more standard collaborative approaches to recommendations, the workshop participants will discuss new recommendation strategies that leverage the retrospective analysis of the user's past actions and behavioral patterns.

- How can change in behaviour be achieved by employing conversational agents? In a case study on alcohol consumption behavior, the benefits of conversational agents to persuade user to control their personal consumption of alcoholic beverages will be illustrated.

- Some contributions to the workshop discuss users' classification, adequate user models for LSCRS, models for motivations and concerns of users, and aspects of context modeling for lifestyle change recommendations.

- A number of application domains for lifestyle change behavior will be presented ranging from recommending meals and meal plans to travel routes under ecological constraints.

3. WORKSHOP FORMAT

The workshop is organized as a half day venue. It includes the presentation of research papers as well as system demonstrations and a plenary discussion and brainstorming session on foundations for lifestyle change recommendation technologies.

4. DISSEMINATION OF RESULTS

All accepted submissions will be collected in proceedings that will be available at http://ceur-ws.org/. Up-to-date information is available on the workshop home page: http://pc57724.uni-regensburg.de/lifestyleChange/.

Personalizing the Local Mobile Experience
Workshop at RecSys 2012

Henriette Cramer
Mobile Life @ SICS
Stockholm, Sweden
henriette@mobilelifecentre.org

Karen Church
Telefonica Research
Barcelone, Spain
karen@tid.es

Neal Lathia
University of Cambridge
Cambridge, UK
neal.lathia@cl.cam.ac.uk

Daniele Quercia
University of Cambridge
Cambridge, UK
daniele.quercia@cl.cam.ac.uk

ABSTRACT

Mobile, local recommendations are on the rise. Surprisingly however, research addressing user perceptions of local recommendations and local differences when interacting with such recommendation services is yet scarce. Location-based recommendation services are mostly evaluated from a 'recommendation systems' standpoint, with limited experiential insights from users and limited focus on local differences that may apply. This workshop focuses on the local, personal user experience, and provides a forum to exchange experiences, insights and strategies in personalizing local mobile applications and generating local recommendations that fit local user needs.

Categories and Subject Descriptors

H.4 [**Information Systems Applications**]: Miscellaneous

General Terms

Design, Experimentation, Human Factors

Keywords

Recommender systems, location-based recommenders, location-based services, mobile computing, mobile recommendations, mobile applications, personalization

1. INTRODUCTION

The rise of location-enabled mobile phones and location based services offers a great opportunity to apply personalization and recommender system technology to people's everyday lives. Location-based services are increasingly popular and used by millions of people around the globe. The millions of local data points generated by these users offer great opportunities for generating local recommendations. Local adaptation of adverts, search results, and recommendations of local places, things (and even nearby people) are now standard practice across many commercial services (e.g. Google search, Foursquare recommendations, Tripadvisor, Yelp etc.).

A variety of digital traces (to name a few: venue/event check-ins and ratings, GPS and GSM traces, transport records, and geo-located photographs) can be used to infer how people move about their city and extract their context and habits. Personalization and recommender systems, potentially merged with the data that people store online (e.g., social networks, web ratings), can then not only be used to recommend new places and events that they may find interesting to attend, but, more broadly, personalize and enhance any service that people find themselves using.

Surprisingly however, the amount of research addressing and understanding user perceptions of local recommendations and local differences when interacting with such recommendation services is yet scarce. The vast majority of location-based recommendation services have been evaluated from a 'recommendation systems' standpoint, with limited subjective insights from users of such systems and limited focus on any local or cultural differences that may apply.

The goal of this workshop is to provide a forum for researchers and developers from both academia and industry to exchange experiences, insights and strategies in: (1) personalizing local mobile applications, and (2) generating local recommendations that fit local user needs. We aim to generate an overview of local considerations that can be applied by both researchers and practitioners in their work, to spur on future research.

2. CHALLENGES

This workshop focuses on the user and their mobile experiences when encountering location-based and personalized recommendations. Designing great user experiences requires a deeper understanding of the information needs, behaviors and underlying motivations of users. As such there are a number of research challenges and open research questions that that arise. For example:

- How do we model mobile users and their behaviors in the context of personalization and recommendations

- How do we handle and design privacy controls and increase awareness?

- How do we engage, persuade and incentivize users while on-the-move?

- What can we learn from traditional web-based recommender system methodologies that can be to these new datasets and mobile scenarios?

- What types of interfaces and interactions can enrich location-based recommendations? Which local (social) factors should be taken into account?

- How can we evaluate mobile recommendation systems in-the-wild? And what are the factors that raise acceptance of such recommendation services?

- How do we measure the success of deployed mobile recommendation systems?

- What new strategies do we need to understand 'the local experience'? For example, what international and intercultural settings need to be considered?

- How can we deal the large data sets/digital traces? And how can these datasets be used in combination with in-depth qualitative research to reveal more insights about mobile users and their recommendation experiences?

3. WORKSHOP GOALS

Our research questions span a variety of fields: pervasive and persuasive technology, smart city/ubiquitous systems, personalization and recommender systems, mobile computing, location-based services, design, social networking, user experience and human-computer interaction. This workshop aims to: (1) build a community of researchers working within this space, who are currently spread across many research fields, such as HCI, Design, Mobile Computing, Information Retrieval, Recommender Systems and Personalization; (2) to promote RecSys as a compelling venue for all kinds of mobile Web related research; and (3) to set an agenda for future research in this important area.

Our goal is to provide a forum for researchers and developers from both academia and industry to exchange experiences, insights and strategies in generating local and personalized mobile recommendations that fit local needs. Themes of interest include (but are not limited to):

- Location, context-based and mobile recommendations
- Local, individual or cultural differences in recommendation, personalization and presentation preferences
- Incentives, behavioural change, usage motivations and persuasion in mobile settings
- Systems using and adapting to local data sources, such as end-user generated content, local services, social media, location-based social networks, etc.
- Mobility modeling and behaviour prediction
- Local search, recommendation on-the-spot and distant planning
- Recommender systems for smart cities/urban environments
- Innovative applications and interfaces for supporting local mobile recommendations
- Design guidelines for location-based and mobile recommender systems
- Security, privacy, reputation and trust issues in mobile recommenders
- Case studies of recommender systems in mobile environments.

4. WORKSHOP ORGANIZATION

This RecSys 2012 workshop explores local mobile experiences from two directions and this is reflected in our work program (See Table 1). The workshop features contributions from designers, developers and researchers from both industry and academia. Beyond a set of position pitches from our participants, this workshop will provide a practical insight in the local experience when encountering personalized recommendations.

We'll spend half a day exploring Dublin using local recommender systems, on-the-spot. We'll be using commercial systems, and any services that participants themselves may have built - and we'll explore what's great, what's missing, and where our research should be heading. The expected outcome is an overview of the challenges in generating local recommendations and strategies for overcoming them, and generating new insights in the factors that influence the local user experience when interacting with local personalization.

Time	Activity
09:00-09:30	Welcome & introductions
09:30-10:30	Paper Session 1
10:30-11:00	Coffee break
11:00-12:00	Paper Session 2
12:00-13:00	Outdoor activity briefing
13:00-15:45	Chasing local recommendations in-the-Dublin-wild
16:00-17:30	Discussion

Table 1: Workshop Program.

5. WORKSHOP ORGANIZERS

Henriette Cramer is a Senior researcher at Mobile Life @ SICS. Her research focuses on mobile (location-based) services, using apps for Research in the Large, and people's interaction with adaptive and autonomous 'things'.

Karen Church is a Researcher at Telefonica Research, Barcelona. Her research focuses on mobile information access behaviours, specifically mobile search, with a view to improving the online experiences of mobile users.

Neal Lathia is Research Associate in the Cambridge University's Computer Laboratory, where he is working on the EPSRC UBhave (Ubiquitous and Social Computing for Positive Behaviour Change) Project. His research falls in the intersection of data mining, mobile systems, ubiquitous/pervasive systems, and personalisation/ recommender systems.

Daniele Quercia is Senior Horizon Research Associate at the Computer Laboratory of the University of Cambridge, where he is exploring the complex relationship between our offline and online worlds. His research lies at the intersection of data mining, social computing, urban informatics, computational social science, and web science.

Author Index

www.ingramcontent.com/pod-product-compliance
Lightning Source LLC
Chambersburg PA
CBHW080713220326

41598CB00033B/5401